D0222419

Applied Solar Energy

AN INTRODUCTION

Applied Solar Energy

AN INTRODUCTION

ADEN B. MEINEL AND MARJORIE P. MEINEL
Optical Sciences Center, University of Arizona

ADDISON-WESLEY PUBLISHING COMPANY
Reading, Massachusetts · Menlo Park, California
London · Amsterdam · Don Mills, Ontario · Sydney

This book is in the
ADDISON-WESLEY SERIES IN PHYSICS

Reproduced by Addison-Wesley
from camera-ready copy prepared by Martha W. Stockton, University of Arizona,
and the authors.

Library of Congress Catalog Card No. 75-40904.

ISBN 0-201-04719-5
ABCDEFGHIJ-MA-79876

This book is dedicated to
the students of today
on whom the success of solar energy rests
tomorrow

Preface

This book is an introduction into the theory that must be mastered in order to successfully engineer and evaluate the performance of solar energy systems. The most important goal in solar energy applications is the ability to calculate output from a proposed design application and thereby establish the value of the energy delivered and a fair price for the system. To this end we carefully build the necessary background and information in successive chapters, culminating in a section on representative applications.

We give some extra emphasis to the topic of solar collector optics, because the optical functioning of collectors is subject to frequent misunderstanding. Since the collector is the most extensive and expensive part of a solar collector system, this emphasis is justified by the hard realities of economics.

Although this book is intended primarily for seniors and graduate students in energy conversion courses, we aim at more. We begin with a brief history of solar energy, with special emphasis on the magnificent achievements made in the late Victorian era. The history of solar energy contains lessons not fully appreciated, and one cannot read this history without admiring those pioneers of the Industrial Revolution who probed the possibilities and tested the barriers that still lie behind the bright facade of solar energy.

We are attempting to satisfy the curiosity and needs of a broad spectrum of readers, from the interested layman to the engineering undergraduate or graduate researcher now probing in these fields. For example, some of the chapters and sections go into considerable detail about results and questions raised by current research. We feel that the simple solar technologies of water and house heating await only the development of a mass market; the real challenges in solar energy lie in the technology applicable to second- and third-generation solar conversion goals, especially the conversion to electrical energy and chemical processing. Here the fields of materials science, solid-state physics, and systems integration have major challenges. We hope our textbook will prove a useful introduction to these fields.

For the interested layman we try to summarize the essentials, and we give a chapter on simple order-of-magnitude calculations for answers to typical simple questions asked by the many potential users of solar energy. We hope the advanced student will not dismiss these simple answers, but will accept the challenge to refine them with the details given in the preceding chapters of this book.

In technical depth we cover only our own area of specialization—thermal conversion. We have, however, touched upon the essence of direct conversion by means of solar cells and summarized a number of related fields.

In some parts of the book we are attempting to examine advanced questions where we found little guidance in the solar energy literature. We have done our best, and it is a challenge to the student to question these parts and thereby begin the refinement process that someday will lead to a definitive text on the topics we barely touch on. One problem we encountered is that engineers speak in different terms than physicists. To a physicist, mass is fundamental and gravity is completely separate. To the engineer, mass can be found in terms of poundals and slugs, where gravity is intertwined with the mass. We have attempted to write equations in which mass, length, and time are the primary variables, writing the gravity term explicitly. One sure way for the student to assess the correctness of equations is to parallel each equation with a "units equation," as we have done in Chapter 11.

We appreciate the patience of both professional and lay readers with our effort to combine these ends in one book. We also thank the reviewers of the manuscript for their valuable comments and suggestions, which we were glad to include in this book, especially Drs. Harry Tabor and Roland Winston.

We would also like to acknowledge the help we have received in preparing this material. Considerable research done by our colleagues at the University of Arizona is represented. Major contributions have been made by B. O. Seraphin, Dean B. McKenney, A. Francis Turner, W. T. Beauchamp, K. D. Masterson, and Walter Meinel. We also acknowledge the contributions to this book by Drs. I. F. Quercia and N. A. Mancini and by the faculty and students at the summer course on solar energy of the International College of Applied Physics, held each summer at some pleasant location in Italy.

This work has been supported indirectly, in part, by a grant from the Mulcahy Foundation to the authors, by the National Science Foundation under grants NSF-GI-30022 and NSF/RANN/GI-36731 at the University of Arizona, grant NSF/RANN/SE/GI-41895 at Helio Associates, and by Helio Associates corporate research funds.

Tucson, Arizona A.B.M.
January 1976 M.P.M.

ὁ μεταδιδοὺς ἐν ἁπλότητι

He that imparts: let him do it in simplicity.
—Rom. 12:8

Contents

Chapter 1
History of Applications

1.0 INTRODUCTION

The history of solar energy stretches back into the dim recesses of prehistory, perhaps as far as the clay-tablet era in Mesopotamia, when the temple priestesses used polished golden vessels to ignite the altar fires. The "sounding statues" of Amenkhotep III (1455–1419 B.C.) of Egypt are cited by Veinberg (1959) as being caused by the escape of air heated by sunlight falling on the statue at the rising of the sun. Veinberg also cites the singing of an artificial bird at the tomb of Zari Memnon, Amenkhotep's son, as being caused by the early morning sunlight. The tales of these solar devices stirred much discussion in the Middle Ages, as is related by Athanasius Kircher (1601–1680), and led to speculation on how the expansion of air in the pedestals of the statues could account for these phenomena.

The great challenge to the science that was to emerge in the 17th Century, however, was the story of Archimedes (287–212 B.C.), the great inventive genius who was chief scientist of the ancient city-state ruled by Heiron II of Syracuse (214–200 B.C.). The story of how he repelled the invading Roman fleet of Marcellus in 212 B.C., as related by Galen (A.D. 130–220), tells of the burning of the fleet by means of solar rays. This story was widely debated and finally relegated to myth, since Livy (59 B.C.–A.D. 17) did not write of this event in describing the Roman invasion. Plutarch (A.D. 46–120) only alluded to it obliquely: ". . . whence the Romans, seeing that indefinite mischief overwhelmed them from no visible means, began to think they were fighting with the gods." Some say he referred to "Greek fire," but this was well known by 212 B.C. and would not have evoked such a statement. The ambiguity posed a problem to Renaissance science.

The basic question was whether or not Archimedes knew enough about the science of optics to devise a simple way to concentrate sunlight to a point where ships could be burned from a distance. Archimedes had written a book "On Burning Mirrors," but no copy has survived to give evidence. This book must have antedated the invasion since he was killed less than a year later when Syracuse finally fell to land attack. The fact that he had written a book on burning mirrors would indicate that he had the basic knowledge to create the barrage of reflectors along the harbor walls to terrify the invading fleet. His name is now attached to any assemblage of mirror reflectors directing their beams of sunlight to a common point.

After Archimedes, we find little progress in solar energy applications until the Renaissance. The last word from the days of the Roman Empire concerning solar energy had been recorded in the 12th Century by Ioanne Zonaras, who told how Proclus repeated the feat of Archimedes using a large number of mirrors to burn the fleet of Vitellius at the siege of Constantinople. Southward across the Mediterranean, science was flourishing in the Arabic world, but little has survived to tell us of any utilization of solar energy through man-made devices. Astronomy, mathematics, and other sciences were being advanced by the learned men of the Moslem empire, but we must forever remain curious about what they must have done to put the sun to work in sunny North Africa. The seeds of science, however, were slowly transplanted in Europe via the resumption of sea commerce stimulated by the Crusades and emerged with vigor in the Renaissance. We finally see evidence of a rebirth of interest in solar energy in the 17th Century.

1.1 THE SEVENTEENTH CENTURY

Eighteen hundred years after Archimedes, Athanasius Kircher (1601–1680) made some experiments to set fire to a woodpile at a distance in order to see whether the story of Archimedes had any scientific validity. Burning glasses had undoubtedly remained the chief means of using solar energy over the intervening centuries, but once more man's curiosity led him to seek the answer to larger questions.

In 1615, Salomon de Caux published a description of a working solar "motor." He used a number of glass lenses mounted in a frame that concentrated the sun's rays on an airtight metal chamber partially filled with water. The sunlight heated the air, which expanded and forced the water out as a small fountain. This was only a play device for the entertainment of royalty—but it did reflect the revival of interest in solar energy.

The next incident involving the sun was in Florence, the city whose spirit had given birth earlier to the renewal of learning—the Renaissance. The experimenters, Averani and Targioni, attempted in 1695 to melt a diamond with a burning mirror. Little is recorded of this experiment, but we can safely predict that, if the mirror was large enough, either the thermal shock shattered the specimen or the specimen began to burn, emitting carbon dioxide and leaving no trace of residue. Diamond does not melt; it sublimes (evaporates without melting). This theme of using solar energy to attack materials that could not be melted by any known flame is a recurring one, and even today large solar furnaces are being used to process or manufacture extremely refractory substances.

More practical uses for solar energy were developed by Ehrenfried von Tschirnhaus (1651–1700), a member of the French National Academy of Science. He used lenses up to 76 cm in diameter to melt ceramic materials. One of his lenses was obtained by Homberg, physician to the Duke of Orleans, who used it to melt gold and silver. A colleague, Geoffroy, later used this lens to smelt iron, copper, tin, and mercury from their ores.

1.2 THE EIGHTEENTH CENTURY

Work using solar energy in the last part of the 17th Century gave impetus to a number of achievements centered in France.

The famed naturalist George Louis Leclerc Buffon (1707–1788) made the first of many spectacular French multiple-mirror solar furnaces, an art that still shines with excellence in France today. His largest consisted of 360 small flat mirrors individually pointed to send sunlight to a common focus. Using a smaller furnace with 168 flat mirrors 6 inches square, he put on a spectacular show in the Royal Gardens in 1747 by igniting a woodpile from a distance of 60 m. Count Buffon concluded that Archimedes' feat was possible and that he had probably worked at a distance of 30–45 m when he set the Roman ships on fire. The ancient harbor of Syracuse would have required 50–60 m; how Buffon arrived at the smaller distance is not explained, but he probably felt that it would have been too difficult to build as large a solar furnace as his, considering the primitive state of technology in 212 B.C. Buffon apparently discounted the possibility that Archimedes understood the principle and was therefore able to use things at hand in the besieged city to produce the effect of a burning glass.

The first experiments relating to ovens for food preparation are described by Nicholas de Saussure (1740–1799). His oven consisted of spaced glass blocks on top of a blackened surface enclosed by an insulated box. The sunlight entered the box through the glass and was absorbed by the black surface. A temperature of 88°C (191°F) was achieved. When a black coating was added to the glass surface, a temperature as high as 160°C (320°F) could be reached.

In 1747, the French astronomer Jacques Cassini of the Paris Observatory constructed a burning glass 112 cm in diameter that was presented to Louis XV. Cassini was able to obtain temperatures in excess of 1000°C (1832°F), which was sufficient to melt an iron rod in a few seconds and to melt silver to such a fluid state that the drippings formed hairlike strands when chilled in water.

Antoine Lavoisier (1743–1794), the founder of modern chemistry, experimented with solar furnaces because they provide the purest heat source possible. His furnace was made of two curved sheets of glass mounted to form a double-convex lens. The space between the plates was filled with alcohol. The liquid lens had a diameter of 130 cm and a focal length of 320 cm. The refracting power of the large liquid lens was insufficient to obtain high temperatures at the focus, so a smaller lens was placed near the focus to make the effective focal length much shorter. With this compound lens system he was able to melt even platinum at 1760°C (3200°F).

1.3 THE NINETEENTH CENTURY

Solar ovens again appear in the literature as described by the English astronomer John Herschel, son of the famous astronomer Sir William Herschel. John Herschel constructed a simple device for practical use on his expedition to the Cape of Good Hope in 1837. His oven was simply a black box that was buried in sand for insulation and had a double-layer glass cover that allowed sunlight to enter and prevented heat from escaping. A temperature of 116°C (240°F) was recorded. The oven was used by Herschel's staff to cook meat and vegetables for the dinner table of the expedition.

Henry Bessemer (1813–1898), of steel manufacturing fame, experimented with smelting of metals, constructing a solar furnace 305 cm in diameter containing many small flat mirrors. Even though it could process copper and zinc, Bessemer soon lost interest in solar furnaces and turned his attention to other thermal sources for refining iron into steel.

The first experiments in which solar energy was used to provide heat inside a vacuum enclosure were by Stock and Heynemann in Germany. Their furnace was made of several glass lenses 76 cm in diameter and 50 cm in focal length. The glass vacuum bulb, without itself being heated, transmitted the solar energy to the specimen, which was held in a magnesia crucible. Silicon, copper, iron, and manganese were among the samples successfully melted.

The earliest attempts to convert solar energy into other forms revolved around the generation of low-pressure steam to operate steam engines. August Mouchot pioneered this field by constructing and operating several solar-powered steam engines between 1864 and 1878. Evaluation of one built at Tours by the French government showed that it was too expensive to be considered feasible if constructed on a scale sufficient for the practical needs of commerce. Another was set up in Algeria, and Mouchot described some of his experiments with it in an 1869 publication.

In 1875, Mouchot made a notable advance in solar collector design by making one in the form of a truncated cone reflector. The spherical or parabolic mirror arrays of his predecessors had focused all the light at one small spot in space where the absorber or specimen was placed. Mouchot's cone, called an *axicon* in modern usage, focused light uniformly along the axis of the cone so that a tube could be used for the energy-absorbing surface, as shown in Figs. 1.1 and 1.2. Since the light was more diluted than when it came to one small spot, the maximum attainable temperature was much lower, but then an engine designer is not interested in melting a hole in his boiler if by accident the water level should drop too low!

Mouchot's axicon consisted of silver-plated metal plates and had a diameter of 540 cm and a collecting area of 18.6 m^2; the moving parts weighed 1400 kg. It allegedly collected and focused 87% of the sun's heat on the boiler enclosure. The steam engine was claimed to deliver 1.5 kW, but this meant it was using less than 3% of the heat received. Conventional coal-fueled steam engines of the day produced work with an efficiency of between 9% and 11%, principally because they could work at a higher boiler temperature.

Abel Pifre was a contemporary of Mouchot who also made solar engines and who had a flair for demonstrating them before the public eye. Pifre's solar collectors were parabolic reflectors made of many small mirrors; they looked rather similar in shape to Mouchot's truncated cones. In 1878 at the Paris Exhibition, Pifre exhibited one of his solar engines operating a printing press, as shown in Fig. 1.3. He had hoped to find

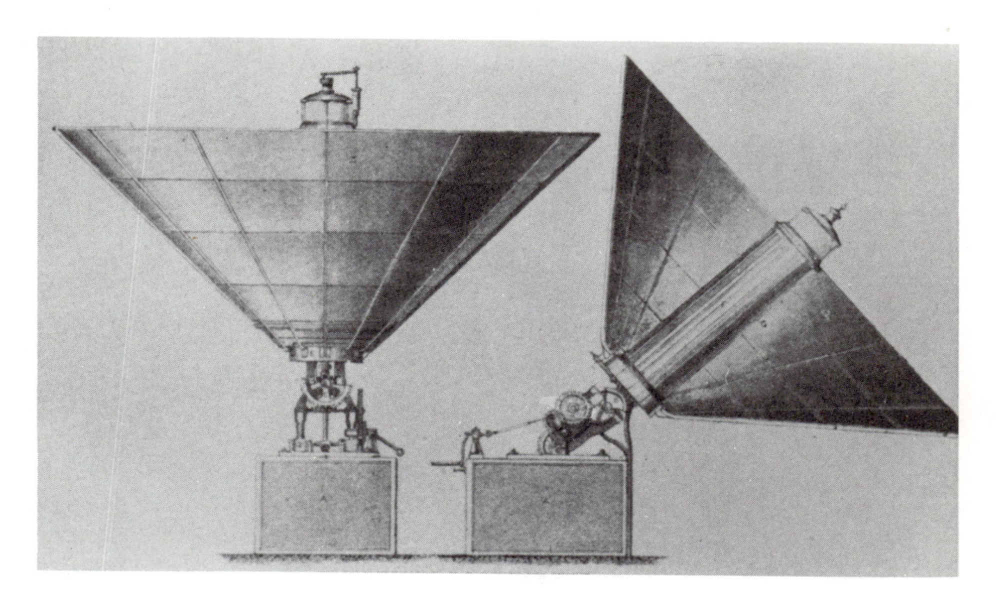

Fig. 1.1 August Mouchot's multiple-tube sun-heat absorber of 1878.

Fig. 1.2 The first large solar collector of the axicon type was exhibited by August Mouchot in 1878.

Fig. 1.3 Pifre's sun-power plant of 1878 driving a printing press.

customers for his solar engine by this means and was no doubt disappointed to find much curiosity but no buyers. Later, in 1882, he set up the engine in the gardens of the Tuileries Palace, where he again demonstrated it operating a printing press.

John Ericsson, of Civil War fame for his role in the development of the warship *Monitor*, also gave his attention to solar-powered engines between 1871 and 1884. He invented the Ericsson-cycle hot-air engine for the conversion of solar heat into motive power, which he used with a paraboloid collector, as shown in Fig. 1.4. His 1883 model, one of seven that he constructed, is shown in Fig. 1.5. It used a cylindrical reflecting surface bent into a parabolic shape, in which thin glass plates, silvered on the rear surface, were bent into shape by iron ribs. The device pivoted about a vertical axis to follow the sun. This device, it was claimed, produced 0.7 kW for 9.3 m^2 of reflecting surface.

Ericsson was able to look critically at the question of practicability of solar engines, perhaps because he was spending his own money! During the course of his experiments he spent some $90,000, a very sizable sum in those days. He concluded after his last engine that solar-powered engines cost 10 times more than conventional engines and that their use could be economically justified only for remote areas of "the sun-burnt regions of our planet."

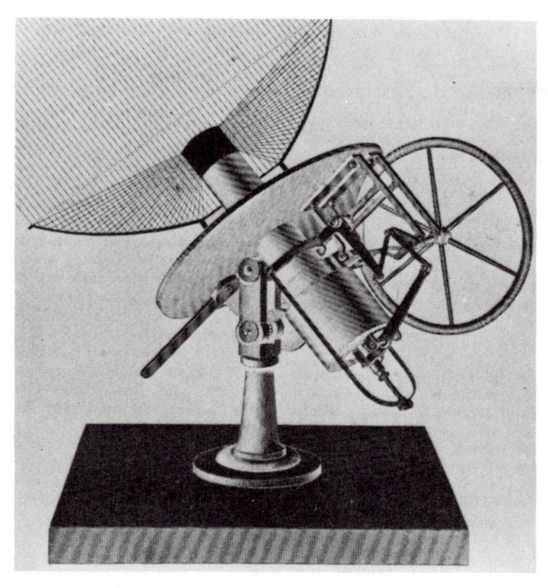

Fig. 1.4 Solar-powered hot-air engine tested by Ericsson nearly 90 years ago.

Fig. 1.5 Ericsson's solar collector of 1883 used a parabolic cylinder to focus on the absorbing tube mounted above the mirror.

W. Adams, an Englishman serving in Bombay, worked on the practical side of making solar engines. In 1878 he wrote a small book, published in India, on his experiments with solar heat. His largest engine consisted of a spherical hollow 12 m in diameter, the inside of which was covered with small sheets of silvered glass. It was successful as an engine, but one of his enthusiastic helpers got the brilliant idea that he could help the machine work even better by putting a brick over the safety valve. The resulting explosion temporarily suspended experiments. A new and stronger boiler then enabled Adams to operate a 2-kW pump on a continuous basis in the compound next to his bungalow at Middle Colaba, in Bombay. Adams also made solar cookers and discovered, as had Mouchot, that direct sunlight on meat creates an offensive odor and taste quite unlike that of fire-barbecued meat.

The solar engines built by Mouchot, along with the increasing popular interest in science in the last two decades of the 19th Century, led, as we have seen, to considerable activity. This activity was also evidenced by the issuance of patents. Even though a patent is supposed to be issued only upon "reduction to practice," we know that completely impractical devices and even technical absurdities get patents. We therefore know little about the merits of some of the solar devices that received patents in these two decades, as the literature gives no information on whether or not they were actually built. Some of the patent descriptions are interesting to read because, like much science fiction, they can describe certain properties of the materials used in their devices without regard to limitations of real materials, such as "a substance that neither conducts heat or electricity." We all wish that such ideal materials existed–in which case many things now impossible would become possible.

Patents for solar engines were issued in India to W. Adams in 1878, in Germany to E. J. Molera and J. C. Cerbrain in 1880 and to C. Tellier in 1885, and in the United States to W. Calver in 1882 and 1883. A general summary of current ideas and work as of 1885 was published in *Scientific American* by Charles H. Pope, who also was an experimenter with solar engines.

Solar engines of the 1880s worked only at the convenience of the sun. Nighttime and cloudy days imposed a real limitation on their usefulness. In 1893, M. L. Severy obtained a patent for a solar engine operating in conjunction with wet storage batteries to enable the user to have 24-hour electrical power. Nothing was said about the economics of such an arrangement, but the cost of storage batteries would only make solar engines less economic than had already been concluded by Ericsson. The proposed system apparently was never built.

A lull in activity followed the advances made by Mouchot, Pifre, Ericsson, and Adams, and it was almost the 20th Century before a new surge of activity began. Ambitious ideas were forming, however, as indicated by a patent issued in 1896 to C. G. O. Barr for a very large solar engine. Barr's engine used a semiparabolic mirror array mounted on railroad cars on a circular track, with a fixed boiler at the focus of the system. The engine was a precursor of some ambitious Soviet designs in the 20th

Century, and also of the United States' "Power Tower" system. The Barr engine was not built, and the Soviet system was abandoned.

One notable exception to the usual mirror system plus steam boiler for translating sunshine into mechanical work was proposed by E. Weston in 1888. He proposed using something rather new, a device called a thermocouple. The solar energy was focused on the thermocouple, which developed a voltage between the hot junction and a cold junction; each junction consisted of a connected pair of wires of different materials, such as iron and nickel. Weston apparently stimulated some other ideas of this type, since thermoelectric solar converters enjoyed considerable activity about 10 years later, again evidenced primarily in the patent office. In 1897 H. C. Reagan, Jr., proposed a solar thermojunction device in which he achieved a cool junction (where thermodynamic heat must be dissipated according to the fundamental laws of thermodynamics) by blowing air on it from a windmill. In 1898 H. F. Cottle obtained a patent on a unique hybrid system in which he would store solar heat by focusing sunlight on a pile of stones and then extracting energy from the stones by an undefined "thermoelectric generator." In 1899 R. H. Dunn obtained a British patent for a "thermopile" solar converter, but the patent description is enough to raise doubts about the reality of the proposed scheme. Nevertheless, all of these inventors were on the track of an effect—thermoelectricity—that in later years was to be reduced to practice and would yield high conversion efficiencies, but at such high system costs that it would become useful only for esoteric applications, as in space vehicles.

Although engines were finding few customers in the 1880s and 1890s, there are some instances of other uses of solar energy worth noting. A solar cooker was described by W. Adams of Bombay in an 1878 article in *Scientific American.* Adams used a conical octagon box lined with silvered glass mirrors that focused light through a cylindrical bell jar into the food container. His device really did work well. His other suggestion for using solar heat "for the cremation of deceased Hindus and others" did not.

Water distillation was another topic of some interest at this time, and a system was designed by J. Harding and built by Charles Wilson at Las Salinas, Chile, in 1874. The still covered 4700 m^2 and produced up to 23,000 l of fresh water per day, in clear sun and at an altitude of 1300 m. This still was operated for 40 years and was abandoned only after a fresh-water pipeline was laid from the Andes Mountains down to Antofagasta. It was said that the cost of water produced by this still, including capital and maintenance costs, was $.001 per liter.

1.4 THE TWENTIETH CENTURY

Significant developments were occurring at the dawn of the 20th Century. Already in 1893 and continuing up to 1909, the patents issued to M. L. Severy gave evidence of new ideas for combining solar steam-powered devices and storage batteries for possible power systems. Because of the efficiency problems that Mouchot and Ericsson had encountered, experimenters were using more sophisticated engines. Some used below-atmospheric pressure engines to help vaporize the working fluid. This new concept was introduced in patents issued to E. H. McHenry in 1900 and 1911 and to E. C. Ketchum in 1905 for engines that used two fluids: water to collect the heat, and a working fluid of lower boiling point than water to drive the engine.

In 1901 several notable patents were issued to A. G. Eneas. He built some very large engines that attracted wide attention. (They were, in fact, not quite as large as the largest made by W. Adams, but Southern California makes a better publicity format than Bombay, India.) In that same year there were also two patents issued to M. M. Baker for solar engines, and one to K.C. Wideen for a solar furnace. In 1905 and 1906, E. P. Brown and Carl Gunther obtained patents on solar steam boilers, as did W. Maier and A. Remshardt in 1907.

H. E. Willsie and John Boyle, Jr., pursued a different approach to solar engines from 1902 to 1918. Instead of using mirror reflectors to concentrate the sunlight, they used what we now call flat-plate collectors. Sunlight entered the system through a glass window and was absorbed by a thin layer of water flowing over a dark bottom. The hot water was then used to vaporize a volatile liquid such as ammonia, ether, or sulfur dioxide. Willsie claimed to be the first to propose a two-fluid system, but McHenry had already patented a two-fluid system in 1900. The first Willsie engine was built at Olney, Illinois, out of rather unsophisticated materials. The absorber was a shallow wooden tank covered with two layers of window glass. The tank was insulated with hay and the bottom lined with black tar paper. With this crude solar collector, temperatures were obtained that even in cold, raw October weather were high enough to vaporize sulfur dioxide for the engine.

The second Willsie engine, built in Arizona near Hardyville, used sand rather than hay as an insulator. The tests were sufficiently encouraging that Willsie decided to form the Willsie Sun Power Company. In 1904 this company built an ammonia-powered system at St. Louis, Missouri, incorporating a 5-kW engine. It is interesting that Willsie prudently provided for the water to be heated on cloudy days by burning fuel. His final solar engines were built by Boyle and Edward Wyman in 1905 at Needles, California, on the hot desert bordering on the Colorado River. This 15-kW system operated an irrigation pump that also supplied cool water for the condenser of the engine. The sulfur dioxide-powered engine developed about 11 kW from a total collecting area of about 186 m^2. Although the Willsie engines were a commercial failure, they were a technical success since they demonstrated that the flat-plate collectors were not as sensitive to cloudiness as the mirror systems used previously.

A large solar furnace was constructed in 1904 by a Portuguese priest, Father Himilaya, for exhibition at the St. Louis World's Fair. This furnace appears quite modern in structure, being a large, off-axis, parabolic horn collector, standing 12.8 m high (see Fig. 1.6). The mirror is covered with many small flat mirrors, each 5 by 10 cm, and the system yields a flux approximately 2000 times solar flux in the furnace at the lower left of the structure. Note how large the structure must be to track the sun

Fig. 1.6 The solar furnace built by Himilaya for smelting metals, exhibited in 1904.

in right ascension and declination; in the face of strict economic limits and the severities of seasonal storms, overcoming the problem of size is a major challenge even today.

Frank Shuman began work in 1906 on a solar engine concept that, like Willsie's, used flat-plate collectors. By 1907 he had completed his first engine. It developed 2.6 kW from a collector of 110 m^2 that heated water, which in turn vaporized ether. In his 1911 system, built at Tacony (near Philadelphia), Shuman added a significant new item to his planar collectors—a flat mirror along the north and south edges that reflected additional sunlight onto the absorber, doubling the energy output. This system, shown in Fig. 1.7, had a total collecting area of 960 m^2 and yielded 32 kW with steam as the working fluid. This unit was sufficiently promising that the Sun Power Company (Eastern Hemisphere, Ltd.) was formed to market the devices. The next project was scheduled to be in Florida, but this did not develop. A more exciting task appeared on the horizon for Shuman and his new company.

Fig. 1.7 General view from the west of the flat-plate collectors with booster mirrors constructed by Shuman at Tacony (Philadelphia) in 1911.

In 1912 Shuman, in collaboration with C. V. Boys, undertook to build the world's largest solar pumping plant in Egypt. The original boiler exploded but was replaced in 1913 by a stronger one. The system at Meadi was placed in operation in 1913 using the engine removed from the Tacony system. The Meadi system departed considerably from the simple Tacony plant, using long parabolic cylinders to focus sunlight on a long absorbing tube, as shown in Figs. 1.8 and 1.9. Each cylinder was 62 m long, and the total area of the several banks of cylinders was 1200 m^2. The cylindrical reflectors were mounted in circular hoops driven in unison to keep the sun focused on the absorbing tube. The Shuman-Boys solar engine developed as much as 37 to 45 kW continuously for a 5-hour period. The project, however, was abandoned in 1915 as a dual casualty of World War I and the competition of cheaper sources of power that became available about that time.

Shuman and Boys obtained a patent on their solar engine in 1917, as did R. A. Fessenden for a vapor engine to operate at lower than atmospheric pressure. New engine ideas were popular during this period: three patents were issued to T. F. Nichols from 1912 to 1915, one to C. E. Burnap in 1914 for an engine operating with ammonia as a working fluid, and one to M. Koller in France in 1913.

Attention also was once more directed to the more sophisticated systems—thermoelectric devices. In Russia in 1908 W. Zerassky built a solar thermoelectric device in which the thermoelectric junction was formed by wires of a zinc-antimony alloy and silver-plated wires. The hot junction was placed in a glass-covered box as the solar absorber. W. Coblentz in 1913 used wires of copper and constantin. In a 1910 article in *Scientific American,* Frank C. Perkins described work in France on a thermoelectric device for converting sunlight into electrical power. A popular article by A. Rordame in 1911 described an unusual thermochemical system for which he quoted an output of 5 W for an input of 8.7 W of sunlight—a fabulous 57% efficiency. Needless to say, there was an error somewhere since even the newest binary mixes of transition element metals yield efficiencies of only 10% to 15%.

The 1920s were rather quiet on the solar energy front, perhaps because everyone had his attention drawn to other aspects of a prosperous and expansive decade. A solar engine idea was patented by W. J. Harvey in 1921, by L. H. Shipman in 1928, and four by Robert H. Goddard, of rocket fame, from 1929 to 1934, the first being issued a week after the stock market crash in 1929. None of these persons built a working engine, however. Cesare Romagnoli in Italy did develop a working engine in 1923, using hot water at 55°C (130°F) to heat a second liquid of lower boiling point, ethyl chloride.

However, the 1920s also saw the development of the interest of C. G. Abbot in solar energy. He had first studied the sun in 1905 and from time to time had personal contact with the experimenters, including Eneas and Willsie. These contacts later included Goddard when the latter was developing solid propellant rockets for the U.S. Army Signal Corps in 1917. Abbot was a frequent visitor to the laboratories of the Mount Wilson Observatory in Pasadena, California, and it is highly probable that the two compared ideas on the topic during these meetings.

In his early years Abbot published little on solar energy, but a long list begins in 1926 and extends until 1973, when he was 100 years old. We have had personal contact with two of his projects. First was the cooker placed in a shed just below the astronomer's dormitory, called the "Monastery," on the steep south face of Mount Wilson. The cooker was still in working order in 1941, but it had long since been displaced by bottled gas as a means of helping to feed the astronomers; nevertheless, it was ready to work when Abbot would visit Pasadena and spend some time on the mountain. The other device was the model of his flash boiler, which he developed in the mid-thirties and built in 1947 at the University of Arizona.

The 1930s saw a remarkable increase in interest in solar energy, but along rather different lines of application. Being a time of global economic depression, the 30s may have seen many people with time to think and dream—and with time to seek a way to get "free" energy. There was remarkable activity in Japan, for example, where 39 patents were issued for solar hot-water heaters. The climate of Japan is far from ideal for solar energy, but energy sources in the volcanic island chain were scarce and the people loved their hot evening bath. The result was a rash of roof-type hot-water heaters that would have 100–200 l of very hot water ready for the evening ablutions. It seems hard to imagine 39 *different* ideas for a solar hot-water heater; nevertheless, 20 more appeared in the decade of the 40s. Japan was one country, moreover, that believed in its solar hot-water heaters, since even in 1960 there were an estimated 250,000 units in operation.

Fig. 1.8 A view from the south of the Shuman-Boys solar collector system constructed at Meadi, Egypt, in 1913.

Fig. 1.9 Detail from the north of the long parabolic-cylinder mirror collectors used in the Shuman-Boys system. Note the hoops on which the mirror was tilted to follow the sun.

A second notable line of interest—house heating—first appeared in the mid 1930s. It became a topic of intense interest in the last half of the 40s. Perhaps this new line of thought was the natural outcome of the intensive interest in hot-water heating, since a good supply of hot water could be fed into a hot-water radiator system of the type already fueled by coal in millions of homes. The pioneering experiments at the Zurich Institute of Technology were published by M. Hottinger in 1935. The M.I.T. experiments by H. C. Hottel and his colleagues soon followed and were described in various national magazines in 1939. Unfortunately the glowing predictions of solar-heated houses did not materialize. In 1955 it was predicted that several million homes would be heated by solar energy by 1970. The actual count in 1970 may, in fact, be lower than it was in 1955. The hard facts of economics defeated the dream of solar-heated homes.

In the area of power experiments, H. Delecourt at d'Oisquerq, France, developed and constructed a solar engine in 1930 that used ethyl chloride. Owing to a series of accidents and breakage, the device never reached the working stage, and it was finally abandoned. In 1932 J. Boisse de Black described a hospital at Colomb-Bechar that heated its water by solar energy. He went on to describe in a very serious manner the potential of solar power in the Sahara Desert—even to the extent of presenting a proposal, with cost estimates, for operating the trains of the ill-fated Trans-Saharan Railroad with solar power. Neither the railroad nor the solar-powered engines were built—so the world missed the unusual sight of a locomotive with a large solar mirror collector teetering on top!

Patents for solar engines were issued in 1930 to C. H. Drane for an ammonia engine; in 1931 to G. W. Dooley; in 1932 to W. L. R. Emmet, whose solar collector, mirrors and all, was inside a vacuum envelope; and in 1935 to F. A. Gill. Solar cells also got a lot of attention when L. Bergmann, in 1932 and 1936, and G. Bail, in 1938, studied the effects under sunlight of semiconductors using transparent gold coatings for electrodes. In 1933 R. Audubert studied various copper oxide cells and cadmium sulfide, while C. G. Fink and M. E. Fogel in 1934 did additional work with copper oxide cells. A new type of semiconductor, lead sulfide, was investigated by Fr. Fischer and B. Godden in 1937 and 1938 and another, thallium sulfide, by C. F. Nix and A. W. Treptow in 1939.

The world entered a period of turmoil with the start of World War II, and the study of such mundane things as the utilization of solar energy was laid aside until the second half of the decade, when it was resumed with renewed hope and vigor as the vision of the endless frontier of science emerged in 1945.

1.5 SOLAR MOTORS IN WESTERN AMERICA

No history of solar power would be complete without the story of the Solar Motor Company of Boston. Its creation at the beginning of the 20th Century represented a determined effort to produce a commercially successful solar engine to meet the unfulfilled need for power in the newly developing California-Arizona desert regions.

In the second half of the 19th Century, more and more Americans were discovering that the Far West had much more to offer than gold. The coastal regions had a moderate climate, pleasant to live in, but curious compared with that of the Midwest and East Coast. Absolutely no rainfall could be counted on for half the year. It was early discovered that the land was very fertile and that as soon as water was added the crops were superb, benefiting from the unbroken sequence of sunny days. The development of irrigation, therefore, was given high priority, and the available surface water supplies were soon fully utilized.

The Imperial Valley was one area where the demand for water was at a premium. This valley was originally formed when the upper portion of the Gulf of California was sealed off by a deposit of silt at the mouth of the Colorado River and by a gentle uplift along the San Andreas fault zone. Millennia later, bare of water, the ancient sea bed was a rich basin of salt surrounded by a vast shield of excellent soil. The soil was fertile and the climate salubrious, but water wells drilled into the valley floor yielded only hot and brackish water that was a handicap to crops. (Today they are geothermal wells!) However, the valley lies some 60 m below sea level, so the problem could be solved by a gravity-flow canal from the Colorado River.

The early efforts at irrigation were not without incident. Some of the early impoundment dams were inadequately designed and built; they failed, resulting in loss of life and destruction of settlements below them. The All-American Canal has a special story of its own, as it successfully challenged the "impassable" sand dune zone west of the Colorado River. Even vehicular roads did not surmount the ceaseless challenge of the shifting dunes until several decades later.

When solar energy entered the picture, the salt beds in the Imperial Valley were still being harvested for salt. A 1901 issue of *Scientific American* described an accident in a rare summer rainstorm when the canal waters burst their banks and briefly flooded the salt works. Four years later, a major break in the same canal, again caused by a summer flash flood, was to cause the permanent submersion of the salt flats and create the Salton Sea. The level of this sea has not stopped rising since that flood in 1905–1907. A gentle flow of Colorado River water continues, now as irrigation runoff, surface and subsurface.

It was in regard to the Salton Sea that we first came into contact with the fascinating topic of solar energy. Since we have digressed from the Solar Motor Company this far, we would like to tell you a story before we return to our subject. We both were working during World War II in C. C. Lauritsen's rocket research group at Cal Tech, since almost all college studies had been suspended for the duration of the war. One of the test ranges for training U.S. Marine troops in rocket devices was situated at about sea level east of the Salton Sea, facing the absolutely barren and rugged Chocolate Mountains. We had noticed in our frequent drives along the Salton Sea that stumps of palm and citrus trees stood submerged in the water, clear evidence of the rising water level. Shortly thereafter, a distinguished friend of ours, William DeWitt Lacey of Pasadena, who was serving the war effort at Cal Tech, invited us to meet Mr. John Burnham, a man of great vision who owned much land in the Imperial Valley and in other parts of southern California, notably Santa Fe Springs. This brother of the famous architect of the Chicago skyline asked us if we could explore the use of solar energy to control the level of the Salton Sea and at the same time produce pure water as a byproduct. The value of land being submerged each year placed a premium on this objective.

Since we both had only recently studied thermodynamics and physical chemistry and since Marjorie's father, Dr. Edison Pettit, was a solar astronomer at Mt. Wilson Observatory and had a solar cooking device in his back yard in Pasadena, we were interested in researching the question. We soon learned a fundamental problem about the distillation of water. The "heat of vaporization" for water is very high, 540 calories per gram at the boiling point. Even though this heat is reemitted when the steam condenses to water, its temperature is exactly the same as that of the boiling water, so it is impossible under these conditions to cause the heat to flow from the condensing steam to the boiling water. We did devise a way to solve this problem—by compressing the water/steam before condensation so that it condensed at a temperature above that of the boiling water in the next stage. We noted that the work done in compressing the water/steam was considerably less than was regained from condensation. This same approach was later used in commercial sea-water distillation units as "pressure-staged distillation."

When we came to the question of economics, we soon learned one of the basic facts that has arrested all attempts to use solar energy: the cost of the equipment installation and maintenance is greater than the cost of doing the same task by cheap fossil fuels. We did, however, come to the novel conclusion that by far the most effective way to control the level of the Salton Sea was to assist nature in the way that she originally got rid of the ocean entrapped in this arm: evaporation into the air. One immediately lost the yield of distilled water by this route, however. We proposed simply to spray the briny water into the air through giant sprinkler systems set offshore in the shallow sea bottom. The energy to pump the water could be of either solar or fossil origin since only a small amount of energy was required compared with that needed for distillation of the same volume of sea water. The energy source to evaporate the water was now the dry heated air of the valley—solar energy collected in a different form. The dry hot desert air almost always blows a moderate breeze over the lake. We found that a long array of spray jets would evaporate water effectively and at the same time drop the air temperature by 10 to 20°C. The breeze would carry this cooled air, which would have a humidity of 50% to 60%, over the eastern shore of the sea, and we would thus manufacture a much more pleasant climate, still with the perpetually clear skies of the area, along the 80 km of barren shoreline.

But back to our story of the Solar Motor Company. In 1900 the Far West was short on energy supplies. Wood and coal were the principal energy sources, and the vehicles of motive power were the steam engine or the steam electric plant that powered the electric trolley system. Oil had not yet appeared in any sizable quantity, though it soon would be found in abundance to delay the application of solar power for almost a century. We can, therefore, begin to appreciate the hope attached to the future of solar power in the Far West that attracted Mr. Aubrey G. Eneas and his Boston backers, Frank J. Post, Charles Sakett, and C. Lansing Haskell, into the field of solar energy.

We read about these hopes in the 16 March 1901 issue of *Scientific American*: "For many years the attention of inventors has been directed to the question of utilizing the direct rays of the sun as a substitute for coal, wood, or other fuel; large burning glasses or reflectors being the general form of the various machines. Especially in France have these been seen." This last sentence reminds one of a similar attempt to spur national incentive in the 1960s by pointing to Soviet achievements in space.

In leading up to the Eneas engine, the article continues:

> It was for a long time difficult to build a concave mirror of very large size, but this was finally overcome by having the surface of the concave mirror covered with small pieces of glass, or mirrors, each of which is so placed that the light or reflection from each side is thrown upon the same spot, the sum total, or the amount of heat centralized, being equivalent to the amount reflected by each glass, multiplied by the number of mirrors. In Europe the early solar glasses were generally of two kinds: that is, the heat was concentrated in two ways–by reflection from polished concave mirrors and by refraction through a convex lens. The earliest use, centuries ago, of such a contrivance was theoretically to dazzle or blind an enemy, metal disks being employed; but nearly all such devices failed to be of any practical value and fell into the category of "curiosities."

It is interesting that solar energy was then, as today, classified as an exotic topic and that the devices that are constructed are still very much in the category of curiosities, except in one very important area: the use of silicon chips, mounted thousands to a panel, used to provide dependable electric power for spacecraft. Their efficiencies are quite respectable, early ones having about 8% conversion efficiency and the current ones 12% to 15%. We must note here that while these efficiencies of conversion of solar heat into power are less than the 35% to 40% for modern steam turbines powered by fossil fuel heat, they are far higher than the 1% to 3% more characteristic of the solar engines of the Eneas era.

Continuing from *Scientific American,* we see that the opportunities and goals remain attractive:

> In Western America within the past twenty years it has been found that there are regions where it is especially desirable to obtain a motor which can be run practically without fuel. Such a region is the Californian desert where vast mining interests have sprung up, and in arid sections where irrigation is necessary, and even in the richest portions of fertile California in connection with the question

of irrigation. On the desert the sun shines almost continuously and in Southern California the percentage of sunshine to cloud is remarkable. These conditions have called attention to the possibility of a practical sun motor, and it is interesting to note that in South Pasadena, California, such a machine has been set up and is successfully accomplishing the work for which it was made—an automatic engine running by the heat of the sun.

This machine was exhibited at the Edwin Cawston Ostrich Farm, and has attracted the attention of a vast number of people, especially as Southern California is now thronged with tourists. In appearance the motor resembles a huge disc of glass, and at some distance might well be taken for a windmill of some kind; but the disc is a reflector thirty-three feet six inches in diameter at the top, and thirteen feet on the bottom. The inner surface is made up of seventeen hundred and eighty eight (1788) small mirrors, all arranged so that they concentrate the sun upon the central or focal point.

In reality the mirror was a section of a cone, so the light, instead of coming to a point "focus," was focused into a line image up the axis of the cone; here Eneas placed the boiler unit.

Here, as shown in the accompanying illustrations [see Figs. 1.10 and 1.11], is suspended the boiler, which is thirteen feet six inches in length, and holds one hundred gallons of water, leaving eight cubic feet for steam. At the time of the writer's visit (Charles F. Holder) to the farm the motor was the subject of no little comment, and the attendant stated, confidentially, that some of the questions asked were remarkable. One man assumed that it had something to do with the incubation of the ostrich eggs; and many asked what made it go, being unable to understand or appreciate the idea . . . .

The amount of heat concentrated in the boiler by the seventeen hundred and odd mirrors cannot be realized, as nothing can be seen but a small cloud of escaping steam; but should a man climb upon the disc and cross it he would literally be burned to a crisp in a few seconds. Copper is melted in a short time here, and a pole of wood thrust into the magic circle flames up like a match. That the motor is a success is seen by the work it is doing—pumping water from a well, illustrating the possibilities of cheap irrigation, and lifting fourteen hundred gallons per minute—equal to one hundred and fifty-five miner's inches. Up to the present time the motor has produced results equal to about ten horse power, but fifteen is claimed for it.

The motor is the result of a number of experiments by a band of Boston capitalists. One of the first productions was a silver reflector, which cost many thousands of dollars, but was abandoned. The next was modeled after the Ericsson machine of 1884; but it was a failure. A third was erected at Longwood, proving also a failure. A fourth attempt was made, this time in Denver, which was

Fig. 1.10 The 33-ft axicon collector of Eneas at the Cawston Ostrich Farm in California in 1901. The boiler is suspended in the center, and the drive mechanism is at the lower left.

Fig. 1.11 Side view of the Eneas solar collector, showing the tracks for seasonal adjustment of the pointing of the mirror and the weight used to drive the mechanism to follow the daily motion of the sun.

fairly successful, doing one-half the work of the Pasadena model. Finally the latter was produced and found to be a success. A duplicate, perhaps improved, will be erected at the Pan-American Exhibition. Dwellers in the East, where rain falls every few days throughout the year, cannot realize what such a perfected motor means to the West, where arid lands await but the flow of water to blossom as the rose. In such regions—and they represent millions of square miles—fuel is usually very scarce, often being so important a factor that the question of it determines the success or failure of the work. This is essentially true of the Californian desert and vast regions in Colorado, Utah, and surrounding States and Territories. Mines and pumping plants are often far from railroads, and in sections where no fuel is in sight, wood and coal being hauled from long distances. In such locations the solar motor is a boon. The skies are comparatively free of clouds, and the machine can begin work an hour after sunrise, possibly earlier, and continue until half an hour before sunset. It is possible that with cheaper methods of storing electricity sufficient power may be stored during the day to run the engine at night, or during the absence of the sun. Inventors are already experimenting upon methods of increasing the effectiveness of the motors, and probably larger ones, and groups of them, will be seen in the near future.

No invention of modern times has given such an impetus to the development of arid lands as the solar motor, and it has been visited by many interested in the question. The development of Lower California has been seriously impeded by the lack of fuel; the country being dry and barren in localities where rich mines are known to exist. The country is cloudless for months—in every sense the land for the solar motor, as water underlies the surface almost everywhere, and when pumped up and sent out upon the soil the region, which was formerly a desert, can be made fertile and literally to blossom as the rose.

Aubrey G. Eneas obtained two patents on his solar engine on 26 March 1901. Things began to look up for the Solar Motor Company of Boston. The company opened an office in Tempe, Arizona, near Phoenix, with J. Murdo Bruns and Clifford S. Estes as its resident representatives.

The first Arizona solar pump was built for Dr. A. J. Chandler to pump water on his ranch a few miles south of Mesa, Arizona, where the city of Chandler is now located. The installation was made in the summer of 1903 by Major W. H. Jacques, formerly an engineer with the Bethlehem Steel Company, but the system did not get much of a chance to demonstrate performance. The motor was started, but after a few days a part that was supporting the boiler at the focus of the collector broke. The part was heavy, and when it fell it damaged a considerable portion of the reflector and cracked the boiler tube. The pump was never rebuilt, and the remains disappeared.

The second (and last!) solar pump was built in 1904 and was originally located adjacent to the railroad tracks of the Santa Fe, Prescott and Phoenix Railroad near

Tempe. It was situated, according to the *Arizona Republican,* "at the wells near the race track, one mile south and one quarter mile west of town." The stated purpose was for testing the design and making minor changes to correct deficiencies, and to publicize the capabilities of the pump prior to offering it for sale to the public.

The Tempe sun motor had a rather dismal inauguration on 21 March 1904, a date no doubt considered auspicious, being the spring equinox. The weather was cloudy that day—and for the next week. We remind the reader that direct sunlight is an absolute necessity for the concentrating type of solar collector.

David Griffiths of the Department of Agriculture came to Tempe from Washington, D.C., to see the engine in operation. He apparently was quite pleased since the Phoenix *Republican* on 28 June 1904 quoted him thus: "The principle has now been established; all that is now necessary to make it a machine of practical use is to overcome a few mechanical imperfections which will undoubtedly be taken care of very soon."

The pump worked well until 24 August 1904, when it was sold to John May, a rancher from southern Arizona. The device was dismantled and moved to Willcox, Arizona, in September 1904. It was sold a second time and moved to the McCall Ranch near Cochise, not far from Willcox, where it met its end in a sudden wind storm. The designers claimed that the open-bottom design would enable the mirror collector to survive a 160-km/h wind. We can state from personal experience that a 160-km/h steady Boston wind is no match for the energetic dust devils that one sees on a calm hot spring day near Cochise, Arizona. Dust devils may be small, only a few yards in diameter at their base, but the rapid whirling wind and suspended dirt can reduce well-built structures to their basic ingredients in short order. We suspect that this may have been the fate of the last of the Eneas machines built for use in the United States.

The Tempe unit had an upper diameter of 11 m and a lower diameter of 5.5 m, slightly larger than the 1901 California model and the 1903 model built for Chandler. The weight of the moving parts was 3800 kg. The sale price was $2500, or $3000 if installed by the manufacturer. The mirrors were of white (clear, not greenish) glass, silvered on the back, made by Chance Brothers of London. The device was operated by winding up a large weight attached to an escapement on the high tower. In the morning the attendant would point the mirror east, toward the sun, wind the weight, and start the escapement functioning, whereupon the mirror would point to the sun without further attention all day. Every few days the attendant would shift the top pivot position to correct for the slow seasonal change of declination of the sun.

Sales of the solar motor were rather disappointing to the Boston enthusiasts—but the performances were also a bit short on fulfillment. One was sold to the Egyptian government for installation in Khartoum, and orders were received for two more, one from Bloemfontein and one for Johannesburg, South Africa. None was sold in the United States. In the end the return was small, and the total expenditures by the company finally totaled $125,000. In spite of this major effort by Eneas and his

Boston backers, almost ten years of effort resulted in nothing but some interesting pictures in the newspapers and journals of the day, now gathering dust in the archives of libraries.

Even though the Eneas solar engines had already gone from the scene, solar energy still caused activity in the Southwest. We find ads in the newspapers of 1907 for the Solar Furnace and Power Company of Phoenix, Arizona. They developed a solar furnace with a hot-air engine, but the purchaser had the option of a steam engine also. Since we find no record of actual working installations, it seems probable that the main product of the Solar Furnace and Power Company was the sale of stock. Some of the ads for stock are quite effusive. Here is an excerpt:

> Any amount of water can be pumped from either deep or shallow wells: no fuel is required, and when a plant is once installed the expense is ended. [We wonder whether they also sold perpetual motion machines as a sideline item!] Stop and try to realize what it means to be able without fire or fuel to produce heat so intense that it is hundreds of times greater than has ever before been developed by the greatest scientific men in their most powerful furnaces [they do not say that their furnace will do this either] . . . to smelt ores and minerals right at the mines without expense of shipping often many hundreds of miles to smelter, to pump water to irrigate millions of acres of land that is now worthless, but with proper irrigation would make homes for thousands of people, to run dynamos and generate electricity to run street cars, light towns and cities, heat and light houses and cook any and all kinds of food.
>
> HOW PROFITS CAN BE MADE
>
> Machines can be sold for pumping and irrigation plants, various kinds of power plants, household use, heating, lighting, mining and hundreds of other purposes too numerous to mention. About a million shares left in the treasury [shares were considerately being offered at $2 each, but they were sure to go to $10 to $15 in a year!] and every share of this stock will in all probability sell readily for more than $2 per share. With part of this money we can buy thousands of acres of land that is now practically worthless at 25¢ to $5 per acre, and by installing Solar Pumping plants can sell it $10 to $100 per acre . . . .

In view of this pitch it is no wonder that solar energy got the reputation it did and that the public quickly put solar energy on that mental shelf reserved for all get-rich-quick schemes.

If we now return to the real world of technical experiments in solar power, can we see a message in the failures of these noble efforts? The lack of cheap energy sources made it attractive to exploit the sun in the Southwest. Electrical energy was available in the cities to a limited extent, but electrical distribution lines were not available in the areas of low population density. It seems hard for us now to realize that even in the early 1930s rural electrification was only a dream. A reason for failure was the discovery and development of oil fields in California and Texas. This new energy source was transportable to rural areas and became cheap because of its abundance. The energy content of a gallon of petroleum fuel means that internal combustion engines could deliver the power to pumps with a very compact device that was easy to operate and maintain (although gasoline engines in the 1920s and 1930s did try the patience on frequent occasions).

Another factor, and one that is very important, is that solar power from the Eneas engines could be obtained only during the daytime on clear days. Power, therefore, was not available at the click of a switch or the turn of a starter crank. The limitation clearly was real to the Boston group, since its members promised that this problem would disappear with the cheap and long-lived storage batteries "just around the corner." Well, 70 year later we still have not turned that corner. Batteries are still expensive and short-lived, and are useful only for limited applications.

The reader may wonder at this point why we persevere in our dream of solar energy as the great hope for power for mankind. Solar engines seem to be dismal failures when matched up to our hopes. Is there still reason to hope for a solution? The promise of ultimate success is, we think, given by the story of the next person on the scene of solar energy, Robert H. Goddard.

Goddard's Vision

On 17 July 1929, near Worcester, Massachusetts, a strange spindly contraption was hauled out of a barn. It consisted of two elongated metal cylinders connected by piping and had a combustion chamber at the topmost part. The men struggling with the thing got it placed vertically between three steel posts. A few minutes later there was a loud roar, a cloud of smoke and dust, and the spindly contraption slowly moved skyward—all the way up to 300 feet, where an explosion terminated the flight. Robert H. Goddard had launched the first liquid-fuel rocket to go above 50 feet. (The first liquid-fuel rocket in history had also been launched by him on 16 March 1926, three years earlier, but with no success.)

In spite of the minuscule altitude reached, one observer, Charles A. Lindbergh, grasped the enormous significance of Goddard's vision and convinced the Guggenheim family to back Goddard with $100,000. Could they really have had any idea that on a July day 40 short years and 3 days later two men would step from a spindly contraption onto the Sea of Tranquillity 360,000 km from the planet Earth?

To return from this digression, what does Professor Goddard have to do with solar energy? The answer is—a great deal. The story goes back to 1917 in Pasadena, California, and to C. G. Abbot and his colleague G. E. Hale, astronomer and founder of observatories. Both were then working on war research at Hale's latest observatory, the Mount Wilson Observatory. Goddard was there making experimental solid-fuel rockets for the Army Signal Corps and organizing his ultimate dream, described in a 1919 publication entitled "A Method of Reaching Extreme Altitudes." When he talked of "extreme altitudes," Goddard had the planets Mars and Venus as well as the moon in mind.

While at the Mount Wilson Laboratories, Goddard and Abbot discussed another challenge to mankind—solar energy. Goddard reacted with enthusiasm and vision, and the first of five patents on solar energy was issued to him on 10 July 1924. Goddard's contribution through this and four subsequent inventions was not in basic new ideas but in engineering innovations designed to render earlier ideas practical. It is apparent that he shared Abbot's views, expressed earlier: "In time, manufacturing will to a great extent follow the sun. The deserts may yet become great industrial areas."

The devices designed by Goddard were supposed to increase the energy utilization from the usual 3% or 4% up to, he claimed, 50%. His approach seemed sound. He would increase the efficiency by increasing the temperature of the working fluid, the Carnot efficiency. We even note that Goddard's patent drawings show nozzles, as on his rockets, for increasing the velocity of the vapors emitted from his "solar accumulators." In 1929, the Sunday *New York Times* carried an article that attracted much attention to Goddard's publication of an article in the October 1929 *Popular Science Monthly* entitled "A New Invention to Harness the Sun." It was the last word from Goddard on solar energy.

From the solar energy point of view it is unfortunate that Lindbergh convinced the Guggenheims to support Goddard's rocket experiments, because Goddard's attention was soon fully occupied with his new rocket test range near Roswell, New Mexico. For space exploration it was an epochal decision, but it leaves the challenge of solar energy unfulfilled as of today.

1.6 POST-WORLD WAR II (1945–1965)

Solar energy came into considerable prominence immediately after World War II, when scientists looked around for new places to apply the talents they had been devoting to wartime programs. One major problem was that of the new nations emerging from the wreckage of the colonies. Most of these nations were energy underprivileged and capital poor. Perhaps solar energy could be useful to them, even if not to the advanced nations. This goal presented a major challenge and was eagerly pursued by many researchers. The advanced nations had not yet come to realize how important solar energy would be in their own future, so the results fell far short of the goal. The disappointment that appeared to settle on the field in the late 1960s should help prepare us to avoid new disappointments now that solar energy has suddenly become a major new energy option of the 1970s.

While the resurgence of interest in the potential of solar energy had focused initially on the new nations, it was soon extended to the advanced nations as well. A survey made in 1952 by the United States government, based upon the enthusiasm evidenced in the magazines of the day, predicted that by 1975, 13,000,000 homes in the United States would be heated and cooled by solar energy. By 1958 there were perhaps two dozen houses heated by solar energy and, in 1970, less than half that number; almost all the rest had been converted to more "practical" heating systems. (One lone figure still living in a home with solar-augmented heating was that pioneer of modern solar energy applications, Dr. George Lof of Fort Collins, Colorado.) Why the prediction missed so badly is a natural question. The answer is twofold: (1) there was no urgent need for solar energy since other cheap fuels seemed abundant, and (2) solar heat was not economically competitive.

In the area of hot-water heating there was more success. In 1960 it was estimated from actual sales records that 25,000 home units were in use in Florida and California. Some 250,000 were then in use in Japan—simple and cheap units to be sure, but capable of providing a hot evening bath in a country short on fuels.

Evidence of the resurgence of hope for the utilization of solar energy was the onset of special conferences dealing with this subject. Let us look at several:

1950—H. C. Hottel organized a symposium on "Space Heating with Solar Energy," which generated much interest in house heating.

1951—The astronomer Harlow Shapley organized a conference on "The Sun in the Service of Man" with emphasis on biological utilization.

1952—A conference was held at Ohio State University on "The Trapping of Solar Energy."

1953—Farrington Daniels organized a symposium on the "Utilization of Solar Energy" sponsored by the National Science Foundation, at that time a new enterprise to foster the development of science in the United States. The conference was broad in scope, and the purpose was to assess the present state of technology relating to solar energy uses and to look toward future uses. This conference was the first of many to cover the entire scope of solar energy applications and did much to cause the great crescendo of activity of the last half of the decade.

Worldwide interest and activity in solar energy was now sufficiently apparent for a World Symposium on Applied Solar Energy to be held in 1955 in Tucson and Phoenix, Arizona. Some 900 persons registered at the symposium, representing industry, government, education, and finance. In addition there were scientists and delegates from 36 countries. A total of 80 solar devices were exhibited, ranging from solar cookers to steam engines, and quite a few electronic devices using the new solar cells.

A look at the categories of interests listed by the Association for Applied Solar Energy in 1955 shows the scope of work going on at that time.

Biological conversion
Concentrating collectors
Cookers
Distillation
Engines and pumps
Flat-plate collectors
Furnaces
Heat pumps
Heat storage systems
House cooling
House heating
Photoelectric
Salt production
Steam generators
Water heaters

The number of countries reporting research in solar energy applications is also impressive:

Algeria
Argentina
Australia
Belgian Congo
Belgium
Brazil
Canada
Cuba
Cyprus
Egypt
England
France
French West Africa
Germany
Holland
India
Israel
Italy
Japan
Kenya
Lebanon
Morocco
Netherlands
New Zealand
South Africa
Switzerland
USSR
USA

The scope of the renewal of interest in solar energy can be seen by looking at the number of publications in each decade. This census of articles excludes "natural" uses such as evaporation for salt, but covers all others, including patents:

1870–1879: 6
1880–1889: 22
1890–1899: 20
1900–1909: 33
1910–1919: 26
1920–1929: 21
1930–1939: 101
1940–1949: 126
1950–1959: 878
1960–1969: 2144

This list clearly shows the surge of interest in the 1900s and 1910s and the resurgence in the 1930s and 1940s, but these are completely eclipsed by the record of the 1950s and 1960s. In these latter years science was rampant and was expected to solve all of the world's problems. In the early 1970s we realized painfully that solutions to problems do not always come automatically upon the application of Big Science. It was at this time that interest in publications waned to a point where there was no longer enough support to maintain either the headquarters of the International Solar Energy Society in Tempe, Arizona, or publication of the journal *Solar Energy*. Headquarters were transferred to Australia, where the flame of interest was kept alive. But the journal actually ceased to publish for long periods in the early 1970s, until the energy crisis descended with abruptness in 1973.

To return to our narration of the history of solar energy, two events loom large in the decade of the 1950s. The first was the announcement in 1954 by the Bell Telephone Laboratories of a solar battery. The second was the impact of the orbiting of the first man-made satellite by the USSR in 1958.

The new photovoltaic cell announced by Bell Telephone scientists D. M. Chapin, C. S. Fuller, and G. L. Pearson created quite a stir. The physical effect was not really new, since Becquerel had discovered the photovoltaic effect in selenium in 1839. The excitement was over the high conversion efficiency of the new silicon cells. The scientists reported 6% efficiency from their first cells and 11% within the next 12 months. Theoretical work indicated that 22% might be achieved.

The predictions concerning the future of solar cells ran the gamut of the imagination. Such things as a house roof paved with solar cells seemed the answer to the problem of converting sunlight into electrical power for domestic uses. There were other problems, to be sure: at that time the cost of covering the roof of a modest dwelling was $2 million!

Solar cells have found practical and important uses in space, where cost is no barrier and no other source of power is available. Three years later, in 1957, the first silicon cells were sent aloft on rockets to demonstrate that they could survive the journey and produce useful power. In 1959 the first successful Vanguard satellite carried 108 chips to power its radio. The growth in the number of silicon chips in space application is shown by the following:

1958	108 chips (0.5 W total)
1962	154,000 chips (5.4 kW total)
1969	3,000,000 chips (105 kW total)

The cost has been reduced drastically in view of the volume required, from an early cost of over $1000/W to about $100/W ($0.35 per chip) in 1970. For comparison, a modern steam turbine electrical generator costs about $100–150/kW, about a thousand times less than solar cells. Even though the steam generator uses fuel during its 20-year lifetime, the target cost before solar cells became attractive for "bulk" power production was in the vicinity of $0.05 per chip; this goal still seems beyond hope of achievement even though much effort has been spent in the past 15 years attempting this breakthrough.

A few years after the Bell Telephone discovery, word came that the USSR had developed a new type of solar cell that did not deteriorate in space under ionic particle bombardment. They had discovered that if the bulk of the silicon was originally doped *p* and then phosphorus atoms instead of boron atoms were diffused into the surface, the surface became *n*. In this case the *n* layer is made thick enough to absorb the sunlight photon and deliver the ejected photon to the *p* substrate. Still, basic costs are so high that only space applications and a very few others can use the silicon cells. The great challenge today is to find a way to greatly reduce the cost of solar cells. In this book we devote little attention to the cells themselves, as they are outside our range of competence. We do, however, discuss solar collectors applicable for use as concentrators for solar cells.

Research in the 1960s resulted in the discovery of other photovoltaic materials, compounds this time, such as gallium arsenide (GaAs). These compounds could operate at higher temperatures than silicon but were much more expensive. Cadmium sulfide (CdS) also was developed, but although it could be made in larger chips, its peak efficiency was low, less than 5%, and it had other problems. As a net result, photovoltaic cells have not emerged as the answer to common electrical power needs.

The rapid development of space capabilities by NASA and the U.S. Air Force during the 1960s made it possible for industry, government, and university research teams to study the entire spectrum of power-generating effects. Since cost factors were not a significant barrier, every known effect was studied. The most important are as follows:

Chemical batteries
Fuel cells
Solar cells (photovoltaic)
Solar engines/turbines
Nuclear and solar thermoelectric cells
Magnetohydrodynamic converters
Thermionic converters
Photoelectric emissive converters

Fuel cells were given publicity through their use in powering the Apollo mission spacecraft to the moon. They are much more efficient in the use of their fuels, hydrogen and oxygen, than would be the case if the fuels were simply burned and the heat used by an engine or turbine to drive the electrical generator. They are more nearly electrical batteries but use high-energy reactants rather than chemical ions as in a chemical battery.

A typical storage battery has an energy capacity in the vicinity of 22–44 Wh/kg of battery. A typical fuel cell, on the other hand, has a capacity on the order of 550 Wh/kg. A thermal energy storage cell, important in solar energy applications where sunlight is intermittent, such as lithium hydride (LiH), has a capacity of 5000 Wh/kg. The latter sounds excellent until we remember that this is heat energy rather than electrical energy, so the 2300 Wh must be multiplied by the efficiency of the "engine" that transforms it into electrical energy. A 10% conversion cycle would make an LiH cell comparable to a fuel cell, but a 1% to 2% efficiency is more likely, in which case it is comparable to a chemical battery.

Nuclear power sources, widely known as SNAP generators, have been used for long-duration space missions. Despite the glamor of the name "nuclear," these generators simply use the crude heat of the disintegrating β-emitting radioisotope, for example polonium 210. Thermoelectric devices convert this heat into electricity.

Thermoelectricity was discovered in 1822 by a physicist named Seebeck, and even though he did not comprehend what he observed it has become known as the "Seebeck effect." A thermocouple consists of two dissimilar wires with intertwined ends. When one of these "couples" is placed in a hot surround and the other in a colder surround, an electrical current flows in the wires. Different metals have different values of the thermoelectric potential, so early research was directed at finding the combinations that gave the largest effects. Net efficiencies as high as 4% were reported by Abram Ioffe in 1956 in the USSR. The most recent SNAP generators yield 5.5%.

Some rather exotic materials have been used for thermocouples in place of the traditional iron-constantin, or platinum and platinum-iridium combinations. In binary compounds we now have lead telluride with zinc antimonide, with a theoretical 10% efficiency. Samarium sulfide is an exotic compound that will permit thermocouple operation to 1100°C. In ternary compounds the mixture of silver, antimony, and tellurium is useful, and there is even one quaternary compound of bismuth, tellurium, selenium, and antimony.

The yield of a thermocouple depends on the temperature difference that can be maintained between the hot and cold ends, so for high efficiencies the materials must be used near their melting points. For solar energy conversion we therefore need concentrating collectors.

Westinghouse Corporation constructed a solar thermoelectric generator for NASA that used a paraboloid 2.0 m in diameter focused on a 32-junction thermocouple and yielded 125 W at 4 V, an efficiency of 3%. They estimated that the potential cost of such a system could be as low as $0.07–0.10/kWh, about 20 times the cost of "raw" electrical power.

The General Atomic Division of General Dynamics Corporation constructed a very light-weight solar thermoelectric collector for NASA that yielded power at the rate of 1 kW for 9 kg.

In 1954 Maria Telkes made thermoelectric solar generators that used several types of metal compounds for the sensitive junctions. She reported that the most sensitive junctions were obtained using one electrode of zinc antimonide doped with other additives and a second electrode of bismuth antimonide in a 91:9 ratio. A "hot-box" collector with this thermoelement had an efficiency of 0.6%, while one with a concentrating collector gave 3.4% efficiency and an output of 0.15 W.

Work in the USSR reported by V. A. Baum included a solar thermoelectric generator consisting of 840 zinc-antimony and constantin junctions heated by a parabolic mirror 2.0 m in diameter. The unit produced 18.9 W at 21 V for an efficiency of 3.4%. At power outputs of 40 W, the junctions quickly deteriorated. In 1960 the researchers at the Krzhizhanovski Power Institute in Moscow made a 1.2-m^2 parabolic mirror and used several tellurium alloy thermocouple junctions that operated at 1.5% efficiency.

The main problem with a thermoelectric generator is in finding materials that combine high thermoelectric properties with durability suitable for a long operating lifetime. A one-year life is a reasonable goal for such materials for space applications, far from the 20- to 40-year life needed to make an economic system for terrestrial high-power applications.

The thermionic effect has also been used to convert heat, as from the sun or other heat source, into electrical energy. When a metal or other suitable element or compound is heated to incandescence in a vacuum, it emits electrons spontaneously. Ingeniously devised triodes can separate these electrons and produce a current between the hot cathode and cold anode. The addition of vapor, such as cesium, into the evacuated tube increases the efficiency of the process. A basic problem, however, is that very high temperatures are needed—1100° to 2600°C. Since the anode must be close to the cathode (only a fraction of a thousandth of an inch), it is hard to maintain a large temperature difference. Since the Carnot efficiency applies here as to all other heat conversion systems, the efficiency depends directly on the ability of a particular design to reject heat from the anode.

General Electric in 1957 produced a 1-W thermionic converter. By 1960, TRW Corporation had constructed a 250-W, 28-V system with a 15% maximum efficiency. In the same year General Electric produced a 500-W solar-heated generator for the Air Force, and in 1962 they had in operation a 105-diode unit heated by a mirror 4.9 m in diameter, at their Phoenix, Arizona, research station. In 1963 the Thermo-Electron Corporation of Waltham, Massachusetts, constructed a 5-diode, 3.5-lb solar-powered unit for NASA that was expected to have a one-year operating life. The lifetime of a thermionic converter is limited since the temperature must be so high that the cathode material, even of ultrarefractory metals like tungsten or tantalum, slowly evaporates and deposits on the window of the vacuum tube. Most recent cesium vapor thermionic converters of molybdenum, copper, and tantalum, when operated at 2200°C, can produce 20 W/cm^2 with an efficiency of 15%. The operating lifetime of such units can now reach two or three years at these power levels.

Several proposals have been made to replace the thermionic cathode with a photoemissive cathode. Photoemissive surfaces are widely used in photocells, photomultipliers, and image tube devices. They do not develop a current or a voltage by themselves, as does the silicon cell, for example. Other techniques must be used to obtain power *out* of the devices, so even though photoemissive surfaces are inexpensive to prepare, no practical devices have been developed owing to these other complicating factors.

Thus far we have talked about photosensitive or heat-sensitive elements and inorganic compounds. What about organic compounds? Many organic molecules are photosensitive, and some have been explored for solar energy conversion. We are familiar with one in everyday life—the molecule chlorophyll. A natural biological system using chlorophyll is quite inefficient even though, over the eons, it made possible the world's coal deposits. In a corn plant the net efficiency is 0.3% and in wheat it is 0.1%. Heavy cellulose-producing crops yield about 1090 kg/ha dry fuel per year. Experiments with algae culture by the A. D. Little Company promised 7000 kg/ha per year at a cost at the "factory" of about $0.55/kg—rather expensive for a fuel (in 1970).

There are modes of using photosensitive molecules other than in biological systems. It is conceivable that the equivalent of a silicon cell might be made at a small cost per square foot. Little progress has been made or appears possible, since the expected efficiencies are low and these molecules decompose readily in sunlight—a fatal flaw in a molecule that must function in sunlight for a long time to be economically attractive.

The Stirling cycle engine, invented in 1816 by the Reverend Robert Stirling, still plays a role in solar energy devices even for space applications. It is a hot-air external "combustion" or heat engine with a regenerator that prevents dissipation of heat between cycles. In 1958 the N. V. Phillips Laboratories at Eindhoven, Holland, made some experimental engines that were reported to be 40% efficient. Although efficiency was achieved, the relatively slow transfer of heat from a hot end of the gas chamber into the gas limits the dynamic performance of the engine. Also in 1958, the Allison Division of General Motors developed a Stirling engine.

In 1960 Farrington Daniels of the University of Wisconsin and T. Finkelstein of the Battelle Memorial Institute made an improvement in the solar-type Stirling engine by replacing the *heat-conducting* head of the cylinder by a *sunlight-transmitting* quartz window. The heat could therefore be directly transmitted into the piston surface. Since the heat still must get into the working gas of the engine, this modification was less effective than it first appeared. One needs an *opaque* working gas plus the quartz head to improve the Stirling cycle enough for practical application.

Thermo-Electron Corporation experimented with closed-cycle Stirling engines, concentrating not on solar energy but on burned fuel as the source of heat. Their engine used thiophene (Monsanto's CP-34) as a working vapor; however, complications are present since thiophene is a highly volatile and toxic liquid. At this moment the project is troubled by the four opponents of economic systems: cost, size, weight, and power-demand inconveniences.

Harry Tabor of the National Physical Laboratory in Israel developed a turbine engine to use solar energy in the 1960s. For a working gas he substituted a heavy hydrocarbon molecule, monochlorobenzene, operating in a closed cycle. His turbine operated at 150°C (302°F), at 18,000 rpm, and produced electricity at 50 cycles per second. He estimated the cost of the power generated to be $0.035/kWh. His collectors consisted of inflated plastic cylinders, clear on top to admit sunlight and aluminized on the bottom to focus the light on a heat-collecting tube. He estimated that 75% of the cost of his system was due to the plastic collector, even though the cost of his collectors represented a breakthrough in cost of focusing collectors.

The SOMOR Corporation of Lecco, Italy, produced on a commercial basis a small electrical generator driven by solar flat-plate absorbers and Shuman-type mirror boosters (see Fig. 1.12). Operation of this unit was very sensitive to heat losses, and the margin between operation and nonoperation was reported as very small. The efficiency of conversion was only a few percent at best, and the cost of electricity produced under favorable conditions was on the order of 100 mills/kWh. SOMOR solar-operated pumps were made in five sizes ranging from 0.1–2.2 kW. The 0.1-kW unit sold for $437 in 1955. SOMOR solar pumps received wide publicity, but sales were disappointing from a commercial standpoint.

Fig. 1.12 Design of the flat-plate collector of the SOMOR pump in Menlo Park, California.

Experiments with a variety of fluids proceeded, in attempts to find ones that would give superior performance in solar engines. Ethyl chloride is one fluid that has received considerable attention. Cesare Romagnoli in 1923 constructed an irrigation pump using hot water from the solar collector at 55°C (130°F) to heat the ethyl chloride working fluid. A modification using a turbine driven by ethyl chloride was used in Libya in 1954 by L. d'Amelio. Another ethyl chloride motor was built by Enzio Carlivari on the island of Ischia that is reported to have developed 3.4 kW.

Several experimental engines were constructed in India by M. L. Khanna and M. L. Ghai at the National Physics Laboratory in New Delhi. These were small, hot-air, open-cycle solar engines, and one was designed to operate at a high Carnot potential. The collector for this engine was a paraboloid and produced temperatures as high as 700°C, but the working parts of the engine could not stand this temperature for more than a short period of operation.

Attempts to utilize solar power with large systems were made by the USSR in the 1950s. F. Molero of the Helio Power Laboratory of the Energy Institute of the USSR Academy of Sciences developed a steam-generating plant in Tashkent, Uzbekistan, using a parabolic collector. This system was used for pumping water for irrigation and livestock, for a refrigeration plant, and for space heating.

A very ambitious project planned in 1957 was described by V. A. Baum, R. R. Aparasi, and B. A. Garf of the Energy Institute of Moscow. This system is reminiscent of the proposal by C. G. O. Barr in 1896. The large boiler of the power generator was fixed, and the mirrors moved about it on railroad tracks. The boiler was on a platform raised 42 m above the ground so that the concentric arcs of the mirrors on the railroad cars could reflect sunlight up to the boiler without mutual interference. The solar collector was to have an area of 19,000 m^2. These mirrors would be mounted on 1293 railroad flatcars, grouped into 23 trains on 23 separate concentric tracks. Each mirror was 3.0 m by 5.0 m, made up from 28 smaller mirrors. As the sun moved during the day, the trains would change position and the angle of the mirrors would be changed so that the solar image always fell on the boiler.

A 1:50 scale model of this large system was built, and complete studies were made of the distribution of solar energy at the focus. The variation of "temperature" in the vicinity of the focus was then utilized in the design of the absorber and various related functions of the boiler system. The full-scale system was never built, owing to the inability of the system to compete with fuel oil, which was then rapidly becoming available in Siberia, the proposed location of the giant system.

One of the most recent suggestions for the generation of electrical power approached the problem from a new angle. If we cannot build a large collector, what about using "natural" collectors—the saline ponds of the world?

In 1963 H. Tabor summarized his studies of this question of large-scale production of electrical power by the relatively new method involving the trapping of heat in large saline ponds. His model for a large-scale system was a solar pond 1–2 m deep with an artificial black bottom. Heat is trapped near the bottom of the pond by water containing a strong density gradient of suitable salts, such as magnesium chloride. The collected heat is withdrawn by conduction to pipes of the collecting fluid laid in the bottom of the pond. The temperature rise in the fluid is small, from about 30°C to 45°C, so the Carnot efficiency is very low, about 10%. Allowing for all losses, Tabor estimated a net efficiency of 1.3% to 1.7%.

Tabor cites the case of a 1-km^2 solar pond that would generate 3400 kW, yielding about 30 million kWh of power per year. Although the pond has some practical problems connected with wind disturbance of the density gradient, extracting of the low-grade heat, evaporation loss, and keeping the pond clean, as of 1962 Tabor felt that it offered the promise of being the cheapest method of extracting solar energy on a large scale.

1.7 SOURCES OF INFORMATION

The preceding survey of solar energy activities between 1945 and 1965 is necessarily brief. The amount of work can be seen by the large number of articles listed in the annual totals, which are still relatively available to interested persons in the reference journals *Solar Energy* and *Applied Solar Energy (Geliotekhnika)* and in a few books published during this period as well as the nontechnical magazine *Sun at Work. Solar Energy* and *Applied Solar Energy* are rapidly becoming the leading media for the reporting of current research and development. A further source of information on current work is provided by the Progress Reports issued under National Science Foundation (NSF) and Energy Research and Development Administration (ERDA) contracts, which are listed in the National Technical Information Service (NTIS) publications. Access to specific projects or types of work is still not easy to obtain, but interested persons will undoubtedly find access to these important reports easier in the future.

The National Science Foundation has prepared a book of abstracts of reports done under the NSF/RANN (Research Applied to National Needs) program. The first covers from 1970 to 1975 (NSF 75–6) and gives the NTIS accession numbers for many projects, making it easy to obtain desired reports. Presumably ERDA will continue this useful service. In the absence of NTIS numbers, a user can request the desired material from NTIS by referencing "key descriptors." Examples of such descriptors are "Solar Energy, Space Heating" or "Solar Energy, Solar Collectors–Flat Plate." The more closely one can define the area of interest by giving descriptors, the quicker will be the response. The cost of each report delivered is $3.00 for a paper copy and $0.95 for a microfiche copy, payable in advance. Requests should be addressed to: National Technical Information Service (NTIS), Document Sales, U.S. Department of Commerce, Springfield VA 22161. Some current reports can be obtained from RANN Document Center, Office of Intergovernmental Science and Research Utilization, 1800 G Street NW, Washington, D.C. 20550.

Part 1
The Energy Resource

Chapter 2
Solar Flux and Weather Data

2.0 INTRODUCTION

This chapter presents in considerable detail the basic resource upon which solar energy utilization depends. One must know the spectral distribution of sunlight in order to evaluate absorbing-surface behavior. We will define the diurnal variation of sunlight for three cases of interest: (1) the *standard solar flux*, (2) a *desert sea-level flux*, and (3) an *urban solar flux*. The standard flux is that generally used in solar energy computations. The desert sea-level flux is slightly higher, in accordance with observations made in the desert of the southwestern United States. The urban flux is significantly lower, approximating the solar flux observed in major urban communities of the midwestern and eastern United States.

The contributions from sunlight scattered by the atmosphere and nearby terrain also differ for the above three cases. The level of this diffuse sunlight is also defined.

Weather significantly modifies the solar flux, and data are presented for the variation of sunlight by month for a number of locations covering a wide range of latitude.

The types of observational instruments used to measure different aspects of solar flux are discussed, with attention to how they differ and the validity of each for specific applications. Because the *nonstandard* solar fluxes, as influenced by weather, are critical to the evaluation of real collector behavior, we give close attention to the types of daily records obtained, using Tucson, Arizona, as an example. The use of modulation transfer function (MTF) to analyze systems under varying solar input is presented in elementary form. Finally, we present various correlations between different types of solar measurements, and the correlation of direct solar flux as a function of wind velocity, an important factor in heliostat design.

It is important to clearly define two terms used in this book. Where *direct* and *diffuse* are usually used for the two components of sunlight important for solar energy collectors, we use *direct* and *scattered* because they are more precise and they simplify notation.

Direct Radiation. Direct radiation refers to the solar flux arriving at the collector without having suffered any scattering in traversal of the atmosphere. This radiation is from the geometrical disc of the sun. It is this component of sunlight that is focused by an optical system into an image of the disc of the sun. We neglect the very small amount of scattered light that is superimposed on the disc of the sun relative to the brightness of the disc of the sun.

Scattered Radiation. The atmosphere produces a diffuse component of scattered light through Rayleigh scattering and dust and aerosol scattering. Clouds also scatter sunlight, which adds to the total scattered flux but is not strictly diffuse in nature. And finally, ground surfaces scatter sunlight, adding to the total flux arriving at the collector. Scattered radiation cannot be focused by any optical system, but it does contribute flux to flat-plate and other nonfocusing collectors.

In this text we generally use the symbol I to denote the solar flux intensity. This symbol is further divided by subscript into I_d and I_s. The first, I_d, denotes the direct component of solar radiation, often denoted simply by the symbol I in other publications. In diagrams, we use the symbol D to denote the direct flux. The symbol I_s denotes the scattered ("diffuse") solar flux. In diagrams, we use S for scattered flux. The notation $D + S$ therefore describes the combined direct and scattered component of solar flux. Where we show $D + S$, other publications may use $I + D$. One should be careful to keep the appropriate definitions in mind when reading any specific article, since standard usage is not well defined. In our text we use the term S to include, when appropriate, other components of scattered sunlight, as from clouds and buildings. We feel this to be appropriate because the original intent of separate designations is to separate the *focusable* component of sunlight from the *nonfocusable* component.

2.1 THE SOLAR CONSTANT

The amount of solar energy received per unit of time per unit of collector area at the mean distance of the earth from the sun on a surface normal to the sun is called the *solar constant.* This quantity is difficult to measure from the surface of the earth because of the uncertainties in correcting the observed flux for the effects of the atmosphere. Measurements from satellites remove much of this difficulty, but they introduce uncertainties of absolute calibration. As a consequence, early satellite measurements resulted in an upward revision of the solar constant from the value determined by C. G. Abbot. The value was later revised downward again, and currently stands close to the Abbot value. The current value of the solar constant in several units is given in Table 2.1.

Above the atmosphere the intensity of sunlight shows a small variation. This variation is caused not by changes in the sun but by changes in the distance of the earth from the sun owing to a small ellipticity of the orbit of the earth with respect to the sun. The seasonal variation amounts to –3.27% at aphelion and +3.42% at perihelion. The amounts by months are shown in Table 2.2.

Table 2.1 The Solar Constant

Extraterrestrial solar constant	*Direct desert sea level*	*Direct standard sea level*
1.3530 kW/m^2	0.970	0.930
1.940 cal/cm^2min	1.39	1.33
4.871 MJ/m^2h	3.492	3.348
0.0324 cal/cm^2sec	0.0232	0.0222
7.16 Btu/ft^2min	5.13	4.92
429.2 Btu/ft^2h	307.7	295.0

Table 2.2 Annual Variation of Solar Flux Due to Orbital Eccentricity

Date	*Departure from mean*	*Solar flux, kW/m^2*
Jan. 1	1.0342	1.438
Feb. 1	1.0296	1.431
Mar. 1	1.0181	1.415
Apr. 1	1.0016	1.392
May 1	0.9848	1.369
June 1	0.9721	1.351
July 1	0.9673	1.345
Aug. 1	0.9716	1.350
Sept. 1	0.9835	1.367
Oct. 1	1.0003	1.390
Nov. 1	1.0172	1.414
Dec. 1	1.0296	1.431

2.2 THE SPECTRUM OF THE SUN

The distribution of the energy from the sun as a function of wavelength is very important in the functioning of solar energy collectors. The variation is basically that of a blackbody at 5800°K, but it is modified by absorption in the solar atmosphere from atomic lines, by the negative hydrogen ion continuum, and by a few molecular bands. The resulting changes in the exoatmospheric spectrum are small and are shown in the region from 0.3 to 0.6 μm in Fig. 2.1.

The temperature of the sun differs according to the particular application. The value quoted above of 5800°K is used to describe the variation of intensity with wavelength for the solar spectrum in the visible and near infrared. The bolometric temperature quoted by Allen (1955) is 6350°K. The bolometric quantity involves (1) the apparent brightness of the sun integrated over all wavelengths, which is the solar constant, (2) the angular diameter of the sun, and (3) the value of the astronomical unit, which is the mean distance from the earth to the sun. Slight changes in the brightness of the sun and in the measured distance to the sun affect the precise value of the bolometric temperature. The solar temperature derived from the changes of intensity with wavelength is less subject to errors of measurement. The spectral temperature of 5800°K is used in evaluating the fraction of sunlight within a given wavelength interval, as with solar cells or selective surfaces. The bolometric temperature of 6350°K is used in evaluating performance of a solar furnace.

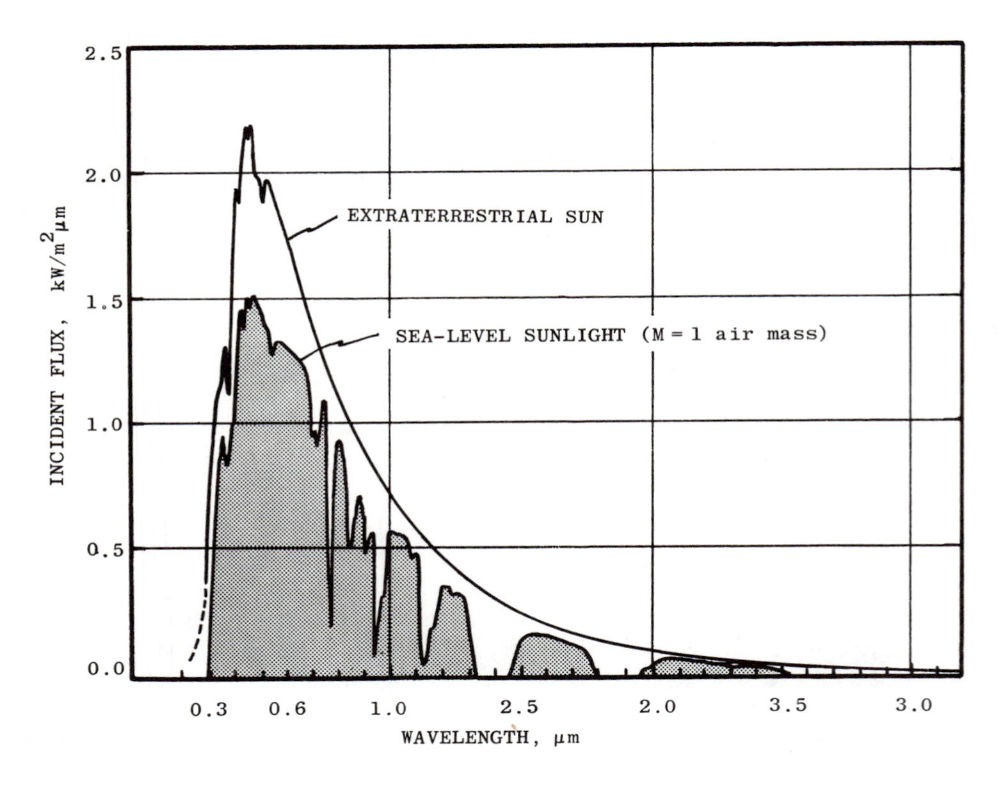

Fig. 2.1 Spectral distribution of energy in sunlight above the atmosphere and after passage through one air mass (zenith) for an atmosphere containing 20 mm precipitable water vapor.

The atmosphere of the earth significantly modifies the exoatmospheric solar spectrum by the presence of absorption bands of water vapor and carbon dioxide, and to a lesser extent by the presence of ozone in the terrestrial atmosphere. Ozone absorption effectively terminates the solar spectrum at 0.300 μm in the ultraviolet, and water vapor terminates it effectively at about 20 μm in the infrared. Between these two extremes there are many other absorption features that modify the spectrum, as shown in Fig. 2.1. The molecular absorption bands in the region where there is significant solar flux are shown more clearly on a logarithmic intensity scale in Fig. 2.2.

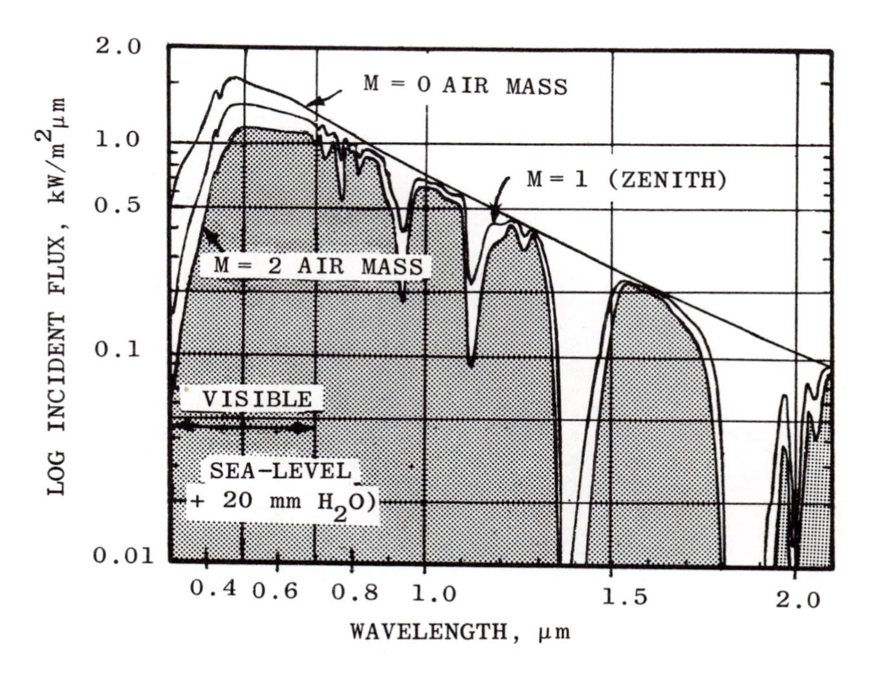

Fig. 2.2 Spectral distribution of energy in sunlight on a logarithmic intensity scale to permit a more accurate evaluation of the effect of the molecular absorption bands in the near infrared.

The standard sea-level atmosphere used in this book is that defined by Moon (1940), in which 20 mm of precipitable water are assumed to exist. The increase in absorption with change from one to two air masses (30° elevation above the horizon) is shown in Fig. 2.2. The percentage of the total energy in the spectrum of the sun longward of a given wavelength is given for three cases in Table 2.3. These cases are: (1) for a 5800°K blackbody, (2) for the exoatmospheric sun, and (3) for sea-level sunlight. Note the wavelengths for which 90% of the solar energy lies at shorter wavelengths. We will use this type of information in Chapter 9.

Table 2.3 Percentage of Solar Energy at Shorter Wavelengths Than a Given Wavelength (λ)

	5800°K	*Sun*		*Sea level*
(μm)	$\Sigma\lambda>0$	$\Sigma\lambda>0$	$\Sigma\lambda>0.3$	$\Sigma\lambda>0.3$
0.30	3.6	1.27	0.0	0.0
0.40	13.6	9.15	8.0	2.3
0.50	25.5	23.5	22.5	14.4
0.60	38.0	37.4	36.5	30.0
0.70	51.3	49.1	48.3	45.1
0.80	60.7	58.3	57.7	57.3
0.90	67.6	65.1	64.6	67.4
1.00	73.4	71.3	70.9	73.4
1.10	77.8	76.1	75.8	80.8
1.20	81.7	80.1	79.8	83.6
1.30	84.5	83.3	83.1	88.6
1.40	86.9	85.9	85.7	89.7
1.50	88.7	88.0	87.8	90.1
1.60	90.1	89.8	89.7	92.8
1.70	91.5	91.2	91.1	95.1
1.80	92.6	92.4	92.3	98.6
1.90	93.5	93.3	93.2	96.1
2.00	94.3	94.1	94.0	96.3
3.00	98.1	98.0	98.0	99.0
4.00	99.2	99.1	99.1	99.5
5.00	99.5	99.5	99.5	99.6
10.00	99.0	99.9	99.9	99.9

2.3 DIURNAL VARIATION OF DIRECT SUNLIGHT

When we measure the brightness of the sun at any particular time of day and then remeasure it sometime later, we usually find a different apparent brightness. This is because of a change in the angular altitude of the sun and a corresponding change in the air mass through which the sunlight travels. These changes must be taken into account in predicting solar collector performance. The changes vary from day to day as atmospheric conditions change, but the variation on a clear day is definable, as we will show. Since most of the research on diurnal variation of sunlight has been done in other than a desert climate, but whereas power and hydrogen conversion from sunlight will most likely be done in a desert climate, we have made an effort to define the difference between the *standard* flux and an appropriate value for the *desert* flux.

In Fig. 2.3 we show some observations made by Laue (1970) in the Mojave Desert of California. His observed data points are indicated by triangles for Table Mountain and by circles for observations from the floor of the desert near Palmdale. From his points we see that the flux was significantly higher than both the Haurwitz (1948) standard solar flux and the data point by Moon (1940) for air mass = 2. From the difference in elevation between the two sets of observations we are able to define the dependence of flux on altitude as shown by the curves in Fig. 2.3. We have defined the curve for zero altitude as *desert sea-level* solar flux (see Table 2.4).

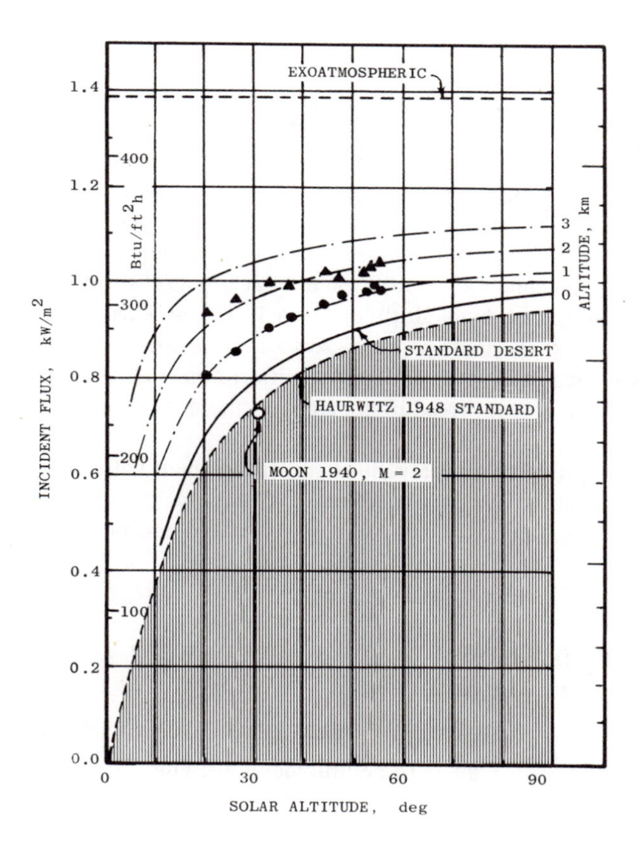

Fig. 2.3 Variation of direct solar (normal) flux with zenith distance and altitude based upon the observations by Laue (1970), and compared to the standard curves by Haurwitz (1948) and Moon (1940).

The reduction of intensity with decreasing altitude (increasing zenith distance) of the sun is generally assumed to be directly proportional to the increase in air mass, an assumption that considers the atmosphere to be unstratified with regard to absorbing or scattering impurities. For a plane-parallel atmosphere the air mass varies with the secant of the zenith distance. For a curved atmosphere a cubic term is added, but the real air mass as determined by astronomical observations does not differ significantly from the secant law for the zenith distance angles of importance in solar energy applications. We find that Laue's data can be fitted with the addition of an exponent s to the $\sec z$ term:

$$I(z) = I_0 e^{-c(\sec z)^s}, \tag{2.1}$$

where I_0 is the exoatmospheric solar flux and z is the zenith distance. The two empirical numerical constants are $c = 0.357$ and $s = 0.678$.

2.4 HEIGHT VARIATION OF DIRECT SUNLIGHT

Solar flux increases with the altitude of the point of observation. The direct sunlight increases as one gets higher because there is less atmosphere to absorb and scatter sunlight. This effect is visually apparent in the increased "blueness" of the sky at high altitudes, the result of the reduction in aerosol scattering and gross dust absorption from the urban atmosphere. The increase in solar flux is also apparent from the severe sunburn one can get at high altitudes, even in the absence of a snow cover. It is important to note that much of the desert of the sunny Southwest is one to two kilometers above sea level, thus adding a significant increase in the solar flux obtained at such sites. The increase is readily apparent in the data from Laue (1970) shown in Fig. 2.3.

We have taken Laue's observations and fitted them with an equation to permit calculation of the solar flux as a function of altitude of the site. This equation, based upon air-mass change and with empirical constants as determined from observations, is

$$I(z,h) = I_0(1 - ah)e^{-c(\sec z)^s} + ahI_0, \tag{2.2}$$

where $I(z,h)$ is the flux at zenith distance z and at a site at elevation h in kilometers, I_0 is the exoatmospheric solar flux, and

$$\begin{aligned} I_0 &= 1.353\ \text{kW/m}^2, \\ a &= 0.14 \text{ per kilometer altitude}, \\ c &= 0.357, \\ s &= 0.678. \end{aligned}$$

The linear term $1 - ah$ in Eq. (2.2), however, means that this equation can be used only for the first few kilometers altitude.

2.5 STANDARD ATMOSPHERE ZENITH DISTANCE FLUX VARIATION

The several standard atmospheres discussed in the preceding sections are used to show the variation of solar flux with zenith distance as shown in Fig. 2.4. There are basically three different cases shown. The upper curves refer to the change in direct solar flux and the total of direct plus scattered sunlight for the desert solar flux. The Haurwitz standard is shown only for direct solar flux, lying about 4% below the desert standard curves. Two curves are shown using solar flux and sky brightness data for an urban location. Note that the amount of scattered sunlight for a clear desert sea-level site is about 8% of the solar flux, but that for the urban location the amount rises to 22%. For some urban locations the amount of scattered light could be significantly higher, so if an urban site is proposed for a solar collector one must take enough observations to establish the appropriate solar flux data.

It is interesting to note from Fig. 2.4 that the large lowering of the direct solar flux is partially compensated for by the increase in scattered sunlight, so the urban $D + S$ curve is not seriously below the standard direct curve. These curves indicate that for an urban scattering day about half the scattered sunlight reaches the earth and half escapes upward.

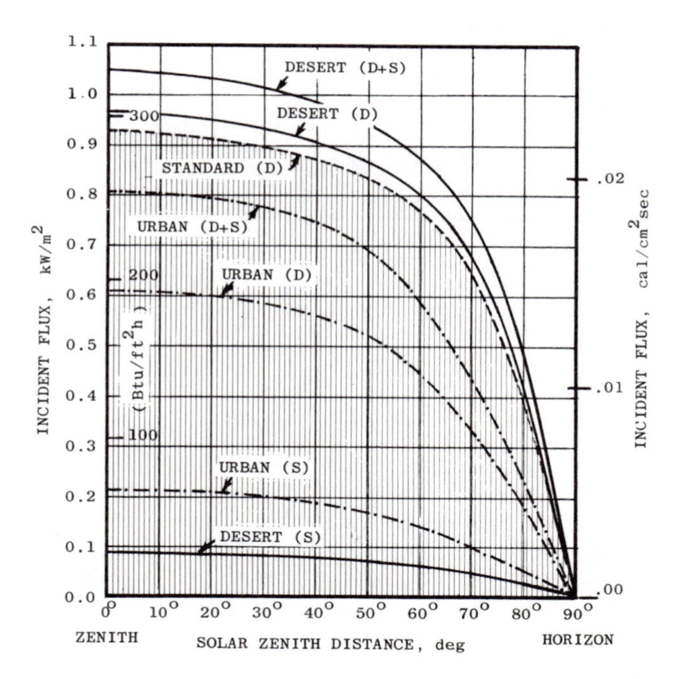

Fig. 2.4 Defining curves for solar flux variation with zenith distance for a desert atmosphere and a standard atmosphere. The curves for urban atmosphere are derived from observations in the eastern seaboard of the United States.

The curves shown in Fig. 2.4 are basic to the sets of curves that will follow. In these curves we assume that the scattered light changes with zenith distance exactly as the direct component changes, so the difference between D and $D + S$ is a constant factor for all values of zenith distance.

2.6 GROUND-SCATTERED SOLAR FLUX

In the preceding section we have shown the characteristics of direct sunlight and sunlight scattered by the atmosphere. This scattered sunlight component is important for flat-plate and other nonconcentrating or low-concentration collectors as an additional source of energy. The ground, and local environmental features like buildings, can also scatter significant amounts of solar energy that a tilted collector can utilize. In order to define these additional sources of solar flux we have taken various observed data and summarized them in Fig. 2.5.

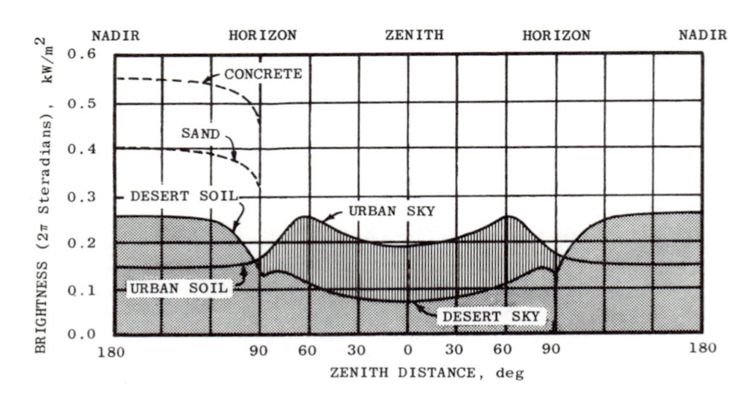

Fig. 2.5 Variation of scattered sunlight as a function of angle, for both the sky and ground under different conditions. Urban sky is brighter than urban soil, whereas desert sky is darker than desert soil.

Figure 2.5 shows the scattered component of sunlight as a function of zenith distance, both for the sky and for the ground. Note that for the desert the sky scatter can be significantly lower than the ground scatter, because the albedo (reflectance) of desert soil is approximately 0.25. For the urban environment the reverse is the case—urban ground cover has an albedo in the vicinity of 0.15—while the sky has an albedo of 0.22. On the average the sky and ground scatter are the same, so as a first approximation one can assume that a tilted collector would receive about the same scattered sunlight component as if it were looking directly up at the sky. If the ground cover is particularly light, as would be the case of concrete or sand, or particularly dark, as of asphalt, one would need to modify this statement accordingly.

For the reader who finds references that specify fluxes for diffuse radiation in units of steradians, it may be useful to note the definition of a steradian shown in Fig. 2.6. A steradian is the solid angle that determines the surface on a sphere having an area equal to the square of the radius of the sphere ($S = R^2$), which is a solid angle of vertex angle 65.4°.

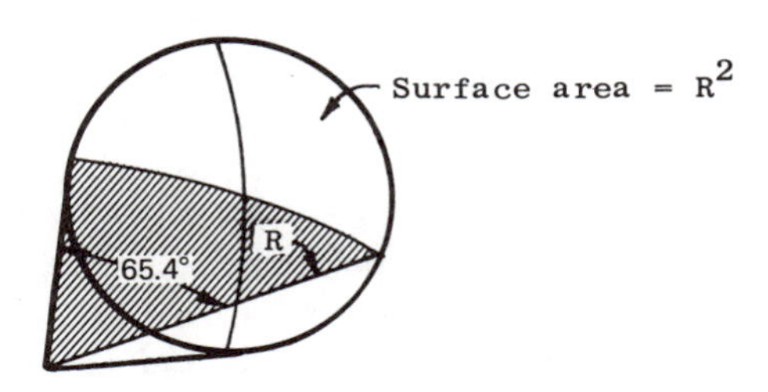

Fig. 2.6 A steradian is the solid angle of a cone with vertex angle 65.4085°, containing 3282.8 square degrees.

2.7 SUMMARY OF ZENITH SOLAR FLUX

On the basis of the preceding discussion, we can summarize the solar flux for zenith sun for the several cases of interest in Table 2.4. A variety of flux units are also provided for the use of the reader. Additional unit conversion tables are provided in Appendix A.

Table 2.4 Zenith Solar Fluxes

Units	*Exo-atmospheric (direct)*	*Desert sea level (direct)*	*Desert sea level (total)*	*Standard sea level (direct)*
kW/m^2	1.353	0.970	1.050	0.930
cal/cm^{2}min	1.940	1.39	1.50	1.33
L/min	1.940	1.39	1.50	1.33
cal/cm^{2}sec	0.0323	0.0232	0.0251	0.0222
cal/m^{2}sec	323.4	232.	251.	222.
W/cm^2	0.1353	0.0970	0.1050	0.0930
J/cm^{2}sec	0.1353	0.0970	0.1050	0.0930
MJ/m^{2}h	4.871	3.492	3.780	3.348
Btu/ft^{2}h	429.2	308.	333.	295.
Btu/ft^{2}min	7.16	5.13	5.55	4.92
hp/yd^2	1.54	1.10	1.19	1.06

2.8 GEOGRAPHICAL DISTRIBUTION OF SUNSHINE

The global distribution of sunshine in the absence of the atmosphere would make the north and south poles almost as sunny as any other latitude, even on an annual basis, but with a very poor annual distribution if continuity of sun were necessary. Weather modifies this geometrical distribution to a major extent, so the sunniest regions actually lie in two latitude bands, at approximately 20–30°N and 20–30°S. These regions are zones of atmospheric subsidence, and the dry descending air results in few clouds and sparse rainfall most of the year. In these zones, which mark the great deserts of the earth as shown in Fig. 2.7, the annual average of solar radiation exceeds 90%. Cloudiness increases and total insolation decreases both north and south of these two bands, the equatorial zone being one of localized convective cloudiness and tropical rainfall and the poleward zones being of polar storm circulation and frontal systems.

In Table 2.5 we give the average daily solar insolation for direct plus scattered light by month for some representative cities in the United States and throughout the

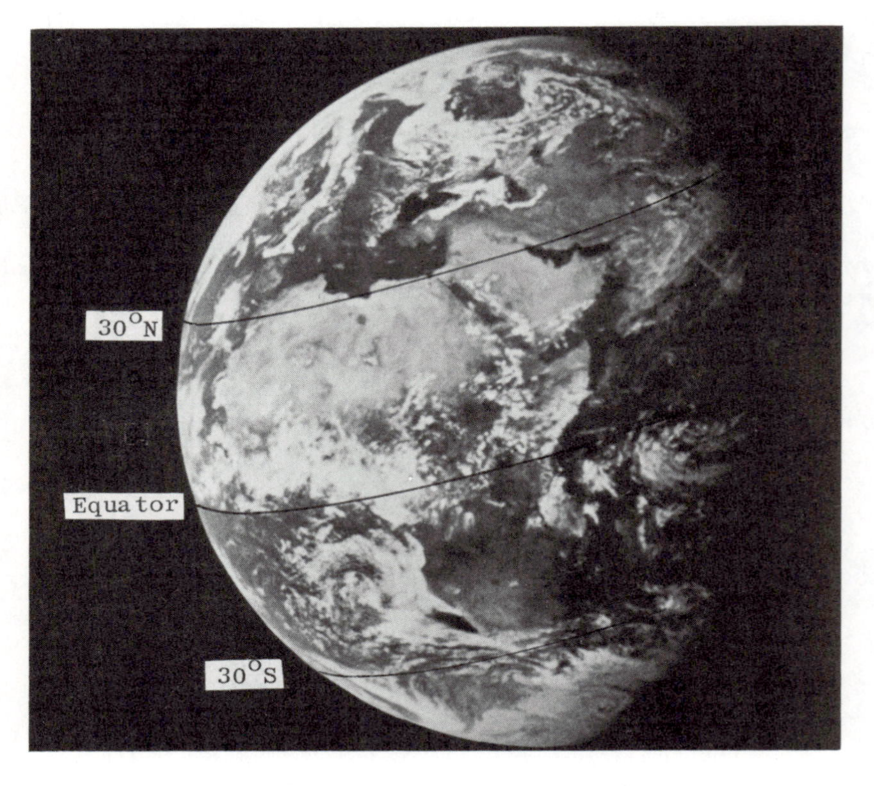

Fig. 2.7 Photograph of the Earth taken from Apollo 11, showing the 30° north and south latitude desert zones in the eastern hemisphere.

Table 2.5 Daily Averages of Solar Energy Received on a Horizontal Surface (Monthly)

Location	°*N* *lat.*	*Jan*	*Feb*	*Mar*	*Apr*	*May*	*Jun*	*Jul*	*Aug*	*Sep*	*Oct*	*Nov*	*Dec*
Yangami, Congo	01	409	450	458	446	438	397	352	361	408	410	423	373
		1507	1659	1688	1644	1616	1464	1298	1331	1504	1512	1559	1375
Dakar, Senegal	15	460	538	633	627	619	580	512	456	464	449	452	470
		1696	1983	2334	2311	2282	2139	1888	1681	1711	1655	1666	1732
Calcutta, India	22	532	617	701	781	784	817	816	800	645	624	557	501
		1961	2275	2584	2879	2890	3012	3008	2949	2378	2300	2053	1847
Tokyo, Japan	36	190	231	274	312	343	303	336	338	254	202	185	169
		700	851	1010	1150	1264	1117	1239	1246	936	745	682	623
Tucson, Ariz.	32.5	315	391	540	655	729	699	626	588	570	442	356	305
		1161	1442	1991	2415	2688	2577	2308	2168	2102	1630	1313	1124
Charleston, S.C.	33	250	334	451	558	620	587	539	528	409	326	308	213
		923	1232	1664	2059	2288	2166	1989	1945	1509	1203	1137	786
Riverside, Calif.	34	264	312	408	494	544	599	592	545	465	358	276	212
		974	1151	1506	1823	2007	2207	2184	2011	1712	1321	1018	782
Albuquerque, N.M.	35	307	367	497	606	676	746	679	624	547	464	348	292
		1133	1345	1834	2236	2494	2749	2502	2299	2018	1712	1284	1085
Santa Maria, Calif.	35.5	290	374	510	610	680	651	659	642	528	432	302	235
		1070	1380	1882	2251	2506	2399	2428	2369	1945	1594	1114	867
Oak Ridge, Tenn.	36	174	231	286	402	570	542	499	463	317	306	204	162
		642	852	1055	1483	2103	2000	1838	1708	1517	1129	753	598
Lincoln, Neb.	41	186	252	338	427	502	557	576	481	409	302	209	166
		686	930	1247	1576	1852	2052	2122	1775	1509	1114	771	613
Salt Lake City, Utah	41	195	246	386	510	546	595	624	509	442	278	179	120
		572	908	1424	1882	2015	2192	2303	2247	1631	1026	661	442
State College, Penn.	41	137	174	275	387	426	500	513	457	358	248	164	115
		506	642	1015	1428	1572	1845	1889	1683	1321	915	605	424
New York, N.Y.	42	122	191	259	363	426	446	439	366	316	243	148	107
		450	705	956	1339	1572	1646	1620	1351	1166	897	546	395
Blue Hill, Mass.	42.5	163	250	324	397	477	518	499	463	364	292	163	135
		601	923	1196	1465	1758	1911	1838	1708	1343	1077	601	498
St. Cloud, Minn.	46	170	238	398	470	561	560	575	452	364	284	175	152
		627	878	1461	1734	2070	2066	2118	1667	1343	1048	646	561
Glasgow, Mont.	48	156	244	392	502	640	767	661	545	378	244	174	122
		576	900	1446	1852	2362	2494	2435	2011	1395	900	642	450
Spokane, Wash.	48	120	198	336	572	483	615	674	581	507	250	133	78
		443	731	1240	2111	1782	2269	2487	2144	1587	923	491	280
Brussels, Belgium	51	56	108	206	346	406	441	406	354	251	158	76	47
		206	398	759	1276	1497	1626	1497	1305	925	582	280	173
Stockholm, Sweden	59	29	78	201	308	467	517	500	392	243	112	32	18
		107	279	741	1135	1722	2138	1843	1445	896	413	118	66

Upper values are in langleys/day (cal/cm^{2}day); lower values are in Btu/ft^{2}day.
To convert from Btu/ft^{2}day to cal/cm^{2}day, multiply by 0.27125.

world to show the typical seasonal variations to be expected. The upper figure is the flux in cal/cm^2 day and the lower figure is in Btu/ft^2 day. It should be noted that these fluxes are subject to calibration errors in different instruments. We have some indication that in the United States data there is a scatter due to instrumental differences that may be as large as 5%. In this table the very high values for Calcutta look suspicious considering that Albuquerque, in the high desert, has 89% of the possible insolation. To check the validity of the maximum readings in Table 2.5 we have used the theoretical maximum daily total insolation from Kondratyev (1954) for a horizontal flat plate located at the given latitude in the absence of an atmosphere. To derive the effect of the atmosphere on the theoretical maximum we have taken the ratio of exoatmospheric solar flux to the desert sea-level total flux from Table 2.4 (1.050/1.353 = 77.6%) to prepare the theoretical maximum curve plotted in Fig. 2.8. On this graph we have plotted the maximum annual daily totals from Table 2.5. The graph shows that the Calcutta maximum is well above the theoretical 100% value. On further examination of the Calcutta data, it appears that almost every month has

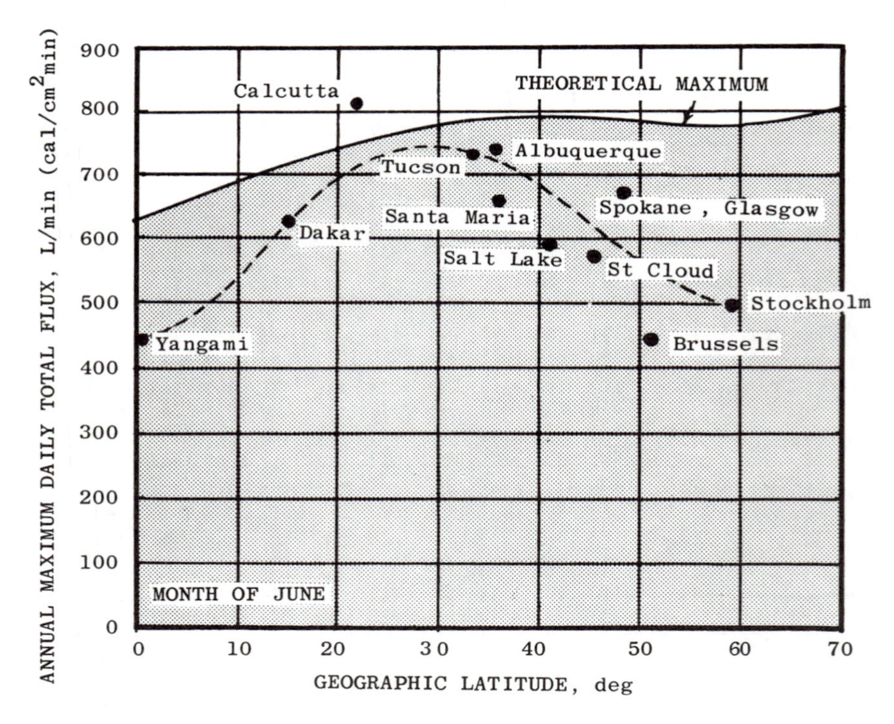

Fig. 2.8 Comparison of the theoretical maximum insolation on a flat horizontal plate as a function of latitude, based on Kondratyev (1954), with observed maxima for different stations. The data point for Calcutta is as quoted in Duffie and Beckman (1974, Table 3.3.1). Note the appearance of the desert zone and statistically clearest skies in the 20–35° latitude zone.

values above the theoretical maximum for that month. This can be explained by calibration error, or by a statistically significant effect of heightened albedo either from clouds to the north of the station (toward the Himalaya range) or from nearby buildings. Calibration errors also affect much of the United States pyranometer data, according to the National Ocean and Atmosphere Administration (NOAA).

The data points in Fig. 2.8 show clearly the presence of a cloud-free arid zone between 20–30° latitude. The theoretical curve rises from 60°N to a maximum at the North Pole, where the daily total flux reaches 85 langleys. The points for Spokane, in the dry northern desert of the western United States, and Brussels, in the wet temperate zone of western Europe, indicate the range of flux variation due to weather in a single latitude zone.

In Figs. 2.9 through 2.12 we show the global isoflux contours for total hemispheric insolation in MJ/m^{2}day for the spring equinox, summer solstice, fall equinox, and winter solstice. It should be noted that these curves are large-scale regional averages, and that the prediction of local insolation must take into account local climatological variations from the mean.

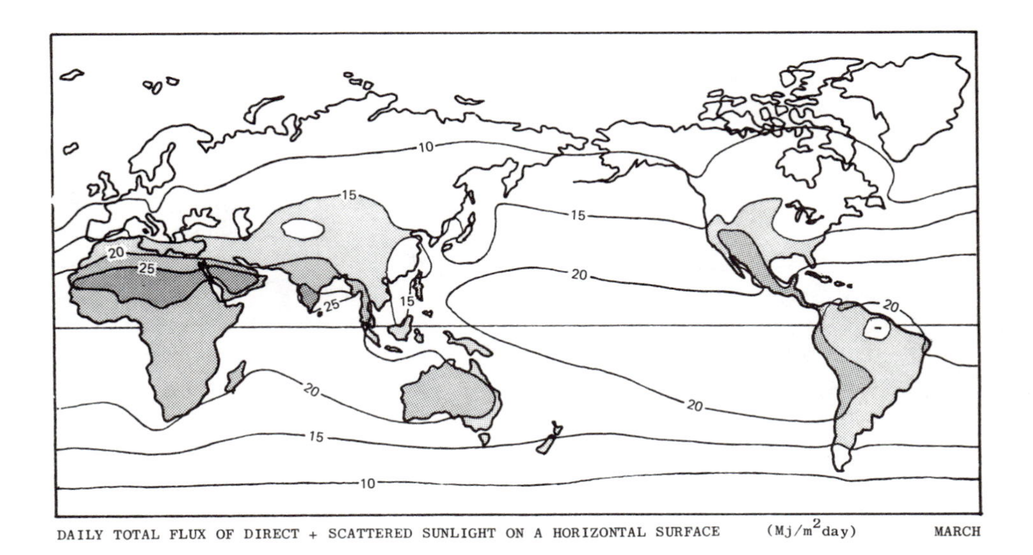

Fig. 2.9 Global isoflux contours for March.

In this and the three figures following, contours are labeled in MJ/m^{2}day of daily total flux of direct plus scattered sunlight on a horizontal surface. A 10-MJ/m^{2}day contour is equivalent to 881 Btu/ft^{2}day.

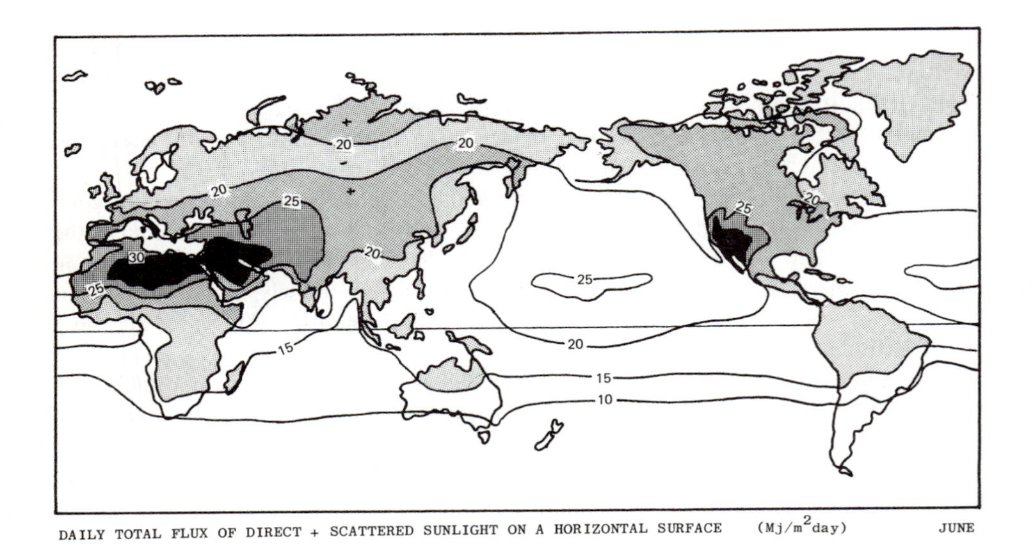

Fig. 2.10 Global isoflux contours for June.

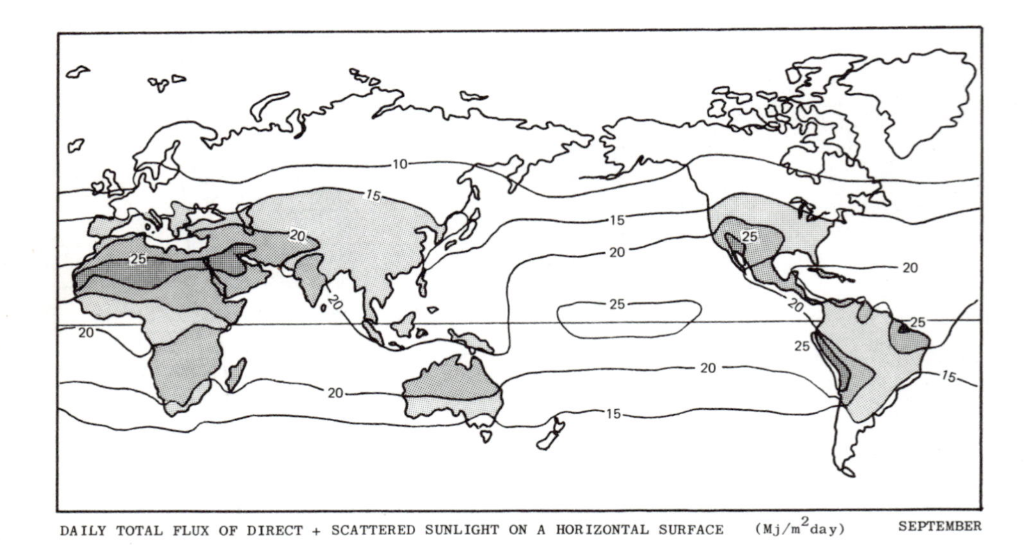

Fig. 2.11 Global isoflux contours for September.

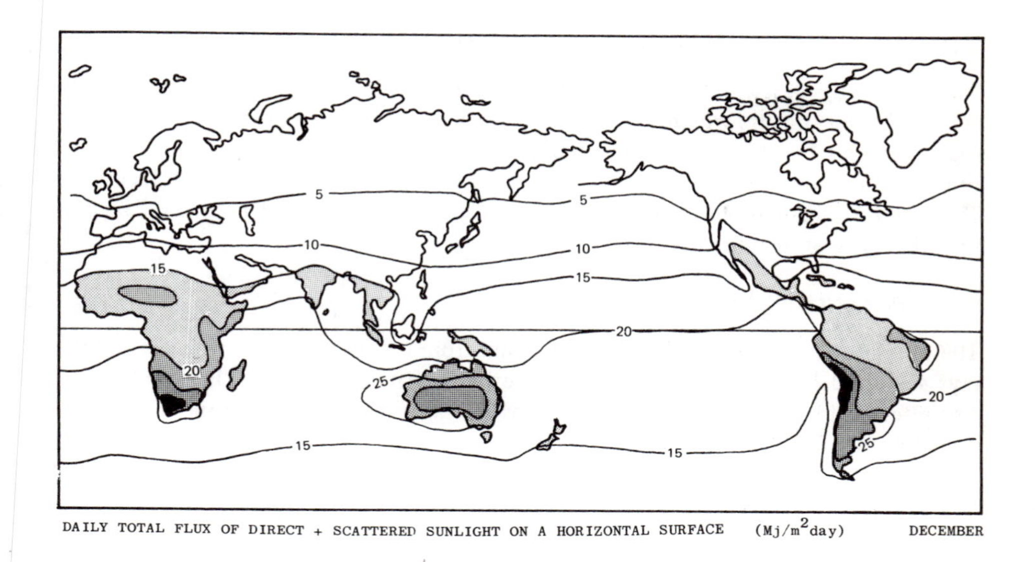

Fig. 2.12 Global isoflux contours for December.

2.9 SOLAR FLUX VARIABILITY FROM WEATHER

The short-duration fluctuations in sunlight caused by clouds are of great importance in evaluation of solar collector performance. One can readily determine the *average* effect of clouds from the type of data presented in the preceding section. If solar thermal collectors were linear transducers of sunlight, then the average flux would predict the average system performance. This is not the case because system losses cause any given system to operate nonlinearly. Below a certain flux level a system will cease to attain a given temperature, but output at this temperature will rapidly improve above this threshold up to design output performance at full solar flux.

An important question is how frequently the sunlight must be sampled to be reliable for thermal system performance evaluation. The most detailed records usually available are hourly averages or integrals. If the system had thermal inertia large compared to an hour, these types of records would be adequate. Most systems, however, have response times of an hour or less; hence hourly insolation figures could lead to inaccurate performance predictions. Let us examine the types of data that are recorded for sunlight.

2.10 SOLAR FLUX OBSERVATION INSTRUMENTATION

We will briefly describe several instruments used to record solar insolation so as to clarify the specific information provided by each. It is important to recognize what each measures because some collectors utilize total flux and others only direct flux. Furthermore, some instruments are used widely and others at only a few locations. The validity of published data is sometimes clouded by the use of imprecise labels as to which type of measurement has been made.

2.11 PYRANOMETER

The pyranometer measures the total solar flux from the direct rays as well as from the rays scattered by the entire hemisphere viewed by the instrument. This total flux is also referred to as the *global flux*. A schematic diagram of one type of pyranometer is shown in Fig. 2.13.

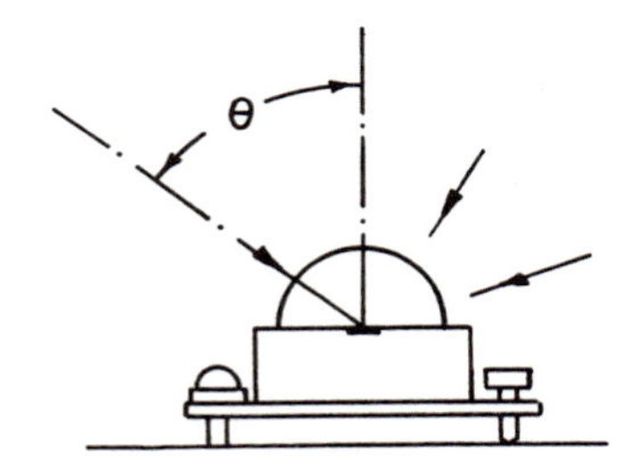

Fig. 2.13 Schematic diagram of an Eppley pyranometer. The detector must have an accurate cosine θ response to properly read the contribution of direct sunlight and of scattered sunlight so that the reading is accurate for *total* intensity.

The pyranometer is widely used because it is a simple instrument, requiring no tracking of the sun. Probably 90% or more of the sunshine data from around the world is gathered using some type of pyranometer.

The pyranometer is generally placed in a horizontal position so that the 2π hemisphere accepted by the instrument covers the entire sky. One must be careful that no buildings or bright reflecting objects are included in this 2π field of view or inaccurate readings will result. The detector must have a cosine response to incident radiation. The detector in the Eppley design uses concentric white and black rings to generate a temperature difference, which is measured by multiple-junction thermocouples. Parson's black for the black and magnesium oxide for the white comparison both provide angular behavior approximating the desired Lambertian, or cosine, characteristic. Long-term accuracy depends upon the stability of the response of the black and white to aging. One or two hemispherical glass windows are provided to protect the detection surfaces from the environment. The glass, however, has a wavelength cutoff in the infrared at about 3 μm. A quartz window or fused silica window would have a cutoff at about 5 μm, so the use of windows eliminates the thermal infrared. Some thermal infrared related to incident sunlight is recorded, however, because the glass envelope acquires the ambient temperature at the site, which is related to solar input, and emits thermal infrared radiation from the envelope to the detector.

The principal pyranometers in use in the past half century of data gathering are:

1. Fuess model 58D, Robitzsch pyranometer
2. Casella model, Robitzsch pyranometer
3. SIAP model, Robitzsch pyranometer
4. Moll-Gorczynski pyranometer
5. Eppley model 1571 pyranometer
6. Eppley model 1220 pyranometer
7. Kipp & Zonen pyranometer

A description of each of these types and a discussion of their calibration characteristics is given by Robinson (1966).

2.12 SHADING-RING PYRANOMETER

The pyranometer is sometimes modified with a shading ring to exclude direct sunlight and thereby enable measurement of the scattered components of sunlight. When this measure of global radiation is subtracted from the reading of a standard pyranometer, the result is a measure of the direct solar flux. The usual shading-ring instrument is a standard pyranometer to which a ring-shaped hoop sunshield is added. To keep the obscuration of the sky small, the ring is made narrow, necessitating changing the position of the ring every few days to keep the shadow centered on the pyranometer.

2.13 PYRHELIOMETER

The pyrheliometer measures the direct solar flux plus a very small amount of sky included in the field of view of the instrument, approximately 5.5° for the Eppley pyrheliometer. This small field of view means that the pyrheliometer must be mounted and driven to track the daily and seasonal motion of the sun. A schematic diagram for a pyrheliometer is shown in Fig. 2.14. Because the instrument always points to the sun, the observations are often termed *normal incidence* insolation measures. This term could be confusing because the pyranometer is sometimes tilted so that it is normal to the noon position of the sun, which also yields what could also be considered as a normal incidence measure.

Pyrheliometer measurement stations are relatively rare compared to pyranometer stations because the instrument needs more attention. The daily drive must be checked, and the instrument must be adjusted to accommodate seasonal changes in

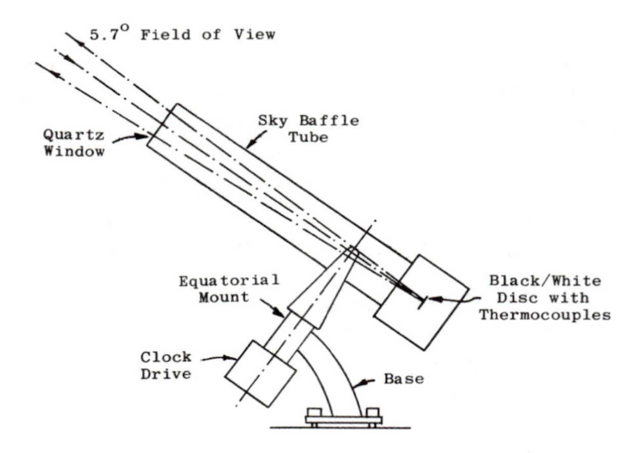

Fig. 2.14 Schematic arrangement of an Eppley pyrheliometer used to measure the normal incidence (direct) intensity of sunlight. The 5.7° are sufficient to reduce the requirement for tracking to several days of operation.

solar declination. In 1975 there were only five official pyrheliometer recording stations in the United States network of weather stations.

Pyrheliometer measurements are essential to the data needed to predict the performance of concentrating solar collectors. In the absence of the pyrheliometer data needed for concentrating collector evaluation, one must reconstruct such data from indirect correlations between direct and total insolation, as is discussed in detail below.

The pyrheliometer generally uses a blackened disc, or in some cases both a blackened disc and a white reference surface. It has been found that a significant number of such instruments over the course of a few decades have deteriorated, the black becoming greenish and the white surfaces yellowed. Consequently, the existing records must be compared with caution.

The principal pyrheliometers in use in the past half century of data gathering are:

1. Water-flow pyrheliometer (Smithsonian)
2. Abbot silver disc pyrheliometer
3. Angstrom pyrheliometer
4. Michelson-Buttner pyrheliometer
5. Linke-Fuessner Armour pyrheliometer
6. Linke-Fuessner Kipp & Zonen pyrheliometer
7. Eppley pyrheliometer
8. Savinov-Yanishevskiy pyrheliometer

A description of each of these types and a discussion of their calibration characteristics is given by Robinson (1966).

2.14 MOVING SHADOW-BAR PYRHELIOMETER

A single instrument can be used to measure direct, scattered, and total insolation. This instrument uses a moving shadow bar to modulate the incoming direct component of sunlight. A schematic diagram is given in Fig. 2.15. The detector is a silicon diode, placed beneath a glass cover that protects the detector and diffuses the incoming sunlight to adjust for exact cosine response. Because the detector is very small, the shadow bar can be narrow and hence close to the detector, making for a compact instrument. A hemispherical glass cover is inserted at half the distance to the shadow bar to keep dust from the smaller diffusing glass surface.

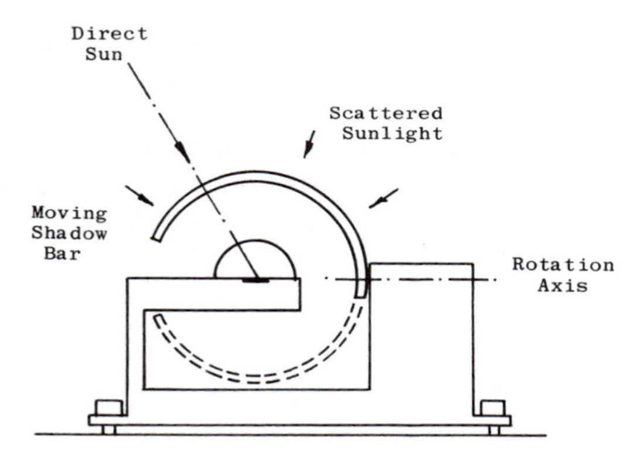

Fig. 2.15 Schematic diagram of a Helio moving shadow-bar pyrheliometer. The detector is alternately exposed to sun plus sky and to sky only, recording thereby both the direct and the direct plus scattered intensities. Some records taken with this type of instrument are shown in Figs. 2.17–2.22.

A silicon diode does not respond to the total spectrum of sunlight arriving at the surface of the earth because silicon has a cutoff beyond approximately 1.6 μm. The spectral response is also not equal at all wavelengths where it does respond to sunlight, as shown in Fig. 15.4. One must therefore calibrate these instruments against an absolute instrument like a pyranometer or pyrheliometer, which measure the total spectrum by converting it to heat by a black surface. A silicon diode is nevertheless a reasonably good secondary detector because clouds and haze have only slight "color" effects between the visible and the infrared.

The moving shadow-bar pyrheliometer is mounted with the axis of rotation of the bar north-south. The bar is a large enough portion of a circle that no seasonal changes are required to keep the sun occulted by the bar. Typical records taken with the Helio moving shadow-bar pyrheliometer are shown in Figs. 2.17–2.22.

2.15 SUNSHINE RECORDER

The U.S. Weather Bureau's development of a *percent of available sunshine* meter has resulted in a new type of sunshine measure now being widely recorded in the United States. This instrument measures the amount of direct sunshine, within certain limitations. It requires no tracking of the sun, but only approximate answers are obtained. The instrument basically uses two detectors, one of which is shadowed from direct sunlight. When the flux *difference* between the two detectors is above a set threshold, a small motor is turned on that records the hours and tenths of sunshine. This quantity and the percentage of possible sunshine it represents are regularly published in the "Local Climatological Data" reports issued by the Department of Commerce, NOAA, Environmental Data Service. A sample of such a data sheet is shown in Fig. 2.16.

This recorder is limited in that its accuracy depends upon the threshold setting for the difference between the two signals and the absolute brightness of the day. It does provide, in our estimation, a better index of the amount of direct sun than can be derived from statistical inferences from pyranometer data. A further limitation from the standpoint of systems analysis is that the sunshine recorder gives no data on the time duration of intervals of sunshine and cloudiness, data that are essential to systems response.

LOCAL CLIMATOLOGICAL DATA
U.S. DEPARTMENT OF COMMERCE
NATIONAL OCEANIC AND ATMOSPHERIC ADMINISTRATION
ENVIRONMENTAL DATA SERVICE

TUCSON, ARIZONA
NATIONAL WEATHER SERVICE OFC
INTERNATIONAL AIRPORT
APRIL 1974

LATITUDE 32° 07′ N LONGITUDE 110° 56′ W ELEVATION (GROUND) 2584 FT. STANDARD TIME USED: MOUNTAIN WBAN #23160

DATE	TEMPERATURE °F							WEATHER TYPES ON DATES OF OCCURRENCE	SNOW, ICE PELLETS OR ICE ON GROUND AT 05AM IN.	PRECIPITATION		AVG. STATION PRESSURE IN. - - - ELEV. 2555 FEET M.S.L.	WIND					SUNSHINE		SKY COVER TENTHS		DATE
	MAXIMUM	MINIMUM	AVERAGE	DEPARTURE FROM NORMAL	AVERAGE DEW POINT	DEGREE DAYS BASE 65° HEATING	DEGREE DAYS BASE 65° COOLING	1 FOG 2 HEAVY FOG X 3 THUNDERSTORM 4 ICE PELLETS 5 HAIL 6 GLAZE 7 DUSTSTORM 8 SMOKE, HAZE 9 BLOWING SNOW		WATER EQUIVALENT IN	SNOW, ICE PELLETS IN.		RESULTANT DIRECTION	RESULTANT SPEED M.P.H.	AVERAGE SPEED M.P.H.	FASTEST MILE SPEED M.P.H.	FASTEST MILE DIRECTION	HOURS AND TENTHS	PERCENT OF POSSIBLE	SUNRISE TO SUNSET	MIDNIGHT TO MIDNIGHT	
1	2	3	4	5	6	7A	7B	8	9	10	11	12	13	14	15	16	17	18	19	20	21	22
1	76	45	61	0	29	4	0		0	0	0	27.27	19	5.6	8.3	17	W	11.9	95	7	5	1
2	73	57	65	3	37	0	0		0	T	0	27.14	25	15.6	17.8	37	W	9.2	73	8	7	2
3	70	48	59	-3	33	6	0		0	0	0	27.23	30	7.4	8.8	29	W	12.6	100	4	2	3
4	74	39*	57	-5	29	8	0		0	0	0	27.46	33	2.8	5.3	13	NW	12.6	100	1	1	4
5	80	51	71	8	31	0	6		0	0	0	27.39	12	9.5	10.4	29	SE	12.6	100	0	0	5
6	86	53	70	7	28	0	5		0	0	0	27.30	24	5.3	11.2	29	W	12.7	100	0	0	6
7	87	50	69	6	32	0	4		0	0	0	27.37	25	.9	7.2	13	NW	12.5	96	3	3	7
8	94*	50	72	9	34	0	7		0	0	0	27.36	15	5.1	7.6	22	SE	12.7	100	4	2	8
9	85	58	72	8	30	0	7		0	0	0	27.19	21	9.9	13.1	28	SW	12.2	96	7	4	9
10	63	47	55*	-9	27	10	0		0	0	0	27.20	28	9.8	10.4	25	W	11.2	88	4	2	10
11	71	44	58	-6	32	7	0		0	0	0	27.27	26	2.3	7.8	22	NW	12.8	100	0	0	11
12	77	47	62	-3	22	3	0		0	0	0	27.27	27	3.7	9.4	20	NW	12.9	100	3	2	12
13	77	49	63	-2	14	2	0		0	0	0	27.32	28	.9	7.9	17	N	12.9	100	0	0	13
14	80	44	62	-3	11	3	0		0	0	0	27.35	27	2.9	7.5	18	NW	12.9	100	0	0	14
15	82	45	64	-1	12	1	0		0	0	0	27.30	02	1.1	6.9	15	W	13.0	100	0	0	15

Fig. 2.16 Sample of the data sheet published by the Department of Commerce giving the measurements made by the "percent of sunshine" meter.

2.16 DIURNAL VARIATIONS OF SUNLIGHT

The diurnal variation of sunlight is seldom represented by an idealized curve (such as is presented in Fig. 2.4). These idealized flux versus time-of-day curves can be seriously modified by cloudiness. To illustrate the range of types of daily records, we will use data gathered with the Helio moving shadow-bar pyrheliometer. Figure 2.17 shows a typical month of data. The scale of these diagrams is small, so we show some typical days to make the information content clear. In Fig. 2.18 we show a typical "clear" day, where the only variation is the cosine projection effect of the fixed horizontal detector. The upper envelope records the total intensity of direct plus scattered sunlight. The depth of the line shading records the direct component, caused by the shadow bar eliminating the direct beam and a small fraction of the sky radiation occulted by the bar. (This small fraction is apparent in Fig. 2.22, where no sunlight was visible and the small modulation was caused by the shadow bar.)

Typical summer cloudiness is caused by discrete clouds occulting the sun, with full sunlight coming through between the clouds. This type of day is shown in Fig. 2.19. Note that, when the sun is added to the sky signal, the total flux goes well above that for a clear day. In winter one can have a significant number of those cloudy days where a continuous thin cloud cover veils the sun, diminishing but not extinguishing its contribution to the total flux. This type of day is shown in Fig. 2.20. Note how the scattered component rises during the latter half of the day, when clouds traverse the location.

Figure 2.21 shows a mostly cloudy day, with heavy overcast resulting in a low scattered component but with occasional breaks in the overcast through which the sun shines. Days of this type show total fluxes that can be a significant fraction of that of a clear day, but with little flux of use in a concentrating collector system. Figure 2.22 shows a completely cloudy day, with no direct sun but with almost 30% of the clear-day flux appearing as scattered sunlight. Note the small modulations in this record caused by the shadow bar.

2.17 CORRELATIONS BETWEEN DIRECT AND TOTAL INSOLATION

Although different solar collectors require different measures of solar flux, as described in the preceding sections, one is often limited in the type of observational records available for a given location. It has therefore been necessary to utilize correlations between different types of measures in order to derive some estimates of the flux required by a specific system. One of the first extensive studies of this type was published by Liu and Jordan (1960) and is widely quoted. Their graphs show correlations with "extraterrestrial daily solar insolation." This quantity is not directly observable. It is interesting to note that some plotted points lie between 0.8 and 1.0 of the extraterrestrial flux. Even a clear desert atmosphere shows a maximum on a clear day at only 0.77.

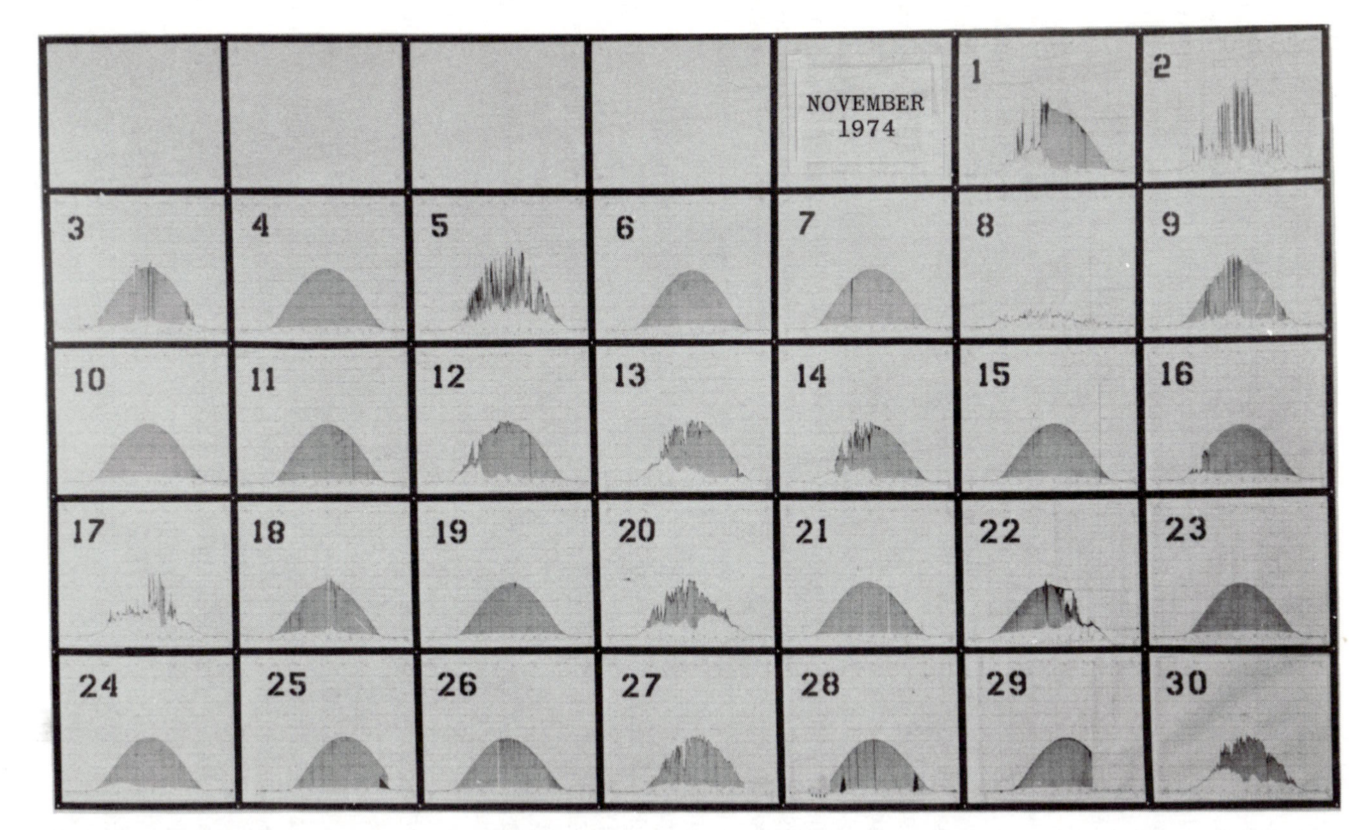

Fig. 2.17 Solar flux data by days for the month of September 1974 for Tucson, Arizona. Recordings were made by McKenney (1974) using a Helio moving shadow-bar pyrheliometer. Note the number of days having considerable total energy but consisting of frequently interrupted direct sun plus enhanced scattered light.

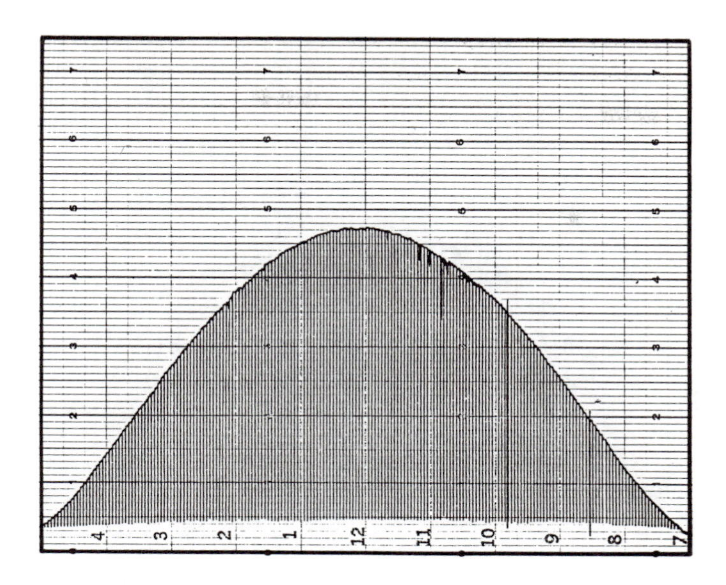

Fig. 2.18 Moving shadow-bar pyrheliometer record for a clear day, 11 November 1974. The time scale runs from morning at the right to evening at the left.

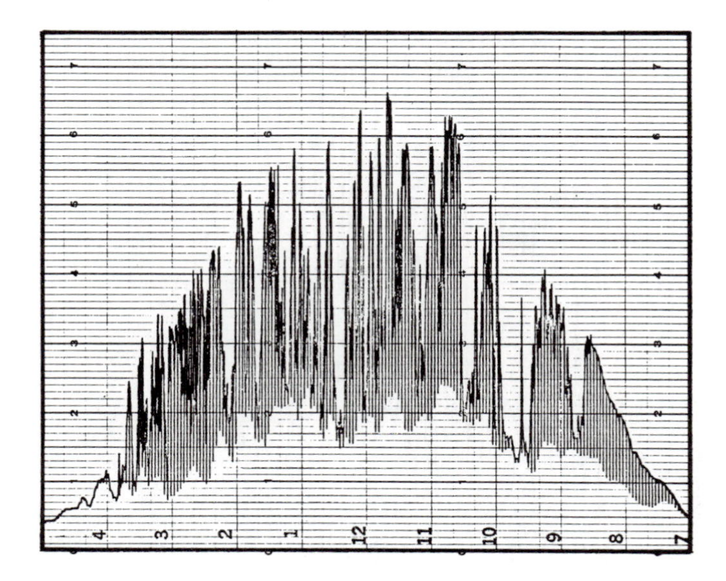

Fig. 2.19 Moving shadow-bar pyrheliometer record for a cloudy day, 5 November 1974. Note that the peak intensity values rise considerably above the total flux for a clear day, but the direct flux is substantially less than that for a clear day at these peaks.

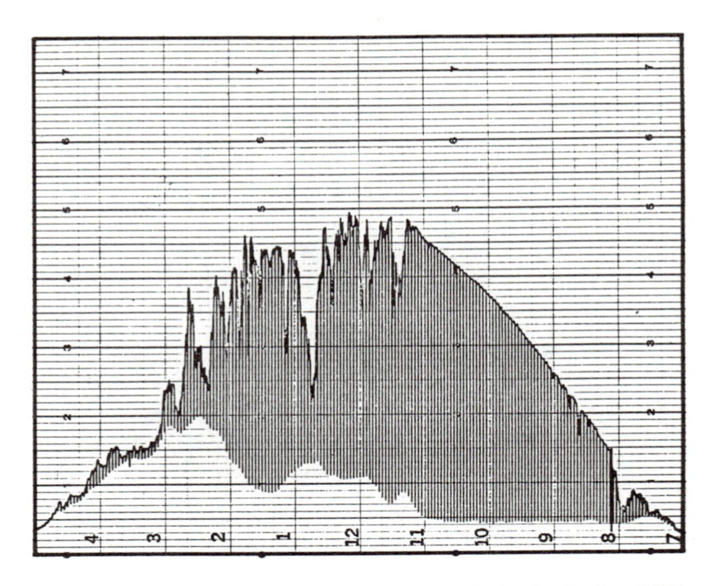

Fig. 2.20 Moving shadow-bar pyrheliometer record for a partially cloudy day, 13 November 1974, where the morning was clear and thin clouds moved over the station in the afternoon, raising the scattered component and diminishing the direct component.

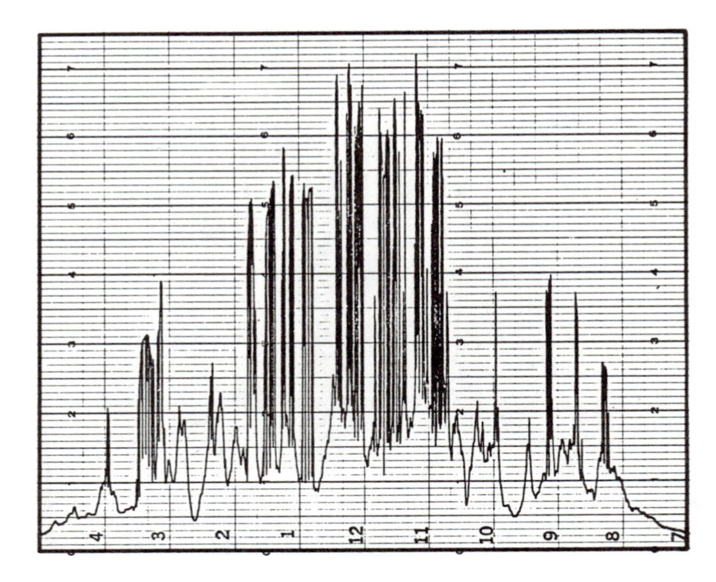

Fig. 2.21 Moving shadow-bar pyrheliometer record for a cloudy day with a few periods of sunshine, 2 November 1974. Such a day would pose serious operating problems for a concentrating system.

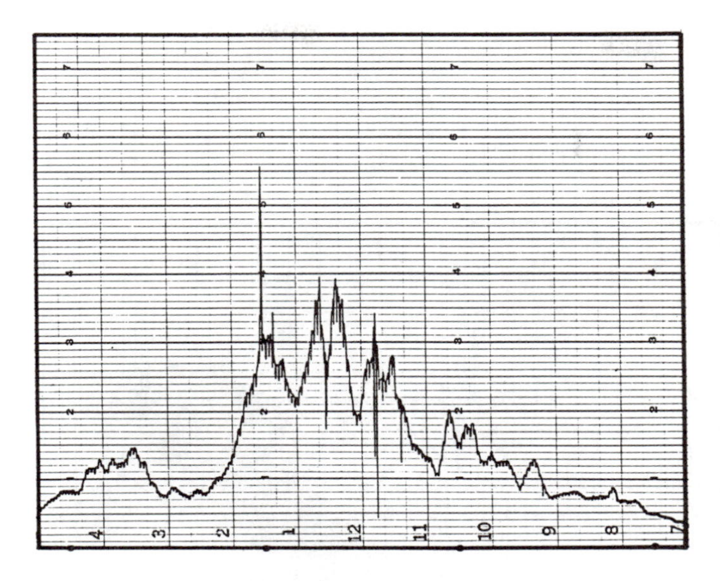

Fig. 2.22 Moving shadow-bar pyrheliometer record for a totally overcast day, 7 October 1974. Direct sun was absent all day except briefly at 1335 h. The small modulations are due to the shadow bar.

A recent study by Bos (1975) shows a good correlation between "percent of possible total insolation" and "normal incidence insolation." The exact observables are not clearly specified, but one would assume that "percent of possible total insolation" refers to the data published for the percent of possible *sunshine,* which is a daily average. On the other hand, the published values of normal incidence insolation are taken only at noon from the pyrheliometer records at Albuquerque, New Mexico; referring to Fig. 2.23, we see quite a scatter, largely due to this difference in the times under consideration.

Correlation of total (global) insolation with direct flux is less well defined, as presented in Fig. 2.24. In this figure we plot data for Tucson, Arizona, for daily horizontal solar flux, as measured by the pyranometer, versus daily total hours of sunshine, which is closely equal to the amount of direct sun. The winter months are plotted, since these are the critical months for performance of a solar energy system, the cloudiness tending to a maximum when the hours of sunshine tend to a minimum. Because seasonal change moves both the total flux and the sunshine upward from midwinter, there is a locus for the seasonal change on clear days. The locus of cloudiness is less well defined but is indicated by the extremity of points to the left of the swath of points.

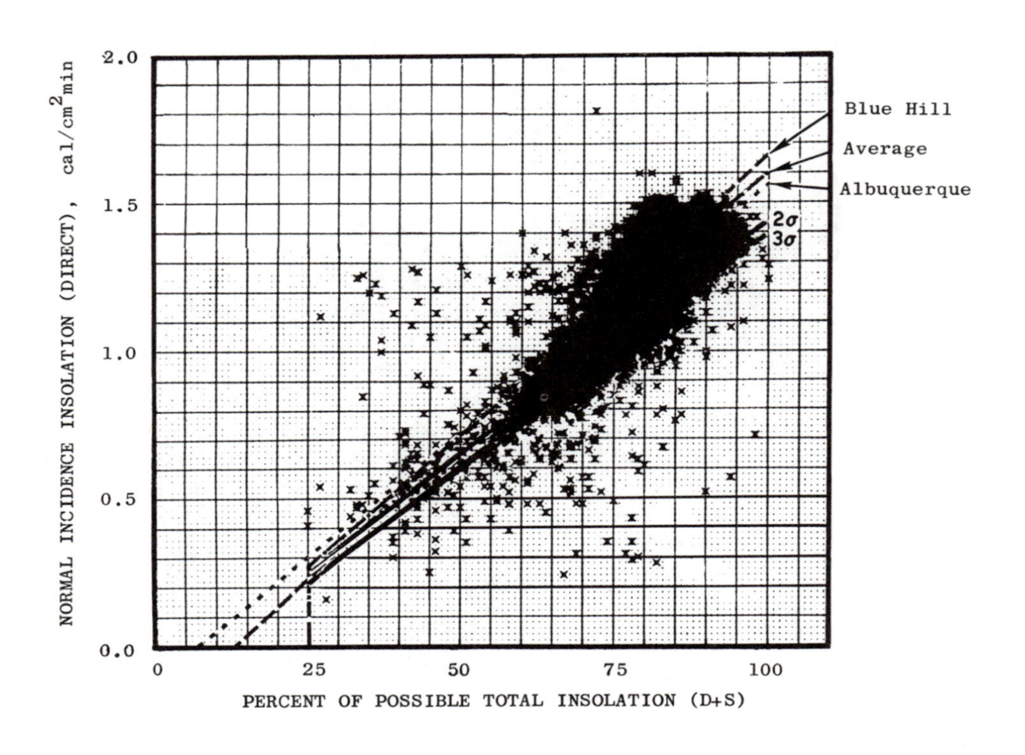

Fig. 2.23 Correlation of normal incidence insolation (direct pyrheliometer readings at noon) versus percentage of possible total insolation (pyranometer horizontal readings), after Bos (1974).

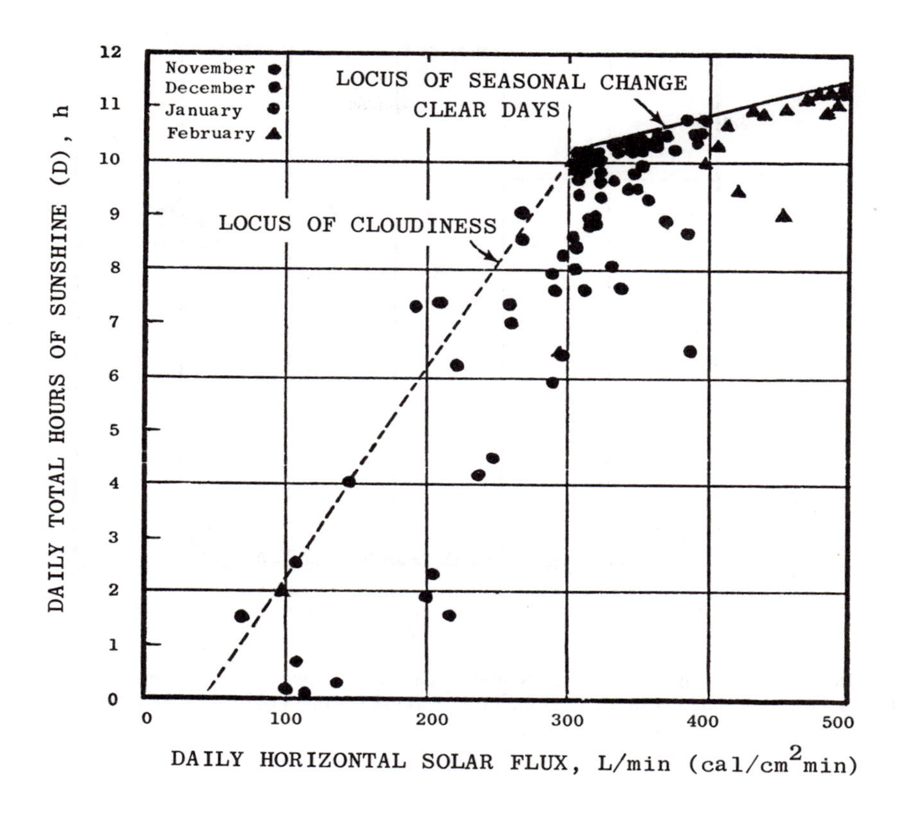

Fig. 2.24 Correlation of daily total hours of sunshine from a sunshine meter (direct) with daily total horizontal flux (pyranometer, direct plus scattered).

Correlation of the published pyrheliometer data for Tucson and total insolation is shown in Fig. 2.25, for the winter months. In this case we see that the direct solar flux is either of full brightness or zero, somewhat as one would expect from visual examination of the daily curves shown in Figs. 2.19, 2.21, and 2.22. A daily average for the direct flux is almost meaningless. When the actual records are consulted it becomes apparent that the published data were intended for other uses than ours. In Fig. 2.26 we show part of the Tucson record, showing many missing data points. These missing points represent readings affected by clouds and are omitted even if the record shows some flux level. A further limitation of this type of data is that the readings are instantaneous at the sun angles indicated, and are not averages over a period of time.

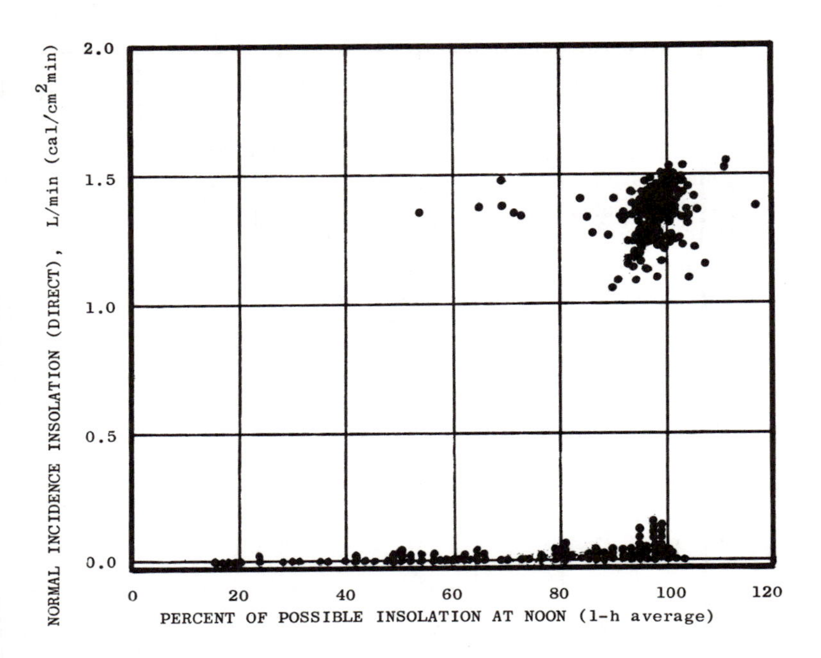

Fig. 2.25 Correlation of Tucson pyrheliometer readings at noon versus percentage of possible total insolation measured with a pyranometer, integrated for one hour (direct plus scattered), for November, December, and January 1974.

WB Form 610-9
(4-57)

U.S. DEPARTMENT OF COMMERCE, WEATHER BUREAU

NORMAL INCIDENCE SOLAR RADIATION INTENSITIES

Tabulated in Langleys Per Minute on a Surface Normal to the Direction of the Sun

Station UNIVERSITY OF ARIZONA, TUCSON Month and year APRIL 1974

SUN'S ZENIT DISTANCE	A. M.				Solar Noon	P. M.			
	78.7°	75.7°	70.7°	60.0°		60.0°	70.7°	75.7°	78.7°
AIR MASS	0.464	0.371	0.278	0.186	x	0.186	0.278	0.371	0.465
1			1.16	1.30			1.19	1.03	0.9
2				1.28		1.24	1.06		
3	0.73	0.84	0.97	1.11	1.38		1.06	0.94	0.85
4	0.99	1.08		1.35	1.53	1.36	1.19	1.07	0.97
5			1.25	1.38	1.56	1.31	1.13	1.01	0.95
6	0.94	1.05	1.17	1.32	1.49	1.32	1.17	1.05	0.94
7	0.89	1.00	1.13	1.30	1.48	1.24	1.05	0.96	0.86
8	0.83	0.95	1.07	1.26	1.37	1.18	1.07	0.91	0.80

Fig. 2.26 Portion of the record published by the U.S. Department of Commerce for pyrheliometer normal incidence solar radiation intensities, for Tucson, Arizona, April 1974.

Perhaps the best correlation between direct insolation and total insolation as it affects collector performance would be obtained from a direct comparison of collector behavior. This type of data is available in the published records of the Desert Sunshine Exposure Tests station at New River, Arizona, near Phoenix. We have taken their data for the integrated daily total flux and the total daily flux for EMMA, a concentrating collector of $X = 10$, from a nonconcentrating collector facing 45° south. The results are plotted in Fig. 2.27. The correlation is excellent, but it should be noted that, even in a desert, 17 days out of two winter months showed no direct flux during the entire day. This good correlation indicates that one could go from total daily measurements to *average* daily direct flux, again being unable to determine the fraction or time distribution of direct solar flux at normal brightness.

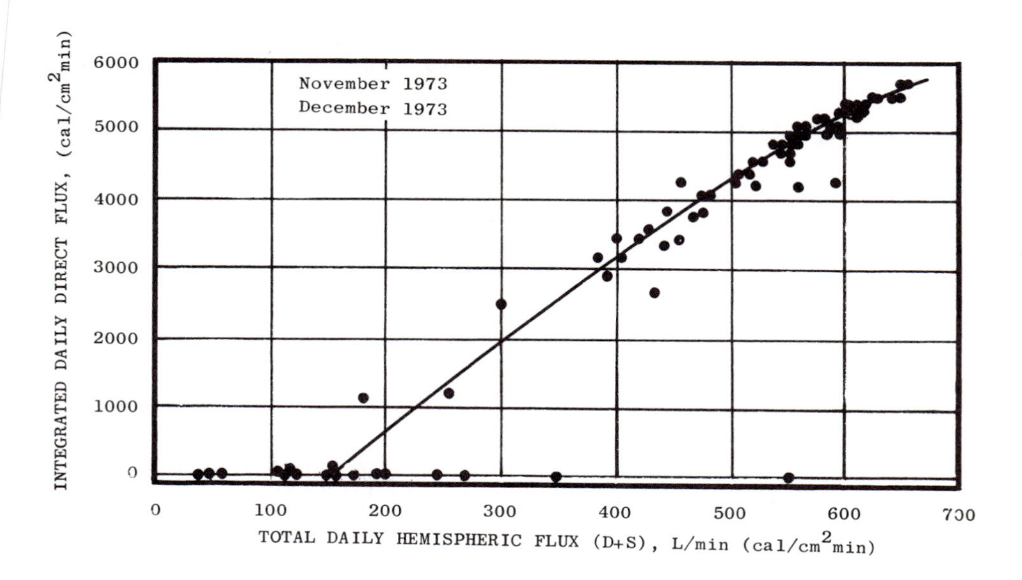

Fig. 2.27 Correlation of observations for a concentrating collector, EMMA, with a pyranometer tilted 45°S, from Desert Sunshine Exposure Tests, New River, Arizona.

A further analysis of the New River data is shown in Figs. 2.28 and 2.29, which present the annual distributions for total solar flux from a nonconcentrating collector (Fig. 2.28) and from a concentrating collector (EMMA, Fig. 2.29). From these curves one could plot the fraction of time during the year that the flux is above a given level, a useful factor in determining the effect of minimum flux cutoffs for a given system.

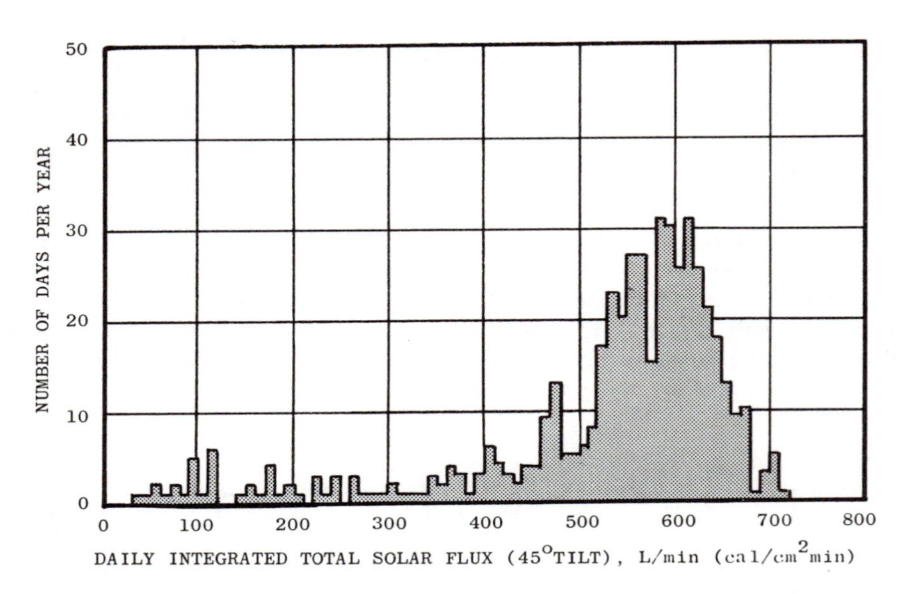

Fig. 2.28 Frequency distribution of daily total integrated flux from a pyranometer for New River, Arizona, for one year.

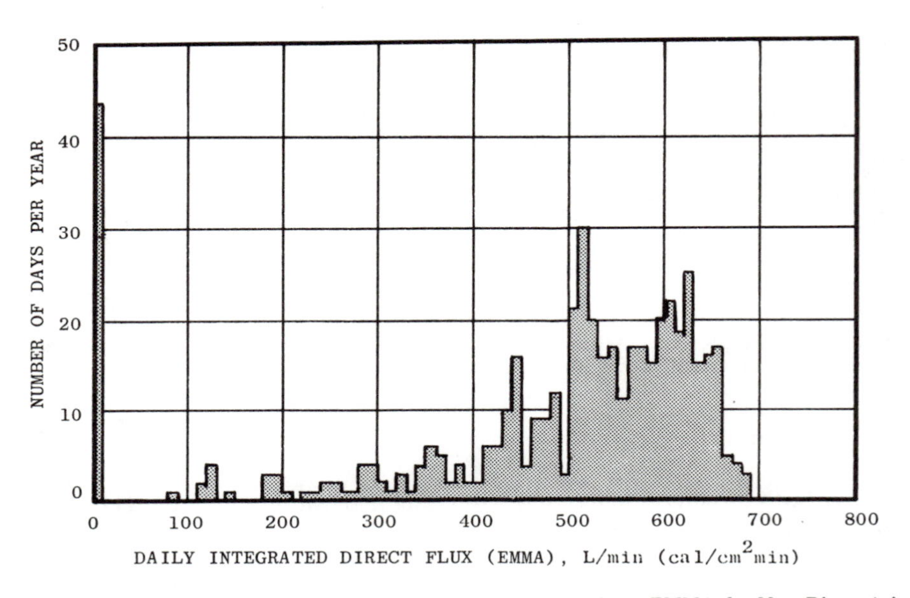

Fig. 2.29 Frequency distribution of daily integrated direct flux from EMMA for New River, Arizona, for one year. Note that 44 days showed no concentratable solar flux.

The preceding discussion shows that there are correlations between the various observables about solar flux. It must be remembered, however, that these observables deal with average values, while system performance deals with the aggregate of instantaneous values. The solar records obtained with the moving shadow-bar pyrheliometer show that in general there are large intensity fluctuations not describable by average values. According to Liu and Jordan (1960), a knowledge of the average intensity of diffuse radiations at different times of the day is needed in many problems dealing with solar radiation. Since solar radiation data are not presented for intervals shorter than one hour, the nearest approach to the true average intensity at an instant obtainable from the solar radiation data commonly available is the hourly average intensity. Again it must be emphasized that extremely variable cloudiness precludes the possibility of obtaining a true instantaneous radiation intensity during cloudy days except from direct experimentation. The same qualification holds even more imperatively for direct radiation.

2.18 SOLAR FLUX VARIATION DYNAMICS

In this section we will discuss the dynamic behavior of solar flux variations based upon digital data obtained using one-minute integrations of the direct solar flux component, the component of interest in concentrating solar collector systems.

McKenney (1974) has used detailed solar flux data to compute the amount of energy available above a given threshold using one-minute integrations. In Fig. 2.30 we show the relationship between flux cutoff and daily accumulated energy for the direct component of sunlight. Note that the accumulated daily amount rises very rapidly with lowered flux cutoff on both the clear and the cloudy day. In Fig. 2.31 we show the curves for the same two days for the total horizontal flux. The effect of a lowered cutoff flux is less steep on the cloudy day than on the clear day. It is interesting to note that on the cloudy day the total horizontal flux was slightly brighter than on the clear day, and that the total at zero cutoff was greater than for the direct component, somewhat over 5000 Wh/m^2.

The specification of the variability of sunlight is complicated by the random nature of its variations. In view of this complexity many authors simply ignore the variability and assume a clear day as the standard for evaluating performance. A steady-state analysis based upon input fluxes of a definite value enables the specification of system performance under a wide range of flux values, but these calculations are valid only when the actual flux remains at the specified value for a period of time that is longer than the system response time. A look at Fig. 2.17 will convince one that days when this condition prevails are not in the majority.

A useful tool in analyzing the performance of a solar thermal system under fluctuating solar input is that of the *modulation transfer function* (MTF). With this tool we utilize the Fourier spectrum of the input solar fluctuations. We begin by taking the variation of direct solar flux intensity, and transform it into the Fourier frequency domain. The steps in the process are outlined in Fig. 2.32. The advantage of

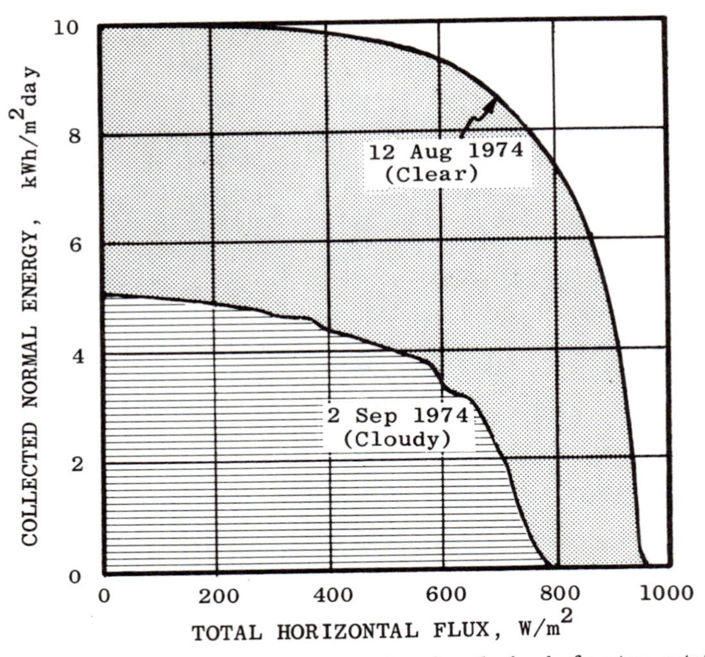

Fig. 2.30 Graph of the amount of solar energy gathered as the level of system cutoff flux is increased, for direct solar flux, for a clear day, 12 August 1974, and a cloudy day, 2 September 1974.

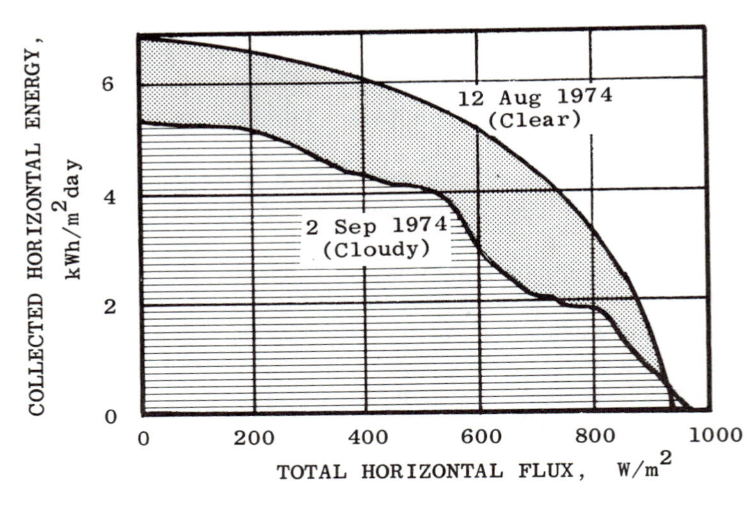

Fig. 2.31 Graph of the amount of solar energy gathered as the level of system cutoff flux is increased, for total solar flux, for 12 August 1974 and 2 September 1974, after McKenney (1974).

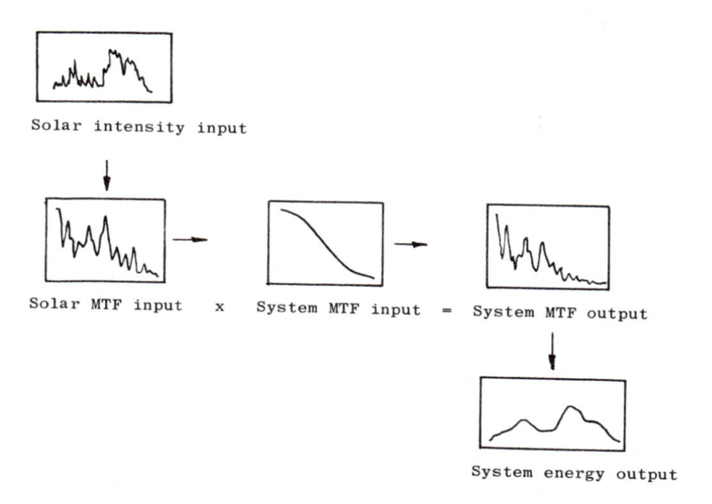

Fig. 2.32 Schematic diagram of the sequence of operations to go from the solar input to a given system to the system output. Note that input and output change from the real amplitude/time domain to the Fourier amplitude/frequency domain for the application of the modulation transfer function operations.

going into the Fourier domain is that each of the components of a system can be generally described by an MTF curve. The system output in terms of its modulation transfer function is then simply the product of the individual MTF functions, from which the actual system output is obtained by the inverse Fourier transform.

If the incident direct flux as input to the system is $I_d(t)$, we assume that it can be transformed to a function of frequency. The numerical processes are well known, detailed theorems being found in most texts on Fourier transforms. The flux can be represented by the summation

$$I_d(t) = \sum_{1}^{\infty} [A_n \cos(n\pi t/\tau) + B_n \sin(n\tau t/\tau)] + A_0/2, \tag{2.3}$$

where A_n and B_n are the Fourier sine and cosine coefficients and τ is the period. The coefficients are determined from the input flux by using the Fourier sine and cosine integrals

$$A_n = 1/\tau \int_{-T}^{T} I_d(t) \cos(n\pi t/\tau)\, dt \tag{2.4}$$

and

$$B_n = 1/\tau \int_{-T}^{T} I_d(t) \sin(n\tau t/\tau)\, dt. \tag{2.5}$$

A preferred expansion in this case is

$$I(t) = A_0/2 + \sum C_n \cos(n\pi t/\tau + \gamma_n), \tag{2.6}$$

where

$$C_n = (A_n^2 + B_n^2)^{1/2} \tag{2.7}$$

and

$$\gamma_n = a \tan(A_n/B_n). \tag{2.8}$$

Here we identify the coefficients A_0 and C_n as being the frequency spectrum of the input solar flux. The larger the value of one of the coefficients, the more energy there is in the flux that can be characterized by that frequency. The numerical value of the frequency is given by n/τ.

To illustrate the use of MTFs we show in Fig. 2.33 the spectra obtained by McKenney (1974) for two days, one clear and the other cloudy. The measurements

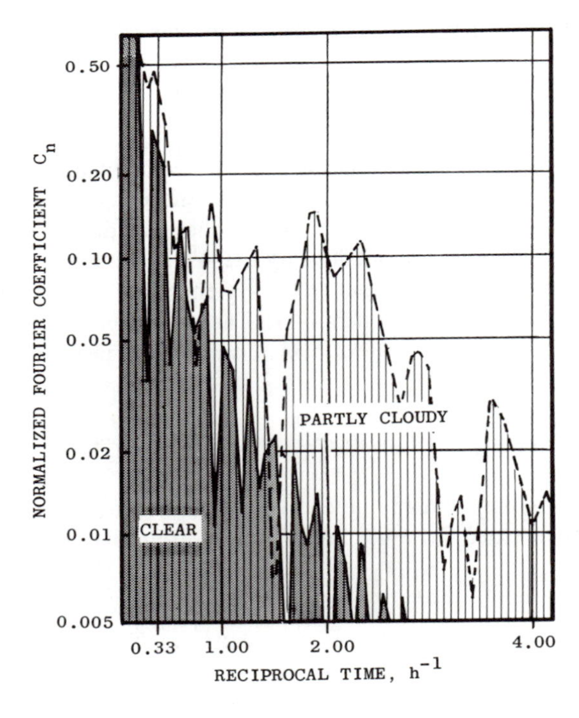

Fig. 2.33 Fourier spectra amplitudes from pyrheliometer records for a clear and a partly cloudy day. The terms for the cloudy day extend to the limit imposed by the one-minute integrations of the original observations.

were obtained with an Eppley pyrheliometer. If horizontal flux were used, the Fourier spectrum would be dominated by the first sine term. The "squareness" of the direct normal measurements, however, results in additional terms having large amplitude, but the spectrum still drops off to negligible values after the first 35 or 40 terms. The spectrum for the cloudy day drops off almost as rapidly for the first few Fourier terms, but then much more slowly, with significant amplitude for 100 to 200 terms depending on the time interval over which the input data are integrated. One would expect to see large Fourier term amplitudes at high frequencies because of the abruptness of the fluctuations shown in Figs. 2.19–2.22.

To indicate the results to be obtained from MTF analysis we show in Fig. 2.34 the cloudy-day tracing used to derive the cloudy-day frequency spectrum in Fig. 2.33. Note that in terms of fluctuation this day is reasonably simple compared to a really cloudy day. For the sample day the solar flux was the same as for a clear day for the first three hours, after which clouds developed. To approximate a using system, McKenney (1974) assumed the system MTF to be unity from the origin out to a given frequency, after which the value dropped to zero. In a real system the response would not change as abruptly but would drop monotonically, as indicated schematically in Fig. 2.32. The result of using all terms up to the frequency corresponding to 2.5 h

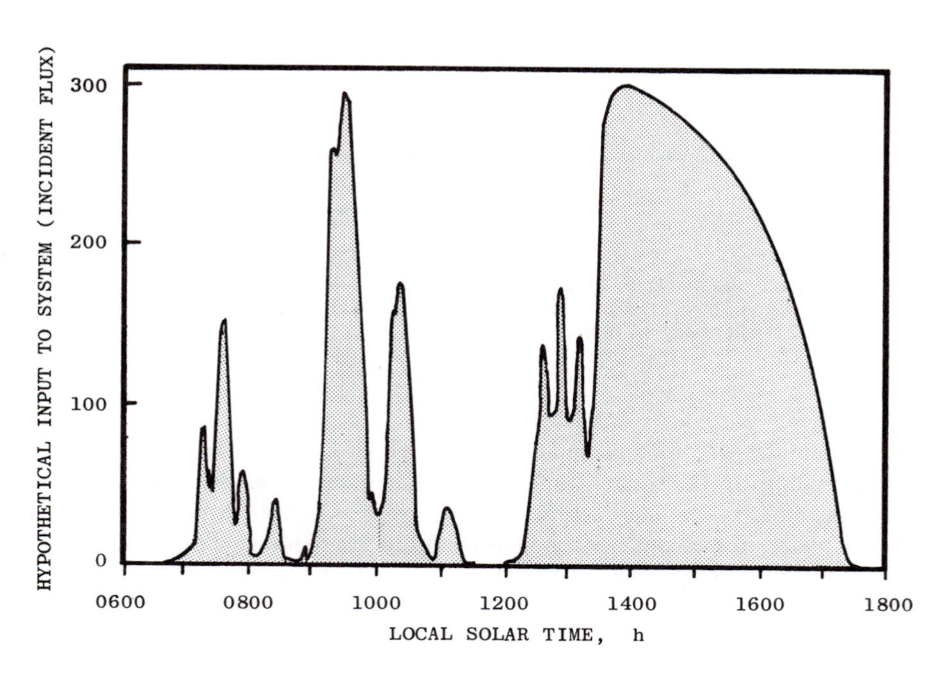

Fig. 2.34 Solar intensity/time trace smoothed from the actual recording to be equivalent to a response time of approximately 2 min.

yielded the curve shown in Fig. 2.35. Note that the three maxima are still clearly observed, with system output dropping to nearly zero in between. This response is approximately what would result if the system had a thermal inertia of about 2.5 h, showing that large thermal inertia and/or storage will be necessary to smooth out the output fluctuations on a day like this one.

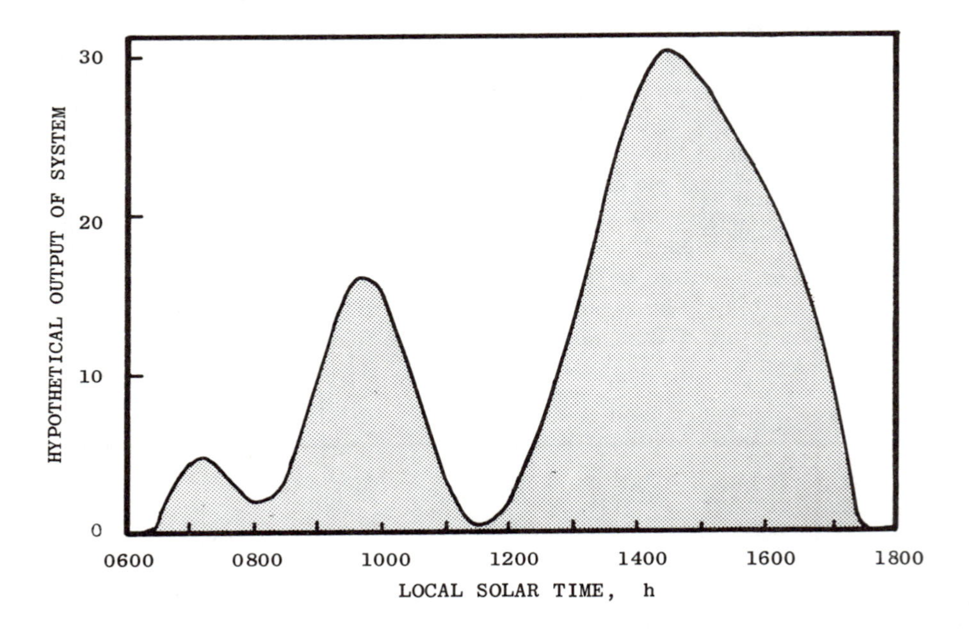

Fig. 2.35 System output versus time for the input spectrum of Fig. 2.32 with a hypothetical system MTF of unity out to 2.5 h (1/2.5 h) and zero for all higher frequencies. This result would approximate a system having 2.5 h of thermal inertia.

To state the MTF for a component of a system in simple terms, it is instructive to imagine an input to the system consisting of a sine wave of a particular frequency and amplitude from zero to unity at the peak input value. The MTF is then the amplitude of the sine wave of the same frequency issuing from the component of the system. It is assumed in the simplest form of the MTF analysis that the attenuated waves for all frequencies of interest issue with the same *phase shift,* a case that is not obvious for a real solar thermal system. Further refinement of the method of modulation transfer functions is necessary for accurate performance evaluation for a system, but the simple form described above is one step better than the constant input value currently widely used.

2.19 CORRELATION OF SUNSHINE WITH WIND VELOCITY

The correlation of direct solar flux and wind velocity is an important consideration in the use of heliostats, which are subject to the upper limit of wind velocity for continued operation. If high winds occur when there is appreciable sunshine, then energy will be lost from the annual output of the system.

In Fig. 2.36 we plot the correlation between the percentage of possible sunshine, a measure of the amount of direct flux, and average wind velocity in knots. Wind observations were taken in the middle of the energy-gathering day rather than averaged for the entire day. The reason for taking hourly averages during midday is that in desert climates a wind springs up during the day, a consequence of the heating of the desert floor. The resultant winds can be significantly higher than the 24-h average winds because desert winds at night are very low.

It can be seen in Fig. 2.36 that a significant number of clear days have hourly average winds in excess of 15 knots (27 km/h). Statistically, cloudy days have lower winds than clear days. It should be noted that the maximum winds during an hourly period are approximately 1.5 times the average wind, so an average wind of 15 knots means gusts of up to 23 knots (40 km/h).

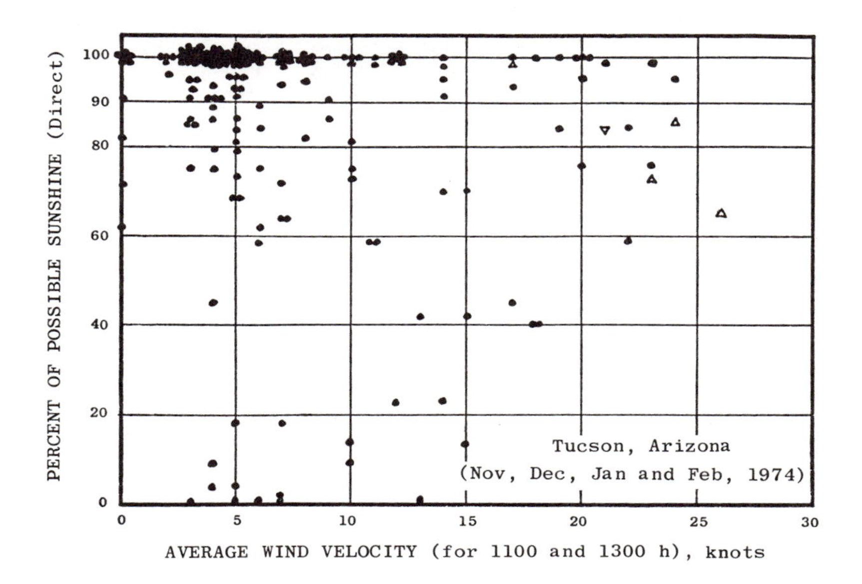

Fig. 2.36 Correlation between average wind speed for 1100 and 1300 h versus percentage of possible sunshine from a sunshine meter (direct). Note that the highest winds tend to occur on the clearest days.

In Fig. 2.37 we show the cumulative percentage of sunshine, equal to the number of occurrences of days with $x\%$ within a 5-knot interval from Fig. 2.36, as a function of the interval of velocity. The cumulative total drops off rapidly with the velocity interval. We can further summarize the wind correlation as shown in Fig. 2.38, where the percentage of energy lost above a given velocity interval is shown as a function of wind velocity. Note that, for example, if the upper limit were a 10-knot average wind, about 27% of the available direct sunshine would be lost from the system.

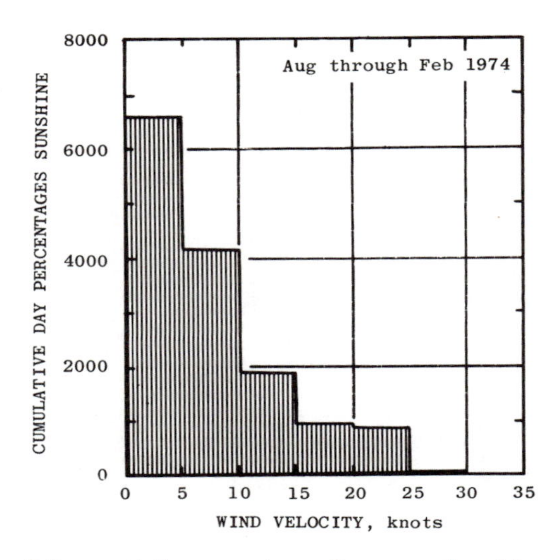

Fig. 2.37 Histogram of the cumulative percentages of energy gathered in each 5-knot interval of wind speed, for winter months.

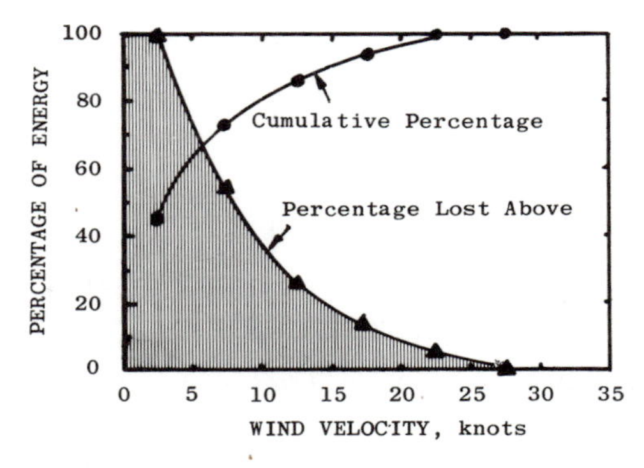

Fig. 2.38 Graphical relationship between the percentage of energy gathered at all velocities lower than a given wind velocity. If the cutoff velocity is 20 knots, then the loss of energy would be approximately 15%.

2.20 ENVIRONMENTAL THERMAL INFRARED FLUX

In discussing collector performance we will refer continually to the thermal infrared flux properties of the collector and the environment. It is convenient to denote thermal infrared flux by a notation to separate the two. We will use TIR to refer to the thermal infrared flux in general and ETIR to refer specifically to the thermal infrared flux from the environment.

The performance of flat-plate collectors is affected by the ETIR. It constitutes an additional source of incoming energy, being as large as one-third the direct solar flux. It is offset, however, by the fact that the collector radiates TIR outward into space.

The environment radiates infrared as a consequence of its temperature, modified by the variation of opacity with the wavelength of the surfaces contributing to this flux. The earth and structures can be assumed to radiate like a gray body in the thermal region, the effective emittance being assigned to the surface involved. To calculate this contribution, one need only know the temperature of the surface and its effective emittance (ϵ). The effective emittances of a number of materials are given in Table 2.6.

Table 2.6 Emittance of Various Materials

Surface	ϵ	*Surface*	ϵ
Asbestos	0.96	Plaster	0.91
Brick	0.93	Roofing paper	0.91
Fireclay	0.75	Water, ice	0.97
Concrete	0.94	Grass	0.95
Glass	0.94	Galvanized iron	0.23
Paints	0.90–0.98	Aluminum, aged	0.19

Table 2.6 shows that most materials encountered in the natural environment radiate nearly as blackbodies in the TIR. Only "bright" new metal surfaces have low thermal emittances. These surfaces generally are "hot" under sunlight and in fact radiate about the same net flux as "cool" surfaces like plaster or grass. This fact simplifies calculation of the ETIR, which can be approximated by assuming the ambient air temperature and an emittance of $\epsilon = 1.00–1.10$.

When we consider the ETIR from the atmosphere, we encounter a much more complex situation than for the land surface. Air has a wide range of opacity with wavelength, further modified by the amount of water vapor in the air. Some spectral regions are transparent to TIR, and a given volume radiates little compared to the same volume at a wavelength centered in one of the absorption bands of water vapor or carbon dioxide. If the air were of infinite extent and at a constant temperature, it too would radiate like a blackbody at the ambient air temperature, T_a. In nature the situation is quite different because a vertical temperature gradient exists within the atmosphere.

Thermal infrared radiation appears to originate from that location where the optical opacity is approximately unity. If air were totally opaque to the TIR, the radiation would originate in proximity to the collector and the TIR spectrum would then, and only then, be that of a blackbody at T_a. The effective path length for unit opacity for air in the ETIR region varies widely, from a few tens of meters in the strong water-absorption bands to kilometers in the transparent regions between the water vapor and carbon dioxide bands. The low flux observed in the 8- to 12-μm region of low atmospheric opacity is due to the atmospheric *lapse rate.* The atmospheric temperature drops with increasing altitude, so the altitude where unit opacity is reached in the transparent 8- to 12-μm region is where the local temperature is very low; hence the ETIR flux spectrum has a minimum in the regions of low atmospheric opacity.

Many authors approximate the effect of the ETIR. The simplest assumption is that the atmosphere acts like a blackbody at ambient temperature. A more appropriate assumption is that the atmosphere acts like a blackbody at reduced temperature, the difference being taken as on the order of 5 C° lower than the ambient temperature. One can determine the effective temperature difference to describe the ETIR from observations at a particular site and use that value to predict performance at that site. One encounters a problem in that this approach is correct only when the humidity of the site is average for that site. In some climates the range of humidity can be large and the effect on a collector significant.

Now that modern infrared measurements, as shown in Fig. 2.39, are available, one can improve the calculation of performance by determining the appropriate ETIR for temperature and humidity conditions. The wavelength variation of the ETIR is, moreover, very important in maximizing such practical matters as sky radiation cooling.

We show two measured examples of the ETIR from the atmosphere in Fig. 2.39. The curve in Fig. 2.39(a) is for a high, dry, cold atmosphere observed at Elk Park, Colorado, at an elevation of 11,000 ft, where the ambient temperature was 8°C. Note the low thermal infrared emission in the 10-μm region compared to that in the absorption bands at 6 and 15 μm. The opacity remains low until the line of sight is near the horizon, where the TIR flux approaches that of a blackbody at 8°C.

The curve in Fig. 2.39(b) is for a warm, humid climate, at Cocoa Beach, Florida, at sea level, with an ambient temperature of 27.5°C. The window at 10 μm is not as transparent and the curve approaches a blackbody more rapidly near the horizon than for the dry atmosphere.

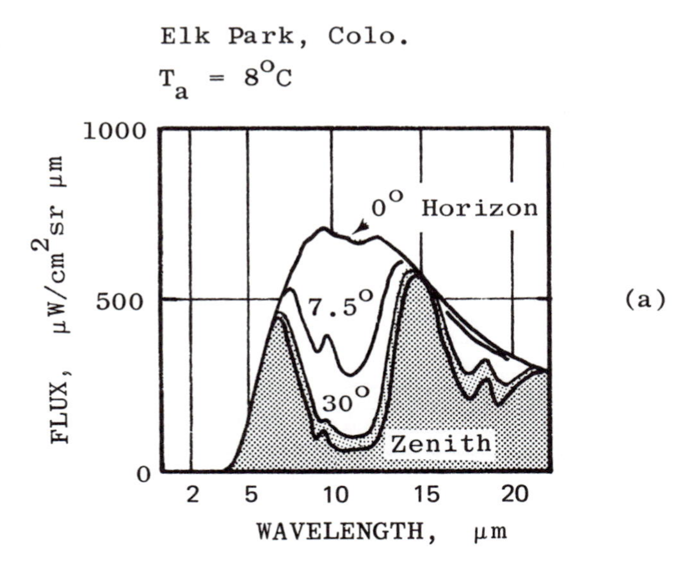

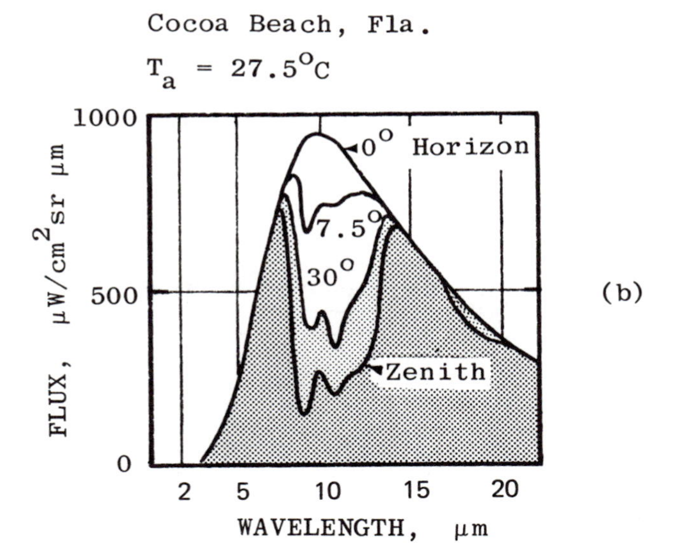

Fig. 2.39 Downwelling sky fluxes at two sites as a function of angle above the horizon (Oetjen, Bell, Young, and Eisner, 1960).

Figure 2.40 shows three high-resolution spectra of the ETIR under what might be called average summer conditions of T_a = 19°C and 74% relative humidity. Note the separation of the 9.6-μm ozone emission bands into two principal branches.

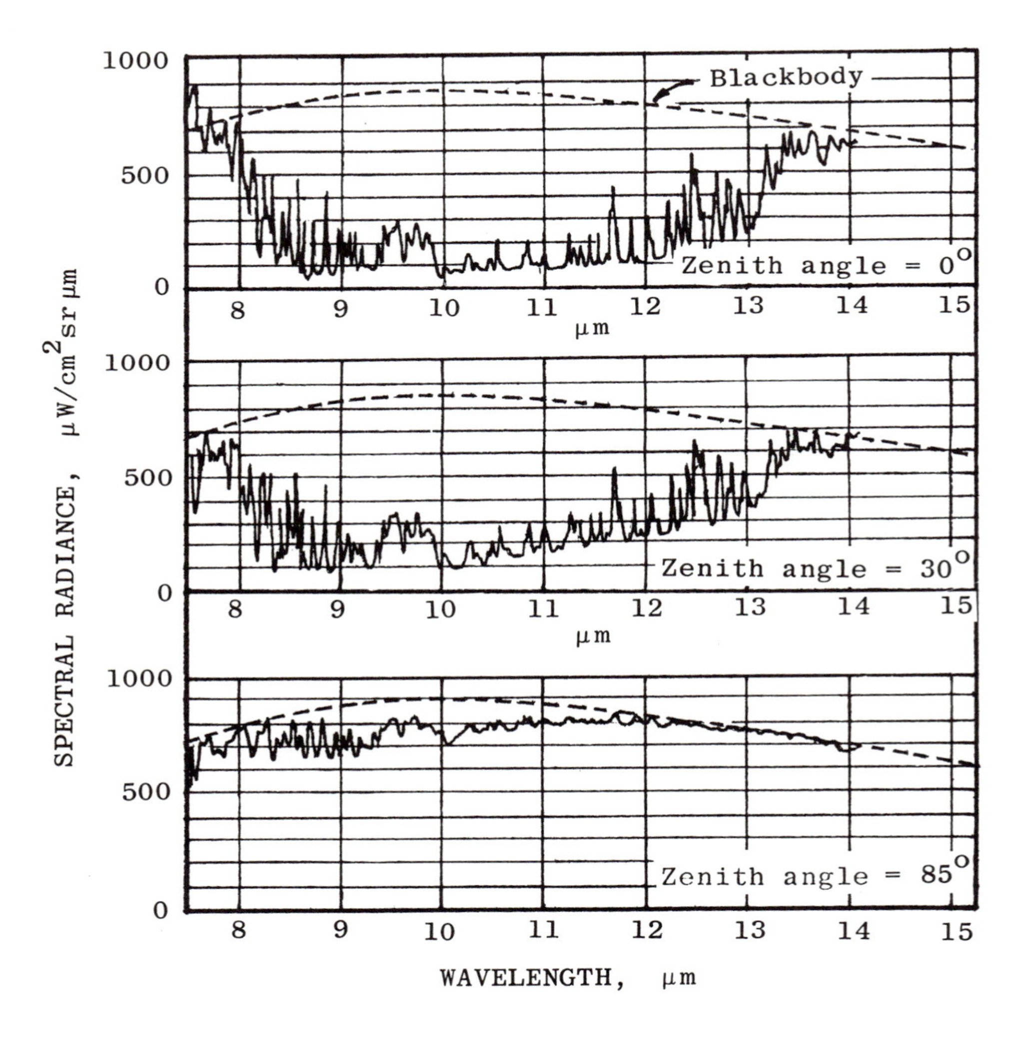

Fig. 2.40 The spectrum of the environmental thermal infrared (ETIR) measured on 24 August 1961 at S. Agata, Italy. T_a = 19°C, 965 mb, 12 g/m^3 water (74% humidity); water in vertical column is 2.1 g/m^3. AFCRL.

If we take the integrated energy under the curves in Fig. 2.39 and extend them as blackbodies beyond 20 μm, we obtain a graph of the total flux, normalized to that for a blackbody at the ambient temperature, shown in Fig. 2.41.

The two curves for Elk Park and Cocoa Beach, plotted in Fig. 2.41, also indicate the hemispherical average fraction. A horizontal collector effectively receives ETIR from the appropriate solid angle of the sky, and because the solid angle subtended by a 1° strip near the horizon is much larger than for one circumzenith, the average flux is strongly weighted toward the horizon. These two curves are informative because they may be reasonably assumed to define the range over which the ETIR can vary, but it would be interesting to see what a more comprehensive observational program would disclose about the range of the fractional blackbody ETIR.

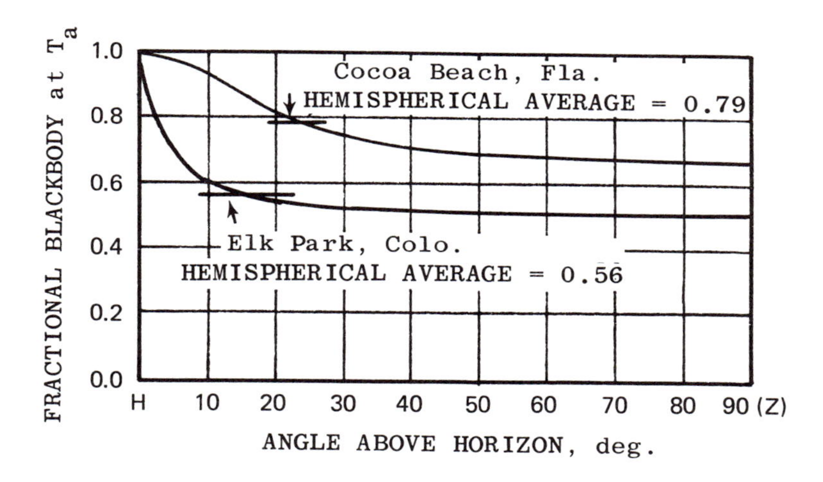

Fig. 2.41 Observed fractional blackbody ETIR as a function of angle above the horizon, averaged over the 5- to 20-μm passband. The hemispherical-weighted averages are indicated.

2.21 ETIR MODEL

In Fig. 2.42 we compile a provisional ETIR model for use in solar energy collector evaluation. The shaded zone represents the anticipated range of flux variation with zenith angle for the wettest and driest conditions. The upper curve corresponds to an ETIR over a 2π solid angle at an ETIR of 0.90 blackbody, typical of the wettest climate likely to be encountered. The lower curve corresponds to an ETIR over a 2π solid angle at an ETIR of 0.60 blackbody, typical of the driest desert climate. If lower

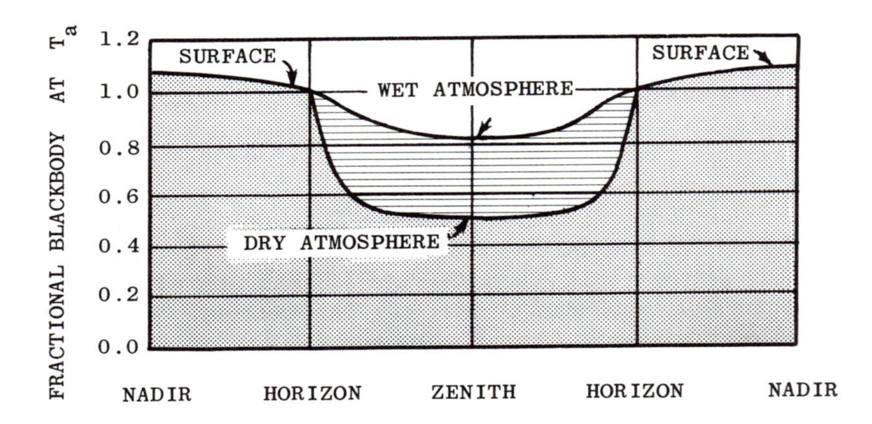

Fig. 2.42 Model proposed by Meinel and McKenney (1975) for the flux of environmental thermal infrared (ETIR) as seen by a solar collector, expressed as the fractional flux of a blackbody at the ambient air temperature.

ETIRs are encountered they will be atypical conditions, such as a high altitude or an exceptionally dry air mass in winter.

The difference between these two models is appreciable. At an ambient temperature of 30°C the ETIR fluxes would be:

$$\begin{aligned} \text{Blackbody } (30°\text{C}) &= 0.46 \text{ kW/m}^2 \text{ (kJ/m}^2\text{sec)}, \\ \text{Wet atmosphere} &= 0.41 \text{ kW/m}^2, \\ \text{Dry atmosphere} &= 0.28 \text{ kW/m}^2, \end{aligned}$$

where the hemispherical-weighted average of Fig. 2.41 is used.

In most flat-plate collectors the windows absorb this ETIR flux, so only part of it is useful, and only insofar as it modifies the flow of heat outward from the absorber. The ETIR flux is very important when one wishes to use radiation to the sky to cool a collector surface, as at night. In this case one is interested in the difference between the blackbody flux and the actual ETIR. On a dry night with sky and collector at the same temperature, one would therefore be able to radiate 0.18 kW/m^2 to the sky.

The calculation of the ETIR flux for a given collector geometry and set of atmospheric conditions involves integration of the model shown in Fig. 2.42 over the solid angle viewed by the collector. An inclined collector will view part of the atmosphere and part of the land surface. It should be remembered that the maximum contribution from a diffuse uniform flux source comes from the conical zone 45° from the normal to the collector, so a collector tilted more than 45° includes a significant portion of the ground ETIR flux.

2.22 PROBLEMS

2.1. In Fig. 2.3, three curves or points are labeled, each referring to the flux of sunlight under certain conditions. Why are these curves different? Since the curves for several altitudes appear to be about equally spaced, how would you expect them to behave for higher altitudes if the atmosphere is assumed to be homogeneous?

2.2. Taking the curves in Fig. 2.4 for standard direct and urban direct and comparing them to the curves for standard scattered and urban scattered, what can you say with regard to the upward scattered sunlight component for a hazy sky? Could you expect the same relationship to hold for cloudy skies? If so, why; if not, why not?

2.3. Assuming the urban curve for direct plus scattered sunlight, graph the relationship between flux on a horizontal plane and zenith distance of the sun.

2.4. In Table 2.4, what is the ratio of exoatmospheric flux to standard sea-level flux? What is the difference? What fraction of the difference appears as scattered sunlight at sea level? Is this ratio significantly different from that encountered in Problem 2.2?

2.5. Derive the relationship between direct, scattered, and total flux on a hazy day, assuming the scattered component escapes half upward and half downward. Plot the curves. If a city has 65% possible total insolation, what does this model predict for the relative amount of direct and scattered solar fluxes?

2.6. If the earth were a cloudless sphere tilted 23.5° to the plane of the sun, and moved in a circular orbit around the sun, what would be the equation for (a) the summer solstice and (b) the annual total of sunlight as a function of latitude? Assume no atmospheric extinction.

2.7. Based upon the model of Problem 2.5, what would be the relative amounts of direct and scattered sunlight for Stockholm and Dakar, based upon the annual average maximum daily total flux for these cities shown in Fig. 2.7?

2.8. Can you convert pyranometer solar data to the flux on other than a horizontal plane? What needs to be known or what model information is necessary to make the transformation to the flux on a plane inclined at an angle equal to the latitude?

2.9. What would be the contribution to scattered sunlight to the reading of a pyrheliometer if the scattered sunlight component were equally bright over the sky? If the sky brightness were ten times the all-sky average within the acceptance angle of the instrument, what would the contribution be?

2.10. If a moving shadow bar pyrheliometer uses a silicon chip solar cell and a glass window over the cell, how much sunlight will not be recorded? Is the difference between this instrument and an absolute thermocouple instrument significant in terms of calibration? If a quartz window is used over a thermocouple instrument, what fraction of the sunlight reaches the detector?

2.11. Estimate the ratio of direct to scattered sunlight for the days whose moving shadow-bar pyrheliometer traces are shown in Figs. 2.19–2.22. What is the total solar insolation on each of these days?

2.12. Explain in your own words the "track" in Fig. 2.24 labeled "locus of cloudiness." How does this graph relate to Fig. 2.23?

2.13. Overlay the curve in Fig. 2.27 onto Fig. 2.23. Use the central line for these two distributions. It is obvious that zeros have been omitted from Fig. 2.23. Are there other significant differences? If so, what is the probable cause?

2.14. In Figs. 2.28 and 2.29, what is the median flux in each case? Draw schematically what you would expect the curve to look like for a cloudier location.

2.15. Is a Fourier series especially appropriate for describing the frequency response differences between a clear and a cloudy day? Can you think of more appropriate functions that would enable a clear day to be described in fewer series terms? If so, how would this function relate to the frequency response of a mechanical system?

2.16. Obtain the weather summary sheets for your city from the Environmental Data Service, NOAA, and graph the sunshine data and wind data as has been done for Tucson in this chapter.

2.17. The ETIR flux shows a large gap in the spectrum from the atmosphere. Would you expect a gap in the ETIR arising from the ground? Under what conditions would the gap be present or absent? Would the effect ever be as large as for the atmosphere?

Chapter 3
Solar Availability

3.0 INTRODUCTION

There are a wide variety of collector orientations that have different advantages and disadvantages. It is important to understand the relationship between these collector orientations and the amount of solar energy each collects so that appropriate tradeoffs can be evaluated. One of the most important of these tradeoffs is the variation of collector efficiency during the year. The earth is inclined with regard to its orbit about the sun by approximately 23.5°. The annual motion of the earth therefore causes the sun to appear to move in declination by 47°, so that in the southern United States the noon sun stands near the zenith in summer but only at about 35° above the southern horizon in winter. This wide range from summer solstice to winter solstice is a major problem that a collector geometry faces. The range of angles for 15°, 30°, and 45° latitude is shown in Fig. 3.1.

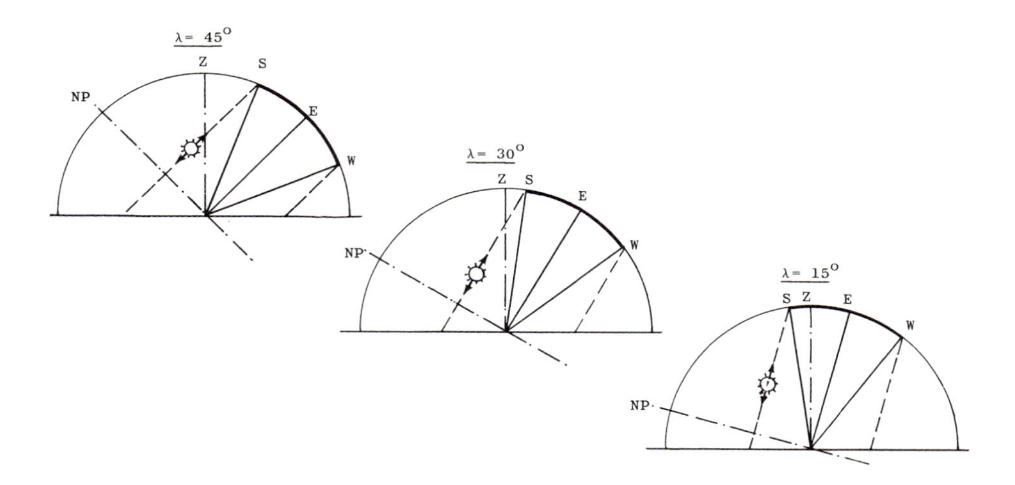

Fig. 3.1 Variation of the position of the sun for the equinox (E), summer solstice (S), and winter solstice (W) as a function of latitude. The dotted line indicates the daily path of the sun on the celestial sphere at the solstices.

3.1 ZENITH DISTANCE VERSUS TIME

The first step in evaluating the availability of sunlight for solar collectors is to determine the relationship between zenith distance and time for the latitude under study. Zenith distance as a function of latitude, solar declination, and time is given by

$$\cos z = \sin\lambda \sin\delta + \cos\lambda \cos\delta \cos t, \tag{3.1}$$

where λ is latitude of the collector site, δ is declination of the sun, and t is the hour angle of the sun, equal to the number of hours that the sun is from the solar meridian (noon). The relationship in graphical terms is shown in Fig. 3.2.

In the following sections we will discuss the various geometries and give the equation relating each to the fundamental variables describing the collector orientation and sun position. The value of the graph in Fig. 3.2 is that one can use it to determine the actual solar flux at the time desired, using the sun-zenith distance models defined in Chapter 2, depending on whether the direct solar flux or total sun plus sky is desired. It also enables one to develop other curves, as for an urban situation. In these curves we use the sea-level desert sun model of Chapter 2, where the zenith direct solar flux is 0.970 kW/m^2.

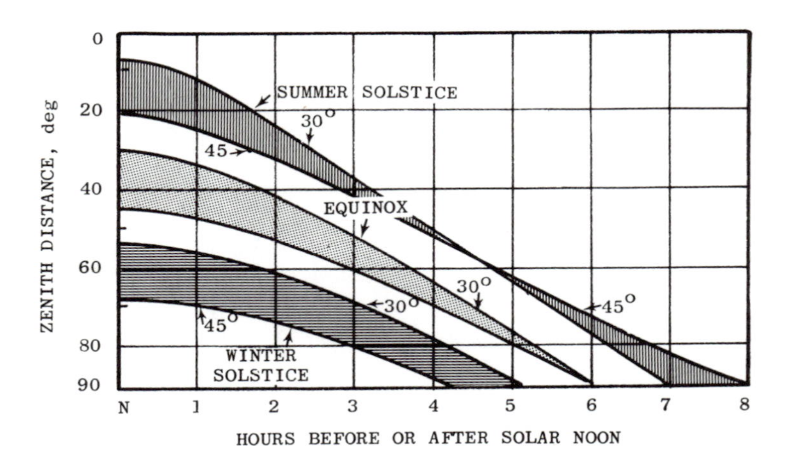

Fig. 3.2 Variation of zenith distance with time of day for latitudes between 30° and 45° as a function of season.

3.2 TIME OF SUNRISE AND SUNSET

The variation of sunrise and sunset times as a function of latitude and time of year is presented in the nomogram in Fig. 3.3. Two lines are needed. The first, shown as a dashed line here, is drawn through the center of the diagram and the latitude of the location, taken as 40° in the sample. A second line is drawn horizontally through the date. The dates of July 1, August 1, and so forth are also shown as horizontal lines. One interpolates between these lines for the date within the month. For guidance, a curve of solar declination versus date is drawn, half on each end of the diagram. In the example, if we want the time of sunrise and sunset at 40°N on August 1, we read sunrise at 0500 h (5:00 a.m.) and sunset at 1900 h (7:00 p.m.). At the summer solstice the times are 0435 and 1925 respectively.

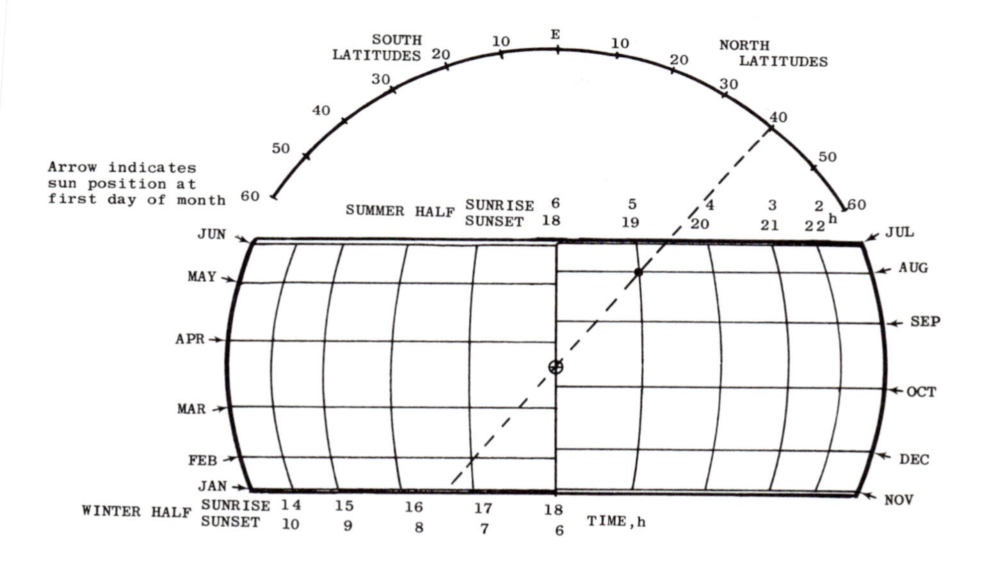

Fig. 3.3 Nomogram for calculating the path of the sun in the sky and the sunrise and sunset times for different times of the year and different latitudes. (Add one hour for Daylight Saving Time.)

Figure 3.3 can also be useful for visualizing the path of the sun during the day as a function of latitude. In Fig. 3.4 we show the ground for the sample location of 40° as a shaded area. If this ground surface is placed horizontally, the path of the sun in the sky is as shown by the lines for different dates. If another line, also dashed, is drawn from the center to a point for a time of day for a given date, the angle above the southern horizon to the sun for an east-west collector is indicated by the angle α.

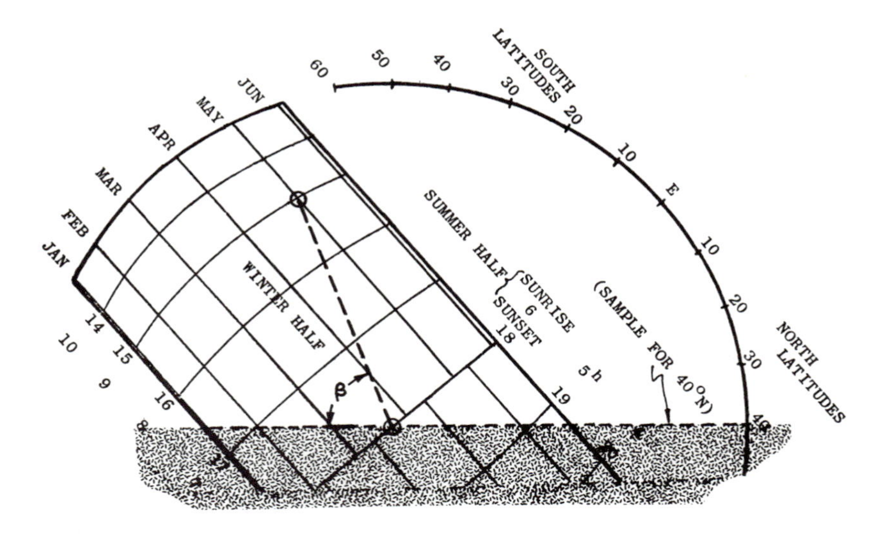

Fig. 3.4 Diagram illustrating the use of the nomogram for determining the vertical angle of the sun from the southern horizon at a given latitude and season and time of day. The example is for 40°N latitude on 1 May at 0900 or 1500 h (Standard Time).

3.3 FULLY TRACKING COLLECTOR

The fully tracking collector is one in which the optical axis of the collector is pointed continuously at the position of the sun in the sky. It is exactly like a telescope in that motion in two coordinates is required. The exact coordinates in which the motion is made are not important. An equatorial mounting where only one uniform motion is required is a redundant case, since one axis of motion (east-west) is made parallel to the axis of rotation of the earth. As a consequence, the sun appears to have no significant daily motion in the transverse coordinate (declination). Any other set of axes of motion, such as altazimuth, requires two motions to track the daily motion of the sun; as a consequence, this type of mounting appears to carry with it a cost increment over simpler mountings, so one is interested in the additional amount of energy the collector gathers each day.

The fully tracking configuration collects the maximum possible sunlight, when one neglects the possible interference of nearby similar collectors when the sun is near the horizon. In the summary tables, therefore, we show this configuration providing 100% for each of the three seasons tabulated.

There are two cases of interest (shown in Figs. 3.5 and 3.6). The principal one (Fig. 3.5) is the variation for direct sunlight (*D*) only, since most fully tracking col-

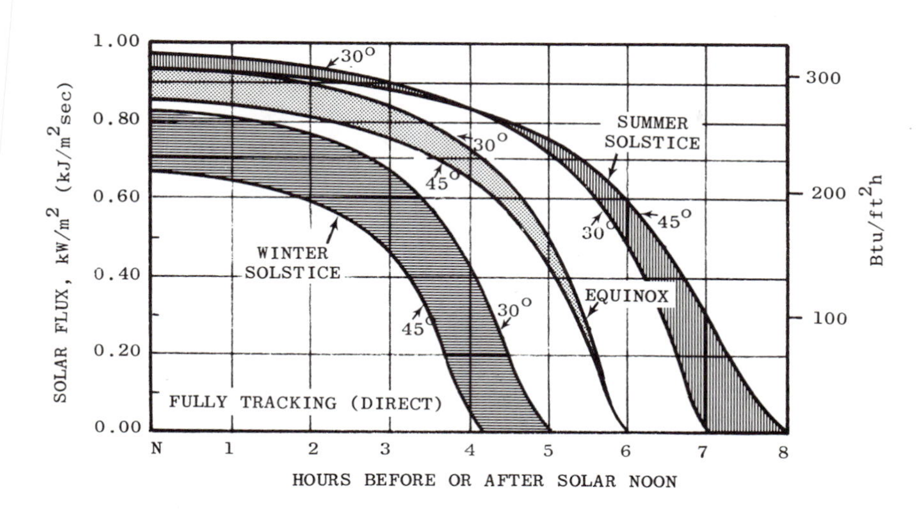

Fig. 3.5 Variation of solar flux for a fully tracking collector using only direct solar rays as a function of time of day and season. The upper edge of the band is for 30° latitude and the lower edge for 45° latitude, except for summer solstice where the curves reverse.

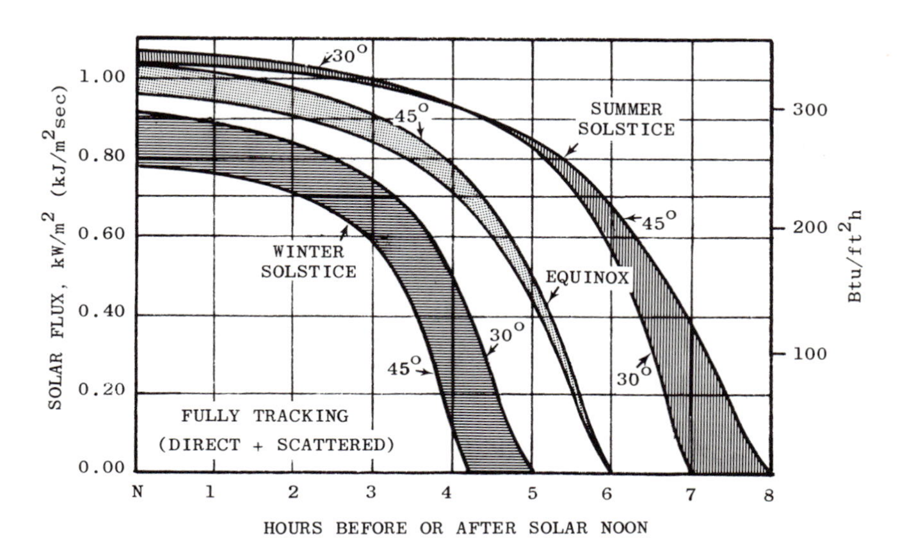

Fig. 3.6 Variation of flux for a fully tracking collector for combined direct and scattered solar flux.

lectors are of the focusing type. The second one (Fig. 3.6), which would apply to a flat plate in a fully tracking mount, is for the variation for direct sun plus scattered sunlight $(D + S)$. Graphs and tables giving the total flux gathered each day for the fully tracking and other collector geometries are presented in Section 3.14.

Note that a characteristic of the fully tracking configuration is that the daily solar flux curve is rather "square'" If there were no atmosphere the curve would be precisely square, the sun being of constant brightness until cut off by the horizon. The rounded profile is due to the atmospheric absorption as assumed in the models derived in Chapter 2.

3.4 VARIATION OF FLUX CURVES WITH LATITUDE AND GEOMETRY

The solar flux curves for different geometries and seasons vary significantly with latitude. Since these curves should be useful over a wide range of latitudes of possible interest to the user, the curves should present enough information to enable their use at a specific latitude. The curves we present are in the form of a band of values, with a separate band for each principal season–summer solstice, equinox, and winter solstice. In general, the upper curve is for a latitude of 30° north or south, and the lower curve is for a latitude of 45° north or south.

To further express the range of possible values we have taken the solar flux versus zenith distance curve as being that for a "desert standard atmosphere sun" for the 30° latitude curves and a "standard atmosphere sun" for the 45° latitude curves, a difference of from 0.97 kW/m^2 for the desert sun to 0.94 kW/m^2 for the standard sun.

The curves can be used for any latitude between 30° and 45° simply by estimating where the curve would be for the specific latitude of interest. One can always quickly derive the noon value of the flux simply by noting the zenith distance of the sun at noon for the given latitude and finding that point on the left-hand ordinate of the curves, constructing the curve from that point following the trend of the illustrated curves.

For the greatest accuracy one can always calculate the value of the solar angle with respect to the collector by means of the equations presented for each geometry.

The general equation for calculating the direct solar flux component on an inclined flat, inclined at an angle θ measured from the horizontal and at an azimuth angle A from the meridian, is given by

$$\begin{aligned} I = I(z)\Big\{&\cos\theta\,(\sin\lambda\,\sin\delta + \cos\lambda\,\cos\delta\,\cos t) \\ &+ \sin\theta\,\cos A\,[\tan\lambda\,(\sin\lambda\,\sin\delta + \cos\lambda\,\cos\delta\,\cos t) \\ &- \sin\delta\,\sec\lambda] + \sin A\,\cos\delta\,\sin t\Big\}. \end{aligned} \tag{3.2}$$

This equation will be simplified in the following specific cases so that the variations caused by different geometries can be readily assessed.

3.5 FIXED HORIZONTAL FLAT PLATE

The simplest case of a fixed collector is the one where the collector is horizontal and directed to the zenith. The solar flux arriving on the collector per unit collector area is given by

$$I(c) = I(z) \cos z, \tag{3.3}$$

where zenith distance (z) is related to time since solar noon (t) by the curves presented in Fig. 3.2. Solar noon can differ from local astronomical time by as much as 15 min owing to the ellipticity of the earth's orbit. Solar noon can also differ from local time when the site is off the center of the time zone.

Note that the daily variation of solar flux no longer has the "squareness" shown in the case of a fully tracking collector. Owing to the cosine of the zenith distance, this change reduces significantly the total amount of energy gathered by a fixed horizontal collector per day, a factor that must be traded off with the cost differences between the two basic types of collectors. Figure 3.7 shows the variation for the direct component of sunlight and Fig. 3.8 for the total of direct plus scattered sunlight.

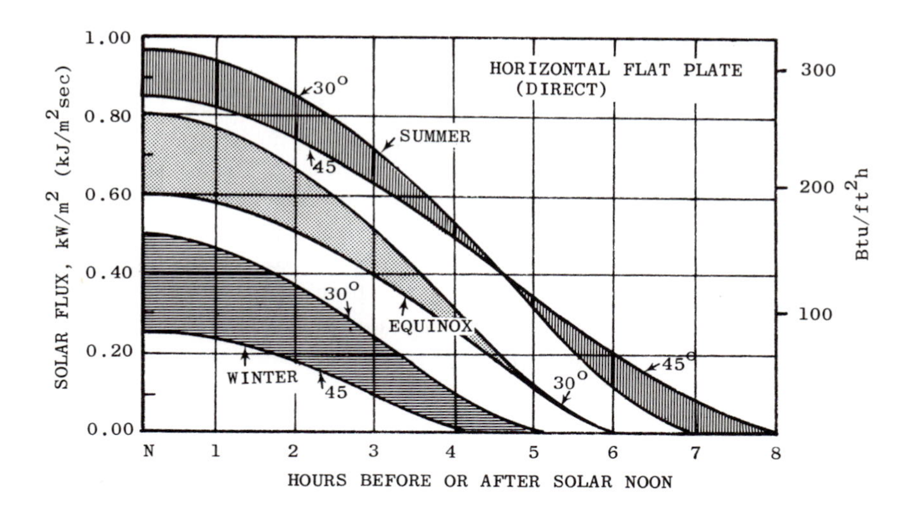

Fig. 3.7 Variation of solar flux (direct sun) on a horizontal flat plate as a function of time of day and season.

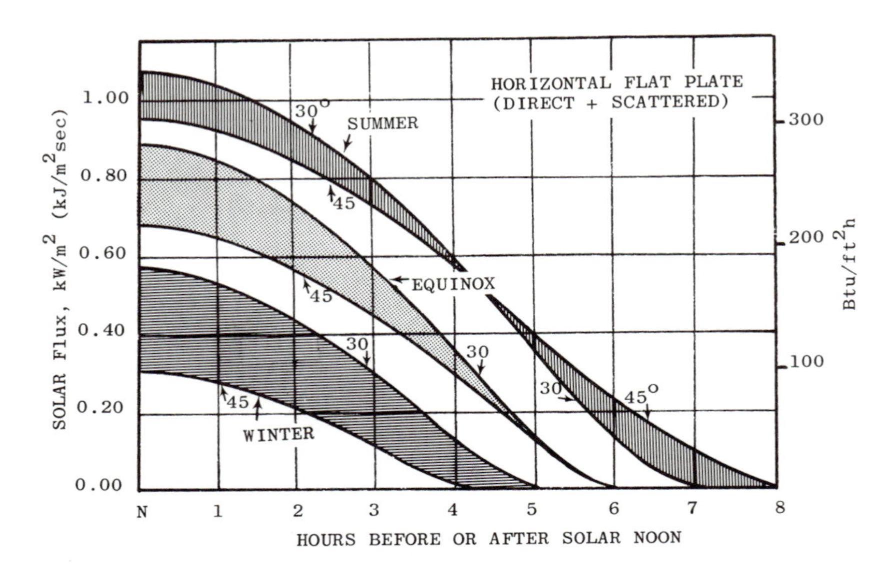

Fig. 3.8 Variation of flux from direct plus scattered sunlight on a horizontal flat plate as a function of time of day and season.

3.6 FIXED LATITUDE TILTED FLAT PLATE

A significant change in seasonal performance can be obtained with a flat-plate collector when it is tilted toward the south. In this case we show the behavior when the collector is tilted at an angle that remains fixed throughout the year. Figure 3.9 shows the case where the collector is tilted at an angle equal to the latitude of the collector. The disadvantage of this angle is that no sunlight is incident on the collector surface during the summer until 6 h before noon; the sunlight is also lost 6 h after noon.

Figure 3.10, which shows one component of the solar zenith as a function of time, is useful for understanding the diurnal position of the sun for the summer solstice, equinox, and winter solstice. We will refer to this diagram later.

When the plate is tilted at an angle θ (measured for the normal to the plate, from the zenith) and where θ is in the meridian, the equation for the angle between the sun and the collector normal is

$$\cos\beta = \sin(\lambda - \theta)\sin\delta + \cos(\lambda - \theta)\cos\delta \cos t. \tag{3.4}$$

In the case illustrated in Fig. 3.9, where the plate is normal to the latitude of the site, $\theta = \lambda$, the equation reduces to

$$\cos\beta = \cos\delta \cos t. \tag{3.5}$$

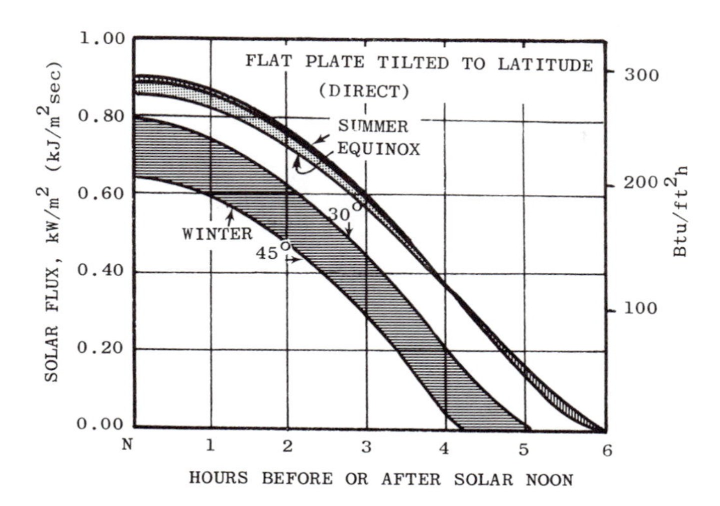

Fig. 3.9 Variation of flux from direct sunlight on a flat plate tilted and fixed at the latitude of the location as a function of time of day and season. Note that the summer and winter performances are equal, and that the tilt eliminates part of the summer sunshine day. For flux for direct plus scattered sunlight, add the sky plus ground albedo as appropriate to the location.

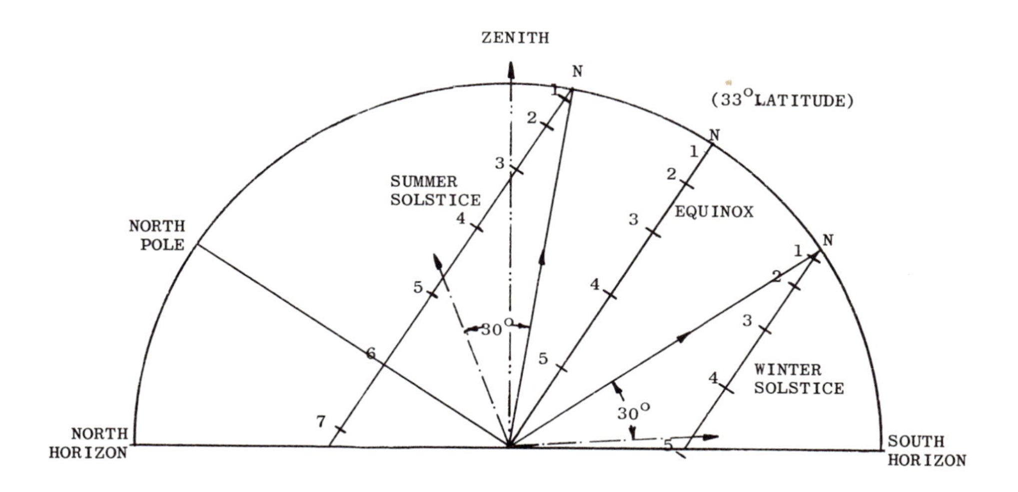

Fig. 3.10 Geometry for the sun as a function of time of day and season for latitude 33°. For an east-west collector a north-south acceptance angle of 30° will accept the direct sunlight for a period of more than 9 h at summer and winter solstice.

Figure 3.9 shows that there is little difference in collector performance for the summer solstice and the equinox, and that the performance at the winter solstice is much improved over that for the horizontal flat plate.

3.7 FIXED LATITUDE, PLUS 15°, TILTED FLAT PLATE

When the tilt of the plate is increased to angles larger than the latitude, the differences between summer and winter performance decrease further. This angle makes the surface almost normal to the winter noon sun, increasing the solar input in winter. The angle for the summer sun is increased so that the cosine factor makes a significant reduction in summer flux; this also reduces the number of hours a day the sun strikes the collector, since the sun is north of the zenith part of the day.

Many researchers recommend that, for optimal seasonal performance for heating the collector, tilt be increased from that of the latitude to the latitude plus 15°. The collector then points 15° below the noon equinox sun position. The resultant curves for 30° and 45° latitude are shown in Fig. 3.11. There is a small increase in winter performance and a significant decrease in summer performance.

The actual best angle for a fixed meridional collector depends in detail upon the cloudiness statistics of the proposed location and the annual demand for energy that

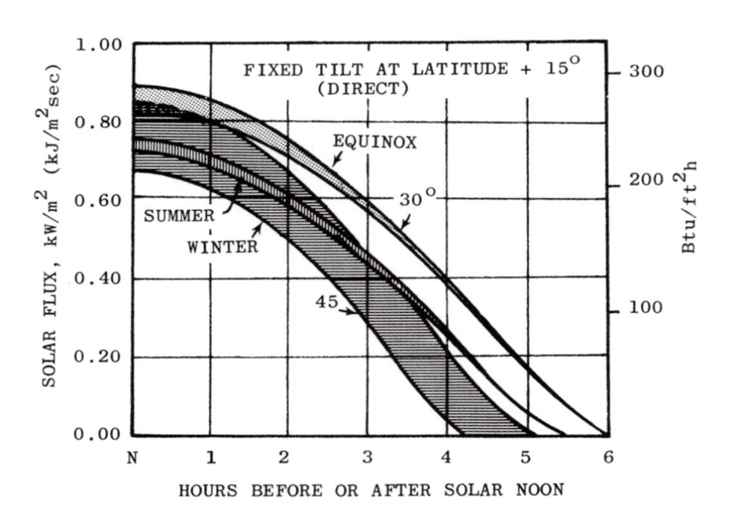

Fig. 3.11 Variation of solar flux on a flat plate tilted at a fixed angle of latitude plus 15° to better face the winter sun, as a function of time of day and season. Note that although winter performance is only slightly improved, the curves move together because summer performance is considerably degraded.

will be placed on the collector system. In Arizona the optimum angle for year-round usage is almost horizontal, since the summer cooling load requires more solar joules than the winter heating load. In northern states the optimum angle will be close to that described above, since the summer cooling load is almost nonexistent and the winter load severe.

It must be remembered that these curves for the flux gathered by the different solar collector geometries apply only to a clear day. Seasonal differences in clarity and cloudiness are important in assessing the capability of a given geometry to satisfy demands. For example, equalizing summer and winter performance can be done through geometrical means, but the probability of cloudiness is higher in winter and will reduce the net monthly yield even though performance on a clear winter day would be the same as on a clear summer day. One also has the additional demand factor to consider. In some locations more energy is demanded in winter; in others, more is demanded in summer.

3.8 VERTICAL SOUTH-FACING FIXED FLAT PLATE

One geometry of particular interest for house heating is that in which the collector faces south and is positioned nearly vertically. When the angle is vertical for latitudes in the 30–35° range, the high angle of the sun in summer results in the collector automatically ceasing to provide heat input; but in winter the vertical wall works efficiently as soon as the morning sun rises in the southeast. The flux curves for this geometry are shown in Fig. 3.12 for the case of the direct component of sunlight. The actual condition for the vertical wall is that significant ground-scattered radiation will be gathered, so the case where the ground has the same scattering efficiency as the atmosphere is shown in Fig. 3.13. Note that although the winter efficiency has been significantly boosted, the total summer flux remains low. One can further boost the winter performance if the ground is snow covered or otherwise whitened.

When the effect of scattered sunlight is taken into account, as was shown in Fig. 3.13, the winter performance is significantly boosted. In this figure we assumed the ground to scatter like the sky—about 15%, typical of a clear day. Note that the summer input is boosted, so such a collector at low latitude only partially turns itself "off" during the summer.

In actual situations where hazy or cloudy skies are common, the effect of scattered sunlight is very important. At high latitudes the amount of sunlight can actually be larger on the north wall of a building. The sunlight gathered on the east and west walls can also be important to the heat balance of a house where hazy skies are common.

To summarize the variation of collector performance as a function of the tilt of the collector we show in Fig. 3.14 the variation for 40° latitude. Note that at the latitude plus 20° the annual curve shows the minimum fluctuations. A general rule applied by many researchers in the United States is that the optimum angle should be $\lambda + 15°$, but Veinberg (1959) states $\lambda + 20°$. On the basis of the curves in Fig. 3.14 the

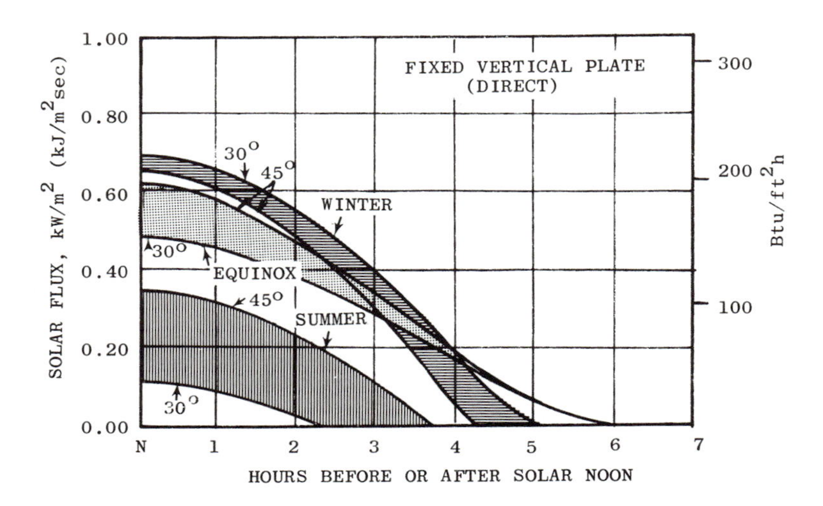

Fig. 3.12 Variation of solar flux on a vertical fixed plate for direct sunlight as a function of time of day and season. Winter performance for 30° and 45° latitude is approximately equal.

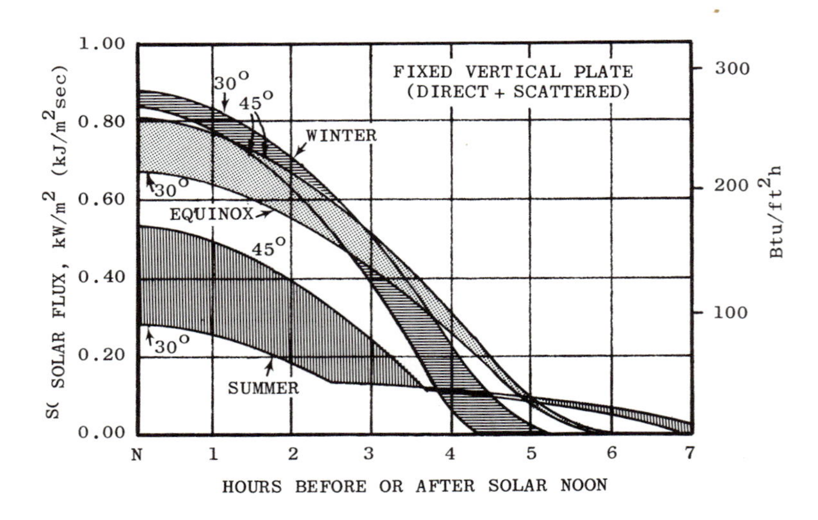

Fig. 3.13 Variation of solar flux on a vertical fixed plate for direct sun plus scattered (diffuse) sunlight scattered off the surrounding landscape and sky.

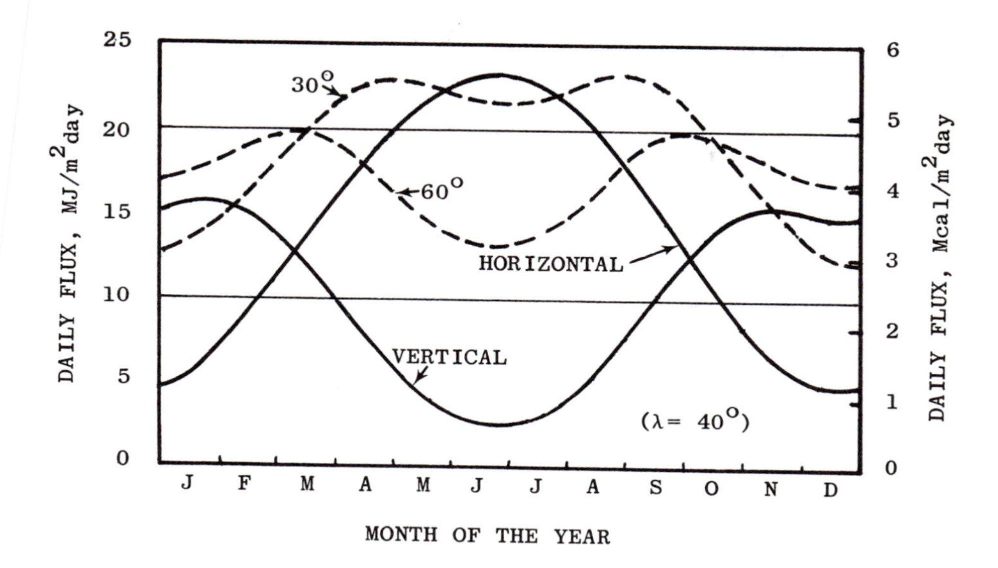

Fig. 3.14 Variation of daily energy input to an inclined flat-plate collector as a function of inclination angle from the horizontal at different times of year for 40° latitude.

optimum would appear to be +20° or slightly more, considering the higher relative amount of cloudiness in winter as compared to summer.

3.9 SEASONALLY TILTED FLAT PLATE

One can further improve the performance of the geometry shown in Figs. 3.9 and 3.11 if the tilt is changed from time to time during the year so that the normal to the collector points to the zenith distance of the sun at noon. This modification is not a great change over the fixed tilted collector because the change need not be continuous, but incremental. Figure 3.15 shows the number of *consecutive* days that the sun stays within a 4° "window" in zenith distance at noon. Note that the sun is essentially close to either the summer solstice or the winter solstice most of the year, moving rapidly between the extremes. This means that for nearly 70 consecutive days the sun is within 4° of an extreme position, spending only 9 days in the 4° window at the equinox. This means that a seasonally tilted collector need be changed only occasionally.

Tilted collectors can cast shadows on adjacent collectors. This means that in terms of land utilization (*fill factor*), tilted collectors lose some of their gains relative to land fully covered with horizontal collectors. We will discuss fill factor in more detail in Section 6.14.

The variation of incident flux as a function of time of day is shown in Fig. 3.16. Note that the difference between winter, equinox, and summer is significantly decreased from the case of a fixed tilted collector shown in Figs. 3.9 and 3.11.

The equation for the angle between the normal to the collector and the sun is

$$\cos\beta = \sin^2\delta + \cos^2\delta \cos t . \tag{3.6}$$

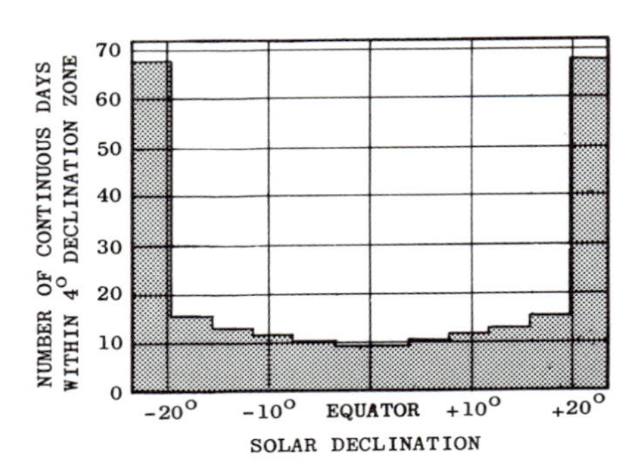

Fig. 3.15 Graph of the number of consecutive days that the sun remains in a given 4° zone of declination, showing that the sun spends most of the time near the extremes of summer and winter solstices.

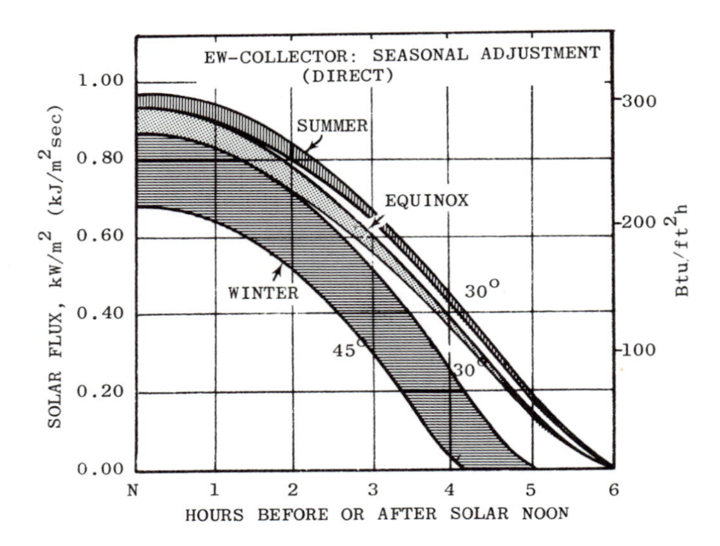

Fig. 3.16 Variation of solar flux for a collector tilted seasonally to point to the elevation of the noon sun, for direct sunlight only, as a function of time of day and season. Because the flat plate is tilted on the meridian, the collector never receives sunlight beyond 6 h.

3.10 NORTH-SOUTH HORIZONTAL, EAST-WEST TRACKING

The basic geometry for a collector placed horizontally with its axis of rotation in a north-south horizontal plane is shown in Fig. 3.17 and the performance curves are shown in Fig. 3.18.

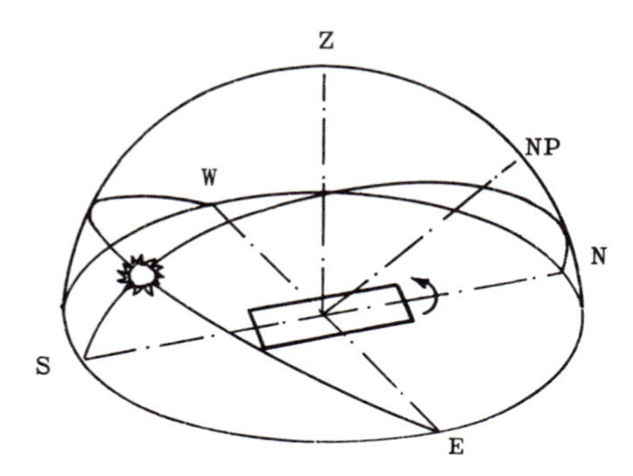

Fig. 3.17 Geometry of a collector having its axis of rotation in a north-south horizontal plane, with east-west daily motion.

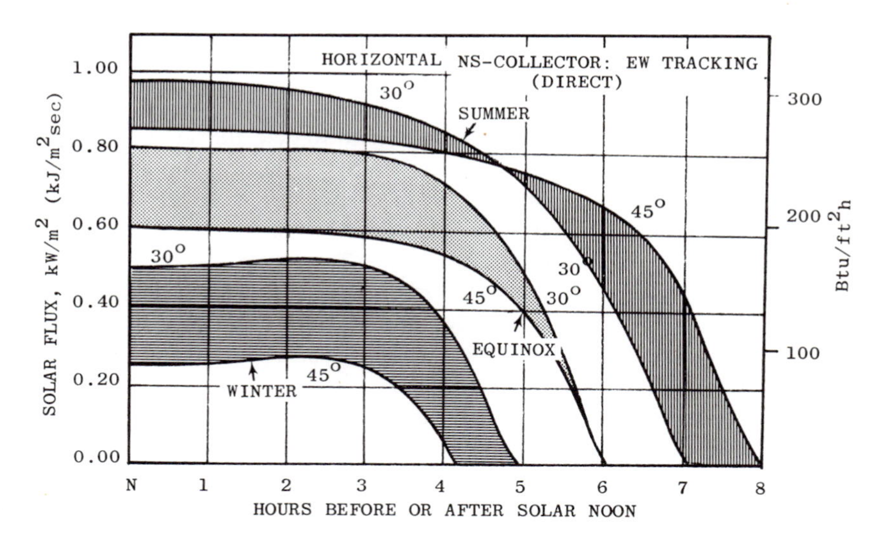

Fig. 3.18 Variation of solar flux for a collector with rotation axis north-south and daily motion east-west as a function of time of day and season, for direct sunlight only. Note that for an array of such collectors, the fill factor will cut the sun off somewhat before it reaches the horizon, the time depending on the latitude.

The axis of a horizontal collector could, if necessary, be placed in any azimuth, but we discuss only the north-south (NS) configuration (this section) and the east-west (EW) configuration (Section 3.12). The NS axis has an advantage in that it approximates the fully tracking collector in summer, but the cosine foreshortening in winter greatly reduces its effectiveness. The two principal variants of the NS axis configuration are (a) axis horizontal and (b) axis parallel to the axis of rotation of the earth.

For the case of a NS horizontal axis, EW tracking system, the equation for the angle between the collector normal and the sun is

$$\cos\beta = \cos z \cos t + \cos\delta \sin^2 t. \tag{3.7}$$

The curves in Fig. 3.18 show that this mount yields a rather "square" profile, ideal for leveling the variation during the day of some collector configurations. The winter curve, however, is seriously depressed relative to the summer curve, so winter performance, when one typically has the worst cloudiness and shortest days, would be poor. The drop in winter performance is even more serious for higher latitudes.

3.11 NORTH-SOUTH POLAR, EAST-WEST TRACKING

One can elect to raise the axis of a NS collector until the axis is aligned with the polar axis of the earth, as shown in Fig. 3.19. This means that the sun is normal to the collector at the equinox and the cosine foreshortening is minimized at the solstices, but with the complication of additional structure. In this case the equation for the angle between the collector normal and the sun is

$$\cos\beta = \cos\delta. \tag{3.8}$$

A disadvantage of the polar axis arrangement is that the upward tilt of the collector causes one collector to shadow another to the north. The arrangement is further reduced in effectiveness in that, when several collectors are deployed, the collectors to the east and west obscure the sun when the hour angle of the sun is large. Shadowing completely blocks access to the summer sun for two to three hours in the morning and again in the afternoon. The result is that the polar mount collector configuration has a relatively poor fill factor, especially if shadowing of collectors during winter is to be minimized to keep up annual performance factors. For single collectors the polar axis configuration is excellent.

The performance curves for the polar axis fixed declination configuration are shown in Fig. 3.20. The equinox performance and summer noon performance are essentially equal, the smaller air mass for summer solstice offsetting the small cosine projection effect. The winter noon value, on the other hand, is reduced because these two effects combine rather than cancel. If one desired to equalize the summer and winter curves, an inclination higher than polar would suffice; but the physical height of such a configuration would be a potential penalty to be traded off in cost effectiveness with the lower structure of the polar mount.

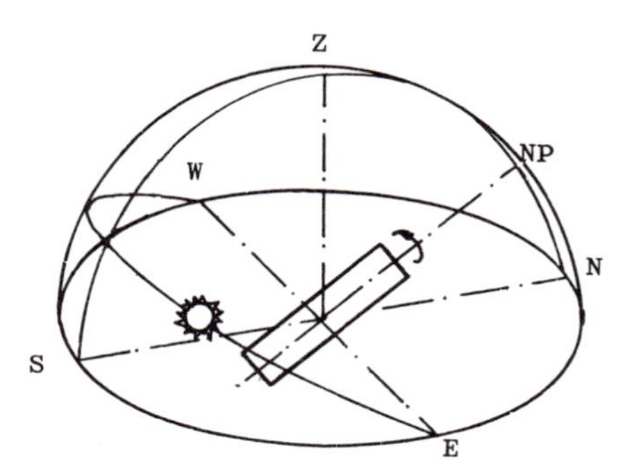

Fig. 3.19 Geometry of a collector having its axis of rotation coincident with the polar axis of the earth, with daily motion.

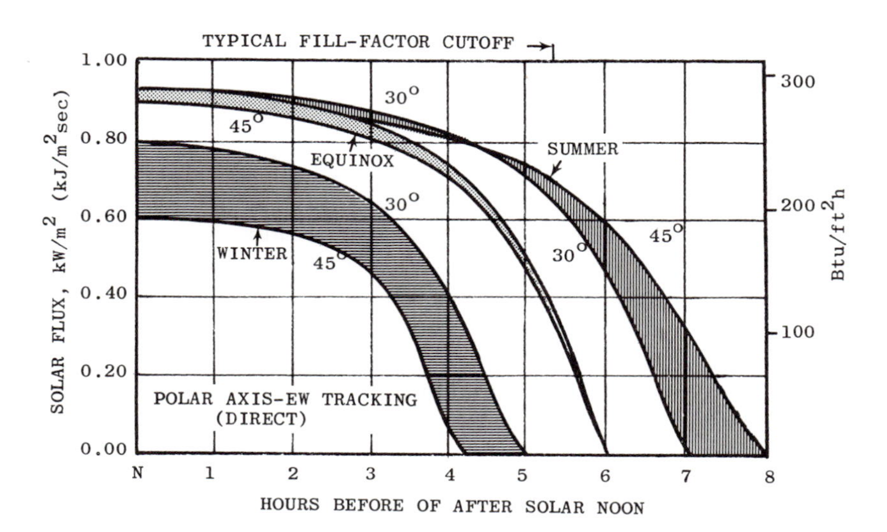

Fig. 3.20 Variation of flux for polar axis collector tracking the sun in the east-west direction, for direct sunlight only, as a function of time of day and season. Note that for an array of such collectors the fill factor effect will cut the sun off somewhat before 6 h.

3.12 EAST-WEST HORIZONTAL, NORTH-SOUTH TRACKING

The basic geometry for a collector with an EW axis and NS tracking motion in one coordinate is shown in Fig. 3.21. The axis of the collector is always made horizontal and in the EW direction. The collector accepts a daily cosine projection factor reduction in performance in exchange for seasonal optimization when only one degree of tracking is permitted. The collectors, being horizontal, minimize the shadowing of adjacent collectors. The principal shadowing is caused when the collector is tipped to a maximum degree south, when the sun casts a shadow toward the collector to the north.

The diurnal north-south motion of the EW horizontal, NS tracking geometry depends on the season. Although no daily motion is required at the equinox, motion is a maximum at the solstices. The sun spends most of the year at the extremities close to solstice (Fig. 3.15); therefore, this configuration involves considerable daily motion most of the year.

The performance curves for the north-south tracking collector are shown in Fig. 3.22. Note that the square profile is not present, the curve being close to a cosine function. The equation for the angle between the collector normal and the sun is

$$\cos\beta = (\sin^2\delta + \cos^2\delta\,\cos^2 t)^{1/2}. \tag{3.9}$$

The daily variation shows the peaked curves typical of the fixed geometries, but the degradation at winter solstice is small and such a collector gives well balanced seasonal performance.

If we refer to Fig. 3.10 we note that the north-south motion is small for a reasonable portion of the day, even at the solstices. In Fig. 3.23 we show the *change* in the meridional angle (the angle being tracked) from the noon zenith distance for the summer and winter solstices. These two curves are symmetrical at about zero, but the winter curve is shortened by the sunset and sunrise lines. It is possible, therefore, to fix the north-south motion of the collector when the acceptance angle of the collector and absorber is such that the image of the sun remains on the absorber during that time. Winston (1974) has used this fact to make a collector having sufficient field of view to accept several hours of motion without the image leaving the absorber surface. In Fig. 3.10 we show two 30° angles, and one notes that with this large an acceptance angle the sun would remain usable for more than nine hours without the necessity of moving the collector in the meridional angle. We later show how the Trombe cusp collector geometry can be modified to provide this acceptance angle and still provide appreciable flux concentration (Section 7.7).

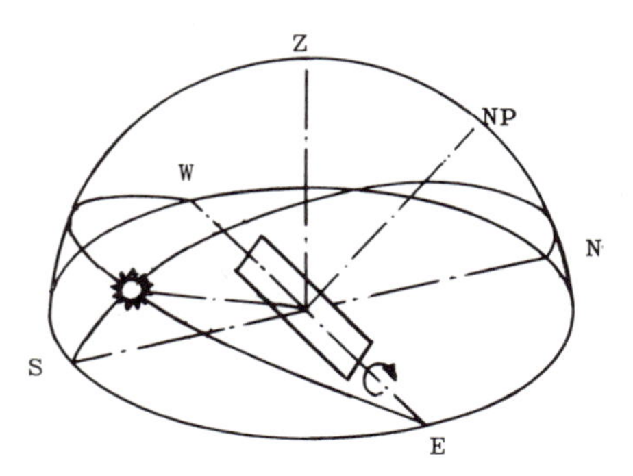

Fig. 3.21 Geometry of a collector having its axis of rotation in an east-west horizontal plane, with north-south daily motion, for direct sunlight only.

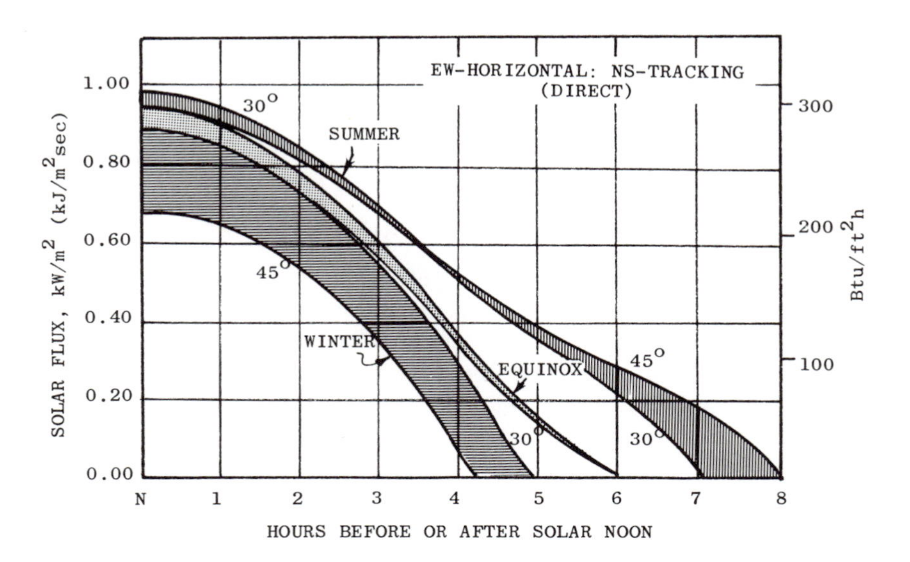

Fig. 3.22 Variation of solar flux for a collector with axis of rotation east-west and daily motion north-south for direct sunlight only, as a function of time of day and season.

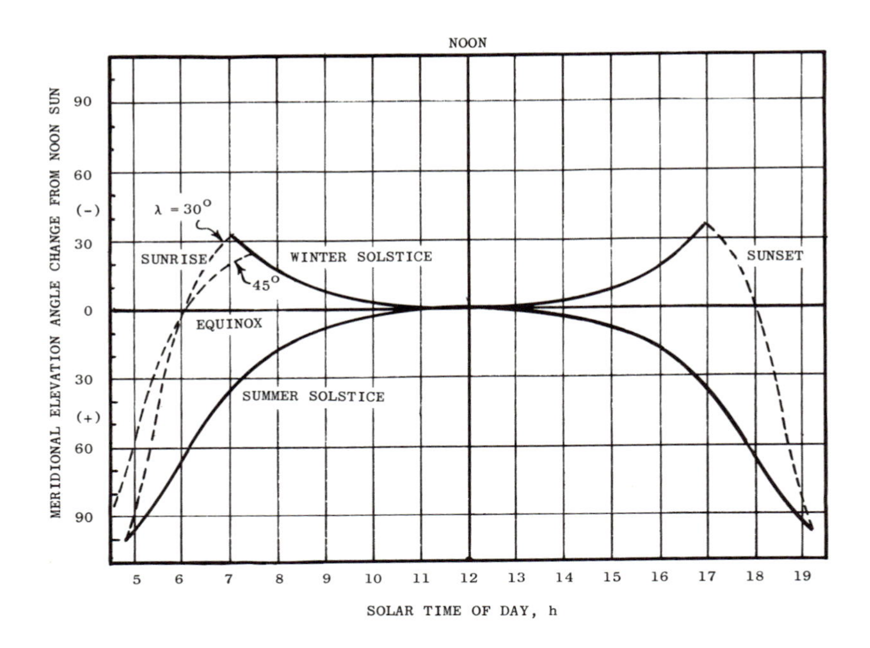

Fig. 3.23 Graph of the north-south angle change for a collector with horizontal east-west axis of rotation with respect to the noon position of the sun.

3.13 COMPARISON OF THEORETICAL CURVES WITH OBSERVATIONS

The theoretical curves for the variation of solar availability as a function of month, shown in the preceding section, are symmetric around the solstices. This symmetry arises solely from the mathematical equations used for the calculation. In Fig. 3.24 we show for comparison an actual case for the very clear desert climate of Tucson, Arizona. The hemispherical fluxes were measured with a pyranometer, data points being taken only on days when the sky was noted as clear and free from any detectable haze.

The interesting feature of the fluxes in Fig. 3.24 is that they are distinctly not symmetrical about the summer solstice. The peak yearly solar flux occurs about 1 June. The clear days are less frequent in summer and fall than in spring, due undoubtedly to the fact that summer atmospheric circulation brings occasional thunderstorms into the region. Nevertheless, one should utilize this departure from the ideal situation in calculating the theoretical "best" performance of a collector in any given region, rather than relying on the idealized flux availability curves.

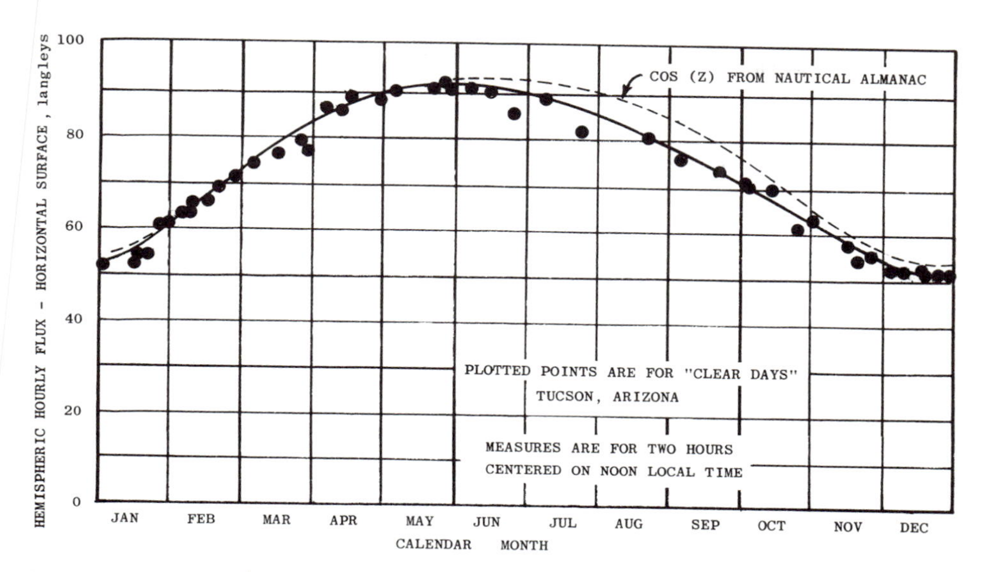

Fig. 3.24 Comparison of clear-day hemispherical fluxes measured with a pyranometer, showing a seasonal asymmetry with respect to the theoretical curves for a horizontal flat plate and the ephemeris values for the solar declination.

3.14 COMPARISON OF DAILY OUTPUTS

The choice among the many geometries discussed in this chapter comes down to one thing: how much energy each collects per day as a function of season, in relation to its complications and cost. In order to compare the relative performances of the different geometries we have summarized the daily output for each for 30° and 45° latitude.

We present the data in two forms. The first, Table 3.1, is a table of values. (Table 3.1 also shows the performance of each geometry relative to that for a fully tracking collector.) Although the tabular form gives numbers, it conveys few of the subtleties of the differences. We therefore also present the information in graphical form in Figs. 3.25 and 3.26. These figures are useful when one wants the average value over a period of several months.

The data in Table 3.1 and Figs. 3.25 and 3.26 were obtained by integrating the area under the curves presented in this chapter rather than by algebraic integrations. The values are therefore of limited accuracy but are sufficient for distinguishing among the various geometrical options and for calculating approximate behavior of specific solar collectors, as is done in Chapter 16.

For purposes of comparison we have divided the cases into (1) those where only the direct sunlight component is collected and (2) those where both the direct and scattered sunlight are collected.

Table 3.1 Collector Performances

		Energy collected				*Percent of fully tracking collector*		
Geometry		*Summer*	*Equinox*	*Winter*		*Sum.*	*Equi.*	*Win.*
Fully tracking (*D*)	(45°)	11.47	7.89	4.40	kWh/m²day	100	100	100
		41.3	28.4	15.7	MJ/m²day			
		3.63	2.50	1.39	kBtu/ft²day			
	(30°)	10.82	8.85	6.24	kWh/m²day	100	100	100
		39.0	32.0	22.4	MJ/m²day			
		3.44	2.81	1.97	kBtu/ft²day			
Fully tracking (*D*+*S*)	(45°)	12.67	8.59	5.10	kWh/m²day	110	110	110
		46.0	30.9	17.9	MJ/m²day			
		4.02	2.73	1.57	kBtu/ft²day			
	(30°)	11.96	9.60	6.50	kWh/m²day	110	110	110
		43.0	34.4	23.8	MJ/m²day			
		3.79	3.05	2.10	kBtu/ft²day			
Horizontal flat plate (*D*)	(45°)	7.54	4.31	1.23	kWh/m²day	64	54	27
		27.0	15.5	4.5	MJ/m²day			
		2.38	1.36	0.42	kBtu/ft²day			
	(30°)	7.85	5.53	3.00	kWh/m²day	72	63	46
		28.5	20.0	10.8	MJ/m²day			
		2.50	1.75	0.95	kBtu/ft²day			
Horiz. flat plate (*D*+*S*)	(45°)	8.55	4.90	1.62	kWh/m²day	73	64	37
		31.0	18.0	5.8	MJ/m²day			
		2.70	1.57	0.51	kBtu/ft²day			
	(30°)	8.90	6.34	3.50	kWh/m²day	82	72	56
		32.0	22.9	12.7	MJ/m²day			
		2.82	2.01	1.10	kBtu/ft²day			
Latitude tilt (*D*)	(45°)	6.48	6.10	3.43	kWh/m²day	56	76	78
		23.3	21.8	12.3	MJ/m²day			
		2.05	1.94	1.09	kBtu/ft²day			
	(30°)	6.47	6.33	4.87	kWh/m²day	59	72	79
		23.3	22.8	17.6	MJ/m²day			
		2.05	2.00	1.53	kBtu/ft²day			
Latitude tilt (*D*+*S*)	(45°)	7.60	7.00	3.95	kWh/m²day	64	86	88
		27.3	25.0	14.2	MJ/m²day			
		2.40	2.21	1.25	kBtu/ft²day			
	(30°)	7.63	7.28	5.56	kWh/m²day	69	82	89
		27.5	26.2	20.0	MJ/m²day			
		2.41	2.30	1.76	kBtu/ft²day			
Latitude tilt + 15° (*D*)	(45°)	4.63	6.16	3.56	kWh/m²day	40	79	81
		16.7	22.2	12.8	MJ/m²day			
		1.47	1.95	1.13	kBtu/ft²day			
	(30°)	4.79	6.49	5.10	kWh/m²day	44	74	81
		17.4	23.4	18.4	MJ/m²day			
		1.52	2.06	1.62	kBtu/ft²day			
Latitude tilt + 15° (*D*+*S*)	(45°)	5.80	7.04	4.08	kWh/m²day	49	89	91
		20.7	25.3	14.6	MJ/m²day			
		1.83	2.23	1.29	kBtu/ft²day			

Table 3.1 (Continued) Collector Performances

Geometry		Energy collected: Summer	Equinox	Winter		Percent of fully tracking collector: Sum.	Equi.	Win.
	(30°)	6.00	7.45	5.80	kWh/m²day	54	84	91
		21.4	26.7	20.8	MJ/m²day			
		1.89	2.35	1.84	kBtu/ft²day			
Vertical (*D*)	(45°)	1.65	3.98	3.57	kWh/m²day	14	52	82
		5.95	14.3	12.9	MJ/m²day			
		0.52	1.26	1.14	kBtu/ft²day			
	(30°)	0.36	3.22	4.40	kWh/m²day	3	36	71
		1.3	11.6	15.8	MJ/m²day			
		0.12	1.03	1.40	kBtu/ft²day			
Vertical (*D+S*)	(45°)	2.40	4.82	4.40	kWh/m²day	22	62	92
		8.8	17.3	15.9	MJ/m²day			
		0.78	1.53	1.40	kBtu/ft²day			
	(30°)	1.27	4.16	5.17	kWh/m²day	12	46	81
		4.6	14.9	19.1	MJ/m²day			
		0.40	1.32	1.68	kBtu/ft²day			
Polar axis (*D*)	(45°)	11.20	7.90	3.97	kWh/m²day	94	100	93
		40.0	28.4	14.3	MJ/m²day			
		3.51	2.50	1.26	kBtu/ft²day			
	(30°)	10.20	8.83	5.91	kWh/m²day	93	100	93
		36.7	31.8	21.6	MJ/m²day			
		3.24	2.80	1.88	kBtu/ft²day			
NS horiz.; E-W track. (*D*)	(45°)	11.21	6.04	1.99	kWh/m²day	95	76	46
		40.2	21.8	7.2	MJ/m²day			
		3.54	1.92	0.63	kBtu/ft²day			
	(30°)	10.80	8.03	4.45	kWh/m²day	97	90	71
		39.0	28.9	16.1	MJ/m²day			
		3.43	2.55	1.41	kBtu/ft²day			
EW horiz.; season (*D*)	(45°)	6.65	6.07	3.72	kWh/m²day	58	77	86
		23.9	21.9	13.4	MJ/m²day			
		2.11	1.93	1.18	kBtu/ft²day			
	(30°)	6.92	6.41	5.50	kWh/m²day	64	73	88
		24.9	23.1	19.8	MJ/m²day			
		2.20	2.03	1.71	kBtu/ft²day			
NW horiz.; N-S track (*D*)	(45°)	8.54	6.11	3.92	kWh/m²day	74	77	91
		30.7	21.4	14.1	MJ/m²day			
		2.69	1.88	1.24	kBtu/ft²day			
	(30°)	8.18	6.50	5.72	kWh/m²day	75	72	91
		29.4	23.4	20.6	MJ/m²day			
		2.59	1.97	1.82	kBtu/ft²day			

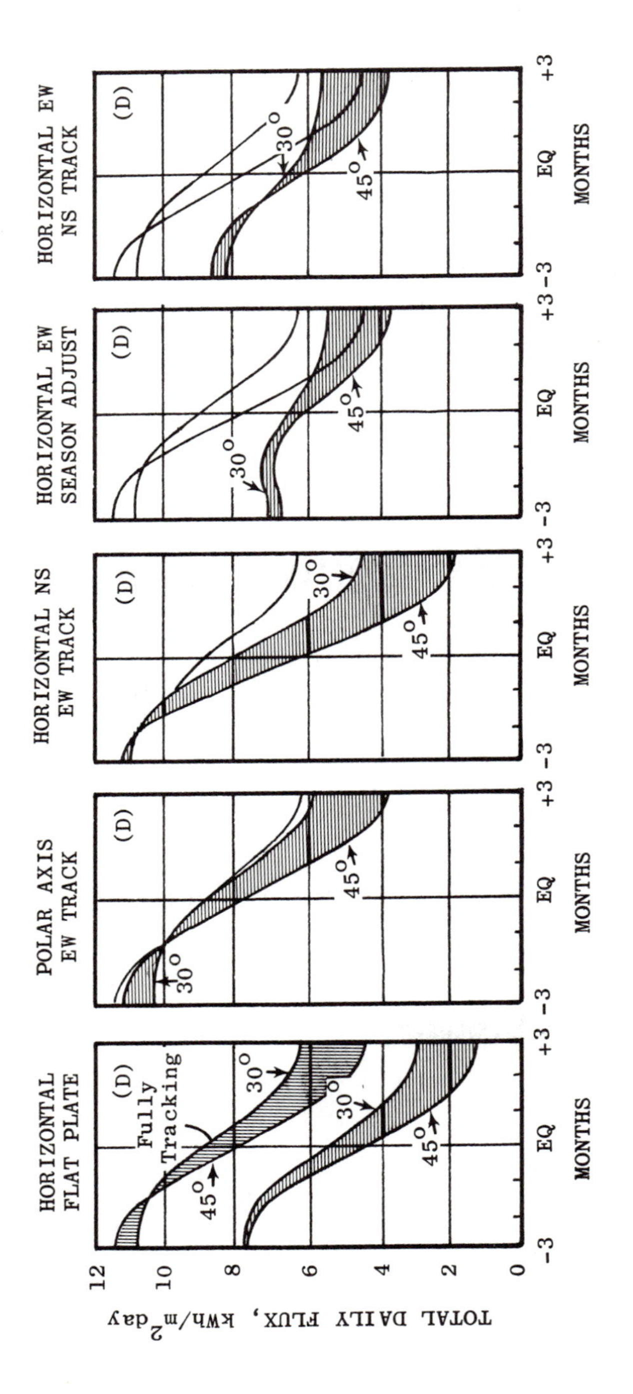

Fig. 3.25 Comparison of daily collector yields in kWh/m^{2}day of collector area for several geometries and with only the direct sun component utilized. The upper swath is for a fully tracking collector also using direct sun only. The limits to the swaths are for latitude 30° and 45° north or south, and are as indicated.

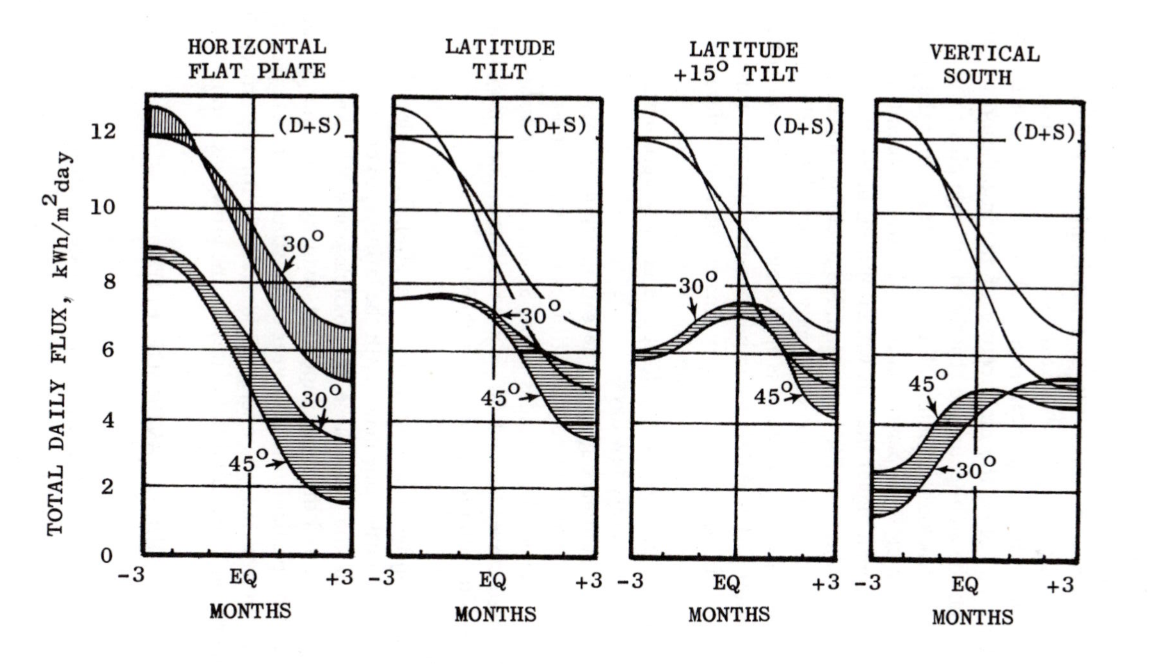

Fig. 3.26 Comparison of daily collector yields in kWh/m^2day of collector area for several geometries and with direct plus scattered sunlight input. The upper swath is for a fully tracking collector also receiving direct plus scattered sunlight.

In each of the figures we show for reference the swath of performance for the latitude zone 30–45° for a fully tracking collector, this being the maximum flux obtainable. Figure 3.25 compares the case of direct sunlight only and Fig. 3.26 direct plus scattered sunlight. Note that in the case of the tilted fixed plate the performance goes through a maximum for a tilt of latitude plus 15° but that the differences between it and a tilt equal to the latitude and the south-facing geometry are not significantly different for winter performance. We would suggest that for heating alone one would therefore allow the architecture of the installation to determine the tilt, rather than stretch the architecture to strictly fit latitude plus 15°.

In the case of tracking collectors (Fig. 3.25), we see that the polar axis EW-tracking collector is very close in performance to the fully tracking collector but that its summer output is much higher than its winter output. In the two cases of EW horizontal, seasonal adjustment, and EW horizontal, NS tracking, the seasonal output is better balanced and the winter output is essentially the same as in the polar axis case. The case of a seasonal adjustable collector is applicable to *fixed-mirror* collectors like those discussed in Chapter 7. The difference in performance between this semiadjustable and the daily north-south tracking case is very small. The cost difference between a semiadjustable and a fully adjustable mounting, however, is also apt to be small. If any means of adjusting is provided, it would appear an easy step to activate the adjustment every hour or so versus every few days.

In using these summary graphs one can interpolate between the edges of the swaths for the latitude of interest. For latitudes below 30° and above 45° one can also extrapolate. The changes for latitudes below 30° are slow, and above 45° are rapid, so extrapolating these curves beyond 50° is not recommended; instead, one should recalculate using the equation presented in this chapter.

The annual yields for 45° and 30° latitude for *perfectly* clear annual weather for several collector configurations of interest is given in Table 3.2. To convert these values to particular locales one needs both the latitude, to enable interpolation for the latitude of use, and the percentage of possible sunshine. Annual averages can be misleading, however, and the more important quantity is the energy yield and percentage sunshine for the particular time of year under study. For example, if bioconversion to trees is being considered, the energy yield available for the process is that amount for the configuration (horizontal flat plate) for the growing season.

3.15 PEAK FLUX VERSUS AVERAGE FLUX

The ratio between peak flux and average flux is frequently a point of miscommunication between persons when discussing average system efficiency or price per watt. In Fig. 3.27 we show the solar flux curves versus time of day for two basic geometries, the fully tracking collector and the fixed collector tilted at the latitude of the site for the equinox. Both curves are for the total of direct flux and scattered flux. In the case of the fully tracking collector the peak flux must be divided by 2.57 to obtain average flux over 24 hours. For the case of the fixed tilted surface the peak flux must be

Table 3.2 **Maximum Possible Annual Energy Yields**

Configuration	*Latitude, degrees*	*Energy yields* 10^3 *kWh/m²yr*	10^6 *kJ/m²yr*	*kBtu/ft²yr*
Fully tracking (*D*)	45	2.85	10.3	910
	30	3.11	11.3	985
Horizontal plate (*D*)	45	1.64	5.9	520
	30	1.92	6.9	610
Horizontal plate (*D* + *S*)	45	1.92	6.9	610
	30	2.26	8.1	710
Fixed (+15°) (*D* + *S*)	45	2.03	7.3	642
	30	2.25	8.1	713
Polar EW tracking (*D*)	45	2.80	10.0	886
	30	3.00	11.0	952
Horizontal EW:NS tracking (*D*)	45	2.30	8.3	728
	30	2.45	8.8	775
Horizontal NS:EW tracking (*D*)	45	1.70	6.1	537
	30	2.51	9.0	790

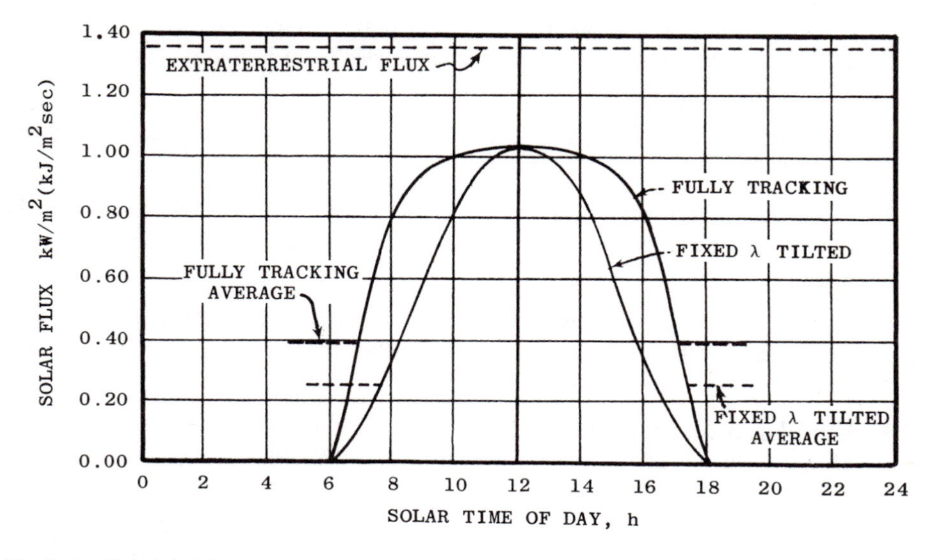

Fig. 3.27 Relationship between the peak intensity and average intensity for a fully tracking collector and a fixed collector tilted at an angle equal to the latitude, for the equinox, and for the sum of direct and scattered sunlight.

divided by 4.20 to obtain average flux over 24 hours. If one is referring to the average flux over 24 hours as compared to the extraterrestrial flux, the fraction is 0.29 for the fully tracking case and 0.18 for the fixed tilted surface. A summary of flux values and ratios is given in Appendix II.

3.16 PROBLEMS

3.1. Prepare a diagram like Fig. 3.1 for your latitude. Mark the hours along the three solar paths: summer and winter solstices, and equinox. What are the times of sunrise and sunset for the solstices at your location?

3.2. Calculate the variation of zenith distance with time of day for your latitude. Plot the results.

3.3. Take the appropriate standard sun curve for your location, from Chapter 2, and calculate the variation of direct solar flux for a fully tracking system for your latitude for the three principal seasons. What is the daily total flux for these three cases?

3.4. Calculate the solar flux on a horizontal flat plate as a function of time of day for your latitude, for each of the three seasons. Use the total of direct plus scattered sunlight.

3.5. Calculate and plot the variation of direct plus scattered sunlight for your latitude on a flat plate inclined at various angles from horizontal to vertical for your location as a function of time of day and season.

3.6. What are the daily total flux collection values for the several cases in Problem 3.5?

3.7. Calculate the flux versus time-of-day curves for a horizontal NS collector tracking the sun east-west for your latitude. Plot the data and calculate the daily total.

3.8. Calculate the flux versus time-of-day curves for a horizontal EW collector tracking the sun north-south for your latitude. Plot the data and calculate the daily total.

3.9. Using Fig. 3.20, what is the length of time that a collector having a field of view of 5° would accept the sun for the horizontal EW collector? At which times of year?

3.10. What is the minimum spacing between collector rows for the EW orientation at winter solstice at noon for your latitude? For latitude of 45°? For latitude of 30°? What is the percentage of shadowing for these cases at 0900 h?

3.11. What is the average flux collected by a flat-plate collector between 0900 and 1500 h for a latitude of 40° for (a) a horizontal collector on 21 June, (b) a horizontal collector on 21 December? What is the average flux collected by a tilted plate at these times when the tilt is equal to the latitude?

3.12. If Q is the amount of heat required to bring a collector from night temperature to day operating temperature, what is the minimum time in minutes after sunrise before the system is ready to operate for a horizontal flat plate on 21 September? Assume (a) the heat required by the collector is $Q = 10$ cal/cm^2 with 50% of the flux being absorbed; (b) $Q = 100$ cal/cm^2.

3.13. What is the additional energy that can be collected with a two-coordinate tracking system compared to the best one-dimensional tracking system? Does this choice of configurations yield the best values at all times of year?

3.14. What is the additional energy that can be collected with a two-dimensional tracking system compared to the best fixed geometry system? Does this choice vary significantly with season? Why? If a best fixed collector on an annual basis were selected, what would be the ratio of energy collected by the tracking collector to that collected by the fixed collector? What does this ratio imply in regard to the allowable additional costs to make the tracking system?

3.15. If the threshold for system operation is a direct flux of 300 W/m^2, what would be the expected output at 40% thermal conversion efficiency for a concentrating collector on 12 August 1974 (Chapter 2)? What would be the output of the same collector for 2 September 1974?

3.16. What is the ratio of daily flux collected for a fully tracking collector at the equinox down to a cutoff point of half the noon flux value compared to a flat plate tilted at your latitude, assuming the fully tracking system to be a concentrating system?

Part 2
The Optics of Solar Collection

Chapter 4
Luminance of Collector Optics

4.0 INTRODUCTION

The utilization of solar energy requires collecting the solar energy and conveying it to the absorber, where it is converted into either electrons or photons. In these five chapters we will discuss the optics involved in collectors and the ways they are utilized in various types of collectors. In this chapter we will discuss the luminance of collector optics. This is an important and frequently misunderstood topic. Luminance refers to the ability of the optical system to increase the solar flux per unit area on the receiver. To understand this function it is necessary to understand some basic properties of optical systems, beginning with definitions of terms used in describing optical systems.

In discussing luminance we use diagrams in which lenses are shown. This is simply for geometrical convenience. A mirror could be substituted with its nodal point coinciding with that shown for the lens, the nodal point being approximately the center of the lens or the center of paraxial curvature of the mirror. The mirror, of course, reverses the direction of the rays and would make a less clear diagram than a lens.

There are basically two types of collectors: (1) the flat-plate and associated non-concentrating collectors and (2) concentrating collectors. The basic difference between the two lies in how they interact with radiation. In Fig. 4.1 we show the basic elements of these two types. The flat-plate collector is simply a photon absorber, converting the photon to a phonon (heat). It is nondirectional in the sense that a photon

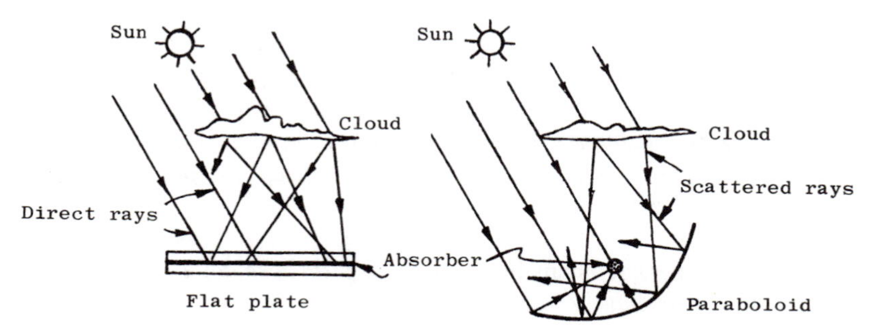

Fig. 4.1 Diagram illustrating the two basic types of solar collectors. The flat-plate collector accepts all photons, regardless of whether they come *directly* from the sun or are *scattered* by clouds. The paraboloid can focus only *direct* sunlight on the absorber.

arriving from any direction can be absorbed; hence energy is absorbed from the sky and environment as well as from the sun. This means that a flat-plate collector is relatively insensitive to sky conditions and operates only on total irradiance upon the absorbing surface. A concentrating collector, on the other hand, focuses the incident flux upon a receiver that is smaller than the aperture of the collector, thereby increasing concentration of flux on the receiver. By the laws of optics discussed below, a concentrating collector cannot concentrate diffuse light from the sky and hence is solely dependent upon the intensity of the *direct* component of sunlight reaching the collector.

4.1 DEFINITIONS

In discussing collectors and their performance it is necessary to define certain terms that are often used interchangeably and sometimes incorrectly. In view of the sometimes subtle differences between terms that appear to have similar meanings, the reader should always refer to the geometry being discussed in order to establish the correct meaning in that context.

Flux. A measure of the power traveling through a system; the units are energy per unit time measured normal to the direction the flux is propagating. Colloquially, flux is also used extensively in this text and elsewhere for radiance or irradiance, which are measures of power per unit time per unit area, outward from a surface or inward toward a surface, respectively. Where we use flux we imply units of kW/m^2, etc.

Luminance. The amount of flux per unit solid angle emitted or scattered per unit *projected* area of surface.

Emittance. The flux emitted by a unit area of surface. *Normal emittance* refers to the flux emitted per unit solid angle normal to the surface.

Hemispherical Emittance. A term used to emphasize that the energy is being emitted into a 2π solid angle, although the term *emittance* is supposed to be sufficient to state this fact. The problem arises in that emittance, as used in discussing selective surfaces, is often measured as 1 minus the reflectance of a surface. Since reflectance is generally measured near normal incidence, one can distinguish between this value of the emittance measure and hemispherical emittance by referring to the former as *normal emittance.*

Intensity. The flux per steradian incident on a surface or reflected from a surface, generally referring to specular reflection.

Irradiance. The flux on a unit area of surface at the angle of incidence of the radiation.

Brightness. The equivalent of luminance, but the term is often used to describe the visual sensation of luminance.

Concentration. A term frequently used to describe the performance of a mirror or lens collector; it is imprecise because there are two distinctly different measures of concentration of importance in describing collector performance.

Brightness Concentration. Denoted by the symbol C, this is the ratio of flux (kW/m^2) arriving at any point on the absorber to the incident flux at the aperture of the system. It is the number of times the brightness is increased in the geometrical image (Gaussian first-order optics) of the solar disc. Gaussian optics are defined as those forming stigmatic images over the entire field of the system being described. For an imperfect optical system the brightness concentration can vary over the blurred image.

Radiation Balance Concentration. Denoted by the symbol X, this is the ratio of the collecting aperture area to the area of the absorber, which is emitting thermal infrared radiation at the absorber temperature. If the area of the absorber is exactly equal to the area of the Gaussian image of the sun, then brightness concentration and radiation balance concentration are identical, but this is seldom the case.

Effective Brightness Concentration. Since a real optical system does not form an image of the size predicted by Gaussian optics, the usual usage of the term *brightness concentration* will be assumed to be the ratio of incident energy per unit area to the energy falling within the actual image of the sun and its area. The actual image is often defined as that area containing a specified fraction ($H\%$) of the flux collected by the optical system. The value of H can vary, depending on the desire of the user. The usual range of values for H is from 80% to 90%. For poor quality optics often found in solar concentrators the upper value may result in an image with very low brightness at the "image" boundary, so lower values are often used when dealing with low quality optics.

Direct Radiation. The term we use to describe that component of the solar flux that arrives at the solar collector without any interaction with the environment. The term *beam radiation* is sometimes found in the literature, referring to the direct solar rays. These rays appear to originate in the solid angle subtended by the disc of the sun, approximately 30 arc min in diameter.

Diffuse Radiation. This term has been used widely in the solar energy literature to refer to the component of scattered sunlight received by a collector. A scientifically more correct term is *scattered radiation.*

Scattered Radiation. The term we use to describe the radiant flux of sunlight scattered by the environment. The sky scatters sunlight, producing a bright hemisphere that diffusely illuminates the solar collector. The ground and nearby buildings also scatter sunlight, producing an additional diffuse illumination. In speaking about the sun and its flux we prefer and use the two terms *direct* and *scattered,* $D + S$.

Principal Plane. The plane where a lens or mirror effectively deviates the incoming rays to the focus, the *principal point* being the place where the principal plane crosses the optical axis.

Paraxial Rays. Rays from the radiant source that enter the lens close to the optical axis, where the imagery thereof can be described by Gaussian optics, that is, where $\sin\theta$ can be replaced by the angle θ.

Chief Rays. Rays from the edge of an object that pass through the principal point of the system, which means that these rays are not deviated in passing through the system. Chief rays define the size of the image in the focal plane of the system.

Field Angle. The angle in the sky subtended by the absorber. It is determined by the linear size of the absorber and the focal length of the optical system, $s/2f = \tan(\phi/2)$.

4.2 LUMINANCE IMAGERY

The geometry of the formation of optical images is basic to understanding collectors. For this purpose we use two distinctly different types of optical rays. For imagery we use *chief rays,* as shown in the upper diagram in Fig. 4.2. In this diagram we show the optical axis of the system as the centerline. The object is located at S, which lies on a field shown by the curved surface AB. The principal plane is shown by the dashed line, which crosses the optical axis at the principal point of the system. The system has a focal plane, generally curved, where object points A, S, and B are imaged at points A', S', and B'. The property of the principal point is that if we draw rays from A and B passing through this point, the rays will be undeviated and will proceed to the focal surface, where points A' and B' are thereby defined.

In a real optical system the principal point may not be the same point on the optical axis for rays from different points in the field, an aberration we call *distortion.* The distance from the principal point to the focal surface is the focal length of the system, and distortion is simply a change in effective focal length with field angle. For the collectors to be discussed herein, the effect of distortion is generally of little importance.

The second type of ray we use to describe an optical system is shown in the lower diagram in Fig. 4.2. The rays shown are *edge rays,* or *rim rays,* and they describe the outer limits of radiant flux in proceeding from the object point to the image point. These rays are deviated by the optical system, the convergence being a measure of the optical power of the system. The meaning of "power" in this context is *not* to be confused with its meaning when used to describe the flux passing through the system.

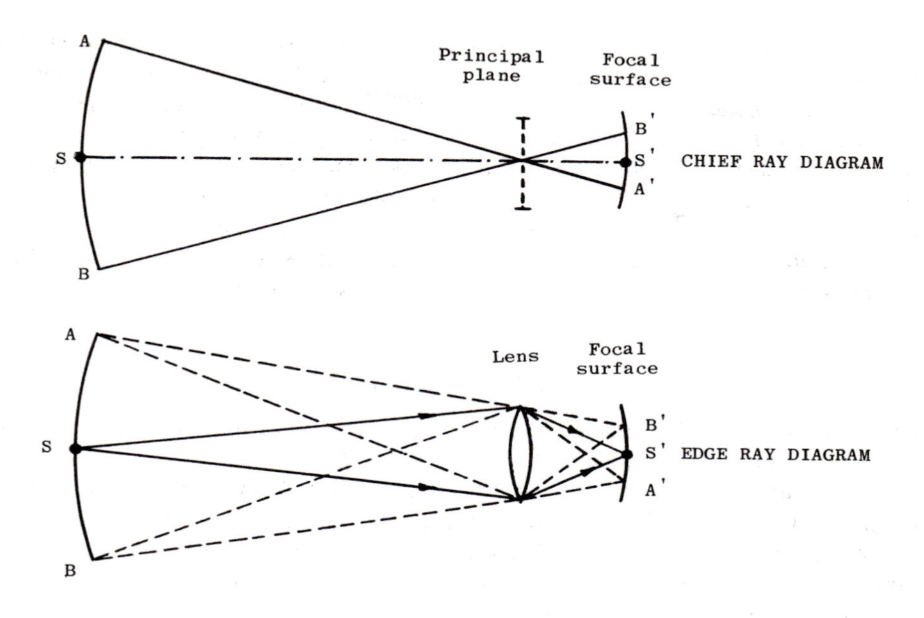

Fig. 4.2 Basic optical ray diagram for an optical system. The chief ray diagram determines *where* the images lie. The edge ray diagram determines the brightness of the image and the aberrations of the image.

There are other rays often used in optics to describe the system, called *paraxial rays.* Paraxial rays resemble those shown in the lower diagram except that they lie close to the chief rays. In most cases the term is used to describe only those rays close to the optical axis. Paraxial rays are rays where the angles involved in transmission of the ray through the system can be replaced by the angle in radians ($\sin\theta$ being replaced by θ and $\cos\theta$ by unity). These rays describe the focal point and the field curvature of the image surface. The edge rays, then, actually define the geometrical aberrations of the system because, in general, they arrive at the focal surface at different points than the paraxial rays do. These optical aberrations are important in many cases of mirror concentrating collectors. For example, a paraboloid is stigmatic only for the ray bundles along the optical axis. For object points off axis a paraboloid shows a *comatic* image, which for fast *f*-ratio paraboloids completely destroys the image within a few millimeters off the axis.

You will note that in this book a diagram often shows both the chief and the edge rays as a quick way of defining the optical behavior of the system. If one is studying one of these diagrams in detail, it is advisable to make a separate sketch of the system for chief rays and for edge rays until one is fully familiar with both.

Before discussing the properties of collectors we must examine *radiometry*, the relationship between the radiant aspects of the object and those of the image. There is a common misconception that a lens can amplify the energy gathered by the area of the lens. A lens only amplifies the irradiance, the energy arriving per unit area of the receiver, and hence the temperature of the receiver, but the total energy received is no more than would be received by a flat-plate collector of the same area. A second common misconception is that a lens or mirror can concentrate diffuse (sky) radiation. Optical elements can actually only decrease the brightness of the image of a diffuse source, in the ultimate limit approaching the brightness of the source.

4.3 SOLID ANGLE OF SCATTERED RADIATION

Before proceeding, it is necessary to define *solid angle*, since it affects the luminance of brightness of the area (da) in terms of the source (dA).

There are different ways of specifying scattered radiation. Since an instrument having a given solid angle of observation is involved, steradians are generally used as part of the unit designation. For clarity of understanding we present a simple derivation of the relationship, which may be useful in collector evaluation.

In Fig. 4.3 we show a hemisphere and a differential annulus of area upon the surface of the sphere. In linear measure the differential area dA is given by

$$dA_L = 2\pi(r\sin\alpha)_L\, ds_L, \tag{4.1}$$

where the subscript L indicates linear measure as contrasted to angular measure, indicated by the subscript o. We wish to express the differential area in angular units:

$$dA_o = 2\pi(r\sin\alpha)_o\, ds_o. \tag{4.2}$$

To convert to radian measure we have

$$2\pi(r\sin\alpha)_o = R(2\pi\sin\alpha)_L, \tag{4.3}$$

$$ds_o = R\, ds_L, \tag{4.4}$$

$$ds_L = r_L\, d\alpha, \tag{4.5}$$

and hence

$$dA_o = R^2\, 2\pi(r^2\sin\alpha)d\alpha. \tag{4.6}$$

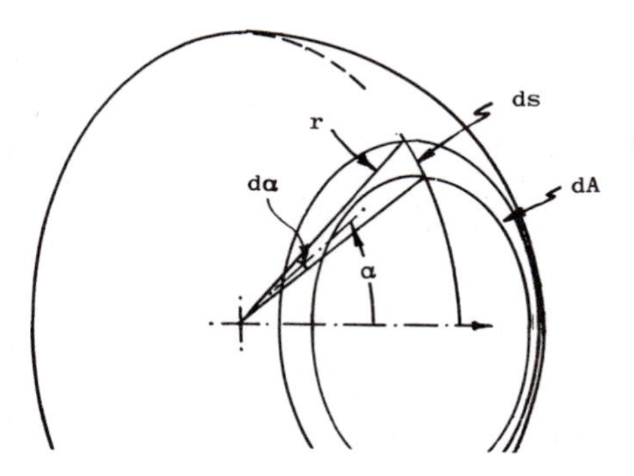

Fig. 4.3 Diagram of a hemisphere of 2π steradians solid angle with the differential angles and areas for useful derivations.

We use a unit sphere $r = 1$ and we have the expression for the number of degrees in a given solid angle as

$$A = 2\pi R^2 \int_0^{\alpha} \sin\alpha \, d\alpha, \tag{4.7}$$

which yields

$$A = 2\pi R^2 (1 - \cos\alpha), \tag{4.8}$$

where $R = 57.2958$ degrees/radian. One steradian is therefore equal to 3282.8 square degrees; and a hemisphere is 2π steradians or 20,626 square degrees. Table 4.1 gives values of the number of square degrees contained in a cone having a given semiangle at its vertex. A graph of this function is shown in Fig. 4.4.

Table 4.1 Number of Square Degrees in a Cone Specified by a Given Zenith Distance or Elevation Angle

Zenith distance	*Elevation angle*	*Square degrees in cone*	*Square degrees outside cone*
0	90	0	20,626
10	80	314	20,312
20	70	1,244	19,382
30	60	2,764	17,862
40	50	4,827	15,799
50	40	7,368	13,258
60	30	10,313	10,313
70	20	13,572	7,054
80	10	17,045	3,581
90	0	20,626	0

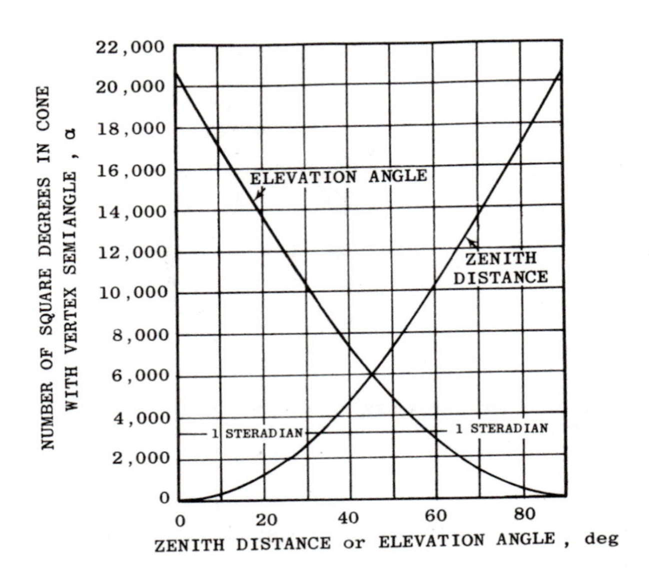

Fig. 4.4 Graph showing the number of square degrees in the cone defined by a zenith distance or elevation angle. The cone containing one steradian has a semiangle of 90° minus 57.2958°, or a cone angle of 65.4084°. One steradian contains 3282.8 square degrees.

4.4 LUMINANCE IN AN OPTICAL IMAGE

To discuss the radiometry of the object and image let us begin with a derivation of some of the important properties of a system. In Fig. 4.5 we show a lens at a distance L from a *point* lying on a small area dA, radiating intensity $B(T)$ in all directions equally. An optical system having an effective aperture of diameter D focuses the object area at a distance F from the principal point of the system into an image area

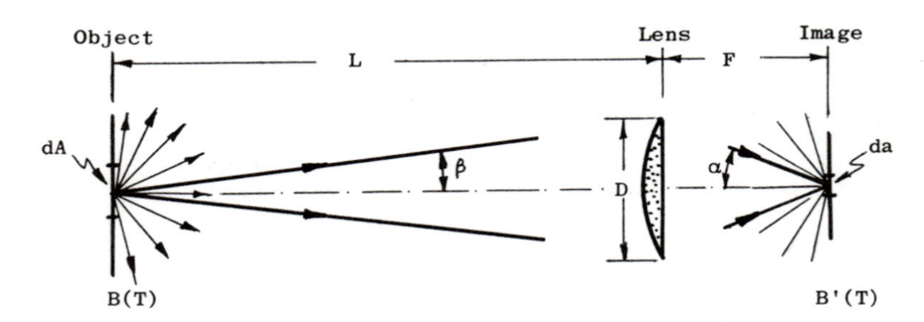

Fig. 4.5 Diagram illustrating the relationship between the angular cone of radiation from the object (sun) accepted by aperture D and focused in the image plane.

da. The edge rays for this configuration accept radiation over a solid angle 2β and focus it into a solid angle 2α.

In Fig. 4.5 the flux radiated into a 2π solid angle is $B(T)$. At the focus of the optical system let the radiated flux that is reflected by a perfect mirror (specular or scattering is immaterial) into a 2π solid angle be $B'(T)$. Then, if an optical aperture D collects flux emitted in a cone of semiangle β, and the flux is converged by the lens in a cone of semiangle α, what is the brightness or luminance at the focus?

First, we should examine the difference between luminance and irradiance. These two quantities are shown diagrammatically in Fig. 4.6. The left-hand diagram illustrates *luminance,* where we consider a radiating point. This point radiates energy equally in all directions, so that a solid angle ω, normal to the surface on which the point is situated, contains exactly as much energy as an equal solid angle ω' taken at some angle θ to the normal. To the eye, such a surface, called *Lambertian,* would appear equally bright from any angle. The *irradiance* of the same surface is shown by the right-hand diagram. In this case we are interested in the energy emitted per unit surface area, and because the surface area is affected by the cosine of the viewing angle θ, less energy is received from large values of θ than at normal incidence.

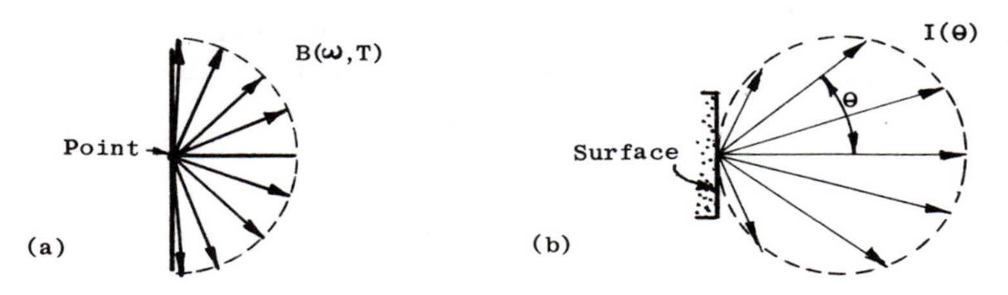

Fig. 4.6 A point that radiates energy equally in all directions (a) and has the same brightness per unit solid angle ω results in a surface (b) that radiates energy per unit surface area, obeying Lambert's law: $I(\theta) = I\cos\theta$.

The expression for the flux I emitted from dA into the solid angle β is obtained from Eq. (4.8) as

$$I_o = B(T)\,dA(1 - \cos\beta), \tag{4.9}$$

and the flux received into *da* is

$$I_1 = B'(T)\,da(1 - \cos\alpha). \tag{4.10}$$

If we have a lens of diameter D, at a distance L from the source and with a back-focal distance F, we have

$$B(T)\,dA(1 - \cos\beta) = B'(T)\,da(1 - \cos\alpha). \tag{4.11}$$

The luminance in the image can be simply evaluated from basic principles of energy conservation. Within the two angular cones, the incremental area in an annular cone of angle d is given by

$$B(T)\cos\beta\, dA\, d\Omega \;=\; B'(T)\cos\alpha\, da\, d\Omega', \tag{4.12}$$

so that

$$B(T)\sin^2\beta_{max}\, dA \;=\; B'(T)\sin^2\alpha_{max}\, da. \tag{4.13}$$

The Abbe sine condition states that

$$dA\,\sin^2\beta_{max} \;=\; da\,\sin^2\alpha_{max}. \tag{4.14}$$

Hence

$$B(T) \;=\; B'(T), \tag{4.15}$$

and the luminance in the image in the absence of absorption in the ray path is equal to the luminance in the object.

Using this equality we can proceed to derive the brightness in the image of the sun. First we must carefully define what we mean by the *f-ratio* of a system in the context of large angles of α and the Abbe sine condition. Let us take the definition of *f*-ratio:

$$f \;=\; 1/(2\sin\alpha_{max}). \tag{4.16}$$

The geometrical interpretation of this definition is shown in Fig. 4.7.

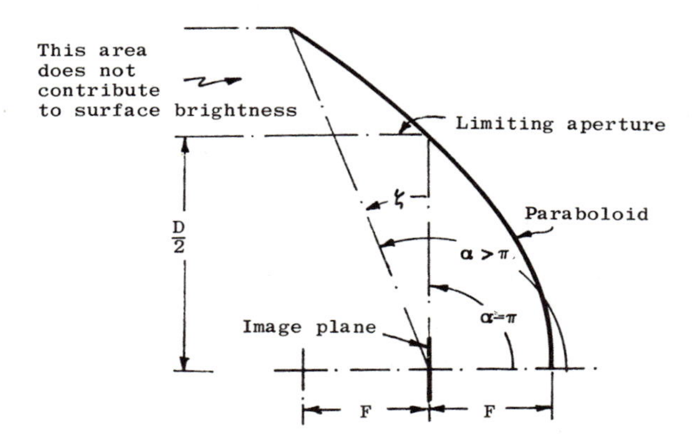

Fig. 4.7 Diagram illustrating an optical system in which α is greater than π. Only the aperture portion $\alpha \leqslant \pi$ arrives at a plane focal surface. When a pipe absorber is used, then a portion 2ζ of the pipe receives the maximum possible brightness.

The definition of brightness of the image in terms of f-ratio is confusing because of the uncertainties in the definition. The classical way of specifying f-ratio is

$$f = F/D, \tag{4.17}$$

where F is the focal length and D is the aperture. For "fast" optical systems the focal length as defined in Fig. 4.7 loses its significance, and it is necessary to resort to the more fundamental definition of f-ratio given in Eq. (4.16). For example, at $f/1.0$ the angle α is 30°, whereas Eq. (4.17) would predict 26.5°.

The aperture of a solar collector is defined by the structure. We can then use Eqs. (4.16) and (4.17) to define the effective focal length F_{eff} as

$$F_{\text{eff}} = D/(2 \sin\alpha_{\max}), \tag{4.18}$$

which is quite different from the usual definition of F, as shown in Fig. 4.7, where $F_{\text{eff}} = D/2$, and where the geometrical focal length is $D/4$.

The maximum value of α in Fig. 4.7 is greater than π, or greater than 90°. A parabolic mirror can appear to have a greater angle, as is also indicated in Fig. 4.7, but the brightness is defined on a plane surface normal to the optical axis, and rays beyond 90° cannot arrive at this surface. A system like that of Fig. 4.7 would require a cylindrical absorbing surface to take advantage of the fact that α is greater than 90°.

Let us consider only the flux emitted in the solid angle β proceeding to the optical system and being converged into solid angle α. We then have the diagrammatic form shown in Fig. 4.8.

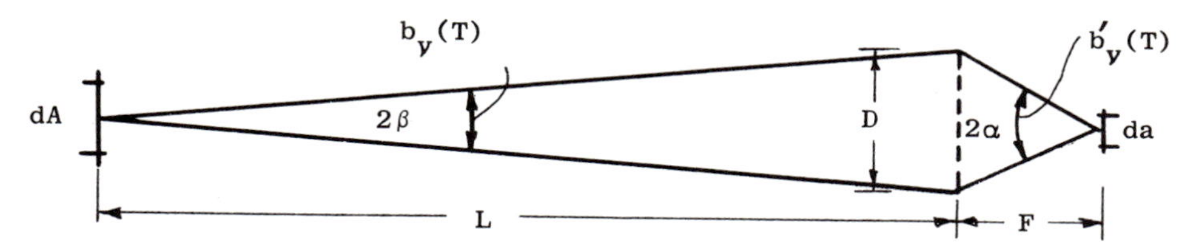

Fig. 4.8 Diagram for the derivation of the actual flux $b'_\nu(T)$ arriving in the focal cone from the optical element of focal length F and diameter D.

The exact relationship between source brightness and image brightness is given by

$$b'(T) = b(T) \sin^2\alpha. \tag{4.19}$$

Inserting the equation for $\sin\alpha_{\max}$ into Eq. (4.16), we obtain the relationship between the brightness of the image and f-ratio of the system:

$$b'(T) = b(T)/4\alpha^2. \tag{4.20}$$

The brightness of the image as a function of angle α is given in graphical form in Fig. 4.9. The relationship is that given by the number of square degrees in each zone of angle multiplied by the cosine of the angle that the zone makes with the surface. Note that even at $F/1$ the surface brightness is only 12% the brightness of the surface of the sun. The fraction rapidly rises until α is on the order of 80°. The last portion of the angle of arrival is not very effective because of the large angle of incidence on the absorber surface.

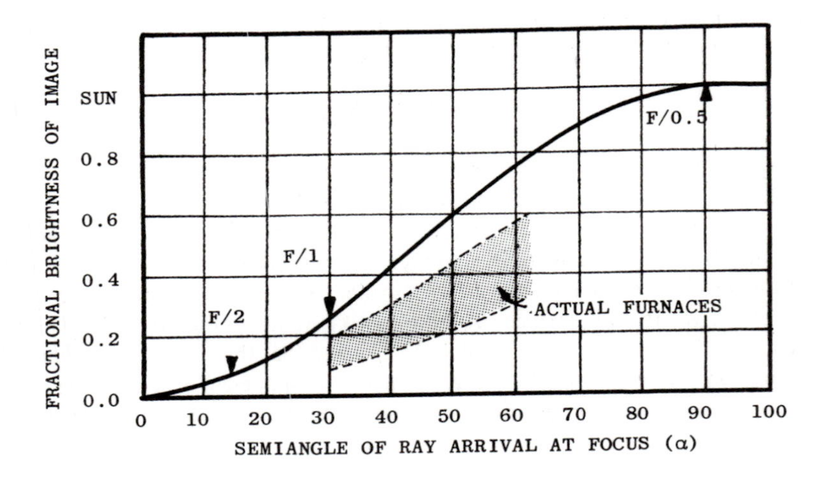

Fig. 4.9 Graph of the surface brightness of the focal image of a perfect optical system as a function of the angle α of arrival at the focal surface. The shaded area shows the range of values for existing solar furnaces.

The effect of image errors in the optical system is to reduce the brightness of the image. If the mean image spread is σ, the image diameter will be increased by approximately σF. The original amount of energy is now spread into a larger image circle, diluting the brightness by the ratio of the square of the perfect image divided by the square of the aberrated image.

In calculations involving brightness for solar furnaces, one should use the geometrical angle α and the fill factor to arrive at the proper brightness without having to define the f-ratio.

In Table 4.2 we show the relationship between the temperature of a blackbody and the flux emitted into a 2π solid angle from a plane surface. An optical system having a 2π solid angle would produce a temperature limited at 6350°K in space, but at only 4365°K at the surface of the earth.

Table 4.2 Radiative Flux Intensities

Temperature, °K	*Flux, W/cm²*	*Temperature, °K*	*Flux, W/cm²*
6000	7360	3100	524.5
(SUN)	6350	3000	460.0
5500	5197	2900	401.7
SUN*	4365	2800	349.1
5000	3549	2700	301.8
4500	2328	2600	259.5
4000	1454	2500	221.8
3900	1314	2400	188.5
3800	1184	2300	159.0
3700	1065	2200	133.1
3600	953.8	2100	110.5
3500	852.2	2000	90.86
3400	758.9		
3300	673.5		
3200	595.5		

*Through one air mass, standard atmosphere.

4.5 INTERPRETATION OF SYSTEM LUMINANCE

It is desirable to become familiar with the results and consequences given in the preceding section. Simple ways of viewing the consequences are presented in Figs. 4.10 and 4.11.

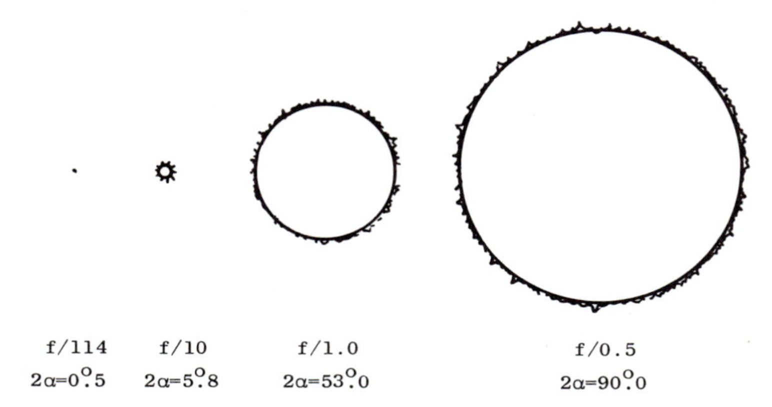

Fig. 4.10 Relative apparent angular size of the sun as viewed at the absorber as a function of the *f*-ratio of the optical system (collector). The *f*/114 optical system is actually the angular size of the sun without optics. To scale, this dot would not be visible in this illustration. The role of concentrating optics is to increase the angular size of the sun as viewed by the absorber.

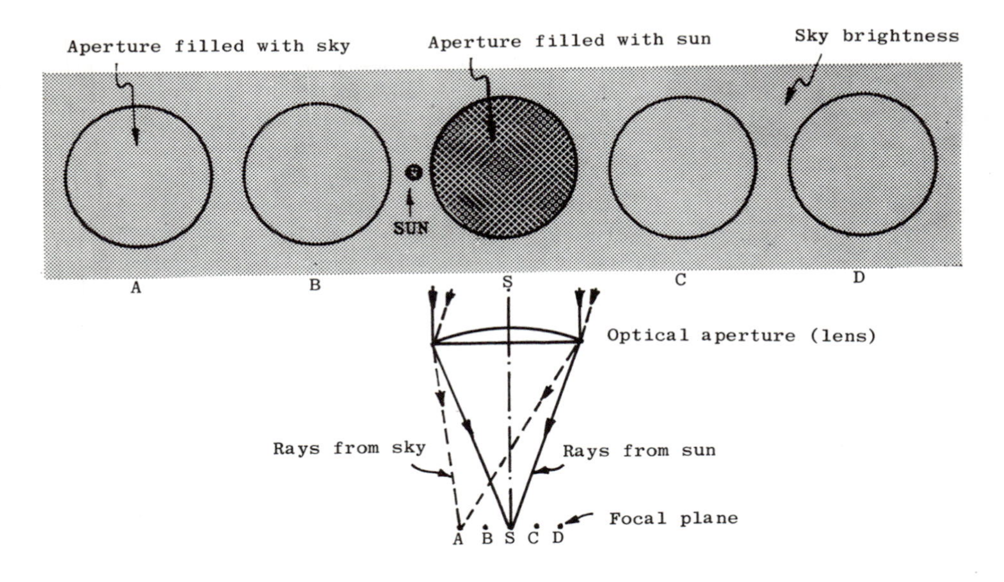

Fig. 4.11 Diagram illustrating the concentrating action of the lens for sunlight by expansion of the apparent angular size of the disc of the sun and its intrinsic surface brightness in contrast to the sky, which already is larger in angular size than the lens, resulting in no net concentration of sky light.

Effect of α

The effect of changing α for an optical system is the same as changing the apparent angular size of the sun. In each of the four diagrams in Fig. 4.10, the brightness or luminance of the sun is unchanged *per steradian.* The flux concentration achieved in the image of the sun is only a consequence of changing the apparent angular size of the sun. No lens means that the effective *f*-ratio of the flat-plate collector is $f/114$, since the sun subtends an angle of approximately ½°.

Nonconcentration of Diffuse Light

The direct consequence of the above interpretation of the equations shows why a lens cannot concentrate diffuse light. Let us take the situation illustrated in Fig. 4.11. In the lower diagram we have a lens focusing the sun at S. If an observer could place his eye at the solar focus (a star is safer and illustrates the point exactly) he would see the lens filled with sunlight, having the luminance (flux per unit solid angle) equal to the surface brightness of the sun. If he moved his eye to other points off the solar image point, he would see only the sky illumination filling the lens. As before, the surface luminance of the sky will not be changed by the lens, so the lens will appear only as bright as the sky, illustrated by the uniform light shading over both lens and sky in the upper part of Fig. 4.11.

Actual Sky Light Luminance at the Focus

In Fig. 4.12 we show three illustrations of what happens to the surface brightness of the image with and without an optical system. Figure 4.12(a) shows the case for a flat-plate collector, where sky luminance arrives from all angles. When a lens and its supporting structure, or simply the mirror (as with a deep parabolic mirror), allows only the luminance defined by the lens angle α to reach the focal surface, the sky flux is significantly reduced by the system. One would get full sky flux if the lens structure were modified so that the sky light from angles greater than α could reach the focal surface. Even though the luminance from a 2π solid angle could reach the focal surface in Fig. 4.12(c), it must be remembered that this is not sufficient for a lens system to utilize sky light as effectively as does a flat-plate collector. The reason is simply that the absorbing surface at the focus in Fig. 4.12(c) is much smaller in total area than is the aperture of the lens, which would be identical to a flat-plate collector receiving equal direct flux from the sun.

The actual sky light luminance at the absorber of a collector can be evaluated by reversing the ray paths. Let us consider any point on the surface of the absorber. Can any ray leaving this point reach the sky? If all rays over 2π solid angle from the absorber can reach the sky, then conversely, the uniform sky brightness reaches the absorber with the same total flux per unit area that a flat-plate collector of the same absorbing surface area would have. In general, lens optics cannot give this equivalence

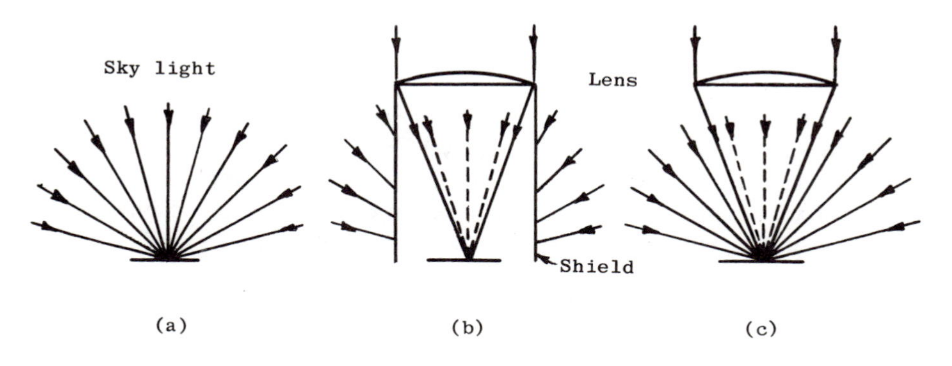

Fig. 4.12 Diagram illustrating the flux of sky light upon absorber surfaces for three cases. In (a), a flat plate receives sky light from a 2π solid angle; in (b) a lens and shield limit sky light to the angular aperture of the lens; in (c), no shield is used and the absorber receives as much sky light as (a) per *unit absorber area.*

with a 2π solid angle of radiation, as illustrated in Fig. 4.12. In Fig. 4.13 we show two cases where one can obtain flux concentration on the absorber and still preserve the total brightness of the sky on the absorber. The case of the parabolic concentrator shows that an arbitrary point does view almost a 2π solid angle of the sky. Only those rays that impinge near the vertex of the mirror return and are intercepted by the absorber. In the case of the boosted flat-plate collector, Section 6.2, the booster mirrors still transfer the rays from an arbitrary point to the sky.

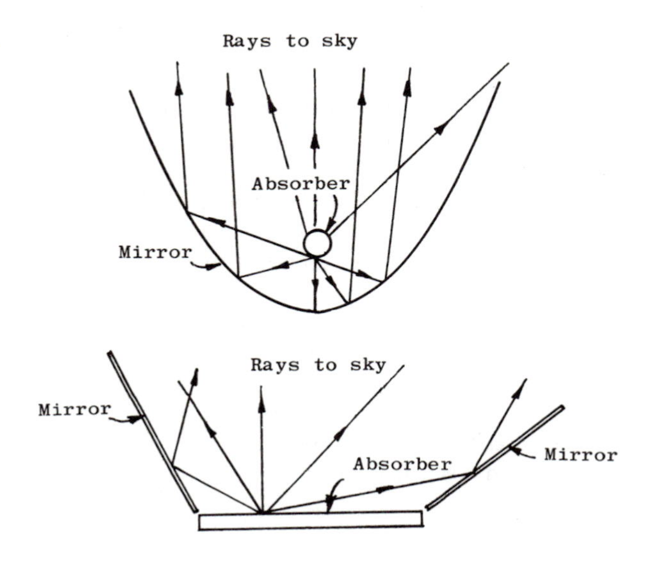

Fig. 4.13 Diagram illustrating that the fraction of scattered sky radiation arriving at the absorber can be evaluated by reversing the ray path.

In the two cases illustrated in Fig. 4.13, it should be noted that although the flux arriving from the sky on the absorber is equivalent to the entire 2π solid angle of the sky (even though the optical acceptance of the absorber is less than the 2π solid angle), the areas of the absorber are *less* than the aperture of the collector. The net result is that the absorbers gather less energy from scattered sunlight than would a flat-plate collector of the *same* optical aperture. The general rule, restated in specific cases to follow, is that the effective contribution of scattered sunlight to a collector is *equal to or less than* $1/X$ that for a flat-plate collector, where X is the radiative balance concentration for the collector-absorber system.

The general rule for the fraction of the scattered or diffuse components utilized by a collector can be summarized by two statements. Both involve the fact that the *upper* limit to the acceptance of sky light on the absorber is set by the 2π solid angle.

The fractional area of the collector aperture covered by the absorber is $1/X$, where X is the radiation balance flux concentration. If the sky light arrives at this absorber surface subtending the 2π solid angle, it is clear that the fraction of sky light accepted by the system, compared to a flat plate, is equal to $1/X$ for a one-dimensional concentrating system, and $1/X^2$ for a two-dimensional concentrating system.

4.6 PROBLEMS

4.1. Prepare a basic optical ray diagram like Fig. 4.2 but for a mirror optical system. Label the chief rays and the edge rays.

4.2. A photometer used to record sky fluxes has an acceptance angle of $f/5$. What fraction of the sky flux will this system collect if the flux is uniformly distributed? By what factor would the measured results be multiplied to convert from observed energy fluxes to flux per steradian?

4.3. A solar concentrator has an effective focal length of 100 cm and an aperture of 150 cm. What is the f-ratio of the system? What is the brightness of the image of the sky relative to the sky itself? What is the solid angle subtended by the radiation arriving at the focus? If the geometrical image of the sun subtends 2° due to mirror errors, what will be the brightness concentration C?

4.4. What is the linear size of the solar image in Problem 4.3? If the mirror were perfect, what would the image size be? If the image is received on a surface 5 cm in diameter, what will be the radiation balance concentration X? State the conditions for your value(s) for X.

4.5. If Eq. (4.16) is correct, how can Eq. (4.17) also be correct? State the arguments in your own words. Can any optical system achieve a brightness concentration of an extended source?

4.6. If $b(T)$ is the flux radiated into a 2π solid angle at the surface of the sun, show that $b(T) = \pi B(T)$.

4.7. A solar furnace is built having an angle of convergence α at the focus of 70°. The mirror is made of many smaller mirrors, the mirrors filling 95% of the aperture. The building housing the focus blocks an additional 10% of the sunlight before it reaches the concave mirror. What is the brightness of the image in terms of the surface brightness of the sun? If all this energy is lost by radiation, what would be the equilibrium temperature? Assume a perfect mirror. Use the radiation table, Table 4.2.

Chapter 5
Refractive Collector Optics

5.0 INTRODUCTION

The optics important in solar energy are generally of two types: the Fresnel lens and the concave mirror. Ordinary lenses are seldom used because their weight and cost exclude them from consideration. The Fresnel lens is important because it can be formed from a thin sheet of transparent material. Fresnel lenses and concave mirrors are also found in both circular symmetry and cylindrical symmetry. We will discuss the optical properties of these two types to the extent that they are useful in solar collectors.

Solar optical systems differ from optical systems in general in that (1) costs limit the optical element to simple surfaces and (2) the need for concentration requires that the aperture-to-focal-length ratio be as large as possible, or, in other words, that the *f*-ratio be as small as possible. The approximations used to describe paraxial optics are not good enough to handle the large geometrical aberrations of solar optics, and recourse to simple, graphical ray-tracing procedures is generally necessary.

Solar energy optics involves some unusual image-forming optical configurations, and also configurations that serve only to collect or redirect sunlight rather than to form an actual image. We will review here most of the designs that have been proposed at one time or another because germs of ideas can arise from these earlier designs that could prove useful in certain situations.

An understanding of optical principles is essential to the appreciation of specific designs and to the recognition of the potential advantages and deficiencies of any particular design.

5.1 REFRACTION OF LIGHT

The operation of lenses depends upon the refraction of light at the interface between the air and the medium of the lens. In solar applications one generally deals only with the transition between a single dielectric material and air, seldom with changes between two media having substantial indices of refraction. The characteristic *index of refraction* is a quantity denoted by n, which is the ratio of the velocity of light in vacuum to that in the medium. In the case of air, the index of refraction is unity to the degree of accuracy needed for system design, the same as it is for vacuum, since the index of air is approximately 1.00029 and the indices of most dielectric materials are in the range of 1.33 to 1.70. Materials of higher index are generally too expensive to be applicable in solar energy collection. Water has a low index of refraction of 1.33.

The law of refraction known as *Snell's law* can be stated as

$$\sin r = (1/n)\sin i, \tag{5.1}$$

where i is the angle between the normal to the refracting surface and the ray entering the surface, and r is the angle between the normal and the refracted ray inside the medium. This geometry is shown in Fig. 5.1. The equation is simply inverted when the ray travels from the higher to the lower index medium.

A convenient graphical construction of refraction is shown in Fig. 5.1, wherein two concentric circles differing in radius by the factor n are used.

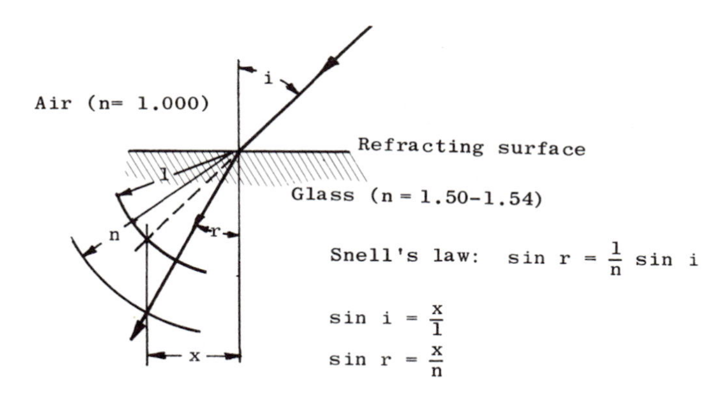

Fig. 5.1 Diagram illustrating Snell's law of refraction with a convenient graphical construction for ray tracing.

5.2 LENSES

The term *thin lens* generally refers to a lens in which the thickness of the lens is small in comparison to the focal length of the system. The basic diagram of a thin lens is shown in Fig. 5.2, where R_1 and R_2 are the radii of curvature for the front and rear surfaces and n is the index of refraction of the material from which the lens is made. The equation for the focal length for a thin lens is

$$1/F = (n-1)(1/R_1 - 1/R_2). \tag{5.2}$$

When one surface of the lens is flat, this expression reduces to

$$F = R/(n-1). \tag{5.3}$$

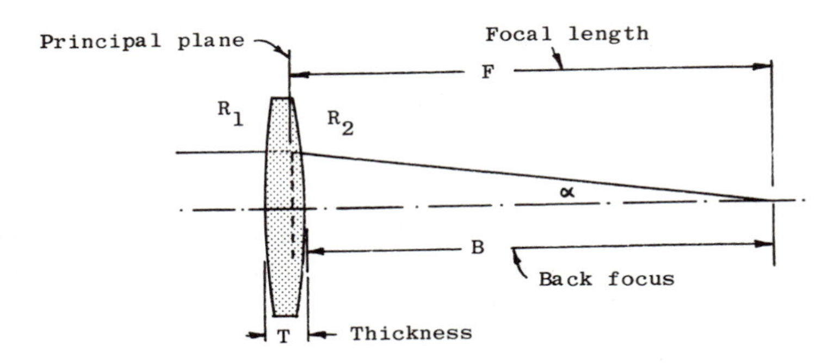

Fig. 5.2 Basic diagram for a lens. Note that R_1 is positive and R_2 is negative as drawn.

The equation for a *thick lens* has one additional term to be added to Eq. (5.2):

$$1/F = (n-1)(1/R_1 - 1/R_2) + [(n-1)/n](T/R_1R_2), \tag{5.4}$$

where T is the thickness of the lens.

The above equations refer to the focal length for the *paraxial rays,* a bundle of rays close to the optical axis. In most solar energy applications the actual lenses have low *f*-values and produce high convergence angles α to the focus. These lenses, therefore, often show focal lengths considerably shorter than predicted by the simple formula. This difference arises from the spherical aberration of the lens, which causes the edge rays to focus closer to the lens than the paraxial rays do. One should *ray trace* the actual edge rays for any proposed solar energy lens application to see where the rays come to a focus.

5.3 RAY-TRACING PROCEDURES

Ray tracing through a refracting system having curved and plane surfaces can be done readily using compass and ruler and following the construction outlined in Figs. 5.3 and 5.4. In Fig. 5.3 the incident ray is projected to the vicinity of the center of curvature of the surface in question. A circle is drawn tangent to the incident ray. In this figure the incident ray is parallel to the optical axis of the system, but this need not be so. If the radius of the tangent circle is h, and if the ray is entering glass from air, one draws a circle of radius h/n. The refracted ray passes tangent to circle h/n. If the ray were instead going from glass to air, in going from left to right the second circle would be drawn with a radius nh, and the ray would diverge tangent to this circle. In general, by progressively applying the construction outlined in Figs. 5.3 and 5.4 one can ray trace any solar energy collecting system with enough accuracy for practical purposes with a standard drafting machine.

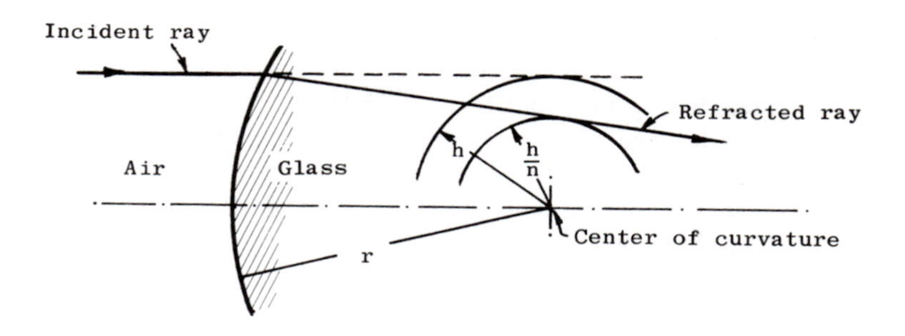

Fig. 5.3 Diagram illustrating a convenient method of tracing rays through a spherical refracting surface.

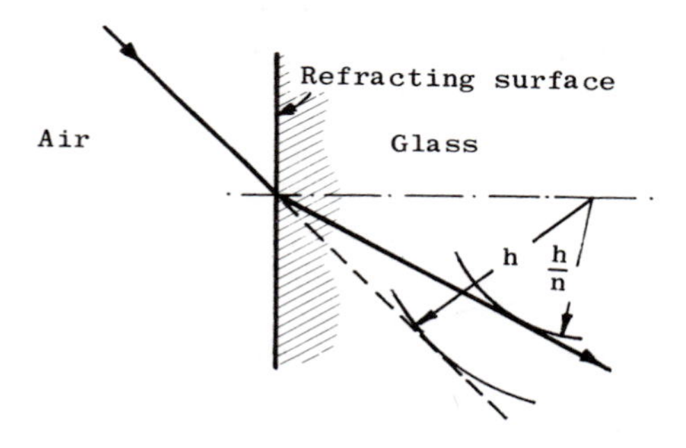

Fig. 5.4 Diagram illustrating Snell's law of refraction with a convenient graphical construction for ray tracing.

Ray tracing through a plane surface involves similar construction, complicated by the fact that the center of curvature is located at infinity. One can, instead, take any point along the normal to the refracting surface where the ray enters and construct the diagram. This alternative rule applies equally well to curved or flat surfaces. At a convenient point along the ray extension one locates the center of the two tangent circles, as shown in Fig. 5.4. A circle of radius h is drawn tangent to the ray extension and a second circle h/n times smaller in radius is drawn. The refracted ray passes tangent to this second circle.

If the ray is *leaving* glass and entering air, the procedure is the same as shown in Figs. 5.3 an 5.4 except that the second circle is drawn with a radius n times the first tangent circle, rather than h/n.

In this chapter we apply this construction in the example of a cylindrical pipe filled with water (see Fig. 5.7). It is a better way of predicting the optical dimensions of the water-filled pipe than is obtained from the conventional paraxial optical equations. The reason is that paraxial equations refer only to rays very close to the optical axis, or to systems having no geometrical aberrations. Almost all of the relatively simple optical systems used for solar energy collectors have significant geometrical aberrations, and since their aperture-to-focal length ratios are large, that is, their *f*-ratios are small, their actual focal positions both on- and off-axis can be significantly different from the paraxial positions.

5.4 FRESNEL LENSES

Optically, the Fresnel lens is closely equivalent to a thin-lens approximation; the relationship is shown schematically in Fig. 5.5. Each of the lens annuli from a plano-convex lens has the angle and curvature of the corresponding portion of the aperture of the lens. In Fresnel lenses for solar energy applications, the faces of the annulus need not be curved, but only tilted at the correct angle to refract the light ray to the focus since the size of the absorbing surface is generally much larger than the width of a Fresnel zone on the lens.

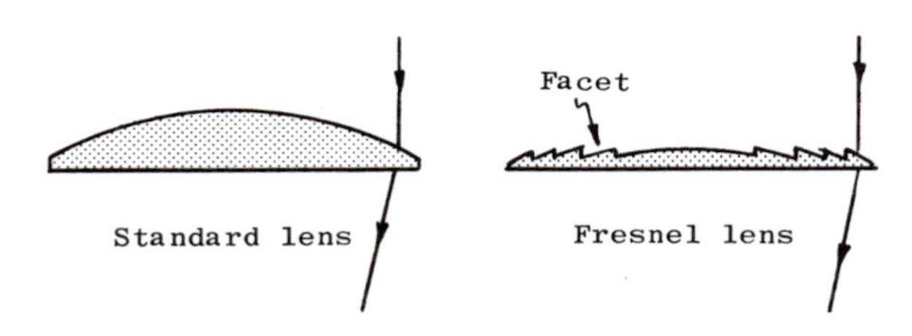

Fig. 5.5 Comparison of a plano-convex lens with the equivalent Fresnel lens. At the same radial zone both lenses produce the same deviation of the incoming solar rays.

The steps of a Fresnel lens produce some error in the precision of the focus. In general there is some rounding of the corners, which scatters light outside the image zone. In addition, for oblique angles from the sun to the lens some sunlight will strike the vertical face of the lens steps and be refracted outside the image zone.

The radius increment for the zones of a Fresnel lens is generally set by the maximum allowable vertical step between lens annuli. This limitation is regulated by the thickness of the lens sheet desired.

Fresnel lenses for solar applications are generally made of plastic although some are made of glass for other applications, such as for use in signal lights or where a long lifetime is required. Plastic is particularly useful because a sheet can be thermally pressed or cast against a die having polished facets. The authors have also made and used cylindrical glass Fresnel lenses for solar tests where the glass facets were milled into the glass sheet and polished with a flexible high-speed lap.

5.5 CYLINDRICAL LENSES

Cylindrical lenses are of potential importance for solar applications because the line focus they produce can be matched to an absorbing pipe structure. The focal length of a cylindrical lens is given by Eqs. (5.2) and (5.3), the same as for a circular lens. A diagram of a cylindrical lens is given in Fig. 5.6.

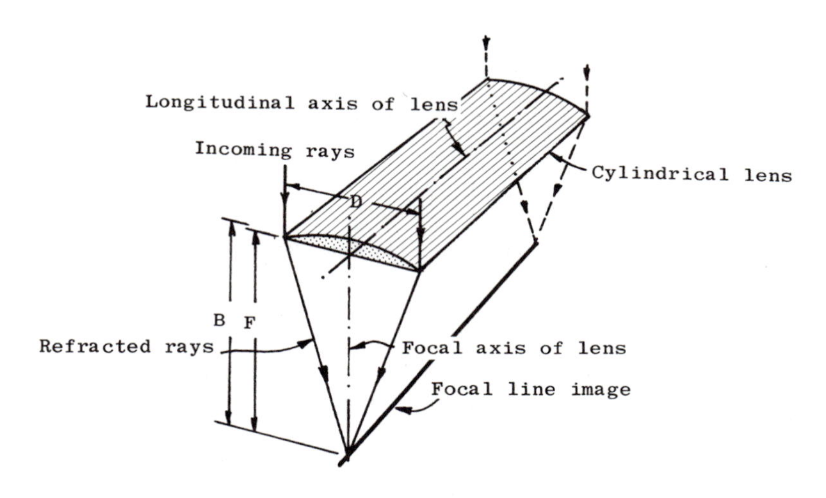

Fig. 5.6 Diagram of a cylindrical lens.

5.6 WATER-FILLED PIPE COLLECTOR

The lens formed by a cylindrical pipe filled with water is of interest because it is inexpensive. We show the derivation of the optical properties of the cylindrical pipe below. In Fig. 5.7 we have a pipe of radius r containing a fluid of index n. The walls of the pipe are considered to be thin in comparison with the diameter of the pipe, so we can neglect the refractive power of the glass walls. The effect of refraction in the glass or plastic walls is canceled to a first approximation because the inner and outer walls approximate a plane-parallel plate that causes no angular deviation of the ray traversing the plate.

The incoming ray, incident at an angle i_1, is refracted and approaches the axis at an angle u_1, striking the second surface nearer the axis. After the second refraction the ray strikes the optical axis, which is defined as parallel to the ray entering the pipe,

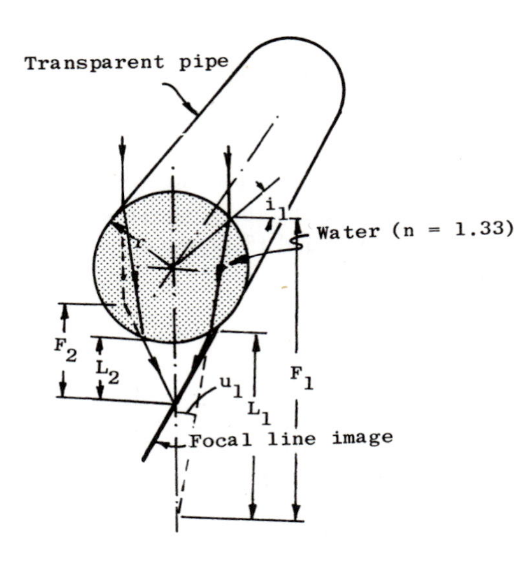

Fig. 5.7 Diagram of a cylindrical lens formed by a pipe filled with water.

and passes through the center of curvature of the pipe, focusing at a distance L_1 behind the pipe. The paraxial ray is refracted by the first surface, and has a focal length F_1 behind the surface of

$$1/F_1 = [(n-1)/n](1/r). \tag{5.5}$$

The location of the focus from the first surface behind the rear surface of the pipe is

$$L_1 = (F_1 - 2r), \tag{5.6}$$

and the distance after the second refraction is

$$L_2 = [(2-n)/2(n-1)]\, r. \tag{5.7}$$

For water, $n = 1.33$, and the distance of the focus behind the pipe is approximately

$$L_2 = 1.01r. \tag{5.8}$$

The *back-focal distance* of $1.01r$ is the focus for the paraxial rays. Rays at a greater distance from the optical axis will be more strongly refracted and focus closer to the pipe. In Fig. 5.8 we show a ray trace through a water-filled pipe and show the optimum location of the absorber. The location of the focus behind a glass sphere, as is used in the burned-track type of solar recorder, is at $0.46r$, or slightly closer, to optimize the energy received in the paper.

Even though the water-filled pipe has considerable spherical aberration, as indicated by the ray trace in Fig. 5.8, it does permit considerable flux concentration to be achieved. If the absorber is made the size of the aberrated bundle of sunlight at the 80% compromise focus, the radiation balance concentration is $X = 10$, while at the 90% focus it is $X = 6.5$. Since the pipe is symmetric about its centerline, the image point for any transverse position of the sun lies at the same back-focal distance, thus defining an image surface that is a concentric cylinder. This concentricity of the image surface means that a simple mechanical geometry can be employed to move the absorber during the day.

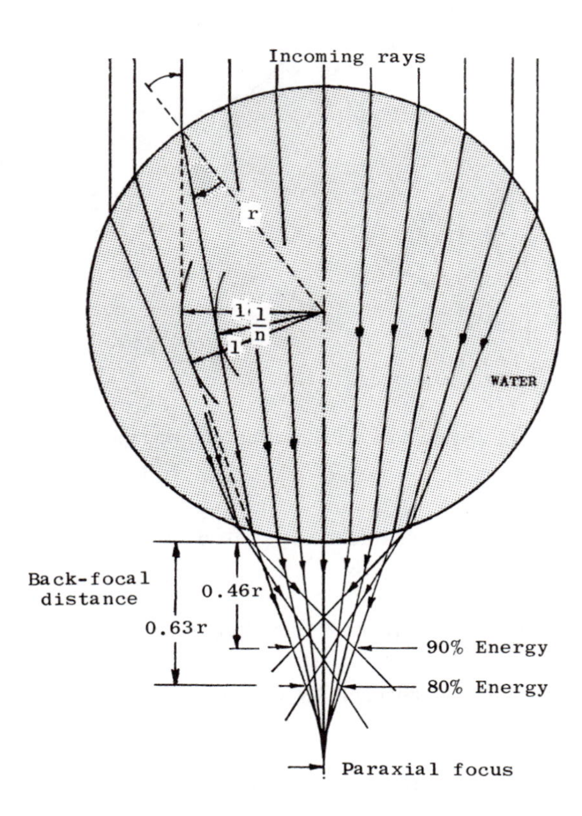

Fig. 5.8 Optical ray trace for a water-filled transparent pipe. Note the simple geometrical procedure for tracing rays graphically.

5.7 OFF-AXIS LONGITUDINAL ϕ-ABERRATIONS

When a lens of any type is used to focus rays arriving off the optical axis of the lens, the focal distance is changed; it occurs closer to the lens than for the on-axis rays. The change in back-focal distance is a complication in the use of lenses that do not track the sun, because the absorber must be moved closer to the lens to reoptimize the focus. The reason the focus shortens is illustrated in Fig. 5.9. The slice through the lens, cylindrical in this case, makes an apparently more powerful lens. The shaded slice is actually the path through the lens after refraction at entry to the lens.

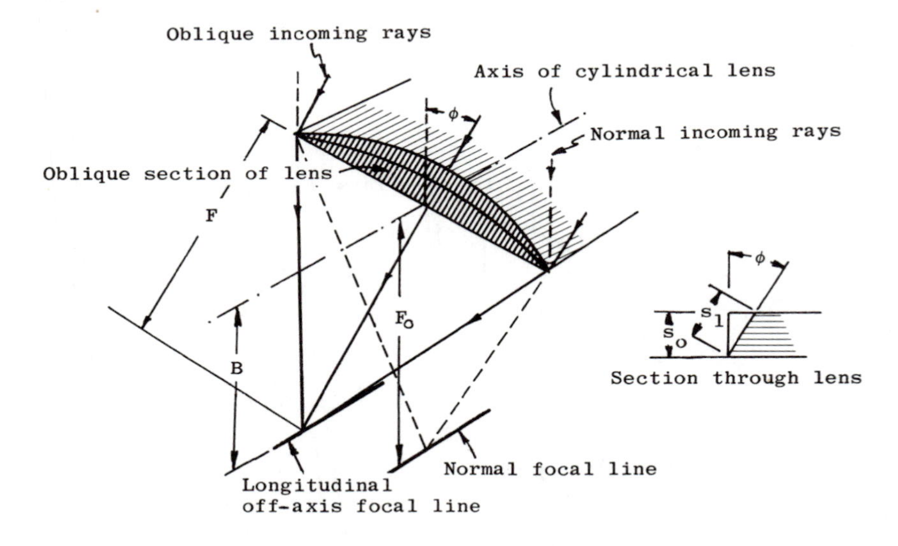

Fig. 5.9 Schematic diagram of a cylindrical lens showing the oblique ray paths.

The "new" lens has a central thickness s_1 given by $s_0 \cos\phi$, which means that the radius of curvature has been changed to

$$r_1 = W^2/8s_1 = W^2 \cos\phi/8s_0 = r_1 \cos\phi. \tag{5.9}$$

Because the rear radius is also changed by this amount ($\cos\phi$), the equation for the focal length of the lens becomes

$$1/F = [(n-1)/\cos\phi](1/R_1 - 1/R_2), \tag{5.10}$$

or

$$F = F_0 \cos\phi, \tag{5.11}$$

but this value of ϕ is the refracted value of the inclined beam. The angle of ϕ_0 outside the lens is given by

$$\sin\phi_0 = n \sin\phi, \tag{5.12}$$

whence

$$\cos\phi_0 = [1 - (1/n^2)\sin^2\phi]^{1/2}, \tag{5.13}$$

which gives us the equation

$$F = F_0 [1 - (1/n^2)\sin^2\phi]^{1/2}. \tag{5.14}$$

(Values of the term in brackets in Eq. (5.14) are tabulated in Table 5.1.) As shown in Fig. 5.10, the radius of curvature of the focal surface is shorter than the axial focal length. The back-focal distance B is then given by $B = F\cos\phi$ or

$$B = F_0\cos\phi[1 - (1/n^2)\sin^2\phi]^{1/2}. \tag{5.15}$$

The relationship between ϕ and the change in back-focal distance predicted by Eq. (5.14) applies to both ordinary cylindrical lenses and cylindrical Fresnel lenses. A

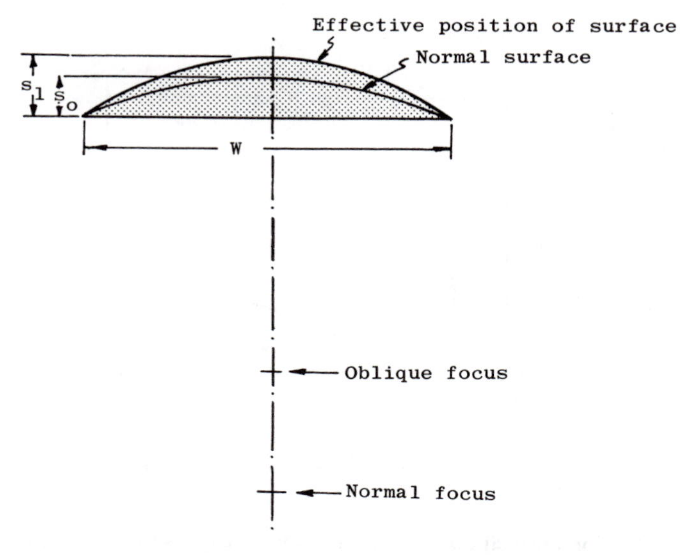

Fig. 5.10 Diagram used in the derivation of the change in back-focal distance of a cylindrical lens for rays inclined with respect to the length of the lens.

comparison of the measured behavior of a Fresnel lens and Eq. (5.14) is presented in Fig. 5.13.

5.8 OFF-AXIS LATERAL θ-ABERRATIONS

The variation of focal length and back focus with angle normal to the length of a cylindrical lens, or for any field angle for a circular lens θ, is different from that just derived for a cylindrical lens ϕ.

Figure 5.11 shows a plano-convex lens with the convex side facing the incoming solar rays. This position for a lens is close to the ideal minimum spherical aberration position, which means that a ray parallel to the axis near the edge passes through the prismatic section near the condition for minimum deviation by a prism. When an inclined ray enters at an angle θ, it emerges from the lens section with a larger deviation θ'. The expression for the curvature $1/R_p$ of the field of a lens is given by the Petzval sum for the system, which for a single lens, and in the absence of significant astigmatism of the lens, is given by

$$1/R_p = 1/nF_0 = 1/nB_0 , \tag{5.16}$$

and the back-focal distance by

$$B = B_0 [1 - n(1 - \cos\delta)] , \tag{5.17}$$

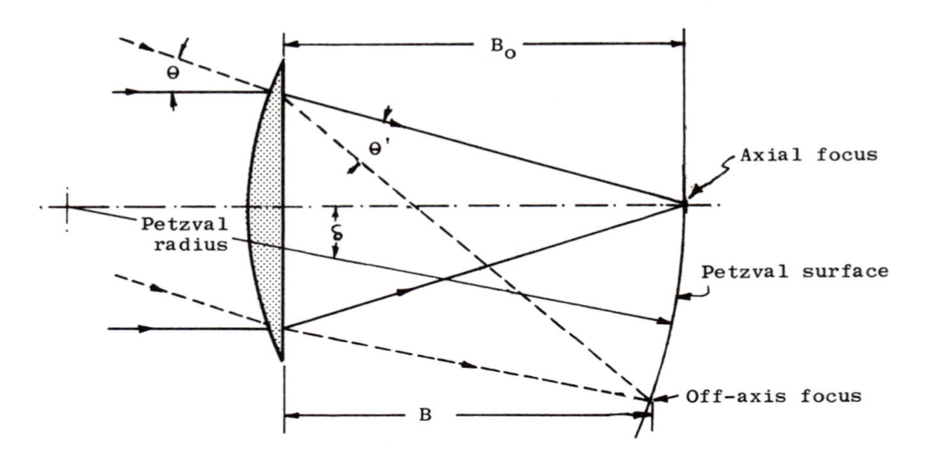

Fig. 5.11 Diagram for a cylindrical lens when the off-axis angle is transverse to the grooves, or for a spherical lens. Note that because the dashed ray passes through the lens at an angle off minimum deviation, θ' is greater than θ.

where δ is as defined in Fig. 5.11. The relationship between θ and δ is approximately given by

$$\delta_{\text{rad}} \cong (1/n)\theta_{\text{rad}}, \tag{5.18}$$

whence we obtain

$$B \cong B_0 [1 - n(1 - \cos\theta/n)]. \tag{5.19}$$

Variations of B with ϕ and θ are tabulated in Table 5.1.

Table 5.1 Variation of Back Focal Distance with Field Angle ($n = 1.50$)

Field angle (ϕ)	*Back-focal distance B*	
	Parallel (ϕ)	*Orthogonal* (θ)
0°	1.000	1.000
10°	0.981	0.990
20°	0.913	0.960
30°	0.817	0.910
45°	0.623	0.799
60°	0.409	0.560
75°	0.198	0.465

Equation (5.19) applies equally well for the change with field angle lying in the plane of the grooves of a Fresnel lens. This is the case, for example, of a linear solar collector lying east-west with the solar angle being ϕ. The line focus would therefore move closer to the lens when the sun angle becomes large.

Simple lens theory says that the field curvature of the lens due to the above effect is given by the Petzval curvature given in Eq. (5.16), shown to scale for $n = 1.5$ in Fig. 5.11. The position of best focus actually lies on the Petzval surface *only* when the lens has no third-order astigmatism. The simple types of lenses, especially Fresnel lenses, have large amounts of astigmatism and the best focus lies on a surface with considerably more curvature (as shown by the measurements for a Fresnel lens shown in Fig. 5.13). One should therefore multiply the values for the θ column in Table 5.1 by a factor of 3.5 to agree with the measurements for the Fresnel lens. This departure of the fast lenses involved in solar energy applications from simple optical theory emphasizes that one should use either ray traces or actual measurements on a small-scale lens before fixing the full-scale design.

5.9 FRESNEL CYLINDRICAL LENS MEASUREMENTS

Some measurements of the oblique focal-length change of a Fresnel lens were made by W. B. Meinel (1972) using one lens of glass and one of fused silica. The results are shown in Fig. 5.12 for angle ϕ measured along the grooves and in Fig. 5.13 for angle θ measured across the grooves. The theoretical change is indicated in each figure by the dashed curve, showing good agreement.

The curves in Figs. 5.12 and 5.13 show that a substantial change in focal position behind the lens will be required to keep the sun focused on the absorber tube. This mechanical complication renders cylindrical lenses less desirable in the EW axis, NS tracking configuration than in a NS axis, EW tracking configuration. In the latter case the diurnal motion of the sun is tracked and only the seasonal motion (±23.5°) causes a focal shift. Referring to Figs. 5.12 and 5.13, one can see that a compromise focus can be selected where the change in focal length between summer and winter solstices is approximately ±10%.

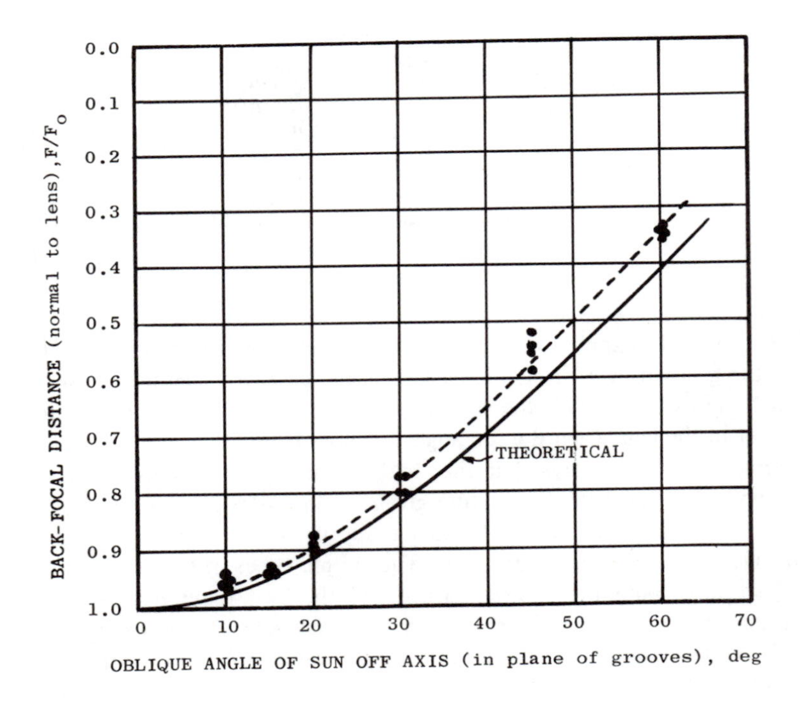

Fig. 5.12 Measured change in back-focal distance for two cylindrical Fresnel lenses, one of fused silica and one of glass, measured both with grooves facing and away from the incident sunlight (W. B. Meinel, 1972).

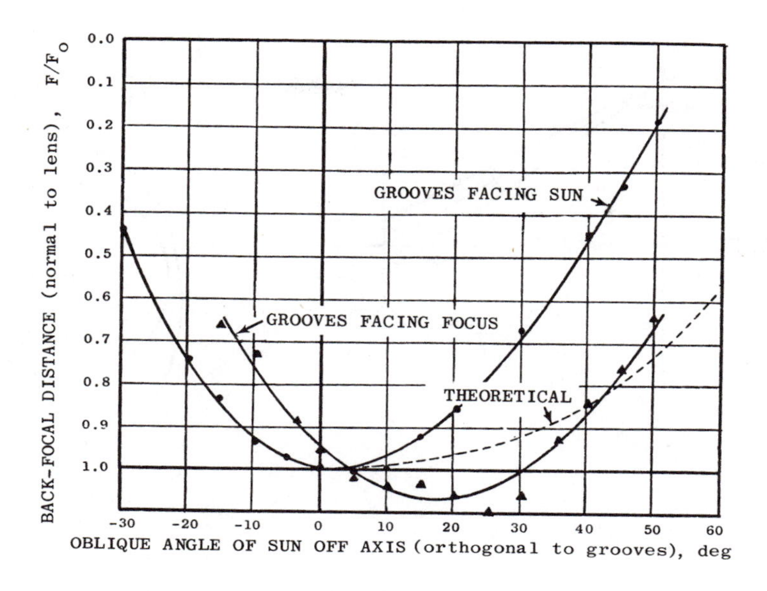

Fig. 5.13 Measured change in back-focal distance for a cylindrical Fresnel lens measured with the grooves facing the sun, the usual mode of use, and facing the focus (W. B. Meinel, 1972).

We see from the experimental data presented in Fig. 5.13 that the simple expression using the Petzval curvature is not at all sufficient to describe the change of focal position with angle transverse to the grooves of a cylindrical Fresnel lens. The data show the effect of *minimum deviation,* since when the grooves face the sun (as when the curved face of a plano-convex lens faces the sun) the variation is approximately symmetrical about the normal to the lens. It should be noted that measurement of the position of best focus is difficult when angles become large or when the grooves face the focus, because the image is poorly defined due to spherical aberration and coma. We recommend using the experimental curves in Figs. 5.12 and 5.13 for engineering design work using cylindrical Fresnel lenses.

5.10 ON-AXIS IMAGE ABERRATIONS

The image formed by a perfect Fresnel lens has significant image defects. The origin of the image defect is shown schematically in Fig. 5.14. In the left-hand diagram we show a Fresnel lens by means of a dashed line normal to the optical axis. A pair of rays from the limb of the sun, where the angle α is exaggerated a factor of 16 times, produces an image of the sun in the focal plane having a diameter A. A pair of rays from the limb of the sun through the edge of the lens also has the same angle α, but since the ray

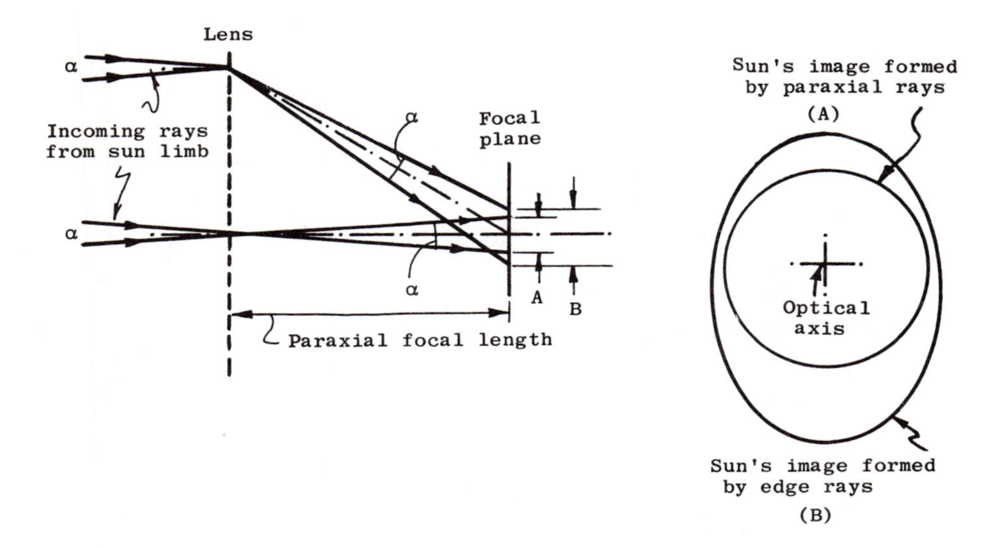

Fig. 5.14 Diagram illustrating the image aberrations due to the oblique focal length being longer than the paraxial focal length for a lens of small *F/D* ratio, an example being a Fresnel lens. An enlargement of the image is at the right.

path is longer to the focal plane and since the ray bundle arrives at the focal plane steeply inclined, the image of the sun is larger and elliptical (*B*). The net result is an image as shown in the diagram on the right. Note that the elliptical image is not exactly centered on the paraxial image. A perfect Fresnel lens, therefore, produces a "soft" image of the sun, softened further by errors of construction.

5.11 LOMONSOV CATADIOPTRIC COLLECTOR

Lomonsov in 1741 developed a solar furnace using a combination of mirrors and lenses. The principle was the same as used by Newton in 1722, wherein seven concave mirrors were angled to direct their reflected solar rays to a common focus. This mirror arrangement produces considerable optical aberration in the images. The Lomonsov design substituted plane mirrors to bend the rays, followed by a series of lenses to focus the sunlight. A diagram of the Lomonsov arrangement is shown in Fig. 5.15, where lenses are indicated as converging the beams over the entire circle. A collector of this type made by Moreau in 1924 is shown in Fig. 5.16.

A different form of the Lomonsov design uses additional lenses ahead of the bending mirrors to allow the final lenses to produce a larger beam convergence angle and a brighter image contribution from each optical system. A system of this type was

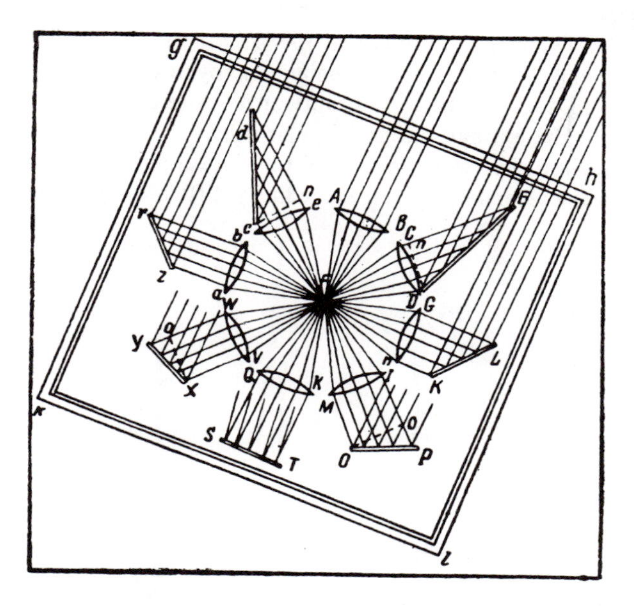

Fig. 5.15 Diagram of the solar furnace designed by Lomonsov in 1741 using multiple lenses to fill a large solid angle (from Veinberg, 1959).

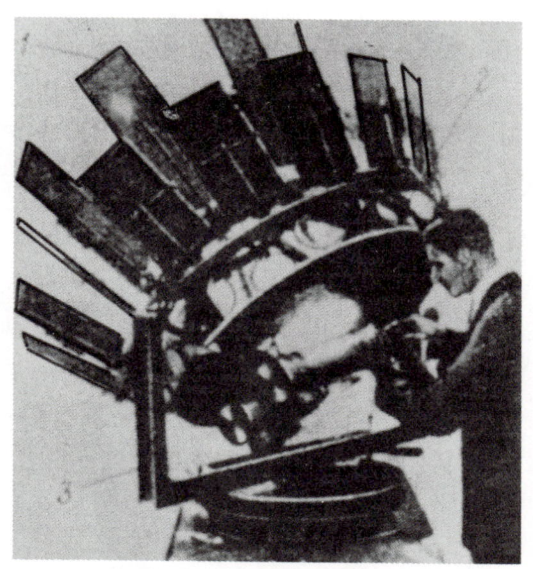

Fig. 5.16 Solar furnace made by Moreau in 1924 using the Lomonsov design (from Veinberg, 1959).

used in Pasadena, California, during the 1930s. An optical diagram is shown in Fig. 5.17. Similar configurations are now being employed in laser fusion experiments, where the large solid angle of the total beam convergence is used to "fill the sky" in the test region with the extreme surface angular brightness of a laser. The additional requirement for laser fusion work is that all the optical systems deliver a wavefront that is in phase for the several beams, an exceedingly difficult requirement.

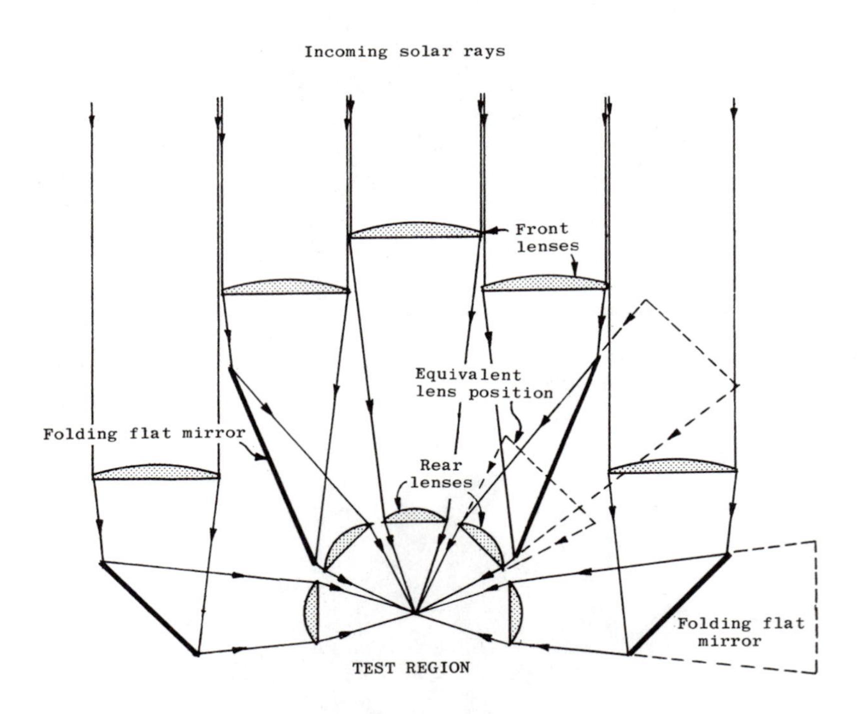

Fig. 5.17 Optical diagram of a multiple-lens solar furnace, creating a large solid angle by use of folding mirrors. Since all wavelengths absorbed by glass have already been absorbed, the sample holder can use glass and not be heated by the solar radiation passing through it. This geometry is similar to that currently being used in laser fusion experiments.

5.12 PROBLEMS

5.1. Draw a diagram like Fig. 5.1 for glass of $n = 1.52$. Draw incoming rays at $i = 15$, 30, 45, 60, 75, and 90°. What is the maximum value of r?

5.2. Draw a plano-convex lens full scale, with the radius of curvature 25 cm and a central thickness of 5 cm. Make rays incident at 1-cm spacings up to the edge of the lens. What is the total spherical aberration of this lens when the lens is made of fused silica (fused quartz) having $n_d = 1.46$? What is the image diameter when the image is intercepted at the paraxial focus? At the position of the smallest circle of light? Light is incident on the curved face.

5.3. Repeat the case in Problem 5.2 with the lens reversed from its normal position, so that light is first incident on the plane surface.

5.4. How would the image sizes and longitudinal spherical aberration change for a Fresnel lens where the facets of the lens have the same curvature at angles as the lenses in Problems 5.2 and 5.3? Give the numerical comparisons.

5.5. Find the index of refraction for a glass cylinder such that the image of the sun would focus on the rear surface of the cylinder.

5.6. A solar collector uses a cylindrical Fresnel lens to focus sunlight on a pipe parallel to the cylindrical lens. The lens has a focal length of 100 cm for incident light normal to the surface of the lens. How much will the pipe (or lens) need to be moved to keep sunlight focused on it over six hours of the day when (a) the lens is placed east-west, (b) the lens is placed north-south, and (c) the polar mount is used for the cylindrical lens and the collector tracks the sun in hour angle?

5.7. Calculate the on-axis image aberration of an $f/0.7$ Fresnel lens of 100-cm focal length. If the lens is circular, calculate approximately the intensity distribution across the solar image, assuming the sun to be a uniform disc 30 arc min in diameter.

5.8. Do the same calculations as in Problem 5.7 except with a cylindrical lens of the same aperture width and focal length. Plot the results in graphical form.

5.9. How many lenses and of what diameter would be needed for the inner lens array of a Lomonsov solar furnace design where a 2π hemisphere is to be optimally filled with radiation? Assume that the sphere on which the lenses are deployed has a radius of 50 cm. A central lens and two rows of lenses are desired. Make a scale drawing.

5.10. If a mirror solar collector produces an $f/1.5$ beam arriving at its focus and one desires to add a lens to increase the effective f-ratio to $f/0.8$, calculate the radius of curvature of a plano-convex lens of fused silica that will accomplish the job. Assume that each surface of the lens reflects 5% of the incident sunlight. Indicate the distance from the furnace focus that you would propose placing this lens. What considerations enter your choices? Is a plano-convex lens the optimum design? If not, what would you suggest?

Chapter 6
Mirror Collector Optics

6.0 INTRODUCTION

Mirrors operate under the same laws as refracting surfaces when the index of refraction is set at $n = -1$. Because mirrors are widely used in solar energy applications we will, however, present the explicit equations applicable to mirrors for the convenience of the user. Mirrors of interest range from flat mirrors, used as boosters for flat-plate collectors, through curved mirrors having cylindrical or circular symmetry. The curved mirrors include conical mirrors, or *axicons* as they are frequently termed in the field of optics. In this section we will discuss the equations relating to mirrors, with some attention to the ones of particular interest in solar energy, but we will reserve the discussion of some aspects of mirrors until we specifically consider collectors.

6.1 PLANE MIRRORS

The basic equations for describing the operation of a plane mirror on a ray, as shown in Fig. 6.1, are simply

$$\sin i = -\sin r \quad \text{or} \quad i = -r. \tag{6.1}$$

When the ray serves as the origin, the reflected ray is deviated through *twice* the angle that the mirror is rotated, as shown schematically in Fig. 6.1. This behavior is important, for example in the case of a solar heliostat. The mirror travels through only 90°

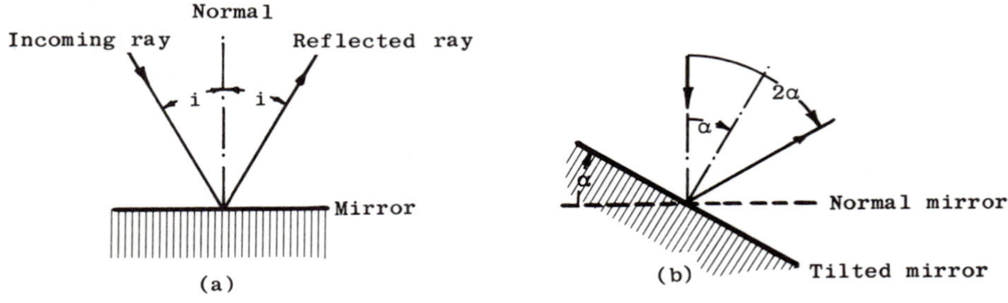

Fig. 6.1 Diagrams for a flat mirror, showing (a) that the angle of reflection is equal to the angle of incidence, and (b) that when a mirror is tilted through an angle α, the reflected beam is tilted through an angle 2α.

to track the sun over 180°. This reduction in motion means that the travel limits and gravity deflections of the structure are significantly reduced from those of a parabolic mirror collector, which tracks the sun.

Pairs of plane mirrors also have important properties for solar applications. In Fig. 6.2 we show two mirrors having an angle β between them. Depending upon the angle, one can have one, two, three, or more reflections between the mirrors, causing the ray to be deviated by multiples of β.

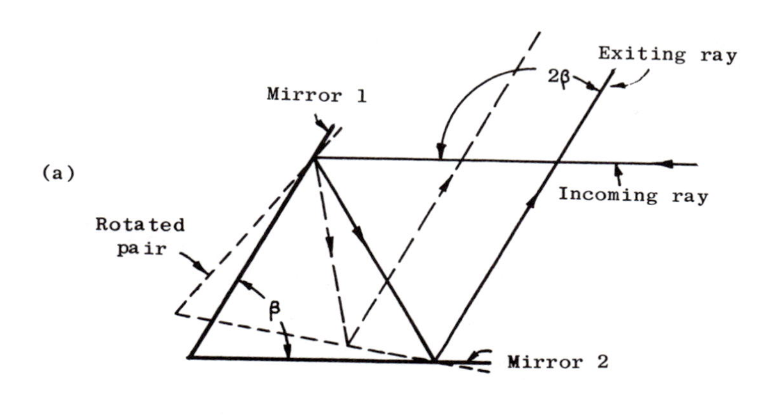

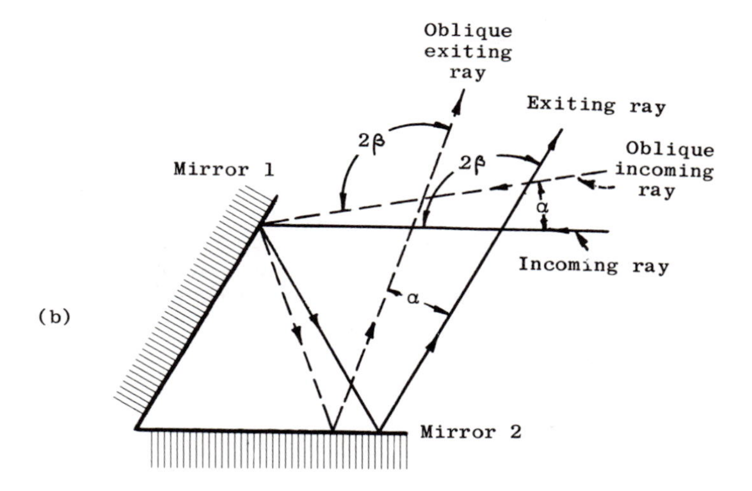

Fig. 6.2 Diagram showing that two mirrors having an angle β between them deviate an incoming ray by an angle 2β. Diagram (a) shows that the angle of deviation does not change as the mirror pair is rotated. Diagram (b) shows that the exiting ray is deviated by the same angle as the incoming ray, but that the angle between incoming and exiting ray remains deviated 2β.

A pair of mirrors having a small vertex angle can act like a concentrating collector. In Fig. 6.3 we show a 30° cone angle mirror pair. The multiply-reflected rays increase their angle with the central axis of the cone until the rays cross this plane normally, whereafter they work their way back out of the cone. In this example the ray from the edge of the aperture reaches its maximum penetration into the cone at the third reflection. When the incoming ray is inclined at the aperture, the depth of penetration into the cone changes, increasing for angles closer to tangency to the mirror wall. In the case shown the flux concentration for the on-axis incoming rays is $X = 2.0$ on a plane absorber bisecting the vertex angle. This absorber must be lengthened to collect all the rays from off-axis angles, reducing the radiative balance flux concentration.

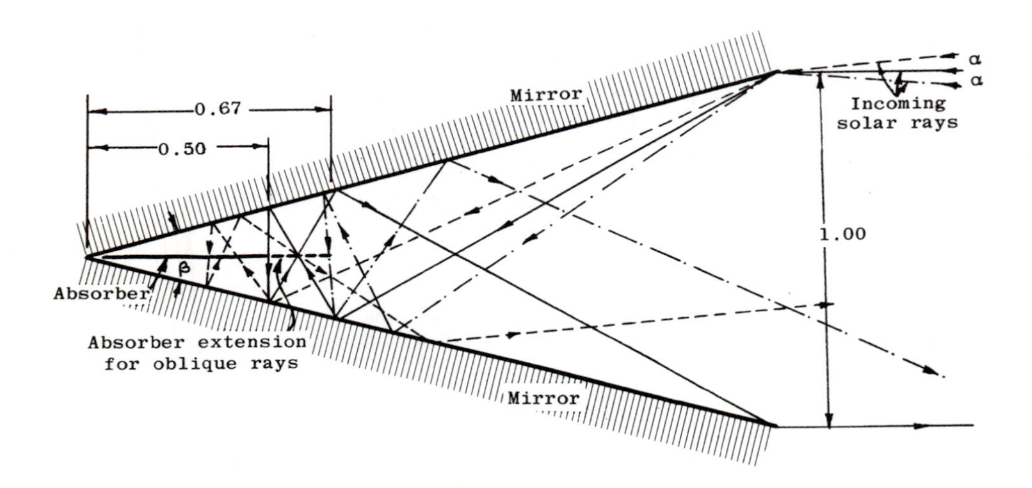

Fig. 6.3 Diagram illustrating the flux concentration effect of a pair of mirrors having a small angle β between them. Note that the absorber surface must be extended (dashed line) to intercept the rays when they arrive off-axis (α). The flux concentration of this 30° wedge is $X = 2.0$ for on-axis and $X = 1.5$ for 5° off-axis rays.

The above geometry for a pair of plane mirrors holds in the case of rotational symmetry, as is formed by an axicon (conical mirror). The flux concentration X for a cone with the geometry in Fig. 6.3 on a conical absorber with a diameter of 0.02 unit at 0.50 length would be approximately 25. Conical mirrors will be discussed in Section 6.3.

6.2 BOOSTER MIRRORS

Flat mirrors have been used to boost performance of flat-plate collectors since they were first introduced in 1911 by Shuman, who used a flat mirror on each side of the absorber in his water-pumping system in Philadelphia. Booster mirrors are useful for increasing efficiency at higher temperatures than those at which a flat-plate collector would usually operate.

The basic booster mirror is shown schematically in Fig. 6.4. The combination has some of the properties of the fixed-mirror collectors discussed in Chapter 7, but requires no curvature for the mirrors.

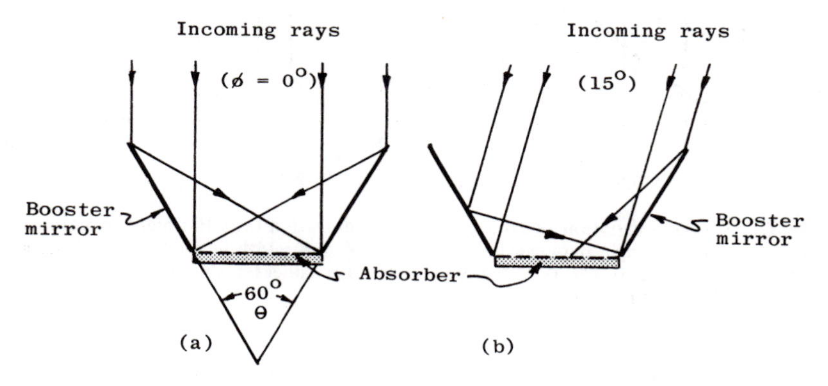

Fig. 6.4 Diagram of a flat-plate collector with two equal-sized booster mirrors. The boosters increase flux near normal but decrease it beyond $\phi = 30°$.

The symmetric booster mirrors in Fig. 6.4 produce a flux concentration (when the sun is normal to the absorber) of 2.0. As the sun angle increases in inclination the mirrors become less effective. When the sun angle exceeds the semiangle θ of the booster mirror, the mirror actually casts a shadow on the absorber, decreasing the system efficiency. A curve of radiation flux balance concentration versus angle is shown in Fig. 6.5. For comparison we show the concentration of the Trombe-Meinel cusp collector (see Section 7.7), which appears somewhat similar in general outline to the boosted flat-plate collector except that the side mirrors are curved to increase the concentration. Note the sharp cutoff of the cusp system compared to the gradual reduction in efficiency of the boosted flat plate.

The efficiency of a boosted flat-plate system can be increased if the angle of the flat mirrors can be changed several times during the year. In Fig. 6.6 we show how the booster mirrors would be changed to optimize the fixed absorber for summer sun angles, which for a considerable portion of the day actually point north of the zenith for an EW collector orientation.

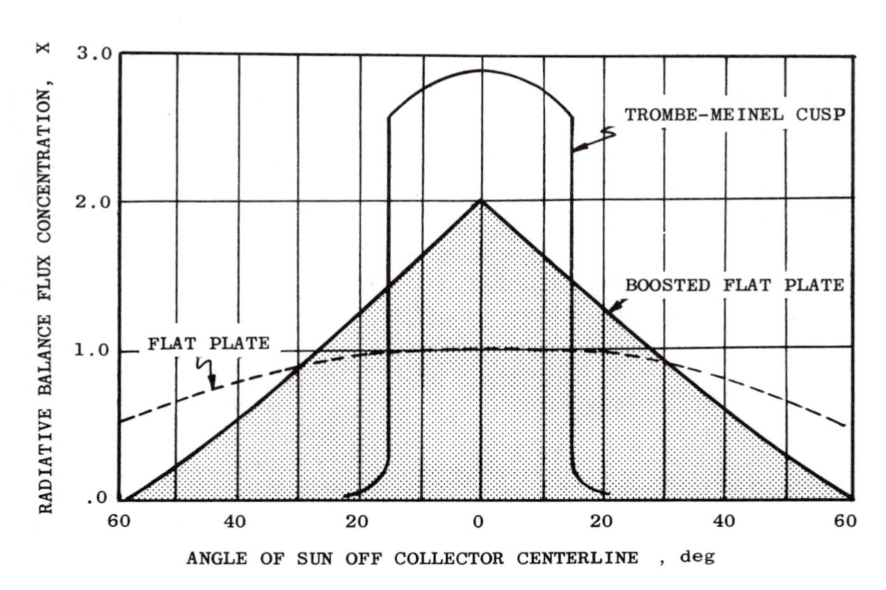

Fig. 6.5 Graph of the variation of effective flux concentration for a flat-plate collector using two side mirrors having an angle between them of $2\theta = 60°$ and width equal to that of the absorber. The high, narrow curve is for the Trombe-Meinel cusp collector shown in Section 7.7, which approximates a $2\theta = 60°$ plane-mirror booster.

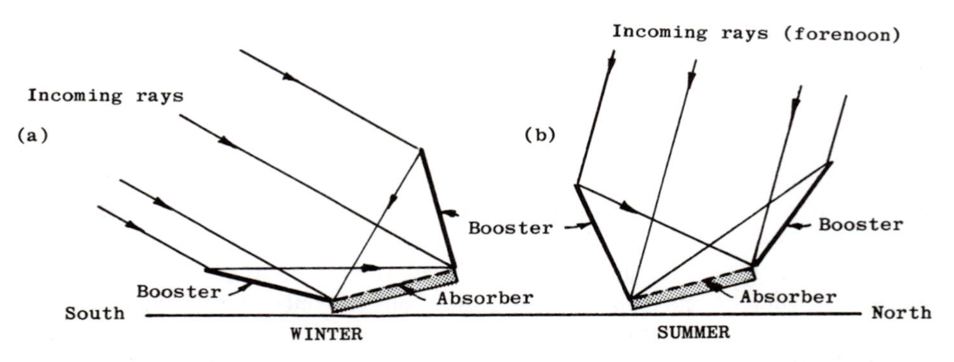

Fig. 6.6 Diagram showing how a pair of booster mirrors can be adjusted to optimize a fixed flat-plate collector for seasonal changes, for an east-west collector orientation and mirrors of equal size to the absorber.

The daily efficiency curve for a boosted flat-plate collector looks much like that for the simple flat-plate collector, with a peak at noon and a cosine drop-off on each side of noon. Tabor (1966) proposed the use of booster mirrors to change this peak at noon. In Fig. 6.7 we show the schematic diagram in which a booster mirror is placed vertically at the western side of the collector. In this position morning sunlight is concentrated on the absorber, but at noon there is no additional boost from the mirror. In the afternoon the mirror is relocated to the eastern side of the absorber, repeating the boost for the afternoon sun. The resulting concentration factor for the collector is shown in Fig. 6.8. The booster causes a peak to occur in this factor at mid-morning and a similar one in mid-afternoon, with a valley in the curve at noon.

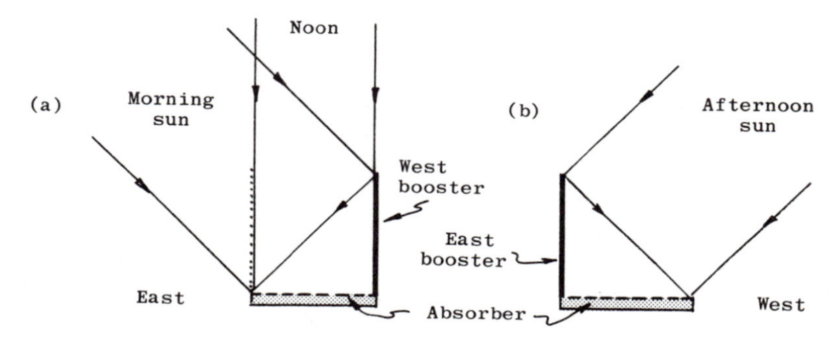

Fig. 6.7 Diagram of the Tabor (1966) east-west booster arrangement. The vertical booster is on the western side of the absorber during the morning and is moved to the opposite side for the afternoon. Booster can be placed slightly off-vertical.

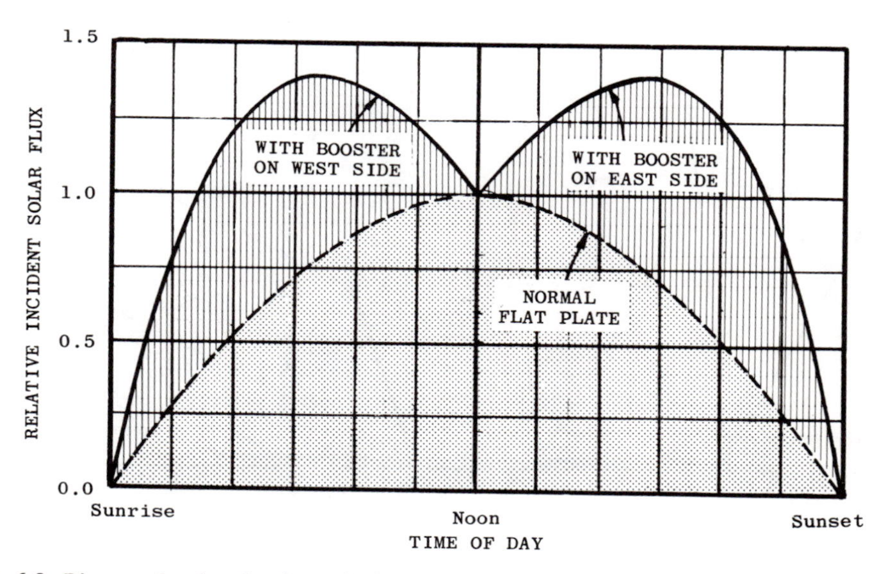

Fig. 6.8 Diagram showing the change in flux with time of day for a flat-plate collector having a removable east-west booster mirror (vertical). The noon dip in the curve can be reduced by inclining the mirrors.

The relative position and profile of the concentration curve near noon can be modified by tilting the flat mirror boosters slightly from the vertical, yielding excellent output characteristics for this design of flat-plate collector. The basic problem faced by this design, however, is the complication of moving the mirror from one side of the collector to the other at noon.

In Fig. 7.7 we show flat booster mirrors used to supplement a cylindrical collector in order to divert back to the absorber the rays that otherwise would miss the absorber. This use is one that, carried to its limit, would yield a Fresnel mirror. If the individual mirrors are smaller than the absorber, one can approximate any shape of mirror by a series of flat mirror strips.

Another option for use of booster mirrors is seasonal application. For example, at low latitudes solar energy collection for cooling purposes during the summer months requires that the collector be approximately horizontal for greatest efficiency. In winter such a horizontal orientation would be very inefficient, but a mirror booster, added for the season, can greatly improve performance. This type of application is illustrated in Fig. 6.9, where the optical cross section for incoming radiation is actually larger during the winter than during the summer. The cross section for the case illustrated is 15% larger for the noon winter sun than for summer sun. The mirror booster can also serve as a cover for the collector during adverse weather, being easily closed when needed. The cover can be left in place during the year if folded flat between adjacent rows of collectors. The simplest mirror booster could, however, be no more than strips of aluminized Mylar stretched between posts spaced along the northern edge of the collectors.

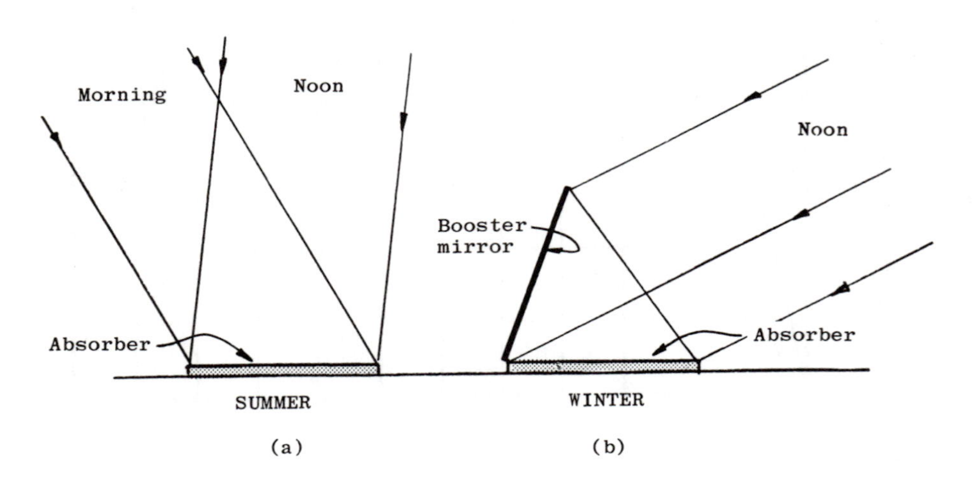

Fig. 6.9 Diagram showing how a horizontal flat-plate collector can be augmented with a plane booster mirror to raise the winter optical cross section above the summer cross section.

6.3 AXICON (CONICAL) MIRRORS

The conical mirror concentrator was first used in 1875 by Mouchot as a simple way of making a large concentrating mirror collector. The basic geometry is shown in Fig. 6.10. In cross section the mirror is made up of straight-line elements, inclined at an angle of 45° to the incoming solar rays. Each line element acts like a plane mirror, simply folding the beam and directing it to the axis of the axicon. (The term *axicon* is employed in the optics field to refer to the conical mirror in order to avoid confusion with the term "conic" mirror, which refers to a mirror having a conic profile as contrasted to those specific conics, the sphere and the paraboloid.)

Since all rays folded by the axicon element cross the axis of the axicon, one obtains a line focal image where the rays cross the axis. One could make the angle of the axicon other than 45°, which simply places the focal line above or below the lip of the mirror.

As shown in Fig. 6.10, the focal surface of the axicon is conical rather than cylindrical, with its vertex coincident with the vertex of the axicon. The angle of the focal absorber cone is 8.0 times the angle of the image of the sun formed by a perfect axicon. The reason for this cone is that the angular size of the sun appears to originate at each point of reflection at the mirror element, and since the mirror elements are farther from the axis at the top of the mirror, the rays from the limb of the sun diverge more in traveling to the focus.

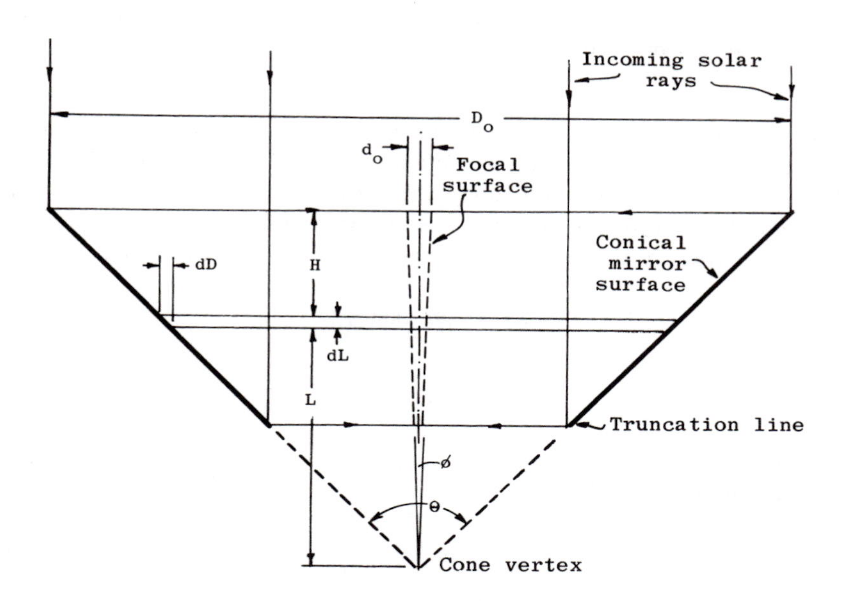

Fig. 6.10 **Basic diagram for an axicon having a 90° vertex angle. Since the meridional section shows only straight lines, this diagram also represents the case of two plane mirrors with $\theta = 90°$.**

In Fig. 6.11 we show a perspective view of an axicon to illustrate the interrelationship between the focal surface and the mirror. It can be seen that the intensity distribution along the length of a conical focal tube will be uniform. The collecting surface area increases with increasing diameter of the conical zone at exactly the same rate as the surface of an element of length along the absorber increases.

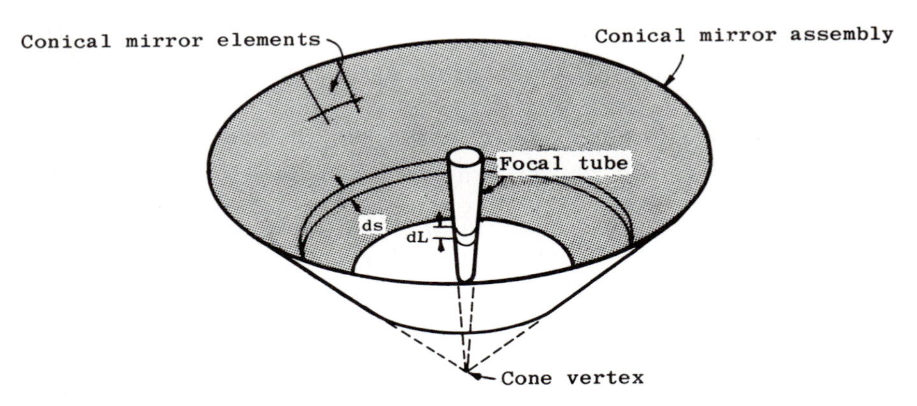

Fig. 6.11 **Conical mirror solar collector first used by Mouchot in 1875. Note that the focal surface for an extended source is a cone having its vertex at the vertex of the conical mirror. Conical mirror segments can be formed from a flat sheet with a bend in only one dimension, a practical simplification over spherical mirror segments.**

The concentration and illumination along the absorber is defined in terms of the vertex angle θ of the axicon and the vertex angle ϕ of the absorber. As shown in Fig. 6.10, the solar illumination along the focal tube will be uniform when the absorber tapers to a vertex common to the mirror vertex. If L is the position up the axicon from the vertex, then the strip dL long contributes a flux from area dA of

$$dA = 2\pi L \tan(\theta/2), \tag{6.2}$$

arriving in a surface area

$$dS = 2\pi L \tan(\phi/2). \tag{6.3}$$

Thus the brightness concentration C, which in this case is also equal to the radiation balance concentration X, is

$$C = \tan(\theta/2)/\tan(\phi/2), \tag{6.4}$$

which, for a perfect axicon having $\theta = 90°$, yields a brightness concentration of $C = 115$.

When a cylindrical absorber is used, the flux concentration will increase with L to a maximum at the top of the absorbing cylinder. In this case we have dS constant with L and the brightness change is given by

$$C = [D_0 - 2H\tan(\theta/2)]\,/d_0\tan(\theta/2), \tag{6.5}$$

where d_0 is the diameter of the absorbing cylinder, D_0 is the diameter of the top of the axicon, and H is the distance along the absorber measured from the top.

The axicon was originally important for a reason that may still be valid. In the manufacturing of mirrors for concentrating mirror systems one faces the task of bending the mirror substrate to the desired curve. In the case of a spherical or paraboloidal mirror, the curvature to be placed on the substrate is in two dimensions. A solid material cannot be curved in two dimensions without either mass flow within the substrate or buckling of the surface. Since the axicon requires the mirror to be curved in one dimension, it can be bent out of sheet metal. A glass mirror with curvature in only one direction is even easier to fabricate.

When the sun is off the axis of the axicon, the image will have aberrations that increase the size of the solar image. In addition, the rays from one side of the axicon will arrive farther up the absorber and those from the opposite side will arrive farther down the absorber. This longitudinal aberration is exactly like that described in the preceding section on plane-mirror pairs. The lateral aberration is shown in Fig. 6.12. The growth of image size due to the lateral aberration is small, indicating that an axicon has a relatively low tolerance for misalignment of the focal tube in terms of image size. It must be remembered, however, that the flux density in the volume near

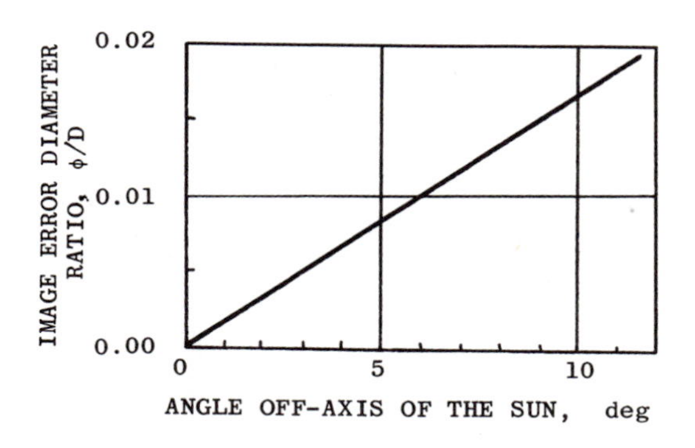

Fig. 6.12 Graph of the growth of image error for an axicon of 90° vertex angle in terms of the fraction of the diameter of the axicon at each focal zone. This growth in image error is small compared to that of a parabolic mirror of the same length-to-diameter ratio.

the focus changes linearly with changes in position from the central axis of the axicon, so although the image quality off the axis of the axicon remains good, the flux drops rapidly when the absorber is off the optical axis of the sun-axicon system.

6.4 AKS COMPOSITE AXICON-LENS COLLECTOR

Aparisi, Kolos, and Shatov (1968) proposed an interesting combination of an axicon and a cylindrical toroidal lens. This design uses the mechanical convenience of a large mirror curved in only one direction, so it can be made of flat sheets of reflective material such as Alcoa Alzak. An optical digram of this design is shown in Fig. 6.13. Figure 6.14 is a photograph of a model; it shows more clearly the nature of the cylindrical toroidal lens, in this case a normal lens, which is shown as a Fresnel lens in the diagram.

Although the AKS design looks like an inconvenience because of the lens, the authors point out that less mirror surface is required than with a parabolic mirror with the same flux concentration. An interesting benefit, suggesting its potential use as a "fixed mirror," is that when the sun moves off the system axis the image moves to a first approximation along a line concentric with the axicon axis. The absorber, if made as a short length of pipe, would accept the solar image for small angles of the sun off axis.

This unusual design could be useful with photovoltaic converters on the periphery of a tube located at the central focus. It also could be used for a sun-powered laser.

The model shown in Fig. 6.14 shows a toroidal lens instead of a Fresnel cylindrical lens. For small sizes this is the easiest way to make the central lens, but for large sizes it would need to be made of a sheet of plastic having the Fresnel grooves impressed in its surface.

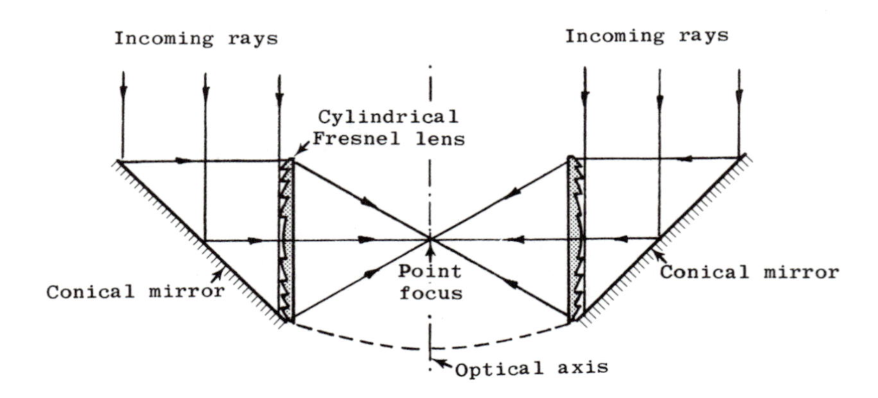

Fig. 6.13 Cross-sectional diagram of the AKS catadioptric collector, using a cylindrical Fresnel lens to reduce an axicon's usual line focus to a point focus.

Fig. 6.14 Photograph of a model of the AKS collector, showing a toroidal lens for the central condensing element.

6.5 SPHERICAL MIRRORS

The basic diagram for a spherical mirror is shown in Fig. 6.15. This diagram applies to cylindrical as well as circular spherical mirrors. An incoming solar ray enters at a height h above the center of curvature. The center of curvature and the line through it parallel to the entering ray define an optical axis. In reality, a spherical mirror has no optical axis, since all axes through the center of curvature are equally usable as temporary optical axes. The ray strikes the mirror at an angle ϕ to the radius R, is reflected by the mirror by a total deviation 2ϕ, and crosses the optical axis close to the paraxial focus point P but on the mirror side of this point because of spherical aberration.

The most useful way to define the part of the ray from the mirror to the focus involves constructing a circle having its center at the center of curvature and drawn tangent to the incoming ray. The reflected ray then proceeds from the mirror to arrive tangent to this circle. Since the angle ϕ is identical at the origin and at the mirror, we have an isosceles triangle with its vertex at the focus point. From this simple geometry we can write

$$L = (R/2)(1/\cos\phi), \tag{6.6}$$

whence

$$S = R - L = R(2 - 1/\cos\phi). \tag{6.7}$$

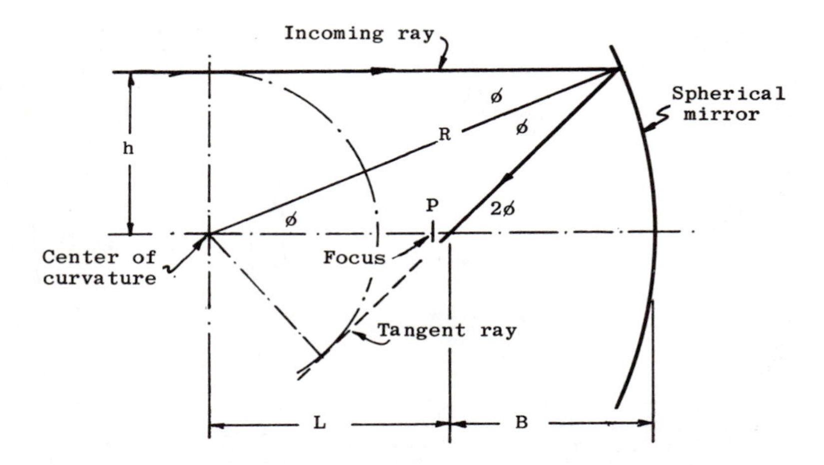

Fig. 6.15 Basic diagram for ray tracing a spherical mirror using the tangent-circle procedure. P denotes the paraxial focus position. Note that L is always larger than B.

The nominal focal length of the mirror is S, but the focal length as defined by chief rays is actually L. Equation (6.7) is exact and can be used if calculations of h values approaching $R/2$ are desired, as in the case of the hemispherical bowl collector discussed later.

In many cases where h is significantly less than $R/2$ we can use a convenient expansion for the value of $\cos\phi$, written

$$1/\cos\phi = 1/(1 - \phi^2 + ...) = 1 + \phi^2 + ..., \tag{6.8}$$

and since the geometry defines ϕ as equal to the angle in radians to the degree of approximation concordant with Eq. (6.8), we have

$$S = R(1 - h^2/R^2), \tag{6.9}$$

which means that S decreases as h increases.

If we place a receptor at the paraxial focus P normal to the optical axis, this change in S will mean that the point at which the ray arrives in the plane of the receptor will change. This error is termed *spherical aberration.* The aberration in S is termed *longitudinal spherical aberration,* and the orthogonal error, lying in the plane of the receptor, is termed *lateral spherical aberration.*

In Fig. 6.16 we show the plane of the receptor, the longitudinal aberration error δ, and the lateral aberration error ϵ. The longitudinal error is given by

$$\delta = L - R/2 = (R/2)[(1/\cos\phi) - 1], \tag{6.10}$$

and the lateral error ϵ is

$$\epsilon = \delta \tan(2\phi) = 2\delta [\tan\phi/(1 - \tan\phi^2)], \tag{6.11}$$

which, when expanded as before, yields

$$\epsilon = h^3/R^2. \tag{6.12}$$

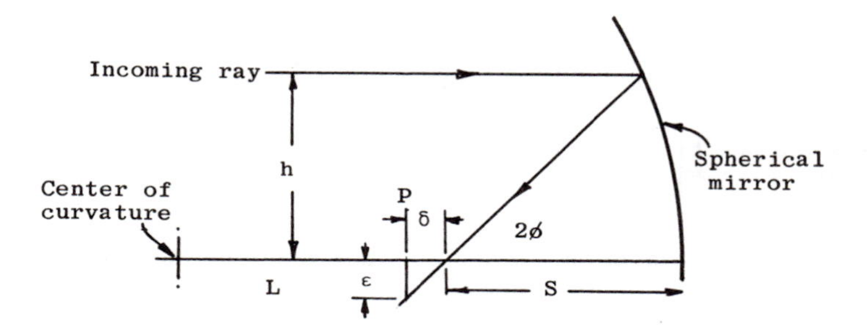

Fig. 6.16 Diagram defining the longitudinal and lateral spherical aberration on a receptor located at the paraxial focus.

6.6 SPHERICAL ABERRATION

When parallel rays enter a hemispherical mirror, the rays after reflection do not proceed to a point but are sent over a wide range of angles, forming a characteristic pattern. In Fig. 6.17 we show such a set of parallel rays incident on half a spherical surface. The ray through the center of curvature forms the paraxial ray, and rays close to it come to a focus at the paraxial image positions, half the radius from the center of curvature. Rays incident on the sphere farther from the axis come to a focus slightly below the paraxial focus, gradually becoming widely spaced along the axis until some rays suffer a second reflection at the mirror surface.

The envelope formed by all the reflected rays is called the *caustic surface*. It is formed by the tangential focus for zones of the mirror, and since it is a focus the

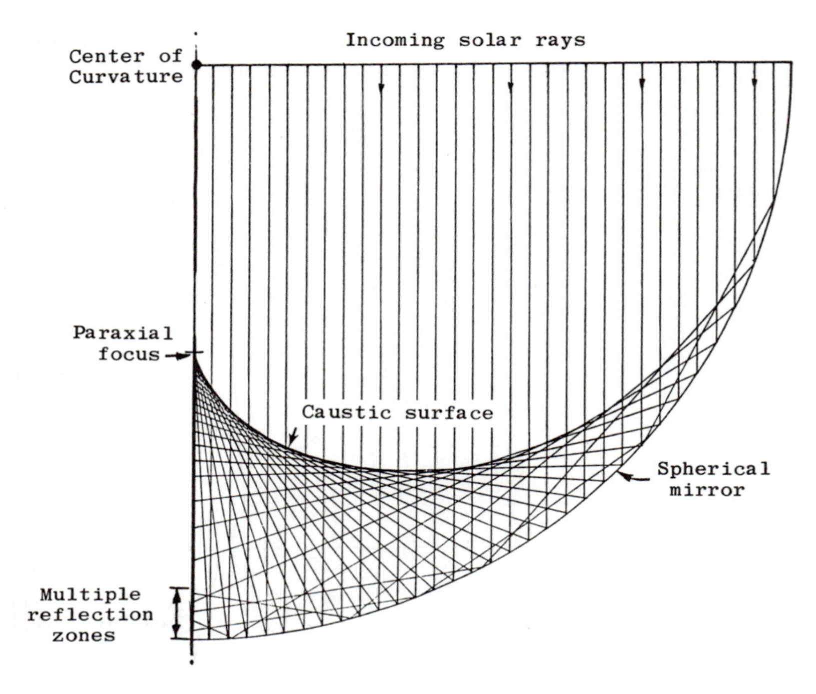

Fig. 6.17 Ray-trace diagram of the spherical aberration of a hemispherical mirror, showing the origin of the "caustic."

brightness is at a maximum on the caustic surface, rapidly shading off for positions between the caustic and the mirror. Some of the optical designs discussed in Chapter 7 utilize the facts about the caustic.

6.7 ASPHERIC MIRRORS

In the preceding section we found that a spherical mirror does not provide a stigmatic focus. We can, however, write the equations slightly differently and find out what surface would yield a stigmatic focus. In Fig. 6.18 we show a generalized surface having its origin at the vertex of the aspheric curve. The distance B is given by

$$B = x + y/\tan(2\phi), \tag{6.13}$$

where $\tan\phi = dx/dy$. Since we can rewrite $\tan(2\phi)$ in terms of $\tan\phi$, we then obtain

$$B = x + y/(2dx/dy) - (y/2)(dx/dy). \tag{6.14}$$

Let us assume that the aspheric surface can be expressed by a power series. Because of the requirement for axial symmetry, this power series can contain only even powers:

$$x = a + by^2 + cy^4 + dy^6 + \ldots . \tag{6.15}$$

Since a is a constant, it refers only to the point of origin for the curve when y equals 0, so we can write $a = 0$. Inserting this expression and its derivative into Eq. (6.14), we obtain

$$B = -cy^4 + 1/4b(1 + 2cy^2). \tag{6.16}$$

Since we desire the curve that yields B constant as a function of y, we see that this is possible only when c equals 0. Hence we find that

$$b = 1/4B, \tag{6.17}$$

and the equation of the mirror surface is

$$x = (1/4B)y^2, \tag{6.18}$$

which is the equation for a parabola whose focal length F is B.

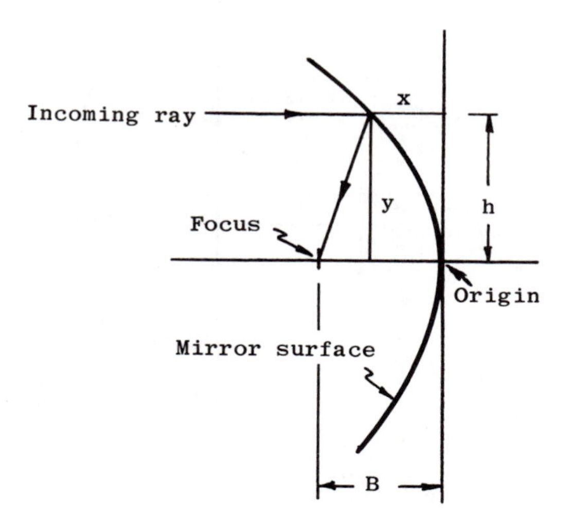

Fig. 6.18 **Diagram for deriving the shape of the mirror surface, which yields B independent of ray height h.**

6.8 PARABOLIC MIRRORS

In the preceding section we showed that a parabolic mirror is the only surface yielding a constant focal distance from the vertex of the mirror. It should be remembered that one is not restricted from allowing the paraxial image to depart from a stigmatic image in order to improve field imagery, but we will now confine our attention to the optical properties of parabolic surfaces, both cylindrical and circular.

The essential geometry of a deep parabolic mirror is shown in Fig. 6.19. A paraboloid has a paraxial center of curvature, where a tangent sphere can be constructed. The paraxial rays come to a focus half the distance between the vertex of the mirror and the center of curvature. In Fig. 6.19 we also show a pair of edge rays, which strike the paraboloid and are deviated 90°, arriving at the focus orthogonal to the paraxial axis. These particular rays also happen to be tangent to the sphere, which is tangent to the central portion of the paraboloid.

The effective center of curvature of the paraboloid, formed by the normal to the paraboloid where the rays strike the surface, is at a position along the axis different from that of the center of curvature for the paraxial region. The distance of this zonal

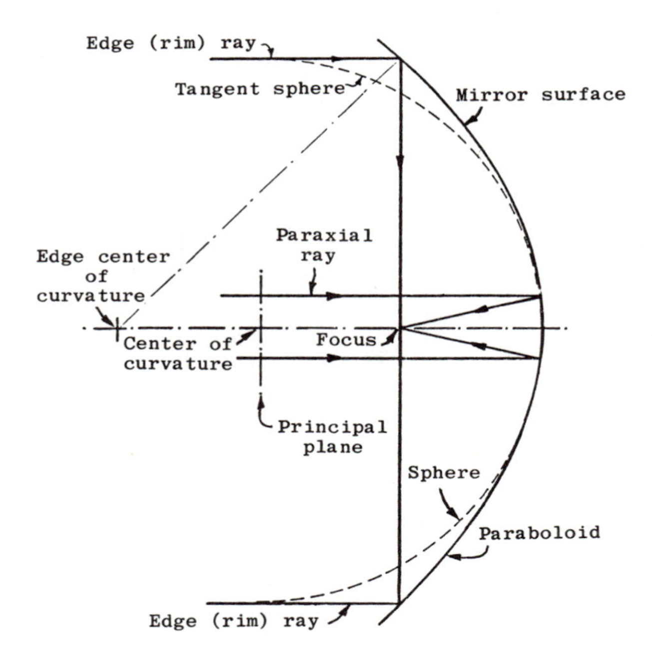

Fig. 6.19 Geometry of an *f*/0.5 parabolic mirror showing the principal elements and the ray paths for the edge rays and the paraxial rays.

center of curvature from the paraxial center of curvature happens to be for a paraboloid *exactly* equal to the sag in the parabolic curve. In Fig. 6.19, the distance C is the same as the distance D.

In the preceding section on radiometry we showed that a convergence cone of radiation of 2π (180°) would result in the brightness of the image equaling the brightness of the source, the sun. It is apparent that one could make a parabolic mirror, as in Fig. 6.19, where the rays approach the focus at a greater total angle than 180° ($\beta > 90°$). This does not mean that the brightness exceeds that of the source, because the flux cannot be incident on a plane surface. The *f*-ratio of 0.5 is the maximum for any optical system in terms of a flat focal surface. For the paraboloid whose convergence angle is greater than 90°, a curved receptor is required and the resulting brightness is equal to or less than that for $f = 0.5$.

The image of a parabolic trough or paraboloid is three-dimensional, even for the sun on the optical axis. The focal volume is approximately ellipsoidal, as shown in Fig. 6.20. In this figure we show the rays from the limb of the sun exaggerated by a factor of 16. We show three mirror elements and the chief ray associated with each element. The angle of divergence of the chief rays from the mirror elements is the same angle α as for the entering rays, but because of the distance the rays travel from the edge element the image size at the focal plane is larger and inclined with respect to the paraxial image. The effect is exactly as described for the Fresnel lens in the preceding case, except it is more severe owing to the large angles of convergence possible with a paraboloid.

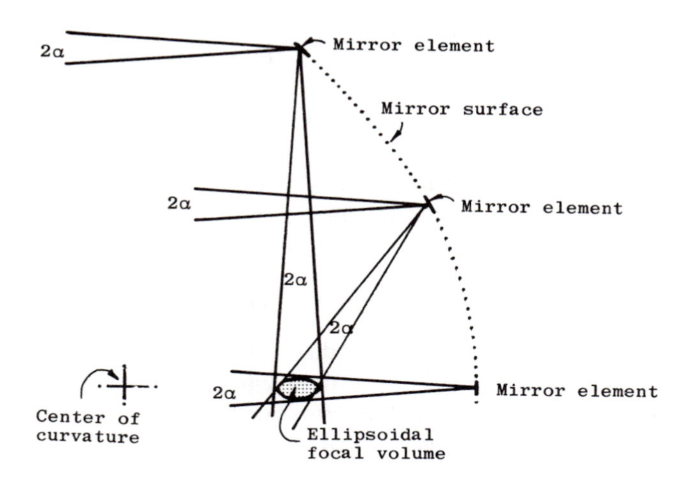

Fig. 6.20 Ray diagram for the solid angle 2α of the sun exaggerated 16 times to show the origin of the ellipsoidal focal volume. For an *f*/0.5 paraboloid the ratio of major to minor axes is 2:1.

In the case illustrated in Fig. 6.20, we have an $f/0.5$ paraboloid, and the ellipsoidal focal volume has a major-to-minor axis ratio of approximately 2:1. The effect of mirror errors in the parabolic surface is to statistically enlarge the focal ellipsoid, preserving the ratio of major to minor axes.

Paraboloids can be made with f-ratios as extreme as one is willing to use, as contrasted to lenses where the maximum convergence angle (minimum f-ratio) is set by the limit of refraction through the edge of the lens, at about $f/0.8$. The resulting concentrations are therefore significantly larger with mirror systems than with lens systems. When errors are taken into account, the maximum flux concentration for a lens is about 10–20, but for a parabolic mirror with similar manufacturing difficulty the maximum is 40–60 for cylindrical optics. For circular optics the corresponding limits are the square of the above values, 100–400 for lenses and 1600–3600 for paraboloids. Higher concentrations can be achieved, but the costs are prohibitive for general solar energy applications.

The limit to the f-ratio for a paraboloid is also set by the cost effectiveness of the mirror surface. For f-ratios faster than 0.5 ($\beta > 90°$), the outer surfaces of the mirror become steeply inclined to the incoming solar rays, and hence their cost effectiveness is reduced. One therefore must take account of the number of square meters of mirror surface required to collect a square meter of solar radiation.

The geometry we have described above applies equally well to spherical and parabolic mirrors that are cylindrical or circular in aperture shape.

Parabolic mirrors yield good images only on axis. When the sun is off axis the image degrades rapidly, primarily because of coma. The amount of coma varies as the square of the f-ratio for third-order coma, and when higher orders are added the rate of growth is much higher when f-ratios of 0.5 and faster are utilized. The actual image error can hardly be recognized as traditional coma in the case of an $f/0.5$ paraboloid. The imagery of an $f/0.5$ mirror is shown in Fig. 6.21. Note that the rays do not form an image, but curl in an arc through the paraxial image position. The extreme rays never cross at a focus but travel parallel to each other. It is clear from this illustration that the collimation of the absorber in the focal line of a parabolic trough or at the focus of a paraboloid is quite critical. It is for this reason that one cannot use an EW parabolic trough in a fixed-mirror arrangement, but must use more sophisticated optical designs as discussed later.

6.9 FIELD ERRORS OF CYLINDRICAL MIRRORS

In the section on Fresnel lenses we saw that the field errors for cylindrical optics create serious complications in the use of these simple concentrating lenses. The same is true with cylindrical mirror optics. For errors due to acceptance angle changes in the plane transverse to their length, the optical image errors are as described above. The geometry is quite different for acceptance angle changes in the plane of the cylindrical axis. This geometry is shown in Fig. 6.22, where we have a cylindrical mirror, for example a circular or a parabolic cylinder, with an axis as shown. This axis for a

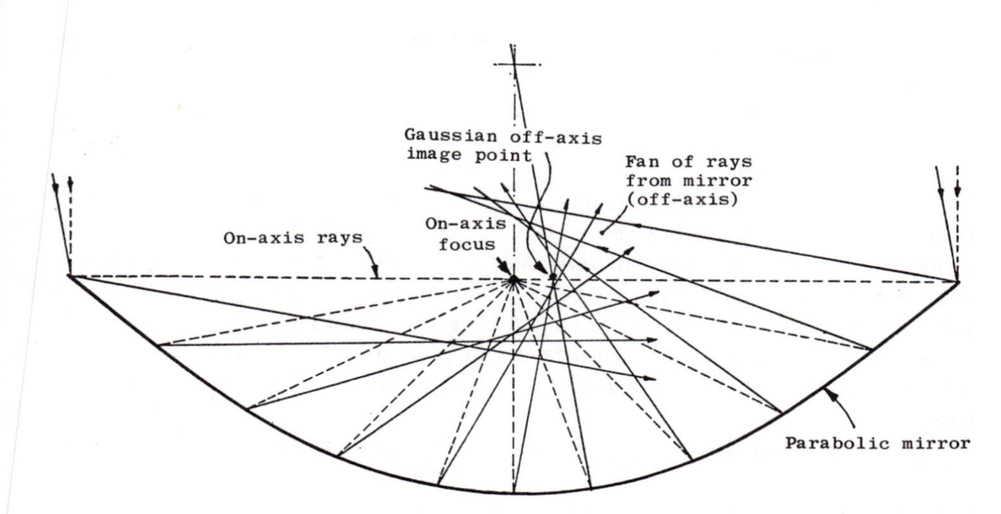

Fig. 6.21 Ray-trace diagram showing the off-axis ray paths for an $f/0.5$ paraboloid. The dots indicate the paraxial image positions.

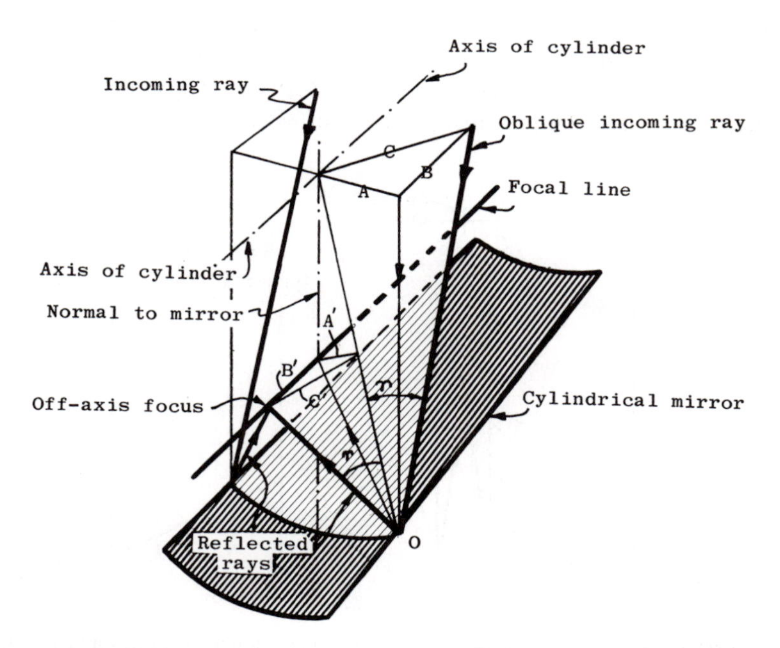

Fig. 6.22 Diagram showing the invariance of the focal line for oblique rays incident on a curved cylinder. The angles γ opposite side C in both triangles are equal and identical to the angle of incidence and reflection of the incoming ray.

parabolic cylinder is the effective center of curvature for the mirror at the distance from the normal to the mirror where the incoming ray strikes the mirror. The angles describing the incoming ray can be resolved into two orthogonal components, making angles A and B with respect to the radius vector to the point of incidence, the angle of incidence being γ, which is proportional to C.

The basic law of reflection and refraction is that the incident and reflected (refracted) rays lie in a plane containing the normal to the surface at the point of incidence, and that the angles of reflection and incidence are equal for a mirror, or obey Snell's law for refraction. We can therefore construct a solid triangle similar to $ABCO$ as $A'B'C'O$, so that all angle components are equal to the corresponding angle components in triangle $ABCO$. This triangle therefore must have its base B' coincident with and parallel to the focal line defined solely by the normal plane triangles having bases A and A'.

It is because of this invariance of the focal-line position that cylindrical mirrors are useful for solar collectors, while cylindrical lenses are complicated by a major shift in position of the focal line for oblique incoming rays.

6.10 HELIOSTAT PARABOLOIDS

In some applications of current interest the need for a large paraboloid and the means to mount it to track the sun are replaced by fixing the focus and reducing the mirror to a large number of discrete mirrors, each steerable so that a point image is kept on the focus. We show the general arrangement of this type of system in Fig. 6.23. It is interesting to note that the principal plane has effectively been transformed into the

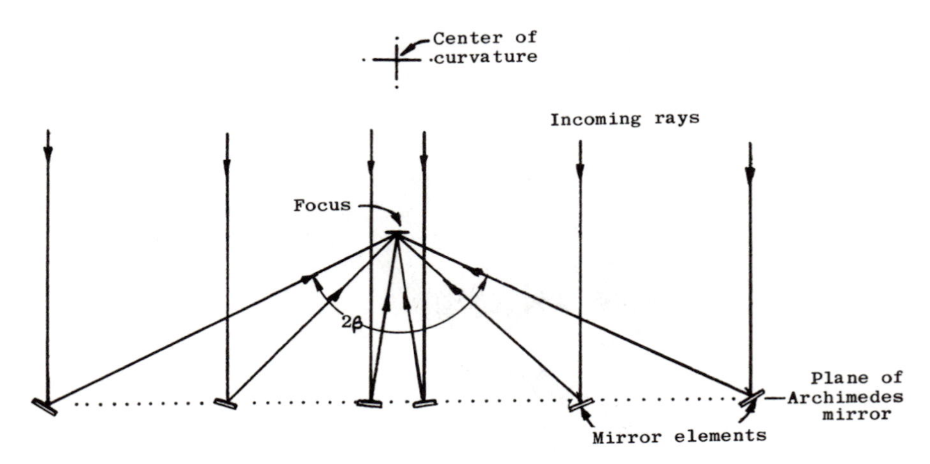

Fig. 6.23 Diagram of the optical equivalent of an *f*/0.50 paraboloid as formed by a multiplicity of heliostats (power tower). Note that in this case the brightness at the focus is set by the convergence angle *from* the mirrors. If the plane of the mirror appeared to be fully filled with mirrors, the brightness would be the same as from an *f*/0.50 paraboloid.

mirror plane, and the surface brightness at the image is given by the angle of convergence of the rays to the focus. In practice one does not achieve this level of brightness because the entire ground is not fully covered with mirrors. One could achieve full coverage or 100% fill factor and obtain the theoretical brightness, but in practice the array would be redundant for most positions of the sun in the sky and the cost effectiveness so low as to be impractical.

The multiple-mirror paraboloid also yields a focal ellipsoid, as is shown in Fig. 6.24. The ellipsoid in this case is eccentric, with a coma-like image error softening the focus at the bottom of the ellipsoid.

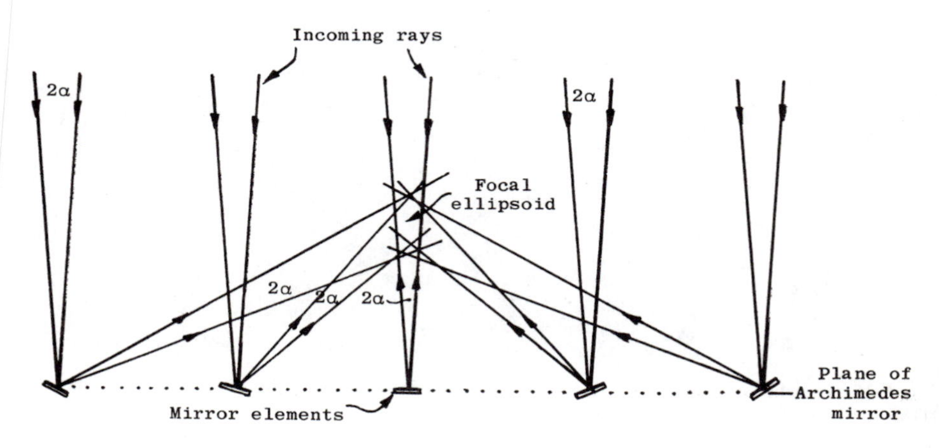

Fig. 6.24 Diagram illustrating the formation of the focal ellipsoid for a Fresnel paraboloid. The angle 2α is exaggerated 16 times the solar disc angle. Note that the focal ellipsoid is asymmetric, with a coma-like error at the bottom.

The multiple-mirror paraboloid has the advantage that a constant focal ellipsoid is formed for any position of the sun in the sky; however, this is at the expense and hazard of having heliostats that must track in two coordinates and be rugged enough to survive, yet inexpensive enough to yield cost-effective benefits.

A drawing of a heliostat field with tower absorber set at one edge is shown in Fig. 6.25. Each heliostat is mounted in an altazimuth mounting so that the angle of the mirrors can be adjusted to keep the reflected solar rays directed toward the tower.

The size of the heliostat mirror is determined by the maximum allowable image spread at the central tower focus. The basic image size is set by the angular size of the sun. This size is further enlarged by the surface errors of the mirrors. The finite size of the heliostat mirror adds the mirror dimension to the blurred solar image. If the mirror size is allowed to double the solar image size, this means that the mirror must appear to subtend the same angular size as the sun. Such a mirror will therefore have an *f*-ratio of 115.

Fig. 6.25 Drawing of a heliostat field with tower absorber designed in 1975 by Martin-Marietta. (Drawing by C. Bennett, Martin-Marietta.)

6.11 CURVED HELIOSTAT MIRRORS

For small heliostat installations the individual heliostat mirror is rather small, and one would like to combine a number of heliostats into a single larger mirror unit. In this case one may elect to provide the mirrors with curvature, so that they focus the sunlight in addition to directing it to the tower. Curved heliostats, however, introduce a new optical problem: the aberrations of an inclined spherical mirror.

The image geometry of an inclined spherical mirror is shown in Fig. 6.26. The basic image aberration that is introduced is astigmatism. This means that the mirror will have one linear focus closer to the heliostat than is the paraxial on-axis focus, and a second linear image oriented at right angles to the first and lying farther from the on-axis focus. Halfway between these two linear images a circle of least confusion is obtained, at approximately the same distance from the heliostat mirror as is the original focus. The additional blur of the solar image on the absorber of the heliostat system is therefore this circle of least confusion. A graph of the diameter of this circle as a function of angle of deflection is shown in Fig. 6.27.

The dependence of the size of the blur circle on the geometry can be derived from Fig. 6.26. The distance between the sagittal and tangential focal positions is denoted by ΔF. If the mirror diameter is D and the normal focal length is F, then the diameter of the circle of least confusion ϵ is given by

$$\epsilon = D\Delta F/2F. \tag{6.19}$$

The difference between the sagittal and tangential foci as a function of the angle of incidence and angle of deflection is given in Fig. 6.27. In this graph we show as the ordinate the blur circle diameter for several f-ratios of the heliostat mirror. To arrive at the value of ΔF from this graph one would multiply the ordinate for the $f/10$ curve by the factor 20. This means that the difference in focal length for a mirror deflection of

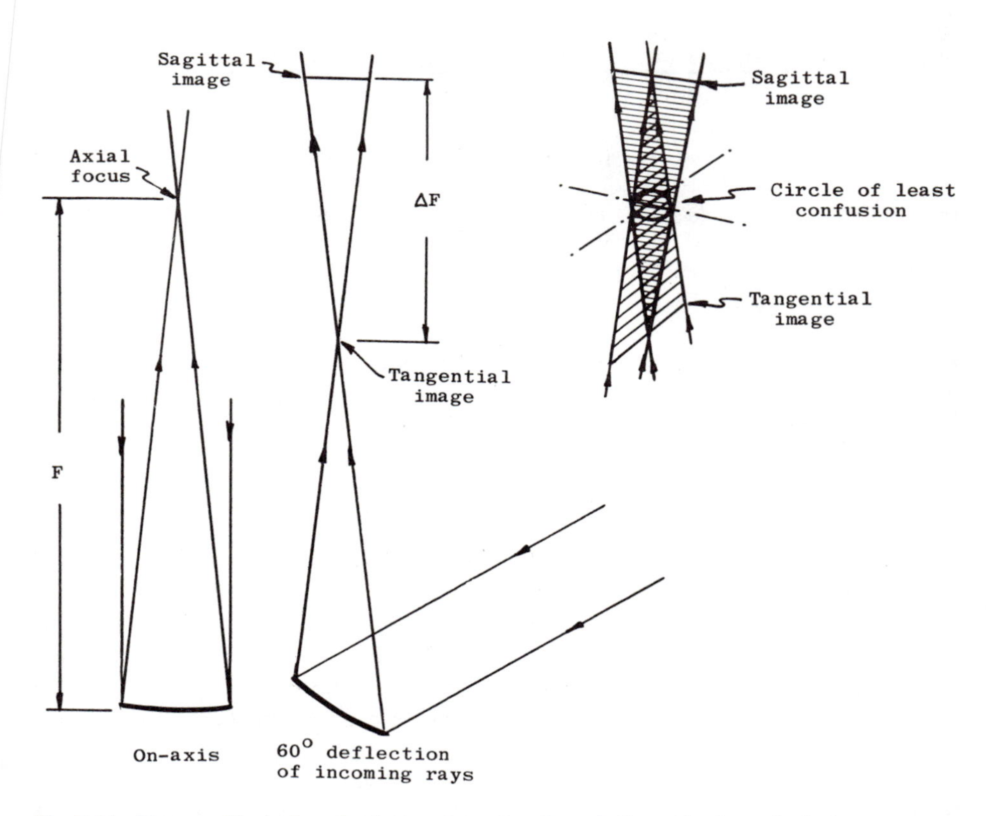

Fig. 6.26 Diagram illustrating the image aberration for a heliostat having spherical power when the mirror is tilted off axis. The inset shows a perspective view of the image, showing the two line foci and the circle of minimum size.

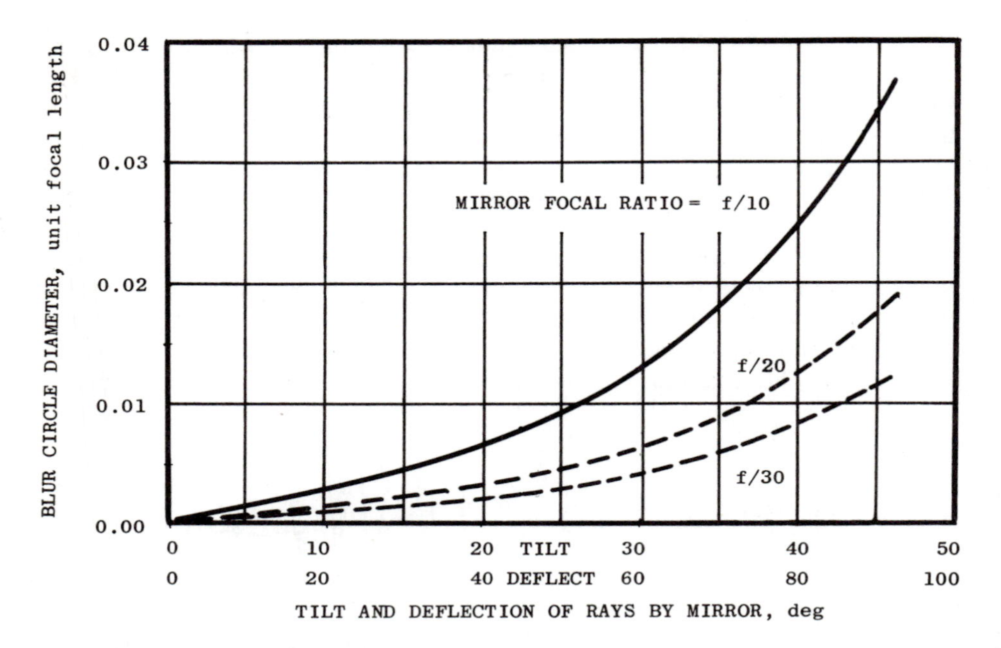

Fig. 6.27 Relationship between the angle of mirror tilt or ray deflection by the heliostat mirror and the diameter of the circle of least confusion. The units are in terms of the focal length of the mirror.

80° would be 0.50*F*, but the circle of least confusion would be only 0.025*F*. A heliostat mirror 8 m in diameter focusing on a central tower 160 m away would add a blur of 1.92 m at an 80° deflection to the normal image from the heliostat, an error of significant magnitude.

6.12 MOVABLE-SLAT CYLINDRICAL COLLECTOR

A simple solar collector adapted especially for cylindrical geometry can be made of a group of mirror slats, as shown in Fig. 6.28. Each of the slats is tilted to direct the reflected rays to a common focal position. For different positions of the sun the slats need to be tilted to different angles to preserve the focal position, as shown in the left-hand diagram. The required rotation of each slat about an axis lying in the plane of the reflective surface or close to the surface can be accomplished by a simple linkage, as indicated.

The slat-type cylindrical mirror can also be used in a fixed configuration, where the entire collector is tilted to follow the sun. Collectors of this type, one of which is shown in Fig. 6.29, have been used extensively at the Desert Sunshine Testing Laboratory in Arizona.

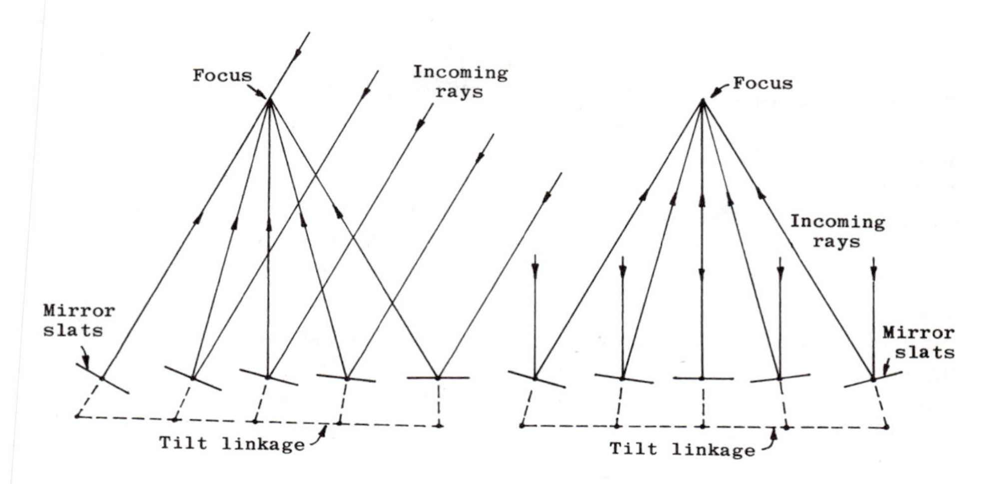

Fig. 6.28 Schematic diagram of a tilting-slat mirror collector suitable for a cylindrical Fresnel paraboloid where the absorber is fixed. If slats are flat, the absorber diameter must be larger than the slat width. If cylindrical curved slats are used, the absorber must accept the astigmatism of the image where solar rays strike the slat at large angles of incidence, as at the edges of the collector.

Fig. 6.29 Photograph of the EMMA solar testing apparatus of Desert Sunshine Testing Laboratory, Arizona, where slat mirrors provide 10X concentration at a linear focus.

Curved slats may be desirable as a means of reducing the number of slats in a collector. The choice of slat size and curvature is set by the relative size of slat and absorber. One desires the directed sunlight to arrive within the cross section of the absorber. This means that the width of a slat must be less than that of the absorber by an amount equal to the angular size of the sun plus surface errors on the slat. The number of slats is therefore somewhat greater than the optical concentration of the collector. If a concentration of $C = 20$ is desired, the number of slats must be greater than 20, on the order of 25 to 30, depending on the tolerances involved.

The introduction of a curved slat reduces the number because the sunlight is actually focused on the absorber rather than simply redirected to the absorber. The width of a single collector is limited by the amount of astigmatism encountered from the curved slats when the sun strikes them at large angles, as at the edge of the collector. This tolerance is in turn related to the size of the absorber relative to the collector aperture. The amounts of astigmatism for curved heliostat mirrors are given in Fig. 6.27.

6.13 VEINBERG PARABOLIC COLLECTOR

Veinberg (1959) describes a number of useful collector designs utilizing parabolic cusps, some of which anticipate the properties of the Winston double parabolic cusp collectors (see Section 7.7). Since Veinberg presents a unified picture of the modes of using parabolic sections, we will summarize them.

The three basic forms of the paraboloid depend upon the orientation of the absorber.

1. Absorber Facing the Mirror Vertex. This is the usual mode of using a paraboloid, illustrated in Fig. 6.30(a). The maximum average solar flux concentration on the absorber has the angles and dimensions indicated on this figure.

2. Absorber Parallel to the Paraboloidal Axis. In this arrangement sunlight arrives on both sides of the absorber for a cylindrical system or on the surface of a cylinder for a circular system, as shown in Fig. 6.30(b).

3. Absorber Facing the Sun. In this arrangement, shown in Fig. 6.30(c), all of the paraboloid lies above the absorber, as in the Winston designs (Section 7.7). Although this arrangement has the same collecting area cross section, it is clear that much more mirror surface is required.

All the paraboloids in the Veinberg family have the absorber cross section equal for the case of a parabolic cylinder. In other words, the surface of Fig. 6.30(b) on both sides of the absorber is equal to the surfaces facing the incoming light in Figs. 6.30(a) and (c). For a circular paraboloid the actual length of the absorber in Fig. 6.30(b) would be lengthened by 1.41 for the absorber discs to have equal area. In each of the

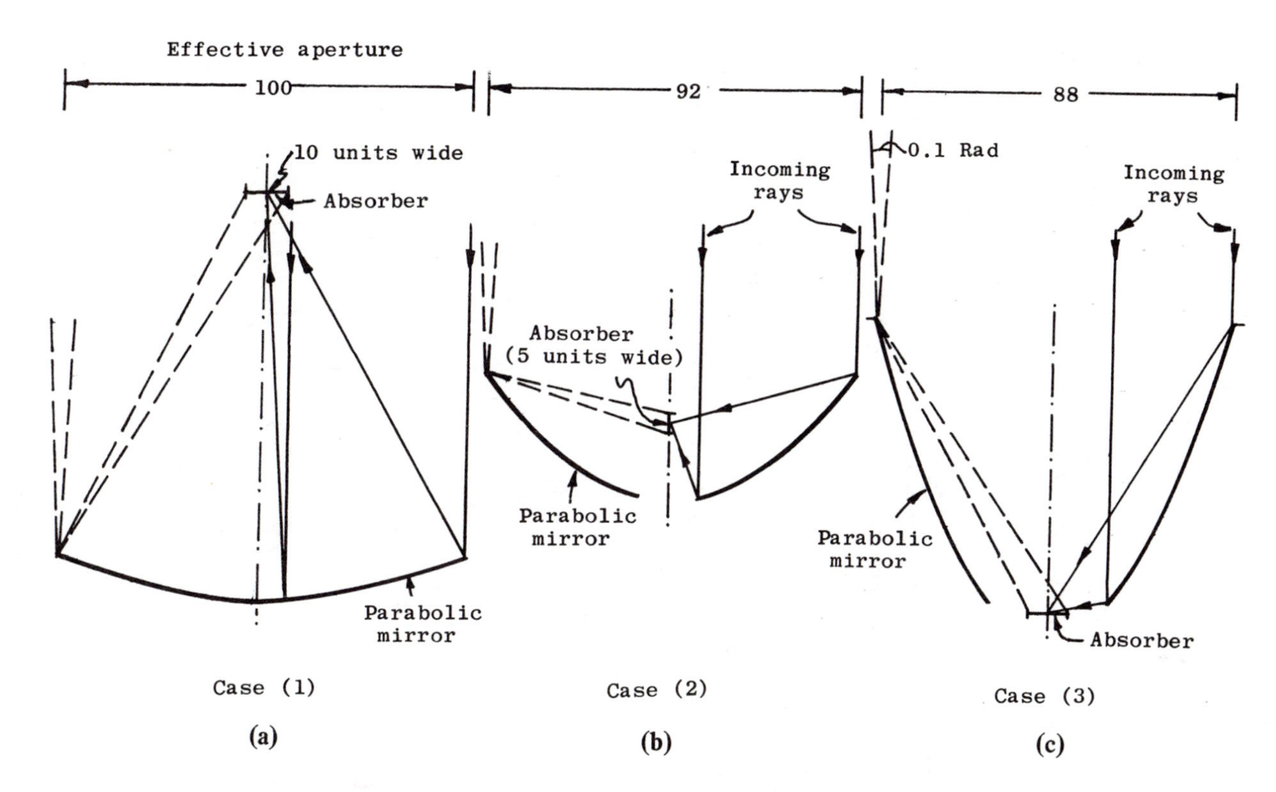

Fig. 6.30 Relative sizes of the three Veinberg parabolic cases shown for equal absorber surface areas and field acceptance angle. Note that the height of the absorber in diagram (b) is half the width of that for diagrams (a) and (c), because sunlight arrives on both surfaces of the absorber and infrared radiation is lost from both.

cases the angular size of the absorber is 0.1 rad, as indicated by the dashed lines. Under these conditions of similarity we can compare the relative apertures of the three cases.

The aperture for Fig. 6.30(a) is 100 units for a 10-unit-wide absorber, which means that the radiative balance concentration is X = 10.0. For Fig. 6.30(b) the aperture is 92 units and X = 9.2, and for Fig. 6.30(c) the aperture is 88 units and X = 8.8, both less the area of the central hole. This means that one loses concentration in Fig. 6.30(c) in exchange for the convenience of having the focus accessible at the bottom of the collectors, and at the added expense of requiring more mirror reflective surface.

Note that Fig. 6.30(c) is very similar in general appearance to the Winston collector, which is discussed in the next chapter. The design also functions similarly, the angle of acceptance being indicated by the dashed lines, 0.1 rad in the illustration. In the Winston designs the two halves of the paraboloid are in effect moved together so that the gap between the absorber and the bottom of the cusps is eliminated. The Veinberg geometry can therefore be used to provide an angular acceptance of 0.1 rad or larger depending on the size of the absorber and at correspondingly lower flux concentrations, as discussed in Chapter 7, or as Veinberg points out, the angle can be used to allow the mirror surfaces to be of low optical figure accuracy.

The design for Fig. 6.30(c) can be extended to include several concentric paraboloidal sections, as shown in Fig. 6.31. This three-cone design gives close to the same brightness concentration in the solar image as a paraboloid of the same cross section,

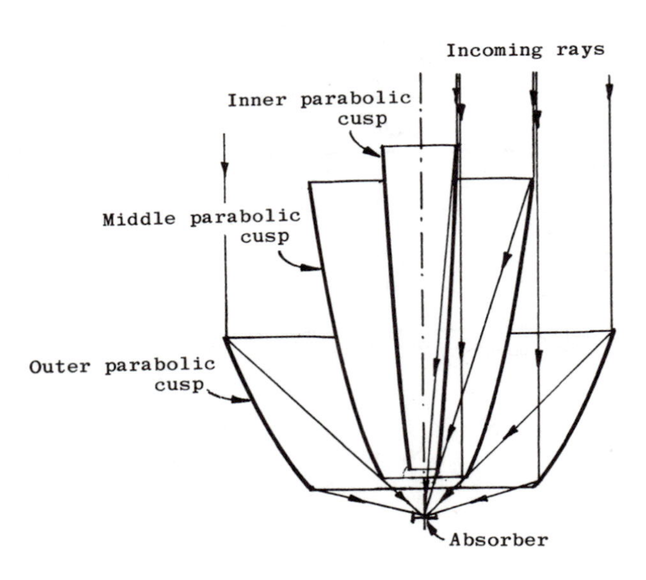

Fig. 6.31 Diagram of a Veinberg triple paraboloidal mirror having the same flux concentration as an $f/0.13$ lens.

the chief advantage of this unusual geometry being that it places the focus in an accessible position with the mouth of the absorber cavity facing upward.

Veinberg also shows several designs where circular sections of off-axis paraboloids are used to focus laterally on both sides of a vertical absorber, as shown in Fig. 7.15.

6.14 FILL FACTOR

The term *fill factor* is used to denote that fraction of the available ground area actually covered with collector surfaces. The only collector geometry that results in a 100% fill factor is the horizontal collector, such as a flat-plate collector. Such a collector works at good efficiency when the sun is near the zenith, but poorly when the sun is near the horizon, as it is for several months in winter. When a collector is tilted from the horizontal it casts a shadow on its neighbors, and one then finds it expeditious to space the neighbors so that the effect of shadowing is minimized. The basic geometry for shadowing is shown in Fig. 6.32.

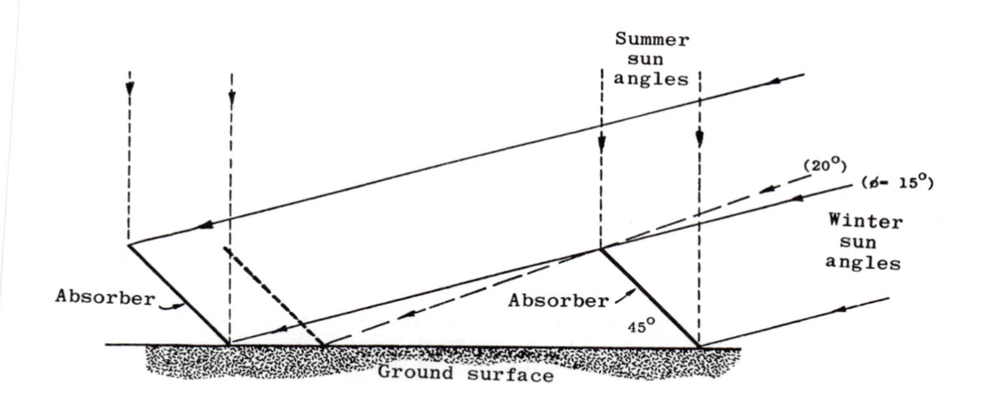

Fig. 6.32 Diagram illustrating the fill-factor geometries for avoidance of shadowing of adjacent east-west rows of solar collectors placed at 45°. The fill factor for no shadowing for a winter sun at 15° elevation is 21%, increasing to 26% for a sun angle of 20° elevation (approximately noon sun at winter solstice at 46.5° latitude).

The amount of spacing required between collectors increases as the angle of the collector increases. The fill factor for an array of collectors is defined as that fraction of the land surface covered with collecting surfaces when the sun is in the zenith. Typical values of fill factor range between 2 and 4, depending upon the system. A vertical collector would actually have a fill factor of zero by the above definition; such collectors are seasonal, however, and one then desires a maximum-effective fill factor during the season of use, the fill factor being differently defined in these special cases.

Fill factor becomes important in discussing the tradeoffs among different large-scale collector systems. For example, the power-tower system uses moving heliostats deployed over a large land area. How close can these heliostats be placed before degrading system performance for low sun? It is obvious that at some angle toward the horizon the foreshortening of the many mirrors will make the field appear completely filled with mirrors. For lower sun angles the total collecting field then begins to act like a horizontal collector, following the cosine curve. The closer the mirrors are placed, the higher the fill factor and the higher the noon output, but the sooner the cosine effect takes over. In Fig. 6.33 we show the effects of fill factor on the daily output of a dilute collector array.

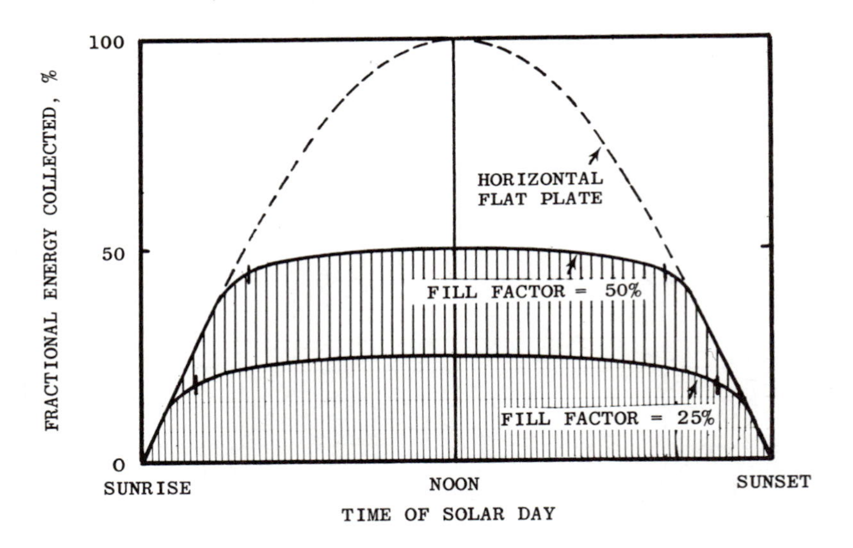

Fig. 6.33 Diagram illustrating the effect of fill factor on solar tracking collector systems. The flat curves are typical of central tower heliostat systems and the cosine curve of the hemispherical bowl collector systems.

Fill factor also affects cylindrical systems. Horizontal rows of collectors must be spaced, whether they are east-west or north-south, so that the nearest neighbor is not in shadow. When the cylindrical collectors are tilted, as in the case of the polar axis cylinder, the shadowing effects are severe at large angles and poor fill factors result. In any event, an array of polar axis collectors cannot follow the sun beyond an hour angle of the sun of 6 h, and the finite angular size of the nearest neighbor means even less coverage without vignetting the beam entering the collectors, as is shown in Fig. 6.34.

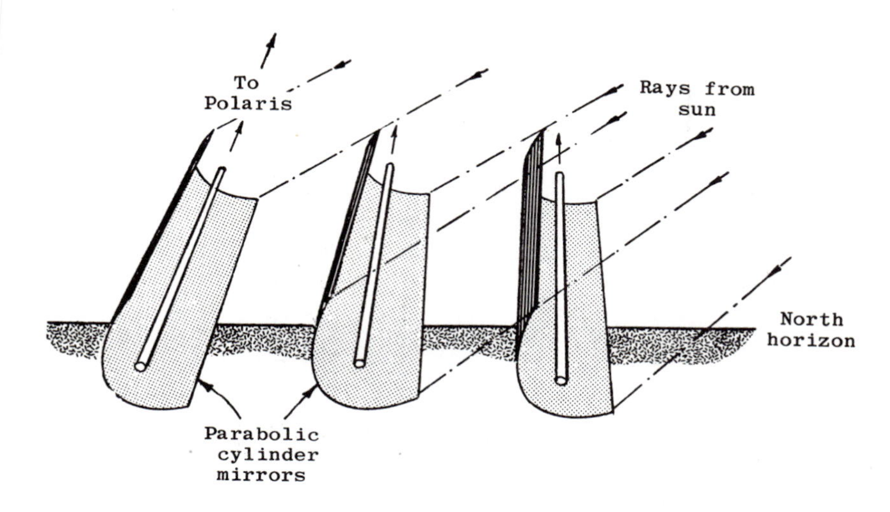

Fig. 6.34 Diagram illustrating the mutual interference of polar-mounted collectors. The collectors are limited to solar hour angles of less than 5 h on each side of solar noon, losing operating hours at summer solstice and even at equinox.

The fill factor of the hemispherical bowl is 100% for all sun angles, but this is because it acts like a flat-plate collector, having the full cosine effect for all angles of the sun, depending upon the direction in which the bowl aperture is directed.

The net consequence of fill factor is to flatten or peak the daily output curve, depending on the value of the fill factor. A flat daily response curve, approaching that of a *single* fully tracking collector, means that the land area is poorly utilized because the collectors must be widely spaced from each other. A peaked daily response curve, on the other hand, means excellent land usage but a large dynamic range over which the heat transfer and utilization system must operate. The tradeoffs are complicated and are under intensive study for the several collector geometries currently being proposed for "solar collector farms."

6.15 INFLATED MIRRORS

Plastic film concentrating mirrors present a more difficult problem than flat film mirrors. Since they offer the potential of very light-weight mirrors at low cost per unit area they may have potential uses in solar energy collection. Taylor (1975) has constructed concentrating mirrors of aluminized Mylar in the form shown in Fig. 6.35. The lower surface is aluminized or silvered, the reflecting surface facing the focus. The upper surface is transparent and can have a nonreflecting coating on both sides. The curvature of the mirror is caused by inflating the space between the surfaces.

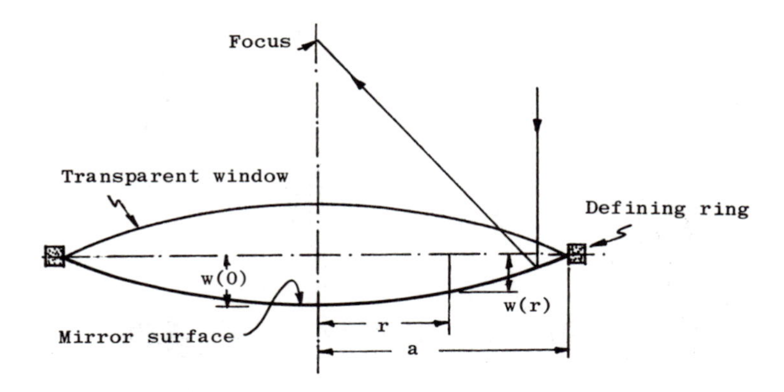

Fig. 6.35 Cross section of an inflated membrane mirror defining the factors shown in Eq. (6.20). The inner surface of the lower membrane is aluminized or silvered to form the mirror.

The deflection of the mirror surface is given by

$$w(r) = w(0)\,[1 - 0.9(r/a)^2 - 0.1(r/a)^4 + \ldots]\,, \tag{6.20}$$

where a is the semiaperture of the mirror, r is the distance outward from the central axis of the mirror to the point where the depth is being calculated, $w(r)$ is the depth at point r, and $w(0)$ is the central depth of the mirror.

This expression holds for modest deflections of the mirror, but should not be used without higher order terms if the mirror focal length exceeds the mirror diameter.

Although Taylor proposed using this mirror in conjunction with a Cassegrain secondary mirror with the secondary mirror compensating for the residual spherical aberration, the primary mirror can be used alone without serious image degradation. If we subtract a paraboloid from the above equation,

$$w(r) = w(0)\,[1 - 1.05(r/a)^2]\,, \tag{6.21}$$

we have the residual error that would give rise to the minimum aberrated image of the sun. The residual depth error is then

$$w(r) = w(0)\,[0.3(r/a^2) - 0.4(r^3/a^4) + \ldots]\,. \tag{6.22}$$

In this equation the slopes at $r = 0.707$ and $r = 1.00$ are equal and opposite in sign and represent the best focus of the inflated mirror. The slope at the edge is found to be 0.02 rad when the depth of the mirror is 0.2 the semidiameter of the mirror. This slope error produces a deviation of the solar ray of 0.04 rad, or 4.4 solar diameters. If a perfect mirror produced a concentration of C = 10,000, this angular error would reduce the value to $C' = 500$, a brightness concentration still useful for some high-temperature applications.

The inflated mirror must be constrained by a ring at the periphery. A metal ring can be used since the compressive forces all lie in the plane of the ring. Taylor has proposed using a tube like a bicycle tire inflated to a high enough pressure to be more rigid than the mirror. Variants of this approach to inflated mirrors have been proposed from time to time. Photographs of one of the Taylor mirrors are shown in Fig. 6.36.

Through front surface. **Rear surface.**

Fig. 6.36 Photographs of a membrane mirror by Taylor (CERN). Note the small amount of scattered light from the transparent plastic membrane, which lowers the contrast of the reflected image.

There are several practical problems to be faced with membrane mirrors. The planeness of the seal between the two halves determines the basic accuracy of the mirror. A rigid support ring facilitates the establishment of this seal line. Uniform film thickness and uniform elasticity of the material are also necessary for an accurate mirror. The flux traversing the transparent plastic film is slightly greater than two solar fluxes, creating materials problems by raising the film temperature and shortening the lifetime of the film. This increase of temperature of the mirror unit also changes the focal length of the mirrors by raising the pressure of the gas inflating the mirror; it expands the material, also shortening the focal length.

6.16 PROBLEMS

6.1. A plane mirror is to be used to send light from a heliostat to a concave furnace mirror placed overhead. The mirror is placed so that the angle from the concave mirror to the plane mirror in the north-south direction is 15°; thus the concave mirror will not shadow the heliostat during summer noon. What is the angle of this plane mirror at noon with respect to the horizontal at summer solstice and at winter solstice? Through what angle has the mirror been moved during the seasonal change? If the mirror acquires the sun at the horizon at the solstices, what is the daily motion in degrees of the heliostat in declination? What is the foreshortening of the aperture at these extreme positions?

6.2. If the heliostat is a slat mirror and fixed at some optimum meridian angle, what would be the optical foreshortening as a function of tilt angle with respect to the equinox noon mirror position? What causes this lowered optical cross section as compared to the moving single-mirror heliostat?

6.3. A flux compressor made of two inclined plane mirrors has an angle between the mirrors of 15°. What is the maximum flux augmentation, and how long an absorbing surface would be required to intercept the rays entering the aperture parallel to the central axis of the wedge? What would be the loss of rays on this absorber for rays incident on the wedge at 5° and 10° off the wedge axis?

6.4. Plot the variation of incident flux on a flat-plate collector placed at an angle λ, facing the noon equinox sun, as a function of solar declination when the angle of the mirror boosters is 50° between the mirrors. Each mirror has a width equal to that of the collector absorber surface.

6.5. Two booster mirrors, as in Fig. 6.4, are to be placed on the north and south sides respectively of an east-west collector. The mirrors can be placed in three positions. Each mirror has a width equal to that of the absorber. Taking into account the duration of sun at different declinations during the year, give the optimum angles for placement of the absorber and the booster mirrors. Make drawings to scale for the three times of year you select.

6.6. Given that the collector shown in Fig. 6.9 has only one fixed angle for the booster mirror, the absorber is horizontal, and the latitude is 35°, select an optimum angle for the booster mirror and plot the annual variation of optical cross section for the collector. What is the optimum angle you have chosen and on what criteria did you base your selection?

6.7. If one placed the absorber at a steep angle, so that hail damage would be less likely, and used a plane mirror (the approximate inverse of Problem 6.6), what angle would you choose for the collector absorber and for the plane mirror for (a) summer and (b) winter? Plot the annual variation of optical cross section for this type of collector.

6.8. Is there any reason why an axicon collector is made with a vertex cone angle of 90°? Draw the optical diagram for an axicon with $\theta = 120°$ and with $\theta = 60°$. Are there any possible advantages or disadvantages to these angles from a practical engineering standpoint? What is the ratio of mirror surface to collector aperture?

6.9. Derive the approximate shape of the optical image formed by the AKS design, where the distance from the focus to the cylindrical Fresnel lens is equal to the radius zone of the entrance aperture, approximately as shown in Fig. 6.13. What is the shape of the image volume? Given that you had a paraboloid of the same nominal image size, is the statement about the AKS having less mirror surface correct? What factor would you state instead?

6.10. Make a scale drawing of a spherical mirror solar collector where the aperture is 1.4 times the paraxial focal length. Draw the mirror having a radius of 25 cm. How large an absorber would be required to intercept the radiation if the absorber were placed at the paraxial focus normal to the axis? How large would the absorber be if placed at the circle of least confusion? Would an absorber of a different shape be better? In what ways? Is the surface of this absorber larger or smaller than the surface of the flat plate at the circle of least confusion? What is the ratio of these two surfaces?

6.11. We have an $f/0.4$ paraboloid for a solar furnace. What is the size and shape of the focal ellipsoid when the mirror has an aperture of 50 m? What is the distance of the focus from the vertex of the mirror? What is the angle of convergence of the rays at the focus?

6.12. We have an $f/0.4$ system of plane heliostat mirrors in a field, yielding this f-ratio, neglecting fill-factor effects. How high is the tower where the focal absorber is placed? What is the shape of the focal ellipsoid, and what are its dimensions when the aperture is 500 m? What is the angle of convergence of the rays at the focus?

6.13. The focal absorber in Problem 6.11 is made twice the size of the focal ellipsoid. What is the approximate field angle (the angular distance the sun can move before the solar image begins to move off the absorber)? How long will it be before the entire solar image is off the absorber? State the exact sun-collector geometry you use to answer this question in the simplest way.

6.14. A heliostat for a solar tower system is made with seven mirrors on a floating disc. The disc has a diameter one-twentieth the distance of the disc from the base of the tower. The system has a tower height equal to the distance of the center of the disc from the base of the tower. What is the astigmatism of the beam formed by these seven mirrors when the angle of deviation of the mirrors to the tower is 90°?

6.15. If the linkages of a tilting-slat collector are of equal length and if the tilts are such that a perfect focus is obtained for the sun in the zenith, what is the image error when (a) the sun is 30° off zenith and (b) the sun is 60° off zenith?

6.16. What are the relative fill factors for an array of east-west cylindrical collectors at 35° latitude, tilted at 35°, with and without side booster mirrors each of width equal to that of the absorber? Draw the geometry involved.

Chapter 7
Fixed-Mirror Collectors

7.0 INTRODUCTION

The cost of moving mirrors and making a solar tracking structure rugged enough to survive and still be within the tight cost limits imposed by the market value of the energy delivered make fixed mirror collectors an important option. The basic problem with a fixed mirror is that the sun moves in declination during the year and in right ascension during the day. The angles through which the sun moves even during the year are considerable. An east-west fixed cylindrical collector, for example, needs a range of about 120° in north-south motion to track the sun from one hour after sunrise at summer solstice to one hour after sunrise at winter solstice. The same angle for daily solar motion in an east-west tracking system would be 150°. It clearly is difficult to accomplish this range of angles with a fixed-mirror system without finding very special geometries for the optics.

Fixed collectors discussed in this chapter comprise two basic classes: (1) fixed mirrors for daily solar motion and (2) fixed mirrors for daily and seasonal motion. The first class involves a limited acceptance angle by the collector. At the equinox an east-west cylindrical collector needs no north-south motion, but at solstices it requires an acceptance angle of 20–30° to yield enough hours of operation for practical use. We discuss a number of options that yield this acceptance angle. The second class requires a very large acceptance angle; we discuss two options with this capability, one using cylindrical optics and the other hemispherical optics.

The ideal semifixed collector would be one where the sun image remained at one point or along a fixed line image regardless of where the sun was in the sky. There are only two cases where this result could be obtained. The first is if the effective focal length of the lens were zero. A lens with a focal length of zero and a finite aperture width would have an *f*-ratio of *f*/0.000, a geometric impossibility. The second case is where the absorber is located at an image point of the objective, an optical possibility discussed below.

Quite a few semifixed collectors have been explored over the years, and the compromises are well understood today. In the following sections we discuss the principal options for this type of collector, which for brevity is generally referred to not as a semifixed collector but as a *fixed collector.* The term *fixed* is understood to mean that the collector is fixed during the day, with annual changes being accepted. We also discuss two mirror designs that are fixed in the absolute sense: the Russell Fresnel cylindrical mirror and the Steward-Meinel hemispherical mirror.

7.1 THE FIELD LENS OPTION

In the second ideal case referred to above, the absorber is located at an image of the objective. This optical configuration can be referred to as the *field lens option* because a second optical surface is required to form the image of the objective; this is analogous to the function of a field lens in traditional optical systems. The optical diagram for a field lens is shown in Fig. 7.1.

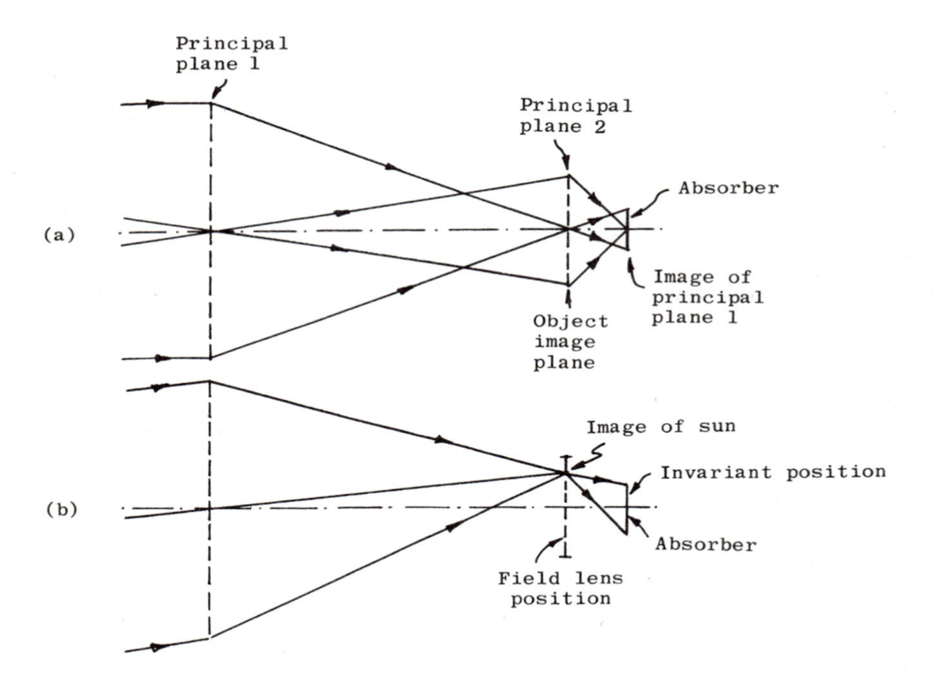

Fig. 7.1 Optical diagram of an objective plus field lens to yield an invariant position of the sun on the absorber.

Diagram (a) in Fig. 7.1 shows the optical relationships involved. No lenses or mirrors are shown for reasons of generality. Instead, the positions of the principal planes are shown. For lenses, this means that physically the lens is located coincident with the principal planes. For mirrors, the mirrors are located at a distance equal to the radius of the mirror from the principal plane (see Chapter 6). The basic function of lens 2 is to form an image of lens 1 on the absorber. In the case of a point source of light, and also with the sun as the source, the distribution of light on the absorber is a uniform disc (or strip, in the case of cylindrical optics). As the image of the sun moves about within the limits of lens 2, the disc of light remains substantially unmoved on the absorber, as is illustrated by the ray paths in diagram (b) of Fig. 7.1.

One problem in the use of this approach to a wide acceptance angle for the sun is the additional cost of the field lens. Another problem is that the field lens must be as large as possible to accept the sun as far off the axis of the system as possible. Furthermore, this lens must bring the rays from the extreme position of the sun to the absorber.

Possible combinations that would achieve this focal invariance are (1) two lenses; (2) two mirrors; (3) lens objective, mirror field lens; and (4) mirror objective, lens field lens. In the past these options have not been used to any extent for solar collector optics, but practical systems of these types appear possible and worthy of exploration.

7.2 LENGTH OF TIME SUN IS ON ABSORBER

The general problem with fixed-mirror collectors is the tradeoff between the acceptance angle of the collector-absorber geometry and the number of hours the sun can be utilized at the solstices. In Fig. 7.2 we show the basic geometry relating the meridional zenith distance of the sun to time. This diagram is a meridional section of the celestial sphere, showing the path of the sun as a function of time of year and time before or after solar noon. At the equinox the sun moves in a great circle and no change is required in the north-south angle for an east-west collector configuration. At the solstices the angle change is at a maximum, and because the sun is within 4° of the solstice positions for 140 days each year, the acceptance angle at this time of year is very critical for a fixed-mirror collector.

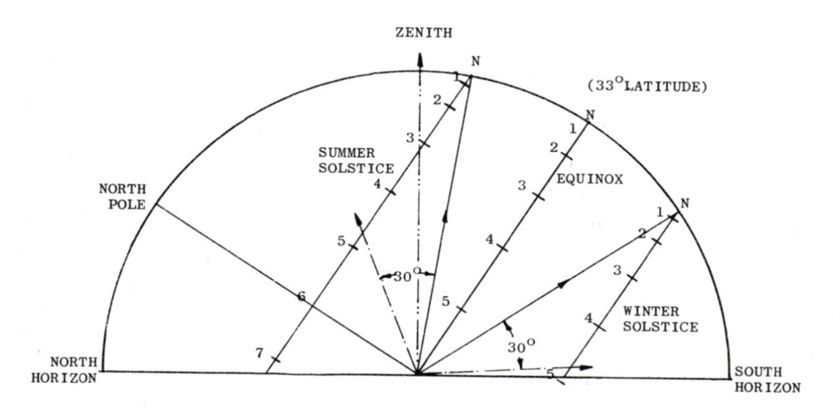

Fig. 7.2 Geometry for the sun as a function of time of day and season for latitude 33°. For an east-west collector a north-south acceptance angle of 30° will accept the direct sunlight for a period of more than 9 h at summer and winter solstice.

Figure 7.3 shows the relationship between acceptance angle and the number of hours of sun collection at the solstices. Because the maximum concentration of fixed-mirror collectors is a linear function of acceptance angle, there is an important trade-off between the two factors and the relationship between concentration and system efficiency. The graph shows that the gain in the number of collection hours increases rapidly at small angles but slowly at larger angles, as between 20° and 30° total acceptance angle. Optical systems having between 15° and 30° appear to be adequate for yielding enough collection hours at the summer and winter solstices.

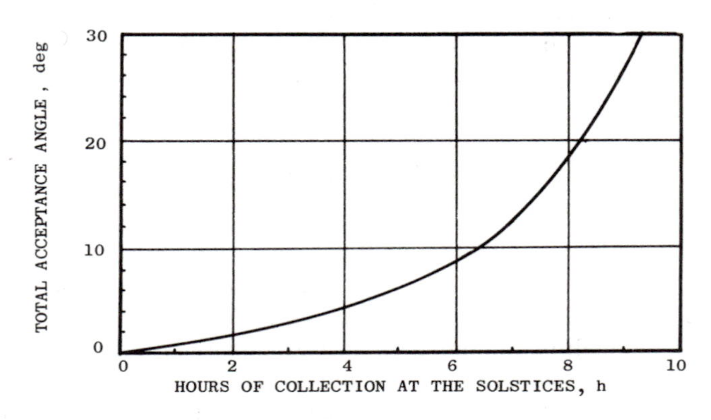

Fig. 7.3 Relationship between field acceptance angle of an east-west cylindrical collector fixed with respect to the north-south angle during the day and the number of hours of collection at the solstices. Note that the number of hours versus angle is independent of latitude.

7.3 CYLINDRICAL IMAGERY

Cylindrical optical configurations are particularly useful when tracking simplification is desired. A number of cylindrical geometries have acceptance angles sufficient to minimize or eliminate daily tracking requirements when the collector is used in the east-west orientation. For the sake of illustration let us look at a cylindrical lens, as shown in Fig. 7.4. This system has an absorber width W larger than the image of the sun. If the width of the optical aperture is D, we immediately have two different flux concentrations: (1) the brightness concentration C, defined by the ratio of the angle of lens convergence 2β to the apparent angular size of the optical image 2α, or $C = \beta/\alpha$, and (2) the radiation balance concentration X, defined by the ratio of the angle of lens convergence to the angle subtended by the absorber (2θ), or $X = \beta/\theta = D/W$. (The angle α, which is the angle subtended by the disc of the sun, is not shown.

The length of time the sun remains within the acceptance angle of a fixed collector and the fraction of diffuse skylight the collector gathers is rendered clear in Fig. 7.5. The lens or mirror can be considered to image the absorber on the sky. The fraction of the 2π solid angle of the sky subtended by the image on the sky is the fraction of the scattered sunlight, relative to a flat-plate collector, that is gathered by the cylindrical optics under evaluation. The length of time the collector gathers sunlight is likewise determined by how long the sun track remains within the image strip in the sky. We will discuss this point in detail under the sections dealing with specific fixed-mirror designs.

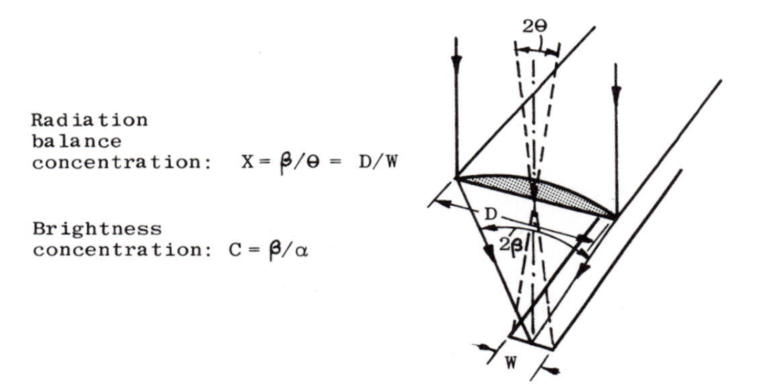

Fig. 7.4 Diagram defining the flux concentration of a cylindrical lens in terms of (1) convergence angle of the lens and (2) angle subtended by the absorber, and of the ratio of lens aperture to absorber width.

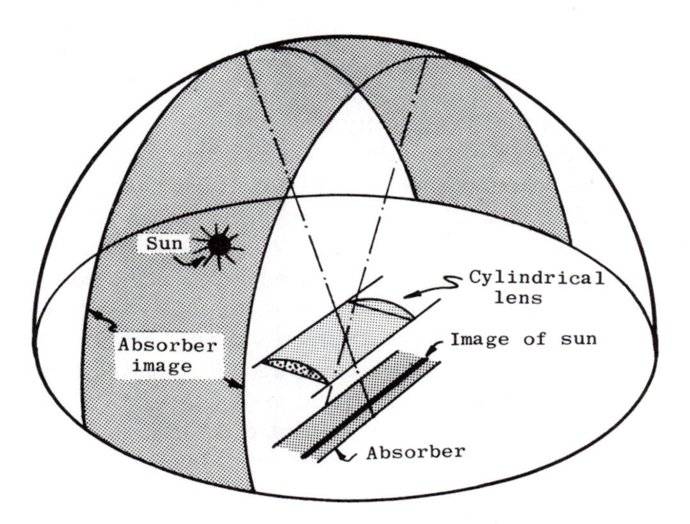

Fig. 7.5 Diagram illustrating the fraction of sky imaged onto the absorber for a cylindrical optical system. The image of the sun remains on the absorber when the sun lies in the strip image of the absorber on the sky.

7.4 TABOR CIRCULAR CYLINDER

Tabor and Zeimer (1961) devised an especially simple cylindrical optical system for application to a medium-temperature system. This optical collector consisted of an inflated plastic cylinder whose lower portion was aluminized to reflect sunlight to a focus near the bottom of the cylinder. The general configuration of this design is shown in Fig. 7.6.

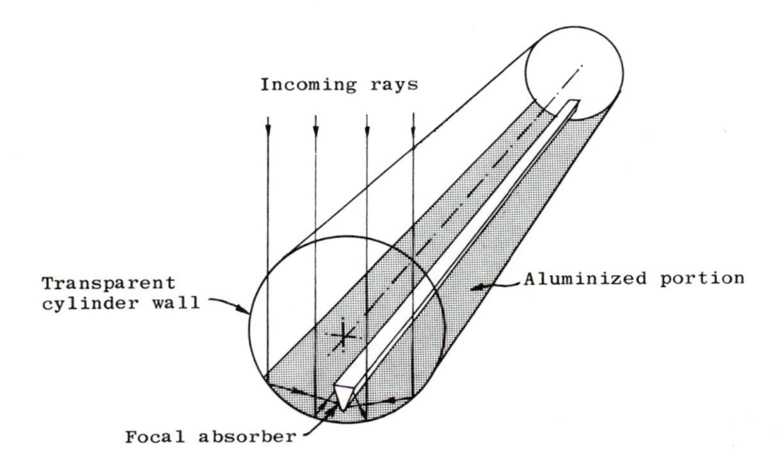

Fig. 7.6 Diagram of a cylindrical collector of simple construction, with a transparent cylinder as the basic collector structure. The side boosters used to extend the useful aperture are not shown in this view.

The optical diagram for this geometry is shown in Fig. 7.7. The spherical aberration of the reflective portion of the cylinder means that a small absorber can collect sunlight only from the central portion. If one chose to make the absorber occupy the entire space between the paraxial focus and the cylinder, then all the rays would be collected, but at the price of a lower radiative balance concentration because both sides of the absorber radiate losses. Tabor selected a triangular absorber of sufficient depth for two plane side mirrors to be added to direct the rays otherwise lost to the absorber. He further noted that if the absorber is slightly oversized the acceptance angle becomes large enough that the collector need not be moved during the day. We find that the size of the absorber required to yield 6.0–8.0 h of collection at the solstices makes the absorber so large that the flux concentration is reduced to about $X = 2$, and other fixed-mirror designs, discussed later, become more attractive. The basic geometry of a cylinder is, however, attractive for a low-cost collector, allowing some cost margin for devising simple ways of moving the cylinder or the absorber during the day. A photograph of this collector installation as set up for demonstration at the 1961 Rome conference is shown in Fig. 7.8.

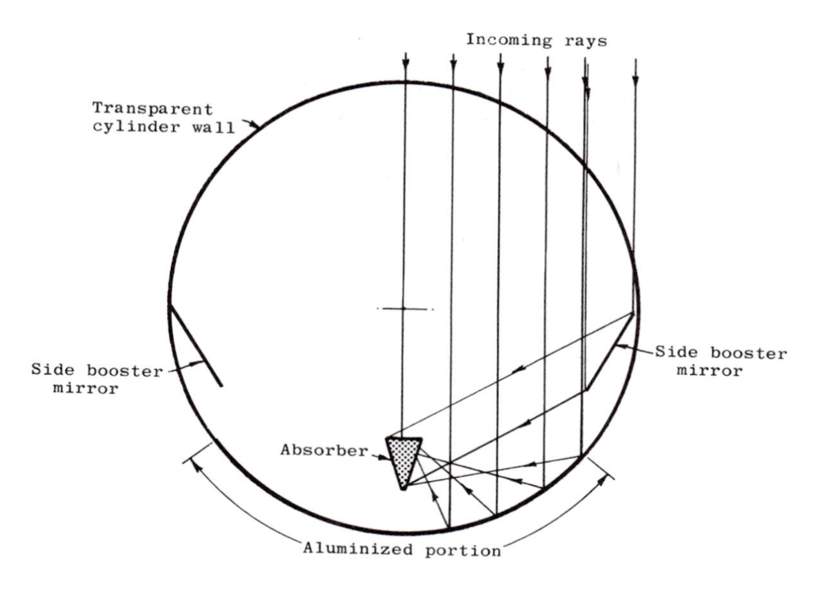

Fig. 7.7 Cross section of a cylindrical tubular collector showing the addition of two plane-mirror boosters to enable use of the fill aperture of the tube, from the solar power unit developed by Tabor and Zeimer (1962), shown in Fig. 7.8.

Fig. 7.8 Test solar-power system using the cylindrical inflatable mirror configuration, Rome, 1961.

7.5 CYLINDRICAL APPROXIMATION TO A PARABOLIC TROUGH

A parabolic trough can be approximated by use of two cylindrical troughs when moderate flux concentration and low-quality mirrors are to be used. When the mirror errors approach a certain level, the perfection of the axial focus of a parabolic trough becomes small compared to the contribution of the errors to the image size. In such cases one can substitute two cylindrical mirrors and obtain reasonable performance.

In Fig. 7.9 we show one cylindrical mirror with rays incident at uniformly spaced intervals showing the generation of the caustic envelope. The minimum bundle of rays for four apertures is indicated by the arrows, showing that the minimum absorber cross sections change rather rapidly as the aperture exceeds $F = 1.0$. The diagram to the left of the ray-trace diagram shows how two cylindrical halves are combined into an approximation of a parabolic trough. Note that the absorber V-section angle changes with aperture, as indicated by the inclination of the arrows in the right-hand diagram. For a fuller discussion, see Tabor (1958).

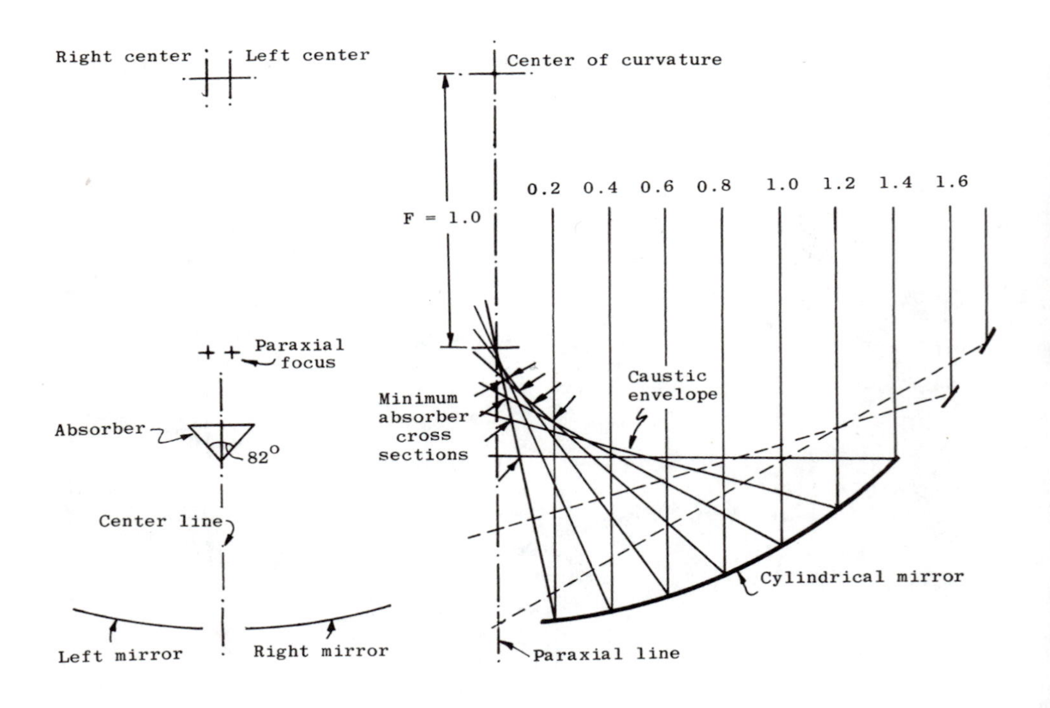

Fig. 7.9 Diagram for optimum utilization of two cylindrical halves in lieu of a parabolic cylinder. Note that the mirror centers are on opposite sides of the center line for the pair from the mirrors.

In Fig. 7.10 we show the effective radiative balance flux concentration for the cylindrical trough as a function of the size of the image and collector aperture. For a practical mirror error, yielding a total image 3.0 times the image of the sun, the optimum concentration is $X = 17$ when the aperture of a mirror half is about 0.8. The effective f-ratio for this optimum is $f/0.63$.

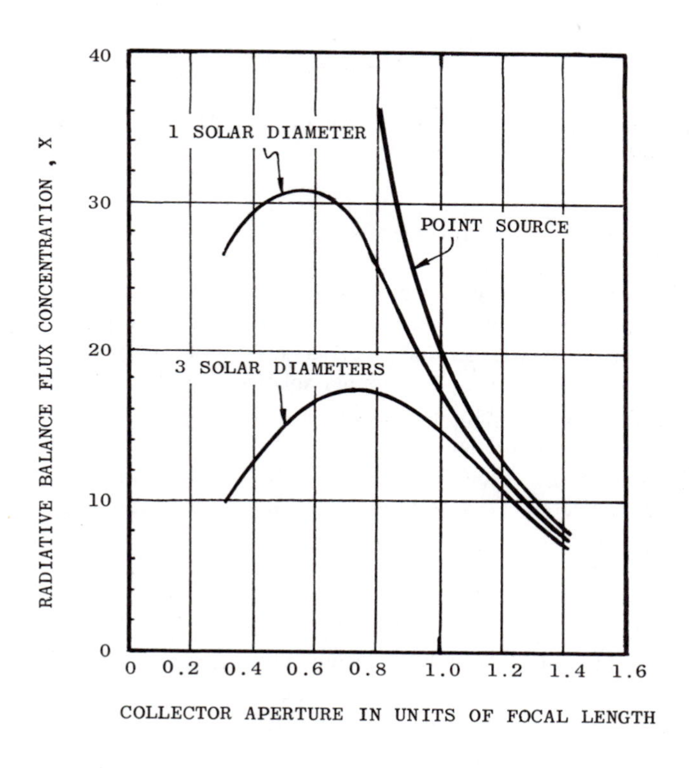

Fig. 7.10 Variation of the radiative balance flux concentration X as a function of the collector aperture for a perfect mirror (1 solar diameter) and a degraded mirror (3 solar diameters), for two spherical cylinders with displaced centers and V-absorber as shown in Fig. 7.9.

7.6 VEINBERG FIXED-MIRROR COLLECTOR

The off-axis aberrated image of a paraboloid can be considered to act like two separate optical systems, as shown in Fig. 7.11, one forming an "image" upward near the optical axis of the paraboloid and the other forming an image downward. This property of a parabolic cusp has been used in two interesting designs, one by Veinberg (discussed in this section) and the other by Winston (see Section 7.8).

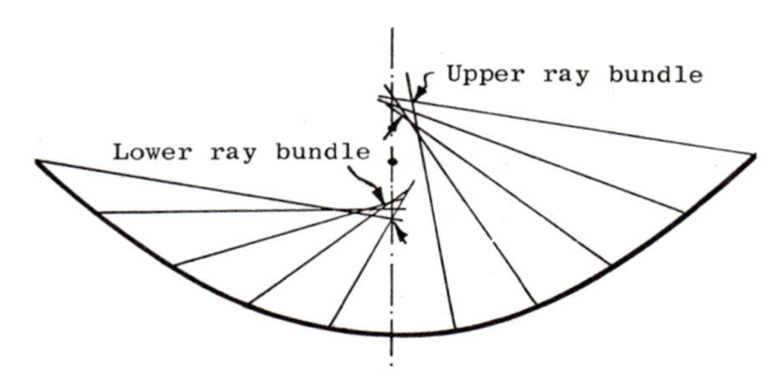

Fig. 7.11 Isolation of the two ray bundles from opposite sides of the paraboloid showing their grouping along the optical axis of the paraboloid.

The Veinberg geometry takes advantage of the property of a parabolic cusp to form a reasonable image over a considerable range of angles of the incoming solar rays. At the equinox the Veinberg collector looks like a parabolic cylinder, as shown in Fig. 7.12. For the geometry at winter solstice, Fig. 7.13, the parabolic cusps are rotated about their extremities so that the solar rays are again brought to a mean focus on the center of the absorber. At summer solstice, Fig. 7.14, the cusp positions are reversed. The chief ray diagram, shown in Fig. 7.15, shows how the solar image position changes as the meridional zenith angle of an east-west collector configuration changes. The size of the absorber needed to collect the sun's rays over a given zenith angle change is indicated. To accommodate a 20° range of sun angles, equal to about 7.5 h at the solstices, the absorber becomes large. Because solar flux arrives on both sides of the absorber, the radiative balance concentration X is approximately equal to 3 in the case shown. If the field acceptance angle is reduced to a total range of 10°, the flux concentration is approximately $X = 5.9$.

The Veinberg configuration has a second advantage in that the absorber as well as the mirrors are fixed. Thus for a moderate concentration system this advantage can be combined with the use of selective surfaces on the absorber and associated windows to yield the equivalent of a collector having an effective concentration 5–30 times the geometrical concentration, or $X(\alpha/\epsilon) = 15–180$.

The Veinberg geometry also looks usable as a collector where the mirrors are moved during the day. The pivot points are at or near the edges of the mirrors, affording maximum security of the system during high winds. Because one needs to be able to make annual changes in the mirrors in any case, the existence of the moving mechanisms offers one the easy option of making the daily motion adjustments several times a day. If this option is elected, the geometrical flux concentration is limited only by the width of the aberrated image at the solstices. In the case shown in Fig. 7.16, this limit is about $X = 9.5$ geometrical, and appropriately higher when selective surfaces are employed for the absorber surface.

Figure 7.17 is a photograph of the ray fans for the Veinberg parabolic cusp collector for winter solstice sun angle.

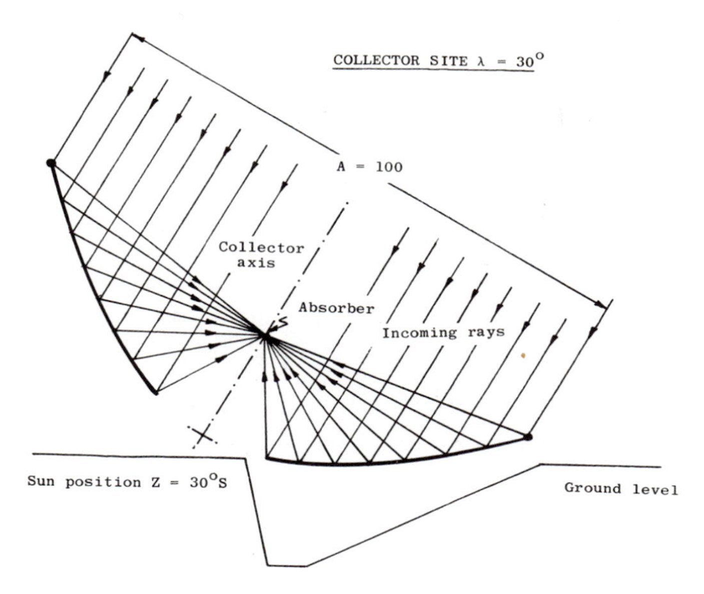

Fig. 7.12 Mirror position and ray diagram for equinox position of the sun. Solar image remains on one point during the day at this time of year.

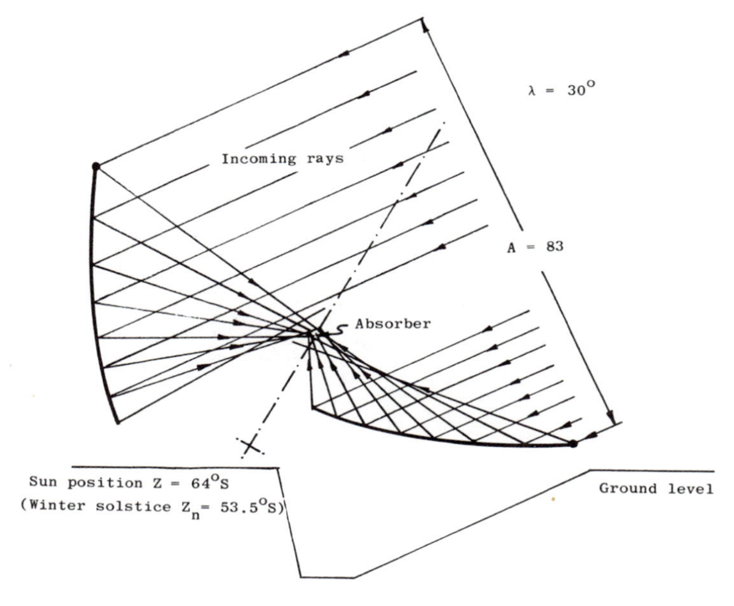

Fig. 7.13 Mirror position and ray diagram for winter solstice. The comatic aberration and daily solar image travel on the absorber are at a maximum at this time of year.

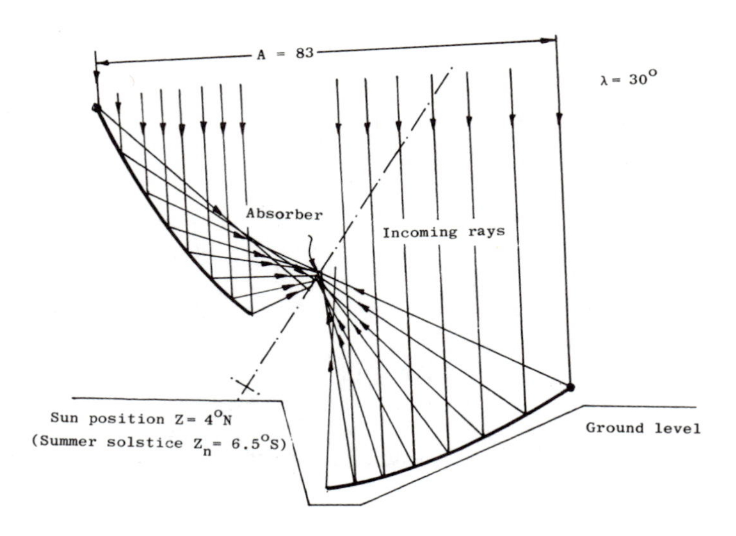

Fig. 7.14 Mirror position and ray diagram for summer solstice.

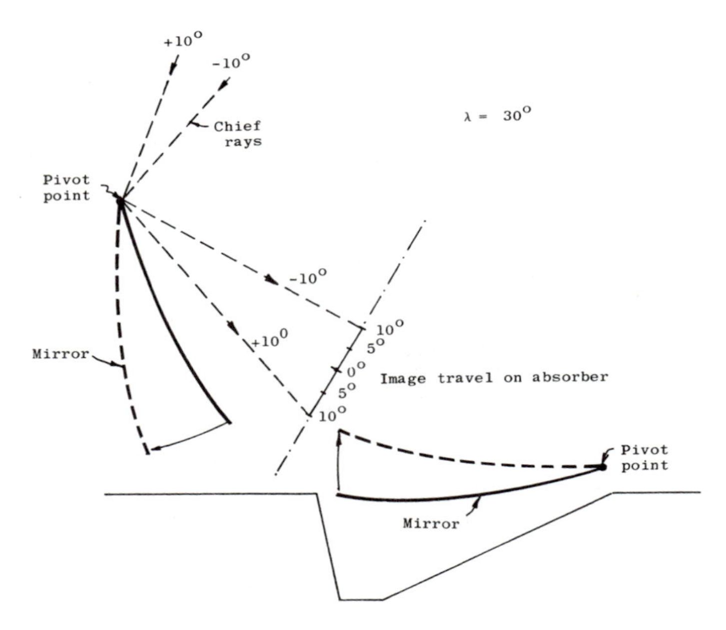

Fig. 7.15 Chief ray diagram showing the travel of the solar image when the mirrors are fixed and the sun travels through a given angle.

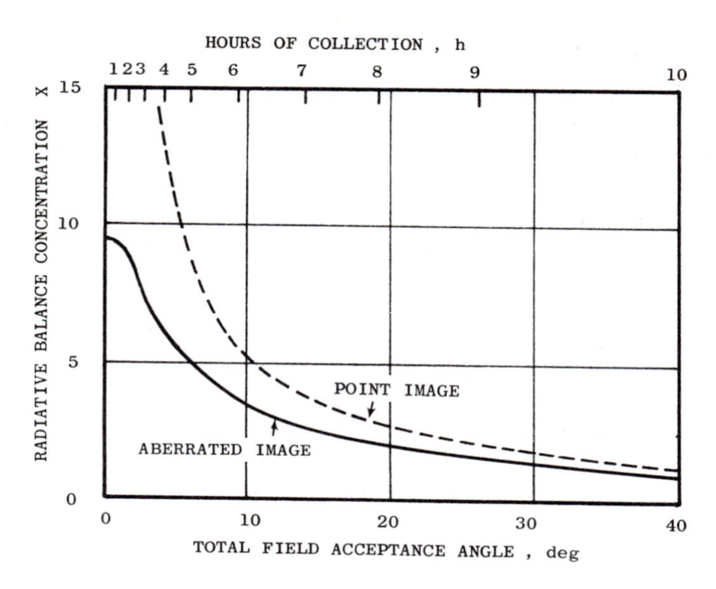

Fig. 7.16 Relationship between total field angle, hours of collection, and effective radiative balance flux concentration X for the Veinberg fixed parabolic cusp collector. The aberrated image size is taken at the solstices and for the sun at half the daily excursion angle from noon position.

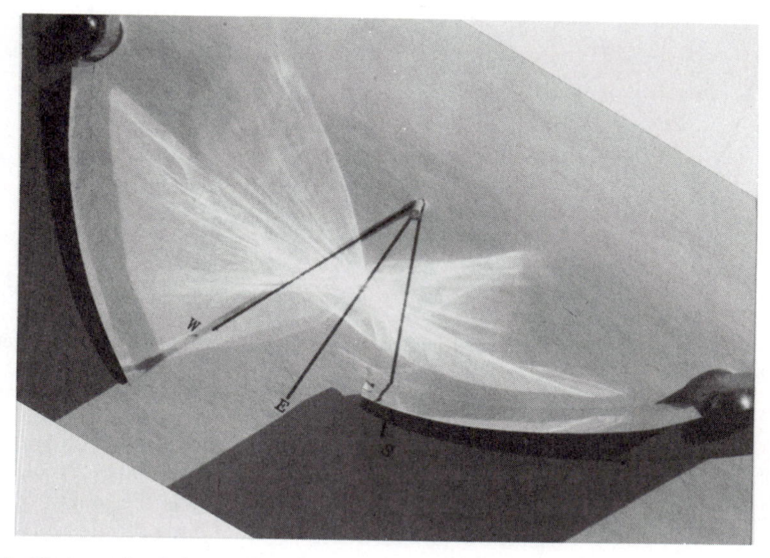

Fig. 7.17 Photograph of the ray fans for the Veinberg parabolic cusp collector for winter solstice sun angle.

7.7 TROMBE-MEINEL CUSP

The 180° acceptance angle cusp collector was first developed by Trombe in 1957 but was never published, being recorded solely as a French patent, now expired. The properties of this cusp were independently discovered by Meinel (1972). The function of the cusp is to form the equivalent of a flat-plate collector, in which the solar flux is optically conveyed to a central absorber pipe containing the heat-transfer fluid, the same mode of energy transmission employed by the central tower system.

The condition that the cusp act like a flat plate requires (1) that the sunlight falling on the aperture reach the absorber from any position of the sun in the sky and (2) that all points on the absorber view a 2π solid angle of the scattered radiation from the sky. The first condition does not require that the incoming flux be distributed uniformly over the absorber; as with the Winston cusp, it can be concentrated to varying degrees on different portions of the pipe at different sun angles. This focusing of radiation on a portion of the absorber pipe is caused by portions of the cusp acting like a parabolic cylinder of poor optical quality, as can be quickly verified by tracing a few additional rays in Fig. 7.18. The second condition does result in a uniform flux of scattered sunlight arriving at the pipe, and optimum performance for a nonimaging collector.

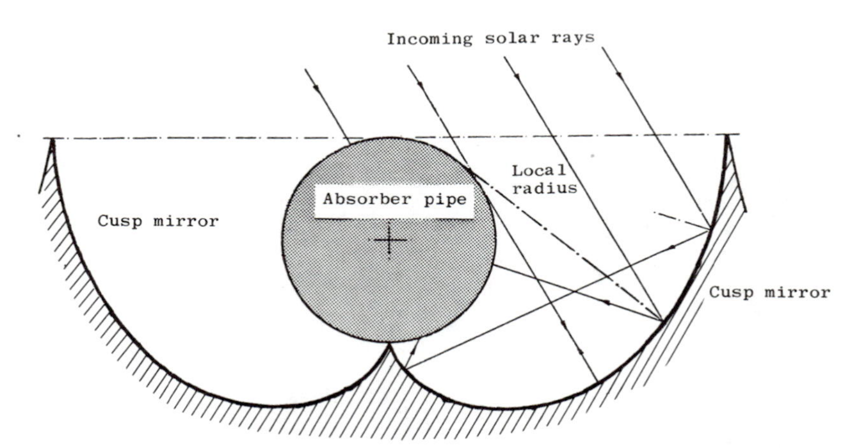

Fig. 7.18 Diagram of the Trombe-Meinel cusp design for 180° acceptance angle, formed as the involute of the circle.

As shown in Fig. 7.18, the basic Trombe-Meinel design is a collector surface that is described by the locus of a string unwrapped from about the pipe. By elementary geometry the instantaneous radius of curvature of the cusp is the length of the unwrapped portion of the string, with the center of curvature lying on the surface of the pipe. All rays missing the pipe will then be deviated by the reflective cusp arriving at the surface of the pipe. The aperture of the cusp is exactly π times the diameter of the absorbing pipe. When a multiplicity of cusps are placed side by side it is clear that one can truncate the cusp short of the vertical slope position and still have all of the incident sunlight arrive at the absorber.

In the symmetric design case shown in Fig. 7.18 the plane of a collector consisting of many cusps side by side is normal to the direction to the central position of the sky. One can also make asymmetric cusps in which the plane is skewed. In this case the length of the cusp is extended on one side and shortened on the other, following the same locus geometry.

Although the original design was for a 180° acceptance angle, it is clear that one can reduce the acceptance angle and gain optical concentration. For example, if one truncates the cusp, the aperture drops below πD, but one can then flatten the cusp to increase the aperture at the expense of losing full capability for sun positions near the horizon. Since the intensity of sunlight drops rapidly as the sun nears the horizon, an optimized cusp can depart from the basic cusp without sacrificing efficiency, still fulfilling the second condition. Condition 2 can be restated as the condition that from any point on the absorber a ray drawn in any direction will escape to the sky either directly or after reflection by the wall of the cusp.

The extension of the cusp for flux concentrations comparable to those of the truncated Winston collector involves adding a tangent curve to the cusp and expanding the width and height of the curve as much as desired. The simplest tangent curve is a parabolic or circular cylinder. When the cylindrical portion becomes large relative to the cusp section, the collector design takes on a different aspect. One can then consider it to be as shown in Fig. 7.19.

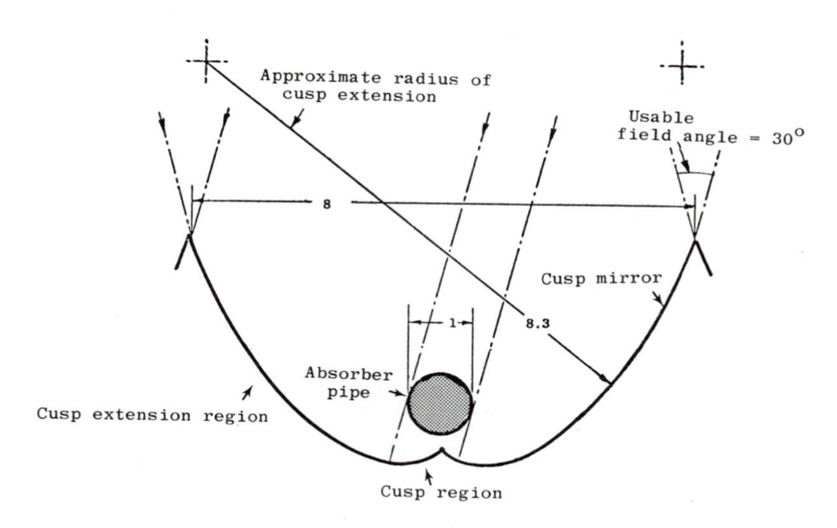

Fig. 7.19 Diagram of a modified Trombe-Meinel cusp for a 30° acceptance angle and a concentration of $X = 2.55$. A triangular absorber can also be used effectively in lieu of the pipe.

In the design of Fig. 7.19 the aperture of the collector is 8.0 times the diameter of the absorbing pipe; this design yields a radiative flux concentration of $X = 2.55$. Other variants are possible and can affect the net concentration. For example, the pipe absorber can be replaced with a flat vertical absorber to reduce the total reradiating surface area. One can also redistribute the cusp portion, placing some of the cusp action at the top of the absorbing plate, as shown in Fig. 7.20. At first glance one would wonder why one would want a reflective surface at the top, thus adding an apparently useless cost; however, in practice one needs a convection-suppressing enclosure around the absorber, and the V-shaped top reflector can be built into the top of the glass enclosure, adding no significant cost and widening the field angle of acceptance of the collector. Photographs of rays for such a design are shown in Fig. 7.20.

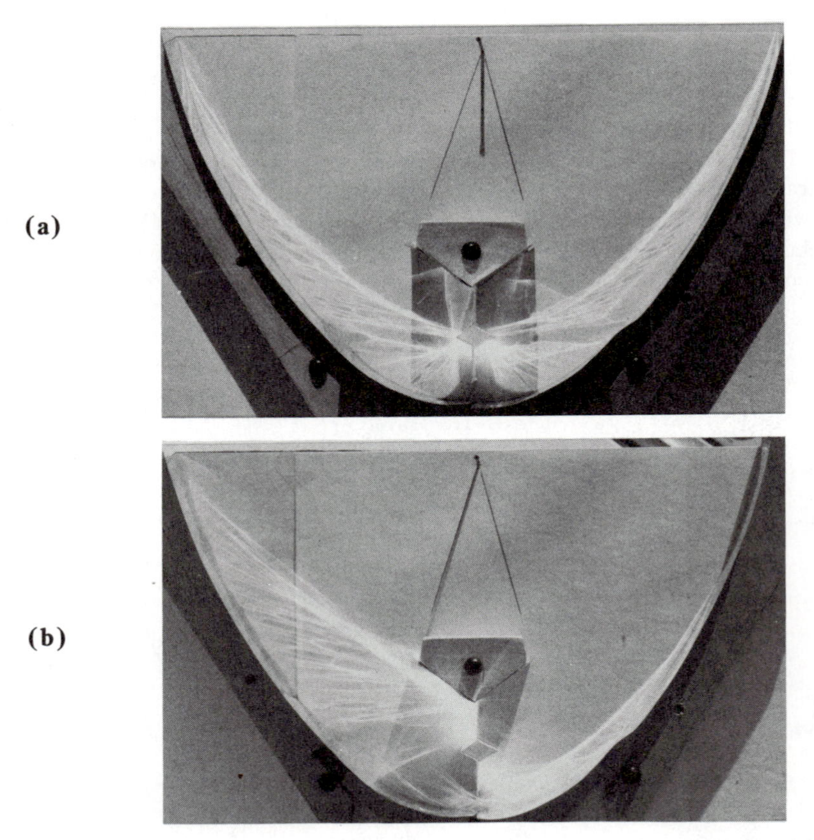

Fig. 7.20 Ray photographs of a cusp collector in which a vertical flat plate is the absorber: (a) sun on collector axis and (b) sun 15° off collector axis. A fluid pipe is shown near the center of the absorber. Most of the rays on the right-hand side of the 15° off-axis case are multiply-reflected in traveling to the absorber.

It should be noted that the Trombe-Meinel cusp family merges imperceptibly into the Winston cusp family; however, the upper limit of efficiency is still yielded by the geometry of the "full" Winston design, as summarized in Fig. 7.21.

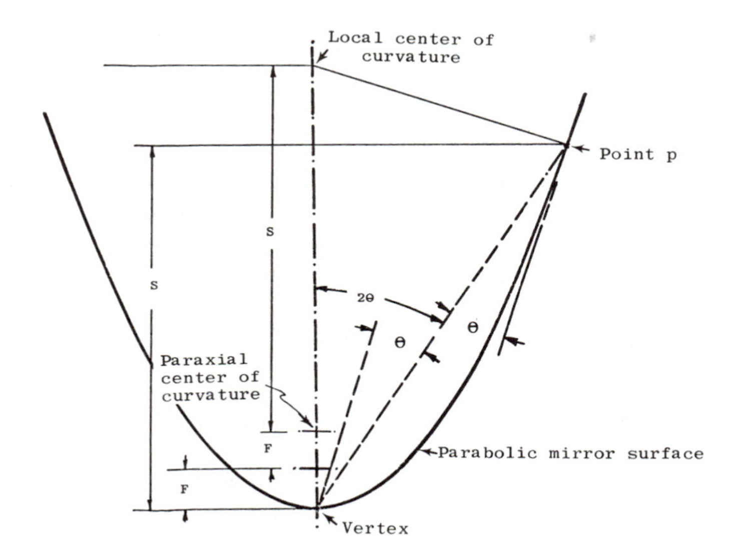

Fig. 7.21 Basic geometry of a paraboloid. Note that the local center of curvature is the same distance from the paraxial center of curvature as point *p* on the parabolic cusp is from the vertex.

7.8 WINSTON PARABOLIC CUSP

Winston (1974) derived via an elegant mathematical argument a class of parabolic cusp collectors that is close to the theoretical upper limit of performance of fixed-mirror collectors. The basic idea is to design the shape of the mirrors so that the rays entering the collector at the maximum field angle are reflected tangent to the surface of the receiver. This means that the maximum angle α is 90° and the effective *f*-ratio is close to 0.5. The reason it is not exactly 0.5 even though α_{max} is 90° is that some of the cone of converging rays is missing from the aperture of the pair of cusps, like a central obscuration in a mirror collector. Since the tangent rays are not efficiently absorbed by a flat absorber, a cavity or curved absorber can be used with little change in performance. Winston has extended the concept to a class of "truncated" Winston collectors that provide modest flux concentrations and an acceptance angle wide enough to enable an east-west cylindrical collector to remain fixed during the day.

The Winston design evolved from the design of Cerenkov's radiation collectors, which used circularly symmetric reflective cusps and followed the same basic geometrical concept as the solar collector design (Winston, 1965). The same class of optical systems was developed by Baranov (1966) for solar energy applications.

The essence of the Winston geometry is to take two symmetrical half-paraboloidal cylinders and fulfill the following two conditions: (1) the focus of one cusp must be placed on the other, and (2) the cusps must be tilted inward until their tops are parallel to the Winston axis and therefore parallel to each other. The first condition defines a whole family of collectors independent of the second condition; the second condition limits the family to one particular set of curves. This second condition means, however, that the upper end of the collector will not function very efficiently because the reflective surface is parallel to the incoming sunlight when the sun is on the axis. For this reason Winston has extended the basic design concept to truncated designs where the upper portion of the cusp is eliminated, reducing the concentration by only a small amount.

The interesting properties of a paraboloid, useful in designing a Winston cusp collector, were summarized in Fig. 7.21. The angle from the focus to any point on the paraboloid is twice the angle of slope of the cusp. If the angle from the focus to the paraboloid is 2θ, then the slope of the paraboloid is θ. A ray arriving at the cusp, therefore, is deviated and leaves the cusp at an angle θ.

In the following derivation let the focal length be 0.5 unit, whereupon the gap between the two foci, the location of the absorber, is approximately unity. The change to achieve the Winston geometry is shown in Fig. 7.22, where θ is parallel to the slope

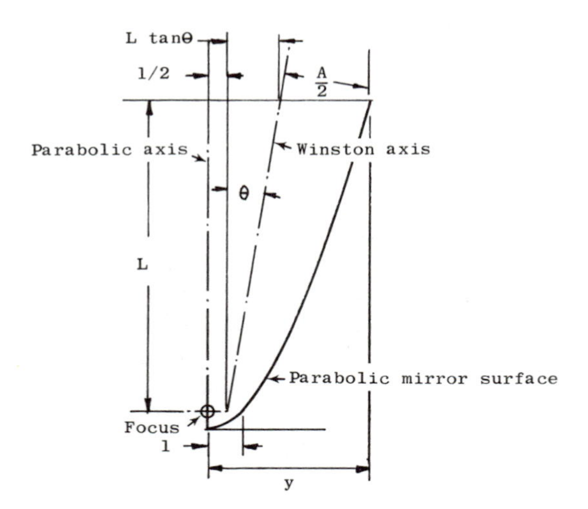

Fig. 7.22 Geometry used to derive the approximate length and aperture for the Winston collector in terms of the angle θ, through which the original axis of the paraboloid is rotated. The horizontal distance from focus to mirror is unity.

of the parabola at the upper end of the cusp for the full Winston design. The aperture of the design is then

$$A = 1/\sin\theta_{max}. \tag{7.1}$$

The radiation balance flux concentration is then approximately

$$X = A. \tag{7.2}$$

The length of the Winston collectors is then given by

$$L = \tfrac{1}{2}(A + 1)\cot\theta_{max}, \tag{7.3}$$

or

$$L = \tfrac{1}{2}[(1 + \sin\theta)/\sin\theta]\ \cot\theta. \tag{7.4}$$

If we take the case for $\theta_{max} = 6°$ having a total field angle transverse to the length of the collector of 12°, then $L = 50$ units and $A = 9.4$ units, making the collector length 4.7 times the aperture and giving it a concentration of $X = 9.4$, a shape illustrated in Fig. 7.23. If the acceptance angle is reduced to $\theta = 3°$, the length increases to 191 units and the aperture to 19 units, with $X = 19$. These values are plotted in Fig. 7.30 and labeled "full Winston."

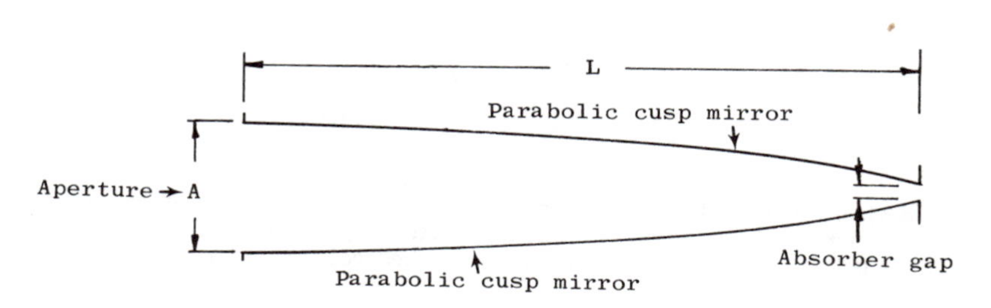

Fig. 7.23 Scale drawing of a Winston cusp collector for $\theta_{max} = 6°$, with a flux concentration of 9.4. The width of the absorber gap is indicated at the right-hand end.

If we examine the geometry of the parabolic cusps, it becomes apparent that two types of families can be defined: (1) collectors of truncated length and (2) collectors of constant length but with changed tilt of the individual parabolic cusp axes. In Fig. 7.24 we show a design using the full Winston cusp with an acceptance angle of $2\theta = 27.5°$. If we cut the cusp in half, the angles remain unchanged but the aperture is

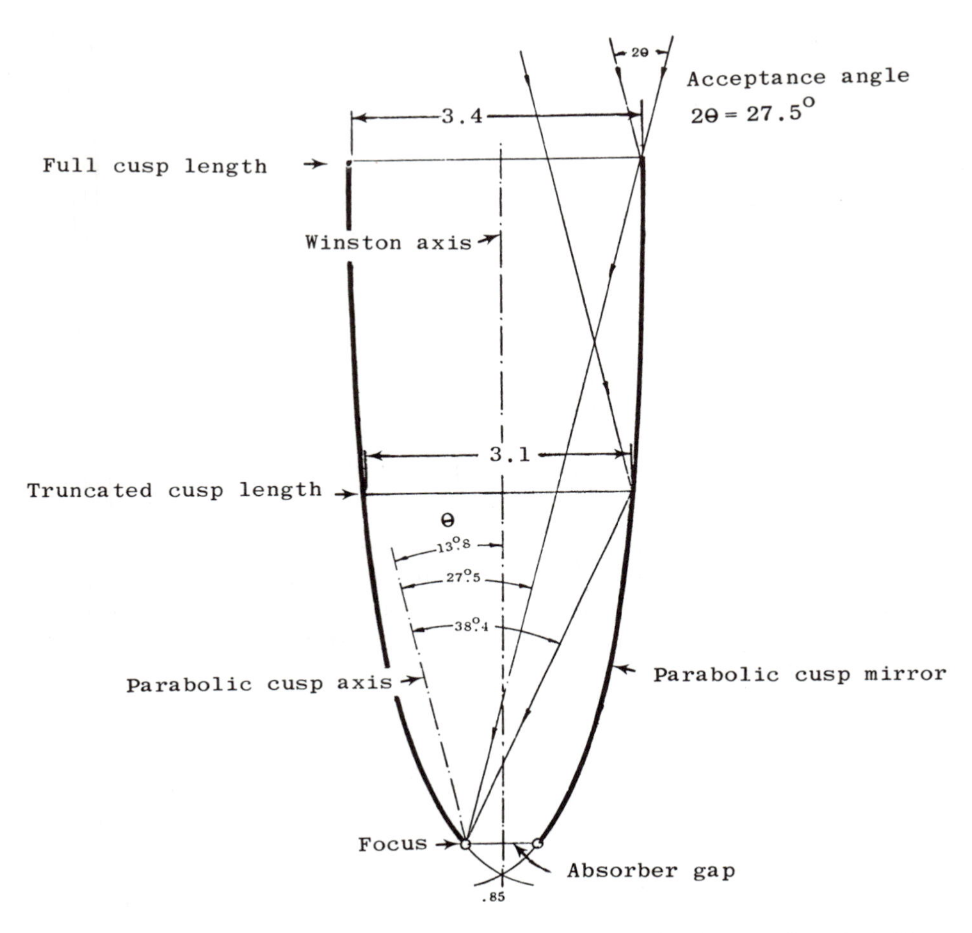

Fig. 7.24 Diagram for the full and the truncated Winston cusp for the case of constant tilt angle between the original parabolic axes and the collector axis ($2\theta_{max}$ = 27.5° in the illustration). For half length, the radiation balance flux concentration X is reduced from 4.2 to 3.6.

reduced relative to the absorber gap, reducing the concentration. (The locus of this change is shown in Fig. 7.30 as the vertical dotted line to the curve marked "truncated Winston.") In Fig. 7.24 we show two lengths for the reflective cusp. The aperture is 3.4 units for the full length but 3.1 units for the shortened cusps, both versions having the same acceptance angle. The amount of reflective surface has been reduced by 50%, but the concentration X has been reduced by only 11%.

In the case of the full Winston design, the field is sharply limited at the acceptance angle $2\theta_{max}$ because the lip of the cusp shields the absorber. When the truncated design is used, some light reaches the absorber gap for larger angles, resulting in a graded decrease in yield when 2θ is exceeded.

Figure 7.25 shows the second case, where the cusp length is kept constant and the angle θ_{max} is reduced to open the aperture. In this case the concentration increases to $X = 5.9$, but the acceptance angle decreases to 13.8°. The shaded portion in Fig. 7.25 shows the angle through which the full aperture sends rays to the absorber. Beyond this angle some rays reach the absorber up to a limit of 3θ, or 42.5°. The locus of constant cusp-length change is also shown in Fig. 7.30.

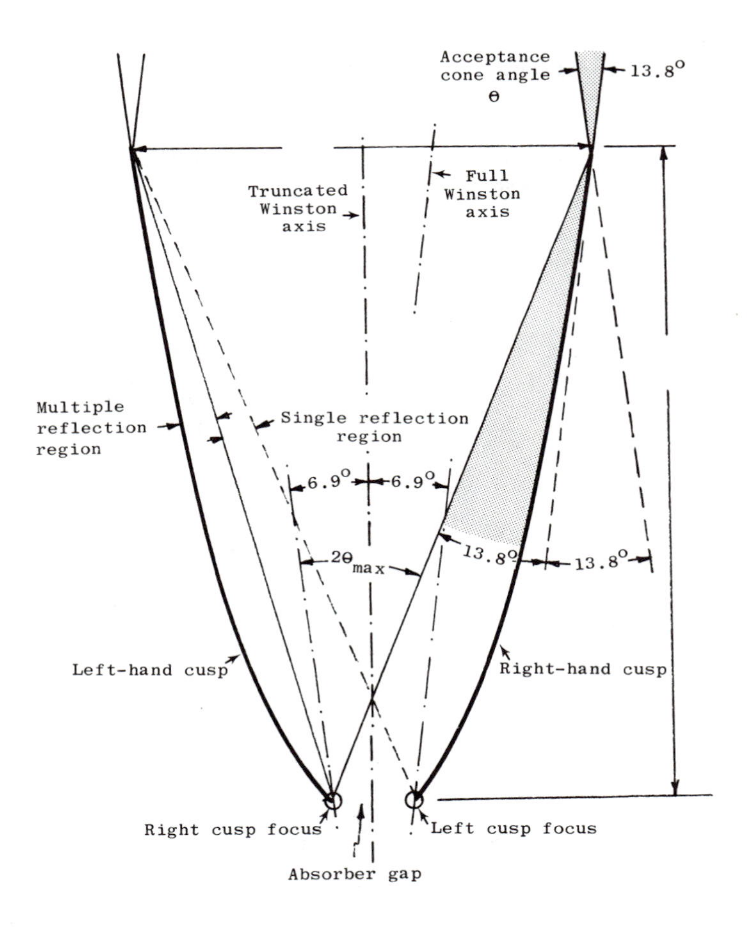

Fig. 7.25 Basic diagram for a double parabolic cusp collector. In this case the original parabolic axis for the right-hand cusp has been rotated 6.9° to the modified axis, half the angle required for the Winston axis. Note that in this case the slope of the parabolic cusp is not parallel to the axis of symmetry of the collector.

The efficiency of the Winston collector in accepting diffuse sky radiation was shown by Winston to be precisely the reciprocal of the concentration factor $1/X$. This is exactly the same relationship as for the maximum sky light accepted by any cylindrical concentrator as shown in Fig. 7.5. The actual behavior of a pair of parabolic cusps under diffuse light is shown in Fig. 7.26. In this diagram we take a point on the surface of the absorber and ask how much of a 2π solid angle the surface views. The reversibility of optical rays theorem indicates how much of the sky sends contributions to the absorber surface. As is diagrammed, all the rays leaving the absorber in a 2π solid angle do proceed either directly or after one reflection to the sky. The absorber therefore acts as if it were exposed directly to the entire sky, and behaves like a flat plate. However, because the absorber in this case is $1/X$ the size of the collecting aperture, this collector collects only $1/X$ of the scattered sunlight component that a flat-plate collector of equal aperture would.

Winston (1974) refers to the double cusp as a *nonimaging* system. Because the double cusp uses a parabolic shape the student may wonder why the term "nonimaging" is used, since a paraboloid does form an optical image. Also, the cusp uses

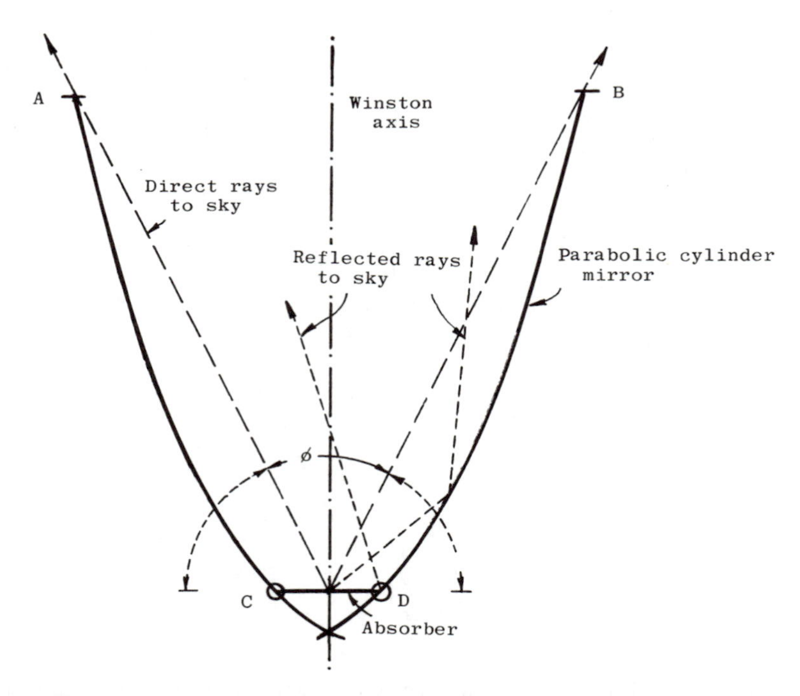

Fig. 7.26 Diagram showing the path of rays from the sky and TIR rays to the sky. The absorber "sees" the 2π solid angle of the sky by either direct or singly reflected rays; however, the absorber receives only $1/X$ of the amount of sky light as a flat plate having an aperture AB.

normal ray optics. Winston has clarified the situation by noting that the Winston cusp and the Trombe-Meinel cusp direct rays only with the requirement that the rays strike *somewhere* on the surface of an absorber. Technically, the definition of an imaging system is that it satisfies the Abbe sine law. By this definition almost all optical systems are nonimaging because they show small departures from the sine condition and hence have aberrated images. The concentrators of the Winston and Trombe-Meinel type obey a weaker thermodynamic condition in that the concentration is

$$X \leqslant 1/\sin\theta_{\max}. \tag{7.5}$$

The soft image formed by a Winston cusp is shown in Fig. 7.27 for the case where the sun is on the optical axis. The missing cone of rays from the aperture is shown, this portion causing the inequality to appear in Eq. (7.5).

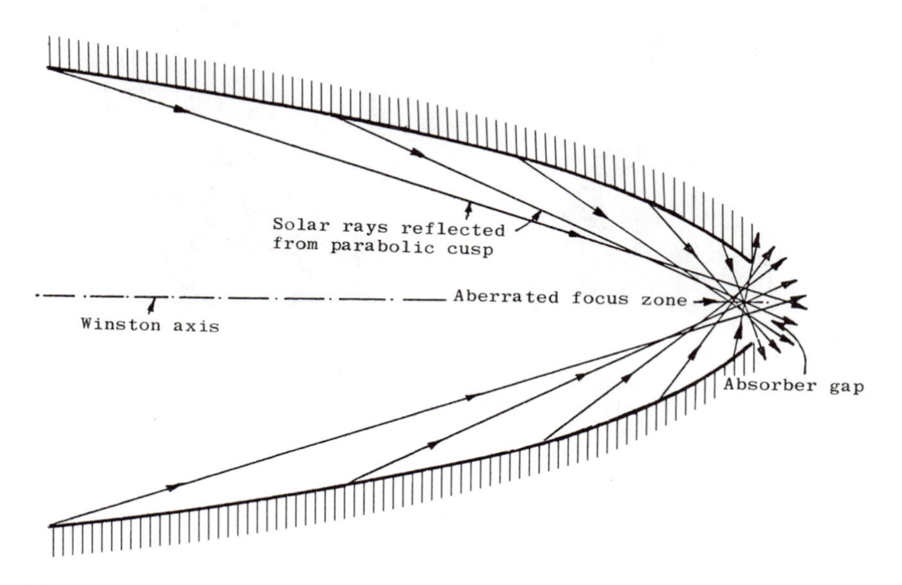

Fig. 7.27 Ray-trace diagram showing the size of the blur image formed when the sun is located on the system axis.

Simple models of cylindrical collector geometries can be made by sawing the profile in a piece of wood and taping reflective Mylar to the curve. The stiffness of the Mylar smooths the sawed profile and approximates what surface inaccuracies can do to the focus concentration. A photograph of rays formed by a model of a cusp collector is shown in Fig. 7.28.

Fig. 7.28 Model of a truncated Winston cusp collector showing the ray paths when the sun is on the axis of the system. The absorber position is approximately the same as that of the horizontal line at the bottom of the cusp.

7.9 COMPARISON OF FIXED-MIRROR DESIGNS

In Fig. 7.29 we show three optical systems, each having the same acceptance angle. The first is a Fresnel lens focusing rays on the upper side of the absorber. The second is a parabolic mirror focusing rays on the under side of the absorber, even though optically the rays act as though they were refracted at the principal plane. Note that in this case the concentration is reduced by obscuration of the parabolic mirror by the absorber, so the concentration is reduced from that of the other two cases. The third case is a truncated Winston cusp.

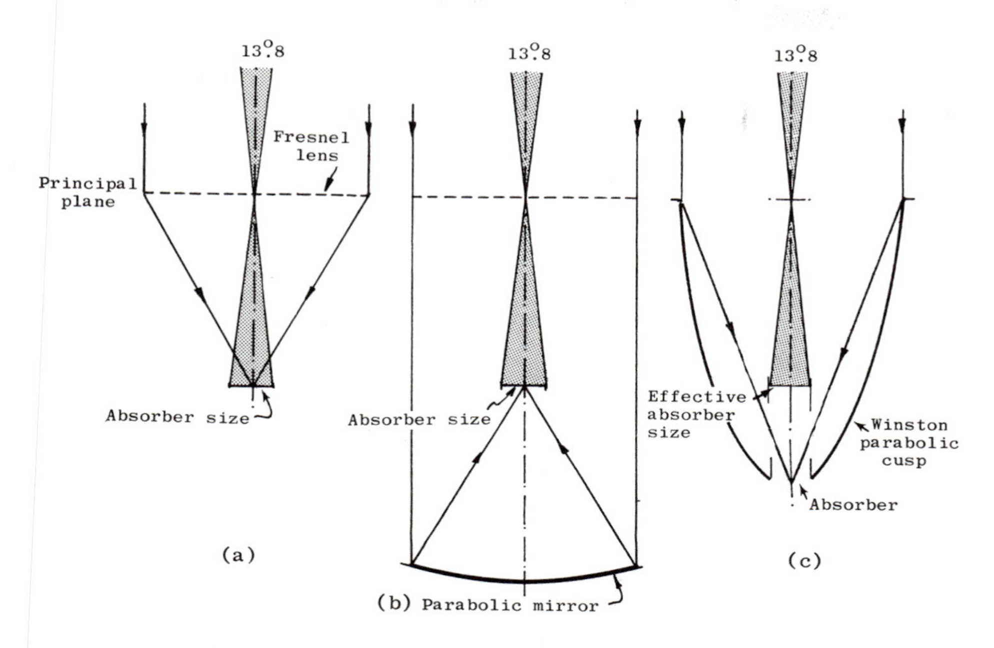

Fig. 7.29 Comparison of three optical configurations, each having the same field acceptance angle and flux concentration.

In Fig. 7.30 we show the quantitative relationships among the several alternatives studied to date. The optimum curve is the upper one for the full Winston design. A whole family of truncated Winston designs (shaded area) lies slightly below this curve. By means of dotted lines we also show the locus of changes, the vertical line being the change caused by truncating the full design and the nearly horizontal line being the locus of changing the angle θ while keeping the cusp length constant.

Three curves from Tabor and Zeimer (1962) in Fig. 7.30 represent a paraboloid (*A*), similar to the one in Fig. 7.29(b); a double-sided parabolic mirror (*D*), a faint beginning of the Winston geometry; and a spherical cylinder having a triangular absorber (*C*). All lie significantly below the Winston curve but require much less mirror surface in the collector. It is important to note that although these three configurations appear substantially equal, in reality they are not. The reason is that the Fresnel lens and the parabolic systems have serious field aberrations, and one loses much efficiency in the fixed-collector mode from these aberrations. See, for example, the change in focal length for a Fresnel lens in Figs. 5.12 and 5.13. As a consequence, flux concentrations are limited to approximately 2 for other than the Winston double-sided parabolic cusp.

The dashed curve in Fig. 7.30 is from a recent study by the authors of extensions of the Trombe cusp to smaller field angles. The Trombe cusp was developed for a 2π solid angle of acceptance and unit flux concentration. When the field is narrowed to 30° the flux concentration can be increased, but the limit at 30° appears to be similar to those of other designs studied by Tabor and Zeimer.

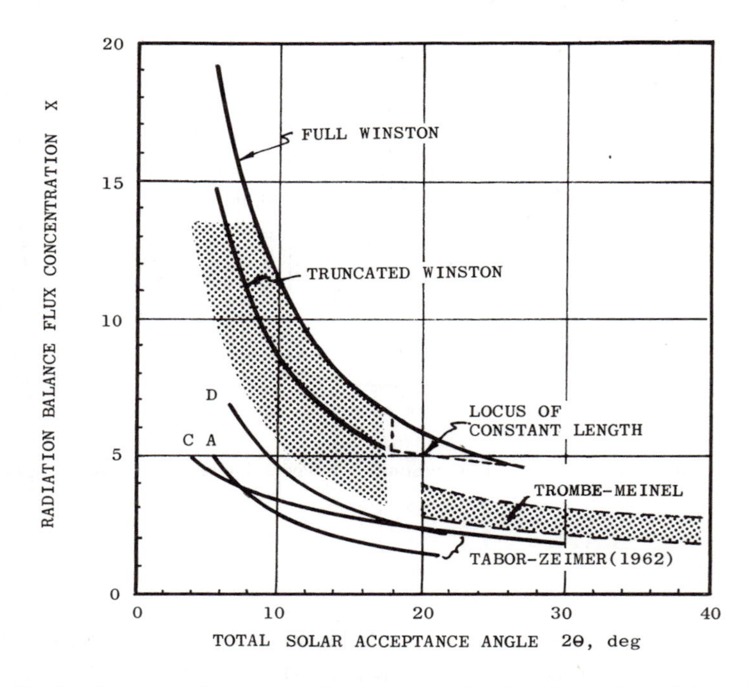

Fig. 7.30 Graph of the radiation balance flux concentration X as a function of field acceptance angle for several fixed-mirror optical systems for low concentration uses. The shaded zones indicate regions of solutions.

We can summarize the performance of the several types of cylindrical fixed-mirror collectors as shown in Fig. 7.31. This graph shows the relationship between the number of hours of continuous operation of an east-west cylindrical collector at the summer and winter solstices as a function of the allowable flux concentration. This performance assumes that the collector is positioned so that at noon the solar image is at one extremity of the field of the collector. The sun then moves out of the collector field of view, the number of hours later as shown. The decision to be made is whether the cost of the additional mirror of the full Winston cusp is worth the additional flux concentration for equal hours of operation of the other options.

The proper functioning of a wide-field collector can be readily ascertained by visual inspection. If you look into such a collector (Fig. 7.32), the mirror will appear

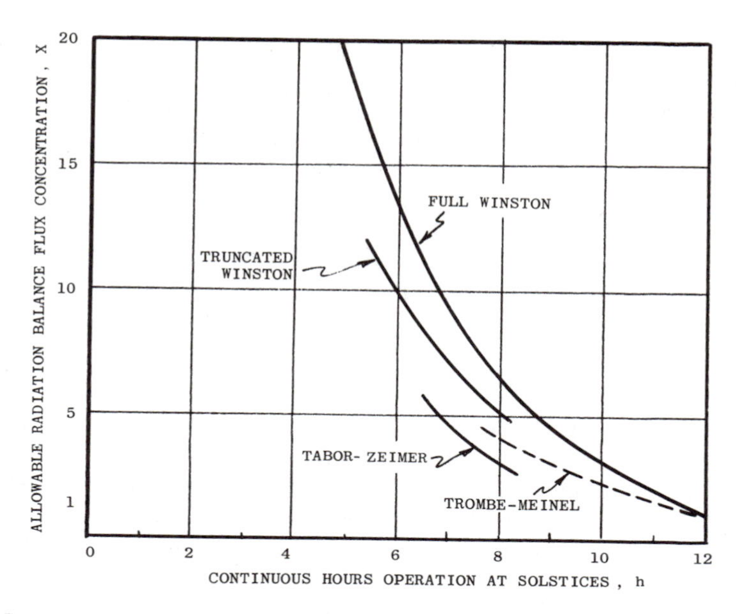

Fig. 7.31 Graph of the number of hours of continuous operation of an east-west cylindrical collector at the solstices as a function of the allowable flux concentration.

Fig. 7.32 Photograph of a cusp type collector showing how the black absorber optically fills the entire aperture of each cusp. Note where the absorber pipe has been removed in two of the cusp sections. When the acceptance angle is exceeded, the cusp portions reflect only sky light to the observer's eye.

black when the optical path from your eye to the mirror ends after reflection at the absorbing surface. If you scan your head until the black disappears from the mirror you have reached the maximum angle of acceptance of the design. You should do this viewing from some distance from the collector; otherwise the ray paths will not be from "infinity." The effect of a finite distance is that the image of the absorber may not appear to make the entire mirror black.

As an example, a Winston collector with accurate cusps and proper location of the absorber will appear as though the errors are velvety black up to the limit of $\pm\theta$, whereupon the errors suddenly turn silvery over their entirety as the ray path misses the absorber and returns to the sky after additional reflection off the cusp.

7.10 OPTICAL CAVITY EFFECTS

An optical cavity can be substituted for a flat absorber when required, such as in the Winston collector, where some of the rays reach the flat absorber at high angles of incidence. In Fig. 7.33 we show the basic geometry of a cavity absorber. The entrance aperture is placed at the focal plane of the basic collector. All rays entering this aperture are intercepted by the absorber pipe located within the cavity, either directly or after reflection from the walls when the central pipe is properly located.

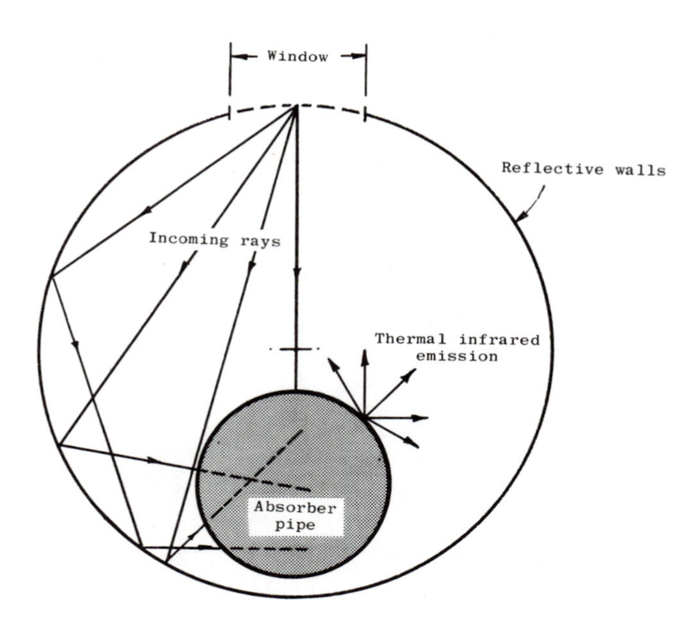

Fig. 7.33 Diagram of an absorber cavity where all the rays accepted by the entrance window arrive at the absorber pipe. Note that if the reflective walls are a perfect cylinder there is a large displacement of the absorber from the center of the cavity. The thermal infrared emission escapes into (is absorbed by) the window either directly or after reflection.

In optical terms the cylindrical reflective walls act like a reimaging system, forming an image of the entrance window between the center of the pipe and the rear reflective wall. This image position is highly modified by spherical aberration, but a pipe on the order of half the diameter of the cavity can collect all of the reflected rays.

The thermal infrared emission radiated from the absorber is also trapped within the cavity. If this reradiation were fully redistributed in the cavity, the amount escaping would be exactly as though the emitting surface were a blackbody at the entrance window of the cavity. When the cavity walls are specular and return the thermal infrared radiation to the absorber after only one or two reflections, the cavity can act as a selective surface having TIR losses less than that of a blackbody, even though the absorber pipe is a blackbody. The effect of a selective surface, however, is to still further reduce TIR losses, as we show below.

The iterative series method can be utilized to calculate the effective emittance of the absorber when it is enclosed in a reflective cavity. For the geometry shown in Fig. 7.33 the absorber is located asymmetrically with regard to the center of the enclosing pipe, so that it occupies the approximate focus of the window area. In this position every ray emitted from the absorber returns to the absorber after one or two reflections off the walls. If the absorber diameter is small compared to the pipe, we can assume that two reflections are the rule and proceed to evaluate the expression for the infinite series that results. The series is infinite because when the ray returns to the absorber some of the incident energy is reflected and some absorbed. The equation that results then need only be modified because some rays (about 20–30%) have only one reflection before returning to the absorber.

If we denote the effect in reducing the effective emittance of the absorber from ϵ to ϵ'

$$\epsilon' = \epsilon(1/\delta), \tag{7.6}$$

the value of the enclosure suppression factor δ is given by

$$\delta = w + [1 - r^2(1-w)]\,(1-w)/G, \tag{7.7}$$

where G is given by

$$G = 1 - (1-w)(1-\epsilon)r^2, \tag{7.8}$$

and where ϵ is the emittance of the absorber surface, w is the fraction of 180° occupied by the window in the cavity, and r is the reflectance of the cavity walls.

The consequences of the above equations are interesting. If the cavity walls are fully absorbing, the effective emittance of the pipe is exactly equal to the surface emittance of the absorber and δ is unity for all values of ϵ. If the cavity walls are fully reflective and the absorber has high emittance, $\epsilon = 1$, the cavity suppression factor is asymptotic to $2/w$. In Fig. 7.34 we show the case where $w = 0.1$ for two values of wall reflectance. The limit when $r = 1.00$ is $\delta = 5.0$. This is explained by the fact that

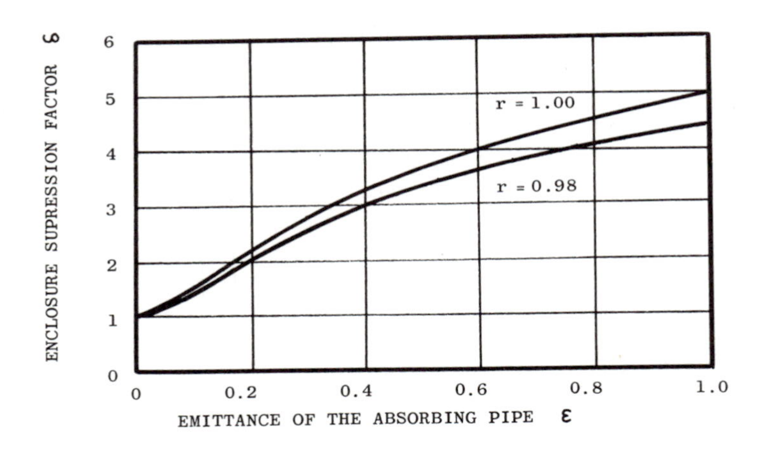

Fig. 7.34 Variation of the enclosure suppression factor δ for a reflective cavity surrounding the absorber, having a window occupying 0.1 of the surface of the cavity wall.

radiation from the absorber has *two* modes of escape through the window, either directly from the absorber or after the first reflection at the wall, the net effective angular size of the window being twice the physical width of the slot.

This mathematical expression should be used with caution, since it is only an approximation to reality for mathematical convenience. It is mainly useful for gaining a feel for the behavior of cavities. Actual design and evaluation of enclosures for use should be done by ray tracing the system.

In summary, one can say that an enclosing cavity is useful when the absorber emittance is high, but is of marginal importance when the absorber employs a highly selective surface having low emittance.

7.11 RUSSELL FRESNEL MIRROR

Russell (1974) showed that a curved Fresnel mirror could produce stigmatic images lying on a circle that is the equivalent of the famous Rowland circle of spectroscopy, except that the conjugate focus is at infinity rather than autocollimated. In Fig. 7.35 we show the essence of the geometry involved. A parabolic mirror of paraxial radius R would normally have an on-axis focus position approximately halfway between the paraxial center of curvature and the mirror, at point A. The off-axis image would show considerable coma. Russell noted that if the cylindrical mirror were broken up into strips and these strips placed on a circle of radius $R/4$, the off-axis images would be stigmatic and also lie on the same circle of radius $R/4$. If cylindrical sections of mirror are used, the centers of curvature of the strips are located slightly inside the paraxial strip. In Fig. 7.35 the centers of strips B, C, and D are located at points on the optical axis marked B, C, and D.

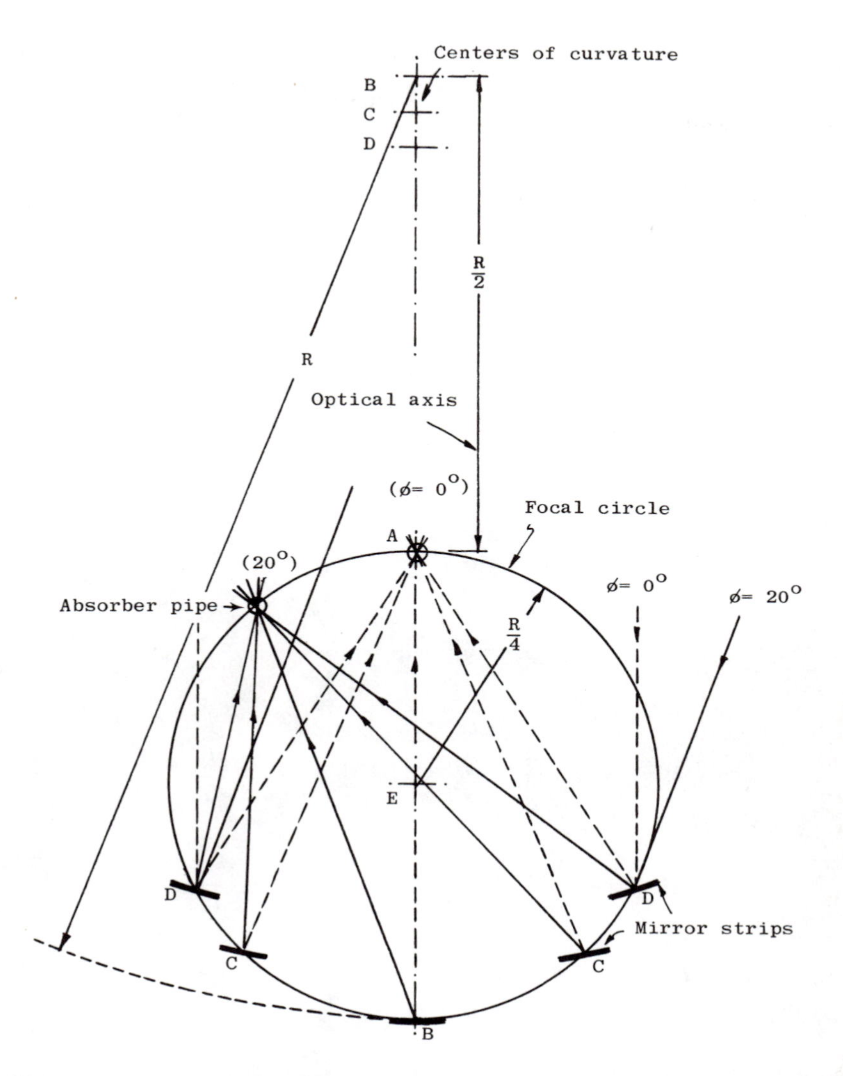

Fig. 7.35 Diagram of the cylindrical Russell geometry for a fixed-mirror moving-focus collector.

The significance of the Russell configuration is that the mirror can be fixed, the only moving part being the absorber, which is pivoted at center *E*. This means that the most vulnerable part of a large-scale solar energy collector system, the mirrors, can be rigidly attached and out of the wind. The chief disadvantage of this design is that the acceptance angle is hardly enough to span the range of angles from early summer sun to early winter sun, about 90°. The geometry is sufficient for such large angles, but the Fresnel strips cast shadows on the adjacent strips when the sun gets far from the optical axis, effectively reducing the aperture below the cosine law limit.

A photograph of a Russell collector built by General Atomics for a group of utilities is shown in Fig. 7.36. Note the narrow width of the individual strips. If the strips are narrow enough, flat reflector strips can be used without significantly affecting the energy concentration at the focus.

Fig. 7.36 Model of the Russell collector built by General Atomics under contract to three utilities. Note the small spacing of the elements of the mirror, enabling flat strips to be used.

7.12 HEMISPHERICAL BOWL FIXED-MIRROR COLLECTOR

The use of a fixed spherical mirror has recently been proposed independently by Steward (1974) and Meinel (1974), but its geometrical elegance was recognized as long ago as 1878 by Adams, who built a 40-ft unit in Bombay, India (see Section 1.3). The hemisphere produces a highly aberrated optical image by usual standards, but because of symmetry, every ray entering the hemisphere must cross the paraxial line at some point between the paraxial focus and the mirror surface. This geometry is illustrated in Fig. 7.37.

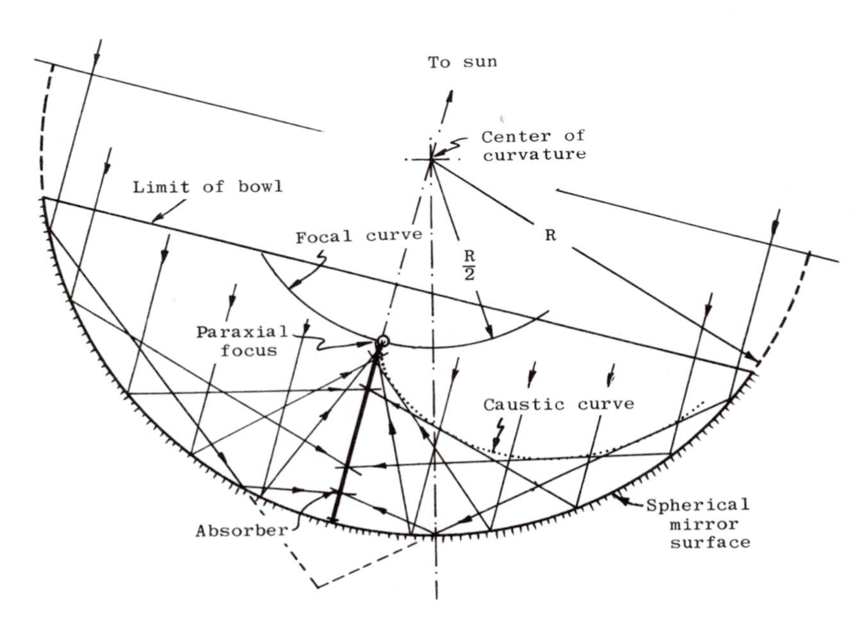

Fig. 7.37 Basic geometry of the hemispherical bowl collector. Note that the angle of arrival of solar rays is nearly normal along much of the length of the absorber cylinder.

The symmetry of the bowl makes it possible to pivot an absorber about the center of curvature of the hemisphere. The absorber intercepts the flux arriving from the mirror at different points along the absorber for different annular zones of the mirrors. The flux brightness varies along this absorber, being bright at the region near the paraxial focus, especially where the caustic intercepts the focal cylinder. The intensity drops along the focal cylinder toward the mirror surface, but rises abruptly where the secondary caustics intersect the cylinder, a rise corresponding to double reflection, triple reflection, and so forth.

The intensity distribution along the focal line of the hemispherical bowl has been studied by McKenney (1975). The variation of the point where the ray intercepts the focal line as a function of aperture is shown in Fig. 7.38. Note that this pattern is truncated by two things: (1) the hemisphere is generally cut short of being a full hemisphere, so typically the total surface of the bowl is between 50% and 75% of the area of a full hemisphere; and (2) when the sun is off the meridian, part of the bowl is foreshortened, eliminating the extreme rays from that side of the bowl but adding them on the opposite side where the solar rays can arrive tangent to the sphere. The intensity distribution shown in Fig. 7.39 is for a focal absorber of a uniform diameter three times the solar image diameter. Allowing for mirror errors contributing one additional solar diameter to the geometrical solar image, this diameter pipe keeps the maximum azimuthal angle of incidence on the absorber under 45°. A photograph of the intensity distribution along a cylinder, showing the brightness gradation and secondary caustics near the junction between the absorber cylinder and the mirror surface, is given in Fig. 7.40.

The design of the absorber for the hemispherical bowl collectors is complicated by the intensity variations along its length. These variations can be accommodated by changing the effective absorber diameter and/or properly directing the flow of the heat-transfer fluid along the length. The flux concentrations that must be handled range from about 200 near the caustics to 50 at the lowest intensity region, the exact amounts depending upon the mirror surface error root-mean-square (RMS) value and the shape factors for the absorber.

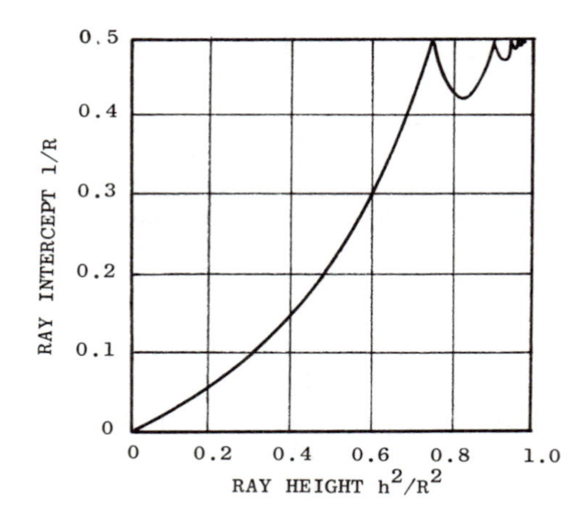

Fig. 7.38 Plot of the axial ray intercepts as a function of the incident ray height for a spherical reflector.

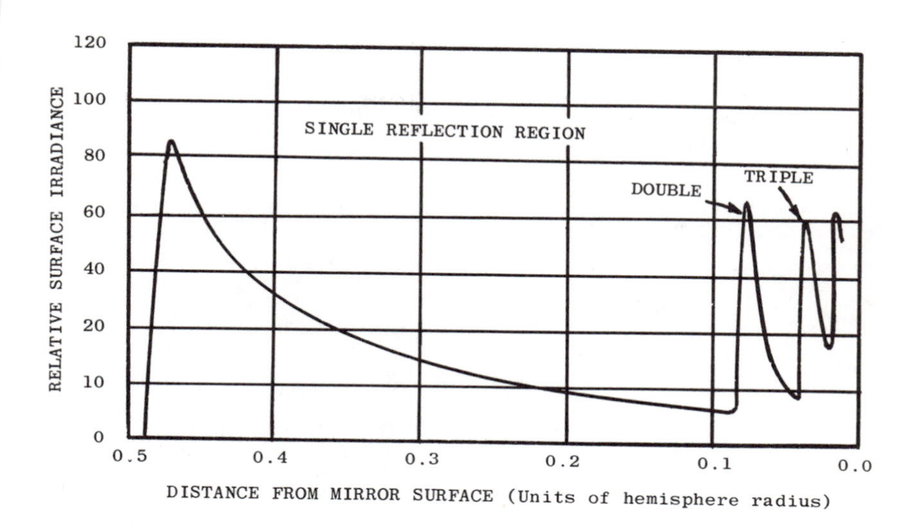

Fig. 7.39 Intensity distribution along a cylindrical absorber having a diameter three times the geometrical solar diameter (McKenney, 1975).

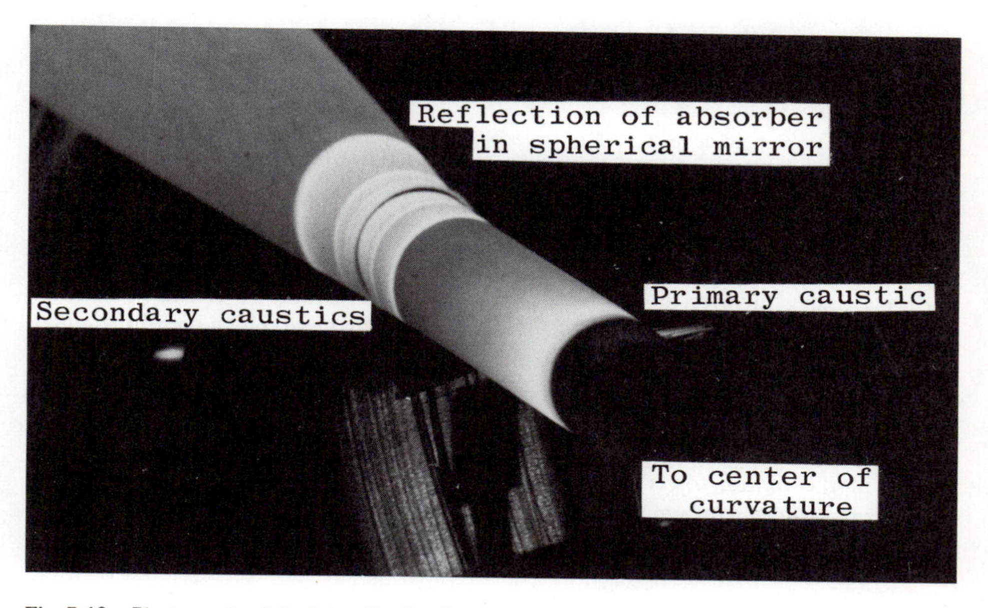

Fig. 7.40 Photograph of the intensity distribution on a cylinder at the line focus of a hemisphere. The single reflection zone is at the right with the multiple reflection zones at left. The image reflected in the hemisphere is at the top left.

The hemispherical *power-bowl* design shown in Fig. 7.41 has been developed as an alternative to the central receiver approach to power systems; the moving heliostats are replaced by the fixed mirrors, and the fixed central tower is replaced by a moving cylindrical distributed-focus absorber.

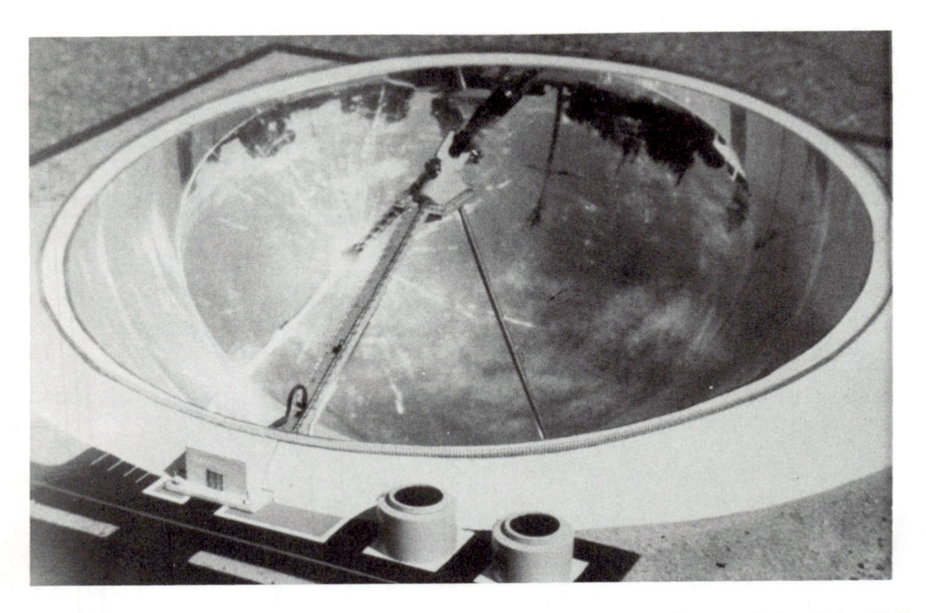

Fig. 7.41 Photograph of a model of a *power-bowl* hemispherical collector, designed by Helio Associates and E-Systems for a modular utility power system. The bowl is 100 m in diameter, and average output for the central 70% of the day is 1 MW_e.

The tradeoffs between the two approaches are significant. The optimization of the bowl design involves the tilt of the structure to the south. If the mirror is horizontal, the flux entering it varies like a horizontal flat plate, being high in summer and low in winter. Tilting the bowl south increases the winter collection efficiency at the expense of summer collection. Optimization further involves the depth of the bowl. Highly inclined mirrors are less cost effective than mirrors nearly normal to the incoming rays. The bowl has mirrors more highly inclined than the heliostat mirrors in the central tower system, but much less than those in the Winston geometry. One would therefore pave the bowl excavation with mirrors only in the regions where maximum performance is obtained. The question of the effect of fill factor on system performance also complicates the direct evaluation of these several collector geometries.

The size of a hemispherical fixed-mirror collector is limited by some maximum diameter. One cannot make the length as great as in the case of the Russell cylindrical geometry, or the surface area as large as with the tower-heliostat geometry. The largest hemispherical bowl covered with mirrors to date is the Arecibo radio telescope, which is 1000 ft across. As a consequence, one is faced with combining the heated working fluid from several bowls or building a turbine for each. Heat losses and moving joints are complications requiring further study, but compact community-sized collectors of the hemispherical type seem to be one option for potential application.

7.13 FIELD-LENS MIRROR CUSPS

Because of the relative compactness of the ray bundles that proceed to the absorber in the case of fixed-mirror collectors, as in the Veingard configuration, it becomes attractive to examine the possibility of constructing a field-lens mirror to reduce the size of the absorber and thereby increase the radiative balance concentration X. Cusp designs of the Winston or Trombe-Meinel type are candidate systems. The design does not demand the same precision as in the case of incident sunlight, because in field-lens usage the ray bundle is not collimated.

In Fig. 7.42 we show a field-lens type of configuration using two parabolic cusps. These particular cusps are not the exact Winston type, but are optimized to redirect

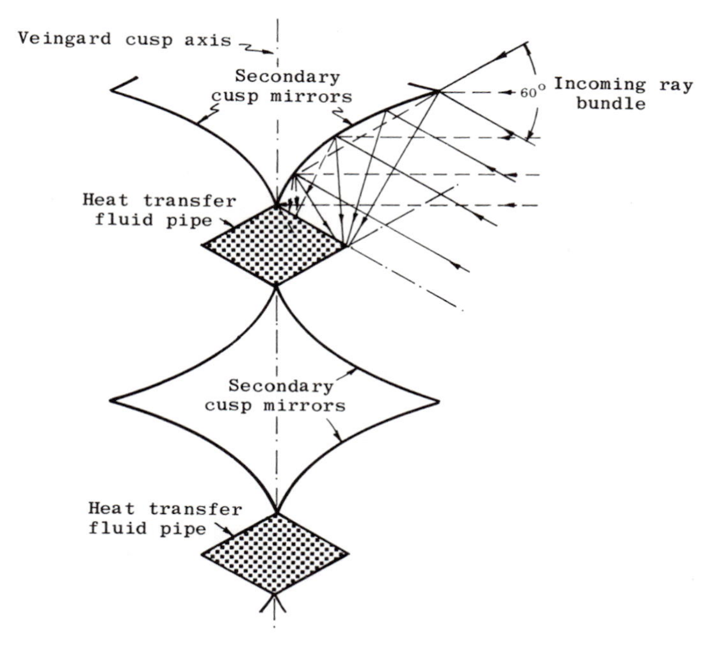

Fig. 7.42 Use of a cusp design field mirror to increase the radiative balance concentration X for a Veingard fixed-mirror configuration. Note that the absorber is now a series of quadrangular pipes instead of a two-sided flat plate.

incoming light from a Veingard collector to a triangular pipe containing the heat-transfer fluid. The triangular pipe geometry is convenient because the third side is a common side. The approximate increase in concentration X is 2.0 for the case illustrated.

There is an upper limit to the amount of additional concentration possible with a field mirror. In the case illustrated, the incoming rays are within a 60° angular cone as they arrive at the cusp aperture. An increase of 2.0 times is equivalent to having a 120° angular cone after operation by the mirror cusps. The absolute limit would be 180° after operation by the field mirror, a case that can be approached only if the cusps of the parabolic sections are extended a great distance, in fact until the opposite cusps become parallel, as in the case of the Winston design.

In Fig. 7.42 the field mirror cusps are shown as being normal to the absorber axis. In actual use the opening of the cusp would be directed toward the center of the incoming angular ray bundle from the mirror. This tilting modifies the shape somewhat; the exact shape must be determined for the exact geometry of the Veingard system on which it is to be used.

A similar field mirror design can be used to advantage with the Steward-Meinel hemispherical bowl design. One of the problems of the absorber design is that there are azimuthal variations in flux brightness, for example on the side where the rays are vignetted owing to a large zenith distance for the sun. A series of cusps, circularly symmetric about the axis shown in Fig. 7.42, would bring the radiation from all azimuths to a single heat-transfer fluid pipe. This type of field mirror is, in fact, easier to do for the hemispherical bowl collector because the incoming ray bundle has a smaller angular spread than the 60° shown in the illustration.

7.14 PROBLEMS

7.1. Draw the optical diagram for a parabolic mirror that uses a field lens to form an image on the absorber. Label the elements of the drawing.

7.2. Draw the optical diagram for a parabolic mirror that uses a field mirror to form an image on the absorber. Suppose that the actual image of the parabolic mirror falls on the absorber when the axis of the paraboloid is pointed at the sun. Relate the angle of acceptance of this system to the relative size of the field mirror.

7.3. Prepare a sun angle diagram for your latitude showing the positions of the sun at the hours of the day for the solstices, the equinox, and the date midway between the equinox and each solstice.

7.4. Compute and graph the fraction of sunlight intercepted by a Tabor cylinder without side boosters as a function of lateral angle of incidence and as a function of the depth of the absorber. What is the effective flux concentration X as a function of absorber size and sun angle?

7.5. Using Figs. 7.12–7.15, calculate the approximate size and position of the minimum-size bundle for an $f/0.25$ paraboloid as a function of off-axis angle of the incident rays. If the total range of angle is 60°, calculate the best position to place the pivot point to convert the paraboloid into a Veinberg double cusp.

7.6. Graph the change in the aberrated image size at a fixed focus of the Veinberg design as a function of angles for the pivot point you have selected in Problem 7.5.

7.7. Draw to scale a Winston cusp 1/3 the length of a full Winston cusp, with a field acceptance angle of 12°. Label the angles involved in defining the cusp. Plot the radiation balance concentration X for the full and the 1/3 Winston design in a diagram like that shown in Fig. 7.30.

7.8. Calculate the self-shadowing of a Russell collector where the steps have a width of 1/20 the aperture of the collector and the depth of the mirror is 0.5 the mirror array radius. The range of angles to be plotted is the range of angles encountered in the seasonal change in position of the sun. Graphical solutions can be used. Plot the results on a graph of effective optical cross section versus sun angle.

7.9. Define the geometry and the significant ratios for the positions in the aperture of a hemispherical bowl where the reflected rays strike the subsolar point after one, two, and three reflections at the mirror surface.

7.10. Calculate the flux concentration of sunlight focused on the surface of the mirror at the subsolar point if all the sunlight is contained uniformly in an angle twice the angular diameter of the sun. What problems does this value suggest?

7.11. Calculate the diurnal variation of optical cross section of a hemispherical bowl collector, tilted 15°S at a latitude of 30°N, when the depth of the bowl is 0.6 of the mirror radius.

7.12. Discuss the engineering tradeoffs regarding a decision to use either (a) a fixed-mirror collector design, making seasonal adjustments, or (b) a similar east-west cylindrical collector in which daily tracking is permitted. What factor would be most important in a decision to build one or the other of these two options?

Chapter 8
Optical Surfaces

8.0 INTRODUCTION

In this chapter we examine the optical aspects of surfaces involved in solar energy collection. Surfaces make the difference between the idealized performance of optical elements as discussed in the preceding chapter and the actual system performance to be discussed later.

The basic problem in achieving optimum performance for a particular design lies in the manufacturing and coating of the surface. The art of manufacturing optical surfaces has become highly advanced in response to the needs imposed by astronomy, scientific research instrumentation, military applications, and so forth. For solar energy applications, however, we are denied recourse to most of the techniques simply because they result in surfaces that are too expensive for the value of the energy delivered by the solar collector. We therefore have to make careful balances among the accuracy of an optical surface, the cost of achieving that accuracy, and resultant performance.

The finished optical surface must be coated and/or protected for operation for long periods of time in an environment that includes dust, moisture, hail, wind, carbon dioxide, and many reactive aerosols. Surface finishes that yield high efficiency in laboratory instruments or telescopes cannot always be relied upon when the optics remain in the open regardless of weather or season. Again, we are restricted by cost to coatings that are inexpensive, or that will become so when produced in the large quantities that will eventually be needed.

8.1 TRANSMISSIVE OPTICAL MATERIALS

The transmissive materials for solar energy applications tend to be limited by cost to ordinary types of glass and plastics. Glass is more durable but it is subject to impact damage. Plastics, on the other hand, tend to be degraded by long exposure to sunlight, the least expensive plastics tending toward most rapid degradation.

Window glass is manufactured by three basic processes. Rolled glass is satisfactory for most window applications. Float glass, where the molten glass is poured onto a bed of molten tin, is of a high surface quality that exceeds the requirements for windows and is good enough for heliostat flat mirrors. Plate glass is polished and is also of high optical quality, but it is more expensive than float glass.

Glass transmits the solar spectrum well, cutting off only that small portion of the ultraviolet shorter than 0.38 μm and longer than 3.0 μm in the infrared. (The transmission curve for glass is shown in Fig. 8.3.) Fused silica transmits both the ultraviolet down to the atmospheric cutoff at 0.30 μm and the infrared out to about 4.5 μm, but is too expensive for use in solar applications other than furnaces. Glass has minor absorption regions in the solar spectrum wavelengths, due primarily to iron impurities in the glass. Glass in "water white" quality can be obtained, but is considerably costlier than ordinary glass. As is noted in Chapter 12, the absorption in glass windows is not a total loss to flat-plate systems because the energy absorbed acts in concert with the normal heat flow out of the collector to impede that heat flow. As a general rule one can consider that *half* the energy absorbed in windows is effectively regained in output from the system. In view of this gain it does not appear cost effective to use other than ordinary float glass for collector windows.

8.2 REFLECTION LOSS

Windows reflect some of the incident light, depending upon the index of refraction of the window material. The reflectivity of glass changes only slowly with index. The reflection loss is about the same for all the materials employed in solar collectors. The dependence of reflection (at normal incidence) on index is given by

$$R = (n-1)^2/(n+1)^2. \tag{8.1}$$

In Fig. 8.1 we show the transmittance of glass windows as a function of number of windows for glass of index $n = 1.51$. The decrease in transmittance with number of windows is due to reflection losses only. There is an additional loss due to absorption, which ranges from less than 1% to 4%, depending on thickness and type of glass. This additional loss should be added to that shown in Fig. 8.1.

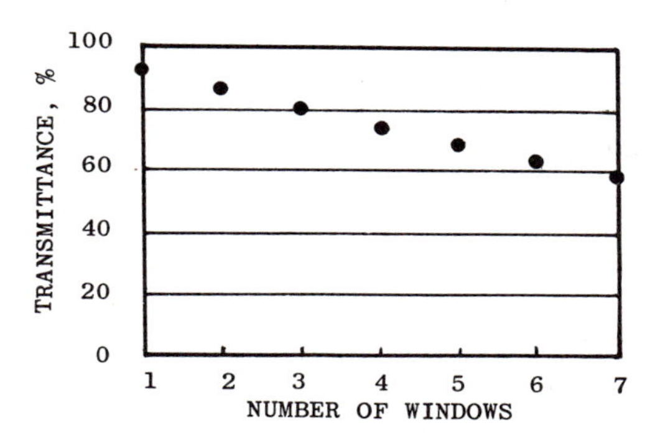

Fig. 8.1 Variation of transmittance through different numbers of windows having an index of refraction of 1.51.

Reflection losses also vary with the angle of incidence of the light. In Fig. 8.2 we show typical curves for one, two, and three glass window plates. For comparison we also show the curves for a single plate of diamond ($n = 2.42$) and one of silicon ($n = 4.0$). Note how rapidly the reflection loss increases with index. Also note that a higher index does not have the same angular dependence as given by the number of glass plates that would equal the normal incidence reflectance of these high-index materials. For example, four glass plates have approximately the same transmission at 0° as diamond, but diamond has better performance at angles above 20°.

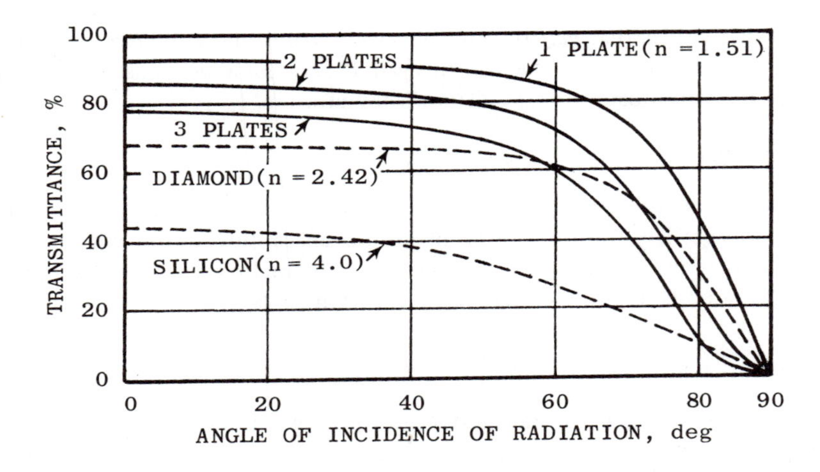

Fig. 8.2 Graph of the relationship between transmittance and angle of incidence of the radiation to one, two, and three glass plates, as compared to other materials of a high index of refraction.

The general expression for reflection as a function of the angle of incidence is given in Eqs. (8.2)–(8.11). E_{0p}^{+} is the electric vector of the wave traveling in the positive direction in the 0th layer, polarized with the electric vector parallel to the plane of incidence upon the medium; E_{0s}^{-} is the electric vector in the negative direction in the 0th layer, polarized perpendicular to the plane of incidence. The Fresnel equations are then:

$$E_{0p}^{-}/E_{0p}^{+} = \frac{n_0 \cos\phi_1 - n_1 \cos\phi_0}{n_0 \cos\phi_1 + n_1 \cos\phi_0} = r_{1p}, \tag{8.2}$$

$$E_{1p}^{+}/E_{0p}^{+} = \frac{2n_0 \cos\phi_0}{n_0 \cos\phi_1 + n_1 \cos\phi_0} = t_{1p}, \tag{8.3}$$

$$E_{0s}^{-}/E_{0s}^{+} = \frac{n_0 \cos\phi_0 - n_1 \cos\phi_1}{n_0 \cos\phi_0 + n_1 \cos\phi_1} = r_{1s}, \quad (8.4)$$

$$E_{1s}^{+}/E_{0s}^{+} = \frac{2n_0 \cos\phi_0}{n_0 \cos\phi_0 + n_1 \cos\phi_1} = t_{1s}. \quad (8.5)$$

An alternative form for amplitude relationships r is

$$r_{1p} = \frac{\tan(\phi_1 - \phi_0)}{\tan(\phi_1 + \phi_0)}, \quad (8.6)$$

$$t_{1p} = \frac{2 \sin\phi_1 \cos\phi_0}{\sin(\phi_1 + \phi_0) \cos(\phi_1 - \phi_0)}, \quad (8.7)$$

$$r_{1s} = \frac{\sin(\phi_1 - \phi_0)}{\sin(\phi_1 + \phi_0)}, \quad (8.8)$$

$$t_{1s} = \frac{2 \sin\phi_1 \cos\phi_0}{\sin(\phi_1 + \phi_0)}. \quad (8.9)$$

The intensity of the reflection is then

$$\rho_p = (E_{0p}^{-})^2/(E_{0p}^{+})^2 = r_{1p}^2$$

and (8.10)

$$\rho_s = (E_{0s}^{-})^2/(E_{0s}^{+})^2 = r_{1s}^2,$$

and the transmittance is

$$\tau_p = n_1(E_{1p}^{+})^2/n_0(E_{0p}^{+})^2 = (n_1/n_0)\,t_{1p}^2$$

and (8.11)

$$\tau_s = n_1(E_{1s}^{+})^2/n_0(E_{0s}^{+})^2 = (n_1/n_0)\,t_{1s}^2.$$

8.3 WAVELENGTH VARIATION OF TRANSMISSION

Plastics are in general more transparent than glass. They absorb in the ultraviolet like glass but, unlike glass, they are damaged by these ultraviolet photons. Plastics have variable transmission in the infrared, depending upon the thickness and the molecular bonds present in the particular plastic. Simple plastics like polyethylene have few absorption bands, whereas complex molecules like Mylar have strong absorption bands in the middle of the environmental thermal emission region.

The infrared absorption of plastics is important in collector behavior. Glass, being opaque to the thermal infrared, traps heat radiation; some plastics, being relatively transparent, allow thermal radiation to escape. If plastic windows are used, the plastic must either be thick enough to absorb the thermal radiation (TIR) or be intrinsically opaque to it. The curves of a number of plastics of importance for solar energy collectors (2.5- to 15.0-μm wavelength region) are shown in Figs. 8.3–8.6.

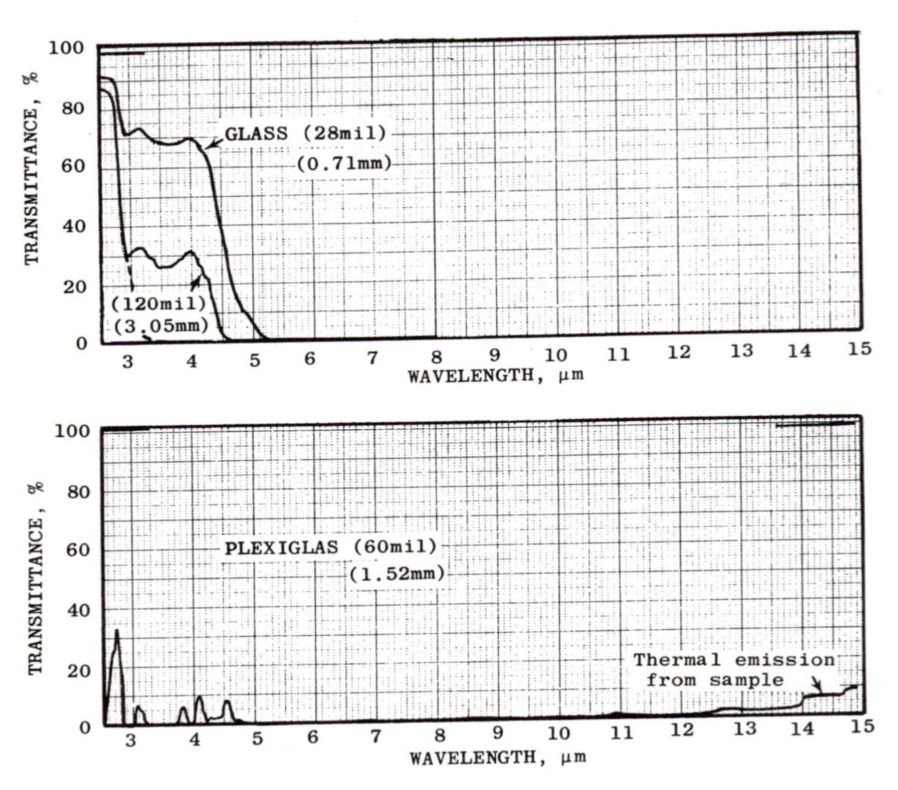

Fig. 8.3 Infrared transmittance curves for glass and Plexiglas. Sample thicknesses are as indicated. Curves include reflection losses; 100% transmittance is close to 100 on the charts.

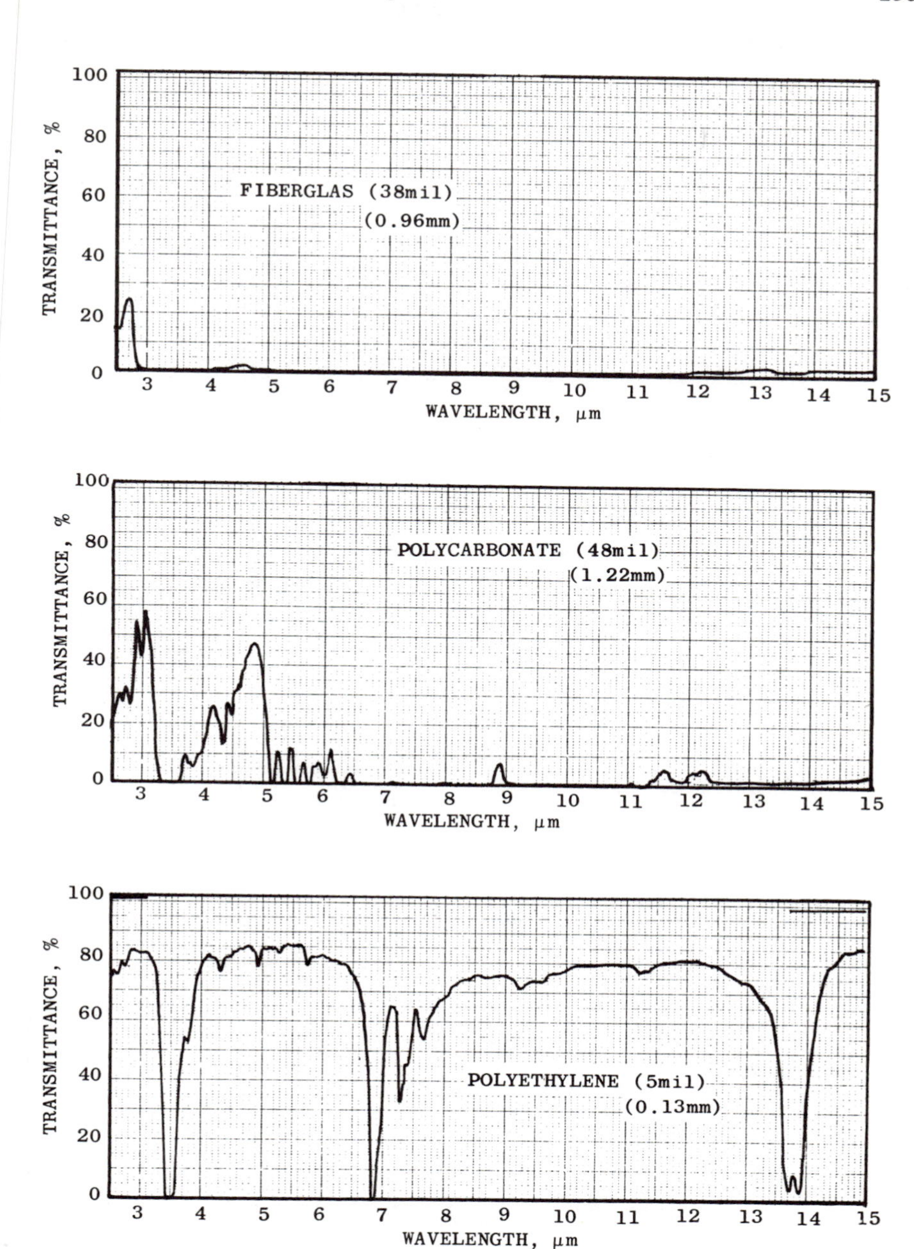

Fig. 8.4 Infrared transmittance curves for Fiberglas, polycarbonate, and polyethylene. Sample thicknesses are as indicated. Curves include reflection losses; 100% transmittance is close to 100 on the charts.

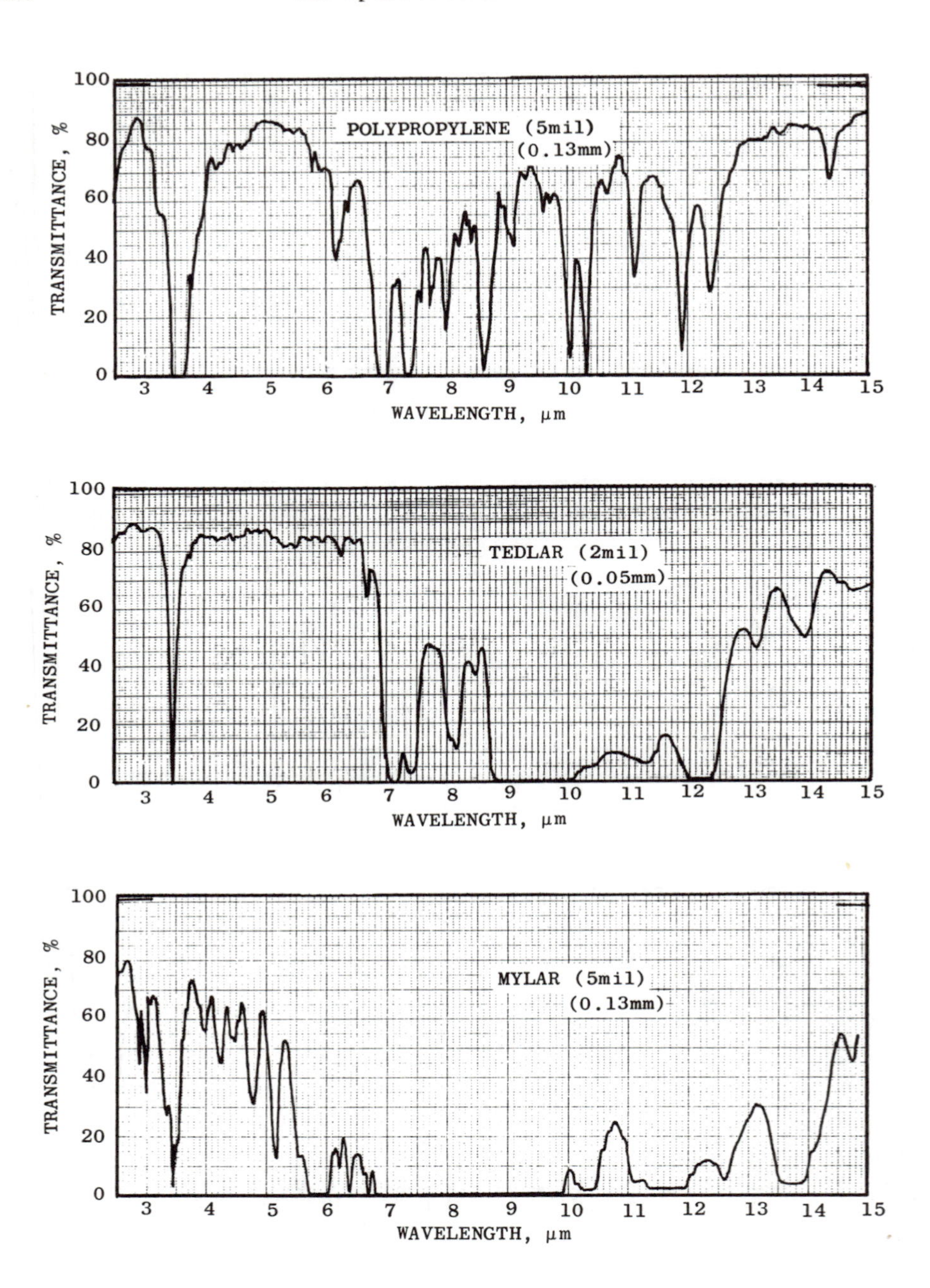

Fig. 8.5 Infrared transmittance curves for polypropylene, Tedlar, and Mylar. Sample thicknesses are as indicated. Curves include reflection losses; 100% transmittance is close to 100 on the charts.

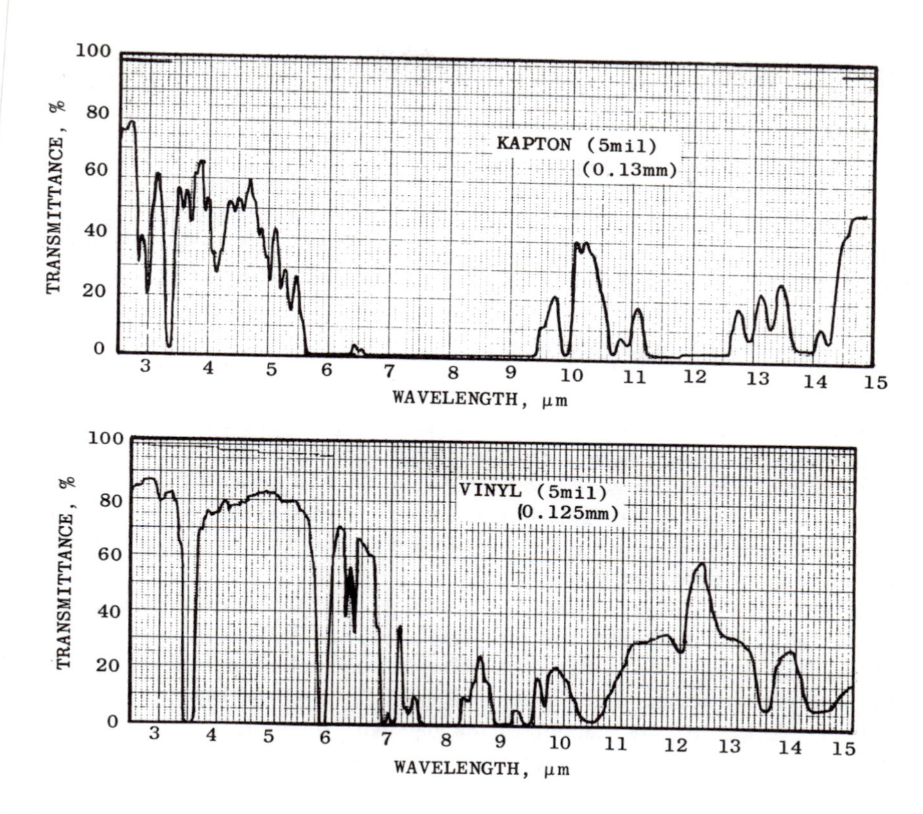

Fig. 8.6 Infrared transmittance curves for Kapton and vinyl. Sample thicknesses are as indicated. Curves include reflection losses; 100% transmittance is close to 100 on the chart.

In the transmittance curves, the thicknesses shown happen to be from typical samples used in solar collector experiments. Plastic *films* are very thin and are used in tension for window coverings. Their thinness tends to make them transparent, whereas the thicker plastics used for rigid window coverings are thick enough to be almost totally opaque in the thermal infrared. Plexiglas and Fiberglas are more opaque than glass, but polycarbonate shows some transmission out to 6 μm.

Polyethylene is the most transparent plastic. The sample tested had been exposed to sunlight for one year, and though it was a "stabilized" type, the net transparency had decreased slightly from that for a new sample. It should be noted that the curves generated in these figures were made with a collimated-beam instrument, so radiation *scattered* by the film can be partially lost to the instrument. In reality, the hemispherical transmittance of the polyethylene film in the solar region remained high, about 90%. The infrared transmittance was about 80%, allowing most of the thermal infrared from a hot surface inside a collector using such a window to escape. Several researchers, including Silvestrini (1975), have taken advantage of this characteristic to make radiative cooling "collectors" for generating cooled air at night.

If one wishes to use plastic film and have high infrared opacity for maximum "greenhouse" effect, one must use a more complex molecular type. Four plastic film materials are shown in order of increasing opacity: polypropylene, Tedlar, Mylar, and Kapton. Kapton is interesting because it has the highest service temperature of any plastic film, up to 400°C, but it also is much more expensive that the poorer ones.

8.4 PRODUCTION PROCEDURES

Transmissive optics include windows and lenses. Windows are frequently used in solar collectors. The materials used are glass and plastics. All of these materials as produced in sheet or film form have surfaces adequate for general solar collector applications. Transmissive surfaces are good enough basically because most of the applications do not require a sharp focus. This insensitivity is further aided by the fact that the deviation caused by a given surface error in the transmission mode is less than for the same error in the reflective mode.

Float glass is a type of glass that combines a relatively low cost with excellent flatness and parallelism, which make it suitable for windows and mirror substrates. It is produced by pouring the molten glass ribbon onto a molten tin surface. Tin is molten over a very wide temperature range, melting at 231.9°C and boiling at 2270°C. Tin oxide is colorless, so any reaction of the glass with the molten tin has minor consequences. The Pilkington process further restricts the diffusion of metal ions in and out of the viscous glass by the application of an electric field. The glass sheet moves along the length of the tin tank, which has a thermal gradient along its length, and when the glass finally becomes "solid" it is still lying on a molten tin surface. The resultant surface is therefore flat on both sides.

As discussed below, glass can be formed into a wide variety of shapes and readily fused into desired structures, such as envelopes to be evacuated.

Plastics come in a variety of forms. Plastic film is blown as long, thin-walled tubes and subsequently slit and rolled in sheet form. For some solar applications, like the crop-drying collector, the plastic tubes are used as supplied by the factory. Sheet plastic can be cast from liquid or injected into molds having the desired faces, flat or otherwise. Since plastics generally become soft when heated, plastic material can be re-formed to the desired shape starting from pellets or sheets.

8.5 SURFACE ACCURACY OF TRANSMISSIVE SUBSTRATES

The surface errors of plate and float flat glass are small, amounting to a few arc seconds, which is negligible for their use in windows in either flat-plate or concentrating collectors. If there is occasion to wonder about the quality of plate or sheet glass, one can quickly evaluate it by shining a bright but small angular diameter light through the sheet, casting a shadow of the errors on a wall some distance away. Typical distances between light source, glass, and wall would be 5 m each.

Transmissive optics also involve lenses made of transparent materials. Two types of lenses can be encountered in solar energy work: simple lenses and Fresnel lenses. Simple lenses tend to be small in solar applications because of the expense and weight of large ones. In general there is no problem with the surface accuracy of either glass or plastic.

Plane Surfaces

The typical manufacturing tolerance for glass simple lenses is a few wavelengths of light for the peak-to-valley, root-mean-square (RMS) surface error. The index homogeneity of ordinary crown glass is sufficient for solar applications, where index variations in the third place ($1.50\underline{0}$) can be accepted. Plastics are also good enough for this application because the inhomogeneities and shrinkage that are problems for precision optics are negligible for plastic simple lenses.

If the surface of a transmitting dielectric has a slope error α, the deviation produced in an optical beam traversing this slope error, as diagrammed in Fig. 8.7, is

$$\delta_t = (n-1)\alpha. \tag{8.12}$$

In the case of reflection the deviation would be

$$\delta_r = 2\alpha. \tag{8.13}$$

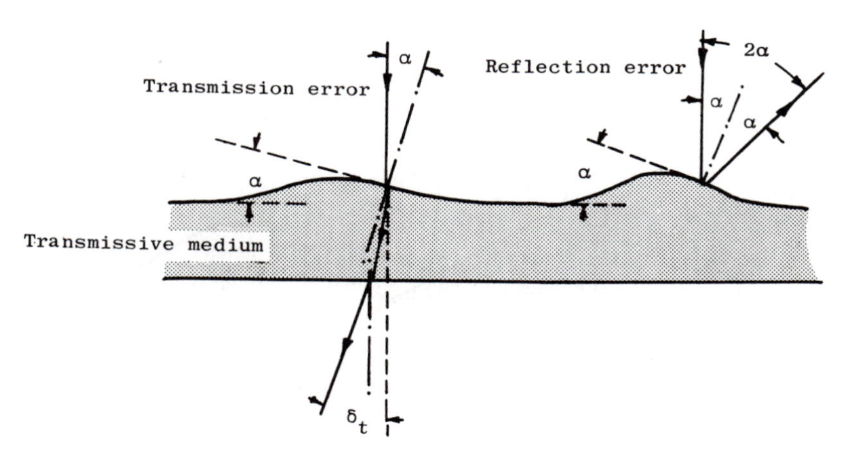

Fig. 8.7 Diagram showing the transmission of a ray through a surface error (left), and reflected by a surface error (right).

Thus the ratio of effects produced via reflection and transmission is

$$\delta_r/\delta_t = 2/(n-1). \tag{8.14}$$

For a typical index of refraction, where $n = 1.50$, the ratio of reflected to refracted error is 4.0:1, a very significant amount.

Fresnel Lenses

Fresnel lenses do introduce tolerance problems, not in the bulk properties of the material but in the precision with which the small lens element faces are made. Some Fresnel lenses are made of glass, but their cost is high unless the required accuracy is *very* low, as with covers for traffic signal lights. W. B. Meinel (1973) has made cylindrical glass and fused silica Fresnel lenses for experiments, but their cost is too high for generalized solar energy use.

The Fresnel lenses important for solar applications are made of plastic. The plastic can be either pressed in a heated mold, or cast or molded from liquid plastic. In either case the accuracy depends upon the accuracy of the mold. As diagrammed in Fig. 8.8, there are two principal sources of error in the resultant Fresnel lens. The first is in the angle of the face of the lens. In general the width of the lens zone is taken so that the face can be made *flat*, avoiding the complication of making curved faces. However, this flat requirement sets an upper limit on the width of the zone, W, in terms of the contribution it makes to the final image blur. The second source of error is in the rounding of the corners of the flat zone. Light entering this curved zone is totally lost to the image-forming aspect of the lens because it is scattered in all directions. The tolerance on the width of the curved zone is set by the width of the desired good portion of the lens and the allowable energy loss by the lens.

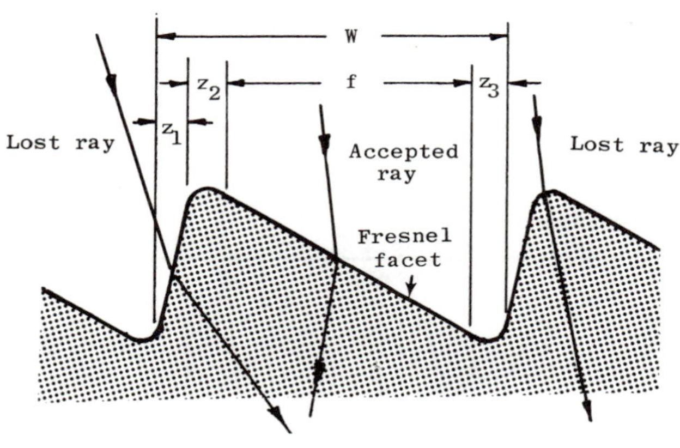

Fig. 8.8 Schematic diagram of a groove of a Fresnel lens showing the three sources of shape factor loss, z_1, z_2, and z_3.

There is a third loss in a Fresnel lens, the loss by oblique rays from the vertical face of the Fresnel grooves, as diagrammed in Fig. 8.8. For applications where the sun is several tens of degrees off the axis of the lens, as could be the case with a semifixed lens collector, this source of energy loss becomes appreciable.

The degree of rounding of the edges of the Fresnel zones is a function of how the lens is made. If the lens is pressed out of glass, the rounding can amount to approximately 1–2 mm. The viscosity of glass is a sharp function of temperature; glass becomes more fluid–less viscous is a better term–with increasing temperature. The lifetime of the mold, however, sharply decreases with increasing glass temperature. Most glass Fresnel lenses are made at as low a temperature as possible and hence have appreciably rounded corners. Even so, the surface facets have poor optical quality owing to problems of maintaining a high polish in the mold. Glass further tends to have rounded corners because of surface tension and because the glass still has reasonable flow after removal from the mold, unless the cost penalty of longer dwell time in the mold is accepted.

Plastic does not have the same problems as glass when lenses are cast directly from liquid or pressed from a heated solid. The low temperatures involved and lack of flow after setting result in Fresnel lenses with sharp corners and highly polished faces. Plastics, in general, have less durability and shorter operational lifetimes than glass.

The surface tolerances in the finished Fresnel lens are given in the equations below and defined in Fig. 8.8. If the loss regions are denoted by z_1, z_2, and z_3 and the width of the useful zone by f, we have the transmitted energy ratio

$$E/E_0 = f/(f + z_1 + z_2 + z_3). \tag{8.15}$$

In Fig. 8.9 we plot the ratio E/E_0 as a function of z/W, assuming that all of the z components are equal. Note that when $z/W = 0.1$ these errors cause the loss of 30% of the energy passing into the Fresnel lens.

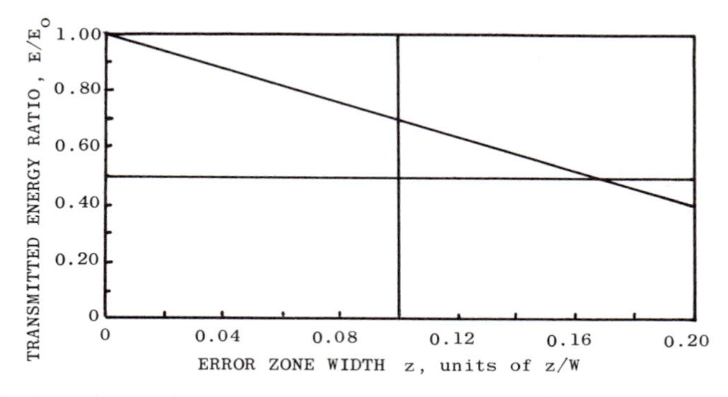

Fig. 8.9 Graph of the change in energy throughput E relative to the available energy E_0 for a Fresnel lens having error zones defined by z, where $\Sigma z = 3z$.

The angular tolerance for the lens mold can also be easily derived in terms of the tolerance placed on the image blur, from the diagram in Fig. 8.10.

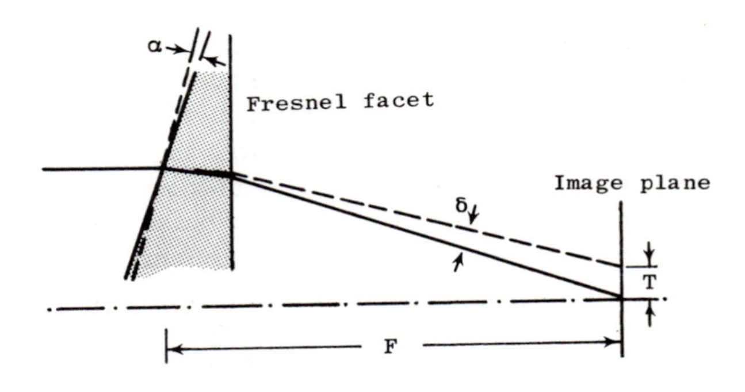

Fig. 8.10 Schematic diagram of the effect of a slope error of a groove of a Fresnel lens.

If the focal length of the lens is F and the tolerance T, the allowable surface error is given by

$$\delta = T/F. \tag{8.16}$$

In terms of the angular error of the surface of the Fresnel element, the error produced in the focal plane is given by

$$\alpha = T/F(n - 1). \tag{8.17}$$

If the allowable ratio of T/F is 0.01 (one centimeter per meter of focal length) and the index of refraction is $n = 1.5$, the allowable surface error, α, is $1.1°$.

We can proceed from the allowable error for the angular tolerance to derive the equivalent error due to face width, assuming that the total error can be equally divided between these two error sources. The spreading of the solar image is a third error contribution, but this amount of spreading would still occur with a perfect lens. We therefore do not include this source of image spread in the spread contribution arising from the use of flat facets on the Fresnel lens. By simple geometry the size of the facet is exactly equal to the tolerance T:

$$W = T. \tag{8.18}$$

Diffraction can add blur due to the finite size of the facet W, but when W is more than 100λ wide, the diffraction blur will be negligible.

8.6 LIFETIME CHARACTERISTICS

The lifetime characteristics of plastics leave much to be desired in solar energy applications, as high solar flux and elevated temperatures combine to make these lifetimes much shorter than in ordinary applications. Plastics discolor and embrittle with age to varying degrees. Polyethylene is the worst in regard to aging; the polycarbonates and acrylics are the best. Neither can stand the elevated temperatures that can be encountered in a flat-plate collector using selective surfaces if the flow of coolant is interrupted long enough for the temperature to reach its upper limit.

Plastics can be stabilized with the addition of small amounts of more complex hydrocarbons. These stabilized forms of polyethylene are made for use in greenhouses. Results of solar collector tests made by the authors of one stabilized form of polyethylene were disappointing because although discoloration and embrittlement were improved, the plastic film sheets still failed after a few months because of "bending fatigue." The plastic film windows were tight at night temperatures but expanded during the day, so vibrations were created by wind even with the windows under positive pressure. Bending fatigue caused the plastic panels to break from their frames around the periphery of the collector and come off in whole sheets. The plastic, however, remained flexible and soft overall.

One should approach plastics in solar collectors with caution. Polypropylene sheet is good for some applications like swimming-pool cover collectors, where temperatures remain about ambient, but its lifetime in higher-temperature applications like house heating remains to be proven, particularly considering the temperature escalation that accompanies cessation of fluid flow in the collectors, as can happen at random times and is the rule during summer.

Glass has disadvantages for use in flat-plate collectors, and it must be much thicker than plastics in order to achieve sufficient durability. Glass, however, presents no hazards from flux and high temperature in the range encountered in flat-plate collectors.

Plastic has similar problems when used for mirror substrates as it does when used for windows, except that changes in transparency are not critical when front-surfaced mirrors are used. The problem of bending fatigue remains, and the aging problem is enhanced in rear-surfaced films. Mylar has excellent properties and aluminum evaporated coatings adhere well, but its cost is several times higher per kilogram than that of polypropylene or polyethylene.

High-temperature plastics, like Kapton, are available and can stand operation at 300°C for a limited time. These high-molecular-weight plastics, however, are not very transparent in the visible, Kapton being deep orange in film 100 μm thick.

The coatability of plastics is a factor to be considered in selecting a mirror substrate material. Plastics containing residual solvents or volatile compounds do not permit coatings with good adhesive quality.

8.7 REFLECTIVE OPTICAL MATERIALS

As stated above, the error in the reflected beam due to an angular error in a reflective surface is 2ϵ. Therefore materials, production methods, and tolerances are significantly different for reflective and refractive optics.

Reflective optics for solar energy applications include a wider range of materials than is possible for transmissive optics because there is no need for them to be transparent. The material needs only a surface with good polish so that the solar rays are specularly reflected with little scattering of light. Mirror materials must, however, be dimensionally stable or constrained to be stable to avoid deterioration of the imaging properties with time.

Metals constitute one source of substrate for mirrors. The metal can be intrinsically reflective, such as a surface of high-purity electro-polished aluminum over a base of aluminum alloy, as in the commercial product Alzak. Sheet metal can be directly used for flat mirrors, or bent into shape for cylindrical mirrors. Making spherical mirrors requires bending the substrate metal in two dimensions, resulting in either bulk plastic flow in the sheet or buckling of the sheet. As a result the optical quality of bulk metal sheets after they are formed into a two-dimensional curve is lower than when they are bent in one dimension.

Metals such as aluminum can be cast into thin mirror shells, as with Tenzalloy, that are reasonably stable and can be optically polished. Optical polishing of metals is extremely tricky when the metal is soft, as is the case with all the highly reflective metals. Overcoating the substrate metal with a hard coating like Kanigen improves the polishability but probably raises the cost beyond the limit acceptable for solar mirrors.

Either clear or opaque plastic can be used for mirror substrates, the reflective surface being emplaced by chemical or vacuum deposition. Plastic film offers the attractive option of making very thin and lightweight mirrors. Such a mirror can be made, as illustrated in Fig. 8.11, by stretching the aluminized film over a defining structure. The weight loss achieved by using plastic film is, however, reduced by the added weight of the defining ring at the periphery of the mirror. The surface quality of the mirror will be no better than the surface planeness of the ring. Plastic film mirrors also suffer from their high thermal expansion relative to the aluminum or steel used for the support ring. If the thermal expansion coefficients are mismatched, a mirror that is under enough tension when cold to be at the deformation limit of the plastic can become limp when the surface heats during the day.

Bulk plastic can also be used for mirror substrates. Such mirrors are very resistant to impact damage, as from stones and hail, but the plastics are soft and easily damaged through abrasion in handling and cleaning. If the mirror surface is protected by being on the second side of the plastic, its optical quality is protected as well, but transparent plastics are expensive and/or deteriorate under the double flux of sunlight passing through them in this arrangement.

Glass or glassy ceramics make excellent mirror substrates. These materials are dimensionally stable and can easily be formed into two-dimensional shapes by thermal

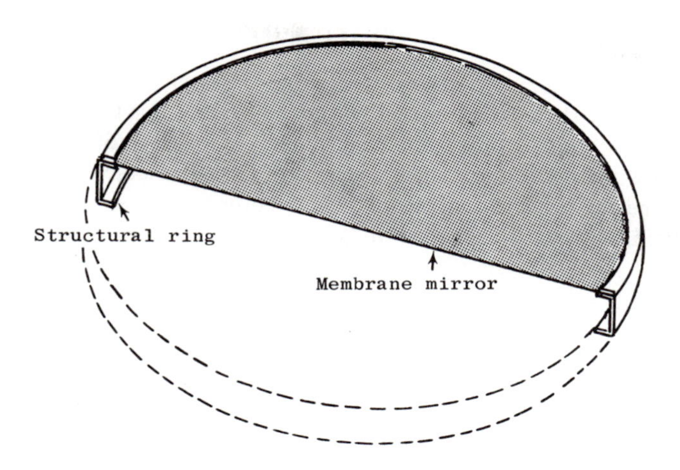

Fig. 8.11 Diagram of a membrane mirror using aluminized Mylar supported by a structural ring.

sagging at temperatures that do not damage the original quality of the external surfaces. The temperature tolerances for sagging are rather critical, however, so sagged glass tends to be more expensive than sagged plastic, even though the basic material is cheaper in terms of dollars per kilogram.

Glass can be front-surface coated by chemical or vacuum processes. Glass that is front-surface coated results in a mirror subject to abrasion and chemical attack. Astronomical mirrors in exposed locations, like the gamma ray telescope of the Smithsonian Astrophysical Observatory on Mt. Hopkins, Arizona, show mirror coating degradation in two or three years. Second-surface glass mirrors are the most durable of all the possible mirror types, except that the glass can be damaged by impact from objects and hail. The thickness of the glass substrate is, therefore, set by the damage tolerance; again, cost is a serious constraint.

The glass for second-surface mirrors must be transparent. The sunlight makes a double pass through the glass, and the weak absorption due to iron impurities in the glass can result in a net loss of 5–7% in glass of typical window quality. One must be careful to avoid unexpected losses when "water white" glass is obtained. White glass can be and often is achieved not by avoiding impurities responsible for absorption but by *adding* chemicals to produce weak absorption in adjoining wavelength regions to achieve a colorless balance to the human eye. The optical transmission and cost of the types of glass being considered must be carefully weighed before any commitment to volume production is made. Improvements in transparency must be assessed as being cost effective to be worth doing.

8.8 SURFACE ACCURACY OF REFLECTIVE SUBSTRATES

How much surface accuracy is needed from a mirror depends largely upon the concentration expected of the collector. For low flux concentrations the surface error can be large compared to that required for high concentrations of solar furnaces. One can perhaps express the desired tolerance in terms of the acceptable flux degradation. Let us take σ as the characteristic width of the image blur in angular measure. As shown in Fig. 8.12, the linear blur at the collector focus from the mirror surface error σ is

$$S_m = 2\sigma F, \tag{8.19}$$

and the total image size to a good order of approximation is then

$$S = S_O + S_m = F(\alpha + 2\sigma). \tag{8.20}$$

If the diameter in the case of a circular collector is D or the width in the case of a cylindrical collector is W, the resultant flux concentration is given by

$$C = D^2/F^2(\alpha + 2\sigma)^2 \quad \text{(for circular optics)} \tag{8.21}$$

and

$$C = W/F(\alpha + 2\sigma) \quad \text{(for cylindrical optics);} \tag{8.22}$$

or in terms of the concentration in the absence of surface errors (C_O),

$$C = C_O[\alpha/(\alpha + 2\sigma)]^2 \quad \text{(for circular optics)} \tag{8.23}$$

and

$$C = C_O[\alpha/(\alpha + 2\sigma)] \quad \text{(for cylindrical optics).} \tag{8.24}$$

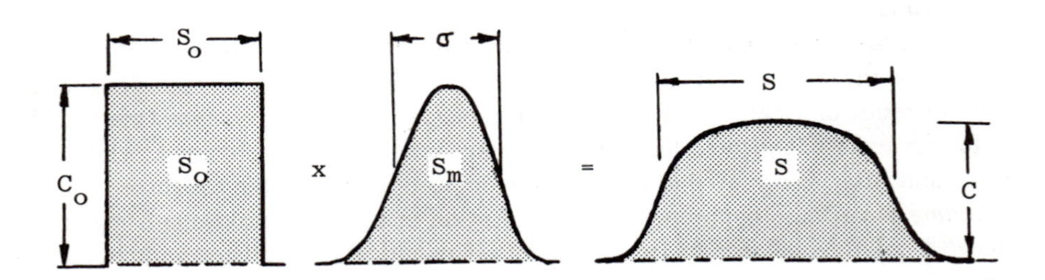

Fig. 8.12 Schematic diagram illustrating the convolution of the solar image (*O*) with the mirror spread function (*M*) to yield the degraded image (*I*).

Equations (8.23) and (8.24) show the basic difference in sensitivity to mirror surface tolerances for the two classes of collector. Cylindrical collectors are less sensitive, while circular furnace-type collectors are much more sensitive. In Fig. 8.13 we show the curves for the ratio C/C_O as a function of σ, where σ is expressed in radian units.

Figure 8.13 indicates that if one were to have the collector reduce the solar brightness flux concentration C by a factor of two, the cylindrical optics would need to have a surface tolerance of $\sigma = 0.005$ rad = 17 arc min, but for the circular optics the tolerance would need to be $\sigma = 0.002$ rad = 7 arc min, almost three times better.

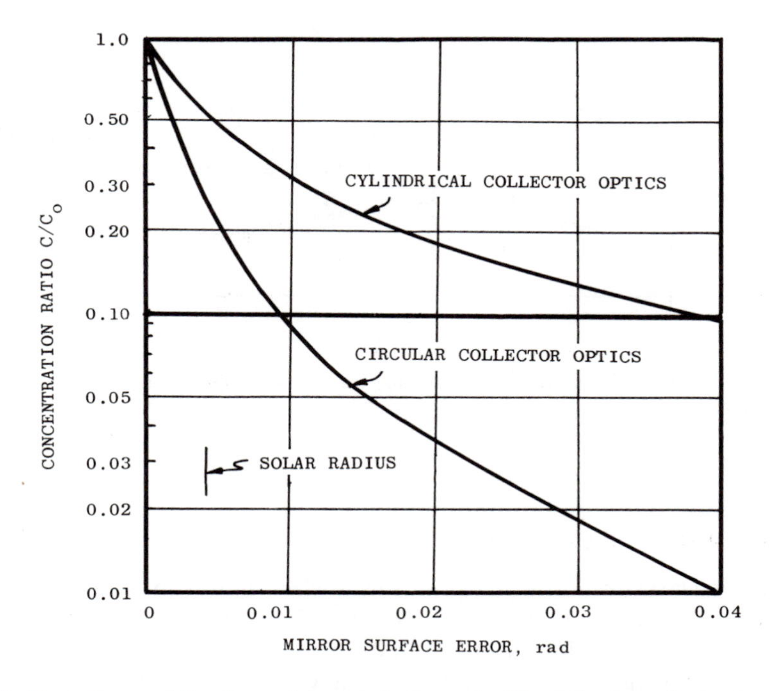

Fig. 8.13 Dependence of the change in concentration ratio (C) relative to perfect optics upon the surface error in the mirror (σ), showing that circular collectors are much more sensitive than cylindrical optics. Note that a mirror error equal to the angle of one solar radius would produce a blur of one solar diameter.

The mirror tolerance is directly related to the cost of forming the mirror. Astronomical mirrors generally have tolerances in the range of 0.04–1.0 arc sec, 500–1000 times better than a typical solar energy collector mirror. Searchlight mirrors have tolerances on the order of 1 arc min and make good solar furnace installations for small-scale experiments. To give a better feel for the size of errors permissible for solar collectors, let us consider the height of a possible error on a mirror. We show such an error in Fig. 8.14. The error is spread over a distance of L cm and has an angular error at the maximum slope of 2σ (so the RMS error is approximately σ). The height of the error zone is therefore given by the expression

$$\epsilon = \sigma L/2. \tag{8.25}$$

For a case where $L = 30$ cm and $\sigma = 0.05$ rad, we find that the height of the error at the halfway point of L is 0.75 mm.

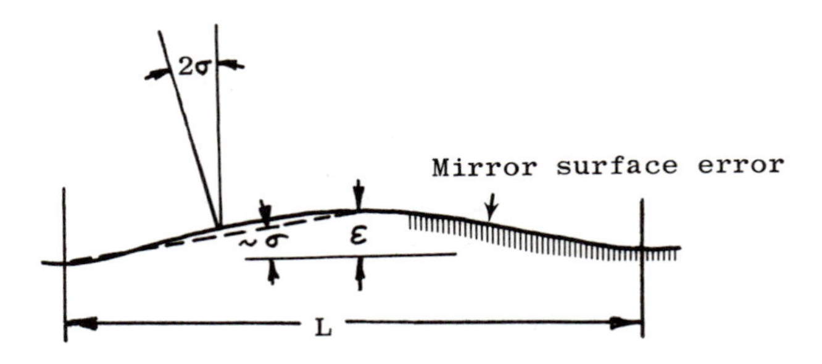

Fig. 8.14 Schematic surface error in a solar collector mirror, where the average slope of the zone is ½ the maximum slope.

8.9 THERMALLY SAGGED MIRRORS

W. B. Meinel (1973) heat-sagged several square glass mirrors in a mold, which were measured by the authors to determine the state of the art of making spherical curved mirrors in this manner. The mirrors were 30.5 cm square, 6 mm thick, and sagged to a radius of curvature of 330 cm. The central portion within a circumscribed circle was of excellent quality, focusing within a circle 10 arc min in diameter, but the portion in the corners of the mirror was much worse—100% of the sunlight was contained within a circle 50 arc min in diameter, but 80% was contained within a circle 30 arc min in diameter. In the sample of three mirrors the corner errors were almost identical, indicating that if the mold had been modified to take into account the errors in the pieces produced, the resultant mirrors would have performed much better, placing 100% of the reflected energy within a circle about 10 arc min in diameter, about 1/3 the solar diameter. It is hoped that subsequent manufacturing experiments will confirm this expected accuracy of sagged mirrors, as this would indicate excellent performance via this avenue for medium-concentration solar collectors ($C = 100$–300).

8.10 LOW-REFLECTANCE COATINGS

All transparent materials reflect some light from their surfaces. This light is generally lost from the solar collector because it travels in the opposite direction from the incoming ray. Any lowering of the surface reflectance of windows and lenses is therefore a direct gain for the collector system.

The amount of light lost by reflection is a function of both the index of refraction of the window material and the angle of incidence of the light. When the ray strikes the surface normal to the surface the equations for reflectance and for transmittance are

$$R = (n - 1)^2/(n + 1)^2 \tag{8.26}$$

and

$$T = 4n/(n + 1)^2 . \tag{8.27}$$

The average reflection from a surface as a function of the angle of incidence was shown for glass in Fig. 8.2. Also shown were the curves for one, two, and three glass plates, and two, four, and six reflections. The maximum contribution from a uniform bright sky to a flat-plate collector comes from a zone of sky 45° from the normal to the absorber, so the light loss at 45° is important. From Fig. 8.2 we note that at 45° the reflection loss is not significantly different from that at normal incidence. The point where window losses begin to be important is beyond a 60° angle of incidence.

Surface reflectance can be reduced by altering the surface index of refraction. The simplest way is to add an evaporated coating one-quarter wavelength of light thick and with an index of refraction approximately equal to the square root of the index of the substrate material. As shown in Fig. 8.15, the quarter-wave layer adds a reflectance amplitude vector equal to the vector from the interface between the surface layer and the substrate. When the two vectors are 180° out of phase, the quarter-wave condition, the vectors cancel and there is no net reflectance.

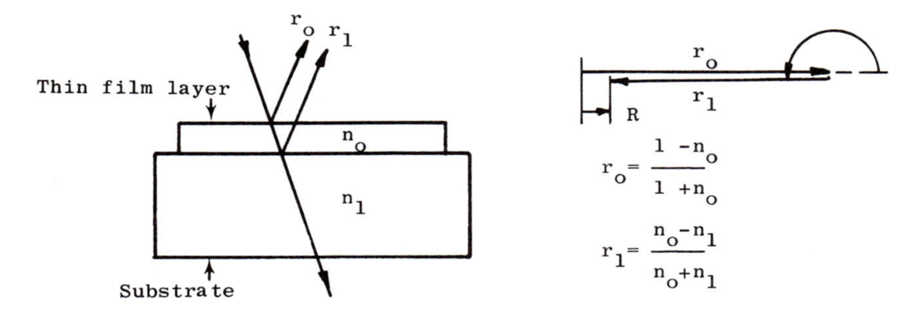

Fig. 8.15 Schematic diagram of the effect of adding a single dielectric layer to a transparent substrate. The condition for zero reflectance is $n_0 = n_1^{1/2}$.

A simple nonreflecting layer is good over only a small range of wavelength because the optical thickness of the layer changes only slowly with wavelength while the wavelength dimension itself changes linearly. Thus the vectors have different phase shifts at different wavelengths and they no longer cancel. The amplitude of the reflectance vectors is half that for each reflection and half that of the original vector from an uncoated surface, so it is clear that the reflectance can build back up to that of an uncoated surface but cannot exceed it, as shown in Fig. 8.16.

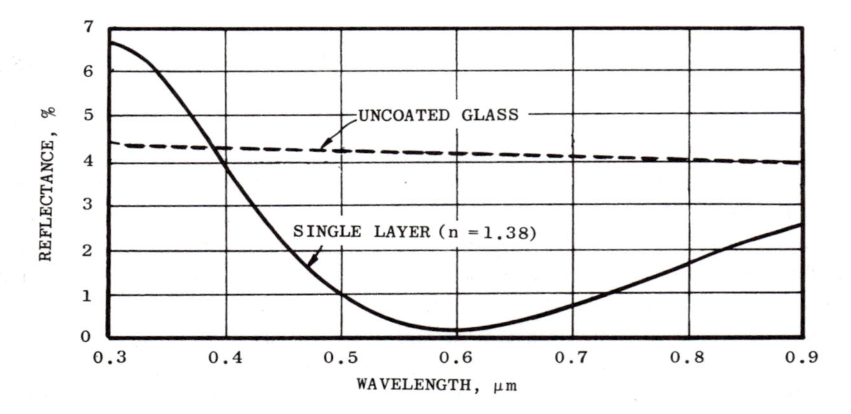

Fig. 8.16 Reflectance versus wavelength for a single layer of MgF_2 over a glass substrate. Note that the reflectance is increased over that for a normal glass surface at short wavelengths. The curve asymptotically approaches glass to the infrared.

The rule is that at half the wavelength of maximum effect for the first minimum the reflectance returns to the original value, followed by another drop for the wavelength one-third the original value, and again at one-fifth. In the infrared the reflectance builds up asymptotically to the normal value for glass at that wavelength. If the minimum is placed in the center of the solar spectrum, at 0.55 μm, the reflectance is unchanged at 0.275 μm in the ultraviolet and is halfway up at about 1.3 μm in the infrared.

With some glass, the effect of nonreflectance can also be obtained by chemically treating the surface with hydrofluoric acid. The process known commercially as Magicoat involves the etching of glass with a solution of hydrofluoric acid fully saturated with silicate ions, so that the silicate part of the glass remains essentially unetched. The removal of the other ions of the glass then leaves the network structure of silicate ionic groups to form a porous and reasonably durable surface. The surface, because of its skeletal structure with many voids, acts like material with a low index of refraction. Thus when the thickness of the etching is properly controlled, the surface of the glass is rendered nonreflecting. One practical drawback to such surfaces is that they readily adsorb impurities such as grease from fingerprints and other hydrocarbons, and hold them so tenaciously that they can hardly be cleaned.

Sophisticated nonreflecting coatings can be applied to glass to broaden the region of low reflectance, achieving values as low as 0.3% over the visible spectrum. In general these coatings are vacuum deposited and consist of at least two layers. The cross section of a typical coating of this type is illustrated in Fig. 8.17. The layer placed directly on the substrate has a higher index than the substrate, giving more latitude for the selection of the upper layer so that durability and compatibility can be optimized without seeking only those materials with the lowest possible index, as are needed for good performance with single-layer coatings. Three-layer coatings, as in Fig. 8.18, provide excellent low-reflection behavior over a wide spectral region.

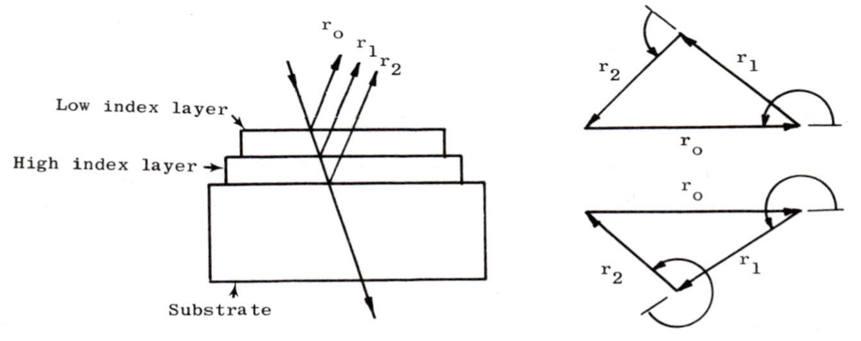

Fig. 8.17 Schematic diagram for two nonreflecting layers. Under ideal conditions one can obtain two zero reflectances, but usually three layers are needed to yield the correct relationships between phase shifts and reflectance amplitudes.

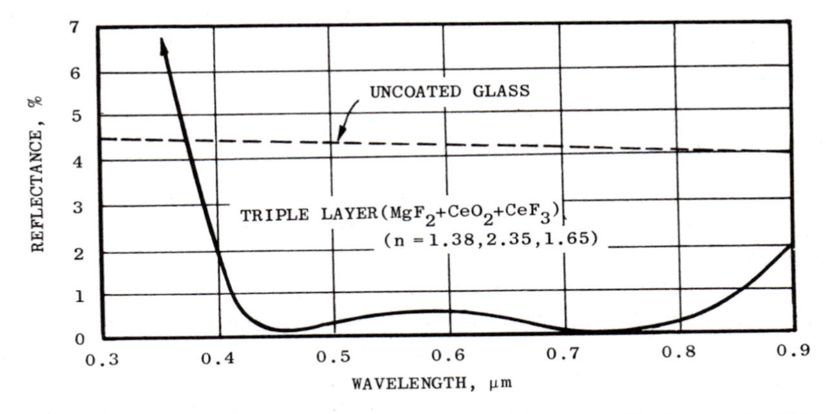

Fig. 8.18 Reflectance versus wavelength for a triple coating on glass. The increased width of the nonreflected region is at the expense of a more rapid rise to high reflectance in the blue region. Infrared reflectance for most coatings remains below normal.

Multilayer nonreflecting coatings are not good over more than a specific range of wavelengths. You will note that a manufacturer never gives the reflectance curve beyond the region in the visible where the coating performs well. You will also note that the reflectance is generally rising rapidly at the ends of the design region, as shown in Fig. 8.18. The reflectance beyond the design region actually rises *above* that for normal glass, and may be two or three times as high in the infrared. If this reflectance were far enough into the infrared, the effect would be beneficial in some important solar applications, such as flat-plate collectors, but it occurs in the near infrared (1.0–3.0 μm) where one would still like to collect the sunlight.

Nonreflecting coatings are used or not used in solar applications strictly on the basis of their cost effectiveness. Multilayer coatings are too expensive for general use. Even the vacuum-deposited single-layer coatings are of questionable value when it comes to cost effectiveness. The chemical etch treatment may be the only process that is economical at present, but very large-volume uses, as might well be possible in solar applications, could make the vacuum-deposited coatings competitive.

8.11 INFRARED REFLECTANCE COATINGS

Infrared reflectance coatings can in principle be useful for inhibiting infrared losses from the absorber in collectors where a window is used. These coatings can consist of a single layer of material that has intrinsically high infrared reflectance, such as tin oxide (SnO_2) or indium oxide (InO_2). We discuss the properties of these selective materials in more detail in Chapter 9. One can also make infrared reflectance surfaces by using multilayer techniques, but only those where the amplitude vectors add rather than cancel. In general, it is too expensive to obtain high infrared reflectance by means of multilayer techniques in solar applications since fifty to several hundred layers may be required. One can, however, benefit from the placement of a simple single layer of low-index material like magnesium fluoride (MgF_2) overlying tin oxide or indium oxide to lower the high intrinsic reflectance from the high index of refraction of these semiconductors.

8.12 BARE METAL REFLECTANCES

Aluminum is the one bare metal that can be considered as suitable for solar collector reflector surfaces because its oxide is relatively transparent. In general, bulk aluminum surfaces are soft and do not yield good lifetime performance without some sort of deliberate overcoating. The best reflective commercial aluminum is marketed under the trade name Alzak. This material is electrolytically polished and then overcoated with a thin layer, about 1 μm, of anodized aluminum (Al_2O_3). This layer does not significantly reduce the solar reflectance, but it does introduce some weak absorption bands in the infrared, as shown in Fig. 8.19. The latter is desirable because the infrared absorption increases thermal emittance and tends to keep the mirror cooler under sunlight.

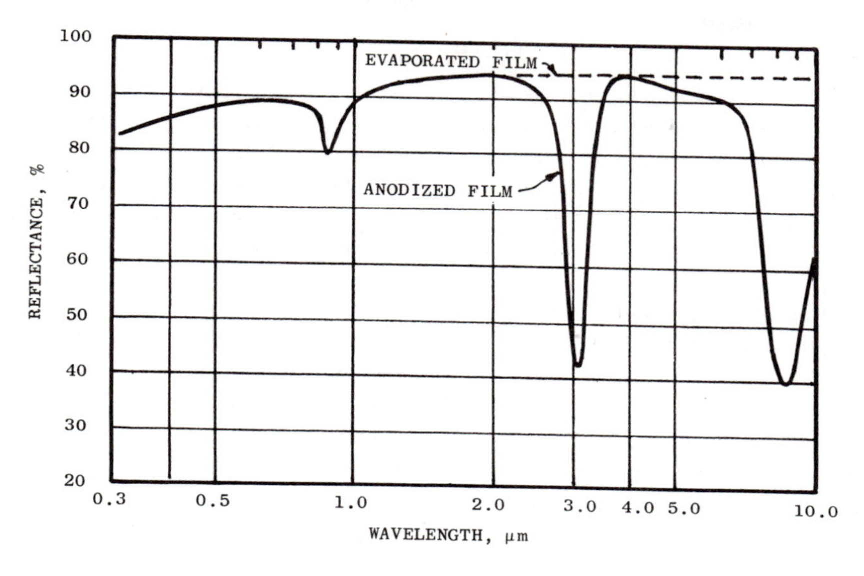

Fig. 8.19 Reflectance curve for commercial polished/anodized aluminum (Alzak) having a 1.4-μm thickness of Al_2O_3.

All other bare metals, except gold, rhodium, platinum, and similar noble metals, lose their new reflectances quickly and are not suitable for solar applications.

8.13 CHEMICAL REFLECTIVE COATINGS

Chemical coatings of silver and aluminum are possible. Silver has been used for years for highly reflective infrared mirrors, but the rapid reduction in solar reflectance of silver due to tarnishing from silver sulfide (AgS) and silver oxide (AgO) reduces its use in solar collectors. Second-surface silvered mirrors are the best possible type of second-surface mirror. When properly protected by back coatings the silver is stable. Some silver second-surface mirrors are electroplated with copper and finished with a good enamel. The problem to avoid is that of water creeping between the silver and the glass or plastic and separating the reflective film. Chemical-deposited silver films, having been laid down in solution, are particularly bad in regard to subsequent separation of the silver film by water because a molecular layer of water apparently remains between the silver and the substrate.

Commercial processes have been developed for placing aluminum on glass and plastic substrates, but proprietary considerations have restricted this process from general application. The substrate in this case must be able to stand considerable heating during the process, whereas the silvering processes, described in every chemical handbook, are done at ambient temperature.

8.14 VACUUM REFLECTIVE COATINGS

Most metals can be deposited by vacuum evaporation or sputtering. Aluminum is particularly easy to deposit by evaporation because of its low melting point and high vapor pressure in liquid form. One can deposit it simply by heating a filament on which small loops of aluminum wire are hung, or by using an electron beam to heat a lump of aluminum in a refractory crucible. Silver is more difficult to evaporate, in part because of its tendency to agglomerate and produce a reflective surface well below the best possible.

The reflectance curves for different types of silver coatings are shown in Fig. 8.20. Silver has the highest solar reflectance of any metal, but unfortunately this high reflectance is not usable because of the tendency of silver to oxidize and to form silver sulfide, and both the oxide and the sulfide are highly absorbent at the solar wavelengths. The infrared reflectance of tarnished silver remains high, and a poor-looking silver mirror can be an excellent infrared mirror.

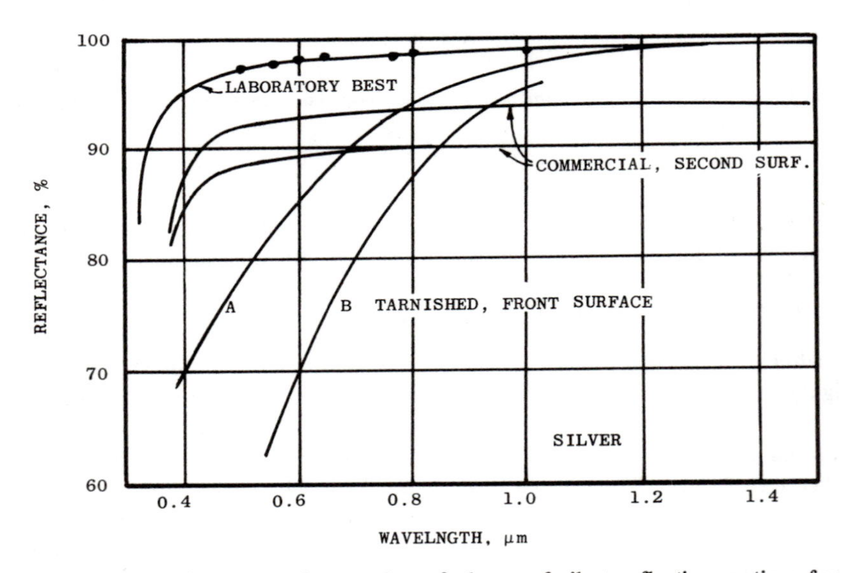

Fig. 8.20 Measured reflectances of a number of classes of silver reflective coatings for solar applications. The initially high visible reflectance of silver quickly diminishes, but second-surface silver is stable. All second-surface coatings have lower reflectance than air surfaces owing to the index of refraction of the substrate on which the coating is placed.

Silver can be protected if one places it on the rear of a transparent substrate like glass or plastic. The glass, plus overcoatings of electrolytic copper and paint on the rear surface, protect the silver film from chemical attack. Front-surface silver mirrors can be overcoated with magnesium fluoride and other transparent oxide and fluorides to slow the tarnishing, but these thin coatings still allow reactive gases to diffuse to the silver, producing tarnish.

Vacuum- or sputter-deposited silver adheres to the glass substrates better than chemical-deposited silver, hence these coatings are not as sensitive to separation by water creeping in from the edges of the mirror.

The reflectance curves for different types of aluminum are shown in Fig. 8.21. One frequently sees curves for these two metals without being cautioned that the results quoted are usually the best that can be achieved under laboratory conditions. In solar applications one must design a system that can use the quality to be expected in commercial manufacture. For this reason we show both the laboratory values and the commercial values. The two vertical bars in Fig. 8.21 are the average reflectances for incandescent light for average evaporated aluminum and commercial Alzak.

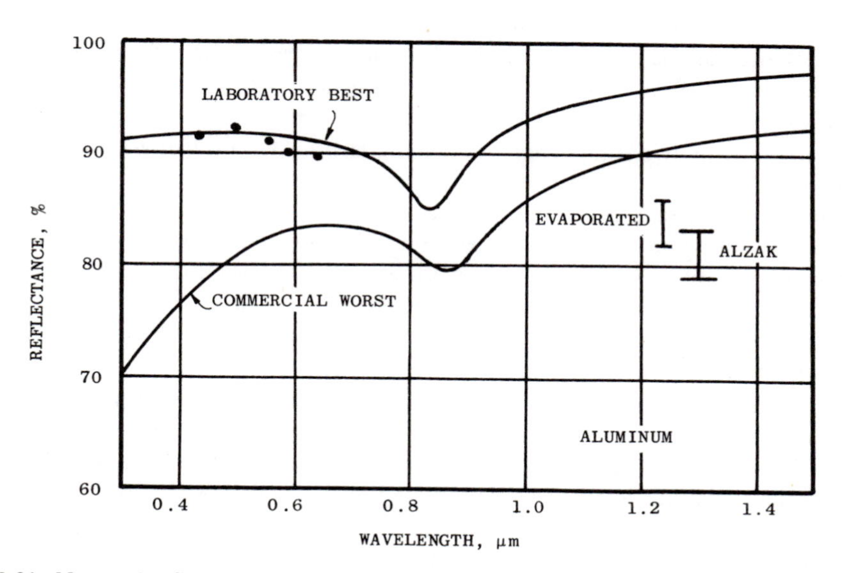

Fig. 8.21 Measured reflectances of a number of types of aluminum reflective coatings for solar applications. Practical reality indicates that a value of 0.83 be used in solar collector performance evaluation. Vertical bars indicate average reflectance over the incandescent light spectrum.

8.15 MIRROR PROTECTIVE OVERCOATINGS

Front-surface mirrors of aluminum must be overcoated to prevent their gradual degradation when exposed to air with its attendant aerosols. In solar energy applications the cost of recoating a mirror every three or four years will be an added financial burden of significant magnitude. One therefore looks to the technique of overcoating a fresh mirror surface to preserve its high reflectance over many years.

The simplest overcoating for aluminum mirrors consists of a half-wavelength of a dielectric material. Magnesium fluoride is used for some applications since it makes a very hard layer, as on camera lenses. It can also be used for protecting mirror surfaces from abrasion during cleaning, but the adhesion is poor unless the substrate mirror can be heated to 250–300°C. Another common single-layer protective coating is silicon monoxide (SiO), but its behavior is sensitive to deposition conditions, especially at the short wavelength end of the solar spectrum.

A good silicon monoxide overcoating must be deposited slowly, exactly the opposite to the conditions for obtaining the best aluminum films. The reason for slow deposition is that the coating should become fully oxidized, so that although it is deposited from a crucible containing silicon monoxide the film is actually closer to silicon dioxide. If fast deposition is applied, the silicon monoxide film looks distinctly yellow (bulk silicon monoxide appears quite black). A good overcoating of SiO_x, where x is between 1.0–2.0, can boost the reflectance of an aluminum mirror by about 6% over the base reflectance, which means increasing the reflectivity of a good coating from 90–96%. The cost of mirror overcoatings is reasonable in view of the extension in operating lifetime that is provided.

An excellent two-component overcoating is provided by half-wave layers of dielectrics of differing index. One type in commercial use is a layer of magnesium fluoride topped with a layer of cerium dioxide (CeO_2), a high index material. This coating combines the low index of magnesium fluoride of 1.38 with the high index of cerium dioxide of 2.3–2.4, depending on evaporation conditions. The reflectance boost is 6–7% and such coatings are very durable. If two stacks of magnesium fluoride and cerium dioxide are added, the net reflectance can approach 99% with a top quality aluminum base film.

Cerium oxide films are also sometimes combined with other low index materials like aluminum oxide (Al_2O_3) ($n = 1.60$), silicon dioxide ($n = 1.46$), and silicon trioxide (Si_2O_3) ($n = 1.55$). Titanium dioxide (TiO_2) ($n = 2.58$) is also a possible substitute for the high index material, as is thorium dioxide (ThO_2) ($n = 2.20$).

In Fig. 8.22 we show a typical reflectance curve for an aluminum film freshly prepared but uncoated, with the same film overcoated with a layer of magnesium fluoride and cerium dioxide. The thickness of the coating is such that the maximum boost of about 6% is obtained at $\lambda = 0.55\ \mu m$. The boost drops off on each side of this wavelength, crossing the original reflectance at 0.4 μm, making the reflectance poorer at shorter wavelengths. Because solar applications need good reflectance over a wide wavelength interval, the boosts given by overcoatings are on the average less than half the maximum boost described above. Thus a coating that boosts the optimum wavelength by 6–7% actually boosts the integrated solar spectrum by less than half that amount, affecting the cost effectiveness of the overcoating. The reflectance of silver, as well as of other metals, is lowered by about 5–10% when the silver is placed as a rear-surface mirror. This lowering is because the reflective layer is in contact with a material of much higher index of refraction than air. One can, if the cost permits, add

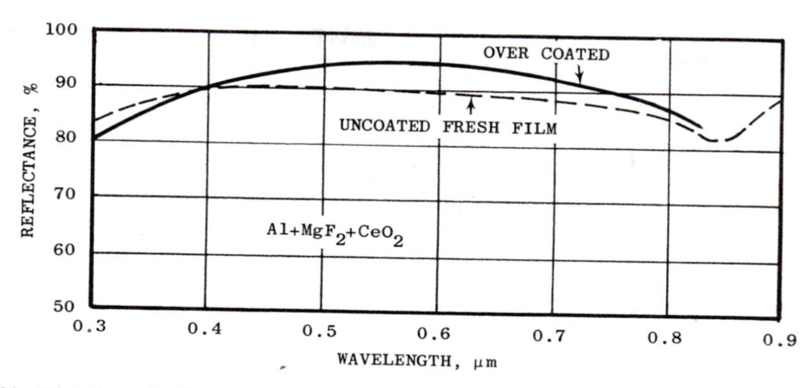

Fig. 8.22 Variation of reflectance with wavelength for a fresh aluminum film and one overcoated with a half-wave of magnesium fluoride and cerium oxide to boost reflectance and provide protection for aging of the aluminum film.

other coatings between the silver and the glass to tune the interface so that the in-air performance of silver is reestablished.

An overcoating is essential in solar applications for reasons other than boost of reflectance. The objective is to lengthen the lifetime of a reflective surface by years. The overcoating is therefore desirable to keep the reactive components of the wet atmosphere from attacking the aluminum film. We are therefore interested in obtaining a coating with maximum freedom from pinholes and micro-defects, which can open avenues for chemical attack of the film. When properly applied, overcoatings can provide a significant degree of protection, but it is harder to get this quality of coating as the size of the piece being coated increases. Manufacturing for solar applications therefore depends upon establishing rigorous quality-control procedures without at the same time driving the cost out of line.

Overcoated mirrors can still show degradation when exposed to the open environment. Laboratory-quality samples can show surprising resistance to weathering, but one must have commercial-quality coatings, cheaply applied, that are resistant. The degradation of the ability of the overcoating to protect the mirror is caused mainly by the micro-defects and pinholes scarcely visible to the unaided eye. These small defects allow the surface film of aqueous solution of aerosols to reach the metal film and both react with the film and creep between the film and the substrate, causing serious degradation to the reflective properties of the coating. These small defects are especially hard to eliminate in any large-volume production without running up the cost of the coating beyond that sustainable by the benefits to be derived from the solar collectors.

McKenney (1973) analyzed aluminum coatings that had been commercially prepared and overcoated with the best techniques available as of 1970. Three years of continuous exposure to the elements in the gamma ray telescope on Mt. Hopkins, Arizona, had degraded their reflectance to about 40%. The reflectance improved to 60%

after they had been carefully cleaned to remove surface deposits. These old coatings had also become semitransparent. Examination under the microscope showed that this transparency was due to myriads of tiny holes where the film had receded, apparently because of chemical reaction of the solutions with the aluminum. One must therefore be alert to the possibility that mirrors in solar collectors may suffer damage that tests of laboratory samples do not reveal.

One of the best overcoatings obtainable is on bulk glass itself. Glass can be evaporated onto metal films, but the same problems arise as when pure dielectrics are used. Bulk glass for second-surface mirrors is excellent and the only safe answer for maintenance of high mirror reflectance over many years. There is, however, a penalty with second-surface mirrors in that the reflectance of the metal-glass interface is lower than that of a metal-air interface. The front-surface reflection of the glass, a nuisance in most optical systems, is not a problem since this energy also reaches the absorber. Second-surface mirrors on plastic substrates can also make a durable mirror. The disadvantage here is that *twice* the solar flux passes through the plastic, tending to accentuate solar damage of the intrinsic plastic material.

Sheldahl Corporation has developed a laminated film mirror consisting of silver deposited on a Mylar substrate and overcoated with flowed-on acrylic plastic, polymerized in place. This mirror exhibits reflectance of about 90% and appears to have a reasonable lifetime as a solar collector mirror. The laminated structure permits the use of a substrate material with desirable mechanical properties and an overcoating with the desired optical lifetime and hardness but without good mechanical properties.

Second-surface mirrors have a property that can be useful in certain applications where the heat load on the mirrors is high. One example of such a heat load is in the case of the hemispherical concave stationary-mirror collector (Section 7.12), where the "line image" of the sun proceeds downward from the paraxial focus until it intersects the mirror surface. The solar brightness concentration can exceed 100 at this point. A second-surface mirror can remain cool under this situation, whereas a front-surface mirror would get very hot. Two effects aid this cooling. First we have the basic advantage that the glass is opaque to the thermal infrared, and hence reradiates readily the energy load it absorbs. A front-surface mirror, on the other hand, has very low infrared emittance and the trapping of absorbed flux results in elevated temperatures. The second advantage, applicable to the hemispherical mirrors of Section 7.12, is that the rays contributing to the heat load have already passed through a double path of glass at the first reflection, absorbing some of the wavelengths that would otherwise add to the heat load at the second point of intercept with the mirror surface.

This prefiltering of the sunlight has been used in the past to reduce subsequent heat loads. The combination lens-mirror collector furnace design by Lomonsov (Section 5.11) uses dielectric material for the lenses so that the same dielectric can be used as a sample container at the focus, the wavelengths that would seriously heat the container having already been absorbed by the lens.

8.16 OPTICALLY SHAPED SURFACES

An optical shape that performs somewhat like an optical surface coating has been proposed by Tabor (1967) to effectively increase the absorptance of an optical surface in a collector. For example, some selective absorber coatings do not have as high a value of absorptance as one would like. If this coating is placed on a substrate with inclined faces, the sunlight reflected from the selective coating would have a second chance to be absorbed.

The range of surface structures offering a second reflection possibility is considerable. A simple "vee" with plane surfaces will accomplish the task for certain angles of arrival of the sun rays. In Fig. 8.23 we show some angles and the range of double reflection. It can be seen that the angle of arrival is then such that the reflection at the first surface is at normal incidence and the ray returns without a second reflection. The best shape is for a curved profile to the groove, the Gothic "vee." This vee is constructed by taking the crest of one groove as the center of curvature for the opposite face. This construction assures that rays arriving from any angle over 180° will have a second reflection.

The basic problem with using structured surfaces to improve absorptance is their cost. A second problem is that the total surface for reradiating infrared heat loss is increased, partially offsetting the low emittance coatings.

Structured surfaces have recently been explored to increase absorptance for use with silicon solar cells. The surface of the silicon crystal is etched to develop a pyramidal structure a few tens of microns in size. The etched surface appears quite dark even without a nonreflecting coating on the silicon. The structure and its regularity are quickly indicated by turning the wafer and noting the regular angles where direct surface reflections are seen.

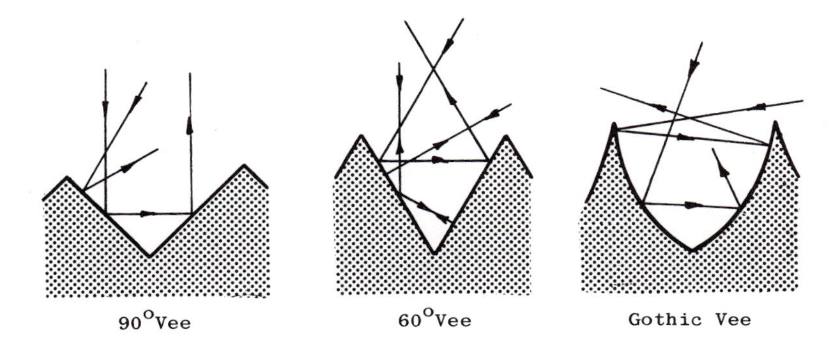

Fig. 8.23 Shaped surfaces on the absorber substrate that increase the net absorptance of the coating through multiple reflections. Note that as the angle increases the angle through which multiple reflections occur increases. The Gothic "vee" ensures two reflections from any angle of incidence.

8.17 PROBLEMS

8.1. Calculate the effective transmission of the window of a flat-plate collector where the index of refraction is 1.52 and the variation of reflection with angle is as given in Fig. 8.2. What is the ratio of the effective loss to the loss at normal incidence? Considering that the absorber is more absorbent at normal incidence than at large angles, how would the ratio change and what would you estimate the change to be?

8.2. Estimate the transmittance of the several materials whose transmission curves are given in Figs. 8.3–8.6 for thermal radiation at a temperature of 400°K.

8.3. Give a concise engineering discussion about the relative reaction of a transmitting and a reflecting surface to the presence of dust defects, such as formed by dirty rainfall. Does the type of absorber or collector design affect the consequences of dust? If so, in what ways?

8.4. The diagram in Fig. 8.17 is schematic only. Is there a further restriction to the lengths of the relative vectors so that a second minimum occurs? Assume that the phase angle is linearly related to the optical thickness of each layer. Discuss qualitatively the case where the upper surface has an index of 1.4 and the lower surface an index of 2.8. Draw the vectors to scale to within the degree of approximation needed to show the nature of the error in Fig. 8.17.

8.5. In making an engineering decision on choice of mirror substrate, reflective coating, and presence or absence of overcoatings, what, in summary form, are the factors you would consider? Would the ease of replacement affect your ordering of the factors and possibly also your recommendation?

Part 3
Absorption, Transfer, and Storage

Chapter 9
Selective Surfaces

9.0 INTRODUCTION

The problem of minimizing heat losses from a solar collector brings us to an examination of the optical properties of the absorber surface and the transparent windows. It is clear that we want as much radiant energy from the sun to reach the absorber as possible while at the same time reducing to a minimum the thermal infrared radiant energy escaping from the hot parts of the collector. Optical properties, which vary widely from one spectral region to another, produce what is termed *selectivity*.

In this chapter we use two terms in a colloquial sense, so it is important to define these terms in both their exact and their colloquial meanings.

Absorptance, α, is, strictly defined, the fraction of incident light of a given wavelength that is absorbed when light strikes an absorbing surface. In the colloquial sense in regard to selective surfaces, absorptance is the average of the absorptance at each wavelength weighted according to the intensity distribution with wavelength of sunlight. The absorptance of the same surface for incandescent light, for example, would be significantly different.

Emittance, ϵ, is, strictly defined, the fraction of the emittance of a perfect blackbody at a given wavelength emitted by a heated surface. In the colloquial sense in regard to selective surfaces, ϵ is the average of the emittance at each wavelength weighted according to the blackbody intensity distribution with wavelength at the temperature of the surface. Selectivity then refers to the wavelength variation of absorptance and emittance over the broad spectral region from 0.3–20 μm.

Many researchers have examined the prospects of selectivity, but it was Tabor (1956) who really brought the capabilities of selective surfaces into practical use. In this section we examine the many ways in which selectivity arises and describe the optical characteristics of selective surfaces. The effects of selective surfaces on solar collector performance will be considered later.

Selectivity in the broadest definition refers to the use of the separation of the input solar spectrum from the thermal infrared emitted by the collector and by the environment to emphasize desired effects. In Fig. 9.1 we reiterate the basic spectral variations that we will be concerned with.

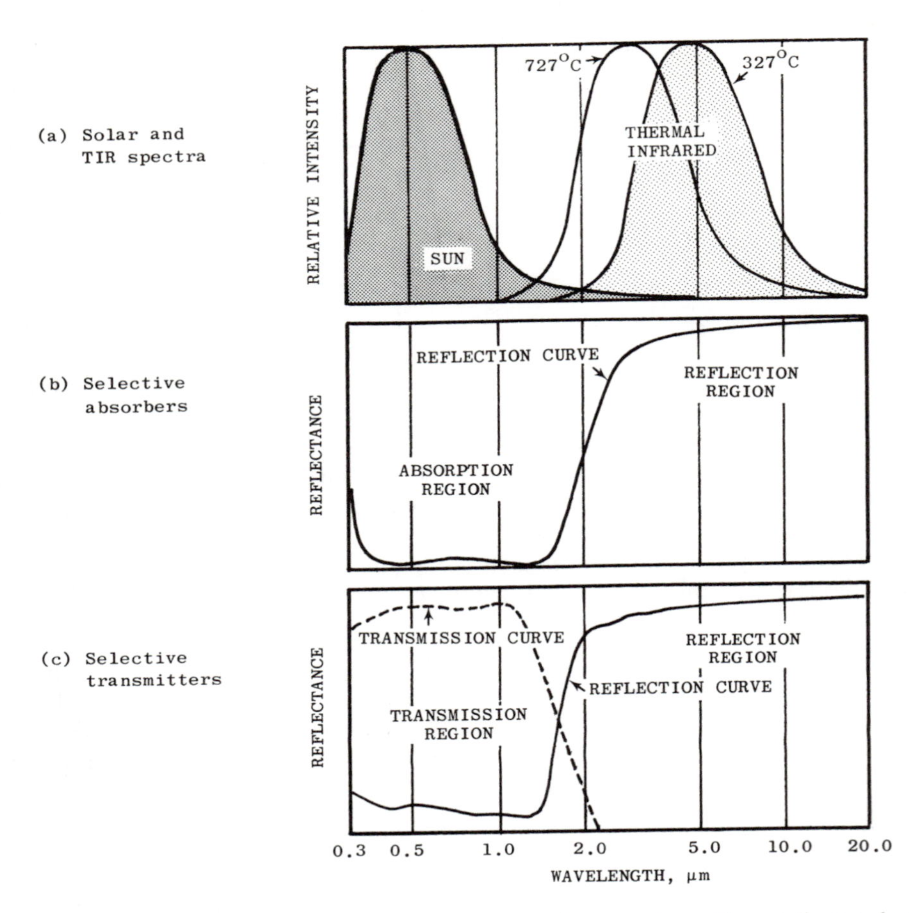

Fig. 9.1 Three diagrams illustrating the basic physics of selective surfaces. The top diagram shows radiant energy curves for the sun and for a hot surface radiating mainly in the thermal infrared; the middle diagram shows a typical curve for a selective absorber; the bottom diagram shows a typical curve for a selective transmitting surface.

There are two basic types of selective surfaces of use in solar collectors:

1. *Selective absorbing surfaces,* where the surface is black to sunlight, making the transition from absorptive to reflective in the region between 1.5 and 3 μm. By Kirchhoff's law a reflective surface is a poor emitter, the value of the emittance being $\epsilon = 1 - r$, where r is the reflectivity of the surface. A highly selective surface is therefore one that has the highest possible reflectance in the thermal infrared. The measure of selectivity is the ratio of the absorptance for sunlight divided by the emittance for thermal infrared at the temperature of the projected

use of the selective surface. This ratio, α/ϵ, can, therefore, vary with temperature, depending on the exact variations of both absorptance and emittance with wavelength.

2. *Selective transmitting surfaces,* where the surface is transparent to sunlight, making the transition from transmissive to reflective in the region between 1.5 and 3 μm. The function of such surfaces is to let sunlight into a collector but to inhibit the loss of thermal infrared (TIR) from the absorber.

There is a third surface of use in solar collectors, but its selectivity is not always in the right direction. This surface, generally referred to as a *nonreflecting film,* is used on the windows of a collector to avoid reflection losses for the incoming sunlight. Nonreflective surfaces lower the reflection at the air-glass interface in only about one octave of the spectrum, often enhancing the reflection in the region to the blue and the red of the nonreflected region, as discussed in Section 8.10.

9.1 LIMITING BEHAVIOR

Selective coatings can never attain the highest levels of absorptance possible with nonselective coatings. This limitation is basically set by the fact that the solar spectrum extends into the thermal infrared with significant amounts of energy out to the 3- to 4-μm region. The spectral distribution of sunlight for one air mass was shown in Fig. 2.1. If we use the values for the sun for two air masses, as tabulated by Moon (1940) and listed in the *AFCRL Handbook* (1965), and assume a cutoff at a given wavelength, we obtain the two curves shown in Fig. 9.2. The continuous curve is for

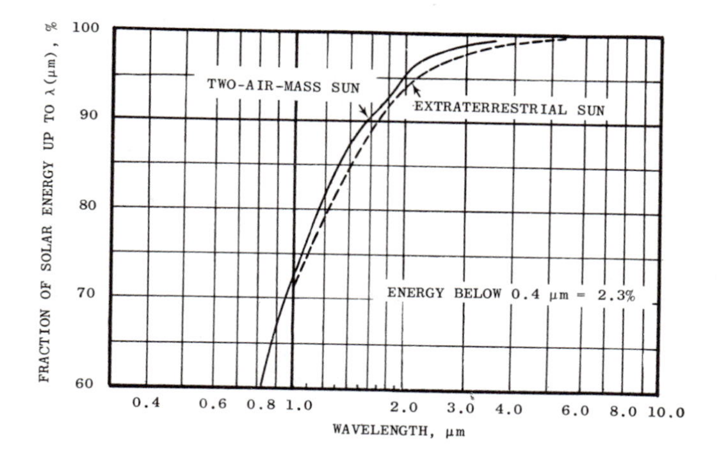

Fig. 9.2 Graph showing the percentage of the energy of sunlight lying to the blue of a given wavelength. The difference between the two curves is caused by atmospheric absorption bands of carbon dioxide and water vapor.

two air masses and differs from the dashed curve by the effect of the infrared water vapor and carbon dioxide absorption features from the atmosphere. This two-air-mass curve shows that if we want to obtain 90% of the solar spectrum the cutoff must be at least at 1.6 μm, in which case the absorptance for all shorter wavelengths must be 100%.

In Fig. 9.3 we show two curves of the type one encounters in selective absorbers, showing how sensitive the total absorptance is to the position and steepness of the change from absorptive to reflective behavior. Figure 9.3 is useful for providing a way of instantly estimating whether the claims for total absorptance and the spectral curve are self-consistent for any particular selective absorbing coating.

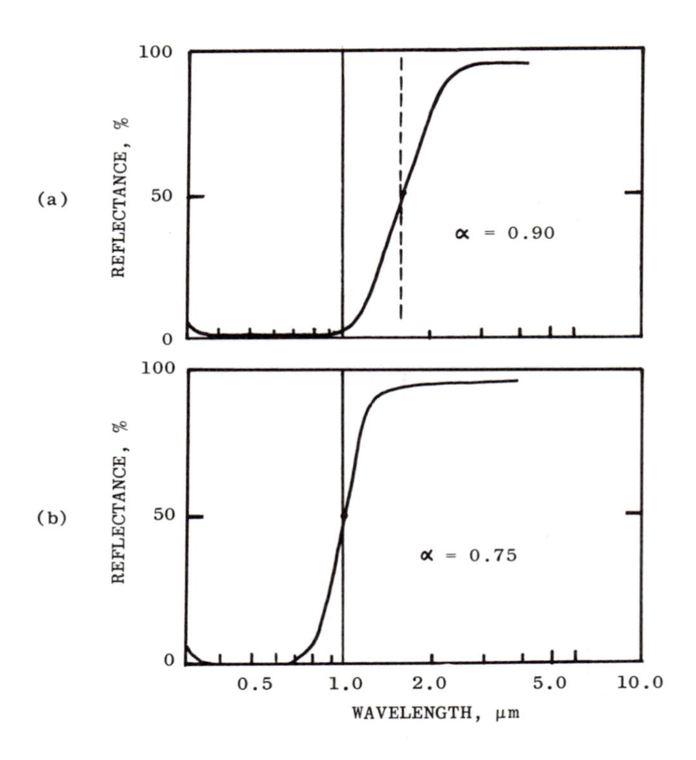

Fig. 9.3 Two possible transitions from absorptive behavior to reflective behavior, (a) at 1.5 μm (dashed line) and (b) at 1.0 μm (solid line), showing how critical the cutoff wavelength is to obtaining a high solar absorptance.

9.2 DISTORTED-WAVELENGTH GRAPHS

Evaluation of the efficiency of a spectral curve for a selective surface is easier to visualize when a distorted-wavelength scale is used for the abscissa of the plot. In this case the wavelength distance interval is proportional to the solar energy falling within that interval. This procedure widens the scale in the visible spectrum and greatly compresses it in the TIR. A typical distorted-λ graph is shown in Fig. 9.4.

It is necessary to use at least *two* distorted-λ graphs for the presentation of selective coating performance. The first graph, in Fig. 9.4, is for the solar spectrum, and is used to estimate the absorptance of the coating. The second graph, in Fig. 9.5, is for the thermal reemission temperature. The sample shown is for 27°C. For most selective surface studies in solar energy the relevant graph would be for a higher temperature, depending on the desired operating temperature of the system that is to use the selective coating.

The fact that two graphs are needed for each selective surface curve is one reason that such curves are not regularly used in daily research activities. They are most convenient in preparing diagrams for nontechnical audiences to emphasize the large values of absorptance and low values of emittance one can obtain through use of selective surface coatings. In this book we retain the logarithmic wavelength scale in all the wavelength diagrams since they can then be translated into any distorted-λ graph as the occasion demands.

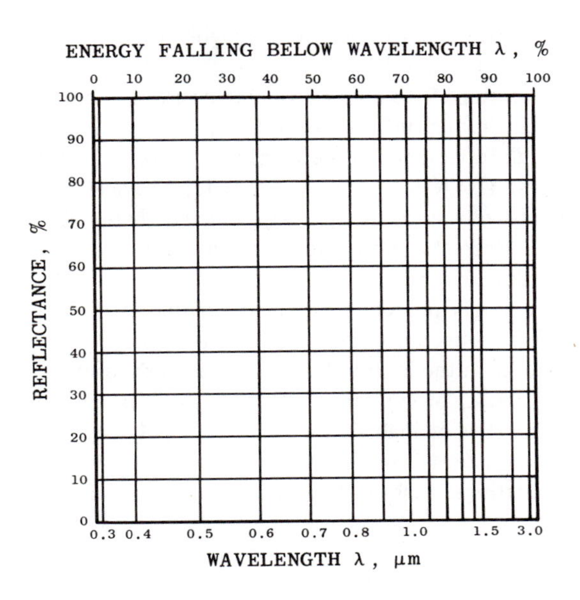

Fig. 9.4 Distorted-λ graph for solar radiation, from Hass (1964).

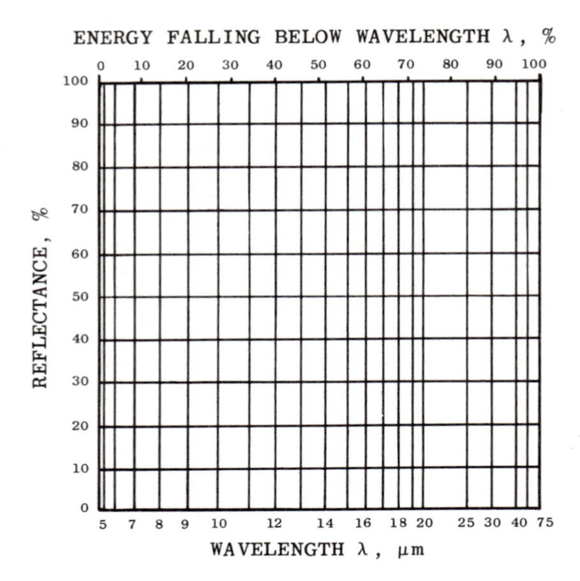

Fig. 9.5 Distorted-λ graph for TIR radiation at 27°C, from Hass (1964).

9.3 METAL REFLECTANCES

The behavior of selective absorber surfaces is highly dependent upon the optical properties of the opaque metal film that serves as the base layer of the interference stack or the tandem stack (Sections 9.13 and 9.16). One desires the highest reflectance, or lowest emittance, in the thermal infrared (TIR). It is only of secondary importance that the metal have a high reflectance in the visible also. Gold (Au), silver (Ag), and copper (Cu) have the highest TIR reflectances, but they tend to agglomerate into small islands of metal on the substrate and, in so doing, change their reflectance and emittance, especially when the effect of agglomeration is enhanced by the superposition of a layer of high refractive index.

The need for a stable reflective film at the temperature of operation makes some metals of lower TIR reflectance important. These metals include aluminum (Al), which is easy to handle in the evaporator and particularly useful when the temperature to be used is under 250°C. Molybdenum (Mo) has been extensively used when the film must retain its properties at high temperature, but only in the absence of air, which causes the film to oxidize. Nickel (Ni) is also used as the base layer in some electroplated multilayer coatings, such as the black nickel selective coatings developed by Tabor (1955).

We present the specular reflectance for the several metals of low melting point and high reflectance in Fig. 9.6, and for the several metals of high melting point and high reflectance in Fig. 9.7.

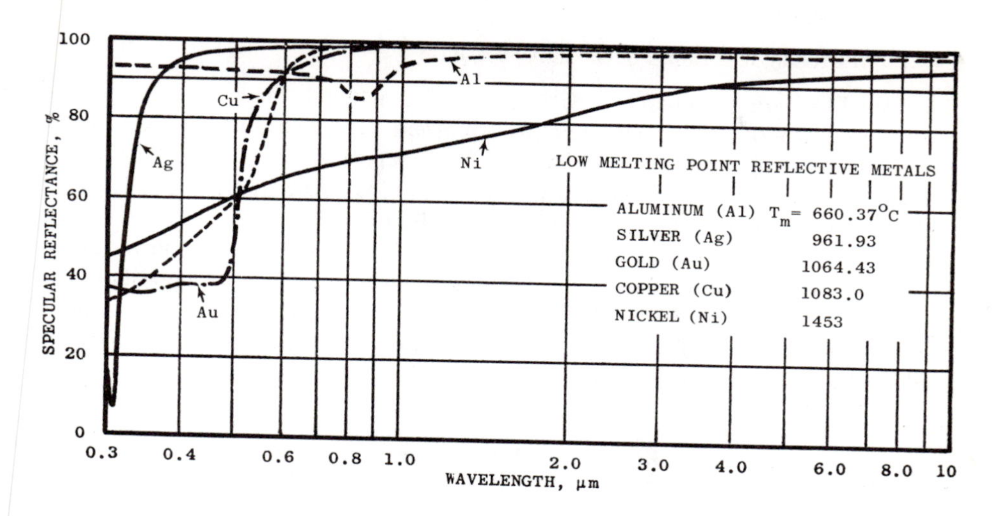

Fig. 9.6 Specular reflectance curves for a number of low-melting-point metal films useful in solar energy applications.

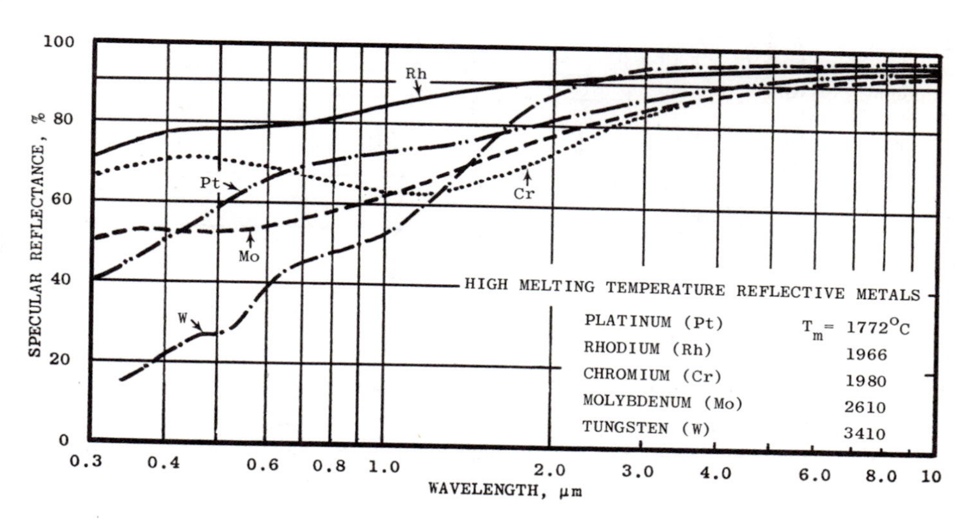

Fig. 9.7 Specular reflectance curves for a number of high-melting-point metal films useful in solar energy applications.

Reflectance is often used to evaluate selective coatings because it is easy to measure. The actual factor of importance is emittance, but emittance is difficult to measure accurately, especially when its wavelength dependence is desired for physical analysis of a coating behavior.

9.4 TEMPERATURE LIMITS OF MATERIALS

The desired operating temperature determines which materials can and cannot be used in a selective absorber stack. On the basis of current experience, we can assign probable limits to temperature as indicated in Table 9.1. The heavy shading indicates temperatures where the metal reflective layer will probably be stable in air. The light shading indicates temperatures where the metal will probably be stable in vacuum. The heavy bar indicates the melting point of the metal. The reason the limit in vacuum is on the order of only 60% of the melting temperature is that the ionic mobility of the metal ions and ions from adjacent layers can readily cause diffusion, altering the basic reflectance of the layer. These limits are not absolute because combinations must be avoided where eutectics, for example, would form.

Table 9.1 Thermal Limits for Several Materials

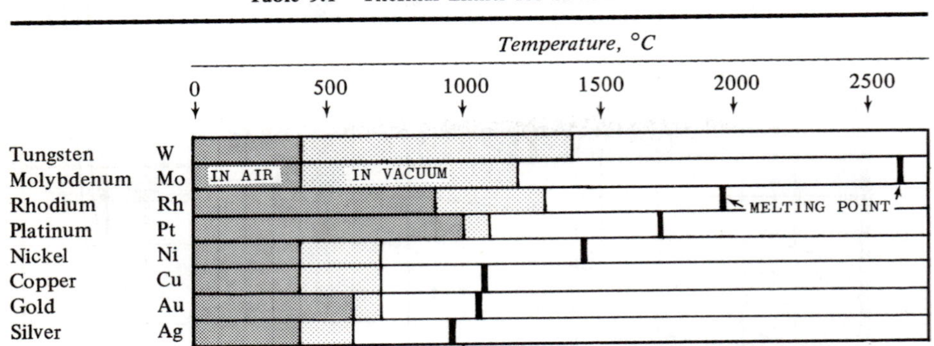

9.5 EMITTANCE OF METAL SURFACES

The emittance of a bare metal surface is of basic importance in the matter of selective surfaces. In the case of both tandem stacks and interference stacks (Sections 9.13 and 9.16), the underlying metal "shows" through in the TIR, and the lowest emittance a selective coating can have is set by the emittance of the bare metal used for the base reflective layer.

The angular variation of emittance of a bare metal (Section 9.6) shows "lobes," where the emittance is higher than at normal incidence. An important question, then, is: How do the overlying layers affect this angular dependence of emittance? We will examine the experimental results later.

The literature on emittance measurements yields data that show a wide scatter. This problem is also encountered when one makes his own measurements of emittance. Emittance is equal to $1 - r$, and when r is very high, small changes in reflectance result in large changes in emittance.

The optical properties of metals vary considerably depending upon the nature of the metal surface. Highly polished bulk metals usually show the highest reflectances and lowest emittances. Vacuum-deposited films and sputtered films often do not reach the reflectance values of the bulk polished material. This is especially true for the refractive metals, like molybdenum and tungsten (W). Aluminum, silver, and gold also show wide ranges in emittance from sample to sample as a result of slightly different deposition conditions. It is necessary in the case of practical applications, like solar energy, to use those values for the metal that will apply to the actual coatings, not necessarily the best values reported in critical tables or handbooks. There is no substitute for actually measuring the optical properties on the base-layer metal one actually will use in the desired application. In the case of aluminum, for example, careful laboratory coatings can show emittances as low as 0.017 at room temperature whereas commercial coatings can show emittances in the 0.05–0.06 range.

With these reservations in mind we show the emittances for a number of metal films, over a range of temperatures of interest for solar energy applications, in Figs. 9.8 and 9.9. Gold and silver have the lowest values; copper is very close but is hard to measure because it readily oxidizes at elevated temperatures. Aluminum has twice the emittance of gold and silver, and nickel and molybdenum lie about an equal distance farther away from gold. Tungsten also has a low emittance when used in bulk form, but it is very difficult to evaporate with an electron beam or to sputter and obtain values within a factor of two of the bulk values. Platinum has the same emittance as molybdenum and nickel, and zinc is close to aluminum.

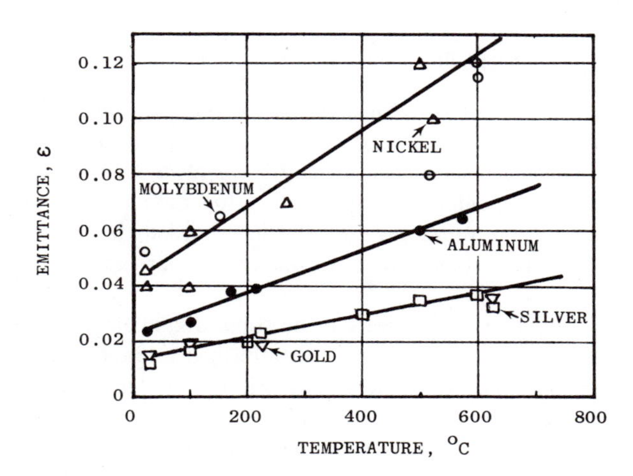

Fig. 9.8 Variation of emittance with temperature for gold, silver, aluminum, nickel, and molybdenum. Data sources: ***Metals Handbook*** **(1949),** ***AIP Handbook*** **(1957),** ***Smithsonian Physical Tables*** **(1934),** ***Engineering Heat Transfer*** **(Welty, 1974), and NASA** ***Conference Report SP-31*** **(1963).**

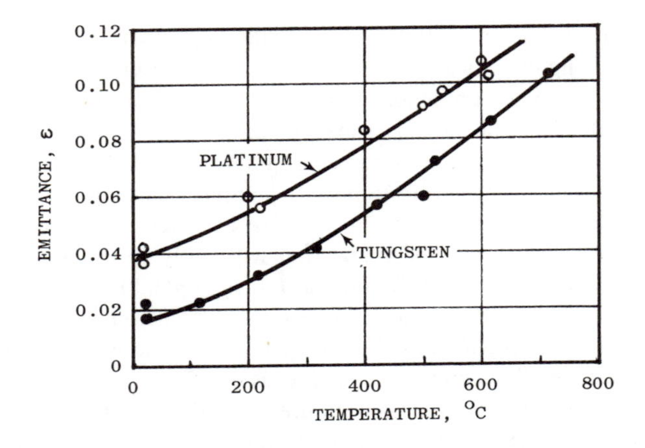

Fig. 9.9 Variation of emittance with temperature for tungsten and platinum. Data sources are the same as for Fig. 9.8.

The property of increase in emittance with rising temperature is very important for the behavior of selective surfaces. The net result is that the selectivity (α/ϵ) measured at room temperature decreases as the temperature increases. It is therefore necessary to measure a selective coating at the temperature at which it will be used, a subject we will discuss later.

9.6 DIRECTIONAL VARIATION OF EMITTANCE

The directional emittance variation for dielectrics and metals is important to an understanding of the properties of the two types of surfaces. According to Kirchhoff's law, the polar plot for a dielectric, shown in Fig. 9.10, is always within a circle tangent to the emittance at normal incidence, dropping to a low value as the reflectance of the surface increases. Reported measures are fully in accord with theory in regard to the angular dependence of reflectance for both planes of polarization, angle, and index of refraction.

In the case of metals there appears to be some disagreement between published polar plots and theory. We show some of the published values on the left-hand side of the top diagram in Fig. 9.10. They are characterized by a large increase in emittance beyond 60° from the normal. What this means is that the reflectance of a metal surface according to these measures *decreases* steadily with increasing angle of incidence. Since any metal can be described in terms of optical behavior by two constants—n, the real component of the index of refraction, and k, the imaginary component—the angular behavior of a metal should follow the Fresnel equation when the experimental values of n and k are used. Bennett and Bennett (1970) have calculated

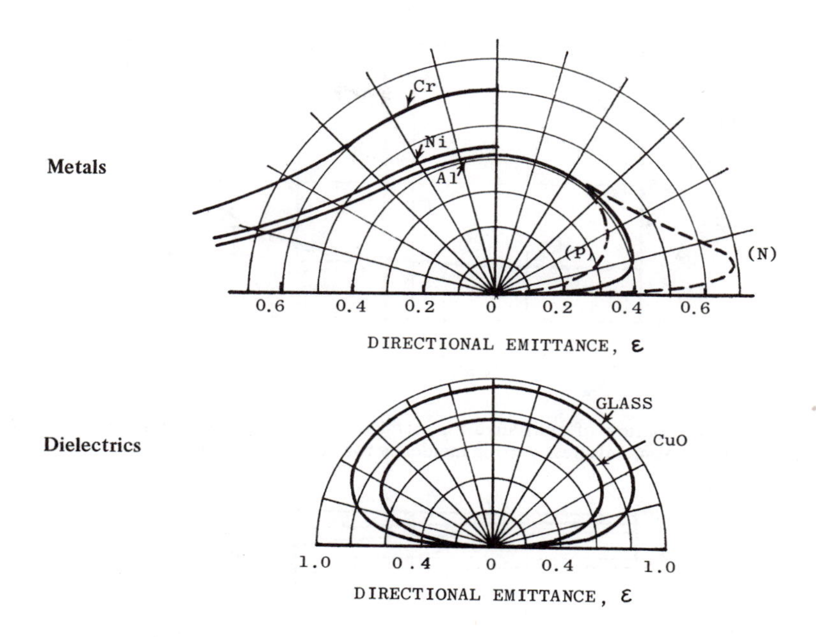

Fig. 9.10 Angular variation of emittance in polar coordinates for metals (top) and dielectrics (bottom). The reported curve for aluminum (top, left) is compared with the theoretical curve (top, right). Also shown are the electric vectors normal (N) and parallel (P) to the surface.

the theoretical variation of reflectance and emittance for metals for polarization parallel to and perpendicular to the metal surface. The types of curves they obtained, translated into a polar diagram, are shown on the right-hand side of the top diagram in Fig. 9.10. There are no combinations of n and k appropriate for common metals in which the average polarization shows a significant increase in emittance at large angles. Only the polarization parallel to the surface has an increase in emittance.

The emittance of metals based upon the Bennetts' work is shown in Fig. 9.11 in rectangular coordinates for the two planes of polarization. The corresponding curves for three dielectrics are shown in Fig. 9.12. The hemispherical total emittance is determined from the polar diagrams or observational data by performing the integral

$$\epsilon = \int_0^{\pi/2} \epsilon_\theta \sin(2\theta)\, d\theta . \tag{9.1}$$

The value this expression yields is significantly different from the emittance obtained from normal incidence measurements. The difference between the earlier values and those of the Bennetts' study is interesting in regard to the practical question of whether or not a metal surface used in solar energy work has an increase in emittance

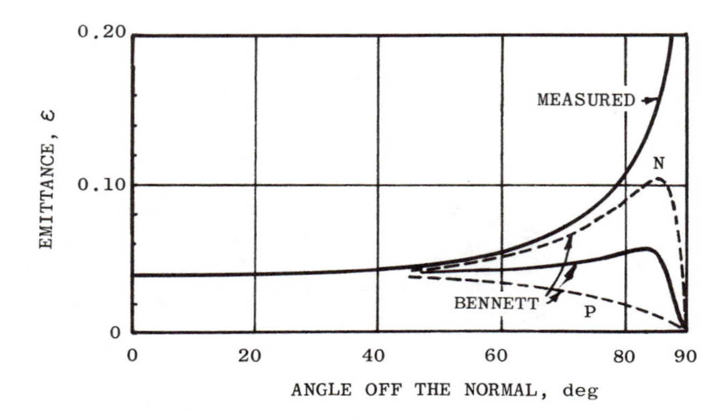

Fig. 9.11 Angular variation of emittance for metal films showing the increase in emittance (decrease in reflectance) of metals near grazing incidence for infrared radiation.

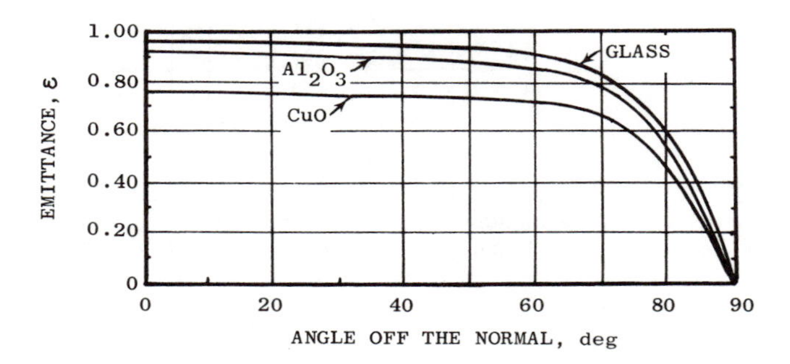

Fig. 9.12 Angular variation of emittance for dielectric surfaces (single surface measurements) showing the large decrease in emittance (increase in reflectance) of dielectrics near grazing incidence for infrared radiation.

at grazing angle. As in the case of vacuum-deposited metal films, we are interested not in the very best laboratory value for reflectance or emittance, but in what a typical surface yields. Two things could be affecting the measures that yield higher grazing emittance. First, the instrument producing the light beam or the measuring instrument might have a partially polarized beam, favoring the component that does show high grazing emittance. Second, the metal surface could have fine structure or contamination that might affect the grazing rays.

It is interesting to note that there is little difference between the two measures we show for the normal emittance and hemispherical emittance for selective interference stacks; this indicates no increase in emittance at grazing angle, which is in better agreement with the Bennetts' theoretical study than with cited observations (Welty, 1974).

9.7 NORMAL VERSUS HEMISPHERICAL EMITTANCE

The usual method of making spectral measurements of reflectance is to mount the sample near normal incidence and calculate the emittance from the relationship $1 - r$. In the emitting mode the surface is emitting into the 2π solid angle, and the hemispherical emittance can be significantly different from the normal emittance. For example, the emittances used to determine the selectivity (α/ϵ) in this book and in most other works on solar selective surfaces are normal emittances. Some values for the ratio of hemispherical to normal emittance are shown in Table 9.2.

Table 9.2 Ratio of Hemispherical to Normal Emittance

Material		*Values*	*Ratio*
Aluminum	(170°C)	0.049/0.039	1.25
Chromium	(150°C)	0.071/0.058	1.22
Nickel	(100°C)	0.053/0.045	1.18
Iron	(150°C)	0.158/0.128	1.23
Glass	(100°C)		0.93
Copper oxide	(150°C)		0.96
Aluminum oxide			0.97

9.8 MEASUREMENT OF WAVELENGTH DEPENDENCE

Measurement of the wavelength dependence of reflectance or emittance is of critical importance for evaluating selective surfaces. Samples currently in use are substrates 5 cm square, which can fit into a standard spectrometer equipped with a reflectance-measuring attachment. The typical two-beam spectrometer then compares the surface reflectance with a comparison beam having no optical surface in the beam, as shown schematically in Fig. 9.13. A standard sample, such as an aluminized mirror, can then be placed in the sample position for comparison of the results with those of a non-selective surface.

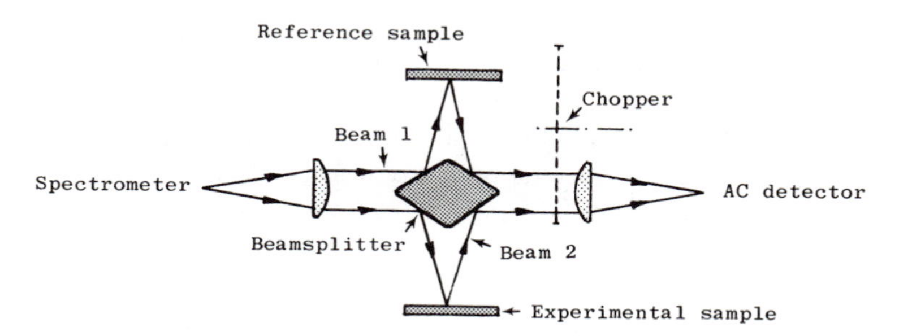

Fig. 9.13 Schematic diagram for the normal incidence spectrometer used to make room-temperature specular reflectance measurements.

In Fig. 9.14 we show a typical tracing for a highly selective coating sample, where the tracing approaches the 100% reflection tracing (upper curve). Note that it is necessary to calibrate the spectrometer runs for such coatings because the quantity of importance is the *difference* between the two curves, and in the illustration this difference is small compared to the wandering of the 100% line.

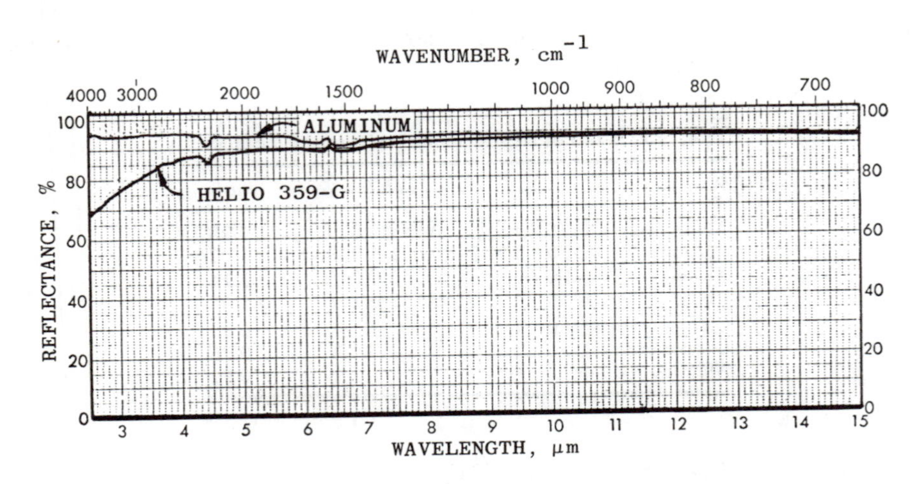

Fig. 9.14 Infrared spectrometer tracing of a highly selective chromium-on-silver design by McKenney *et al.* (1974) compared to a tracing with an aluminum mirror in place of the sample. Traces of the atmospheric absorption bands can be seen as a result of the difference in the two optical paths. This particular coating has absorptance of 0.92 and emittance of 0.04.

The small difference between the above two curves also illustrates the uncertainties incurred in measuring highly selective samples by the reflection method.

The basic problem of measuring samples in a spectrometer is that it measures *specular* reflectance, and only at near normal incidence. The substrate samples must have a near optical quality polish. This means that the substrate surface is considerably different from that encountered in the real world of solar energy, where polished substrates are too expensive. It also means that if the selective surface sample has a roughened or structured surface, the optical measuring beam will be scattered out of the acceptance angle of the spectrometer optics, resulting in deceptively high absorptance readings.

9.9 HIGH-TEMPERATURE REFLECTANCE MEASUREMENTS

Since the emittance of metals and dielectrics changes with temperature, one cannot be certain that measurements made at room temperature will be applicable when the selective surface is to be used at high temperature. It is therefore necessary to make measurements at operating temperatures to correctly determine the emittance of the coating. These types of measurements are complicated by numerous practical problems of making measurements in vacuum and at high temperatures.

The high-temperature spectrometer used to measure the change of reflectance with temperature is shown in Fig. 9.15. The calibration and maintenance of such an instrument is important when reliable results are needed. The time consumed in making a set of spectral runs at elevated temperature is much greater than for samples with room-temperature spectrometers. For this reason it is important to calibrate the changes in reflectance (or emittance) for enough samples to obtain a good calibration curve. In this manner one can use room-temperature measurements for the bulk of any development program and save much time in the laboratory.

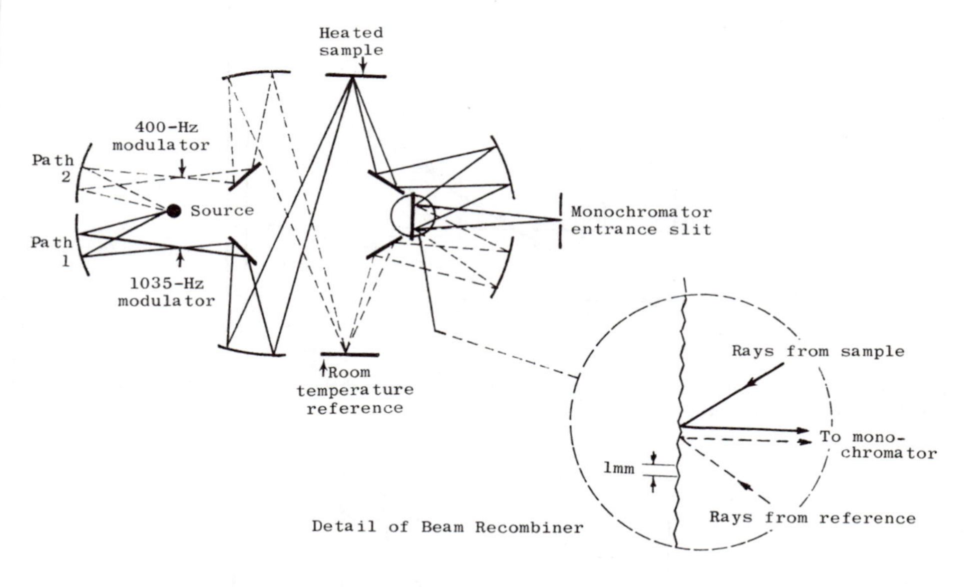

Fig. 9.15 Schematic diagram of the high-temperature reflectometer developed by Seraphin (1974).

In Fig. 9.16 we show some experimental points measured for the emittance of selective coatings, normalized to unity at room temperature. The clear circles show measurements made with the first spectrometer built by Seraphin (1974), which showed nonlinearity above 400°C, illustrating a common problem with instruments in which large thermal gradients and TIR fluxes are encountered. The dark points are measurements made after the causes of the nonlinearity were identified and removed, yielding the expected linear change of emittance with temperature. Note the considerable range in the emittance variation for these two recent samples. In general, however, the emittance variation lies within the spread shown by these two samples.

It is interesting to compare the change of emittance with temperature of the selective surfaces with that of bare metal surfaces. In Fig. 9.17 we show the fractional increase of emittance for several metal films compared to the two slope lines for selective coatings from Fig. 9.16. Clearly, the change of emittance with temperature for these interference selective coatings is much lower than for a bare metal, an apparent

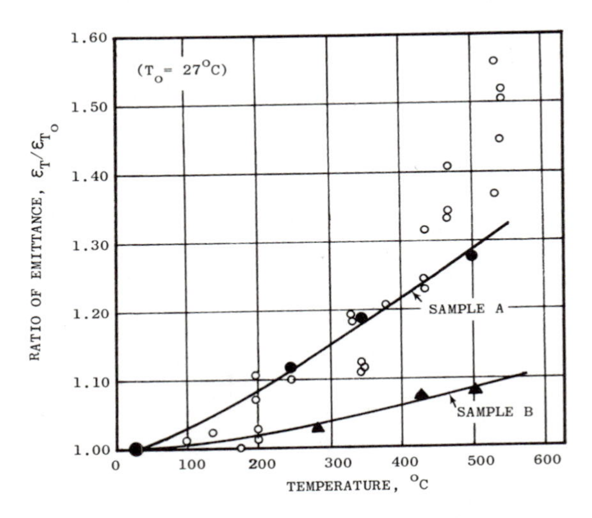

Fig. 9.16 Variation of emittance with temperature for several selective interference stacks measured with a high-temperature spectrometer.

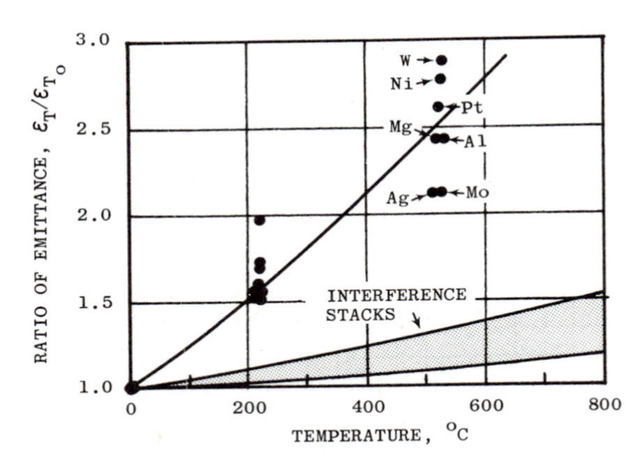

Fig. 9.17 Comparison of the increase of emittance with temperature of metal films compared to preliminary results for some selective interference stacks.

consequence of the fact that the metal layers are overlain by a layer of dielectric material. This fact is important for research in interference-type films in that relatively few measurements at high temperature are sufficient to verify that nothing unusual is being encountered in a development program. At present, we have no similar measurements made with selective coatings of other types to establish whether or not this low change of emittance with temperature is a characteristic of the interference stacks alone or is true for selective absorbing surfaces in general.

9.10 DIRECT MEASUREMENT OF ABSORPTANCE AND EMITTANCE

The spectrometer methods of measuring absorptance and emittance are limited by the requirement for specularity of the sample and for a reference surface, such as an aluminum mirror. Samples placed on pipes and sheet as directly received from a factory are seldom specular, and the cost of polishing each substrate would be prohibitive. Further, the deposition or the heat treatment after deposition can render a coating on a specular substrate nonspecular.

McKenney and Beauchamp (1975) have developed a simple calorimetric instrument for measuring directly the solar absorptance and thermal emittance. The instrument is shown in schematic form in Fig. 9.18. A sample is placed on a heat sink block of copper whose temperature can be varied, the sample being interfaced with the block by means of a thermal jelly. High-conductivity lubricants can be obtained from standard laboratory supply houses. The surroundings of the chamber are maintained at ambient temperature. A heat-flow sensor (like Keithley Model 860 or successor instruments) is placed between the sample and the heated block. Such a meter can read heat flows as low as 10^{-5} cal/cm^{2}sec. A window is also provided to allow solar flux to enter and impinge on the sample.

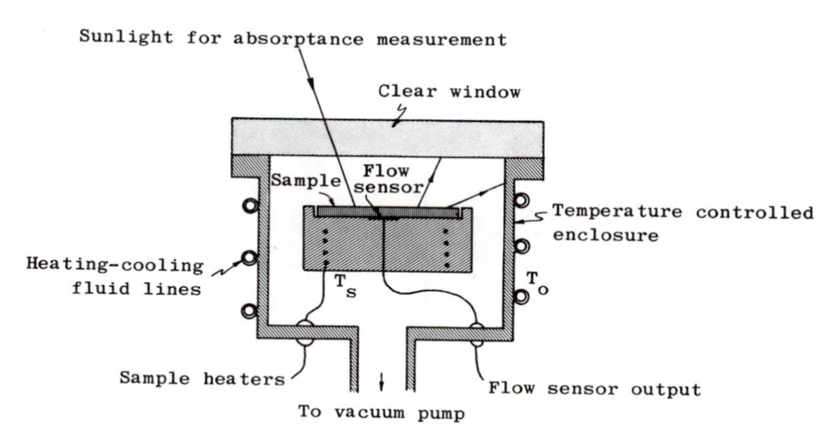

Fig. 9.18 Hemispheric emittance and solar absorptance instrument for measuring selective surface samples (50 X 50 mm).

In use, sunlight is first allowed to enter the chamber, whereupon the sample rises to an equilibrium temperature, determined by adjusting the rate of water flow through the block to maintain the desired temperature:

$$\alpha I = \epsilon\sigma(T_s^4 - T_0^4) + Q(\text{water}). \tag{9.2}$$

If we specifically adjust the flow to hold the sample at the same temperature as the enclosure, then the emittance of the sample can be neglected, and α can be measured directly from the rate at which heat flows from the sample into the block.

To measure ϵ, one blocks the solar flux from entering and heats the block and sample to a higher temperature than the enclosure walls; thus heat radiates out from the sample. The emittance is then determined from the rate at which heat flows from the block into the sample.

Some results obtained with this instrument are shown in Fig. 9.19. Graph (a) is for absorptance α, where solar flux is limited to the same angle of the sky as is measured by an accompanying pyrheliometer. The observed calorimetric α-values agree well with those calculated from laboratory spectrometer curves for the sample used, an expected result because sunlight is well collimated. The graph of interest is graph (b), showing the correlation between normal incidence emittances, calculated from normal incidence spectrometer curves, and hemispherical emittances, the factor of real importance in evaluating collector performance. This curve shows that the spectrometer values predict the hemispherical values accurately when the samples are specular in the infrared. This result shows that the difference sometimes observed for metals does not apply to selective coatings, a point in agreement with the fact that selective coatings do not show the same increases in emittance with temperature as do metals.

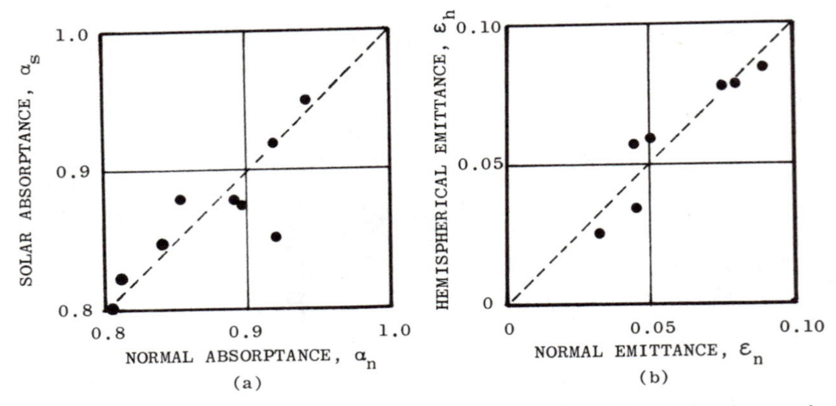

Fig. 9.19 Diagrammatic comparisons of α_s to α_n and ϵ_h to ϵ_n for the best absorber coating designs. α_s and ϵ_h were determined calorimetrically as discussed above. α_n and ϵ_n are calculated from the normal spectral reflectance. Examples cited are a lead sulfide coating and a black nickel coating after 15 h of heating at 500°C.

9.11 BASIC MODES OF OBTAINING SELECTIVITY

The methods of obtaining selectivity are through use of:

1. intrinsic materials,
2. tandem stacks of a semiconductor overlying a reflector,
3. interference stacks of alternating dielectrics and metals,
4. wavefront-discriminating surface roughness,
5. dispersion of metal droplets in a metal or dielectric matrix, and
6. quantum size effects.

We shall describe these methods in Sections 9.12–9.21.

9.12 INTRINSIC MATERIALS

Intrinsic materials are substances that have the desired selectivity naturally, without the need for any other material to augment the wavelength behavior. Hafnium carbide (HfC), the reflectance curve of which is shown in Fig. 9.20, is a material that naturally has high reflectance in the thermal infrared and high absorptance in the region of the solar spectrum. The TIR reflectance is on the order of 0.90 ($\epsilon = 0.10$), but the visible absorptance is only about 0.70. Efficient use of hafnium carbide would actually require some means of increasing its visible absorptance, such as overlaying it with a

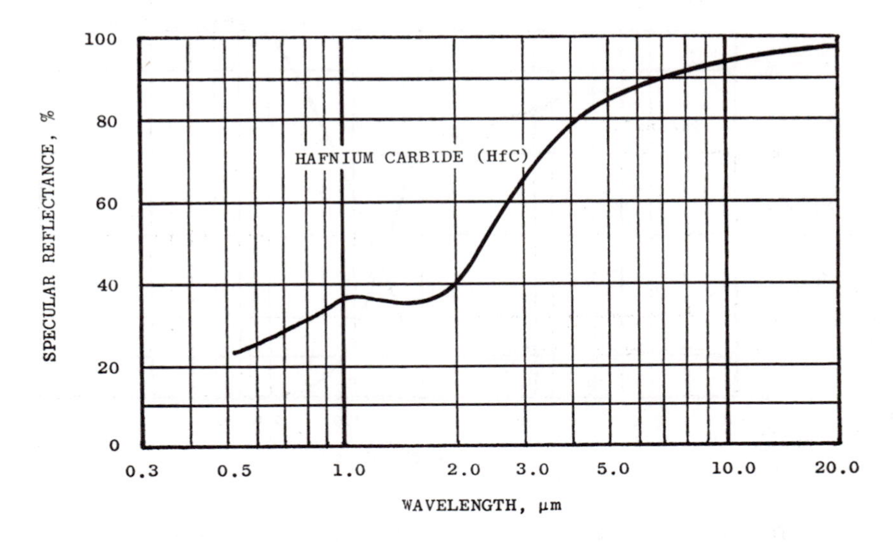

Fig. 9.20 Reflectance curve for an intrinsically selective material, hafnium carbide (HfC). Many other carbides have similar behavior.

quarter wave of a dielectric material, or arranging the absorber so that *two* reflections occur for incident sunlight. Its high melting point, the highest of any compound, also would make it attractive for the absorbing surface of a high-temperature solar collector.

Tin oxide (SnO_2) is also a naturally selective material, but it is used mainly for window applications. It is transparent to the visible, with transmittance of 0.75, but reflective in the TIR with $r = 0.70$. The reflectance curve for tin oxide is shown in Fig. 9.21. Indium oxide (In_2O_3) is selective in the same way, and is often combined with tin oxide in commercial selective or conductive windows. Both tin oxide and indium oxide are placed on the windows as a solution that is baked to provide final adherence. These materials have a high refractive index in the visible region, resulting in high reflection losses. They also have intrinsic internal absorption, as implied in Fig. 9.21. The actual extent of internal absorption is not well determined because the use of these materials for conductive windows also includes the effect of nucleants added to improve conduction. For strictly solar energy applications one does not need conductive enhancement and could use the pure compounds, in which case the internal absorption might be significantly lower than that usually cited.

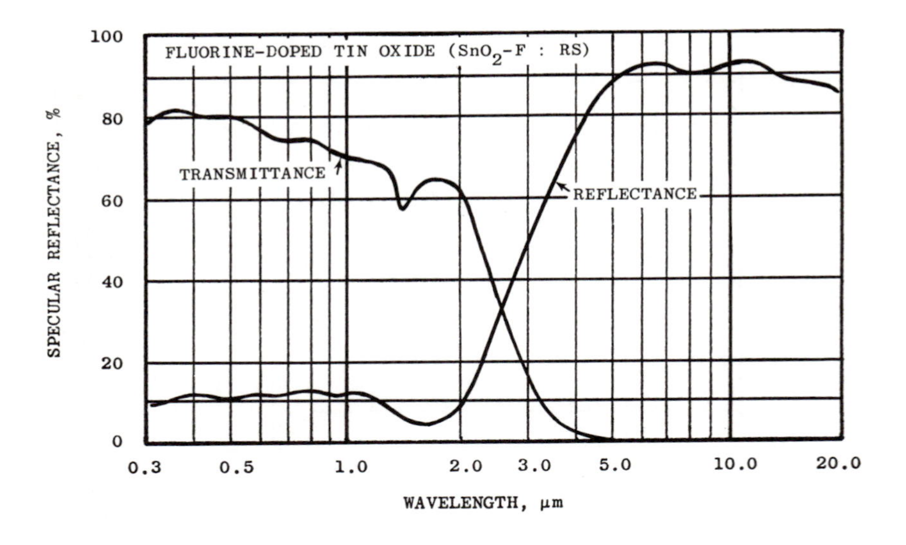

Fig. 9.21 Transmittance curves for tin oxide (SnO_2) with fluorine doping, after Rekant and Sheklein (1969). These curves vary considerably from sample to sample. The SnO_2 is on glass in this example.

Excellent transmitting selective surfaces have been made by Blandenet, Lagarde, and Spitz (1975), using indium oxide doped with a small percentage of tin oxide. The coatings are prepared by pyrolysis from an aerosol generated by ultrasonic vaporization of a solution of organic and inorganic compounds by the Pyrosol® process. A typical curve for such a doped film in comparison to an undoped tin oxide and an indium oxide film is shown in Fig. 9.22. The transition from transmission to reflection is abrupt at about 1.8 μm. The transmission in the visible range is 92%, or about 90% of the solar spectrum, with reflection in the thermal infrared of about 85%. This high transmittance and reflectance compared to traditional tin oxide films opens the way to utilization of this type of selectivity for windows in solar collectors. The previous performance of tin oxide films did not yield a cost-effective gain because of the internal absorptance of the films in the visible and the lower infrared reflectance. The question of the cost effectiveness of the new indium oxide films should be reexamined.

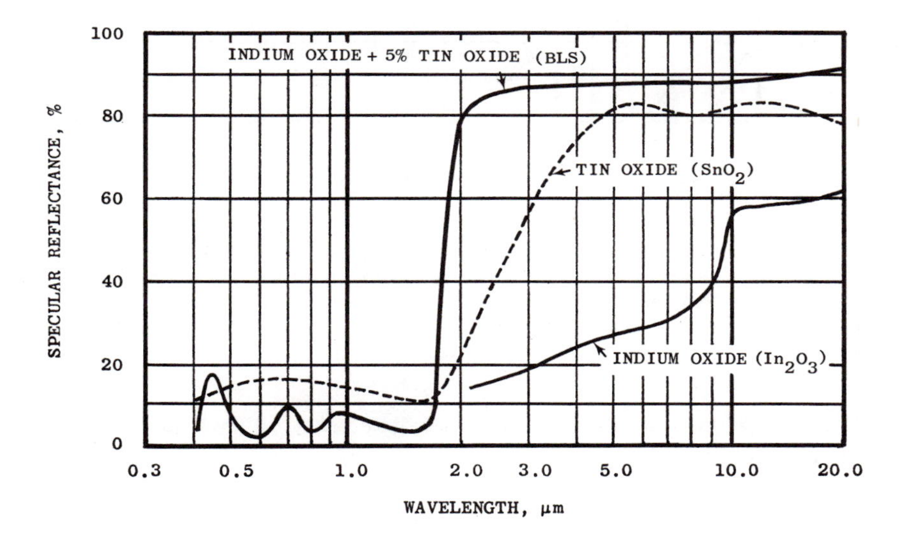

Fig. 9.22 Reflectance curve for indium oxide with 5% doping of tin oxide compared to undoped indium oxide and tin oxide, from Blandenet *et al.* (1975).

Lanthanum hexaboride (LaB_6) is also a selective window material having a TIR reflectance close to 0.90 and a transmittance for the visible of 0.85 when an overcoating of dielectric is used to reduce visible reflectance. The transition from highly reflective to highly transparent occurs in a remarkably short wavelength interval, raising the possibility that other rare-earth hexaborides might offer transitions that are more favorably located to the red of the lanthanum hexaboride transition. The optical characteristics for lanthanum hexaboride are shown in Fig. 9.23.

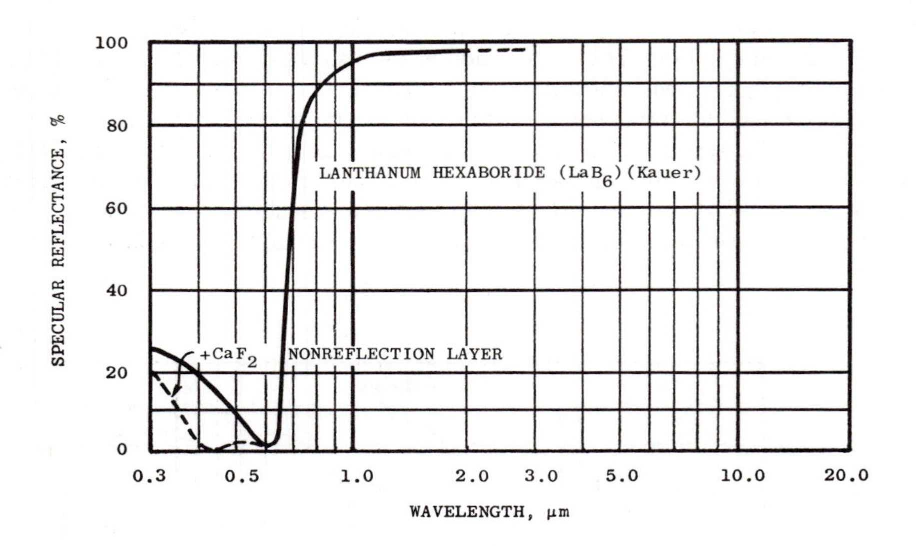

Fig. 9.23 Transmittance and reflectance curves for lanthanum hexaboride on glass, developed by Kauer (1966), with the effect of an added layer of calcium fluoride to nonreflect the visible. This example is to show that the rare-earth hexaborides have unusual optical properties and deserve further study.

The semiconductors offering transmissive selectivity all have high indices of refraction for sunlight. It is therefore necessary to overcoat them with nonreflecting coatings when the best performance is required. In Chapter 12 we examine the consequences in terms of sytem performance so that the necessary cost effectiveness of refinements like nonreflecting selective windows can be evaluated.

9.13 TANDEM STACKS

Tandem stacks are two optically active materials combined in two discrete layers to get the desired net optical effect, as shown in Fig. 9.24. Tandem stacks were first used by Hass (1964) to make a "black" searchlight mirror, black in the visible and highly reflective in the TIR. Hass used a layer of germanium on top of a standard mirror surface, the germanium being transparent to TIR. The germanium, however, quickly disappeared into the aluminum mirror surface. Seraphin (1974) has recently developed

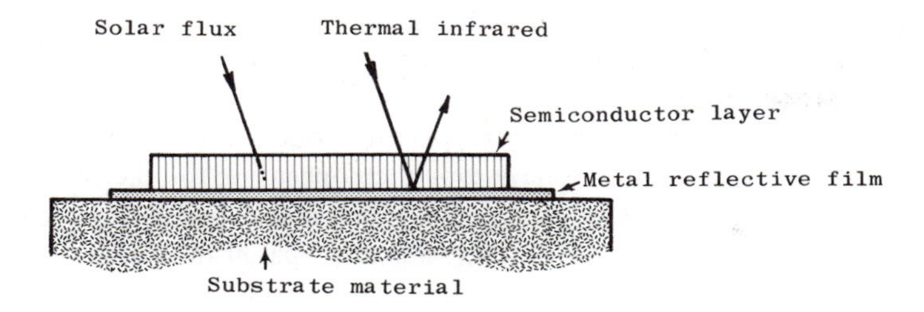

Fig. 9.24 Basic concept for a tandem stack consisting of a semiconductor for the solar absorptive layer and a base metal film for high reflectance in the TIR, where the semiconductor becomes transparent.

this approach to a point where practical utilization appears promising. Williams, Lappin, and Duffie (1963) also have used lead sulfide (another semiconductor) to produce selective tandem stacks, a method recently advanced by McMahon and Jasperson (1974). The basic concept of the two layers is maintained, but additional layers are generally necessary for physical reasons, so a durable, efficient tandem stack in reality becomes much more complex than two materials. The basic optically active component of the tandem stack absorber is a semiconductor. The wavelength region of transition from transparent to reflective depends upon the band gap of the semiconductor. The thickness of the semiconductor required for proper opacity in the visible depends upon the strength of the absorption, which for silicon appears to be in the range of 2–5 μm, but Seraphin has found a way to use layers as thin as 1.5 μm by fully utilizing interference effects in the stack design.

Tandem stacks cover a wide range of selective coatings, ranging from very simple ones to very complex ones that have many layers but functionally act like a tandem layer. In principle, one would like to have a selective "paint." Most of the selective coatings we describe here are in reality rather difficult to make, with methods ranging from simple heating in air to chemical vapor deposition and vacuum evaporation of multilayers. Paints, however, involve two materials–the pigment and the binder. One could find several types of selective materials to use for the pigment, but few choices for a binder. Binders are generally organic in nature, and most organic materials have strong absorption bands in the thermal infrared, so although the pigment may have low emittance in the infrared, the binder will not. It is much easier to make a "cool" paint, the exact opposite of the selective surfaces we desire for solar energy conversion. The combination of titanium dioxide (TiO_2) with a suitable binder makes a paint with low solar absorptance and high infrared emittance.

A very simple form of tandem selective surface can be made by lightly "smoking" a mirror surface. Carbon particles are opaque to visible wavelengths but transparent to the thermal infrared. The basic problem of these selective surfaces is that they are very fragile (Veinberg, 1959), the carbon particles adhering only weakly to the metal surface of the mirror. The presence of a binder to hold the carbon would modify the basic optical properties of the carbon particles with the opacity of the binder. Some binders, like polyethylene, would be transparent enough to preserve the selectivity of the carbon, but most paint binders have too much opacity in the thermal infrared.

Electrochemical selective coatings like black nickel can also be considered as tandem stacks. The simplest form of black nickel prepared in this way is described in the ASHRAE booklet listed in the references. The coating is produced by varying the plating current, but considerable development work is necessary to obtain the excellent uniformity of behavior produced by Miromit in Israel and Honeywell in the United States.

Oxidized Metals

The simplest tandem stack is represented by the modification of the substrate material to introduce selective effects. These types of coatings, as measured by Edwards, Gier, Nelson, and Roddick (1961), are shown in Figs. 9.25 and 9.26. In Fig. 9.25 we show the change in Type 410 stainless steel after it is heated in air at 750°C. The selectivity α/ϵ is small, about 3, and the solar absorptance is low, about 0.75, but the process is simple and inexpensive. This means that if some selectivity is desired, as for concen-

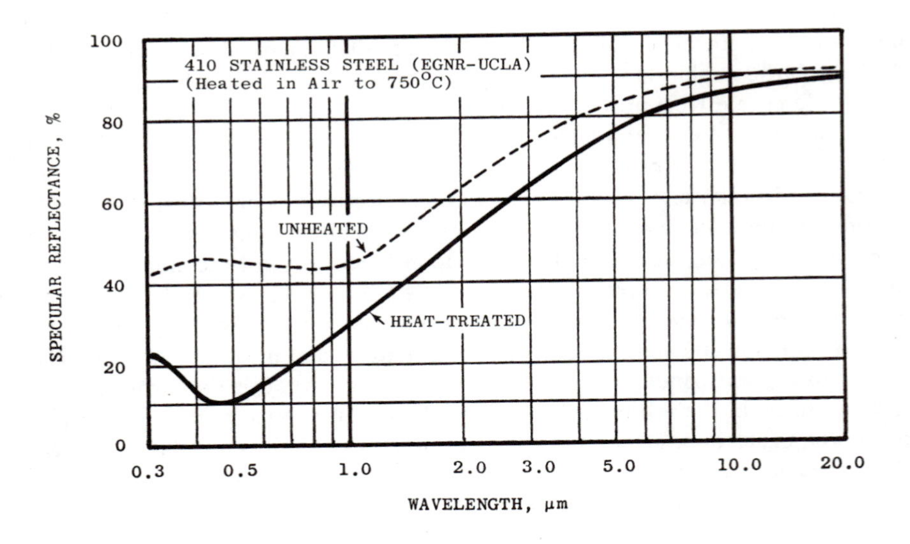

Fig. 9.25 Curves for commercial Type 410 stainless steel as received and after heat treating to form natural oxides, from Edwards *et al.* (1961).

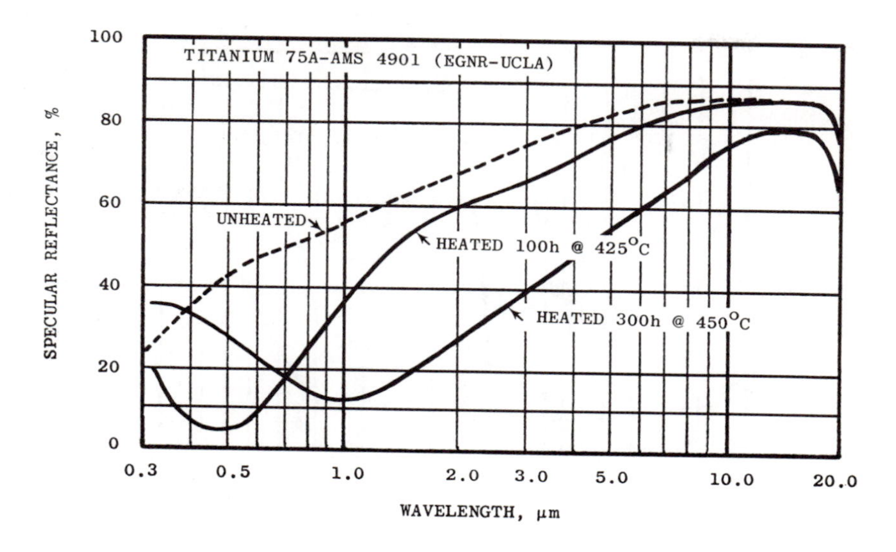

Fig. 9.26 Curves for commercial Type 75A-AMS 4901 titanium sheet as received and after heat treating in air to form natural oxides, from Edwards *et al.* (1961).

trating collectors, then the oxidation of stainless steel could be useful. A second form of simple oxidation selectivity is shown in Fig. 9.26, where titanium (Ti) is heat-treated to produce moderate selectivity.

Silicon/Germanium

The second type of tandem stack, produced by placing a different material over the base film, is shown by the curves in Fig. 9.27. Here the additive material is silicon (Si). The simplest type is shown by the upper curve, by Edwards *et al.* (1961), where silicon is placed over an aluminum substrate, the bulk silicon in this case being 0.5 mm thick. The reflectance in the visible is reduced to the level of bulk silicon, which is further lowered by adding a nonreflecting layer of silicon dioxide (SiO_2). For comparison we show the curve for a commercial silicon solar cell by the IRC Company, in which the silicon wafer is attached to a nickel substrate. The infrared reflectance beyond 5.0 μm is due to the transparency of the silicon, which allows the nickel to show through. The curve denoted BOS-UA is a theoretical curve for a sophisticated multilayer stack that uses both silicon and germanium (Ge) to produce a curve that is close to the ideal for a selective absorber (Seraphin, 1974). In this case the underlying base film is silver and the silicon/germanium layer is "nonreflected" with a composite layer of silicon nitride and silicon dioxide (Si_3N_4/SiO_2).

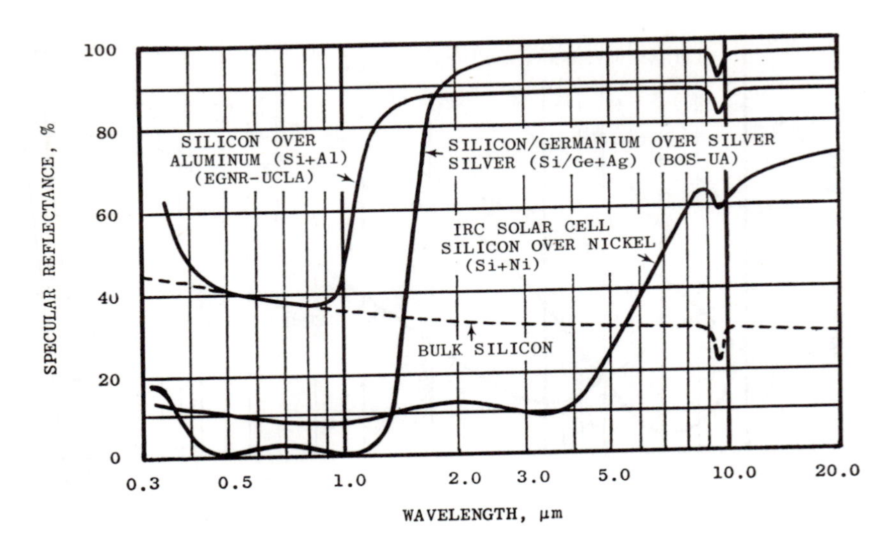

Fig. 9.27 Basic curves for silicon tandem and interference stacks, compared to a silicon solar cell and bulk silicon. Note that this theoretical Seraphin-type tandem stack develops the optical characteristic of a double minimum when the silicon is overcoated with a nonreflecting layer.

The current work by Seraphin (1974) at the University of Arizona is concerned with the development of tandem stacks utilizing chemical vapor deposition (CVD). Rather than using thick layers of silicon over aluminum it uses optically thin layers. The transition from opaque to transparent occurs at 1.0 μm.

The basic transmittance of a very thin layer of silicon is shown in Fig. 9.28. The layer of silicon 1.5 μm thick was placed on a glass substrate, which accounts for the cutoff of measurements at 2.5 μm. The interference fringes due to reflections from the two silicon interfaces are conspicuous. The lower curve in this figure represents the transmission of a 0.7-μm-thick film of germanium, also showing interference fringes. The curve showing the combined effects of the silicon and germanium layers over silver are shown in Fig. 9.29, where the interference fringes are not shown. The solid curve in this case is the average of the maxima and minima.

Actual CVD coatings made to date are more like the curves shown in Fig. 9.29. The pairs of curves shown by dots and dashes represent the envelope of the interference fringes, shown explicitly between 2.5 and 20.0 μm. This coating uses only silicon, and has a nonreflection double layer of silicon nitride (Si_3N_4) and silicon dioxide. The dotted curve shows how the absorption band shifts to the red with increasing temperature.

There are two different temperature effects for a semiconductor stack: (1) a redward shift of the absorption edge with increasing temperature, and (2) a small

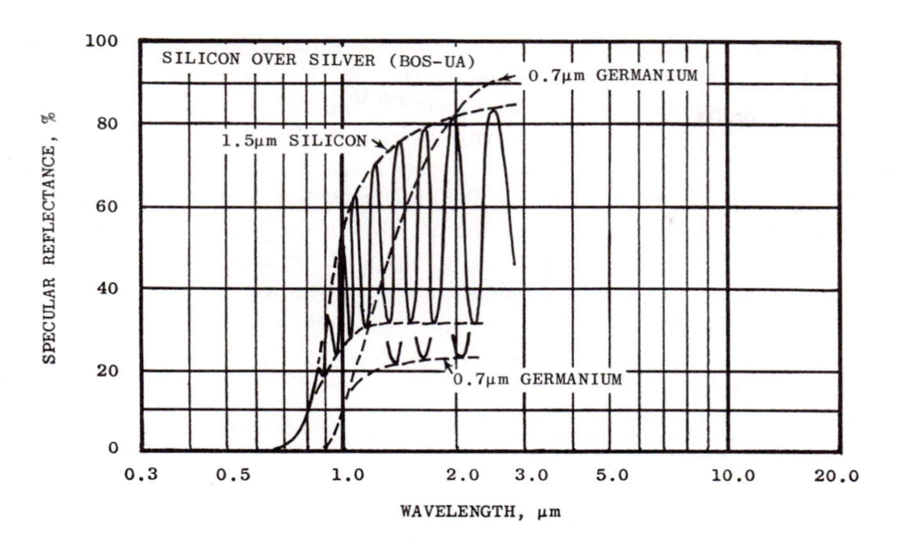

Fig. 9.28 Transmittance curves for samples of silicon (Si) 1.5 μm thick and germanium (Ge) 0.7 μm thick, showing the difference in wavelength onset of transmission and interference fringes (Seraphin, 1974).

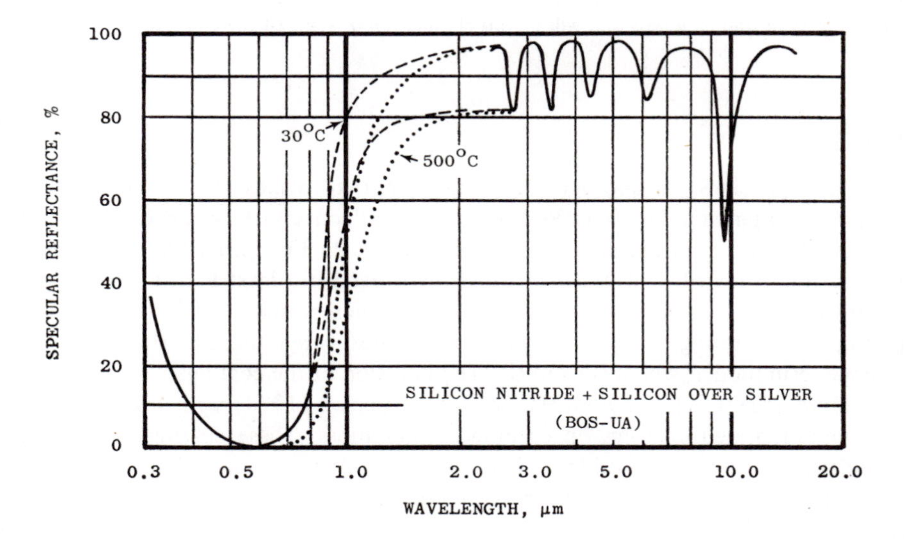

Fig. 9.29 Variation of the reflectance curves for a tandem stack, by Seraphin (1974), showing the effects of increased temperature. The dotted curves show the envelope of the interference fringes, shown explicitly only in the TIR region.

increase in the free-carrier absorption within the silicon. The first effect is clearly shown in Fig. 9.29 by the two curves, one for 30°C and the other for 500°C. The second effect is small, shown only as a slight lowering of the depth of the absorption fringes in the TIR, which increases with wavelength. The extent of the free-carrier absorption that deepens these fringes is dependent upon the purity of the silicon. The small magnitude of the effect in Fig. 9.29 indicates that the CVD silicon is of satisfactory purity to avoid degradation of the selective surface with temperature.

A tandem stack using silicon actually requires more than simply the silicon over the reflective substrate. A fully developed stack is as shown in Fig. 9.30. The reasons for these additional layers will be discussed in Section 9.23.

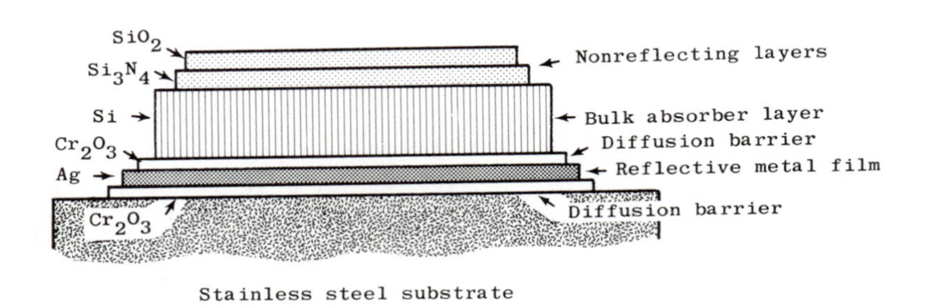

Fig. 9.30 **Schematic cross section of a fully evolved tandem stack selective absorber. The challenge now is to find materials that will permit elimination of some of the above layers.**

Lead Sulfide

Lead sulfide (PbS) is another semiconductor that makes a good tandem stack. Williams, Lappin, and Duffie (1963) reported on selective surfaces made of lead sulfide crystals deposited on aluminum substrates, shown in Fig. 9.31 (WLD-UW). Electron micrographs showed that these coatings were dendritic in structure, so the optical absorption was a combination of the intrinsic absorption of the lead sulfide and particulate scattering. Their lead sulfide crystallites were bonded to the substrate by a silicone resin binder, which adds to the TIR emission of the coatings.

More recent work by McMahon and Jasperson (1974) with lead sulfide used vacuum-evaporated coatings on evaporated aluminum. These coatings did not show specular reflection in the visible, and electron micrographs showed that the lead sulfide layer had surface structure, which was effective in destroying visible reflection but was specular in the TIR. The reflectance curve shown in Fig. 9.31 shows that the TIR reflectance is very high, close to that for pure aluminum, but it does not agree with the calculated reflectance curves, shown as dashed lines. The reason for this departure from calculated behavior is attributed by McMahon and Jasperson to the fact that the front-surface reflection of the lead sulfide does not obey the Fresnel equation for a single dielectric-air interface because of the surface structure of the lead sulfide film. The McMahon and Jasperson coating used approximately 1200 Å of lead sulfide, with no overcoating of dielectric to lower its surface reflection.

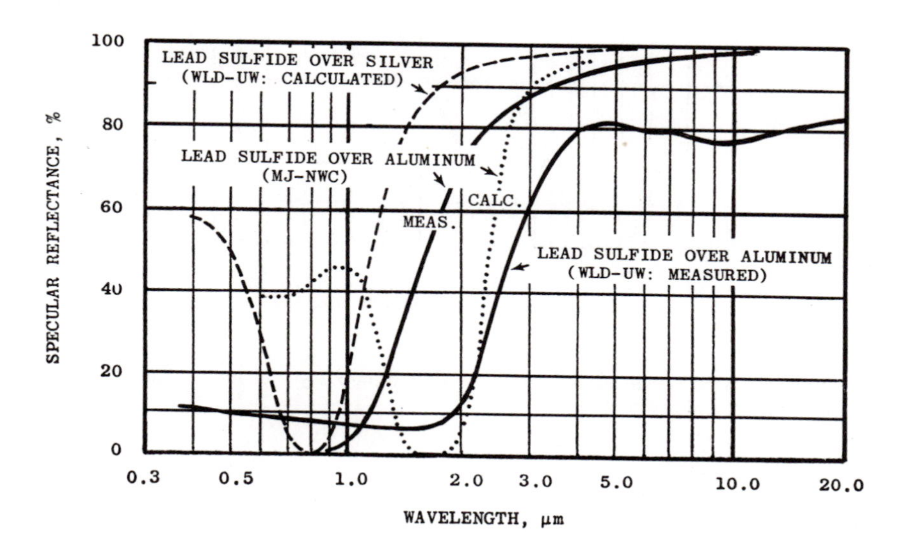

Fig. 9.31 Reflectance curves for lead sulfide (PbS) selective coatings. The calculated curve by Williams *et al.* (WLD-UW) is for a layer of lead sulfide 340 Å thick on a silver substrate; the curve by McMahon and Jasperson (MJ-NWC) is for a layer 1200 Å thick on an aluminum substrate. Both measured curves depart from the calculated curves, probably owing to the effects of surface roughness of the lead sulfide.

9.14 INVERSE TANDEM STACKS

Goldner and Haskal (1975) have proposed a tandem stack selective surface where the TIR emission is suppressed not by an underlying silver layer but by overcoating silicon with a selective transmitter like tin oxide or indium oxide. They report that sprayed-on coatings of indium oxide plus tin oxide of 0.1–1.0 μm thickness yield absorptance of 0.85 and emittance of 0.07. They further suggest that a combination of a reflective metal layer beneath the silicon and an overlying transmitting selective layer could further increase absorptance and lower emittance. The transmission curve for their overcoated silicon is shown in Fig. 9.32.

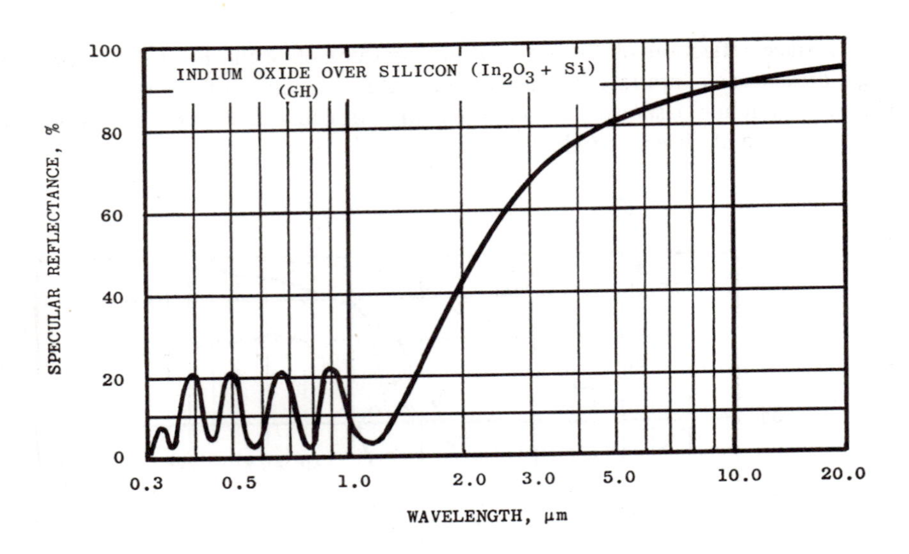

Fig. 9.32 Reflectance versus wavelength for a 0.6-μm-thick layer of indium-tin oxide (ITO) over silicon, by Goldner and Haskal (1975).

9.15 INTERFERENCE STACKS

Interference stacks produce the desired optical effects by optical interference between alternate layers of metal and dielectric. In the case of naturally selective materials the effect is produced by a single pass through the optically active medium, or the return pass after reflection by the underlying mirror surface. In the case of the interference stack the desired effect is the net result of a multiplicity of passes through the dielectric portion of the stack lying between the two reflective surfaces, the upper one of which is partially transparent. Careful tuning of the layer thicknesses and variation of optical constants with wavelength is necessary to get a good broad-band selective surface. The basic concept for a four-layer interference stack is shown in Fig. 9.33. Note that dielectric layer 1 need not have any intrinsic absorptance in the solar spectrum but could constructively have some to complement the natural properties of the metal films.

The general characteristics of an interference stack are illustrated in Fig. 9.34. The first curve is for the opaque reflective metal layer, where the reflectivity is high in the TIR and gradually drops in the visible. When the first layer of dielectric is added, it tends to reduce the reflection at the metal interface, the position of which depends upon the thickness of the dielectric. The selective effect is not as strong as one desires

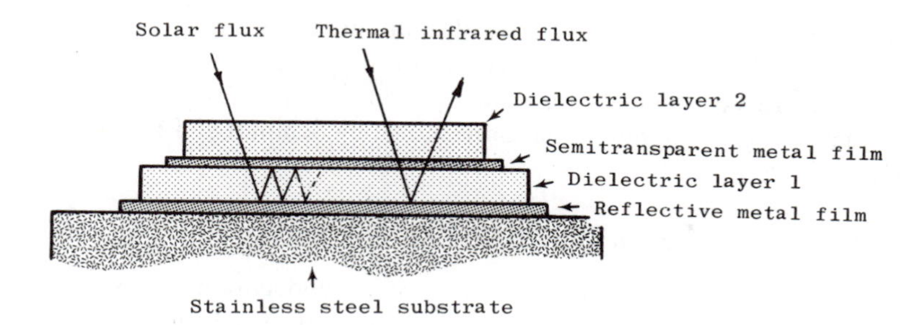

Fig. 9.33 Basic concept for a four-layer interference stack consisting of two dielectric quarter-wave layers separated by a thin, semitransparent metal film. The dielectric layers need not have any intrinsic absorptance in the solar spectral region.

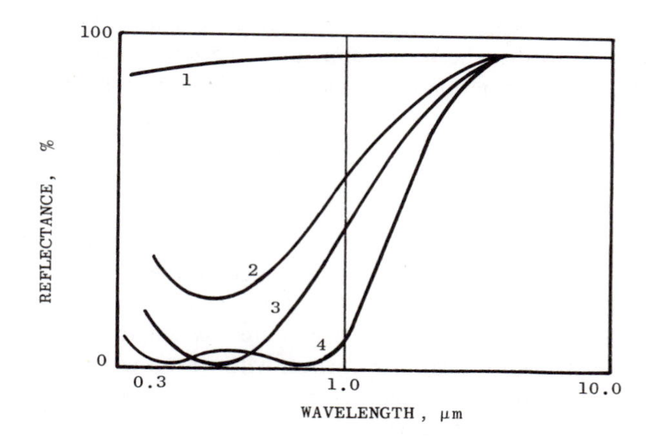

Fig. 9.34 Schematic variation of reflectance with wavelength for a four-layer interference stack as successive layers are added.

because, in general, the reflectivity of the dielectric is weak at its air interface, so the interference effect it produces is weak. Some dielectrics can be selected, such as chromium oxide, that do provide good interface reflection plus some internal absorption, to make a reasonable two-layer interference stack. In general, however, one needs to add a second reflective layer, partially transparent as in the traditional Fabry-Perot interferometer, to strengthen the reflected wave and maximize the internal interference in the dielectric of layer 2. Layer 3 is generally thin, about 50 Å, so its optical properties are quite different from those of the bulk metal. This property of layer 3 is deliberately enhanced to increase its effect. Considerable speculation exists as to whether this thin metal layer is really pure metal or some mixture of metal atoms and impurities from the vacuum chamber. Vacuum art plays a role in getting good layer-3 performance. The final layer of a typical four-layer interference stack is a dielectric, and its role is to broaden the wavelength region of high absorption. Multiples of the basic stack can be used, but, as has been shown by Schmidt and Park (1965), the gains are not cost effective.

9.16 BASIC PHYSICS OF A MULTILAYER STACK

The behavior of a multilayer stack can best be understood by examination of the changes that occur as new layers are added on top of a highly reflecting substrate metal film. Each new layer produces an interface between two materials of different optical constants. These interfaces generate amplitude vectors whose magnitude depends on the optical constants for the interface under the influence of multiple reflections in the case of the lowest reflective vector. These amplitude vectors are turned through an angle that depends upon the optical thickness of the layer. The thicknesses are therefore chosen so that the vector sums are made zero at the desired wavelengths. The phase angles ϕ are a function of wavelength λ; the optical thickness is wavelength dependent:

$$\phi = (2\pi/\lambda)nt. \tag{9.3}$$

The vector diagrams that apply to a four-layer stack are shown in Fig. 9.35 for the cases where zeros occur. There are three vectors, and regardless of the length of the vectors there are two sets of phase angles (wavelengths) where minimum or zero vector sums occur. In Fig. 9.35(b), ϕ_1 is less than 180°; in Fig. 9.35(c), ϕ_1 is greater than 180°. It should be noted that the phase angles are wavelength dependent. This means that the optical thickness of each layer changes with wavelength. The result is that the phase angles for each vector are considerably different at the blue wavelength than they were at the red wavelength, resulting in the two vector diagrams indicating the minimum vector sums in Fig. 9.35. This vector sum relationship is why the typical interference stack has two minima in the visible region.

The actual situation for a stack is more complicated than is implied in Fig. 9.35. The length of the vectors changes with wavelength, and the changes in the three vectors are not strictly determined by the pure materials forming the four layers. The variation of reflectivity of the base metal with wavelength, in the case of copper, produces a high maximum between the two minima. When silver or aluminum is used, the intermediate maximum can be effectively suppressed.

The greatest unknown in stack performance is posed by the thin metal layer, shown in Fig. 9.33. This layer is only 40–60 Å thick, and when measured separately has optical constants considerably different from those of the opaque metal film formed in the same deposition sequence. This layer has a major impact on film stack behavior, having significant effects for both vector $\mathbf{A}_2$ and vector $\mathbf{A}_3$ in Fig. 9.35. Most of the optical absorption of the stack in the visible appears to be due to attenuation of the incoming wave in this thin, semitransparent metal film.

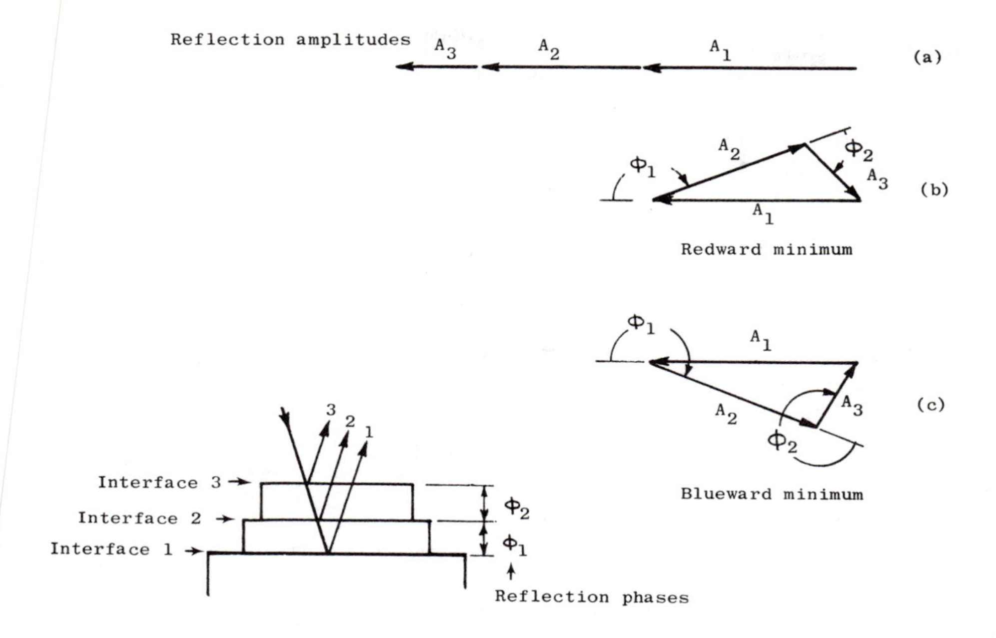

Fig. 9.35 Vector diagrams showing how the double minimum arises for selective absorbers when three reflecting interfaces are involved. The semitransparent metal film is assumed only to increase the magnitude of the amplitude vector A_2.

The vector diagram of Fig. 9.35 is modified in the case of tandem stacks by the fact that the phase angle ϕ_2 is $m2\pi$, where m is large. As a consequence, vector $\mathbf{A}_3$ goes through many revolutions to each revolution of ϕ_1, producing the double minimum in the absorptance versus wavelength shown in curve 4 of Fig. 9.34.

9.17 TYPICAL INTERFERENCE STACKS

In this section we show the optical behavior of a number of interference stacks of different types. Some of these stacks are of historical importance and some are from current research, but all are useful in demonstrating the variety of optical behavior that can be obtained.

In dealing with the older literature it is sometimes difficult to assess what chemical species are being utilized. In some cases the only way of knowing would be to have an Auger spectroscopy analysis (Section 9.23). This is especially true with the sulfides and oxides. Copper oxide can be either Cu_2O or CuO. The former is a semiconductor and has good intrinsic selectivity. The latter is only slightly selective. CuO decomposes to Cu_2O at 1026°C. Likewise, copper sulfide can be either Cu_2S or CuS. In general the more stable form is Cu_2S, as CuS decomposes to Cu_2S at 103°C. A thin film of any of these over bright copper yields selectivity, but if the underlying bright metal slowly reacts with oxygen or sulfur, the selectivity can disappear.

Selective surfaces of copper oxide can be formed in two ways. The first is by deposition of droplets of black copper oxide over anodized aluminum, as studied by Hottel and Unger (1959). The second is by treatment of a bright copper sheet with a 2:1 solution of NaOH and $NaClO_2$ for 3–10 min. The selectivity of both types is shown in Fig. 9.36, the curve for Cu_2O on copper being from Watson-Munro and Horwitz (1975). For comparison we show the curve for one of the early interference selective surfaces of Hass, Schroeder, and Turner (1956) and the curve for bulk CuO. Note that the interference stack can increase the rapidity of the transition from absorptance to reflectance, an important quality for surfaces hot enough to radiate TIR in the range from 3–5 μm. It also shows the basic disadvantage of the multilayer stacks in the rise in reflectance between 0.4 and 0.3 μm. For this reason a selective stack does not reach the high absorptance of the copper oxide/aluminum coatings.

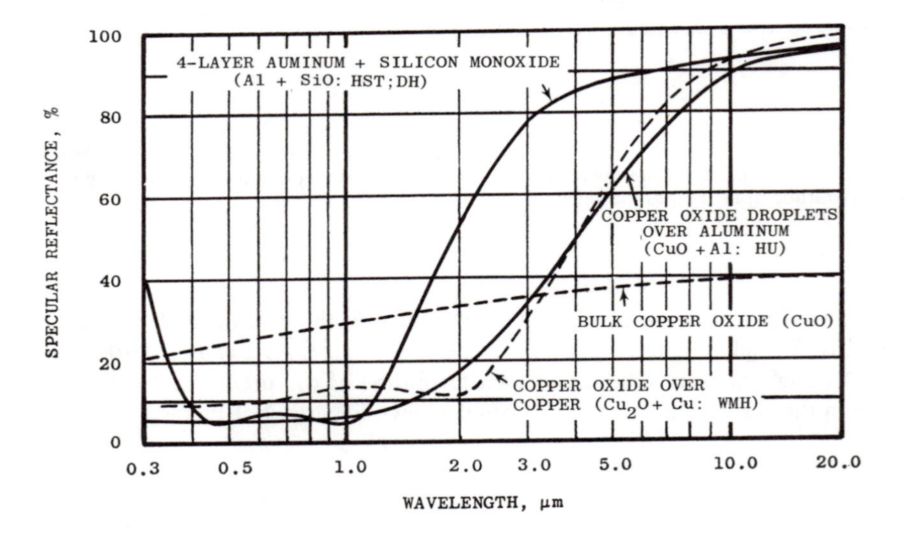

Fig. 9.36 Comparison of an early multilayer stack designed by Hass *et al.* (1956) with two copper oxide coatings. Both the particulate CuO over aluminum (Hottel and Unger, 1959) and the Cu_2O over copper (Watson-Munro and Horwitz, 1975) have a slow transition between absorptance and reflectance. A curve for bulk CuO is also shown.

The electroplated selective black nickel coatings developed by Tabor (1956) are shown in Fig. 9.37. The characteristic double maximum of a multilayer stack is apparent. These coatings are bright nickel on copper substrate, with layering from zinc sulfide. The actual sequence of currents in the plating bath containing the nickel and zinc is alternated so that the multilayer effect is obtained. The upper curve in Fig. 9.37 differs from the lower curve in that a thinner layer of zinc sulfide is used.

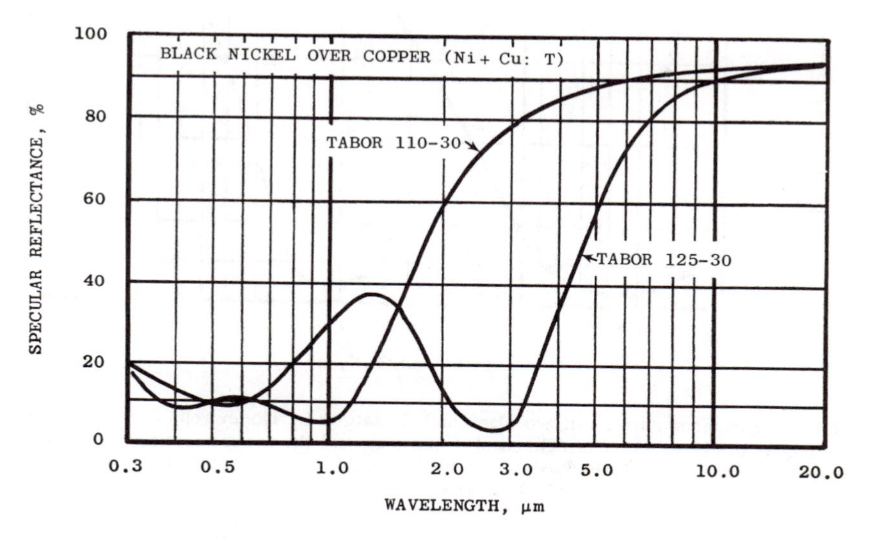

Fig. 9.37 Curves for the black nickel coatings developed by Tabor (1956), showing a typical interference stack behavior of a double minimum. These coatings are electroplated from a bath containing both the nickel and zinc ions necessary for the layers. The top curve is for the thicker layer of zinc sulfide over the base layer of reflective nickel.

In Fig. 9.38 we show two curves for a more recent type of black nickel produced by Honeywell. The two curves show the changes in the coating behavior due to postplating heat treatment. The upper curve is for a coating after production, showing a moderately high absorptance and low emittance. The lower curve shows the same coating after heat treatment at 500°C for 15 h. The transition from absorption to reflection has been moved to the red, increasing the solar absorptance to about 0.98. The net emittance has risen to about 23% for a blackbody at 200°C even though the emittance between 10 and 20 μm is still very low. This effect demonstrates that the cutoff can be controlled by heat treating such a coating far in excess of its normal anticipated operating temperature, which for black nickel is usually assumed to be less than 150°C. When the thickness of the zinc sulfide layer is increased, the second minimum shifts redward to beyond 2.0 μm and the intermediate maximum rises. Schmidt and his coworkers at Honeywell have perfected the black nickel process so that absorptances in excess of 0.90 are reported. These coatings are commercially produced in large sizes for several solar heating and cooling experimental installations in the United States.

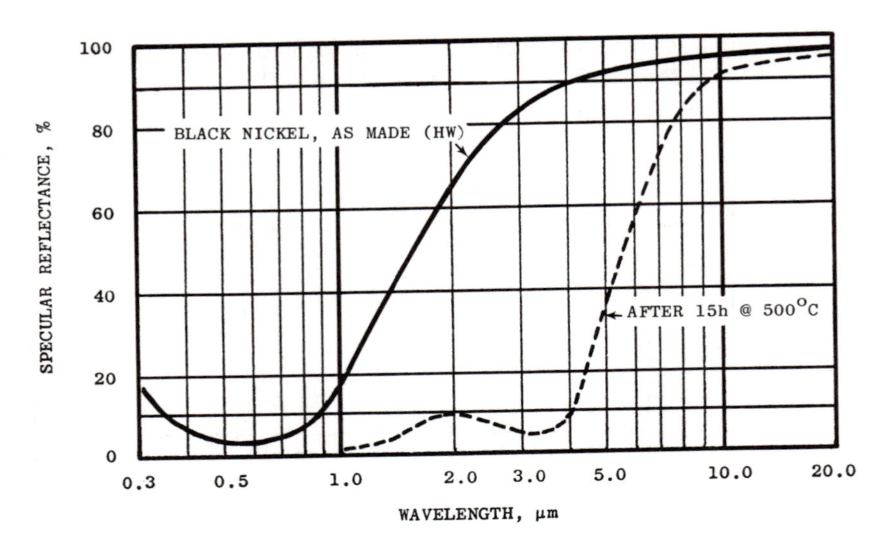

Fig. 9.38 Reflectance curves for two black nickel coatings by Honeywell. The upper curve is for the coating as produced by electrochemical deposition, and the lower curve is after heat treating, showing the shift in the transition wavelength made possible by this procedure.

In Fig. 9.39 the lower curve is the molybdenum multilayer stack performance developed by Schmidt at Honeywell for high-temperature space applications, where the multilayer consists of either aluminum oxide (Al_2O_3) or cerium oxide (CeO), or both, as the dielectric layers. The transition at 2.0 μm is steep, but the TIR reflectance is below 0.90 because of the limitations of evaporated or sputtered molybdenum. The upper curve is a modification of the basic molybdenum stack using a thin layer of silver atop the molybdenum. Gold also could be used to boost the reflectance of the base film, but gold diffuses readily into the molybdenum, whereupon the reflectance returns almost to the pure molybdenum values. Silver appears to be much more durable, and its diffusion can be inhibited by the use of a very thin layer of dielectric between the silver and molybdenum.

Tungsten also appears to be a good high-temperature material to use for the base film, having a reflectance in the TIR approximately equal to that of aluminum. The problem with tungsten is that when it is evaporated or sputtered its reflectance is significantly lower than that of the bulk material. CVD deposition of tungsten appears, however, to be potentially satisfactory for base films.

Figure 9.40 shows a number of stacks made by McKenney and Beauchamp (1974) at Helio Associates, comparing two stacks using silver as the base layer and one using copper. The simpler stack using silver (Helio 07) has good absorptance over a small range of the solar spectrum but too rapid a rise in reflectance at 1.0 μm; hence the net solar absorptance is low, on the order of 0.70–0.75. One can change the curves

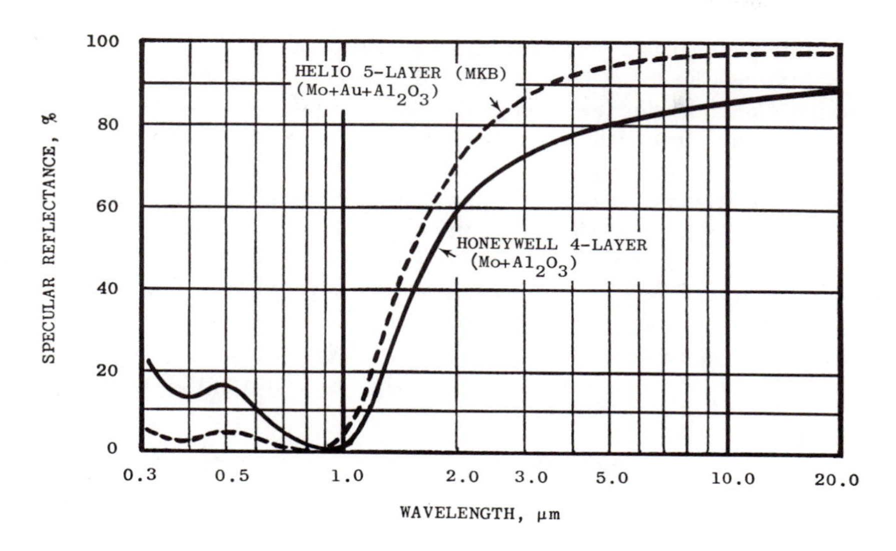

Fig. 9.39 Comparison of two interference stacks. The one by Schmidt (Honeywell) is the standard molybdenum/aluminum oxide four-layer stack, and the one by McKenney and Beauchamp (1974) is a stack with a base reflective layer of gold.

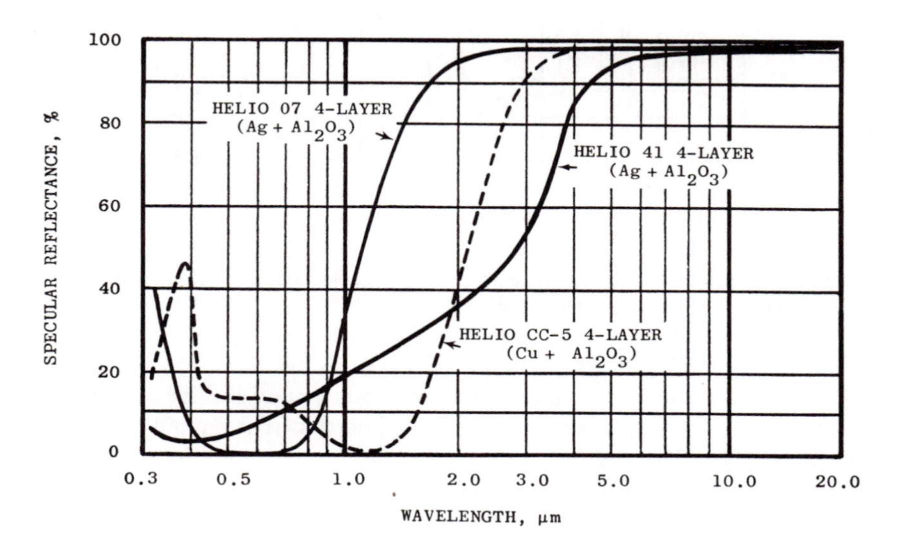

Fig. 9.40 Curves for a number of selective interference stacks by McKenney and Beauchamp (1974) using silver- and copper-based reflective films, showing the variety of curves possible.

significantly by changing thickness and deposition conditions to yield the curve Helio 41. In this case the transition is shifted to where it is desired, about 2.0 μm, but a reflectance peak develops at 0.4 μm, providing little improvement in net absorptance over Helio 07. In curve Helio CC-5, copper is substituted for the base layer and the solar absorptance is increased to more than 0.80.

Several interference stack curves in which copper is the base-layer film are shown in Fig. 9.41. Different stack designs result in significant changes in the behavior of the secondary maxima and minima. Copper is an interesting base-layer film, having high TIR reflectance and low cost, but it can be chemically attacked at elevated temperatures in air, forming copper oxide (CuO), which destroys the selectivity of the film. It is not known at present whether or not stopper layers of dielectrics can adequately protect the underlying copper film to yield satisfactory lifetime of operation.

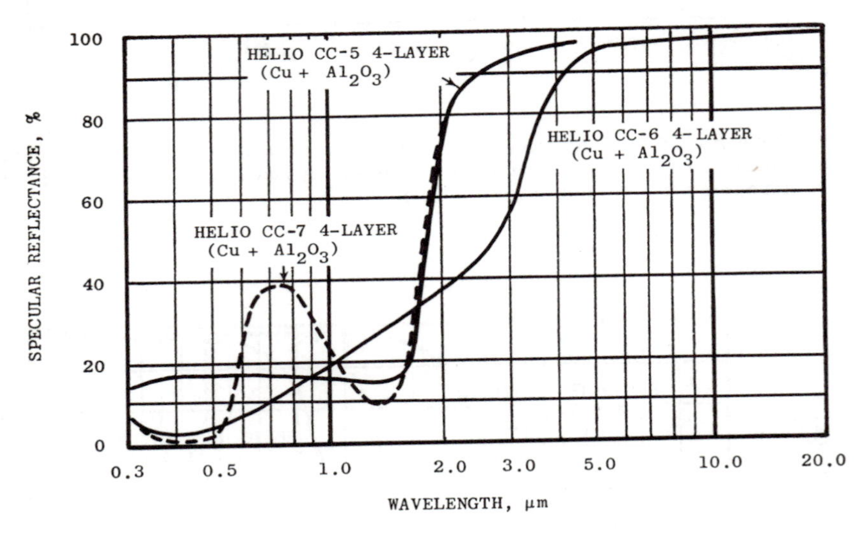

Fig. 9.41 Curves for three interference stacks on copper base film, by McKenney and Beauchamp (1974), showing the variation that can be obtained.

Figure 9.42 shows how one can combine a semiconductor with a metal to obtain an interference stack with excellent selective properties. The dashed curve is for the base layer of silver. The bulk copper sulfide (Cu_2S) curve shows the wavelength variation of reflectance for this semiconductor material. When the two curves are combined into a stack with the copper sulfide between the two silver layers, one gets an excellent interference stack with its transition at 1.6 μm.

Chromium, like nickel, produces excellent selective absorber surfaces. Several curves for different types of black chrome are shown in Fig. 9.43. Black chrome

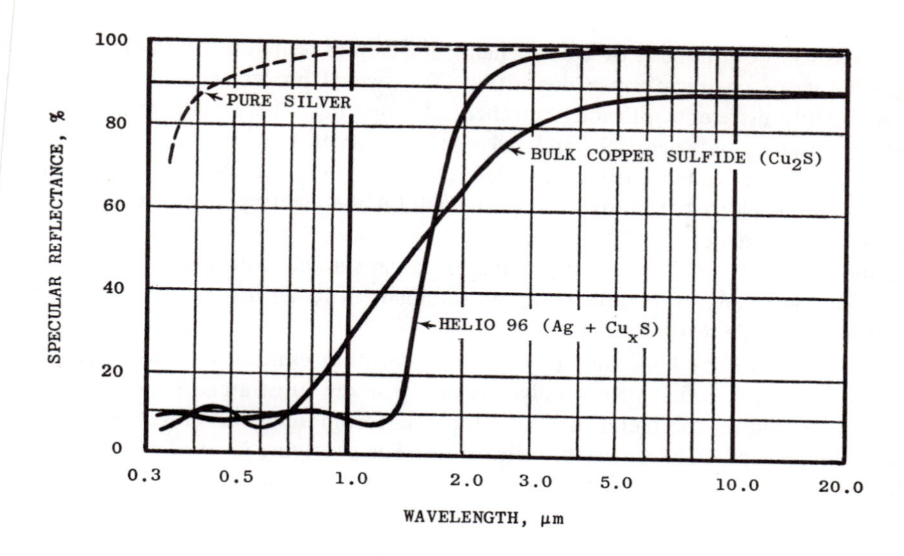

Fig. 9.42 Curves for a multilayer stack using a semiconductor for the dielectric between the two metal layers, with silver as the base reflective film (McKenney and Beauchamp, 1974).

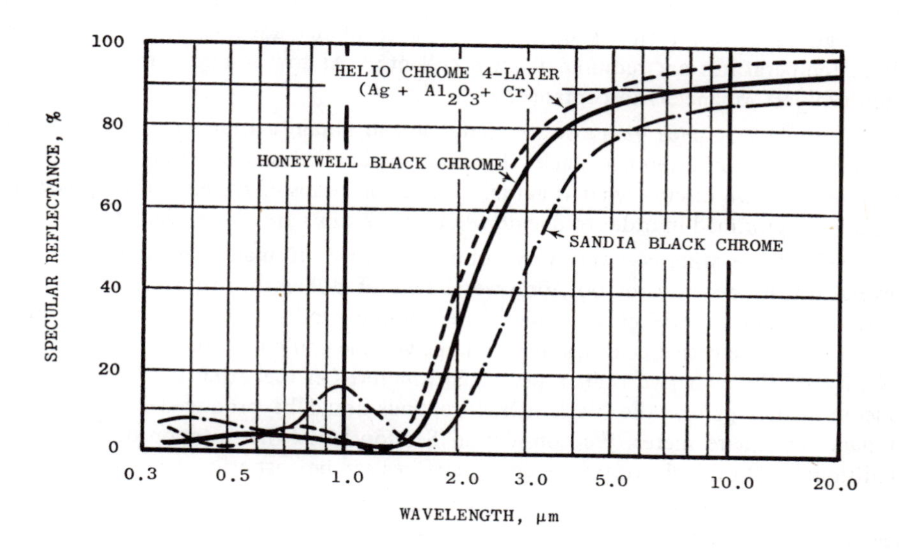

Fig. 9.43 Variation of reflectance with wavelength for several current types of black chrome coatings. The Helio coating (McKenney and Beauchamp, 1974) uses a multilayer stack of aluminum oxide and chromium over a base layer of stabilized silver. Curves for Honeywell and Sandia coatings are by Masterson and Seraphin (1975).

surfaces can be produced by electrochemical means as well as by vacuum evaporation, sputtering, or chemical vapor deposition. The several processes produce coatings of significantly different optical and mechanical properties. These coatings generally have an absorptance between 0.93 and 0.97 and emittance between 0.03 and 0.25. The Honeywell and Sandia coatings are electrochemical, and the Helio coating is vacuum deposited. Under electron-microscope resolution the electroplated coatings generally show surface structure, the size being less than a wavelength of visible light for the Honeywell coatings. The Sandia coating has larger surface structure and is nonspecular at visible wavelengths, complicating measurement of its absorptance with a spectrometer in the visible region.

The coating by McKenney (Helio) shows a lower emittance than the other coatings because the reflectance of silver is used for the substrates. An opaque layer of silver is placed on the metal substrate, separated from the metal substrate by a barrier layer to prevent diffusion of the silver ions. The overlying layer of aluminum oxide further stabilizes the silver from agglomeration and diffusion.

Note that there is considerable latitude for the placement of the transition from high absorptance to high reflectance for the black chrome coatings. This tradeoff can be used in the optimization of the solar collector to use these selective coatings where in some cases a slightly higher α is preferable to a lower ϵ. Note that for the Helio and Honeywell curves the Honeywell designs have higher α but about twice the emittance at 10 μm. This optimization will be discussed in detail in Chapter 12, but in general the coatings of lower emittance and lower absorptance are effective only when medium temperatures are required from a collector. At low temperatures the primary factor in performance is absorptance.

The above curves demonstrate that one can obtain a wide variation of detailed properties from interference stacks that could be applied to solar energy collection. In later sections we discuss what is actually gained in terms of system performance. This must be determined in order to decide whether the gains are cost effective.

Selective coating costs are very volume sensitive. In small quantities the costs are several dollars per square foot, or several tens of dollars per square meter, and upwards. Current experience with moderate volume production in continuous coaters indicates that selective coatings in very large volume could be produced at costs in the vicinity of three to five dollars per square meter. On the other hand, electroplated selective coatings like the black nickel coatings could be produced at two to three dollars per square meter. We consider in Section 16.19 the question of whether the additional cost of sophisticated selective surfaces can be cost effective.

Auger analyses of black nickel and black chrome coatings made by electrochemical deposition show that the layers are indistinct, whereas the layering for the typical interference stack coating has distinct layers with almost indistinguishable results. Interference phenomena in *graded* films are well known, but in multilayer vacuum technology they are seldom used because the same effect can be obtained with discrete layers of varying index.

9.18 WAVEFRONT-DISCRIMINATING MATERIALS

Wavefront-discriminating materials use physical surface roughness or particulate diameters to produce different optical effects in the visible and infrared. A surface that is rough to visible wavelengths can be a good mirror for the TIR simply because the wavelength of the TIR is greater than the dimension of the surface roughness. In the case of small particles the Mie effect operates, and the particles scatter strongly at short wavelengths and weakly at long.

One can make surfaces deliberately rough to emphasize the selectivity or use materials that naturally result in rough surfaces. An example of a naturally rough surface with highly selective performance is shown in Fig. 9.44, a surface formed by chemical vapor deposition of rhenium (Re). Quite a few metals, including tungsten, when their deposited thickness is great, show structural roughness of a selective nature. In these cases the tapered whiskers of metal are highly absorbent at short wavelengths but good reflectors at wavelengths longer than the whisker spacing.

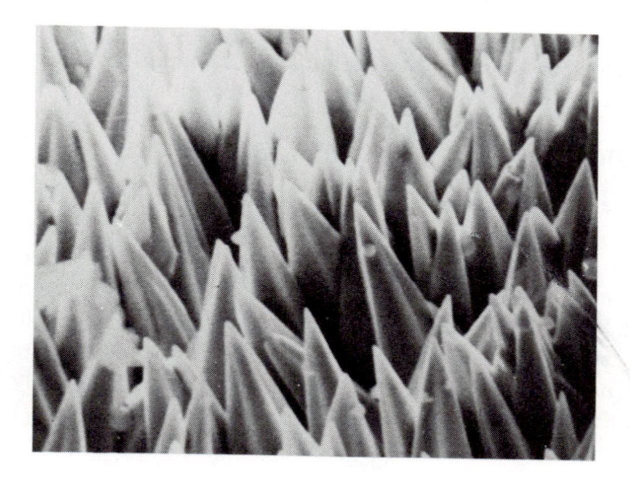

Fig. 9.44 Scanning electron micrograph of a thick layer of chemical-vapor-deposited (CVD) rhenium, displaying conical needles that are highly absorbent for near-normal-incidence solar wavelengths and that are reflective of TIR wavelengths.

We note that one can easily roughen a surface by fine grinding or sand-blasting so that deposition of a selective coating on the roughened surface results in high visible absorptance and high infrared reflectance and low emittance.

The deposition of some materials on metal substrates produces surface roughness of the type to enhance selectivity. The deposition of copper oxide (CuO) on metal substrates has been used by Kokoropoulos, Salam, and Daniels (1959) and lead sulfide on aluminum by Williams, Lappin, and Duffie (1963). A continuous layer of copper oxide (CuO) is not selective, as shown in Fig. 9.36.

9.19 EFFECTS OF SURFACE ROUGHNESS

There are two effects of surface roughness. The first concerns the measurement of performance. A nonspecular surface cannot be directly measured in the typical spectrometer because the reflected rays are scattered outside the acceptance angle of the spectrometer optics. The coatings using lead sulfide on aluminum are of this type. Even though the substrate is optically polished, the lead sulfide coating is particulate or dendritic and hence leads to scattering in the visible.

The second effect of roughness is that the emittance of a metal film on a rough substrate is increased over that on a smooth substrate. This effect has been extensively studied by Bennett (1974). The amount of increased emittance appears to be related to the micro-roughness and crystallite size. The net effects of micro-roughness on emittance appear to be complex, emphasizing that the measurements of relevance to solar energy applications should be done on the same substrate quality that must be used in the actual application. We have begun a series of such experiments and note that the emittance does rise when the surfaces are placed directly on commercial quality substrates. The wide variation of emittances observed in practice, therefore, can result from several effects, including (1) crystallite size, (2) impurities incorporated into the film during coating, and (3) surface roughness of the substrate.

9.20 DISPERSION OF METALS IN METALS

When precipitates of metals are made with impurities or as metal droplets in other metals, one can get broad-band absorption. A dispersion of droplets of vanadium (V), calcium (Ca), or niobium (Nb) in copper produces a broad resonance in the visible region of the spectrum but retains high reflectance in the TIR (Sievers, 1973).

Engelhard Industries has developed a commercial black in which small percentages of oxides cause the basic high reflectance of the principal metal to be lowered in the visible but not affected in the TIR. The coating most frequently cited has the following formula:

	Weight percent in solution	*Weight percent in film*
Au	8.630	89.5
Rh	0.039	0.4
Bi_2O_3	0.430	4.5
Cr_2O_3	0.020	0.2
SiO_2	0.167	1.7
BaO	0.360	3.7

The reflectance of one of these fired coatings is shown in Fig. 9.45. These coatings are applied as an organometallic solution and fired at 300–600°C to obtain the final coating. The organic portion of the solutions is based upon carboxylates, alcoholates, or mercaptides. The relationship between solution composition and optical behavior is

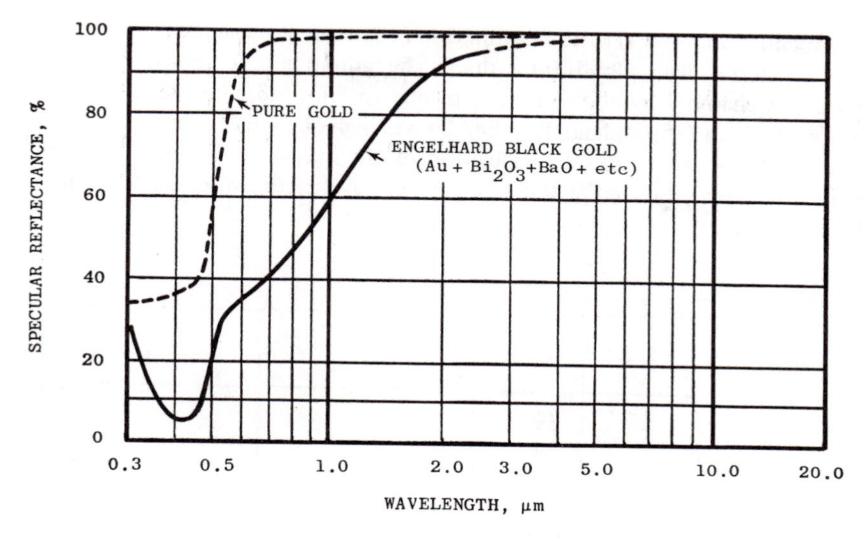

Fig. 9.45 Curves for pure gold and gold modified by the addition of oxides of a number of metals, prepared by Engelhard Industries using the fired-on-ceramics process (Langley, 1974). TIR emittance is low, but the visible absorptance is only about 0.5.

quite empirical. The film thicknesses after firing are about 1000 Å or less according to Langley (1974), who presents an excellent review of the ancient art of firing gold on ceramic or glass substrates. He notes that the addition of rhodium to the gold film results in the formation of rhodium oxide during firing and that without the rhodium the gold film would tend to agglomerate when heated above 300°C. Since the agglomeration of the most reflective metals–gold, silver, and copper–tends to be a problem in keeping the TIR emittance minimized, considerable progress may be possible in stabilizing the thin metal films at elevated temperatures. Another method for inhibiting agglomeration was discovered by Seraphin (1974), as reported in Section 9.13.

9.21 SELECTIVITY BY QUANTUM SIZE EFFECTS

Mancini and his coworkers (Burrafato *et al.*, 1975) have shown that quantum size effects (QSE) in ultrathin films produce selectivity with high visible absorptance and low infrared emittance. When a tandem stack with a highly reflective metal substrate is also used, the combined effects can make a good selective absorber for solar energy conversion. The critical thickness for QSE to be relevant is on the order of 20–30 Å for a metal because the free electron concentration is on the order of 10^{22} electrons/cm^3. For a degenerate semiconductor the free carrier concentration is lower, on the order of 10^{16} electrons/cm^3, and QSE becomes important in the 100–500 Å thickness domain.

Mancini and coworkers used a thin film of indium antimonide (InSb) as the semiconductor atop a silver film 1000 Å thick. The indium antimonide film was deposited in a vacuum chamber by allowing the particles to fall onto a heated surface, the temperature of which (approximately 1000°C) was high enough to evaporate the material but not high enough to cause dissociation of the indium antimonide molecules. An experimental film measured at 0.8- and 1.6-μm wavelength for several film thicknesses is shown in Fig. 9.46.

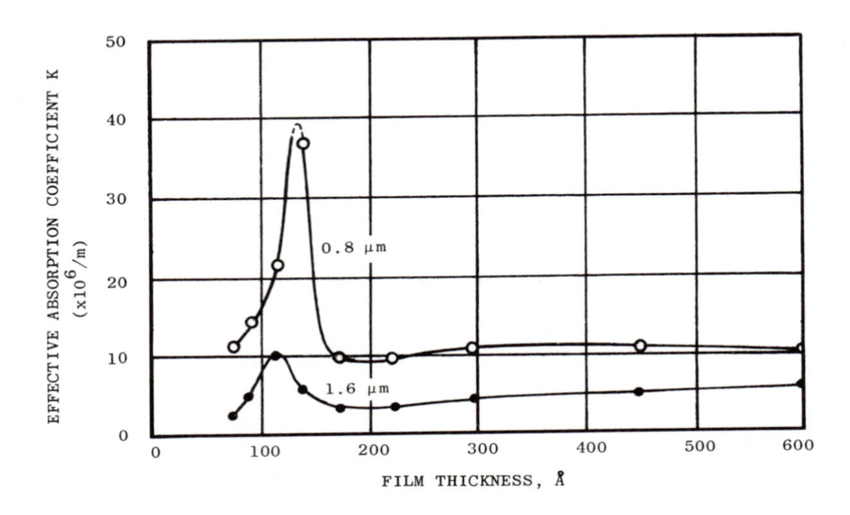

Fig. 9.46 Experimental curves for effective absorption coefficient *K* for evaporated thin films of indium antimonide over a reflecting layer of silver (Burrafato *et al.*, 1975).

Although indium antimonide was used to show a strong QSE at a thickness of 100–150 Å, its melting point is too low to allow good utilization in high-temperature solar collectors. It is important that similar QSE effects be explored in other semiconductors having sharp conduction band minima, which means a low value of the reduced electron mass, m_e/m_o, where m_e is the effective mass of the electron in the thin film and m_o is the free mass.

It is interesting to speculate that multilayer interference coatings have benefited from QSE effects. The ultrathin metal layer between the two dielectric layers is about 40 Å thick for molybdenum. It is this layer that yields the major absorption effect for the stack. It is well known that the apparent optical constants n and k for such thin layers differ widely from those of the bulk material. Part of this effect has been ascribed to the fact that such ultrathin films are not smooth and homogeneous but are made up of small islands of material. Ultrathin films are also variable in characteristics, depending on deposition conditions, interpreted usually as being due to absorption of gases and impurities in the deposited film. Mancini's work opens anew the study of the exact mechanism of light absorption in multilayer stacks.

9.22 REFLECTIVE SURFACE AGGLOMERATION

A basic problem of thin metal films is that they tend to agglomerate when heated. Silver is particularly bad in this respect, the film breaking up into small droplets, as shown in Fig. 9.47. To the eye an agglomerated film looks only slightly hazy under strong light and a dark background, but when a high-index material like silicon is superimposed, the scattering becomes serious. The reason for this behavior is that the wavelength of light is shortened by the index of refraction of the silicon, so 1.0-μm radiation "looks" like 0.25-μm radiation, and the scattering is greatly increased.

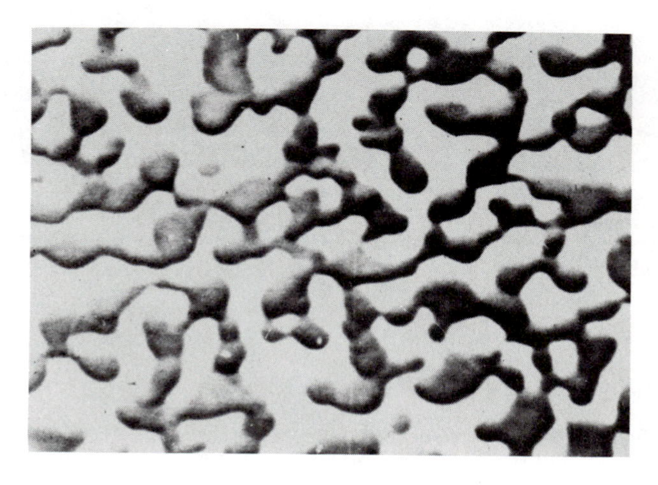

Fig. 9.47 Scanning electron micrograph of a silver surface after short heating at 600°C showing the agglomeration of the silver film into approximately 1-μm droplets.

Agglomeration was found to be particularly serious in the research done by Seraphin (1974) in preparing CVD silicon tandem stacks. The silver film could be prepared on the substrate material, but when placed into the CVD reactor and raised to the required temperature of 700–900°C, the silver would quickly agglomerate, resulting in poor stack performance. Seraphin and his coworkers discovered that the agglomeration could be prevented by placing a "stopper" layer over the silver before introducing the sample into the CVD chamber. Fully oxidized chromium (Cr_2O_3) was found to act both as a stopper layer and as a diffusion barrier.

In Fig. 9.48 we show a micrograph of a sample of silver overcoated with CVD silicon. The left-hand portion has been stabilized by a thin layer of chromium oxide, but the right-hand portion is not stabilized. The agglomeration is indicated by the appearance of many defects.

Fig. 9.48 Micrograph of a section of silicon deposited over silver after CVD processing. The left side has a layer of fully oxidized chromium (Cr_2O_3) and shows no defects, whereas the right side is without the stabilizing layer and shows many surface defects.

9.23 DIFFUSION OF IONS

One of the major questions concerning selective stacks is their operating lifetime under high-temperature conditions. The phenomenon of diffusion of ions is well known, but the extent to which it will limit lifetimes of typical selective surfaces is not known. Diffusion can become evident as a widespread but dilute dispersal of one metal ion into the neighboring metal ion or by the loss of interface optical characteristics. For example, if silicon were placed on a gold base film and the stack were heated, the gold would disappear by eutectic formation into the silicon and the interface would disappear. In the case of silver the eutectic temperature (870°C) is higher than anticipated operating temperatures (the eutectic temperature for gold is 370°C), so the physical layer remains intact. Silver, however, diffuses readily into silicon, and this diffusion would raise the infrared emittance of the silicon and degrade the stack.

The diffusion of ions is readily studied by use of Auger spectroscopy. The effectiveness of the chromium oxide (Cr_2O_3) stopper layer is quite apparent in the spectra by Wehner (1975) in the sharp drop in ionic species at the interface (Figs. 9.49 and 9.50). In these graphs the ordinate is relative number of atoms, and the abscissa is time of ionic erosion of the surface, related in an approximate way to depth in the sample. Figure 9.49 is for a sample of CVD silicon over a silver film placed on an oxidized silicon substrate. A layer of chromium oxide, underoxidized, is placed between the silver and the silicon. Note that the stoichiometric ratio of chromium to oxygen is about 1:1, as indicated by the equality of the chromium and oxygen peaks. In this case the barrier has not been effective because silicon has diffused into the silver.

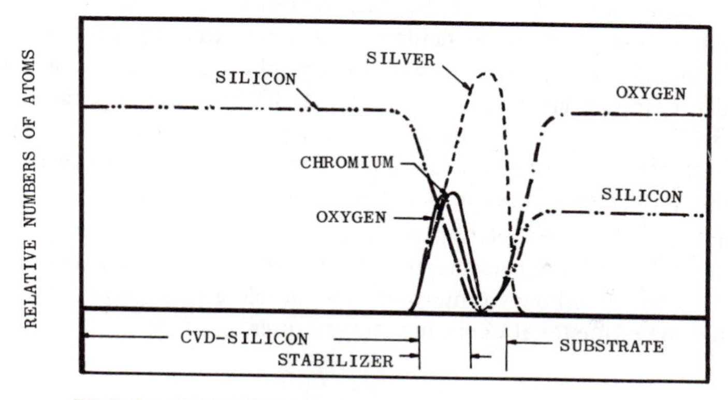

Fig. 9.49 Auger spectra of silicon deposited onto stabilized silver films on oxidized silicon substrates. This sample had an underoxidized stabilizer layer of chromium oxide, and no heat cycling other than in the CVD process (Wehner, 1975).

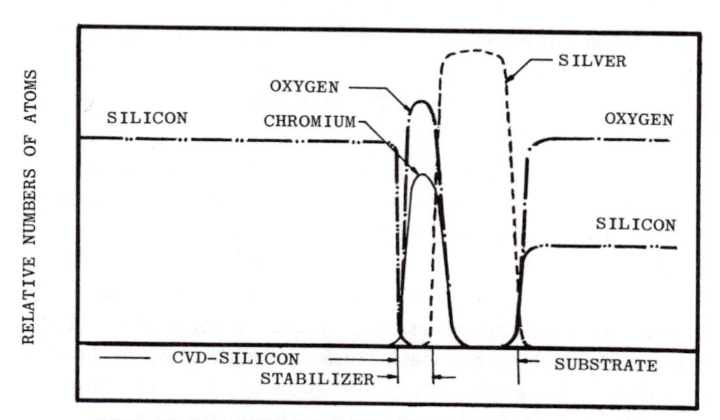

Fig. 9.50 Auger spectra of fully oxidized chromium. The sample was heat cycled for 150 h at 540°C and shows excellent containment of the silicon (Wehner, 1975).

Figure 9.50 is for a sample similar to that in Fig. 9.49 except that the chromium oxide barrier layer is now fully oxidized, indicating that all the chromium atoms are linked to oxygen. In this case the barrier is effective in preventing the migration of silicon into the silver. The steepness of the cutoff for silicon at the chromium oxide layer is at the resolution limit of the Auger technique.

The seriousness of diffusion increases with operating temperature of the multilayer stack. At low temperatures the effects may be negligible, but as one approaches 0.5 of the melting point of any of the stack materials, diffusion can become important. The case of the silicon/silver tandem stack shows that diffusion and agglomeration can be handled successfully, but with the complications and cost increases caused by the need for stopper layers, as shown in Fig. 9.30. The basic tandem stack now has become a five- or six-layer interference stack, depending upon how many layers are used in the nonreflection portion of the stack. The question is yet to be answered as to whether the improvements in performance will be cost effective.

An important question relating to coating lifetime is how to make accelerated tests. It is easy to find a temperature at which the stack will degrade in days, hours, or minutes. The problem is how to interpret these accelerated tests at elevated temperatures into lifetime estimates at operating temperatures.

9.24 LIFETIME TESTING OF COATINGS

In the figures below we show the behavior of some coatings before and after temperature cycling. Since all of these tests were at temperatures in excess of planned operating temperatures, on the order of 600°C compared to 500°C, it is difficult to assess the expected lifetimes. The tests also were of short duration compared to expected lifetimes. In one sense our experience has shown that, if degradation is going to occur, it tends to occur soon after heat treating. These "catastrophic" types of failure can be easily identified and avoided. The major problem is to identify degradations that occur slowly over years of operating lifetime.

In Fig. 9.51 we show a standard molybdenum/aluminum oxide selective coating before and after heat treating in a vacuum, showing that there is no degradation after exposure to 600°C for 48 h. In Fig. 9.52 we show the same type of stack but with a booster layer of gold over the base metal film, added to raise the TIR reflectance and improve the selectivity of the coating. Note the rapid disappearance of the gold into the molybdenum, the stack acting essentially like an unboosted stack after 48 h. In Fig. 9.53 we show the same type of basic stack, but with a booster layer of silver. Note that after 48 h there is little evidence of diffusion of the silver into the molybdenum. Although these tests are short, they serve to indicate that physical effects like diffusion are sensitive to the materials used, and that the proper choice of materials is essential to long operating lifetime at the temperature required for thermal conversion of solar energy into power, on the order of 500–600°C.

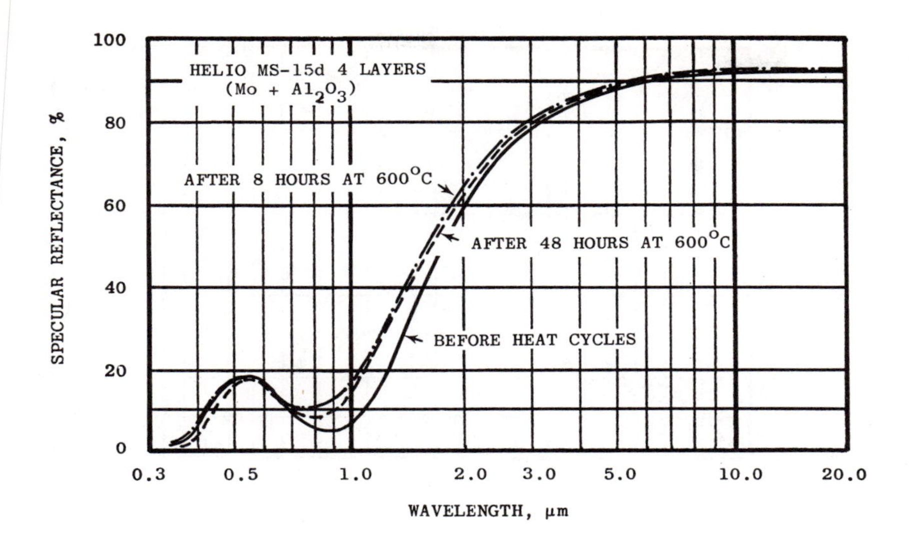

Fig. 9.51 Thermal aging curves for a standard molybdenum/aluminum oxide stack showing no changes over this period of time (McKenney and Beauchamp, 1974).

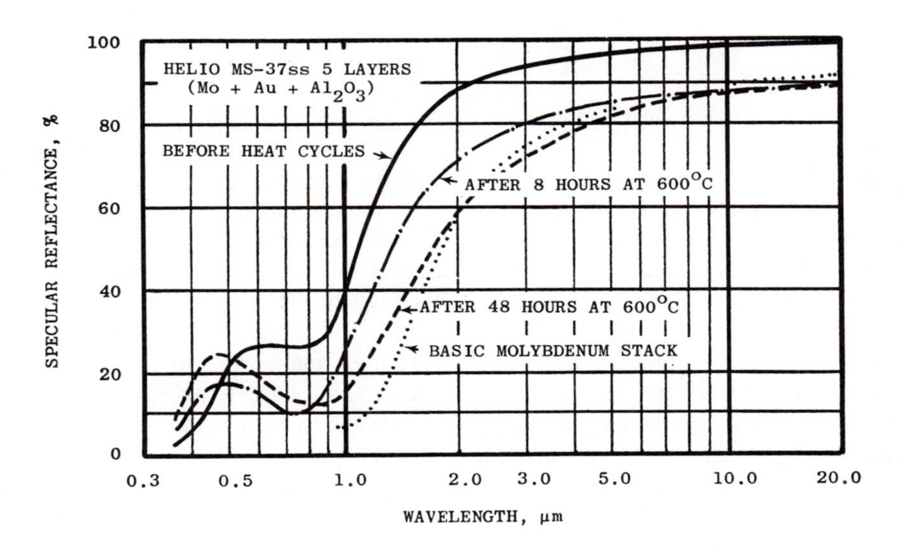

Fig. 9.52 Thermal aging curves for a gold-boosted molybdenum stack showing rapid reduction in reflectance with duration at 600°C, approaching that for the basic molybdenum stack (McKenney and Beauchamp, 1974).

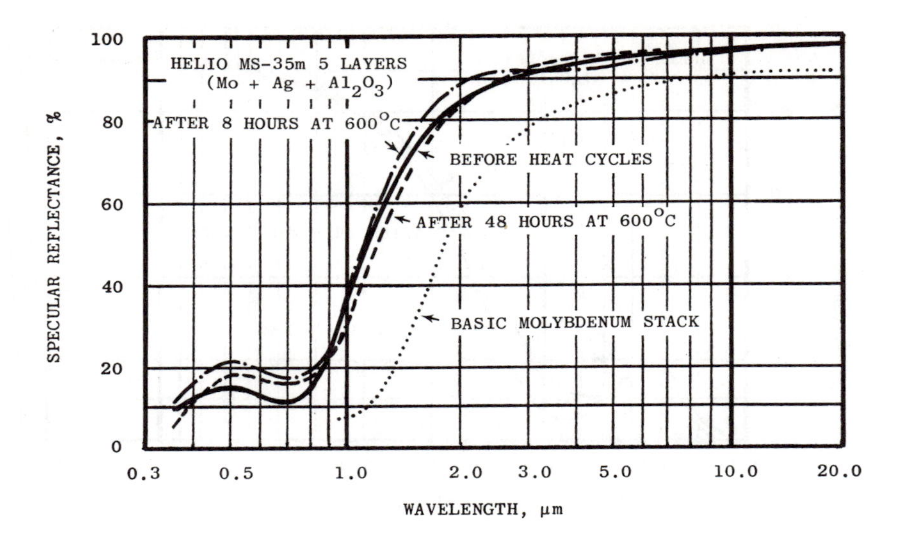

Fig. 9.53 Thermal aging curves for a silver-boosted molybdenum stack showing no reduction in reflectance after 48 h at 600°C (McKenney and Beauchamp, 1974).

A study of thermal aging effects in selective coatings by Masterson and Seraphin (1975) has shown some interesting effects and possibilities associated with thermal changes in optical behavior. A black nickel coating supplied by Honeywell (Fig. 9.54) is reasonably stable up to about 500°C, but above this temperature the cutoff wavelength rapidly shifts redward. A black chrome coating supplied by Sandia (Fig. 9.55) shows orderly and regular changes but little net shift in the cutoff wavelength. Precisely what is occurring in the structure of these coatings is not understood at present.

It is necessary to distinguish between failure for physical reasons and failure for chemical reasons. A physical change involves the mutual interaction of the layers. A chemical change involves interaction of the layers with an external source, such as is provided by the atmosphere. If a selective coating is to be operated in air, one must remember that natural air includes water vapor, carbon dioxide, and a wide variety of chemical aerosols. For example, telescope mirrors exposed to the environment show pitting and stains due to the joint action of moisture and sulfur compounds in the aerosol class, forming among other species sulfuric acid droplets of mild acidity on the mirror surface. It is therefore necessary after physical stability has been demonstrated to further test a selective coating under the most severe environment that it could experience.

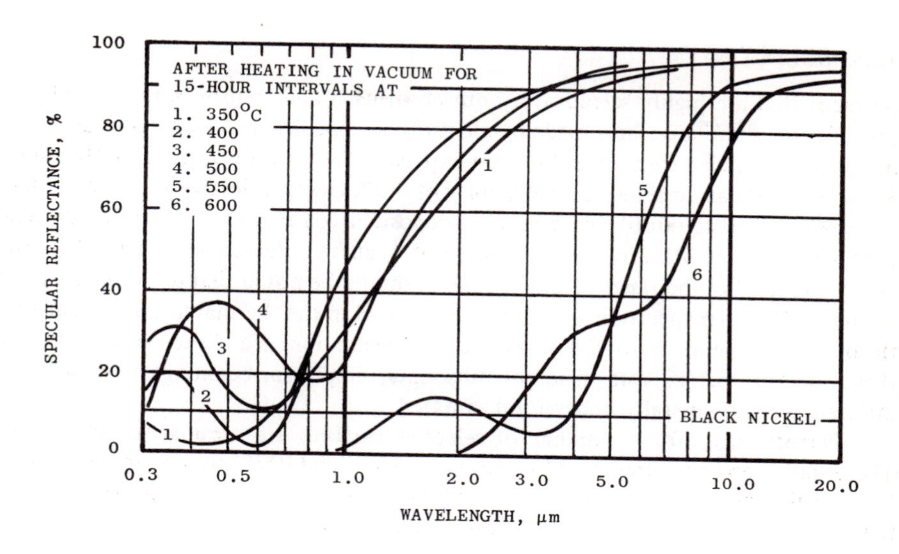

Fig. 9.54 Changes in specular reflectance (absolute) for a black nickel coating after samples are heated in vacuum for the times indicated (Masterson and Seraphin, 1975). Curves 5 and 6 are not extended into the visible because the surfaces were no longer specular.

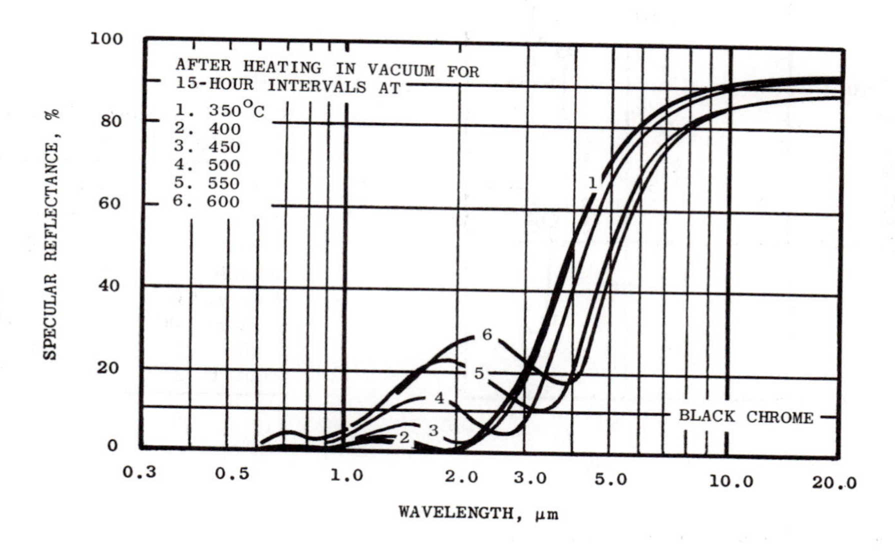

Fig. 9.55 Changes in specular reflectance (absolute) for a black chrome surface after heating in vacuum (Masterson and Seraphin, 1975).

9.25 SELECTIVITY

Surfaces that are highly selective in both absorption and transmission are technically possible. Selective absorbers made by electrolytic deposition of nickel plus zinc sulfide in layered or graded form offer the least expensive form for selectivities in the range of 8–12 and with absorptance in the vicinity of 0.92. Higher selectivities can be obtained by means of either multilayered vacuum-deposited coatings or chemical vapor deposition, where selectivities in the range of 15–50 can be obtained, but where the absorptance is limited to the range of 0.85–0.92. It should be noted that α is evaluated in the region of the solar spectrum whereas ϵ is evaluated in the thermal infrared, characteristic of the temperature of the surface under evaluation. We chose not to distinguish these spectral regions with different subscripts, simply for economy of notation, the difference being generally understood as stated.

Performances for a number of selective surfaces are summarized in Fig. 9.56. These data were measured by Masterson and Seraphin (1975) for coatings supplied by Honeywell, Inc., Sandia Corporation, and Helio Associates. The selectivity ratios for these several coatings at 150°C is as estimated in Table 9.3.

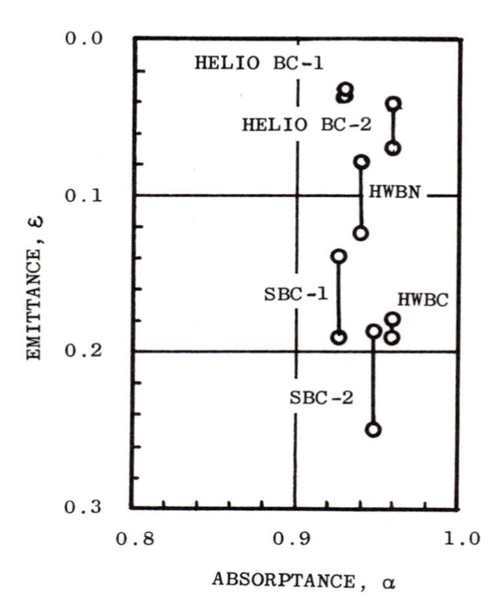

Fig. 9.56 Summary of absorptances and emittances for several selective surfaces with absorptances exceeding 0.90. Upper circles are for room temperature and lower circles for 300°C.

Table 9.3 Selectivity Ratios for Coatings Shown in Fig. 9.56

Type	α/ϵ	α
Helio BC-1	26	0.93
Helio BC-2	18	0.96
Honeywell BN	9.4	0.94
Honeywell BC	5.3	0.96
Sandia BC-1	5.7	0.93
Sandia BC-2	4.4	0.95

9.26 PROBLEMS

9.1. Estimate the emittance of a perfectly sharp wavelength cutoff selective absorber for a surface at 500°K when the solar absorptance ranges from 80–95% absorptance. Show the graphs you prepare to estimate this emittance, and plot the results in graphical form.

9.2. Prepare a distorted-λ graph for 100°C.

9.3. Estimate the net solar reflectance and absorptance of mirrors using aluminum, gold, copper, and molybdenum.

9.4. How much do the absorptances of Problem 9.1 increase when the emittances of gold and aluminum are added?

9.5. Plot a selective surface on a distorted-λ graph for the solar absorptance and for the TIR emittance for 100°C. Use the graph prepared in Problem 9.2. Use any selective surface you wish. (See Appendix J: Distorted-λ Graph.)

9.6. Discuss the criteria you would weigh in choosing a selective surface for the absorber in a collector designed for low-temperature applications. What additional factors become significant when high temperatures are to be used?

9.7. Discuss the problems raised in evaluating the temperature properties of coatings under normal and under emergency operations.

9.8. What factors enter in the discussion of the desirability and relevancy of accelerated testing of selective surfaces?

9.9. A reflective layer of silver is used for a selective surface, and its thickness is 1000 Å. The price of silver is \$5.00 per ounce. What is the value of the silver per square meter of absorber? If gold is to be substituted, below what must its price be in order to keep the value less than \$0.50 per square meter?

9.10. Why does the superposition of a layer of high index of refraction on top of an agglomerated (roughened) surface accentuate the scattering of that surface by reflected light? Make one or more diagrams to show the reasoning.

Chapter 10
Basic Elements of Heat Transfer

10.0 INTRODUCTION

In this chapter we give a concise review of aspects of thermodynamics and heat transfer that will be used and expanded upon when we discuss specific topics. The differential equations needed to describe transient conditions caused by the temporal variations of the source energy are highly complex, so in general we will limit the material to steady-state conditions. We will, however, discuss thermal inertia, which is one transient condition that can be estimated and is important in calculation of collector performance.

Limiting ourselves to steady-state conditions means that the performance calculated using these equations is the *best* to be expected from the system. In view of a common tendency to overestimate solar collector performance, these equations will at least provide a means of estimating the validity of claimed performance. One can realistically expect actual performance within 80–90% of the steady-state predictions if the systems analysis has been properly done using the steady-state equations.

10.1 MODES OF HEAT TRANSFER

Heat transfer occurs by three basic mechanisms and the consequences thereof: *radiation, conduction,* and *convection.* The mechanism through which the transfer occurs is the *photon* in the case of radiation and the *phonon* in the case of conduction. The laws governing each are simple and occur frequently in solar energy systems analysis. The third mode of heat transfer, convection, is in reality only a form of conduction followed by buoyancy effects of the heated fluid.

The several aspects of heat transfer that we will examine are:

1. radiation,
2. conduction,
3. natural convection,
4. forced convection, and
5. wind.

10.2 RADIATION

All heated bodies emit thermal electromagnetic radiation whose wavelengths and intensities are dependent upon the temperature of the body and its optical characteristics. The optical characteristics of significance are emittance (ϵ), absorptance (α), and reflectance (r). These three quantities are related by simple laws:

$$\text{emittance} = 1 - \text{reflectance}$$

$$\epsilon_\lambda = 1 - r_\lambda \tag{10.1}$$

$$\text{absorptance} = 1 - \text{reflectance}$$

$$\alpha_\lambda = 1 - r_\lambda. \tag{10.2}$$

Hence we have Kirchhoff's law:

$$\alpha_\lambda = \epsilon_\lambda. \tag{10.3}$$

The subscript λ is important to note because α, ϵ, and r change widely over the wavelength range of interest in solar energy systems for most materials. The few materials in which they do not vary with λ are termed *gray bodies,* and those with $\alpha =$ 1.00 for all wavelengths are termed *blackbodies.*

In the preceding chapter we discussed the usefulness in solar energy of selective media, where these properties can have widely different values and where $\alpha/\epsilon \neq 1$. When we refer to a selective surface with $\alpha/\epsilon \neq 1$, it does not mean that Kirchhoff's law is violated; it only means that α_λ is evaluated at a different λ than ϵ_λ.

The quantity of radiant thermal emission from a unit area of surface into a 2π steradian solid angle is given by the Stefan-Boltzmann equation:

$$(Q/A)_{\text{rad}} = \epsilon \sigma T^4, \tag{10.4}$$

where

$$\begin{aligned} \sigma &= 5.72 \times 10^{-5}\ \text{erg/cm}^2\text{deg}^4\text{sec} \\ &= 5.72 \times 10^{-8}\ \text{J/m}^2\text{deg}^4\text{sec} \\ &= 5.72 \times 10^{-8}\ \text{W/cm}^2\text{deg}^4 \end{aligned}$$

and T is measured in degrees Kelvin.

In general we will consider individual surfaces, and the transfer between surfaces is then as noted in Fig. 10.1. Surfaces 1 and 2 have no back-surface losses, so the radiant flux is emitted only from surfaces facing the gap.

Radiation from surface 1 is then

$$\text{downward} = \epsilon_1 \sigma T_1^4,$$

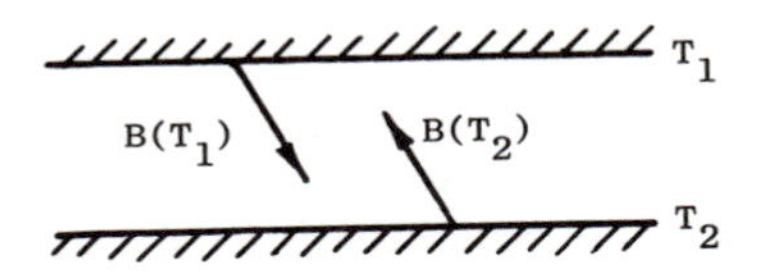

Fig. 10.1 Diagram illustrating the radiative exchange between two surfaces at different temperatures. The flux vectors B(T) represent hemispherical fluxes.

and from surface 2,

$$\text{upward} = \epsilon_2 \sigma T_2^4 .$$

The net gain or loss of surface 1 to surface 2 is then

$$\Delta Q = \sigma(\epsilon_1 T_1^4 - \epsilon_2 T_2^4). \qquad (10.5)$$

If surfaces 1 and 2 have back sides, the radiative exchange must then take into account the radiation field from their surroundings, and the losses or gains are calculated in the same manner.

A useful plot of the hemispherical radiant flux from a blackbody as a function of temperature is shown in Fig. 10.2 and Table 10.1.

The energy within a given wavelength interval varies with temperature for a blackbody radiator. The spectrum emitted is given by Planck's law:

$$J_\lambda = c_1 \lambda^{-5} [\exp(c_2/\lambda T) - 1]^{-1} d\lambda . \qquad (10.6)$$

Values of the Planck function are shown in Fig. 10.3. The integral under the Planck function yields the total hemispherical flux as given by the Stefan-Boltzmann equation. For a real surface the emission is given by multiplying Eq. (10.6) by the emittance ϵ_λ at each wavelength.

Note that although the amplitude of the emittance varies with temperature of the surface and the peak emittance shifts with temperature, at *no* wavelength does a cooler blackbody emit more energy per unit solid angle than a hotter body. This fact is important when we deal with thermal reemission from a heated surface.

The wavelength of maximum emission shifts redward with decreasing temperature. The value of λ_{max} is given by Wein's equation,

$$\lambda_{max} = b/T, \qquad (10.7)$$

where $b = 2.88 \times 10^3$ μm°K.

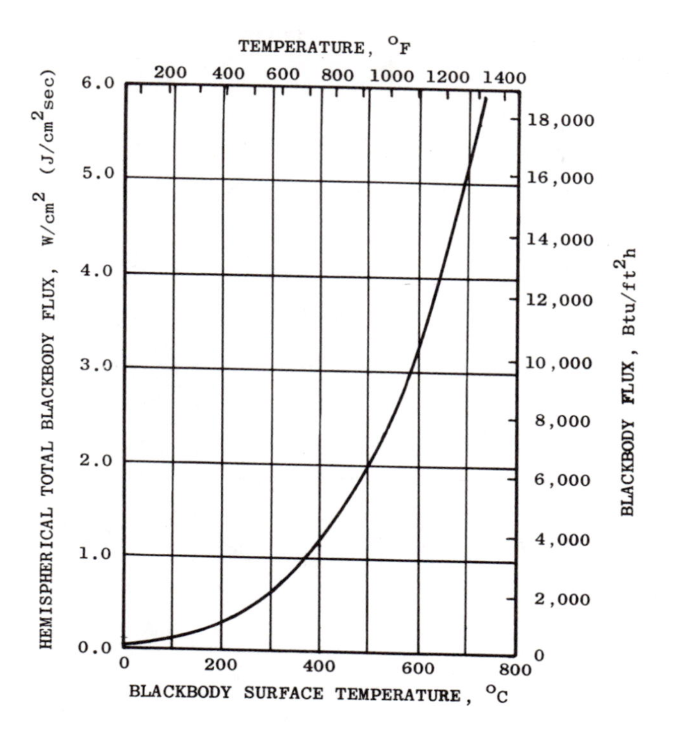

Fig. 10.2 Variation of hemispherical total blackbody flux as a function of blackbody surface temperature, in metric and English units.

Table 10.1 Radiative Thermal Flux from a Blackbody

Temperature			λ_{max},	Flux at λ_{max},	*Hemispherical total flux*		
°K	°C	°F	μm	W/cm²μm	W/cm²	kW/m²	Btu/ft²h
5800	5523	9974	0.50	9500	6420	64,200	2.03×10^6
1200	923	1693	2.4	3.209	11.80	118.0	37,400
1000	723	1333	2.9	1.290	5.679	56.79	18,010
800	523	973	3.6	0.423	2.326	23.26	7378
700	423	793	4.1	0.217	1.364	13.64	4327
600	323	613	4.8	0.100	0.736	7.36	2335
500	223	433	5.8	0.0403	0.355	3.55	1126
400	123	253	7.2	0.0132	0.145	1.45	460
300	23	73	9.6	0.0031	0.046	0.46	146
273	0	32	10.5	0.0019	0.030	0.30	96
Solar flux, sea level, desert					0.097	0.97	308

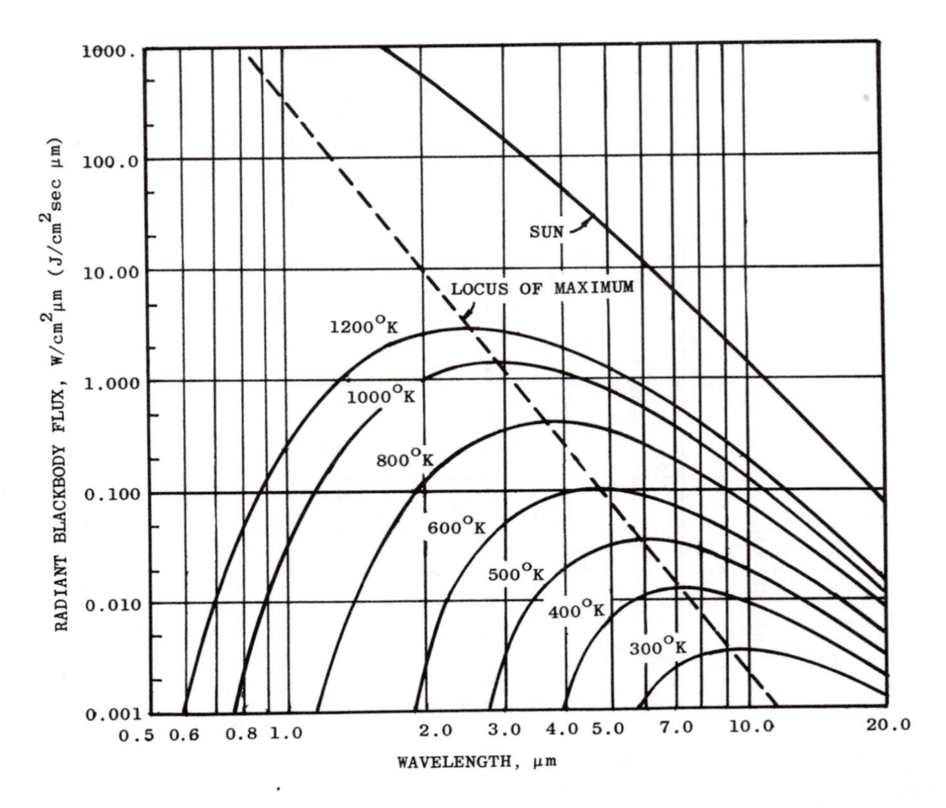

Fig. 10.3 Relationship between radiant flux and wavelength for blackbody radiator. Energy scale is in units of W/cm^2 per micrometer of wavelength and radiated into a 2π steradian solid angle.

10.3 CONDUCTION

Whereas radiative transfer is a function of absolute temperature, conduction is a function of temperature gradient dT/ds. The relationship between the quantity of heat transferred per unit cross-sectional area per unit time is given by

$$(Q/A)_{\text{cond}} = k(dT/ds), \tag{10.8}$$

where k is the coefficient of heat conductivity of the medium. For a slab of material of thickness s, and geometry as shown in Fig. 10.4, the equation becomes

$$\text{Slab:} \qquad (Q/A)_{\text{cond}} = Q_{\text{cond}} = (A/s)k(T_2 - T_1). \tag{10.9}$$

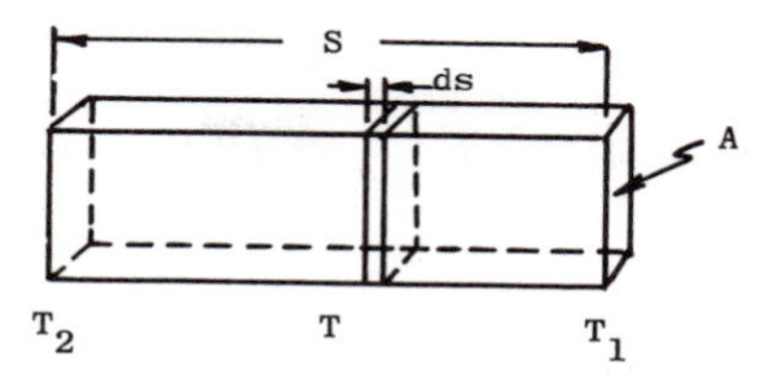

Fig. 10.4 Diagram showing the geometrical relationships for conductive heat transfer in a parallel slab of cross section A.

Conduction is a function of the shape of the body losing heat and the configuration of the surroundings. Two shapes other than plates, shown in Fig. 10.5, are of practical use in solar applications: the insulated pipe and the buried pipe. The relevant equations are

$$\text{Concentric pipe:} \quad Q_{\text{cond}} = \frac{2\pi L}{\ln(r_2/r_1)} k(T_2 - T_1), \tag{10.10}$$

where r_1 is the inner diameter and r_2 is the outer diameter, and

$$\text{Buried pipe:} \quad Q_{\text{cond}} = \frac{2\pi L}{\ln(2d/r)} k(T_2 - T_1), \tag{10.11}$$

where d is the depth at which the pipe is buried and r is the diameter of the pipe. Since pipes and buried pipes occur frequently in the analysis of solar energy systems, we tabulate for convenience in Table 10.2 the values of the natural logarithms occurring in Eqs. (10.10) and (10.11).

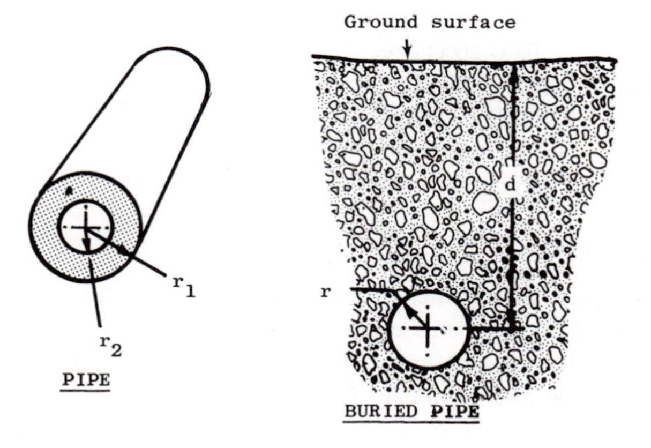

Fig. 10.5 Definition of relevant quantities for the cases of a pipe and a buried pipe of negligible wall thickness. When the buried pipe has a significant wall thickness both equations are used.

Table 10.2 is useful in that it conveys a feeling for the importance of changing either the thickness of the insulation around a pipe or the depth at which a pipe is buried. A change in depth of a factor of 10, from $2d/r = 20$ to 200, barely reduces the conduction heat loss by a factor of 2.

Table 10.2 Logarithmic Ratios

r_2/r_1	$\ln(r_2/r_1)$	$2d/r$	$\ln(2d/r)$
10.0	2.303	200	5.298
8.0	2.079	100	4.605
6.0	1.792	80	4.382
4.0	1.386	60	4.094
2.0	0.693	40	3.689
1.8	0.588	20	2.996
1.6	0.470	18	2.890
1.4	0.336	16	2.773
1.2	0.182	14	2.639
1.10	0.095	12	2.485
1.05	0.049	11	2.398
1.00	0.000	10	2.303

The quantity preceding $k(T_2 - T_1)$ in Eqs. (10.9)–(10.11) is called the *shape factor*; tables of these factors can be found in heat-transfer handbooks.

Conductivity k varies widely for different materials, being high for metals, low for nonmetals, and very low for gases. Conductivity also varies slowly with temperature, but for most solar energy applications a single value is adequate to yield accurate results for net heat losses for a system. The range of conductivities is

Metals	=	1.00–0.10 cal/sec C°cm
Nonmetals	=	0.005–0.0001
Gases	=	0.0003–0.00002.

Thermal conductivity for various materials is given in Table 10.3.

10.4 CONVECTION

Convection is a direct consequence of conduction. In most treatments in engineering applications, convection is considered as an entity, combining the conductive term for the gas or liquid with the true convective term. We will discuss both the combined approach and the separated approach. The combined approach yields equations useful in practical engineering of solar collectors in air. If we wish to examine the dependence of convection on linear dimensions when different gases than air are involved, it becomes necessary to distinguish the conduction term from the true convection term.

Table 10.3 Thermal Conductivity of Various Materials

Material	Temperature, °C	Thermal conductivity: *cal cm/sec C°cm²*	*W cm/C°m²*	*Btu in./h F°ft²*
Metals				
Aluminum	100	0.49	20500	1130
	200	0.55	23000	1590
	400	0.76	31800	2210
	600	1.01	42300	2932
Brass	0	0.25	10400	720
Copper	20	0.934	39300	2711
	100	0.908	38100	2640
	300	0.89	37200	2580
Magnesium	20	0.37	15400	1070
Steel	18	0.115	4800	330
	100	0.107	4500	310
Nonmetals				
Asbestos fiber	500	0.00019	8.0	0.55
Brick	20	0.0015	62	4.3
Concrete	20	0.00071	30	2.0
Cotton wool	20	0.000043	1.8	0.12
Diatomaceous earth	20	0.00013	5.4	0.38
Granite	20	0.0045	188	13
Glass	20	0.0017 –0.0025	71 –100	5.0 –7.2
Glass wool	20	0.000081	3.4	0.24
Infusorial earth	100	0.00034	14.0	1.0
	300	0.00040	17.0	1.2
Magnesia brick	500	0.0050	210	14.0
Plaster of Paris	20	0.00070	29.0	2.0
Sand, dry	20	0.00093	39.0	2.6
Soil, dry	20	0.00033	14.0	0.96
wet	20	0.0010	42.0	2.9
Wood, across grain	20	0.00010 –0.00030	4.2 –12.6	0.29 –0.87
Gases				
Air	20	0.000057	2.4	0.13
Carbon dioxide	20	0.000031	1.3	0.07
Helium	20	0.000339	14.2	0.79
Hydrogen	100	0.000369	15.4	0.86

When a heated body is in contact with a gas or liquid medium having a lower temperature, or the converse, heat is transferred between the surface and the fluid. The heat change causes a change in density of the fluid. This change is large for a gas, being expressed by the gas laws, so the positive (or negative) buoyancy of the gas adjacent to the heated (cooled) surface causes it to move. A new quantity of fluid then comes into contact with the surface and the process continues. This mass motion of the fluid adjacent to the surface, shown schematically in Fig. 10.6, is termed *convection.* In the case of liquids the density change is less, but convection is readily induced in liquids also. An important practical result of buoyancy in liquids is the *thermosiphon,* used to move water through solar water-heater collectors.

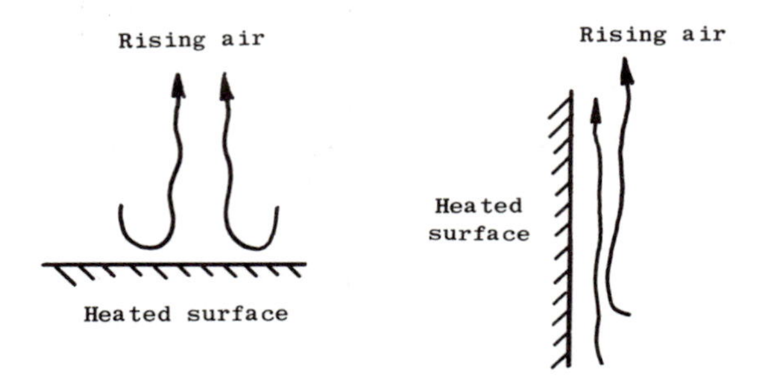

Fig. 10.6 Basic development of convective flow from air in contact with a heated surface. If the heated surface faces downward then no stream of moving air is created, except for edge effects. Loss is then only by conduction.

In Fig. 10.7 we show schematically how convection acts in the transfer of heat from one solid body to one that is nearby but separated by a gap of fluid, gas, or liquid. Heat is conveyed by direct conduction as well as by mass motion of the fluid. In situations where convection is inhibited, such as by honeycomb baffles or the lowering of the pressure of the gas, conduction still operates. A honeycomb adds a new path for conduction. A gas at low pressure still conducts heat as at atmospheric pressure, a point that we will examine in detail later.

Steady-state convection, when there is no external driving force like wind or pumped flow, means a balance between the buoyancy effects and the viscosity of the fluid medium. The force generated by the density difference is exactly balanced by the force generated by the viscosity, the velocity of convection being the variable generating the viscosity force. In the mathematical description of convection the usual treatment is to use *dimensional analysis* to arrive at the appropriate mathematical forms, allowing arbitrary constants to yield the correct answer in a limited domain of applicability.

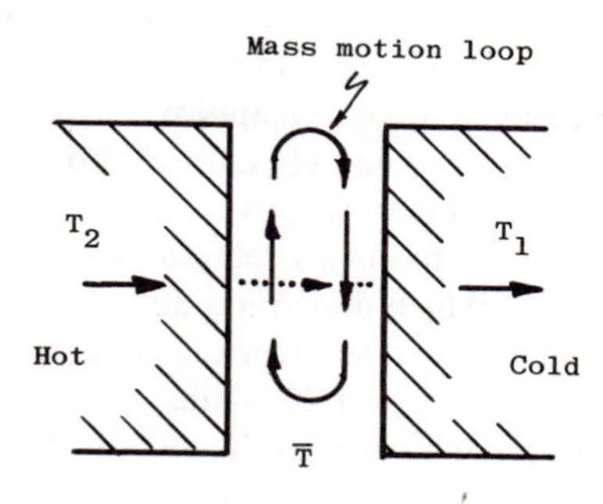

Fig. 10.7 Diagram illustrating convective heat transfer between two surfaces immersed in a fluid. The dotted arrow indicates a conductance term that is added to the mass motion term.

10.5 DIMENSIONAL ANALYSIS

Dimensional analysis is a powerful tool in the analysis of convective loss, but it is blind in regard to details of the physical processes involved. It is, for example, quite useful in predicting performance of systems having approximately similar characteristics but can be inadequate in examining the behavior of a system over a wide dynamic range of conditions. A case where this deficiency is evident is in the examination of the changes of convective heat loss in a system as the pressure is reduced from atmospheric pressure to vacuum. We will therefore examine the basics of convective heat loss to see how to modify the usual dimensionless equations so that they are applicable over a wide change in individual variables involved in the dimensionless "numbers."

Dimensional analysis states that two systems having the same dimensionless numbers are identical in behavior with regard to those particular numbers. Convection involves nine physical quantities:

Dimension	L
Thermal conductivity	k
Fluid velocity	v
Fluid density	ρ
Dynamic viscosity	μ
Specific heat	c_p
Heat transfer coefficient	h
Volume expansivity of the gas	y
Temperature difference	ΔT.

The above terms are all well-defined physical quantities except for L. The dimension L is a matter of some potential confusion because it is not strictly a dimension in the usual sense. This dimensional quantity appears in several of the dimensionless numbers, in each case being somewhat loosely defined. For example, in the case of the Reynolds number as applied to performance of an airplane it is the "characteristic length" of an airplane. An airplane has many relevant lengths, such as the chord of the wing or the length of the body. Which dimension do you use? The answer is that you use the correct length to yield the right numerical value of the quantity being calculated. Turned around, the effective length is that value of L that gives the right numerical magnitude of the matter being evaluated. In dimensional analysis we are primarily concerned with the exponent on L, and when L appears as L^3, as it does for convection, the cubic factor is exact while the numerical value of L is not exact. In the case of convection the closest association of L with reality is to specify L as the *minimum dimension between the hot and cold walls of the convecting system.* When tubes are considered, L is generally taken as the tube diameter. If a surface is convecting into free space, the recommended value for L is about 60 cm (2 ft), according to Tabor (1955, 1956).

The usual dimensionless groups used for convection studies are:

$$\begin{array}{ll} \text{Nusselt number (Nu)} & h(L/k) \\ \text{Reynolds number (Re)} & \rho v L/\mu \\ \text{Prandtl number (Pr)} & \mu c_p/k \\ \text{Grasshof number (Gr)} & g\gamma \Delta T L^3 \rho^2/\mu^2 . \end{array} \tag{10.12}$$

We will first restrict our attention to the traditional method of using these equations and then proceed to use a derivation based on the Rayleigh number as suggested to us by Tabor. The Rayleigh number is

$$\text{Rayleigh number (Ra)} \quad (\rho^2 g\gamma \Delta T L^3/\mu k) c_p . \tag{10.13}$$

The student should note that there are three Rayleigh numbers used in physics, and this particular form is the correct one to use for flat surfaces. A second form, for cylinders, is used in Section 11.25, Eq. (11.81).

It can be noted that the Rayleigh number actually is the product of

$$\mathrm{Ra} = \mathrm{Gr} \cdot \mathrm{Pr}. \tag{10.14}$$

The above dimensionless numbers do have physical meanings that are useful to keep in mind:

Nu = nondimensional heat transfer coefficient,
Re = ratio of inertial to viscous forces, used in the description of forced convection, where the fluid has an initial velocity with respect to the heated surface,
Pr = ratio of molecular diffusivities of momentum with respect to heat,
Gr = ratio of buoyant to viscous force; replaces Re in the cases of natural convection,
Ra = ratio of thermal buoyance to viscous inertia.

Dimensionless numbers are useful in describing the similarities between systems of different physical natures. Two systems having identical values for a particular number will behave convectively in the same way with regard to the characteristic described by that number. This means, for example, that a system having a gaseous heat-transfer medium will behave like one having a liquid heat-transfer medium when the characteristic under examination has the same number. It also means that if one selects the proper physical parameters one can exactly scale one system by means of a smaller system.

The key number we seek to determine for a system is the Nusselt number, because the heat loss is determined by Newton's equation

$$Q/A = h\,\Delta T, \tag{10.15}$$

where h appears in the Nusselt number as

$$h = (\mathrm{Nu})(k/L). \tag{10.16}$$

In using dimensionless numbers we seek an equation that relates what we seek, the value of the Nusselt number, in terms of the variable number. For forced convection this variable number is the Reynolds number (Re). For thermally induced convection it is the product of the Grasshof number and the Prandtl number, which is equal to the Rayleigh number:

$$\begin{aligned} &\text{forced convection:} \quad \mathrm{Nu} = f(\mathrm{Re}) \\ \text{or} \\ &\text{natural convection:} \quad \mathrm{Nu} = f(\mathrm{Gr})\cdot g(\mathrm{Pr}). \end{aligned} \tag{10.17}$$

The functional form of Eq. (10.17) can be determined in two ways: (1) by physical theory or (2) by experimentation. The first avenue is involved. A strong argument is gained from experiments. In Fig. 10.8 we show the experimental curve for a wide variety of fluids (gaseous and liquid), shapes, velocities, and so forth, correlated against the product of the Grasshof number and the Prandtl number. The required function for the Nusselt number, and hence for the heat transfer coefficient h, is obtained by fitting this curve with an equation.

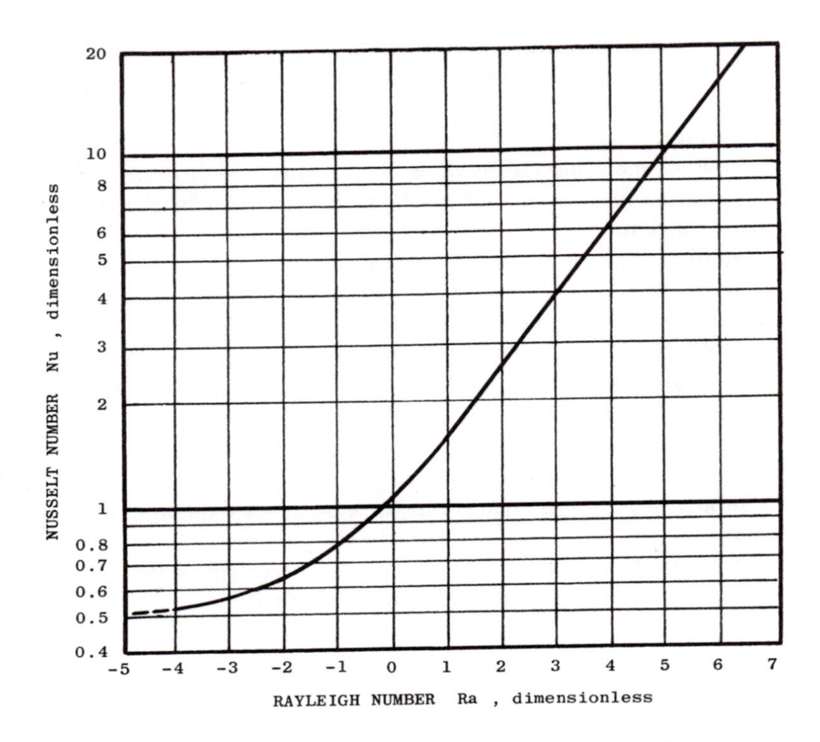

Fig. 10.8 Mean experimental Nusselt numbers versus Rayleigh number for natural convection from a horizontal cylinder, where $L = D$ (from McAdams, 1954).

The usual procedure is to fit the experimental curve with an exponent and a multiplicative constant, defining the range of Gr·Pr, or Re, to which the equation applies. We need to find a procedure with wider applicability in order to examine what happens to h when, for example, we take a system from atmospheric pressure to vacuum. To obtain this relationship in greater generality we need to use physical theory to establish the mathematical form for $f(\mathrm{Gr}) \cdot g(\mathrm{Pr})$ in the limit. We can approach this goal by applying physical theory: in particular, the kinetic theory of gases.

One of the great triumphs of the kinetic theory of gases was the prediction of certain effects later confirmed by experiment. The one factor we wish to draw upon at this time is the prediction that thermal *conductance* in a gas is constant, not varying as the pressure of the gas is changed. This statement is fully confirmed by experiment when the pressure is greater than that wherein the mean free path of a gas molecule is longer than the dimensions of the system under examination. In the case of solar collector dimensions this means that for pressures greater than 10^{-2} gm/cm^2 (1 atm = 1000 gm/cm^2), the conductance is constant.

In Fig. 10.9 we show the relationship between Nu and Re in linear form. It is clear that for low values of Re the value of Nu asymptotically approaches the value of ½. This asymptote defines the limit that we wish to associate with the invariance of heat conductivity with pressure, noted above. Fitting a curve of exponential form, we obtain

$$\mathrm{Nu} = \tfrac{1}{2} + \exp[\tfrac{1}{2}\log(\mathrm{Re} - 1)], \tag{10.18}$$

which fits the curve over a wide range, departing only at high values of the product. This indicates the need for higher-order terms, or a readjustment of the constants if a system in the higher range is to be evaluated. The exponent to a logarithm is readily reduced to a power term, yielding the equation

$$\mathrm{Nu} = \tfrac{1}{2} + 0.61(\mathrm{Re})^{0.217}. \tag{10.19}$$

Because Nu is given by Eq. (10.12), we see that

$$h = k/2L = 0.61k(\mathrm{Re})^{0.217}. \tag{10.20}$$

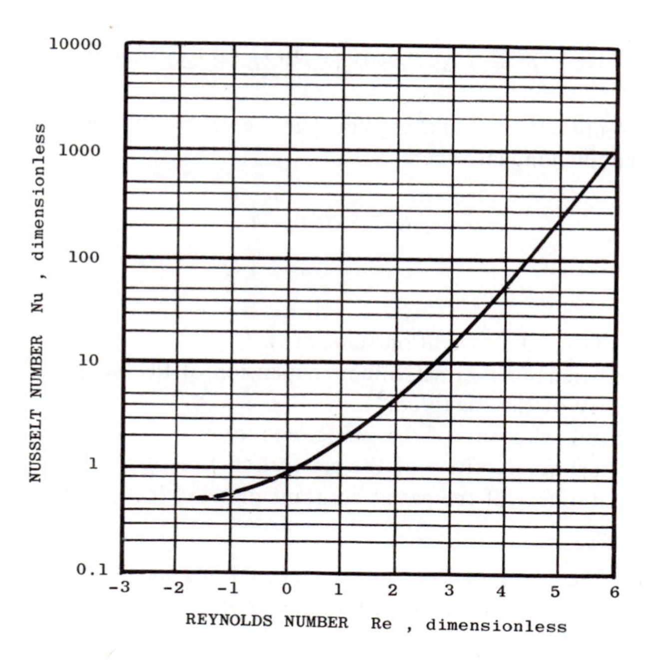

Fig. 10.9 Mean experimental Nusselt numbers versus Reynolds number for forced flow normal to single cylinders (from McAdams, 1954).

We note that in this equation the first term is *half* the conductance between two surfaces separated a distance L at the temperature difference ΔT. This apparent difference is because the experimental data are for loss from the surface of a cylinder into free space. It can be interpreted as indicating that the *effective* distance for conduction is twice the pipe diameter. With this interpretation we obtain the proper separation of convection from total heat loss.

10.6 PARALLEL SURFACES

The importance of separating the pure convection from the pure conduction term becomes apparent when we wish to examine how Nu varies with changes in the system, such as a change in surface separation L, temperature difference ΔT, absolute temperature T, gas species, or gas pressure.

If we define a quantity a as the modulus of the gas, we can express the heat-transfer coefficient h by an equation of the form

$$h = k\Delta T/L + \text{const} \cdot ka^n L^{1-3n} \Delta T^{1+n}, \tag{10.21}$$

where n is in the range of ¼–½ and is often taken as ¼ over the range of frequent usage for solar collectors. In this case we have, as an example,

$$h = k\Delta T/L + 0.5ka^{0.25}L^{-0.25}\Delta T^{1.25}, \tag{10.22}$$

where a is a combination of the physical parameters of the medium involved in the convection phenomenon, and where

$$a = gy\rho^2 c_p/\mu k, \tag{10.23}$$

where

g = gravitational acceleration (cm/sec^2),
y = coefficient of thermal volume expansion of the gas (cm^3/cm^{3}C°),
ρ = density of the gas (g/cm^3),
μ = dynamic viscosity of the gas (g/cm sec),
k = thermal conductivity of the gas (cal/cm C°sec),
c_p = specific heat of the gas at constant pressure (cal/g C°).

In the usual handling of the convection equations one usually finds the term ka^n even though the term k appears within a. The product ka^n is often referred to as the *convectivity* of the fluid.

In Table 10.4 (page 338) we tabulate the quantities appearing in the modulus a of the gas for a number of gases.

The use of the Rayleigh number to describe system behavior is based upon the fact that it contains the terms that appear in both the Grasshof and Prandtl numbers. We do not have the flexibility of having different powers for the Gr and Pr numbers as implied by Eq. (10.17). The basic reason we can simplify the relationship is that Pr does not vary significantly over system variables, such as temperature. We therefore will proceed in this section to utilize Tabor's suggestion and use the Rayleigh number.

The basic experimental data summarized by Tabor (1958) are shown in Fig. 10.10, where the horizontal axis is Ra. One could take these data and apply analytical analysis to derive an exponent for ΔT and L to fit the data for specific ranges of interest. We would prefer instead to use the experimental curve as the fundamental datum and develop the procedures for calculation using this type of curve instead of power formulas.

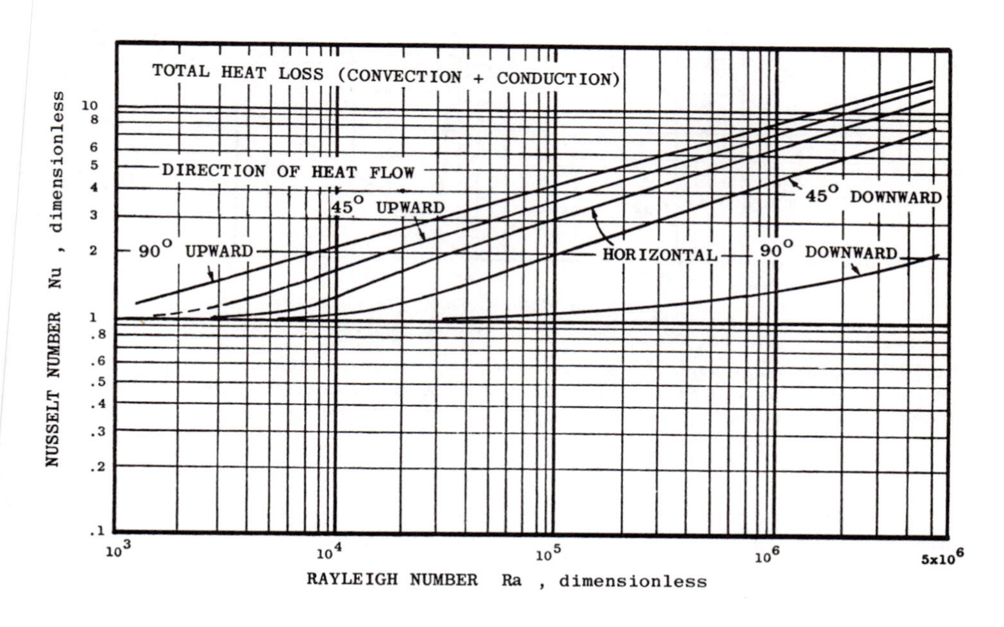

Fig. 10.10 Generalized graph of heat flow between parallel plates, from Tabor (1958). Directions of heat flow from top to bottom curve are, respectively: 90° upward; 45° upward; horizontal; 45° downward; 90° downward.

The basic point of departure in our approach to convection is *first* to separate the true conduction term from the total observed heat loss. The remaining term is the true convection term. Taking the Tabor data and subtracting the conduction term, we are pleased to find that the resultant curves are more homogeneous than the raw data. We are then prepared to do analytical studies of great depth on the variation of convection with input parameters. The curves for pure convection are shown in Fig. 10.11 for five orientations of the pair of parallel plates.

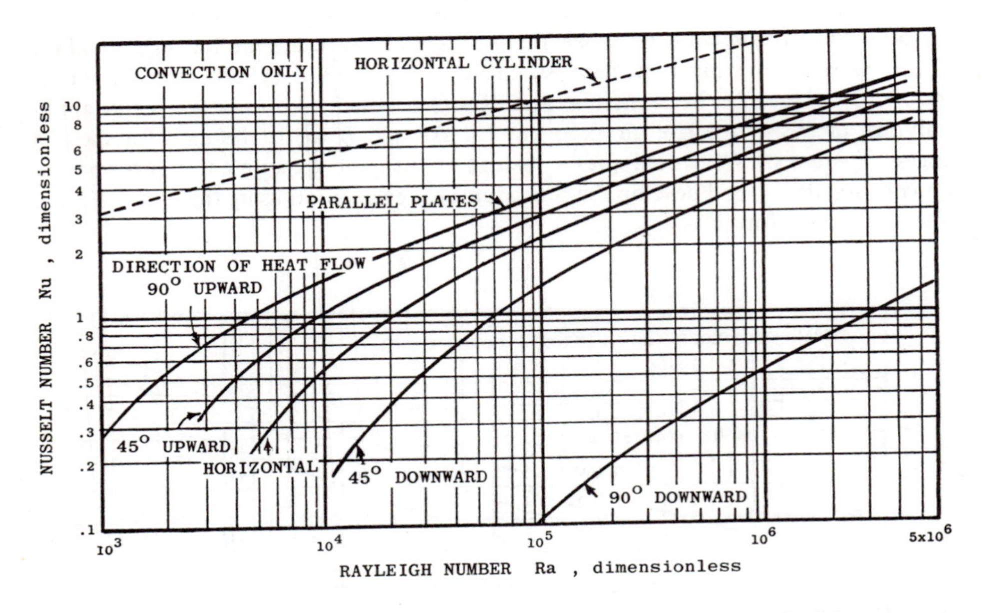

Fig. 10.11 Generalized graph of heat flow between parallel plates after subtracting the conduction loss, from Tabor (1958). Directions of heat flow from top curve to bottom curve are, respectively: 90° upward; 45° upward; horizontal; 45° downward; 90° downward.

The correctness of the separation of the conduction term from the experimental curves is confirmed if we do the same for the McAdams (1954) data. Subtracting Nu = 0.5 from the data curve and replotting it, we have the straight line shown in Fig. 10.12. Note that this curve remains straight at low Reynolds numbers, whereas the parallel plate curves drop significantly at low Rayleigh numbers. The reason for this is that there is a minimum value of Ra below which there is *no* convection, but because the Re data refer to *forced* airflow there is an airflow even at low values of Re. The distinction between laminar and turbulent flow for low values of Re is not noticeable at the scale of accuracy displayed by the mean curve in the McAdams data.

There is a further significance in these two reduced curves. The straight-line relationship for forced convection means that a single power law equation is sufficient to express the value of the Nusselt number over a wide range of Reynolds numbers. The curve in the relationship for thermally induced convection prohibits the use of a power law equation except over a limited range of Rayleigh numbers. For this reason

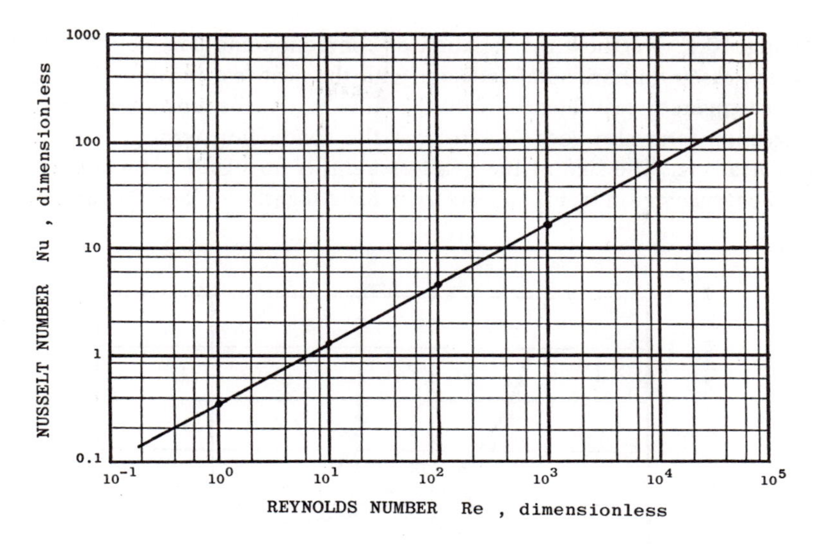

Fig. 10.12 Graph of the pure convection term from the data by McAdams (1954) for heat loss from circular cylinders in a forced flow of air.

the several power law fits as summarized by Duffie and Beckman (1974) are required. We will use this experimental calibration curve directly. If one prefers to use analytical forms to calculate Nu, we recommend that new fits be used to describe the true convection term separately from the combined conduction term.

10.7 VARIATION OF CONVECTIVE LOSS WITH SURFACE SEPARATION

To explore how both the conduction and convection terms vary with surface separation we will not use the equation with an exponent, as in Eq. (10.21), but go directly to the curves based on the Tabor (1958) data. If we were to use Eq. (10.22), the term $L^{-0.25}$ power would have the convection term going to infinity as L goes to zero. This is not the experimental case, as is implied in the fact that as the Rayleigh number goes below a critical value, about 1700, thermal convection ceases. If we use the Tabor data we note that convection is well behaved as L goes to zero.

To calculate the convection and conduction loss for a specific surface spacing or temperature difference it is necessary to repeat the several steps. These steps in summary are:

1. Calculate the value of the Rayleigh number: $\text{Ra} = aL^3\Delta T$.
2. Determine from the graph in Fig. 10.11 the value of the Nusselt number (Nu).
3. Calculate $\text{Nu}\cdot k$.
4. Add the conduction term k to obtain the total loss coefficient $\text{Nu}\cdot k + k$.
5. Multiply by $\Delta T/L$ to obtain the heat loss coefficient h.

In Fig. 10.13 we show the variation of convection and conduction with separation for the case of a horizontal collector with the heat loss flowing directly upward and mean temperature difference of 40 C°. Note that as the separation L goes to zero the convection term also goes to zero, but the conduction term goes to very large values of heat loss. The sum of the two losses shows no significant reduction beyond about 2.5-cm spacing—in fact it appears to increase slightly for larger separations.

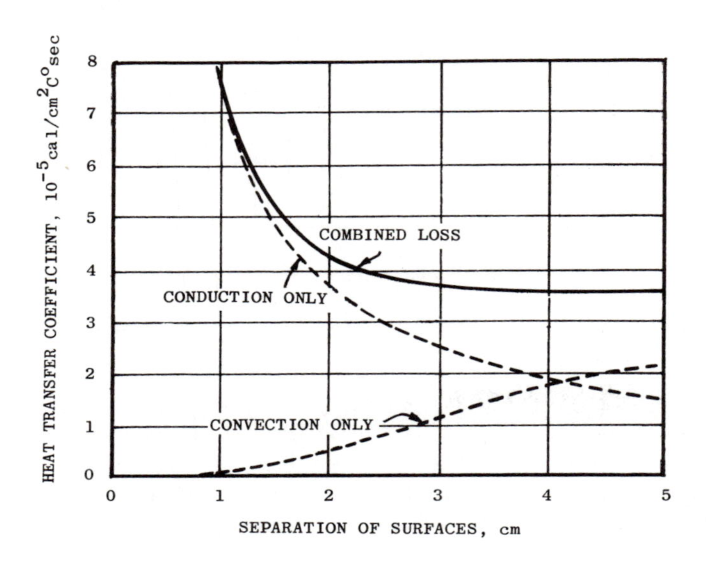

Fig. 10.13 Relationship between heat-transfer coefficient h and the separation of two horizontal surfaces L, for the case where Ra = 8000 for L = 5.0 cm.

10.8 VARIATION OF CONVECTIVE LOSS WITH TEMPERATURE

The modulus of a gas changes with temperature of the gas, and since the modulus appears in the Rayleigh number as

$$\mathrm{Ra} = a\Delta T L^3, \tag{10.24}$$

it is important to have an understanding of how it varies with temperature. In calculating the curve shown in Fig. 10.14 we used values for the different terms in a as tabulated in the *Handbook of Chemistry and Physics,* 51st edition, and the *AIP Handbook,* with some data extrapolated to 100°C.

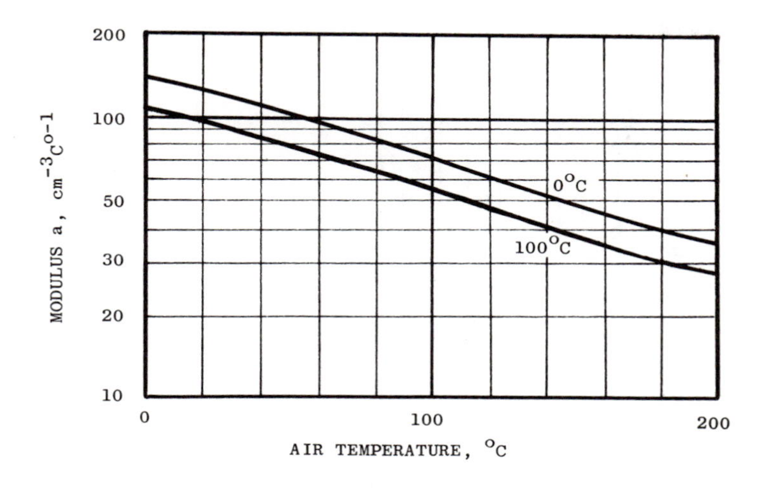

Fig. 10.14 Variation of the modulus of air as a function of temperature.

10.9 VARIATION OF CONVECTIVE LOSS WITH GAS SPECIES

The major challenge with medium-temperature nonconcentrating and low-concentrating collectors is how to raise the efficiency of energy extraction. In general, there is little to be gained by increasing the absorbance α of a collector since present values are between 0.90 and 0.98. One seeks the major gains through reducing the heat losses. We have already seen that selective surfaces can reduce radiation losses from the absorber. We have also seen that honeycomb structures also can suppress convection, but have certain problems that detract from their universal application as convection suppressors. We have also shown that selective windows can reduce either radiation or convection, or both, to some degree. In this section we will consider the general question of how high-molecular-weight gases can reduce convection. The final part of this section will deal with gases at reduced pressure, concluding with the behavior of the collector when fully evacuated.

The numerical values for the physical properties of different gases arranged according to molecular weight are shown in graphical form in Figs. 10.15–10.17. Figure 10.15 shows the variation of viscosity with molecular weight, but whereas kinetic gas theory does relate molecular weight to viscosity, this graph mixes molecules of different numbers of atoms, which causes the expected trends to be less obvious.

In Fig. 10.16 we show the variation of specific heat as a function of molecular weight. The downward trend of c_p with increasing molecular weight is evident, but the scatter is considerable. The species with the lowest specific heats are the noble gases, the exception being bromine.

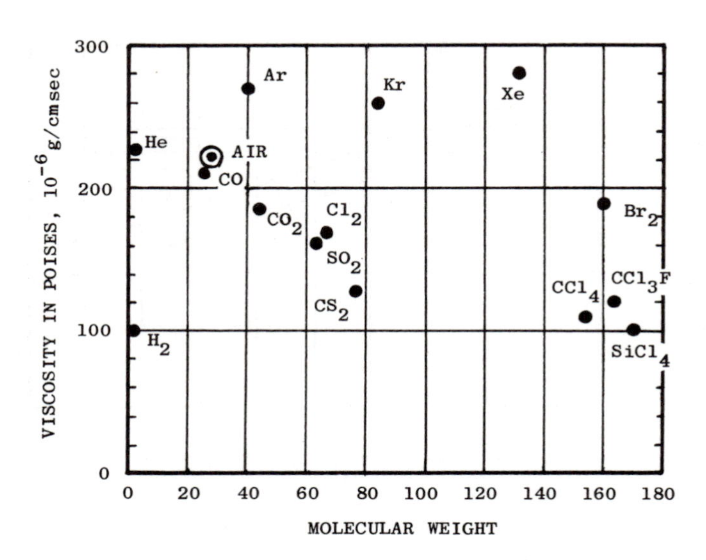

Fig. 10.15 Graph of the relationship between viscosity and molecular weight of the gas, evaluated at $T = 100°C$.

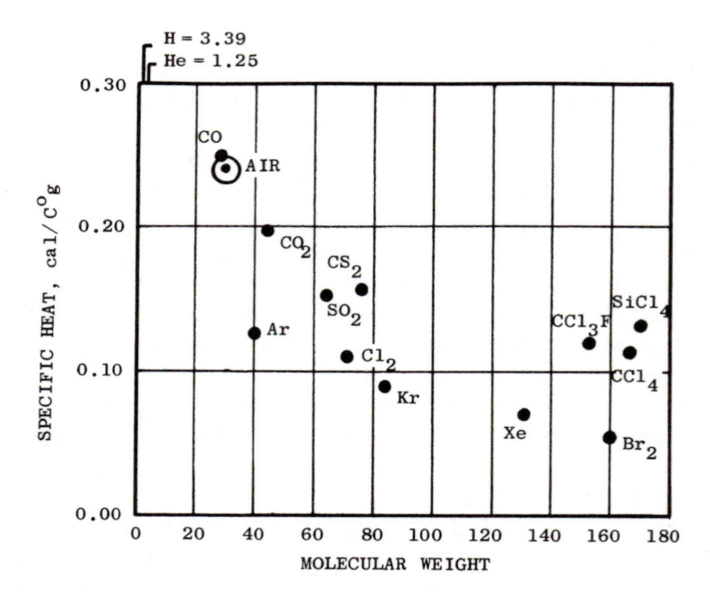

Fig. 10.16 Graph of the relationship between specific heat of a gas and molecular weight, evaluated at $T = 100°C$.

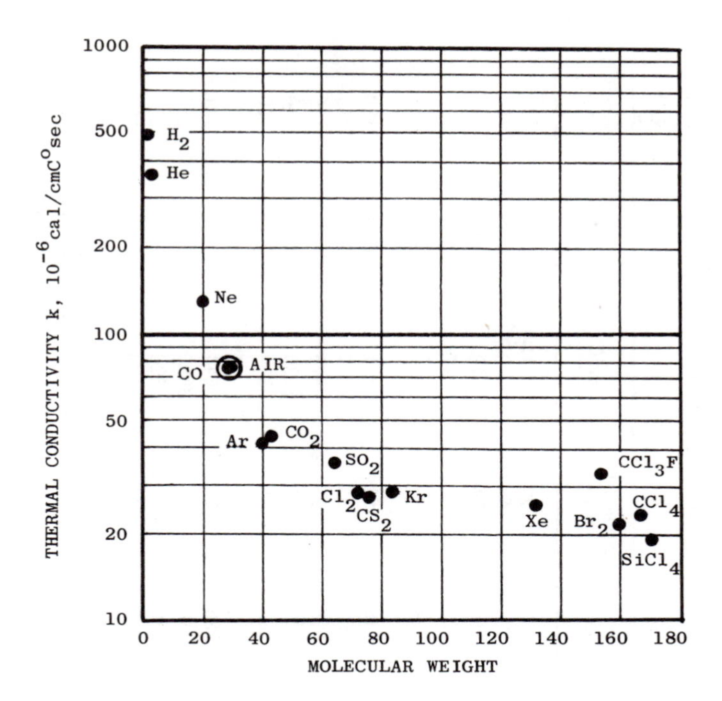

Fig. 10.17 Graph of the observed values for the thermal conductivity of a number of gases.

In Fig. 10.17 we show the variation of thermal conductivity k with molecular weight. Here the trend is well defined and there is little scatter, the high-molecular-weight gases having very low thermal conductivities.

The values of the relevant constants for the different gases of interest in this chapter are summarized in Tables 10.4 and 10.7 (page 345).

To determine the effect of different gases we proceed to calculate the modulus for each gas and the Rayleigh number, and then find the value of the Nusselt number from Fig. 10.11. The values for the Rayleigh number, the resultant Nusselt number, and the heat losses for several gases are presented in Table 10.5. The variation of Nusselt number with molecular weight of the gas is shown in Fig. 10.18. The monoatomic gases have the lowest values of Nu, forming a lower bound for the distribution of points in Fig. 10.18.

In examining the values for the individual terms entering the expression for the convective modulus a, we note that the reduction in conduction for high-molecular-weight gases is almost completely offset by the increase in convectivity. In Table 10.5 we see the separate terms for convection and combined convection and conduction for a number of gases included in Table 10.4, as taken for the specific case of L and ΔT

Table 10.4 Convective Properties of Gases

Gas	Molecular weight M	Density ρ	Specific heat c_p	Thermal expansivity y	Viscosity (poises) μ	Thermal conductivity k	Modulus a	Modulus ratio a/a_{air}	Convectivity $ka^{1/4}$
	g/mole	$\frac{g}{cm^3}$, (10^{-3})	cal/g C°	cm³/cm³C°, (10^{-3})	g/cm sec, (10^{-6})	cal/cm C°sec, (10^{-6})	1/cm³C°		cal/cm²C°, (10^{-6})
H_2	2	0.066	3.39	3.66	98	490	1.1	0.23	499
He	4	0.130	1.25	3.66	228	360	0.9	0.19	352
CO	28	0.915	0.25	3.67	210	74	48	1.02	194
Air	29	0.946	0.24	3.67	220	75	47	1.00	195
Ar	40	1.306	0.125	3.68	269	52	55	1.17	141
CO_2	44	1.447	0.199	3.72	186	54	151	3.21	189
SO_2	64	2.142	0.152	3.90	161	35	477	10.1	163
Cl_2	71	2.352	0.11	3.90	168	27	513	10.9	128
CS_2	76	2.479	0.157	3.90	126	26	1130	24.0	150
Kr	84	2.740	0.09	3.68	260	28	334	7.1	120
Xe	131	4.316	0.07	3.68	280	26	644	13.7	131
CCl_3F	154	5.02	0.12	3.9	110	34	3090	65.7	254
Br_2	160	5.22	0.055	3.9	188	22.5	1320	28.1	136
CCl_4	166	5.41	0.115	3.9	120	24	4470	95.1	196
$SiCl_4$	170	5.54	0.132	3.9	100	19	7640	123.0	178

All data at 100°C.

Table 10.5 Convective and Conductive Losses of Gases

Gas	*Rayleigh number (Ra)*	*Nusselt number (Nu)*	*Convective term (Nu•k)*	*Total term (Nu•k + k)*
H_2	1.84×10^3	0.5	2.45×10^{-4}	7.35×10^{-4}
He	1.52×10^3	0.35	1.25×10^{-4}	4.94×10^{-4}
CO	8.16×10^3	1.45	1.07×10^{-4}	1.81×10^{-4}
*Air**	8.00×10^3	1.4	1.05×10^{-4}	1.80×10^{-4}
Ar	9.36×10^3	1.5	0.78×10^{-4}	1.30×10^{-4}
CO_2	2.57×10^4	2.2	1.18×10^{-4}	1.72×10^{-4}
SO_2	8.08×10^4	3.5	1.22×10^{-4}	1.57×10^{-4}
Cl_2	8.72×10^4	3.5	0.95×10^{-4}	1.22×10^{-4}
CS_2	1.92×10^5	4.6	1.20×10^{-4}	1.45×10^{-4}
Kr	5.70×10^4	3.0	0.84×10^{-4}	1.12×10^{-4}
Xe	1.10×10^5	3.8	0.99×10^{-4}	1.25×10^{-4}
CCl_3F	5.26×10^5	6.4	2.18×10^{-4}	2.52×10^{-4}
Br_2	2.25×10^5	4.7	1.06×10^{-4}	1.28×10^{-4}
CCl_4	7.60×10^5	7.0	1.68×10^{-4}	1.92×10^{-4}
$SiCl_4$	9.84×10^5	7.7	1.46×10^{-4}	1.65×10^{-4}

*Based on $Ra_{air} = 8000$; Nu from Fig. 10.11.

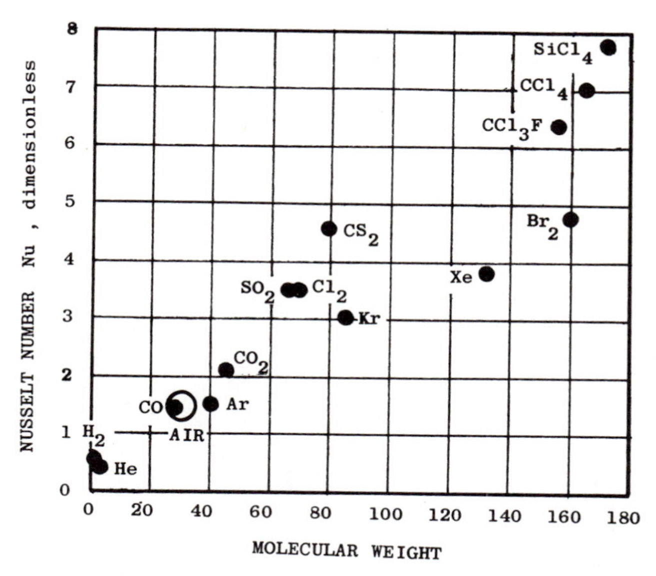

Fig. 10.18 Variation of the Nusselt number (Nu) for different gases for the case of a pair of upward-facing surfaces of 5.0-cm spacing at a mean temperature of 65.5°C and temperature difference of 100 C°.

used to obtain the experimental data shown in Fig. 10.11. The values for the convective term are plotted in Fig. 10.19. Note that the noble gases yield a reduced convective loss, but that the high-molecular-weight molecules actually are worse than air. This same feature also is evident for the total heat loss, as shown in the last column in Table 10.5.

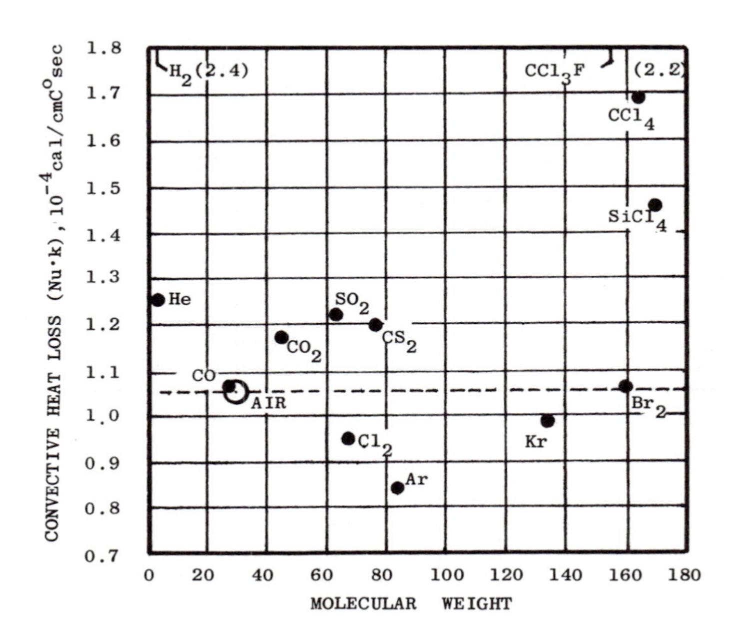

Fig. 10.19 Variation of the convection term Nu·*k* for horizontal parallel surfaces when Ra = 8000 for air, as a function of molecular weight of the gas between the two surfaces.

10.10 VARIATION WITH SEPARATION FOR OTHER GASES

To examine how the total convective heat loss *h* varies with surface separation, we present some data in Fig. 10.20 for gases better than air and a high-molecular gas worse than air. In these examples we start with the case where the separation is 5 cm (2 in.). The major factor to watch is the reduction in the value of the Rayleigh number. If the Rayleigh number happens to be less than 1707 in any specific case, the configuration is stable and no convection occurs. The exact behavior of the convective loss coefficient as Ra approaches 1707 is not explicitly defined in references we have examined, so we assume that the data in regard to the Rayleigh number from the experiments cited by Tabor hold in all cases where the system is sufficiently close to the baseline configuration.

In Table 10.5 we note that while the convection term for the heavy gas CCl_4 is greater than for air (Nu·k = 1.68 versus 1.05), when we add k to get the total loss the difference is small. Since we have L^3 entering the Rayleigh number and only L entering the conduction term, it is possible for different gases to have different variations of total loss with separation of the convecting surfaces. In Fig. 10.20 we show the variation of the total heat-loss coefficient h with separation, for horizontal flat plates having a separation of from 1 to 5 cm at the same temperature conditions as previously used for Fig. 10.13.

In Fig. 10.20 we note that the high-molecular-weight gases gain relative to air for small spacings. This is due primarily to the fact that the Rayleigh numbers are very small and the gas therefore relatively nonconvecting. This is one reason that plastic foams containing residual high-molecular-weight gases are excellent insulators: low convectivity and low conductivity are combined. The differences for practical spacings for solar collectors are so small, however, as to be of little practical importance.

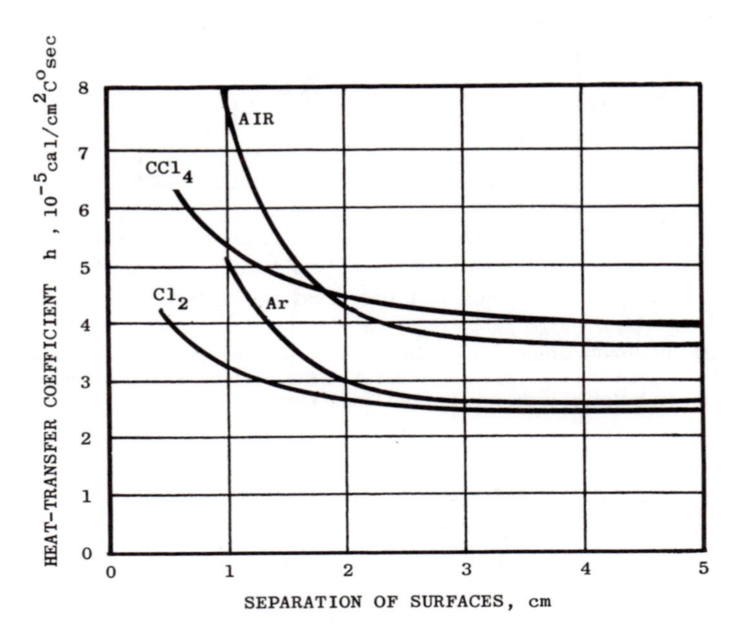

Fig. 10.20 Variation of heat-transfer coefficient h with separation of two horizontal surfaces L, with four different gases between the surfaces, for Ra = 8000 at L = 5 cm.

10.11 VARIATION OF CONVECTIVE LOSS WITH PRESSURE

We can utilize Eq. (10.19) to determine the variation of heat loss when the gas between the convecting surfaces is reduced in pressure. There are three separate pressure domains in regard to heat loss: (1) the convection range, (2) the conduction range, and (3) the Piriani range.

The convection range is that range of pressures near, and greater than, atmospheric, where the dominant loss mechanism is convection. As the pressure is reduced the convection loss drops to a plateau determined by the conductive loss. In accordance with the kinetic theory of gases, conduction remains constant with pressure, the plateau being formed when the convection loss is fully suppressed. The third range begins when the mean free path of the gas molecules begins to be equal to the dimension between the heat-losing surfaces. As shown in Fig. 10.21, the conduction loss drops as the onset of the low-pressure (Piriani range) region is entered. When there is no gas remaining in the space between the surfaces there is no loss of heat, other than by radiation.

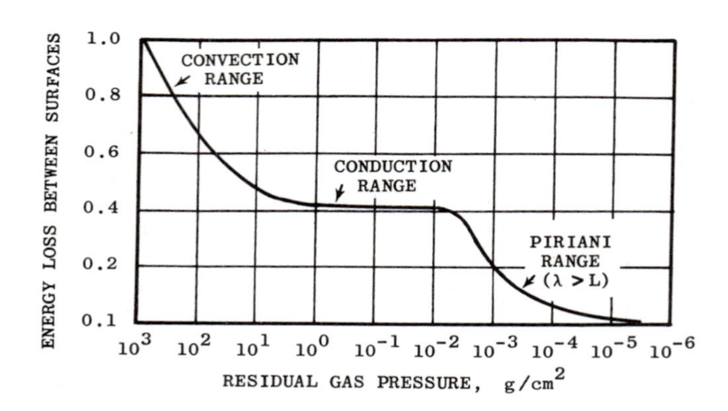

Fig. 10.21 Qualitative variation of energy loss between surfaces as a function of reduced pressure. Note in particular that the existence and duration of the conduction range is dependent upon the geometry of the collector, as is the onset of the Piriani range of pressures.

In the practical sense of designing efficient solar collectors one should have numerical comparisons so that cost effectiveness evaluations can be made. Sealing a system to contain a gas at nominally atmospheric pressure already adds a complication. To further contain the gas at subatmospheric pressures becomes at least an order of magnitude more expensive, so the exact amount to be gained is of importance.

The expression for the heat-loss coefficient in terms of the Rayleigh number makes it easy to evaluate the behavior of convecting surfaces when the pressure is reduced. Further application of the curves for true convection versus Ra enable one to actually make a transition from the convection range into the conduction range. The procedure in determining the variation is exactly as before: compute the change in Ra, then read Nu from the graph.

In the analytical case where a power equation is used, such as when $a^{0.25}$ is encountered, the variation is simple, varying as the 0.5 power of the ratio of pressures, where the value of Nu is obtained from Fig. 10.11:

$$Nu' = Nu(p/p_0)^{0.5}. \tag{10.25}$$

In Fig. 10.22 we have followed the exact procedure of using the graphical relationship between Ra and Nu for a number of gases.

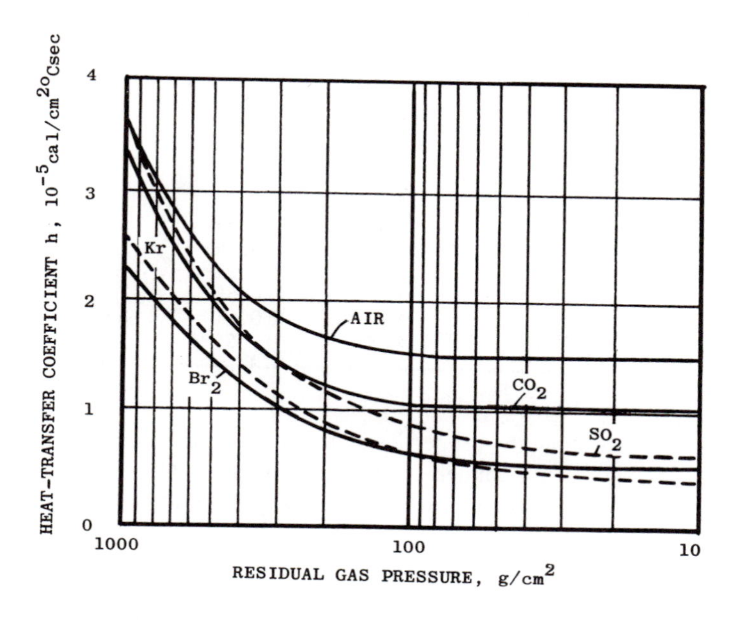

Fig. 10.22 Dependence of the heat-transfer coefficient for two horizontal parallel plates where Ra = 8000 for L = 5 cm for air.

The important quantities are the value of the Rayleigh number used for a baseline, 8000 as used in Fig. 10.20, and the sum of $k(\mathrm{Nu} + 1)$. Fig. 10.22 shows that there are small differences in the rate at which the convection loss decreases with pressure drop. Most gases have reached the conduction plateau at a pressure of 100 g/cm^2, about one-tenth atmosphere. The loss factor remains approximately constant until very low pressures are achieved, the Piriani range.

In the Piriani range of pressure the heat loss is given by

$$Q/A = \tfrac{1}{4}[(\gamma+1)/(\gamma-1)]\,(2R/\pi M)^{1/2}\,(b\,\Delta T/T^{1/2})p, \qquad (10.26)$$

where γ is the ratio of the specific heats, M is the molecular weight of the gas, R is the universal gas constant, and b is a correction factor for the gas type, ranging from a low of 0.3 for helium and being approximately unity for the gases we are considering. The pressures where the Piriani range begins for several gases are shown in Table 10.6.

Table 10.6 Pressures for Onset of Free Molecular Flow

Gas	*Molecular diameter, 10^{-8} cm*	*Pressure for $\lambda = 1$ cm, g/cm²*
He	2.18	0.0285
H_2	2.47	0.0183
Ar	3.36	0.0100
O_2	3.39	0.0100
N_2	3.50	0.0094
CO	3.50	0.0093
NO	3.40	0.0091
CO_2	4.18	0.0066
N_2O	4.27	0.0061
C_2H_4	4.55	0.0055

In Table 10.7 we show the properties of a number of additional gases that may be of use. For gases that would condense during the night we show the vapor pressure of the gas at $T = -10°C$ and also the heat of vaporization per gram of gas.

10.12 FORCED CONVECTION AND WIND LOSS

One encounters forced convection in air-heater collectors and also wind-induced losses from the front surface of collectors. In terms of dimensional analysis the change from natural to forced convection means the substitution of the Reynolds number for the Rayleigh number. One then uses a correlation curve for the relationship between the Nusselt number and the Reynolds number, of the type shown in Fig. 10.9. The Reynolds number is given by

$$\mathrm{Re} = vL/\nu = vL\rho/\mu, \qquad (10.27)$$

where μ is dynamic viscosity (poises) (or absolute viscosity) and ν is kinematic viscosity (stokes). Units are g/cm·sec for dynamic viscosity and g/cm·sec·density for kinematic viscosity.

Table 10.7 Gases of Interest for Convection Suppression

	Molecular weight M	*Boiling point* T_b	*Pressure* $p(-10°C)$	*Heat of vaporization* H
Gas	*g*	°*C*	g/cm^2	*cal/g*
CO	28	−192		
CO_2	44	−78		
CS_2	76	+46	66	19
Ar	40	−186		
Cl_2	71	−34		
Br_2	160	+59		44
$SiCl_4$	170	+96	13	11
Si_2H_6	62	−14		
SiF_4	104	−86		
$SiClF_3$	120	−70		
SF_6	193	−34		
SO_2F_2	102	−55		
SeF_6	193	−34		
COS	60	−48		
CCl_4	159	+77	13	11
CCl_3F	137	+24	144	10
ClF_3	92	+11	295	14
POF_3	104	−40		
NOF	49	−56		
CCl_2F_2 (F12)	121	−30		
$CClF_3$ (F13)	104	−81		
$CHClF_2$ (F22)	86	−41		
C_2ClF_5	154	−38		
C_2NF_7	171	−35		
$GeCl_2F_2$	181	−3		
$GeClF_3$	165	−21		
C_2F_6	138	−79		

One has the option of using power equations for the heat-transfer coefficient for forced convection and, as remarked earlier, the relationship for pure convection is a straight line and hence can be fitted with a power equation. Various equations are actually met in the literature. The original Hottel and Woertz (1942) data have been generally expressed in an equation of the form

$$h_w = 1.00 + 0.304v, \tag{10.28}$$

where v is velocity in mph and h_w is in Btu/ft²h F°.

The literature has quite a number of ways in which the function $f(\mathrm{Re})$ is stated. We will review them so that their differences can be appreciated. The authors advocating each method are also shown:

$$\begin{aligned} \mathrm{Nu} &= 0.0158\,(\mathrm{Re})^{0.80} && \text{(Duffie and Beckman)} \\ \mathrm{Nu} &= 0.023\,(\mathrm{Re})^{0.80} && \text{(Kreith)} \\ \mathrm{Nu} &= 0.170\,(\mathrm{Re})^{0.47} && \text{(Welty-1: Re = 40–4000)} \\ \mathrm{Nu} &= 0.062\,(\mathrm{Re})^{0.62} && \text{(Welty-2: Re = 4000–40,000).} \end{aligned} \tag{10.29}$$

When these expressions are evaluated in terms of physical variables and solved for Q/A, we have

$$\begin{aligned} Q/A &= 2.54 \times 10^{-6}\, v^{0.80} L^{-0.20}\, \Delta T && \text{(D and B),} \\ Q/A &= 6.06 \times 10^{-6}\, v^{0.80} L^{-0.20}\, \Delta T && \text{(K),} \\ Q/A &= 9.09 \times 10^{-5}\, v^{0.47} L^{-0.53}\, \Delta T && \text{(W-1),} \\ Q/A &= 3.25 \times 10^{-6}\, v^{0.62} L^{-0.38}\, \Delta T && \text{(W-2).} \end{aligned} \tag{10.30}$$

These equations appear quite different but are not as different when evaluated. The above equations are written for v in cm/sec, ΔT in C°, and L in cm, yielding Q/A in cal/cm²sec. As an example let us take $v = 100$ cm/sec, $\Delta T = 40$ C°, and $L = 10$ cm. We obtain

$$\begin{aligned} Q/A &= 0.00165 \text{ cal/cm}^2\text{sec} &&= 22 \text{ Btu/ft}^2\text{h} && \text{(D and B),} \\ Q/A &= 0.00384 &&= 51 && \text{(K),} \\ Q/A &= 0.0117 &&= 124 && \text{(W-1),} \\ Q/A &= 0.00909 &&= 125 && \text{(W-2).} \end{aligned} \tag{10.31}$$

Comparing this set of values as inserted in the Hottel and Woertz equation, we find

$$Q/A = 0.0091 \text{ cal/cm}^2\text{sec} = 121 \text{ Btu/ft}^2\text{h} \qquad \text{(H and W).}$$

It is interesting to plot the variation of Q/A as a function of velocity in order to see the differences between the several equations cited above. We note that some of the equations lead to unrealistically low estimates of convection loss compared to the Hottel and Woertz value, which was directly obtained from observing losses from solar collectors. The equations we recommend for use are

$$Q/A = 0.000136\,(1 + 0.0068v)\,\Delta T \qquad \text{cal/cm}^2\,\text{sec C}^\circ, \tag{10.32}$$

$$Q/A = 0.000567\,(1 + 0.0068v)\,\Delta T \qquad \text{W/cm}^2\,\text{C}^\circ, \tag{10.33}$$

where velocity is in cm/sec and temperature in C°, with the result being in cal/cm² sec. To convert this equation to yield Btu/ft² h, multiply the coefficient by 1.33×10^4. The above equations are a linear function fitted to the Hottel and Woertz data. The equation is not dimensionally representative of convection as defined by the above equations, but it agrees well with the slope of the Welty equations, shown in Fig. 10.23, and it has a finite value at zero wind velocity, representing the effect of natural convection.

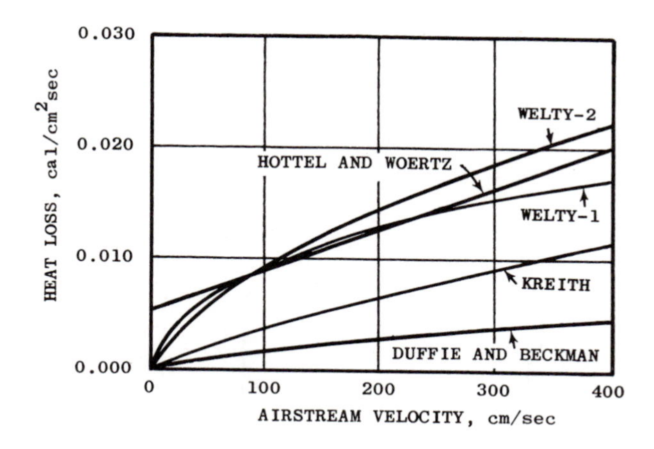

Fig. 10.23 Heat loss under forced convection evaluated for ΔT = 40 C° and a characteristic dimension of L = 10 cm, as a function of temperature for the several equations presented in the literature.

For many applications the input data on airflow are given in cubic feet per minute (ft³/min or cfm). To get the equivalent velocity in cm/sec for ducts of cross section A in ft² and V in ft³/min, the equation to use is

$$v\ (\text{cm/sec}) = 0.508\ \frac{V\ (\text{ft}^3/\text{min})}{A\ (\text{ft}^2)}\ . \tag{10.34}$$

10.13 CONVECTIVE SELF-FLOW THERMOSIPHON

The density differential created by temperature gradients is used to cause the fluid being heated to flow without any external power source other than sunlight. The effect of convective self-flow is generally termed the *thermosiphon effect.* The magnitude of the effect and the velocity of fluid flow can be calculated on the basis of simple physical principles.

In Fig. 10.24 we show a U-shaped tube containing fluid of a total depth h. If the tube is inclined with respect to the vertical, the value of h is the elevation difference, or $l\cos\theta$, where l is the length of the tube and θ is the angle of tilt measured from the zenith. If one side of the tube is heated with respect to the other, the density of the fluid in the heated column will be lowered, but because the two columns are self-balancing, the length of the column in the hot side will be increased a distance dh for the weights in the two columns to be equal. We thus have

$$\rho_2 < \rho_l \quad \text{and} \quad h_2 = h_i + dh. \tag{10.35}$$

The incremental column dh represents a pressure head, and if the tube is cut off on the hot side at height h this force increment will accelerate the entire loop of material.

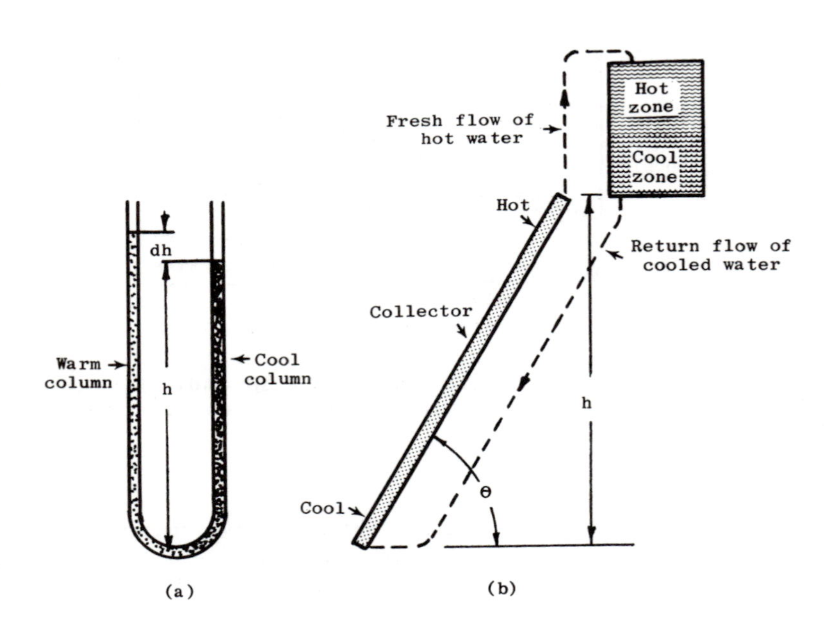

Fig. 10.24 Schematic diagram of the thermosiphon system for hot water. For proper operation the storage tank must be placed above the top of the solar collector.

The force causing the velocity of flow to arise is

$$F = \rho dhAg = ma, \tag{10.36}$$

where A is the cross-sectional area of the tube, g is its gravitational acceleration, and a is the acceleration induced by the force. Substituting $m = 2Ah$ for the total mass of the fluid to be placed in motion, we solve the equation for the acceleration and find

$$a = \rho gdh/2h, \tag{10.37}$$

whence the velocity of flow by the equation $v^2 = 2as$ $(s = dh)$ is

$$v_{\text{flow}}^2 = \rho gdh^2/h. \tag{10.38}$$

But because $dh = kh\Delta T$, we have as the final equation for the velocity of flow

$$v_{\text{flow}} \leqslant (\rho gh)^{1/2}k\Delta T. \tag{10.39}$$

The inequality in Eq. (10.39) means that this velocity is that which would occur if there were no impeding forces in the fluid flow. For low velocities the frictional forces impeding the flow will be small, but one must also include the opposing pressure head

$$dp = f(v^2\rho/2g)(L/D). \tag{10.40}$$

The velocity of flow will be reasonably approximated by Eq. (10.39) or (10.41) until the term for dp reaches a value of approximately 0.1 of the thermal pressure head. See Chapter 11 for a fuller discussion of frictional pressure head.

It should be noted that in an actual solar collector the temperature of the water ranges from cool water entering at the bottom to hot water exiting at the top. Because the column is not therefore isothermally hot, the head dh is approximately one-half that for the isothermal case, and we have

$$v_{\text{flow}} \leqslant (\rho gh/2)^{1/2}k\Delta T. \tag{10.41}$$

The practical application of the thermosiphon to hot-water heaters is shown schematically in Fig. 10.24(b). The location of the storage tank *above* the collector is a requirement, even though this location is not aesthetically attractive. Also, the cold water makeup for the hot water used from the tank must be at the bottom of the tank to provide the proper head of dense cool water.

Gaudenzi (1975) has produced solar collectors in which the inverse thermosiphon effect has been used to store cold water, which is cooled by nocturnal sky radiation from the collector. His circuit for a heating/cooling collector is shown in Fig. 10.25. The bypass sections allow the hot section to be gravitationally isolated at night and the same is true for the cold section during the day. Note that one has the option of emphasizing either heating or cooling by the number and type of windows on the collector, the windows generally aiding the heating process and reducing the cooling process by their opacity to the thermal infrared.

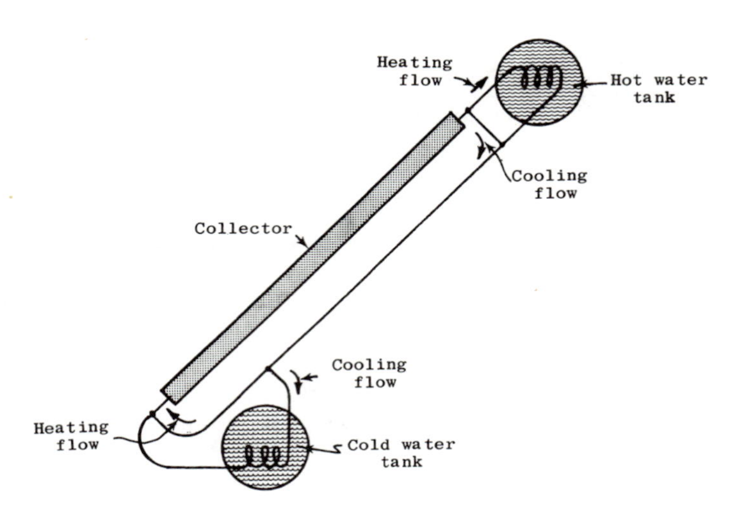

Fig. 10.25 Diagram of the Gaudenzi (1975) heating/cooling solar collector using the direct and inverse thermosiphon effect. The heat transfer medium is Freon 12 or an equivalent fluid.

Self-convective flow is also useful in constructing air heating collectors. A basic type described by Pope (1903) has recently been revived by Trombe (1974) for home-heating applications. It is frequently termed the *Trombe wall.* The configuration is shown in Fig. 10.26 in an orientation well adapted for south walls of buildings. The sunlight is admitted to the absorbing surface through a double glass window. The absorbing wall is vertical and spaced from the glass window by a gap of 100–150 cm to provide a channel for the air to flow through. The surface of the absorbing wall becomes heated, causing the air adjacent to the wall to convect upward. The warm air enters the room when the upper gap is open, but is retained by the collector when the gap is closed, thus enabling heat demands by the house to be met. The heating of the wall, especially when the upper gap is closed, causes a heat pulse to enter the mass of the wall. When the wall is made of massive material having good thermal storage properties (Section 13.11), the wall becomes the thermal storage medium for the house. At night the double window insulates the heated wall, and the wall continues

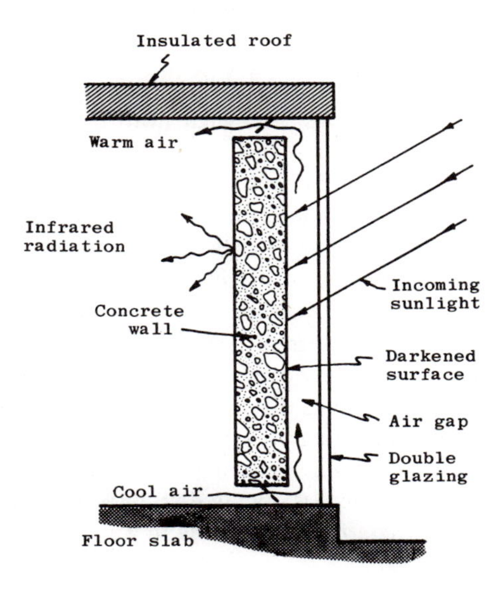

Fig. 10.26 Configuration of a self-convective flow solar air heating collector with inherent thermal storage. This design is called the *Trombe wall.*

to convectively heat cold air from the floor of the room, injecting it back at the top of the room.

Trombe has noted that the wall can be painted other colors than black and still operate with acceptable efficiency. Dark colors in any hue can be used.

10.14 HEAT TRANSFER TO LIQUIDS

In the preceding section we found that heat transfer from surfaces to air is low, in the range of ½ to 5 Btu/ft^{2}h F° or 2×10^{-5} to 20×10^{-5} cal/cm^{2}sec C°. This rate is still high enough to make convection into air a significant loss mechanism for solar collectors, even though it is inefficient as a mechanism to transfer heat into gas as a working fluid. Liquids in general make much better heat-transfer media than air, having heat-transfer coefficients (h) hundreds of times larger. As a consequence, heat can be transferred to liquids with only small temperature differences between the surface and the liquid. It should be noted, however, that gases under high pressures become reasonable heat-transfer media, especially hydrogen and helium. In the graphs in the preceding section the properties of both helium and hydrogen were generally off the top of the graph because their thermal conductivity was so high. Pressurized helium could be considered as a potential heat-transfer medium for solar collector applications if the complications of containing a high-pressure gas are acceptable.

Water

Water is the least expensive heat-transfer medium. Organic liquids, salt compounds, and liquid metals are also candidates with high heat-transfer coefficients. None of these liquids is the ideal heat-transfer medium, each having one or more practical problems. Water acts as a medium to cause mass transfer of the container materials from hot to cold regions of the system, as do the other liquid media. Water has the additional drawback that it boils. The change of phase from water to water vapor precipitates instabilities in heat transfer, because a system can suddenly go from the heat-transfer coefficient for a liquid to that of a gas. Thus when water loses liquid contact with a heated surface transferring heat from a fire, nuclear pile, or sun, the heated surface temperature can rapidly escalate to a temperature where the metal wall fails. The only water system that avoids this potential "burnout" problem is in pressurized water systems where the operating pressure is above the critical pressure of water, 3206 psi or 225 kg/cm^2. This pressure is so high that one then encounters materials strength problems when such a system is operated at 500–600°C.

The various heat-transfer coefficients relating to water and steam are summarized in Table 10.8.

Table 10.8 Typical Heat Transfer Values

	Overall transfer coefficient	
	W/m² C°	*Btu/ft² h F°*
Steam to gases	30–280	5–50
Steam to fluids	550–3500	100–600
Steam to organics	300–3500	50–600
Water to air (high pressure)	60–200	10–30
(Surfaces to air)	3–25	½–5
Water to Freon 12	450–850	80–150
Water to oil	120–350	20–60

The separate regimes of water heat transfer are, briefly, as follows:

1. Free convection.
2. Nucleate boiling, where the bubbles are detached from the heat-transfer surface and quickly absorbed within the water medium.
3. Nucleate boiling, where the bubbles reach the surface.

4. Unstable film nucleate boiling, where the entire transfer surface is briefly free of water transfer, the steam layer collapsing and re-forming but still transferring heat effectively.

5. Stable film nucleate boiling, where no water reaches the transfer surface, the surface being insulated by the steam layer and the heat-transfer rate dropping abruptly, precipitating the condition for burnout.

6. Radiation exchange only, where temperatures escalate until radiation exchange is capable of handling the system heat transfer load.

Salt Eutectics

Salts and salt compounds are also a heat-transfer medium of potential importance, but with certain major problems. Salts are thermally stable materials, unlike organics, which tend to decompose at elevated temperatures. The major disadvantage of salts is that they are generally solid at room temperature, posing a night problem when used in solar collectors. One of the most common salt heat-transfer media is sold under the trade designation Hitec or HTS. It is a eutectic mixture of $NaNO_3$, $NaNO_2$, and KNO_3 and has the following properties:

Melting point	143°C
Boiling point	(decomposes)
Density	2 g/cm^3 (at M.P.) 1.7 g/cm^3 (at 550°C)
Specific heat	0.37 Btu/lb
Heat-transfer coefficient (depending on velocity and temperature)	200–2000 Btu/h ft^2 F°
Heat of fusion	35 Btu/lb
Viscosity	19 centipoise (at M.P.) 1.1 (at 550°C)
Operating temperature range	400–1000°F 202–540°C

Other low-melting-point dielectric eutectics are:

Eutectic	Melting point
$KCl + AlCl_3$	140°C
$NaCl + AlCl_3$	110°C
$NaCl + FeCl_3$	156°C
$TeCl_4 + AlCl_3$	108°C
$KI + HgI$	125°C
$NH_4Cl + CuCl$	130°C

Salt eutectics also offer interesting possibilities for use as thermal storage media, where the heat of fusion is used to store considerable thermal energy at constant temperature. We will discuss them in detail when we consider thermal storage (Section 13.10).

Hitec can be kept from freezing by the addition of some water, forming a slurry at below the normal freezing point. This water, however, evaporates when the eutectic is reheated to operating temperature.

Organics

Organic liquids can be used when the system temperature is compatible with the upper temperature limit for the organic material. These liquids are expensive and have a major change in viscosity over the temperature range that can be encountered from room temperature to operating temperature, presenting startup problems.

One of the widely used organics is Dowtherm A, a mixture of diphenyl and diphenyl oxide. Its properties are:

Melting point	13°C
Boiling point	252°C
	505°C (at 500 psi)
Specific heat	0.65 Btu/lb
Heat of vaporization	120 Btu/lb
Thermal conductivity	0.10 Btu ft/h ft^2 F°
Heat-transfer coefficient	200 Btu/h ft^2 F°
Viscosity	5.0 centipoise (at M.P.)
	0.4 centipoise (at 350°C)
Operating temperature	$<$ 400°C

The polychlorinated biphenyls (PCBs) are also good organic fluids for heat transfer. At one time they were widely used, but PCB is known as an environmentally persistent chemical, and its use is curtailed because of its potential for poisoning the environment. Considering the large amount of fluid needed for heat transfer in solar collectors in terms of net energy production, organics are not particularly attractive, especially if one desires to use the heat-transfer fluid as the thermal energy storage material also. High-temperature organics cost in the vicinity of $1000 per barrel.

Liquid Metals

Liquid metals have gained attention in recent years through their application to the fast-breeder-reactor programs. They have extremely high heat-transfer coefficients and no problems from change of phase in the desired range of operating temperatures. Sodium is the principal metal being used, but others can also be used in appropriate situations. The range of metals explored as heat transfer media include:

Sodium
Sodium plus potassium (NaK)
Potassium
Lithium
Bismuth
Lead plus magnesium
Lead plus bismuth
Mercury
Gallium

The properties of sodium are as follows:

Melting point	97.8°C
Boiling point	877°C
Density	0.97 (at M.P.)
	0.80 (at 600°C)
Specific heat	0.31 cal/g
Heat-transfer coefficient	4000–10,000 Btu/h ft^2 F°
Viscosity	0.41 centipoise (at M.P.)
	0.21 (at 400°C)
Vapor pressure	35.7 mm at 620°C

Sodium becomes reactive with steel containers above 600°C. There is extensive literature on corrosion of sodium because it presents one of the serious materials compatibility problems of the liquid-metal fast breeder reactor. Below 550°C, sodium can be contained adequately in Croalloy steel, according to Oak Ridge National Laboratories (ORNL). Also, in view of the ample wall thicknesses and clearances in the liquid-metal system in the solar energy applications, no serious lifetime problems should be encountered.

The principal corrosion problem with sodium is the penetration of metals by the dissolving of the oxygen-rich grain boundaries of the steel alloys. One normally cold-traps sodium to keep the oxygen content on the order of 10 ppm. Dissolved metals redeposit within the sodium loop: iron in the hot regions, nickel and manganese in the cold regions, and chromium throughout the system. Carbon tends to form sodium carbide and to be deposited in regions of close dimensional fit.

Sodium is inexpensive, about $0.16/lb for ordinary grade (1974). Reactor grade is much more expensive because the calcium impurity must be kept low to avoid precipitation of insoluble CaO, which can plug a nuclear system. The solar transfer loop can adequately use ordinary grade sodium plus a cold trap.

The heat-transfer properties of sodium and a few other fluids are shown in Figs. 10.27–10.29.

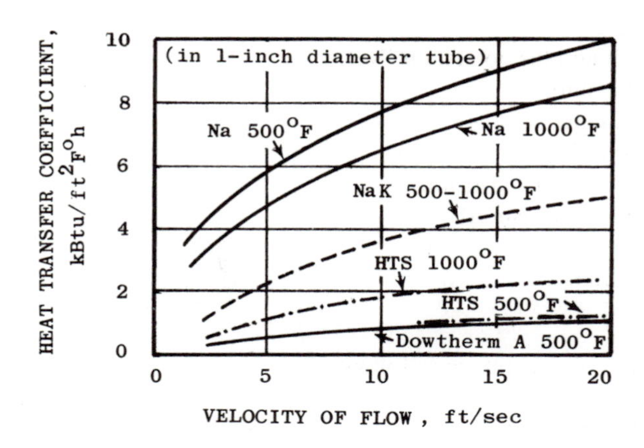

Fig. 10.27 Variation of the heat-transfer coefficient for several heat-transfer fluids as a function of velocity in a pipe 25 mm in diameter.

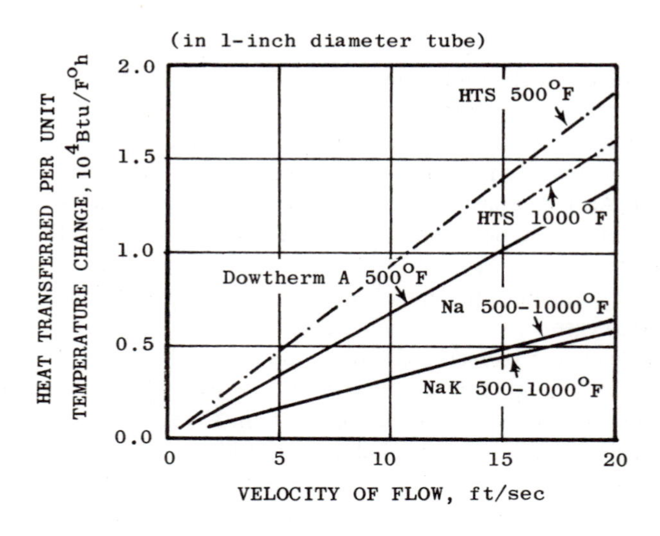

Fig. 10.28 Heat transferred per unit temperature difference for several heat-transfer fluids as a function of velocity in a pipe 25 mm in diameter.

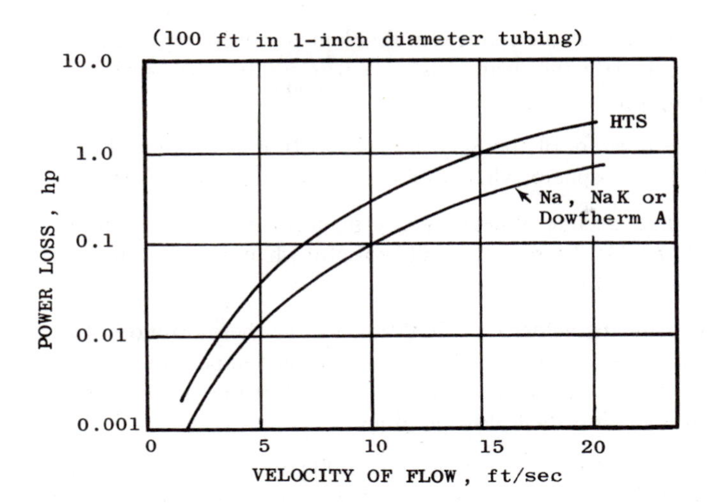

Fig. 10.29 Pumping power required per 30 m of pipe length for a pipe 25 m in diameter as a function of fluid velocity for several heat-transfer fluids.

10.15 PROBLEMS

10.1. Plot the net radiative exchange between two facing surfaces for the case where the lower surface is at 500°K and the upper surface ranges from a low of 300°K up to 500°K. Assume the emittance of both surfaces to be as a blackbody.

10.2. Plot the net radiative exchange between two facing surfaces as above, but with the lower surface being a selective surface with emittance at all temperatures of 0.10, and the upper surface a blackbody.

10.3. Taking Fig. 10.3, plot the flux curves for 400°K, 600°K, 800°K, and 1200°K on a graph where the vertical scale is linear and normalized to unity at the wavelength of maximum radiated flux.

10.4. Given that the relative conduction loss is proportional to the inverse ratio of the logarithmic ratios, recalculate Table 10.2 in a form where the change in heat loss is more directly visible from inspection of the table.

10.5. Calculate the heat-transfer coefficient for air and a plate separation of 1.5 cm when the temperature difference is 80 C°. Plot this point in a graph like that in Fig. 10.10.

10.6. Assuming that the surfaces in Problem 10.5 were separated by SO_2 rather than air, calculate the heat-transfer coefficient for the same spacing and temperature.

10.7. If the pressure of the SO_2 in Problem 10.6 were reduced from atmospheric to 1/100 atmosphere, what would the heat-transfer coefficients be reduced to?

10.8. Calculate a table of heat loss coefficients for wind velocities using Eqs. (10.32) and (10.33). These tables will be useful when you are calculating collector heat balances in Chapter 12.

10.9. Calculate the velocity of fluid flow via the thermosiphon when the hydrostatic head is 200 cm, the fluid is water, and the temperature difference between the top and bottom of the collector is 50 C°. If the pipe has a diameter of 1.0 cm, how many liters will flow per hour?

10.10. If Archimedes had at his disposal 1000 gold shields 1.2 m high and 0.6 m wide, plus 4000 bronze shields 1.1 m high and 0.55 m wide, and if these were deployed around the semicircular harbor walls of Syracuse, what would be the flux concentration at the center of the harbor 100 m from the shore if the mirrors directed the sunlight with four solar diameters accuracy and half of the beams happened to fall uniformly within a circle of 10 m in diameter at the harbor midpoint? If the absorptance of the ships were 0.5 and the only heat loss were radiation, what would be the equilibrium temperature of the surface of the ships? State the assumptions and the values used in arriving at your answer. Would this flux level possibly account for the reputed story?

Chapter 11
Heat Transfer in Solar Collectors

11.0 INTRODUCTION

In this chapter we analyze the thermal behavior of several types of solar collector systems. The objective is to determine the response of given systems to changes in the input variables. Typical questions addressed are the dependence of collector output on heat transfer fluid type, velocity of flow, absorptance of the collector surfaces, pressure drop in the piping, amount of power required to pump the transfer fluid through the collector, difference between air heating collectors and fluid heating collectors, and so forth. We do not cover all aspects of this extensive topic, but simply give enough information and examples to assist the student in solving other specific system conditions.

Heat transfer involves the dynamics of getting the solar energy from the heated surface of the collector into a working heat-transfer fluid of use to the system. Heat transfer involves the properties of the heat-transfer fluid, the rate of injection of heat, and the velocity of fluid flow in the collectors. When the collector system is large, the diameter and length of fluid pipes become important in determining the power expended in pumping the fluid through the system.

There are two cases we wish to examine as illustrations: (1) transfer in a system utilizing a liquid, and (2) transfer in a system utilizing a gas. The differences are significant. In the first case we can ignore to a first approximation the temperature drop between collector surface and fluid. In the second case this temperature drop becomes of major importance in the heat transfer involved.

A third case of heat transfer in solar collector systems is also of interest: where the absorber surface is located in a vacuum. This case is especially simple because the only heat-transfer mechanism of importance is radiation, and elegant behavior is obtained, representing the limiting case where conduction and convection are reduced to zero.

11.1 HEAT LOSSES IN A DISTRIBUTED COLLECTOR SYSTEM

The distributed system of small collector modules is shown schematically in Fig. 11.1. A pair of main trunk lines serves the entire area, with a pair of feeders reaching each pair of columns of collector modules. Within each module are additional absorber lines. Because each class of heat-transfer line will have a different diameter, velocity,

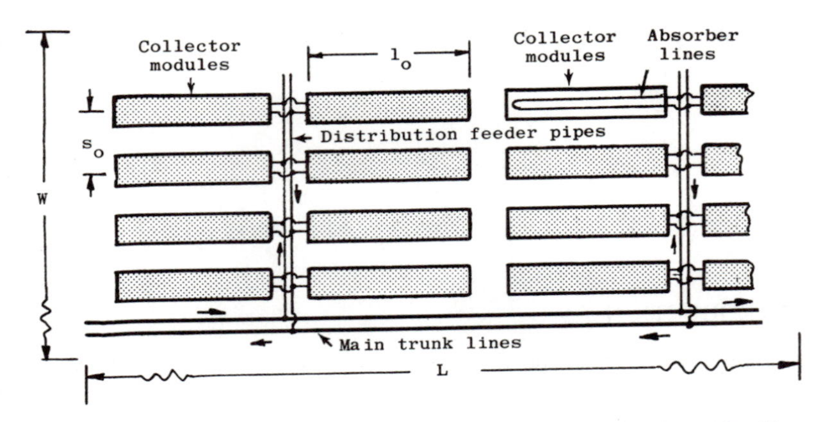

Fig. 11.1 Schematic diagram of a distributed system of collectors wherein hot and cold conduits must be distributed to each collector module.

and degree of insulation, we write the expression for the total length of piping in three terms. Assuming that the space between the modules is small in the direction of length, so that l_0 is equal to L divided by the number of columns of modules, we have

$$\begin{array}{lll} \text{Length in trunks} & L_1 & = 2L \\ \text{Length in feeders} & L_2 & = WL/l_0 \\ \text{Length in absorbers} & L_3 & = 2\,WL/s_0 . \end{array} \tag{11.1}$$

The total area of piping losing heat is therefore

$$A = 2\pi D_1 L + \pi(WL/l_0)D_2, \tag{11.2}$$

where D_i is the diameter of the appropriate pipe. The absorber lines are not included in the loss area because they are presumably gathering heat.

If one compares the total area of heat-losing pipe in the case of the distributed system to the area applicable to the case of a single long module per row of length $W/2$,

$$A = 2\pi D_1 L, \tag{11.3}$$

the reduction of loss that one can achieve by eliminating the headers is considerable, as L/l_0 can be on the order of 1000 to 10,000.

11.2 THE LIQUID TRANSFER MODULE SYSTEM

In the following discussion we will assume that no change of phase occurs in the heat-transfer fluid. The liquid flows at constant velocity along a pipe of uniform cross section. We want the fluid to reach a given temperature after traveling the distance of the collector. The total path length of travel of the fluid in the collector will be denoted by L, which can be subdivided into a path $L/2$ outbound through the collectors and a similar path $L/2$ inbound, gathering energy the entire length of the path.

To avoid complicating the equations defining the transfer mechanism of getting the heat from the absorber surface into the fluid, we define a transfer efficiency coefficient η, where η is the ratio of energy output from a collector to the available energy at the top surface or entrance aperture of the collector. We will see later how to determine η for a given collector from detailed heat-balance equations.

For the sake of specificity we will discuss the heat-transfer relationships for a solar collector in the context of a specific design. The equations we develop are, however, general in that they apply to any collector length, velocity, and so forth.

The *long module*, discussed herein, addresses one of the often-cited reasons why flat-plate or any distributed collector systems are impractical for large-size "farms": their high loss of thermal energy in transporting the heat-transfer fluid from distant collectors to a central location. We feel that this objection is not necessarily valid if a geometrical configuration is possible wherein all portions of the farm—collectors and feeder lines—are collecting energy over their entire extent. We define the term *long module* as a collector in which the unit is gathering energy continuously over an outbound leg $L/2$ long and an inbound leg $L/2$ long, for a total path length of L. A layout of this type of module is shown in Fig. 11.2. The question is: What are the equations that describe the performance of such a long collector, and are the required parameters practical?

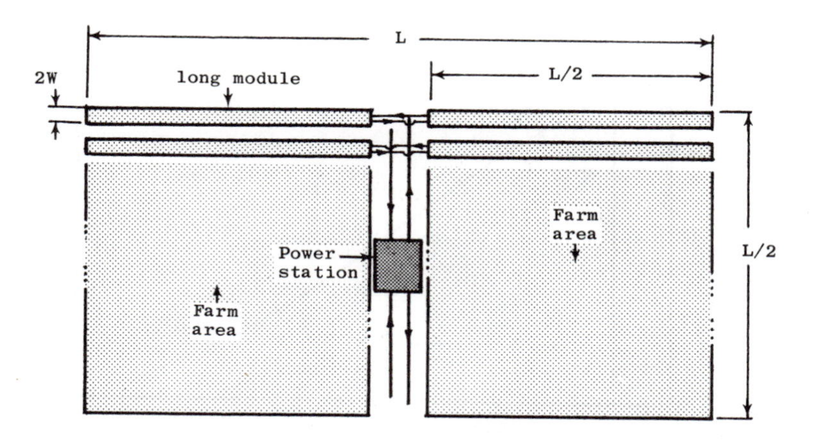

Fig. 11.2 Basic layout of a large solar power farm employing the long-module concept wherein the heat-transfer fluid is being heated along the entire path from the central corridor to the farm periphery and back to the central corridor. The width $2W$ of the module is greatly exaggerated.

11.3 SOLAR HEAT AVAILABILITY

In the collector we assume that the heat-transfer fluid travels along a cylindrical pipe of constant cross section, the fluid therefore flowing at a constant velocity along the entire distance L. The fluid enters at a temperature T_1 and exits at a temperature T_2, experiencing a temperature rise of ΔT. The heat-transfer fluid has a density ρ and heat capacity c. The fluid is contained in a pipe of diameter D and flows with a velocity v. The solar flux I is collected by a system having a net efficiency of extraction $\bar{\eta}$, where $\bar{\eta}$ is defined as the mean value of η over the temperature interval involved. Using the mean value is a simplification, but it raises the question of whether a constant rate of heat transfer is appropriate. Let us examine this important point.

In Fig. 11.3 we show a schematic diagram of the variation of η with distance along a collector. A real collector will have a higher value of η at the entering temperature T_1, the value decreasing with increasing temperature of the heat-transfer fluid. In this figure we also mark temperatures along the curve, showing that a longer distance is required to add an equal increment of temperature as the temperature rises. We also show a mean value $\bar{\eta}$, defined such that fluid traveling a distance L will reach T_2 as though a constant heat flow had occurred along L.

Although we have gone to some length to justify the use of a mean value for η, it will be noted that the expression for the ratio of power output to pumping power input, Eq. (11.33), is independent of the collector parameters. The collector behavior does, however, affect the ratio of length of module to fluid pipe diameter.

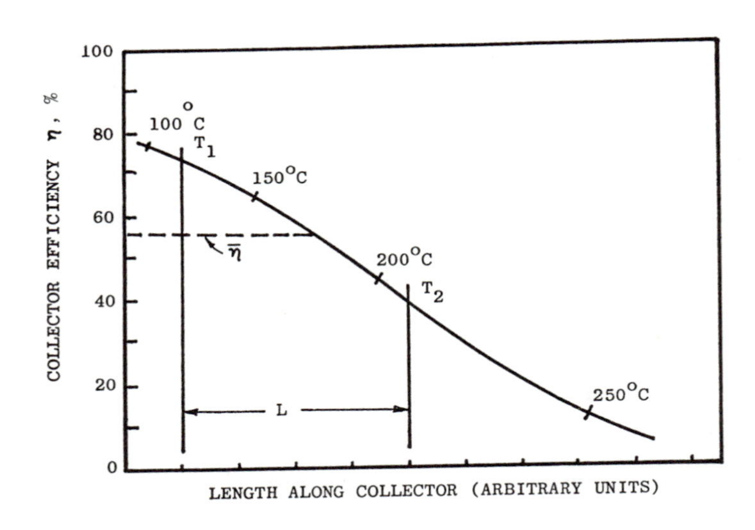

Fig. 11.3 Schematic representation of the variation of collector efficiency η with distance along a linear collector and with temperature. The mean value $\bar{\eta}$ is defined such that fluid traveling a distance L will reach T_2 as though a constant heat flow had occurred along L.

In Chapter 12 we show in detail how the efficiency η is calculated, but at this point in the consideration of system behavior we will assume η to be known. It should be noted that η contains the absorptance α of the collector, one of the variables of considerable interest in optimizing collector efficiency. The factor η also includes losses by reflection and absorption in the sunlight getting to the absorber.

We also find it convenient to define another efficiency factor, that of extraction of *absorbed* solar flux, P. In the case of an evacuated collector this efficiency factor enables one to see clearly the roles of α and ϵ in determining system performance. P and η are related by

$$P = \frac{Q_{\text{out}}}{I\alpha(1-r_1)(1-r_2)\dots(1-a_1)(1-a_2)\dots} \tag{11.4}$$

and

$$\eta = Q_{\text{out}}/I. \tag{11.5}$$

where r and a are the reflectances and absorptances of windows.

There can be a considerable difference between these two definitions of efficiency, so keeping them clearly separate is important. In the first example cited in the model calculations in Section 12.5, the solar flux entering the collector is 940 W/m^2, but the flux absorbed is only 727 W/m^2. The efficiency factor $\overline{\eta}$ is therefore 0.66, while the efficiency of extraction of absorbed solar flux P is 0.85.

In the section below we will generally use the factor η, in which case the absorptance α does not appear.

The solar energy is collected over several times the diameter D of the fluid pipe, with effective width of WD. The quantity W is dimensionless and is a measure of the accumulation of energy in the heat-transfer fluid by means of either optical concentration from the collector or conduction, as in the case of a flat-plate collector. In the following equations it is not necessary to distinguish between these two forms of energy concentration. We also combine all the collector losses into a single factor η, which is the efficiency of heat extraction of heat arriving at the collector by the heat-transfer fluid. The net thermal energy available to heat the fluid is therefore

$$Q/t = WD\overline{\eta}I \qquad \text{(solar input).} \tag{11.6}$$

In traversing a total length of collector L, the fluid rises in temperature ΔT to the desired output temperature T_2, sustaining at the same time a pressure drop Δp due to pumping energy losses in the pipe. We are interested in finding the dependence of Δp on the parameters of the collector, in order to ascertain the maximum allowable collector path length L, which determines how large a collecting area we can use in which no unproductive feeder lines are required. We would then be able to make a farm module with a total width L and a length on the order of L using only a central corridor of feeder lines of diameter large enough to make the surface losses negligible compared to the thermal energy transported.

11.4 FLUID MECHANICS

If we return to elementary fluid mechanics we can then derive the required system parameters to satisfy the desired performance conditions. The basic Bernoulli equation can be written as

$$-dp/dx = fv^2\rho/2gD + (\rho/g)v(dv/dx) \quad \text{g/cm}^3, \tag{11.7}$$

where f is the hydraulic friction coefficient, which is a function of pipe roughness and Reynolds number (Re). When water is the heat-transfer fluid, the range of practical interest for smooth pipes is approximately $f = 3 \times 10^{-2}$. For other fluids one must calculate the Reynolds number:

$$\text{Re} = vD/\nu, \tag{11.8}$$

where v is the velocity of flow, D is the characteristic dimension, equal in this case to the pipe diameter, and ν is the kinematic viscosity of the fluid. The relationship between the hydraulic friction coefficient f and Re for several pipe roughnesses is shown in Fig. 11.4.

In each of the equations in this chapter we will include a representative set of units. The reason for this, which will become obvious as we proceed, is that thermal energy and work energy are not identical, but are related by Joule's constant. One cannot equate the two without proper recognition of the differences. These differences persist in any unit system—metric, SI, English, and so forth—so one must be careful of the units in each of the equations.

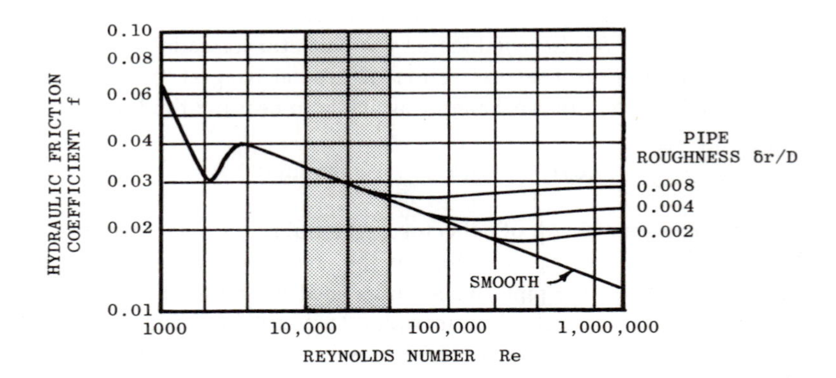

Fig. 11.4 Diagram of the dependence of the hydraulic friction coefficient f on the Reynolds number (Re) and pipe roughness $\delta r/D$. The shaded area represents the region of Re for water with conditions applicable to the long-module collector.

For an incompressible fluid ρ is constant, and in a pipe of constant diameter $dv/dx = 0$, and hence the equation reduces to

$$dp = -(fv^2\rho/2gD) \int_0^L dx. \tag{11.9}$$

Since the velocity is a constant along L, we have

$$\Delta p = -(fv^2\rho/2g)(L/D) \quad \text{g/cm}^2. \tag{11.10}$$

In the case under consideration, shown in Fig. 11.5, we have a disc of fluid of diameter D, 1 cm in length, that travels a total distance L at a velocity v, absorbing thermal energy the entire time. The disc will absorb heat more readily when it enters the pipe than when it leaves, owing to the difference in temperature of the disc, but this variation is incorporated in the factor denoting the percentage of the available energy that is absorbed by the disc, F.

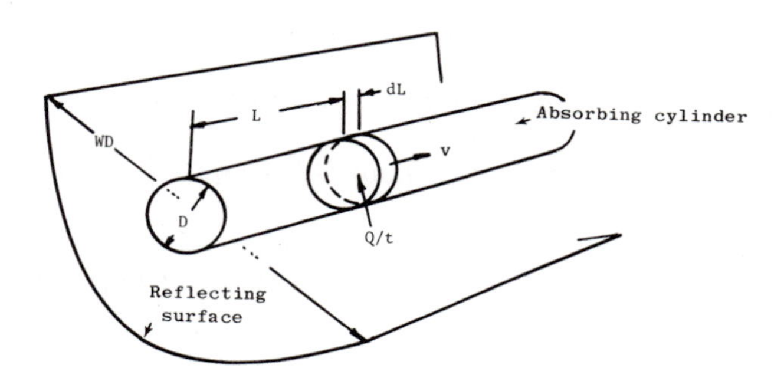

Fig. 11.5 The disc-shaped section of fluid receives heat input as it travels the length of the collector. For a rectangular duct the area elements in the equations are replaced with *ab*, the cross-sectional area of the duct.

The mass of fluid in the disc is

$$M = \rho V = \rho\pi D^2/4 \quad \text{g/cm}. \tag{11.11}$$

The amount of heat acquired per centimeter of disc length is

$$Q = \rho\pi c D^2 \Delta T/4 \quad \text{cal}, \tag{11.12}$$

and the time in the pipe while gathering this amount of heat is

$$t = L/v \quad \text{sec.} \tag{11.13}$$

The solar heat transferred to the fluid can be written as

$$Q = W\bar{\eta} I\, Dt \qquad \text{cal/cm}$$

or

$$Q = W\bar{\eta} I(LD/v) \quad \text{cal/cm}, \tag{11.14}$$

where I is the solar flux arriving at the collector and $\bar{\eta}$ is the appropriate average fraction of the available energy actually transferred into the working fluid. In the equations to follow, we will eliminate the average value notation over $\bar{\eta}$ in order to simplify the notation. The rise in temperature of the fluid can then be written

$$\Delta T = (4W\eta I/\pi c\rho v)(L/D) \quad \text{C}^\circ, \tag{11.15}$$

and solving for v,

$$v = (4W\eta I/\pi c\rho\Delta T)(L/D) \quad \text{cm/sec}, \tag{11.16}$$

and inserting it into the expression for the pressure drop, we have

$$\Delta p = (8\rho f/\pi^2 g)(W\eta I/c\rho)^2 (L/D)^3 (1/\Delta T)^2 \quad \text{g/cm}^2. \tag{11.17}$$

The question now becomes: For realistic values of the several factors, what are the resultant L/D ratios as a function of allowable pressure drops Δp? Solving for L/D, we obtain

$$\frac{L}{D} = \underbrace{\left(\frac{\pi^2 g}{8f}\right)^{1/3}}_{\text{constant}} \underbrace{\left(\rho c^2\right)^{1/3}}_{\text{fluid}} \underbrace{\left(\frac{1}{W\eta I}\right)^{2/3}}_{\text{solar input to fluid}} \underbrace{(\Delta p)^{1/3}}_{\text{pressure drop}} \underbrace{(\Delta T)^{2/3}}_{\text{temperature rise}}. \tag{11.18}$$

It is appropriate to pause at this point to examine the consequences of this general equation. One goal of the long-module concept could be to make as long a collector as possible for one central power plant—in other words to maximize L/D. Let us look at each factor in the expression.

11.5 FLUID PROPERTIES

The quantity $(\rho c^2)^{1/3}$ is a function of the heat-transfer fluid, and one would like the largest value of this relationship. If we examine a number of candidate fluids we find that water is the best for both performance and cost. We are, however, limited by the maximum temperature rise in a water system because pressure is a sharp function of temperature if one is to prevent a phase change to steam. The minimum temperature, on the other hand, is limited by the environmental temperature. In practical terms we arbitrarily limit the allowable ΔT as discussed below. The general trend of available fluid factors is given in Table 11.1. The conclusion from this table is that water is the best heat-transfer fluid, and that one obtains significantly smaller L/D ratios with other materials when the temperature rise is kept constant.

Table 11.1 Fluid Heat-Transfer Factors

Fluid	Specific heat c (cal/g C°)	Density ρ (g/cm^3)	$(\rho c^2)^{1/3}$
Water	1.00	1.00	1.00
Organics	0.30–0.60	1.00	0.45–0.51
Salts	0.25	2.50	0.54
Sodium	0.30	0.97	0.44
Gallium	0.08	5.90	0.34

11.6 TEMPERATURE RISE

The temperature rise permitted depends upon the working fluid used, and one would like to maximize ΔT. We find that economics dictates water to be the best choice, subject to temperature limitations. The pressure must be kept low enough to avoid engineering and materials problems. One would prefer the collector and thermal storage tanks to operate at the same pressure and temperature. Because of the rapid rise in pressure with temperature of water above 100°C, the practical limit appears to be in the range of 150–200°C for a water system.

Other fluids listed in Table 11.1 allow higher temperatures, but at significant cost increases. It is interesting to examine the engineering performance of these fluids in order to learn whether they might be cost effective.

From Table 11.2 we see that the higher temperatures of other fluids boost their performance almost to a par with water. This means that the same L/D ratios can be used. It also implies that the higher Carnot efficiencies associated with higher operating temperatures will tend to favor the use of organics or sodium; however, in each case it is clear that the higher cost per pound of the heat-transfer fluid will more than offset the additional Carnot gain.

Table 11.2 Thermal Properties of Fluids

Fluid	ΔT_{max}, C°	$\Delta T^{2/3}$	$(\rho c^2)^{1/3}(\Delta T)^{2/3}$
Water	150	28	28
	100	21	21.5
Organics	300	45	22–23
	200	34	15–17
Salts	500	63	34
Sodium	500	63	28
	400	55	24
Gallium	600	71	24

11.7 SOLAR FLUX

If the amount of solar flux transferred per unit length is reduced, a greater distance can obviously be traveled before a given ΔT is achieved, but this option is self-defeating because we want the maximum energy gain per unit collector area. The concentration ratio W can be adjusted, but even for a flat-plate collector W is on the order of 3 to 6, the ratio of the diameter of the fluid line to the plate area connected to the fluid line. To reduce W to unity, the absolute limit, one would have a collector with adjacent fluid lines, but the gain is small since the power 2/3 occurs. The magnitude of the dependence of performance on increase of solar flux by concentration W is given in Table 11.3.

Table 11.3 Flux Concentration Factor

Concentration W	$(1/W)^{2/3}$
1	1.00
2	0.63
3	0.48
4	0.40
6	0.30
10	0.22

11.8 PRESSURE DROP

The largest sustainable pressure drops are desirable from the standpoint of the L/D ratio when all other consequences are ignored. Again, the cost of the energy needed to repressurize the fluid to return it to the collectors becomes an important question. We examine the influence of Δp on L/D in Table 11.4.

Table 11.4 Pressure Drop Factors

Pressure drop Δp, g/cm^2	$(\Delta p)^{1/3}$
100	4.6
1,000	10.0
2,000	12.6
4,000	16.0
10,000	21.7

11.9 L/D RATIOS

To illustrate the approximate size of L/D for a typical set of parameters, we take the following values:

$$\begin{aligned}
f &= 2 \times 10^{-2}, \\
g &= 980 \text{ cm/sec}^2, \\
c &= 1.0 \text{ cal/g C}^\circ, \\
\rho &= 1.0 \text{ g/cm}^3, \\
\Delta p &= 1000 \text{ g/cm}^2 \quad (p \approx 10{,}000 \text{ g/cm}^2), \\
\overline{\eta} I &= 0.010 \text{ cal/cm}^2 \text{ sec}, \\
\Delta T &= 100 \text{ C}^\circ.
\end{aligned}$$

We then have

$$\begin{aligned}
(\pi^2 g/8f)^{1/3} &= 39, \\
(\rho c^2)^{1/3}(\Delta T)^{2/3} &= 21.5, \\
(\Delta p)^{1/3} &= 10, \\
(1/\eta I)^{2/3} &= 22.
\end{aligned}$$

Hence we have

$$L/D = 180{,}000\,(1/W)^{2/3}.$$

With a ratio of collector width to pipe diameter of $W = 4$, we have

$$L/D = 72{,}000.$$

This means that for a collector with a fluid pipe 2.5 cm in diameter, the total path length L will be

$$L = 180{,}000 \text{ cm} \quad \text{or} \quad 1.80 \text{ km}.$$

The velocity of heat-transfer fluid flow in the pipes for the above case is

$$\begin{aligned} v &= (4/\pi)(W\eta I)(L/D)(1/\Delta T)(1/c\rho) \\ &= 1.28\,(4 \times 10^{-2})(72 \times 10^{5})(10^{-2}) \\ &= 3.7 \times 10^{1} = 37 \text{ cm/sec}. \end{aligned} \tag{11.19}$$

The numerical values obtained for both the field size and flow velocity are encouraging to the concept that one could collect the solar energy over a considerable area without encountering unusual pressure drops.

The time the fluid is traveling through the collector is then

$$t = L/v = (1.8 \times 10^{5})/37 = 4860 \text{ sec} = 81 \text{ min}.$$

11.10 REYNOLDS NUMBER

We can take these provisional values and calculate the value of the Reynolds number (Re) for the above flow velocity and correct our first estimate of the hydraulic friction coefficient f. We have

$$\text{Re} = vD/\nu = v\rho D/\mu, \tag{11.20}$$

where ν is the kinematic viscosity (stokes) of the fluid, μ is the absolute or dynamic viscosity (poises), and ρ is the density of the fluid.

Substituting the provisional value for v, with $D = 2.5$ cm, we find that at 100°C we have

$$\text{Re} = 32{,}900.$$

We note that this value of Re is in the range of turbulent flow in Fig. 11.4, but at a point where the hydraulic friction coefficient f is about 0.03, significantly higher than the assumed value of $f = 0.02$. Two things become apparent: (1) that we underestimated f, which will lower the L/D ratio and velocity, and (2) that the range of f

is insensitive to pipe roughness in the range up to 100 μm per cm of pipe diameter (0.010 in. per in. of pipe diameter). Because the value of f affects L/D as $f^{-1/3}$, the change in f from 2 to 3 changes the calculated L/D ratio by

$$(L/D)' = 0.87(L/D).$$

The final set of values for the long module in the above example is therefore

Fluid velocity = 32 cm/sec,
Total module length = 1.57 km,
Time of fluid travel = 81 min.

These dimensions are shown in Fig. 11.6.

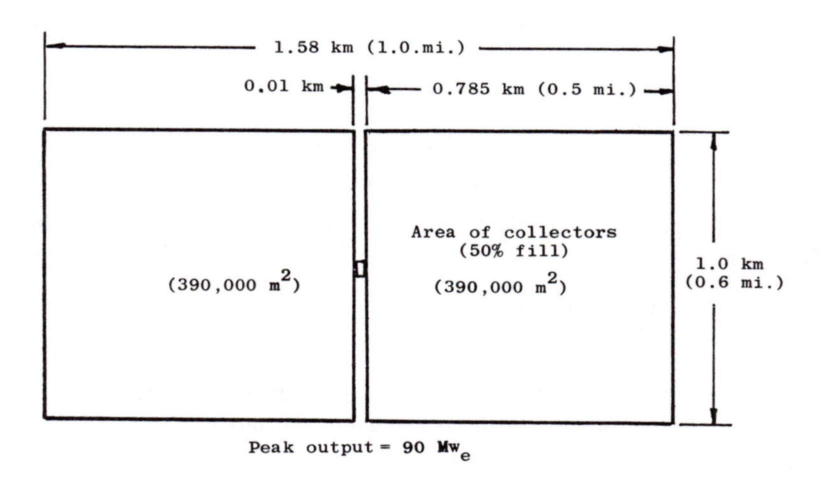

Fig. 11.6 Layout of the solar power farm for $W = 4$ and Δp = 1000 g/cm² with ΔT = 100 C°. With a net conversion efficiency of $\eta = 0.12$ the peak electrical output of this farm unit would be approximately 90 MW_e.

11.11 RATIO OF POWER EXPENDED TO POWER GENERATED

The ultimate factor determining the size of a power farm long module is the amount of energy expended in pumping the fluid through the collector at the required velocity. There are two power terms to be considered: (1) power to sustain the velocity of fluid flow, and (2) power to overcome the friction pressure drop.

The basic equation for the power input to an incompressible fluid is

$$P = \rho VH \quad \text{g cm/sec}, \tag{11.21}$$

where V is the volume rate of flow of fluid and H is the absolute "effective" height through which the flow is moved. We elect to use g cm/sec for power units because they are the individual terms in Eq. (11.21). One could use J/sec, W, or grams force rather than grams mass, but then the conversion factor could be confusing. It is, moreover, important to keep units equations parallel to the symbol equations. At the end of the equation development we can transfer units as we wish without risk of a misstep. For example, the difference between grams mass and grams force appears in the conversion factor applied between Eqs. (11.31) and (11.33).

In Eq. (11.21), V is given by

$$V = vA = v\pi D^2/4 \quad \text{cm}^3\text{/sec}, \tag{11.22}$$

where v is the fluid velocity, and H is given by

$$H = v^2/2g + \Delta p/\rho + z \quad \text{cm}, \tag{11.23}$$

where z is the geometric height. In a closed system we are interested in the change in effective height:

$$\Delta H = \Delta(v^2)/2g + \Delta p/\rho + \Delta z. \tag{11.24}$$

Since $\Delta(v^2)$ and Δz are both zero in a closed system, the term of importance is the pressure drop term. If, on the other hand, the heat-transfer fluid circuit had a storage tank where the fluid remained quiescent before returning to the collectors, then $\Delta(v^2)$ would in fact equal v^2. If one proceeds to evaluate the dynamic term relative to the static terms, the difference is academic in the case of solar collector parameters. For the sake of argument we will proceed as though the dynamic term were applicable. We therefore have two terms: $v^2/2g$, which is the dynamic height, and $\Delta p/\rho$, which is the static pressure height due to the pressure drop in the fluid line.

Substituting these expressions in the power input equation, we obtain

$$P = \rho v(\pi D^2/4)(v^2/2g + \Delta p/\rho) \quad \text{g cm/sec}. \tag{11.25}$$

If we now substitute the expression for v derived in Section 11.9,

$$v = (4/\pi)(W\eta I)(L/D)(1/\Delta T)(1/c\rho), \tag{11.26}$$

we obtain

$$\eta_p P_p = \underbrace{\frac{8}{g\pi^2} \frac{(W\eta I)^3}{\Delta T^3 c^3 \rho^2} \left(\frac{L}{D}\right)^3 D^2}_{\text{dynamic power term}} + \underbrace{\frac{W\eta I}{\Delta T c\rho} \frac{L}{D} D^2 \Delta p}_{\text{pressure drop term}} \quad \text{g cm/sec}, \tag{11.27}$$

where η_p is the pump net efficiency, on the order of 0.5, and P_p is pumping power.

Taking the case previously evaluated for a farm module with $L = 1.8$ km and $v = 37$ cm/sec, we find the relative magnitude of the two terms in Eq. (11.27):

$$\begin{aligned} \text{Dynamic power term} &= 370 \text{ g cm/sec}, \\ \text{Pressure drop term} &= 17{,}900 \text{ g cm/sec}. \end{aligned}$$

It is therefore reasonable to neglect the dynamic power term when the velocity is sufficiently low, in which case the net power requirement reduces to

$$\eta_p P_p = (1/c\rho)(W\eta I/\Delta T)(L/D)D^2 \Delta p \quad \text{g cm/sec}. \tag{11.28}$$

The key quantity in determining the ratio of power generated to power expended is the amount of solar energy collected by the module that can be converted into electrical power. In the derivation form used in this book the amount of energy usefully extracted per unit time is

$$Q/t = W\eta I(DL), \tag{11.29}$$

where WD is the effective collector width and L the collector length. The actual power generated by the collector P_c requires definition of the system conversion efficiency for the heated working fluid delivered to the power plant. We will denote the overall efficiency of conversion by η_c. We then have

$$P_c = \eta_c W\eta I(DL) \quad \text{cal/sec}. \tag{11.30}$$

The ratio of power generated by the collector module to power expended in pumping the heat transfer fluid through the collector pipe is then

$$\frac{P_c}{P_p} = \frac{\eta_p \eta_c}{1/c\rho} \frac{W\eta I(DL)}{W\eta I(L/D)(1/\Delta T)D^2 \Delta p} \quad \frac{\text{cal/sec}}{\text{g cm/sec}} \tag{11.31}$$

or

$$P_c/P_p = \eta_p \eta_c (c\rho)(\Delta T/\Delta p). \tag{11.32}$$

Because the units of power are not identical, we need to multiply by the conversion factor between g cm/sec and cal/sec, or

$$P_c/P_p = 42{,}664\, \eta_p \eta_c (c\rho)(\Delta T/\Delta p), \tag{11.33}$$

where ΔT is in C° and Δp is in g/cm^2.

11.12 MAGNITUDE OF POWER OUTPUT/INPUT RATIO

The expression for the ratio of power generated by the collector to power expended in pumping the heat-transfer fluid through the collector pipe is remarkably simple. It says that the characteristics of the collector do not matter. The only relevant quantities are the properties of the heat-transfer fluid $c\rho$, the temperature rise ΔT, and the *allowable* pressure drop Δp. This means that we can set the ratio of P_c/P_p at a reasonable value and then find the allowable pressure drop, or the reverse.

In the example we have already evaluated, if we take the system thermodynamic conversion efficiency η_c to be

$$\eta_c = 0.20,$$

and ΔT to be

$$\Delta T = 100 \text{ C°},$$

for a pressure drop of

$$\Delta p = 1000 \text{ g/cm}^2 \quad (\sim 1 \text{ atm}),$$

we find the ratio of power generated to power expended to be

$$P_c/P_p = 427.$$

11.13 PARAMETRIC RELATIONSHIPS FOR FLUID TRANSFER

The relationships among module length, flux concentration, flow velocity, dwell time in the collector, and pressure drop are presented in Figs. 11.7–11.10.

11.14 VARIATION OF OUTPUT/INPUT RATIO WITH SOLAR FLUX

In Section 11.11 we derived a very simple equation when the principal variables are pressure drop and temperature rise. In actual use another situation will prevail because the length of linear travel L in the collector will be a fixed quantity; the solar flux, however, will vary.

When one specifies the temperature rise and the pressure drop, the ratio of power output to power input, P_c/P_p, is constant regardless of the solar input and collector absorbtance and energy concentration ratio W. This means, for example, that when the solar flux is less than optimum the fluid velocity is lowered and the path length in the collector is increased so that the pressure drop is the same; then the distance L required for the fluid temperature to rise by an increment ΔT increases so that the ratio P_c/P_p is unchanged.

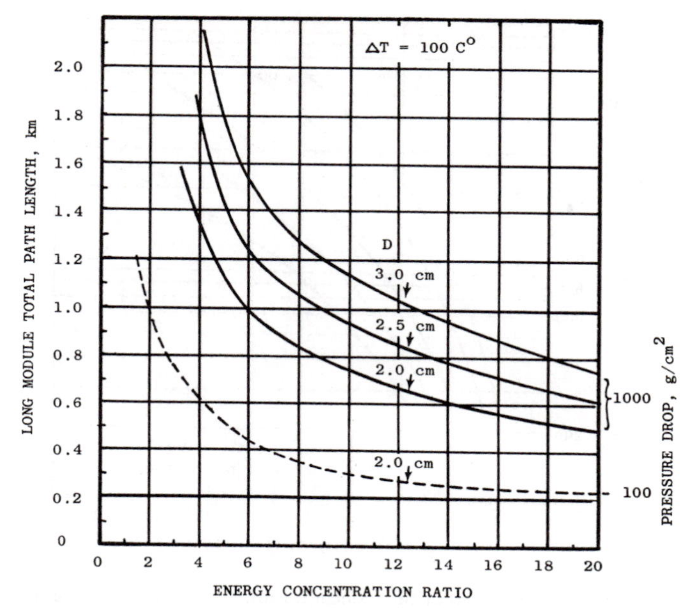

Fig. 11.7 Dependence of the long module total path length L upon pressure drop Δp and fluid pipe diameter D. A Δp of 1000 g/cm^2 is approximately 1 atm or 15 psi.

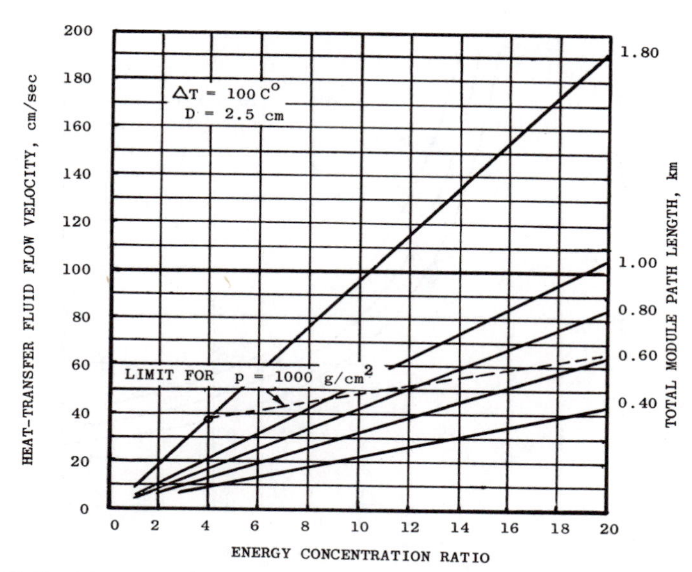

Fig. 11.8 Relationship among flow velocity v, flux concentration factor W, and total module path length for D = 2.5 cm and temperature rise ΔT = 100 C°. The limit where Δp = 1000 g/cm^2 is indicated.

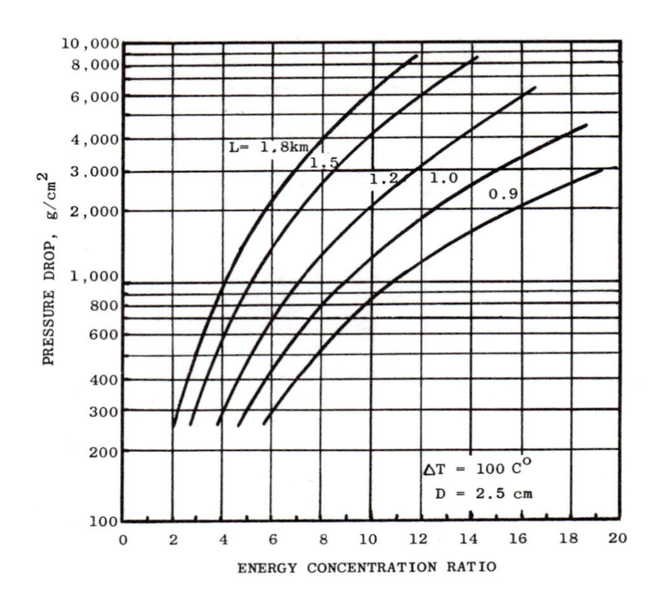

Fig. 11.9 Pressure drop as a function of flux concentration W for different module lengths L for a net temperature rise of $\Delta T = 100$ C° and a pipe diameter D of 2.5 cm.

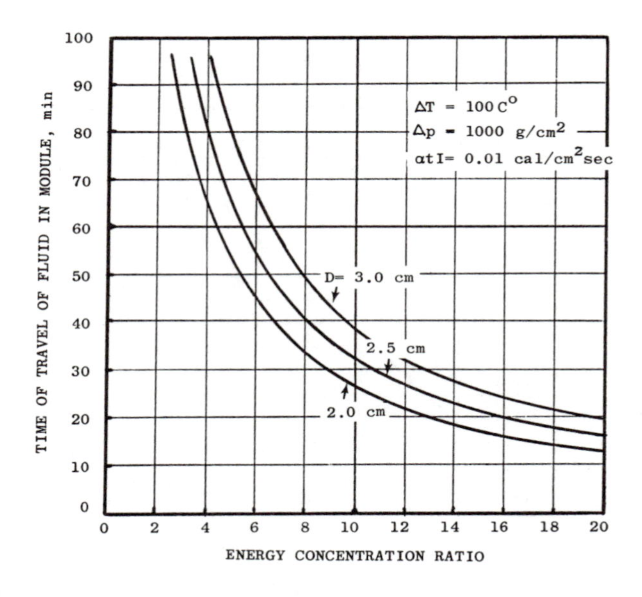

Fig. 11.10 Time of travel in collector as a function of flux concentration factor W for different pipe diameters D for a net solar flux extracted by the fluid of $\alpha tI = 0.01$ cal/cm^{2}sec.

In a typical collector situation, one would like to know how P_c/P_p changes as the solar flux I changes, wherein the path length in the collector is a constant. In this case of lowered solar flux the velocity of flow is reduced so that the fluid takes more time to arrive at the end of the collector, arriving with the desired temperature rise ΔT. In this case the equation becomes

$$P_c/P_p = 42{,}700\,\eta_p\eta_c(\rho c)\Delta T\,\frac{\pi^2 g}{8f}\,(\rho c^2)\,\frac{1}{(W\eta I)^2}\,(D/L)^3(\Delta T)^2$$

$$= 5.11\times 10^7\left(\frac{\eta_p\eta_c}{f}\right)\rho^2c^3\,\frac{1}{(W\eta I)^2}\,(D/L)^3(\Delta T)^3, \qquad (11.34)$$

which can be written

$$P_c/P_p = 5.11\times 10^7\left(\frac{\eta_p\eta_c}{f}\right)\rho^2c^3\,\frac{1}{(W\eta)^2}\,(D/L)^3(\Delta T)^3(1/I)^2. \qquad (11.35)$$

When the above equation is normalized to summer noon clear-day solar flux I_0, the expression for the change in net output to input ratio becomes

$$(P_c/P_p)/(P_c/P_p)_0 = Y = (\Delta T)^3 I_0^2/(\Delta T)_0^3 I^2.$$

The change in Y with reduction in solar flux is plotted in Fig. 11.11. Because the absolute value of P_c/P_p is so large, the solar flux can drop to very low values before the power used in pumping exceeds the power output of the collector module.

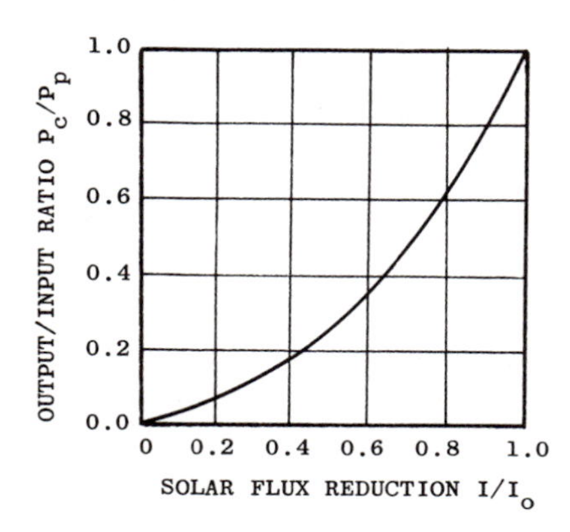

Fig. 11.11 Variation of the output/input ratio P_c/P_p for a long-module collector for constant ΔT output as a function of the reduction in solar flux input.

11.15 AIR-TRANSFER SYSTEMS

Air-transfer systems are basically identical to liquid systems except that the input parameters are generally expressed differently. For the sake of clarity we will take an example of a simple air system having a rectangular duct, as represented in schematic form in Fig. 11.12. The width of the duct will be *a* and the depth *b*, yielding a cross-sectional area *ab*. For a circular duct one can replace *ab* by $\pi D^2/4$.

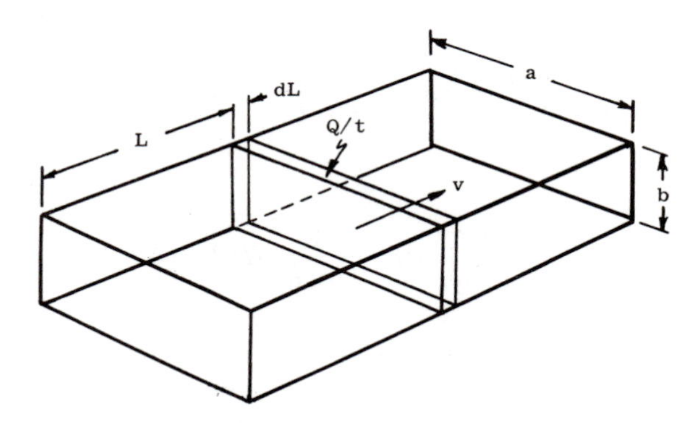

Fig. 11.12 Schematic diagram for a rectangular duct air heater. At this point we do not examine details of how the input energy Q/t is injected into the fluid stream.

The mass of air in the unit length dL of the duct is

$$dM = \rho\, ab\, dL \quad \text{g/cm}^3, \tag{11.36}$$

and the heat capacity of this mass of gas is

$$dQ = \rho c ab\, dL\, dT \quad \text{cal.} \tag{11.37}$$

The solar heat input to this small section of the duct, on the other hand, is

$$dQ = \eta I W a\, dL\, dt \quad \text{cal,} \tag{11.38}$$

where W is the flux concentration into the collector, defined as the ratio of the optical aperture of the collector divided by the width a of the absorber. For a flat-plate collector, W equals 1. Equating these two expressions, we have

$$\rho c ab\, dL\, dT = \eta I W a\, dL\, dt,$$

whence

$$dT = (\eta I/\rho c)(W/b)\, dt \quad \text{C}^\circ, \tag{11.39}$$

and because the time the air is being heated is L/v, the differential time element can be replaced by $dt = dL/v$; hence, we obtain the integral equation

$$\Delta T = \frac{I}{\rho c v b} \int_0^L \eta(I,\Delta T)\, dL \quad \text{C}^\circ, \tag{11.40}$$

where

ΔT = rise in temperature upon traveling a distance L,
ρ = density of air, the mean value over the range of ΔT being acceptable for the accuracy generally desired,
c = specific heat of air at constant pressure, which can vary depending upon the water content of the air,
b = depth of the duct carrying the air,
I = solar flux available to the collector absorber,
η = efficiency with which the solar flux I is transferred to the flowing airstream,
v = velocity of the airflow.

We cannot readily integrate this expression since η is a function of the accrued temperature difference between the ambient air and the air in the tube, which is a function of distance along the duct.

To take an example, we have calculated the relationship between η and ΔT for a solar grain dryer that consists of a tube of black polyethylene surrounded by another tube of clear polyethylene. Air is blown through either the inner black tube or it and the space between the two tubes. The efficiency of this type of collector is not high, but the cost is very low, yielding a good value for the cost of heat delivered into the grain bin. The relationship between η and ΔT is as shown in Fig. 11.13.

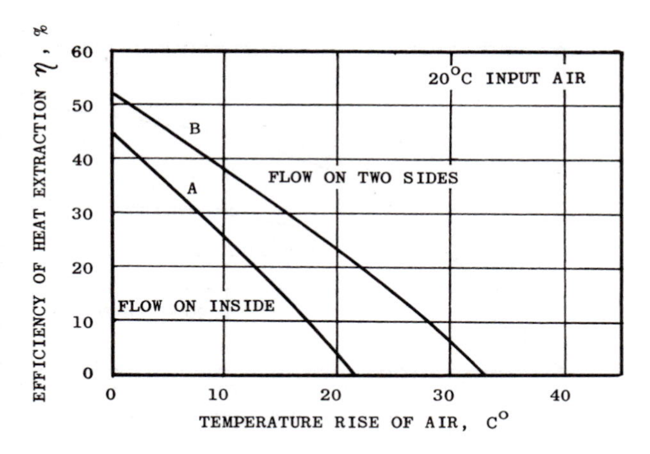

Fig. 11.13 Variation of the efficiency of heat injection into the airstream flowing inside a tubular plastic absorber for a tube absorbance of 0.90 and a solar input of 0.015 cal/cm²sec.

In evaluating the integral of Eq. (11.40) we have to assume definite values for the several quantities appearing in the equation. The effective value of the solar flux is not the flux at the normal point on the cylinder, but the average. The normal solar flux is therefore divided by $\pi/2$. The value used for the actual average flux is the direct solar flux divided by $\pi/2$ plus the diffuse component, 0.10 direct flux. The value of $I = 0.13$ cal/cm^{2}sec is therefore about the noon brightness of sunlight in Kansas in November. The values are

$$\begin{aligned} D &= 100 \text{ cm}, \\ I &= 0.013 \text{ cal/cm}^2 \text{ sec}, \\ c &= 0.24 \text{ cal/g C}^\circ \text{ (at 30°C)}, \\ \rho &= 0.00116 \text{ g/cm}^3 \text{ (at 30°C)}, \\ v &= 34 \text{ cm/sec}, \end{aligned}$$

whence

$$\Delta T = 1.61(\eta) \quad \text{C}^\circ\text{/meter of collector.}$$

In Fig. 11.14 we integrate the temperature rise along the length of the absorber tube for the above example. As one would expect, if the tube is very long the temperature approaches the limit where the curves in Fig. 11.13 reach zero heat transfer. Curve A is for airflow strictly through the central black tube. Curve B is for the case where the dead air space is drawn into the inner tube after picking up some heat from the outer surface of the black tube. Since the two tubes would be decentered, the inner tube lying on the bottom of the outer tube, with the largest air gap at the top, the purging airflow is mainly over the hottest portion of the absorbing tube surface.

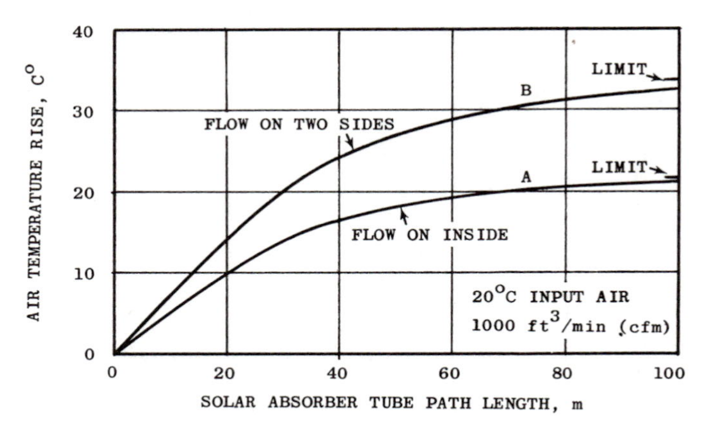

Fig. 11.14 Temperature rise as a function of tube length for (A) airflow on the inside of the absorbing tube only and (B) airflow on two sides of the absorbing tube.

11.16 AIR HEAT TRANSFER IN TERMS OF VOLUME RATE OF FLOW

Air heating systems are generally rated in terms of the volume rate of flow in ft^3/min (cfm), and building engineers are accustomed to specifying heating systems in terms of the amount of heating of the air desired and of the volume rate of flow required to service a building. It is therefore useful to change the preceding equations to accommodate these terms. To take a specific example, let us assume that the collector is made with the absorber suspended in the middle of an air duct. Sunlight is incident from one side, but the airstream surrounds the absorber. What then is the temperature rise to be expected in terms of absorber area and air volume?

The relationship between η and temperature rise at a given velocity is represented by the curve shown in Fig. 11.15. This curve is not for a specific case, but for illustrative purposes only. Curves of this type should be prepared for any given collector before a precise evaluation of the effect of changing airflow volume rate V can be made. Several examples are given in Sections 12.10 and 12.11.

The basic equation to begin with is Eq. (11.40). We now wish to substitute an analytic expression for the dependence of η on velocity v and eventually V. Let us assume that the change can be expressed by the function

$$\eta = \eta_0 (\underset{\text{convection term}}{1} + \underset{\text{forced flow term}}{gv}). \tag{11.41}$$

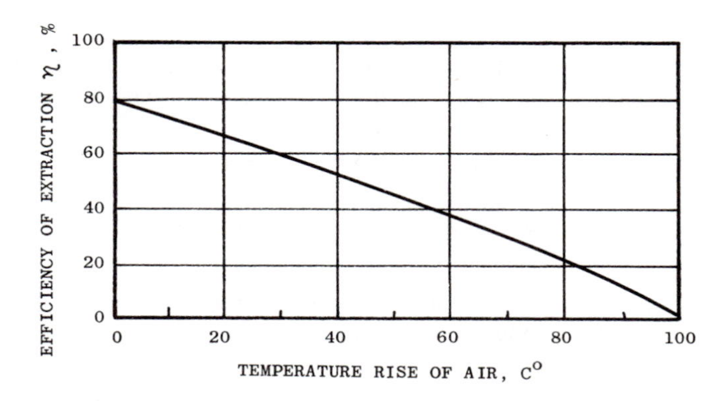

Fig. 11.15 Variation of efficiency of heat extraction as a function of incident solar energy with temperature rise in the airstream. This case assumes zero wind.

Inserting this expression for η into Eq. (11.40) and expressing velocity in terms of V, we obtain

$$\Delta T = (a/\rho c V) \int_0^L I\eta_0 (1 + gV/ab)\, dL, \tag{11.42}$$

which can be reduced to two terms:

$$\Delta T = \underbrace{(a/\rho c V) \int_0^L I\eta_0 dL}_{\text{natural convection}} + \underbrace{(g/\rho c b) \int_0^L I\eta_0 dL}_{\text{forced flow}}. \tag{11.43}$$

We want to minimize V to reduce the sum, as it has actually dropped out of the forced flow term. We also want to maximize the collector width and minimize the duct depth.

The desired collector output is not the temperature of the air alone. This is an important fact to recognize, and one that often clouds the issue of collector performance. *Any* solar collector can yield hot air, and the unsuspecting customer thinks this looks great until he tries to get quantity from it. The specifications for a solar heater should therefore have two quantities: (1) the temperature of the air exiting from the collector, and (2) the volume of air or kilograms of air per second issuing from the collector.

The volume of air defines the quantity of heat delivered to the user by the relationship

$$\Delta TV = (a/\rho c) \int_0^L I\eta_0 dL + (gV/\rho c b) \int_0^L I\eta_0 dL. \tag{11.44}$$

To meet the specifications, the values for both Q and ΔT must be met simultaneously. Denoting the integral by $\overline{\eta}L$, we have the two conditions, with $Q = \Delta T V \rho c$,

$$Q = a(\overline{\eta}L) + g(V/b)(\overline{\eta}L)$$

and

$$\Delta T = (a/\rho c V)(\overline{\eta}L) + (g/\rho c b)(\overline{\eta}L). \tag{11.45}$$

Because volume flow rate appears to minimize one desirable quantity and maximize the other, the exact dependence of system performance on volume V must be determined by actual calculations.

To show the performance of a simple collector as the airflow velocity is changed, we assume a collector as shown in Fig. 11.16, with the absorber located in the center of the airflow duct. This collector has one window and perfect insulation on the rear surface. The available solar flux is 330 Btu/ft^{2}h or 0.104 W/cm^2, of which we have assumed 0.095 W/cm^2 is absorbed, a rather good efficiency. In Fig. 11.17 we show the performance as the airflow velocity is changed. Note that the efficiency increases rapidly for low velocities, but then levels off rapidly. As an example, if the collector had a cross section 2.0 m wide and 0.2 m deep, with an area of 0.4 m^2, and if the fan moved 28 m^3/min (1000 ft^3/min), the airflow velocity would be 1.17 m/sec at the point indicated by the circle in Fig. 11.17.

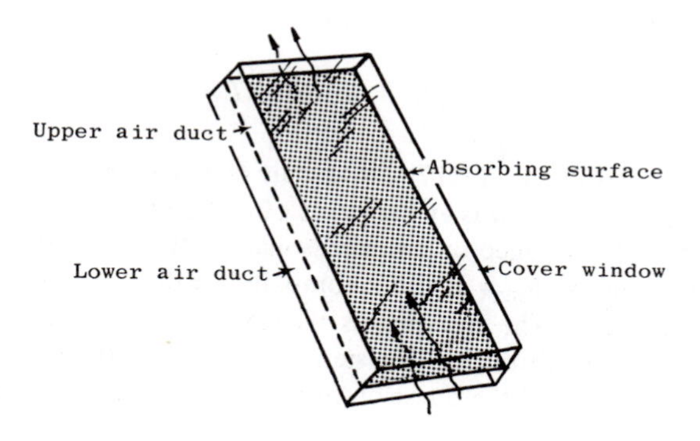

Fig. 11.16 Basic geometry of the example cited. The central absorber can be a solid plate, mesh, or louvered, each case yielding approximately equal performance.

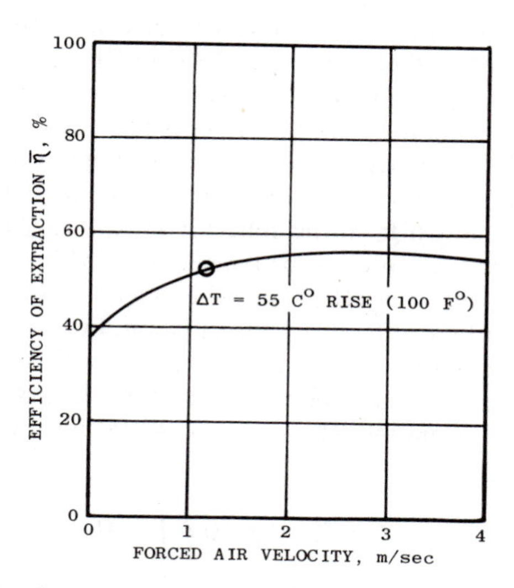

Fig. 11.17 Example of the change of extraction efficiency η with airflow velocity, for a collector having the absorber located in the middle of the airflow duct, with one cover window and perfect rear insulation.

If we take a high velocity to maximize the mass flowing per unit time through the collector, we note that the collector becomes very long for the air to reach 55°C, and the pumping losses will soon begin to affect the net system efficiency. The question becomes: What is the effective efficiency per unit length of collector in achieving a given ΔT? We must multiply ΔTV by the efficiency of extraction η and divide by the length of the collector L. The amount of heat collected per unit length can be written in two forms,

$$Q/L = \text{const}\cdot(\Delta TV\overline{\eta}/L),$$

or

$$Q/L = \text{const}\cdot\Delta T\overline{\eta}, \tag{11.46}$$

where $\overline{\eta}$ is obtained from Fig. 11.17. This result says that the most efficient operating point for a forced air collector is at the point of maximum $\overline{\eta}$. On the other hand, even a simple convection flow collector has a reasonable efficiency. In the example shown in Fig. 11.17 the value of $\overline{\eta}_{max}$ is 0.56, compared to the value of $\overline{\eta} = 0.37$ for natural convection heat transfer. The question is whether the additional efficiency is worth the expense of forcing air through the system, compared to the use of gravity flow.

We save for Chapter 12 the question of how to calculate extraction efficiency for collector designs.

11.17 TYPICAL EVALUATION SITUATION

To illustrate the procedure for determining the length of pass of the air through the collector, let us take the case where the outside air temperature is 0°C (32°F) and the air from the house enters the collector at 20°C (68°F) and is to be heated to 60°C (140°F) before being returned to the house via a rock storage bin. The airflow is 28 m^3/min (1000 ft^3/min) and is sent to three banks of collectors having a duct cross section of 1.2 m^2. How far must the air travel to acquire this temperature rise?

Referring to the diagram in Fig. 11.18, we show the path of the air in terms of efficiency of heat extraction from the collector as the heavy line from 20°C to 60°C, the difference between the entering air temperature and ambient and the exiting air temperature and ambient. The efficiency of extraction varies considerably over this range. Let us examine two methods of computation. First we will use the η curve and do a numerical integration of Eq. (11.40). Second we will assume the mean value of η for the temperature range of this problem and see how the answers compare.

To perform the integration let us take dL = 100 cm. Evaluating the expression, we have

$$T = 8.50 \times 10^{-2} \int_0^L \eta \, dL. \tag{11.47}$$

Because we need to use the value of η at mid-interval, we first calculate the temperature rise for one interval assuming the value of η at the beginning of the first interval, which, from Fig. 11.15, is $\eta = 0.66$. The typical temperature rise per 100 cm is then 5.61 C°. Taking a value of η at 22.8°C and carrying the integration from there, we have the values shown in Table 11.5. We note that we have achieved the desired temperature rise between 8 and 9 m of travel.

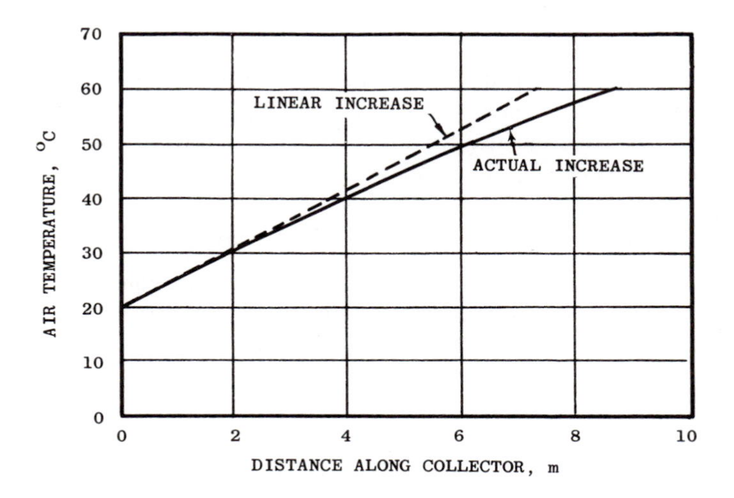

Fig. 11.18 Sample integration of airflow along a collector having a dependence of η on ΔT as shown in Fig. 11.15.

Table 11.5 Sample Integration for an Air Heater Collector

Interval	η	ΔT, C°	T, °C	T, °F
0	–	–	20.00	68.0
100	0.64	5.44	25.44	77.8
200	0.62	5.27	30.71	87.3
300	0.60	5.10	35.81	96.5
400	0.58	4.93	40.74	105.3
500	0.55	4.67	45.41	113.7
600	0.52	4.42	49.83	121.7
700	0.49	4.17	54.00	129.2
800	0.45	3.92	57.92	136.2
900	0.41	3.85	61.77	143.2

Now let us compare the result if we use a mean value of η for the entire temperature rise interval. The value of $\overline{\eta}$ is 0.53, and inserting it into the equation

$$\Delta T = I\overline{\eta}L/\rho cvb \tag{11.48}$$

and solving for L, we have

$$L = 11.76\,\Delta T/\overline{\eta}.$$

which equals 887 cm of travel distance in the collector. We see that the substitution of a mean value for η yields answers to a satisfactory degree of accuracy for practical use.

The actual temperature rise for this example is shown in Fig. 11.18. Note the slowing of the rise caused by the decrease in η.

11.18 ALTERNATIVE FORMS OF THE HEAT-RISE EQUATION

Using the information given above, we can write the heat-rise equation in a number of simple ways avoiding integrals—ways that are useful in different calculations as well as in understanding the tradeoffs with regard to collector size, flow rates, and so forth. The basic equation for collector performance is then

$$\text{Velocity:}\quad \Delta T = I\overline{\eta}L/\rho cvb, \tag{11.49}$$

where b is the depth of the duct carrying the air (or other heat-transfer fluid). In the preceding example we decided on an input value for the volume rate of flow V, then selected the number of collector modules into which this flow would be partitioned, thus defining the velocity v. If we retain volume rate as the input parameter, we have

$$\text{Volume rate:}\quad \Delta T = I\overline{\eta}aL/\rho cV, \tag{11.50}$$

where $\overline{\eta}$ is the value appropriate to the actual velocity flowing in the collectors. If the flow is divided into many modules, the value of $\overline{\eta}$ will be lowered. If $\overline{\eta}$ is not a function of velocity, then according to these equations one can divide the flow and velocity in any way one likes as long as the total collector area is maintained. Since aL equals A, the area of the collector, we have

$$\text{Collector area:}\quad \Delta T = I\overline{\eta}A/\rho cV. \tag{11.51}$$

One can obviously solve these equations for the desired parameter.

If we examine the above equations it becomes evident that the real variable of importance is not length, width, area, or velocity, but the *time* a given volume of fluid remains in the collector:

$$\text{Time:}\quad \Delta T = I\overline{\eta}t/\rho cb. \tag{11.52}$$

This last equation shows that in addition to time as the chief variable, the depth b of the duct is important.

11.19 EFFECT OF CHANGING HEAT-TRANSFER FLUID

A basic difference between heat-transfer fluids is apparent when Eq. (11.40) is evaluated for different fluids. A given volume of collector duct holds a very small mass of fluid when that fluid is gas, compared to when that fluid is liquid. The net result is that a gas-filled collector reaches the desired ΔT in a very short time. Liquid can remain in the collector a much longer time before reaching the same ΔT. This means, for example, that a fluid velocity can be much lower for a given collector length, or a liquid collector can be made very long for the same fluid velocity. These are important system tradeoffs to be considered.

Let us compare air and water. Solving for t, we have

$$t = (\rho cb/\eta I)(\Delta T). \tag{11.53}$$

For water and a duct depth (averaged over the collector width) of 0.1 cm, we have $\eta = 0.50$,

$$t = 10\,\Delta T,$$

and for air, with a duct depth of 10 cm and the same value for η, we have

$$t = 0.31\,\Delta T,$$

a difference of a factor of 30. In terms of the principal variable for heat-transfer fluids, ρc, the difference is 3200 times larger for a liquid than for a gas, assuming equal channel or duct depths.

11.20 HEAT TRANSFER IN EVACUATED COLLECTORS

One of the major heat-loss terms for a solar collector is *radiative.* The other major loss term is *convective.* In Chapter 12 we will discuss aspects of collector performance when convection is the major heat-loss mechanism. In this section we discuss the idealized case where convection is assumed to be negligible, as would be the case if the collector were evacuated to a high vacuum. The definition of collector performance becomes especially simple under this assumption. We also are able to see the full effect of selective coatings on the performance of the system. At this point we will neglect the very formidable problems encountered in the engineering of a collector for high vacuum operation.

The simple mechanical model that we will consider in this analysis is that of a linear collector. The model consists of a long pipe into which solar energy is injected continuously along its length. The pipe is in general associated with some form of optical concentration, where X is the concentration of sunlight on the absorbing pipe surface. A value of $X = 1$ would be the case where sunlight fell uniformly on the total surface of the absorbing pipe. If one were to consider only the optical cross section of the pipe D, where D is the diameter of the pipe, the cross-sectional concentration factor C would be πX. The reason we take the quantity X as the factor for the mathematical derivation of performance is that the pipe *radiates* thermal infrared (TIR) from all of its surface, and we are interested in the ratio of incoming to outgoing energy for the pipe.

In Fig. 11.19 we show the basic collector diagram defining the flux concentration factor X. The diameter of the absorber is D and the aperture of the collector, w, is equal to πDX. The optical concentration generally used by authors would be

$$C = w/D = \pi X. \tag{11.54}$$

We prefer to use X as the concentration, where

$$X = w/\pi D. \tag{11.55}$$

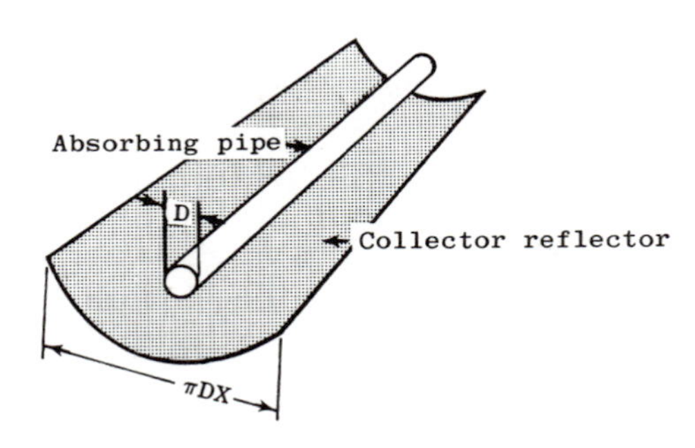

Fig. 11.19 Definition of the flux concentration factor X as used in this section. X is the ratio of the collector aperture to the *surface* of the absorber. If part of the absorber is insulated, then the thermal radiating surface of the absorber is used in forming the ratio X.

The solar input model can be considered as either continuous along the length of the pipe or injected at discrete points along the pipe, as long as the total number of points is reasonably large, on the order of 10 or more. The continuous distribution occurs when the solar flux is directly focused on the pipe by means of a cylindrical mirror or lens. The discrete distribution occurs when the solar flux is focused by a circularly symmetric optical mirror or lens at points along the pipe, or the energy is added by means of an intermediate transfer system such as a heat pipe. In the latter case of the discrete input source distribution, we assume that the entire length of the pipe is losing TIR uniformly along its length. If the section between input points were insulated for heat losses, then the expressions derived herein would not be applicable.

Let us consider the simplest model of a continuous uniform distribution of energy inputs along the length of the collector. We further assume that the pipe has a thermal gradient along its length small enough that conduction in the pipe material or the heat-transfer fluid does not alter the gradient. We allow a heat-transfer fluid to flow down the pipe, entering at a low temperature and gradually becoming heated as it proceeds along the pipe, arriving at the end of the collector with a desired ΔT. At each point the heat-transfer fluid is assumed to be substantially equal to that of the pipe. We further assume that the pipe is enclosed in a vacuum envelope such that the pipe loses energy only by radiative loss.

The energy balance of the pipe and collector is given by

$$\underset{\text{energy input}}{\delta \alpha X t I} = \underset{\text{radiative loss}}{\epsilon \sigma T^4} + \underset{\text{energy extracted}}{Q}, \tag{11.56}$$

where

δ = suppression of infrared losses by geometrical means, such as antiradiant cavities surrounding the absorber,

α = absorptance of the pipe surface over the spectral passband of the energy input flux,

X = surface optical flux concentration of flux (tI = the flux intensity) reaching the absorber pipe,

t = collector transmittance, and

ϵ = thermal infrared emittance of the pipe surface.

When Q is zero, no energy is being extracted from the system and we have the stagnation temperature. *Stagnation temperature* is a term that is usually used in aeronautics, being the temperature at a point on the aircraft where there is no flow of air over the surface and therefore minimum heat removal. We apply the term to solar collectors with a significant difference: it refers to the case where there is no flow of heat-transfer fluid through the absorber. Because no heat is removed by the heat-transfer fluid at zero flow rate, we have a condition of stagnation. If one prefers, this condition could also be referred to as a *zero flow temperature* or a *loss of coolant temperature.* It is nevertheless an important temperature for engineering reasons, because if the system loses flow capability for any reason, the surface of the absorber will rise steadily up to the stagnation temperature:

$$T_S = [(\alpha/\epsilon)(1/\sigma)\delta X t I]^{1/4}. \tag{11.57}$$

The stagnation temperature must be given considerable attention when one selects materials for the absorber pipe and its optical coating. The materials must withstand long-term exposure to daily operating temperatures and *also* a reasonable exposure to emergency temperatures caused by insufficiency of flow of the heat-transfer coolant through the absorber.

A simple substitution in Eq. (11.56) makes the radiative balance equation easier to evaluate and understand. Let us express Q as a fraction of the *absorbed* energy that is usefully extracted from the system. If this fraction is denoted by P, we have

$$Q = (\alpha\delta XtI)P. \tag{11.58}$$

We then have

$$\delta\alpha XtI = \epsilon\sigma T^4 + \delta\alpha XtIP. \tag{11.59}$$

Hence

$$\epsilon\sigma T^4 = \delta\alpha XtI(1 - P),$$

or

$$(1 - P) = (\epsilon/\alpha)(\sigma/\delta)(T^4/XtI), \tag{11.60}$$

or

$$T = [(\alpha/\epsilon)(\delta XtI/\sigma)(1 - P)]^{1/4}. \tag{11.61}$$

Each of the above equations is useful, depending upon the quantity to be calculated. The most important thing to note is that now we have temperature of the absorber as the sole function of the selectivity parameter α/ϵ. Absorptance α is still important, as are the reflection and internal absorption loss, because the absolute amount of incident solar energy Q utilized by the system is $I\alpha(1-r_1)(1-r_2)...(1-a_1)(1-a_2)...$ multiplied by P, as in Eq. (11.4).

In the cases we will treat in this section we will have no cavity suppression of thermal infrared emission; hence δ is unity and we will drop it from the equations following.

Figure 11.20 shows the variation of temperature as a function of the fraction of the absorbed energy removed by the heat-transfer fluid. We show the experimental line published by Speyer (1965) for a simple evacuated collector of low optical concentration. In the Speyer collector the values were $\alpha = 0.9$, $\epsilon = 0.4$, and $X = 0.67$. The concentration factor was low for the Speyer design because the absorbing tube area was 1.5 times the collecting area; hence the value of $X(\alpha/\epsilon)$ is 1.5. The slope of the line is steeper than the family of theoretical curves. The explanation is that for such low values of $X(\alpha/\epsilon)$ the efficiency is quite sensitive to the temperature of the glass vacuum envelope, and when the absorber heats, the envelope temperature also rises. In general the Speyer data confirm the prediction one would make from the family of curves in Fig. 11.20.

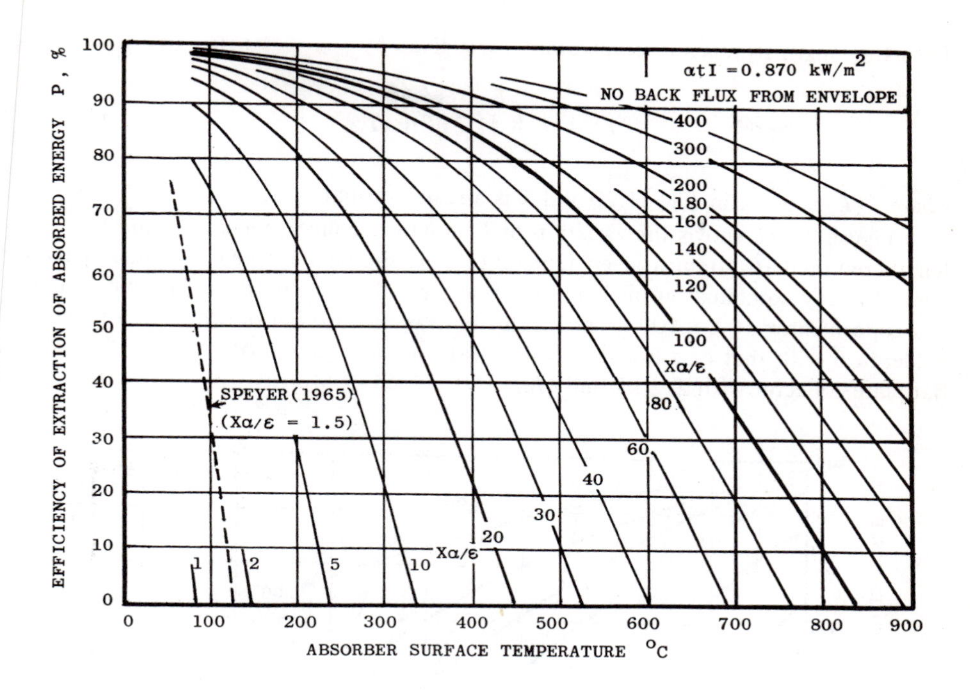

Fig. 11.20 Graph of the variation of the efficiency of energy extraction as a function of temperature, for values of $X(\alpha/\epsilon)$, when the solar input to the absorber αtI is 0.870 kW/m².

The stagnation temperature for each value of $X(\alpha/\epsilon)$ in Fig. 11.20 is indicated by the intersection of the curve with the abscissa, where P is zero. The curves in Fig. 11.20 were computed using the value for the absorbed flux αtI of 0.087 W/cm^2 or 0.0206 cal/cm^2 sec. If a different solar absorbed flux is used, then the effective value of $X(\alpha/\epsilon)$ is given by

$$X(\alpha/\epsilon) = (X\alpha/\epsilon)_0 (I'/I). \tag{11.62}$$

We are in general interested in the efficiency of use of the collector P over a given temperature rise. To determine this quantity we first need to find P as a function of collector length and flow velocity, or the time the fluid is being heated in the collector of a given length L. If we start with a temperature T_1 at the input of the absorber pipe, then it will change with length according to

$$T + \Delta T = T_1 + (dT/dL)\Delta L. \tag{11.63}$$

Then the length to reach a temperature T_2 is given by

$$L = k \int_{T_1}^{T_2} [1/P(T)] \, dT, \tag{11.64}$$

where $P(T)$ is given in Fig. 11.20 and k is a scale constant.

Figure 11.21 shows the variation of P along the collector length L and also the length required to attain a given temperature. It should be noted that these curves apply for any combination of X and α/ϵ that yield the same product. This means that if a concentrator of $X = 100$ is used with a blackbody absorber of $\alpha/\epsilon = 1.0$, the result is identical with that of a low-flux concentrator of $X = 10$ and $\alpha/\epsilon = 10$, or that of a flat-plate collector with $X = 1.0$ and $\alpha/\epsilon = 100$.

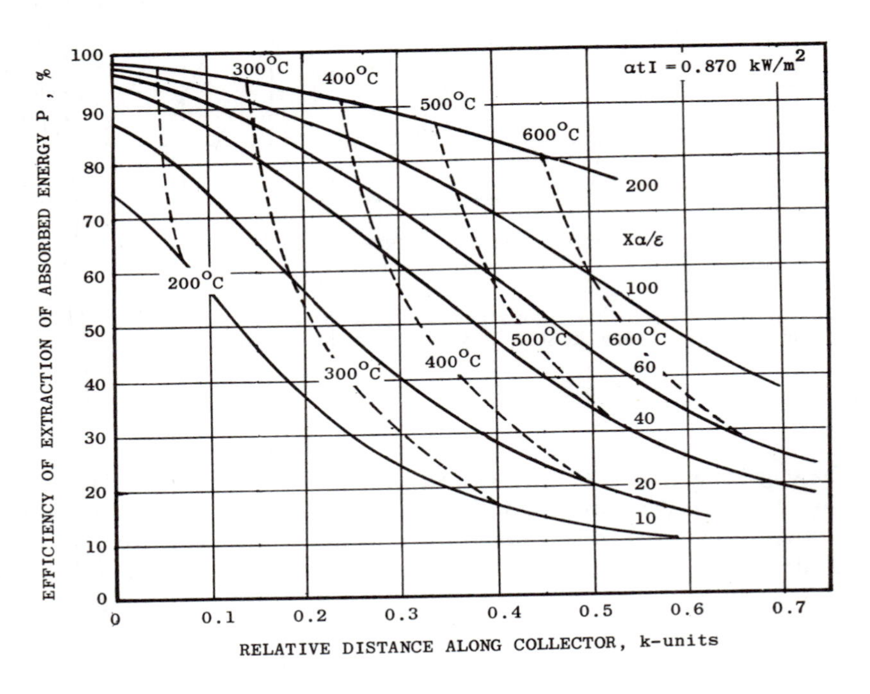

Fig. 11.21 Variation of the efficiency of extraction of absorbed energy P as a function of distance along the collector in units defined by $k\Delta T$. The dashed lines are lines of constant temperature for the exiting heat-transfer fluid.

Figure 11.22 shows the variation of temperature with length along the collector (or with time in the collector) as a function of $X(\alpha/\epsilon)$. It is clear that for operational temperatures in the vicinity of 500°C there is little to be gained in exceeding

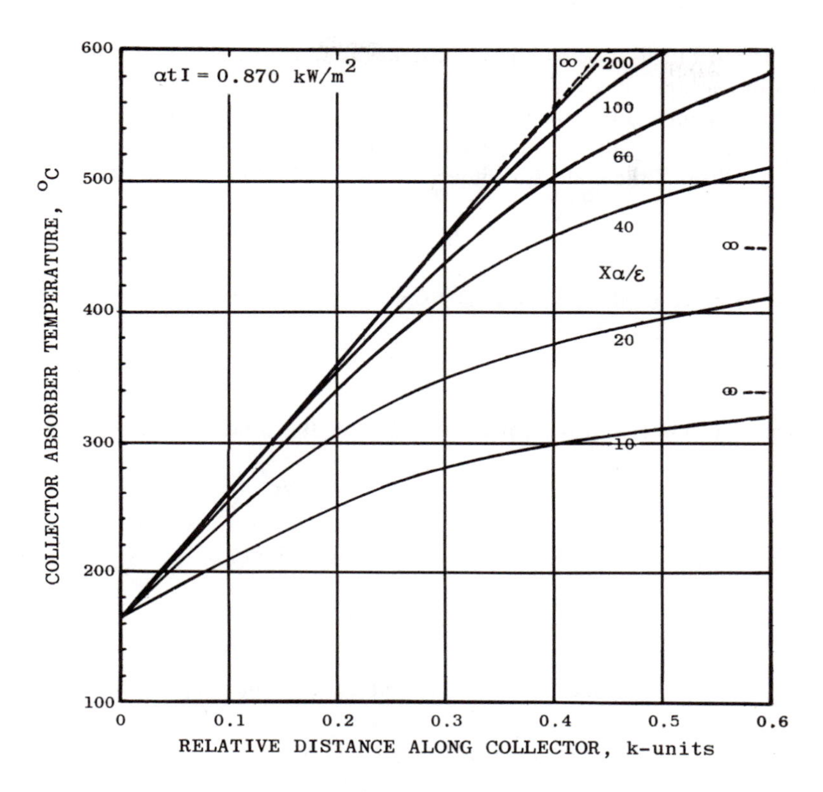

Fig. 11.22 Variation of absorber temperature as a function of distance along the collector for an input temperature of T_i = 167°C. Note that selectivity/concentration ratios $X(\alpha/\epsilon)$ above 100 are little different from ∞, showing that selective absorbers are of little benefit on high-concentration systems.

$X(\alpha/\epsilon) = 100$. This means that selective coatings are not useful in conserving energy in concentrating collectors when X exceeds several hundred. The only use of selective surfaces in the case of high optical concentrating collectors is when one needs to minimize the infrared flux from the absorber on nearby structures.

It should be noted that in the graphs shown in Figs. 11.20–11.22 we have considered that there is no infrared back flux from the envelope of the absorber. Even though the glass vacuum envelope will heat up to some degree, we have neglected this thermal infrared because the selective surfaces are assumed to be highly reflecting in the infrared, and hence have low absorbance for the cover wavelengths. When we use the curves for very low values of $X(\alpha/\epsilon)$, this back radiation may become detectable, so these curves are conservative in this region because the addition of some absorbed infrared would raise the thermal performance of the actual selective surface.

Equation (11.64) yields relative distances along a collector of uniform diameter fluid absorber pipe. If we wish to determine absolute distance we need to write the equations in slightly different form.

Let us consider a disc of heat-transfer fluid in the absorber pipe having a diameter D. The aperture of the collector, indicated by Fig. 11.19, is W, where

$$W = \pi D X. \tag{11.65}$$

The heat capacity of the fluid disc is given by

$$dQ = (\pi/4) D^2 \rho c \, dl \, dT. \tag{11.66}$$

The solar heat input to this element of length of the fluid is

$$dQ = \pi D X (\alpha P t I) \, dl \, dt. \tag{11.67}$$

Equating Eqs. (11.66) and (11.67), and replacing dt by dl/v,

$$dT = (4 X \alpha P t I / \rho c D v) \, dl. \tag{11.68}$$

Hence the length to achieve a given temperature difference is given by

$$L = (\rho c D v / 4 X \alpha t I) \int_{T_1}^{T_2} (1/P) \, dT. \tag{11.69}$$

The constant of porportionality in Eq. (11.64) is therefore given by

$$k = \rho c D v / 4 X \alpha t I. \tag{11.70}$$

We note that as the fluid velocity rises, the length to reach a given ΔT increases linearly. Also, when the diameter is increased, the length to reach a given ΔT increases linearly. Increased absorbtance α, flux concentration X, and solar flux I decrease the required length linearly.

11.21 THERMODYNAMIC UTILIZATION OF COLLECTED ENERGY

In general, one is interested in maximizing the power that can be extracted from the collector system by means of a thermodynamic cycle. One could compute the Rankine efficiency for the use of the collected heat, but this involves the use of specific temperatures and the Mollier diagram of enthalpy versus entropy. We show how this is done in Chapter 14. In this section we will, for simplicity, assume a Carnot cycle as an aid in understanding the role of selective surfaces in improving system performance.

The Carnot potential of a given amount of heat supplied to a Carnot cycle at a temperature T_2 is given by

$$C = (T_2 - T_1)/T_2, \tag{11.71}$$

where T_1 is the system condenser temperature. We note in Figs. 11.20–11.22 that the efficiency of extraction of heat from the collector is high when the temperature is low. Equation (11.71), however, shows that when the temperature rise is low in the collector, T_2 is close to T_1 and the efficiency of conversion of heat to work is low. The values of Eq. (11.71) are plotted in Fig. 11.23.

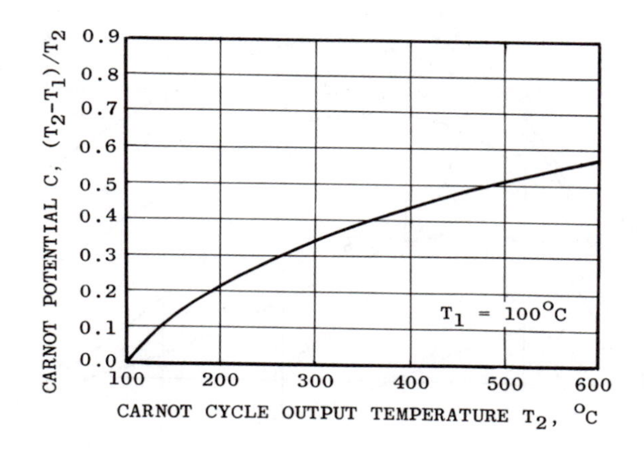

Fig. 11.23 Variation of the Carnot potential as a function of temperature difference, when the condenser temperature is T_1 = 100°C.

The relative distance shown in the preceding figures is in units of k, where k is as defined by Eq. (11.70). To obtain linear distances s, one must calculate

$$s = kT = (\rho c D v / 4 A \alpha t I)(\Delta T). \tag{11.72}$$

The ability to convert absorbed heat into work is given by

$$W = CP. \tag{11.73}$$

Using the values for P from Fig. 11.21, we obtain the graph for CP shown in Fig. 11.24. The temperature assumed for the condenser is T_1 = 100°C. The value of P in this figure is the value at the exiting fluid temperature T_2. This means that the fluid is not significantly heated in passage through the collector. We take this "limiting case" because it is instructive in assessing the value of heating the heat-transfer fluid over a temperature difference. It is equivalent to assuming that P for the whole collector is the worst possible value.

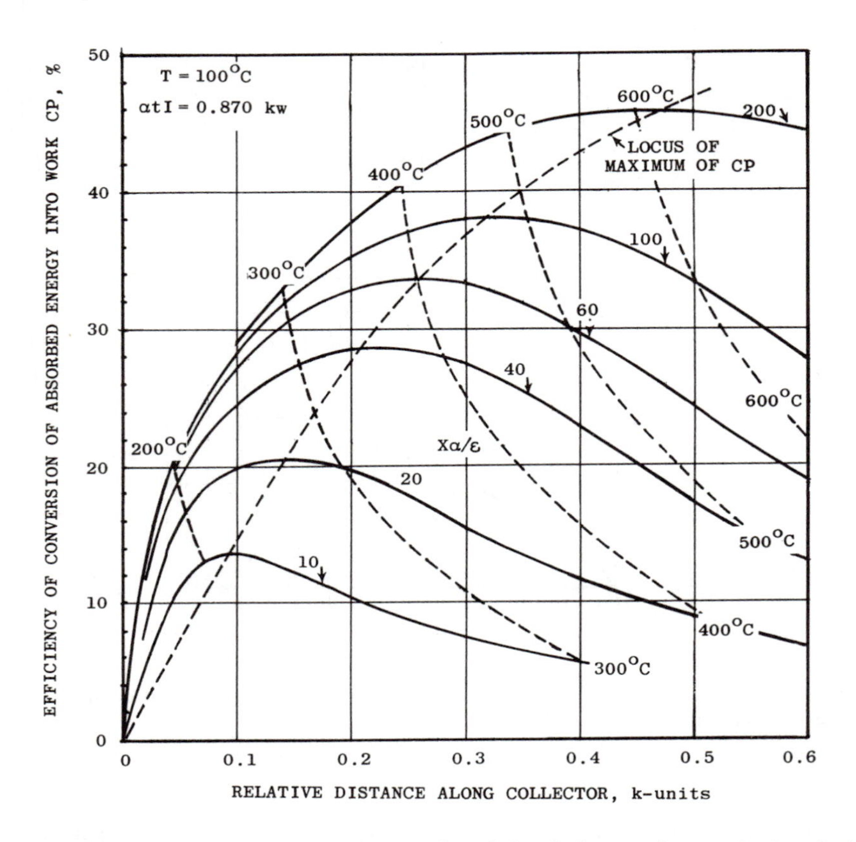

Fig. 11.24 Variation of the efficiency of conversion of absorbed energy into work when the heat-transfer fluid enters and leaves the collector at T_2, a case where P is at the minimum value, taken at T_2.

To give a numerical example of CP, let us take $X(\alpha/\epsilon) = 20$. We wish the value of CP at a distance along the collector of $k = 0.2$ units. At this distance we find from Fig. 11.21 a value of $P = 0.57$ and a temperature of $T_2 = 305°\text{C}$. The value for CP is therefore 0.199.

To take a more practical case, let us assume that the heat-transfer fluid enters the collector at $T_i = 167°\text{C}$, but with the condenser temperature still $T_1 = 100°\text{C}$. In this case we must determine $\overline{P}$, the average value of P over the interval from $T = 167°\text{C}$ to T_2. The resulting values for $C\overline{P}$ are plotted in Fig. 11.25.

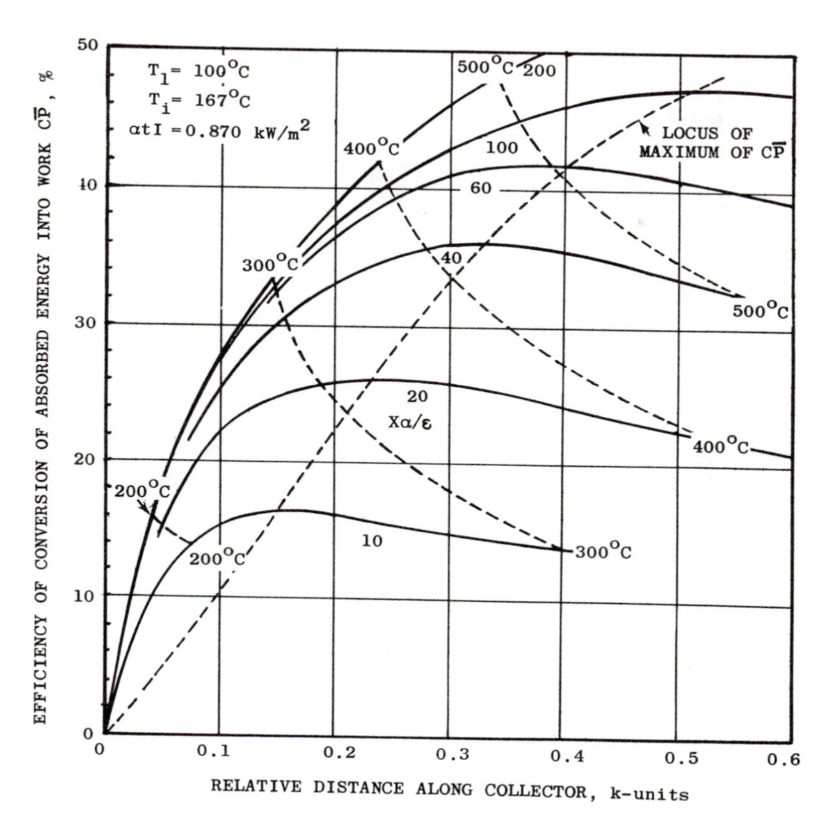

Fig. 11.25 Variation of the efficiency of conversion of absorbed energy into work when the heat-transfer fluid rises from an input temperature of T_i = 167°C to exit temperature T_2.

To give the numerical equivalent to the conditions cited above, let us take $X(\alpha/\epsilon) = 20$. The value of $\overline{P}$ from Fig. 11.21 is 0.73, at a distance along the collector of k = 0.2 units. Since $C = 0.45$ in this case also, the value for the efficiency of conversion of absorbed energy into work is $C\overline{P} = 0.255$, an increase of 0.056 in efficiency.

To show the net changes between the case of zero temperature rise and that of finite rise, we show the differences between Figs. 11.24 and 11.25 in Fig. 11.26.

The reason for showing these two cases is that one must make an efficiency tradeoff upon occasions like the following. The extraction of work from a thermodynamic cycle might not sufficiently cool down the heat-transfer fluid for the maximum net system efficiency. In such a case the fluid returning to the collector would be too hot for best performance. The curves in Fig. 11.26 indicate that one always wants to extract as much energy, per pass, as possible before returning the fluid to the solar collectors, in order to get the best value for $\overline{P}$.

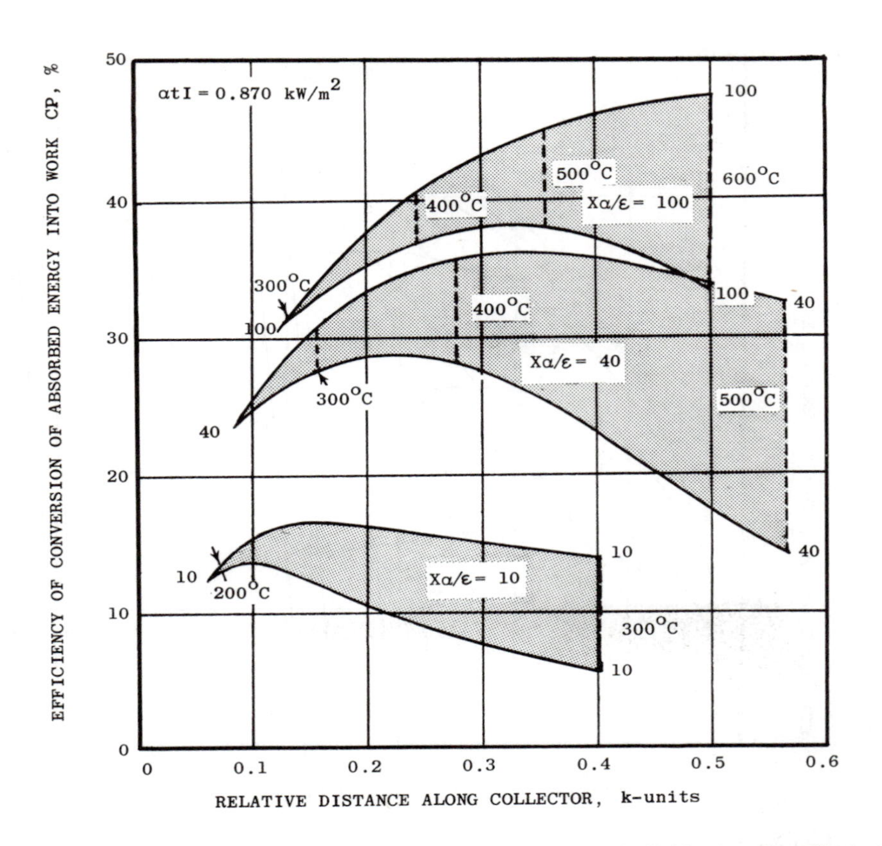

Fig. 11.26 Comparison of the net efficiency of the collector when the fluid enters the collector at essentially T_2 and at an inlet temperature of T_i = 167°C. Note that significant gains in energy conversion efficiency are achieved by operating the heat-transfer fluid over a wide temperature increase in the collector.

One can use Figs. 11.20–11.26 to obtain absolute lengths by evaluating k from Eq. (11.68) and using the k-unit scales in these figures. The temperature increment used in these figures was a finite one of 33 C°. A sum of distances required to achieve this increment was used in lieu of the integral.

To evaluate a given collector configuration operating between T_1 and T_2, we need to determine the average efficiency of extraction $\overline{P}$ over the temperature interval. Figure 11.21 is useful for this purpose. If the average value of P is denoted by $\overline{P}$, the actual quantity of heat available for the Carnot cycle is

$$d\overline{Q} = \pi DX(\alpha\overline{P}tI)dl,$$

and the amount of work that can be extracted by the Carnot cycle is

$$dW = \pi DX(\alpha\bar{P}tI)[(T_2 - T_1)/T_2]\ dl. \tag{11.74}$$

A graph of this function is shown in Fig. 11.24. Note that the curve for each $X(\alpha/\epsilon)$ value has a maximum, and although use of a longer collector length will result in an increased heat-transfer fluid temperature, the system efficiency drops off. This is because the added length adds heat inefficiently, causing $\bar{P}$ to drop more rapidly than the Carnot term increases.

While we are interested in the value of α/ϵ because of its influence on the fluid temperature and the percent of absorbed energy extracted, we are also interested in the absolute value of the absorptance α because we want to absorb *and* extract the maximum amount of the incident energy. The latter factor is indicated by the appearance of α in Eq. (11.74). As an example of the efficiency to be expected, let us take a system having $\alpha = 0.90$, $X(\alpha/\epsilon) = 100$, and $T_2 = 500°C$. The system will collect and transfer into work

$$dW = \pi DXtI \times 0.45 \times 0.90 = 0.405(\pi DXtI) = 40\%$$

of the solar flux arriving at the absorber. To obtain the actual system performance one would need to multiply 40% by the percentage of Carnot for the Rankine cycle and the turbine efficiency factor, which combine in a typical case to total about 75%. The above system would therefore have an actual conversion efficiency, subject to the limiting conditions specified in the beginning of this section, of approximately 30%.

It should be noted that the curves presented in this section are for a particular set of input values. If one desires to calculate a different system, one will need to evaluate the appropriate graphs, beginning with Fig. 11.20.

11.22 EVACUATED COLLECTOR TRADEOFFS

The preceding graphs show that selective surfaces can yield high efficiencies when operated in vacuum and with unit and low flux concentration collecting systems. If α/ϵ can be made as high as 40–60, already demonstrated in the Helio development program, good performance can be obtained with unit flux concentration at temperatures in the 300–400°C range, and if small optical concentration is added, on the order of $X = 2$ or 3, then good performance can be obtained between 500–600°C. The basic problem facing the use of highly selective absorbers is in the engineering and cost of the evacuated envelope for the absorber. A secondary problem is obtaining α greater than 0.95. One can either accept a loss in α or use a "Gothic vee" surface (Section 8.16) for the absorber so that two reflections are made at the absorbing surface, as suggested by Tabor, thus raising the effective α to more than 0.96–0.98.

11.23 LINEAR ABSORBER WITH AIR

In the preceding case we examined the performance of a collector in which the absorber is isolated inside an evacuated enclosure. We are now interested in the changes when air or another gas is used in the enclosure surrounding the absorber, or when the absorber is in the open air. In these cases there is an additional loss of energy from the absorber due to combined conduction and convection. The net effect of this additional energy loss is to remove additional energy from the absorber and to heat the surrounding envelope. This heating of the surrounding envelope, a point neglected in the preceding analysis, causes the envelope to radiate enough thermal infrared, TIR, to affect the temperature of the absorber. If highly selective coatings are used on the absorber, this additional back flux of TIR is still negligible. The additional loss from the absorber by convection is, however, not negligible.

The basic equation for heat balance is modified to

$$\underset{\text{energy input}}{\alpha_0 XtI} = \underset{\substack{\text{radiative loss}\\ \text{(surface 0)}}}{\epsilon_0 \sigma T_0^4} - \underset{\substack{\text{radiative back flux}\\ \text{(surface 1)}}}{\alpha_0' \epsilon_1 \sigma T_1^4} + \underset{\text{convection + conduction}}{(\text{Nu}+1)(k/S)(T_0 - T_1)} + \underset{\text{useful energy}}{Q}, \tag{11.75}$$

where α_0' is the absorptance of the absorber for thermal infrared radiation from the envelope, a different value in general from the absorptance of the absorber for sunlight. In this expression we have followed the procedure of separating conduction from convection, as outlined in Section 10.6. The convective loss is obtained by calculating the value of the Rayleigh number (Ra) for the absorber structure and finding the value of the Nusselt number (Nu) from Fig. 10.11. Since the Nusselt number is the size of the convective loss relative to the conductive loss, the combined loss is Nu + 1. The net loss can then be treated as a conduction loss with the thermal conductivity of the medium k being modified by Nu + 1, resulting in the form of the loss term in Eq. (11.75). The shape factor S for simple conduction is the distance L over which the heat is conducted, but for other shapes, as for a cylinder, the factor is different, as discussed in Section 12.7.

To calculate the losses in this equation it is necessary to know the temperature of the envelope and of the absorber. The envelope temperature depends on the input of heat from the sun, generally small, and the input from the absorber, appreciable when air is in the system. The envelope temperature then rises until the amount of heat it loses to the environment balances the heat inputs, a topic discussed in detail in Sections 12.5 and 12.7. At this point we will assume that the envelope temperature T_1 is known.

Proceeding as before, let us substitute

$$Q = \alpha_0 XtIP. \tag{11.76}$$

Then, solving for the fraction of the absorbed energy extracted from the system, P, we have

$$P = 1 - \frac{\sigma}{(\alpha_0/\epsilon_0)XtI} T_0^4 \left(1 - \frac{\alpha_0' \epsilon_1}{\epsilon_0} \frac{T_1^4}{T_0^4}\right) - \frac{\mathrm{Nu}+1}{\alpha_0 XtI} \frac{k}{S} (T_0 - T_1). \tag{11.77}$$

Since the absorptance of the absorber in the TIR, α_0', is equal to the emittance of the absorber in the same spectral range, ϵ_0, the equation simplifies to

$$P = 1 - \frac{\sigma}{(\alpha_0/\epsilon_0)XtI} T_0^4 \left(1 - \epsilon_1 \frac{T_1^4}{T_0^4}\right) - \frac{\mathrm{Nu}+1}{\alpha_0 XtI} \frac{k}{S} (T_0 - T_1). \tag{11.78}$$

We have selected a typical case to illustrate the effect of adding the conduction and convection losses to the vacuum case. In Fig. 11.27 we show the variation of efficiency of extraction of the absorbed solar energy P with absorber temperature for $X(\alpha/\epsilon) = 40$. The upper curve is for radiative loss only. The dashed curve is for assumed conduction and convection loss; for actual geometries of interest to the student this curve must be calculated as described above. The third curve, for combined loss, is reduced from the vacuum case by the amount of the conduction plus convection. For comparison we have shown the ordinate on the right-hand side of this figure as the efficiency of extraction of useful energy in terms of the available solar energy η. This shows a loss of 28% in transmission of sunlight and absorption by the absorber.

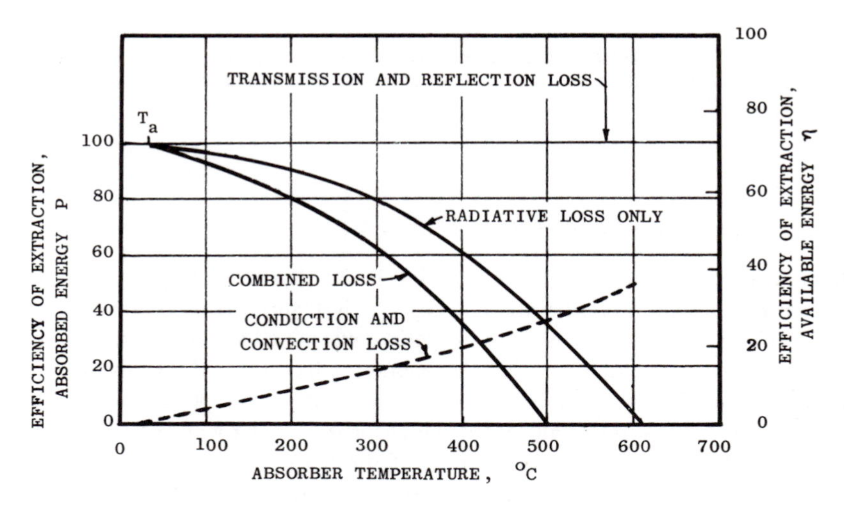

Fig. 11.27 Graph of the relationship between the efficiency of heat extraction of absorbed energy P and temperature for radiative loss only, for conduction plus convection, and for combined losses for $X(\alpha/\epsilon) = 40$. A typical scale of efficiency η in terms of the available sunlight arriving at the collector is also shown.

In Fig. 11.28 we show the result in terms of temperature rise along a collector in relative units. Note that while P is not reduced by a large quantity, the length of collector (or time of stay of the heat transfer fluid in the collector) increases significantly for 400°C fluid temperature. The temperature of the entering fluid is taken as 100°C. One uses a curve of this type to derive the average value of the efficiency, $\overline{\eta}$, needed to calculate system efficiencies in Chapter 16.

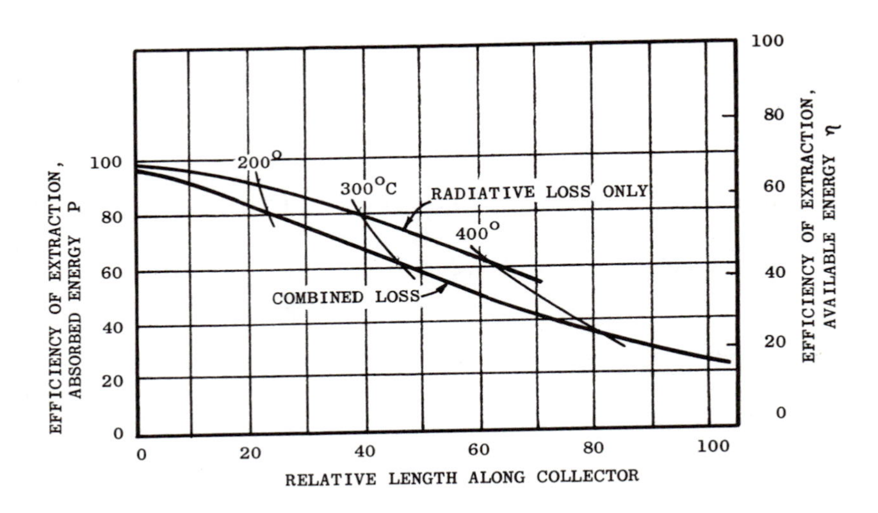

Fig. 11.28 Relationship between temperature rise and distance along a linear collector for $X(\alpha/\epsilon) = 40$ for the case of radiation loss only and for radiation plus conduction plus convection loss, as given in Fig. 11.27.

11.24 RADIATION SUPPRESSION USING HONEYCOMBS

A honeycomb structure, as shown in Fig. 11.29, has certain properties useful in suppressing thermal infrared radiation from a hot surface, as originally shown by Hottel (1927). Such structures when properly constructed act very much like selective surfaces and have potential uses in solar energy collectors. If the walls of the honeycomb are "black" to thermal infrared, then the 2π solid angle of thermal radiation emitted from the absorber will largely be absorbed by the walls of the honeycomb. Only the solid angle of the upper opening of the honeycomb will permit radiation to escape directly to the sky. The radiation absorbed by the honeycomb will be reemitted at a new temperature, that of the wall of the honeycomb. Secondary emission from the walls near the bottom of the honeycomb is also reabsorbed by the walls, with little escaping directly out the aperture of the honeycomb. As a consequence of these

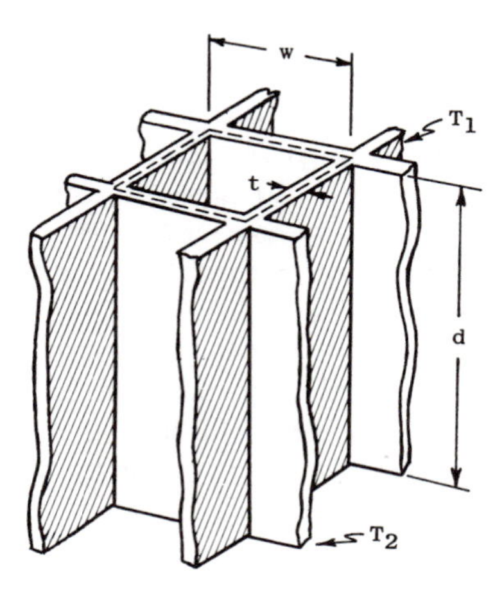

Fig. 11.29 Schematic diagram of a honeycomb mesh for the suppression of thermal radiation loss from the absorber and convection in the air gap between absorber and window surface.

multiple absorptions and emissions, the thermal radiation actually emerging from the honeycomb is effectively at a much lower temperature than the absorber.

Radiation-suppressing meshes for solar applications were first explored by Veinberg and Veinberg (1929). Francia (1961) improved the method of construction for a series of solar power plant designs. The first designs were of opaque mesh material, so sunlight had to be directed down the length of the channels. This requirement did not present economic problems when the honeycomb was used at the receptor of a solar furnace, but the requirement of tracking made it of no advantage for flat-plate collectors. The use of transparent materials for the honeycomb or aluminized material removed some of this limitation.

The ideal honeycomb material would be transparent, absorbing in the infrared. The walls of this material would allow sunlight to pass obliquely to the absorber surface, yet they would successively absorb thermal radiation and act as a selective surface. Glass and plastic would be such materials, but cost and fragility in the first case and inadequate thermal properties in the second make them less than ideal. Aluminizing a honeycomb allows sunlight to enter at oblique angles, with the disadvantage of absorption in the multiple reflections occurring for oblique rays. The aluminized surface, however, is also infrared reflective and the mesh does not significantly reduce the infrared emission of the absorbing surface because reflectance of aluminum is high in the infrared. If one used a selective surface coating that is highly reflective in the visible and absorbing in the infrared—the exact opposite of the selective surfaces discussed in Chapter 9—the honeycomb would function properly.

The basic problem with honeycombs, aside from physical function and durability of the wall material, is the large amount of surface area required per square meter of collector surface. If the ratio of length to width is typically a factor of 10, the total surface of the honeycomb is 30 times the cross section of the collector. If glass or plastic were used, the cost would be 30 times the cost of a simple window on the collector of the same material. When you consider that you can barely afford the cost of single glazing on a flat-plate collector, the factor of 30 looms large. Honeycomb for use with concentrating collectors does not have this same limitation.

A second problem that must be met in selection of the material for the honeycomb is the thermal conductance through the material. If convection is suppressed, one still has thermal conductance losses through the air and the medium of the honeycomb. If the wall thickness of the honeycomb is t, the width is w, and the depth is d, the heat loss per unit collector area is

$$Q/A = (wt/d)k\Delta T. \tag{11.79}$$

Because the conductivity of most materials is 100 to 1000 times that of stagnant air, the total area of the ribs must also be this ratio less than the opening. For a 1-cm square, honeycomb ribs of glass ($k = 0.0025$) would need to be

$$d = (w^2/2w)(k_{\text{air}}/k_{\text{wall}}), \tag{11.80}$$

where for glass $d = 0.0128$ cm, and for aluminum ($k = 0.5$), $d = 0.000064$ cm (640 Å), about the thickness of an opaque evaporated film of aluminum on a mirror.

11.25 CONVECTION SUPPRESSION USING HONEYCOMBS

A second area of potential use of honeycombs for solar applications is in the suppression of convection in collectors. When a vertical channel is made small, it has been found that vertical convection does not begin until a certain finite temperature difference between the top and bottom of the channel is exceeded. A certain minimum Rayleigh number (Ra) must be exceeded for the onset of convection, where Rayleigh number is defined as

$$\text{Ra} = qL^3\rho^2 gyc_p/\mu k^2 x, \tag{11.81}$$

where

q = heat flux per unit time (cal/sec),
L = characteristic dimension, the width of the channel (cm),
y = volumetric expansion of the fluid ($\text{cm}^3/\text{cm}^3\,\text{C}^\circ$),
c_p = specific heat of the fluid at constant pressure (cal/g C°),
μ = dynamic viscosity (poises) (g/cm sec),
k = thermal coefficient of conductivity (cal/cm C°sec), and
x = length of the channel (cm).

The necessity for a minimum temperature gradient for the onset of natural convection can be readily demonstrated, as Lord Rayleigh did, by heating a pan of viscous materials on a stove. The onset of natural convection, indicated by the establishment of hexagonal convection cells, is not as soon as heat is applied but only after a minimum heating of the bottom layers has occurred.

The theoretical minimum value for the onset of convection is when the Grasshof or Rayleigh number is equal to 1707. The presence of the small channels of the mesh effectively increases the value of Gr or Ra for the surface from 1707 up to approximately 200,000. These properties of meshes were reported by Hollands (1965), and experiments were reported by Hollands and Edwards (1967). However, the early results of experiments did not appear to be substantiated by Charters and Peterson (1972). It appears from this conflicting evidence that natural convection in air is only marginally suppressed and is easily upset by external conditions. In any event, the suppression of convection would be effective only for small temperature differences between the bottom and top of the honeycomb. If the absorber temperature rose beyond a certain minimum, then the value of q would cause the Ra value to exceed the limit and convection would begin.

The expected variation of convective heat loss with increasing temperature difference for a honeycomb is shown in Fig. 11.30. Convection normally grows approximately linearly with temperature $T^{5/4}$ and is shown by the dashed line. When the Rayleigh number is sufficiently low, and when either the cell spacing or temperature difference is small, the convection will be fully suppressed. When the Rayleigh number reaches about 2000, the convection begins, and it rises rapidly to reach the normal convection level.

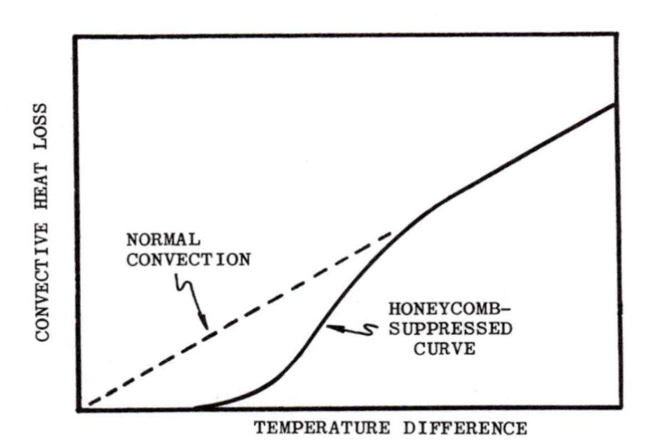

Fig. 11.30 Schematic representation of the effect of convection suppression with honeycombs as a function of increasing temperature difference. The temperature difference at the knee of the curve depends upon the size of the opening of the honeycomb cell.

11.26 HEAT PIPES

The heat pipe is a relatively new device for transferring large amounts of heat with a small temperature drop between heat input and heat output. The basic heat pipe is shown in Fig. 11.31. The pipe generally has a length-to-diameter ratio in the range of 10 to 100, and is designed to remove heat a greater distance from input to output than could be done by simple conduction. The pipe is evacuated and sealed with a working fluid inside. A wick is generally necessary to transport the condensed fluid back to the boiler region and to insure good wetting of the walls of the boiler section.

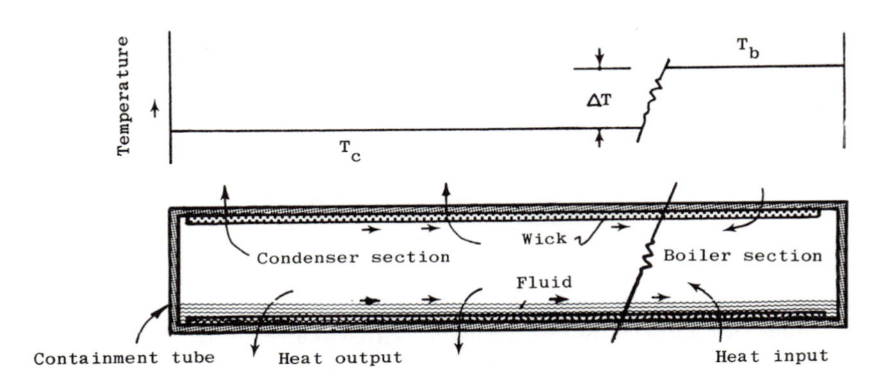

Fig. 11.31 Schematic diagram of a heat pipe. The pipe can be in any orientation if a good wicking material is used. It is evacuated and contains only the fluid and vapor of the working material.

In operation, the fluid wetting the walls of the boiler evaporates. The condenser end of the device is at a lower temperature so that the vapor condenses on the walls of the tube. Since the pressure inside the tube is essentially constant along its length, the temperature of the boiling fluid is identical to that of the condensing fluid. A temperature difference is required, however, between the boiling fluid at the heat input end and the condensing section, sufficient to drive the required number of kilowatts through the pipe walls. Another temperature difference is required between the condensing vapor and the outside of the heat pipe in order to remove the heat of vaporization being released in condensation. The temperature differences are small, being the difference required by ordinary conduction formulas appropriate to the wall thickness of the tube. The large heat-transfer capability of the heat pipe then resolves to that of the walls of the tube, but is effective over the length of the tube. A heat pipe therefore carries approximately as many times the heat load in conduction as the ratio of the length to the wall thickness, about 1000–10,000 times that conducted by a solid bar of the same metal of equal length.

The heat pipe is generally used to remove heat from a small region and dump it into a large region of the heat pipe. For solar applications one would want to have the opposite situation: to gather it over most of its length and dump it into a small region. This means a small temperature difference through the boiler walls and a large difference through the condenser walls. Typical temperature differences will be on the order of 10–30 C° between the surface temperature of the heat pipe in the input length and that at the output length, but along the length of the input section and the output section the temperature will be approximately constant.

Some practical problems are associated with heat pipes. One concerns the fluid used. Water can be used over a medium-temperature range, and liquid metals can be used over a high-temperature range. Organics also can be used as long as the temperature does not exceed the range for chemical stability, whereupon the generation of gases would cause the heat pipe action to be degraded. Any fluid, moreover, will dissolve some of the containment pipe material. This solubility is generally a function of temperature. The result is that some materials whose solubility rises with temperature, removed by solution from the hot end, will be permanently deposited at the cool end. Materials that have the opposite solubility curves will be removed from the cool end and be deposited in the hot end. This mass transport will eventually foul the wick and even puncture the heat pipe.

Heat pipes also have dynamic problems with "start up" and "choking" when the heat flow conditions are critical. The engineering literature on heat pipes is rather extensive, but attempts to use heat pipes for the transport of solar heat have not been fully successful. The series of reports by Kemme *et al.* (1969) is a good starting point for additional information.

11.27 HEAT TRANSFER ALONG THIN SHEETS

An important factor in getting the solar heat into the heat-transfer fluid that is sometimes overlooked in home-built solar flat-plate collectors is the transfer along a thin sheet. A typical water-heating collector is shown in Fig. 11.32, where two fluid pipes are shown separated a distance s. The metal sheet has a thickness t and is made of a material having thermal conductivity k. In order to transfer an absorbed flux of Q/A along the sheet, a temperature gradient must be established. If the thermal impedance to this flow is high, such as when the sheet has low thermal conductivity (as with plastics) or is thin, this temperature gradient can significantly lower the performance of the collector.

The quantity of solar heat collected by the sheet from the center line to a position x is given by

$$Q = \alpha t I x, \tag{11.82}$$

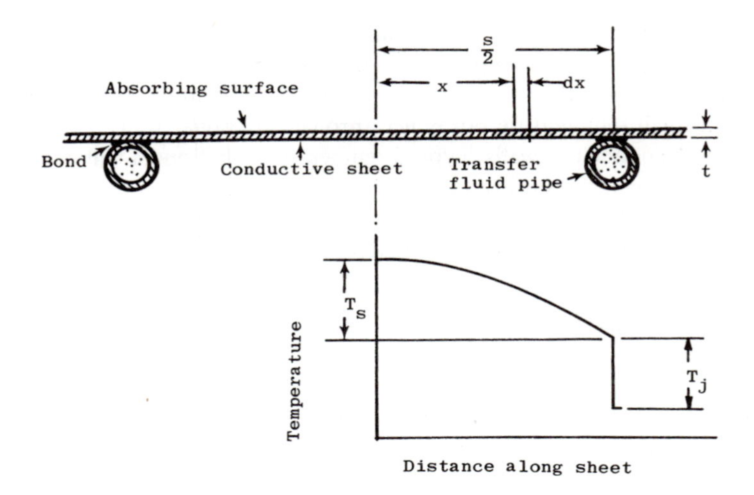

Fig. 11.32 Diagram showing the temperature distribution along a thin sheet absorber with separated heat-transfer fluid pipes. Note that a significant temperature jump can occur if the thermal flow is impeded by a small cross section at the junction of the fluid pipe and sheet.

and the gradient necessary to flow a quantity of heat Q is

$$Q = kA(dT/dx), \tag{11.83}$$

where A is the cross section of the sheet, equal to the thickness t per unit length along the sheet in the transverse direction. Equating these two expressions, we obtain after integration

$$\Delta T = (\alpha tI/2kt)x^2, \tag{11.84}$$

which says that the temperature distribution is parabolic, with the vertex of the parabola at the midpoint of the sheet, as indicated in Fig. 11.32, where the total temperature difference is obtained by setting $x = s/2$.

We also show a second important point in Fig. 11.32. The fluid pipes are shown as not integral with the sheet absorbing the solar flux. We therefore encounter another temperature difference set by the impedance, or bond conductance, of the junction between the pipe and the sheet. The heat flow across this junction is as set by the heat conduction equation, where the area is the area of the actual junction. If the junction is brazed, then the cross-sectional area will be appreciable and the value of ΔT_j small. If, on the other hand, the pipe is simply clamped to the sheet, the area of contact can be very small and the temperature difference large. One should be careful to minimize the thermal impedance of this junction in order to use the collected solar heat at maximum efficiency.

If one takes a set of values for the above case of $s = 10$ cm, $t = 0.1$ cm, and $k = 0.11$ cal/cm sec C°, and a solar flux $\alpha tI = 0.20$ cal/cm^2 sec, the value of ΔT_S is 22.5 C°. This difference is significant, and indicates that the heat losses from the center of the sheet will be significantly larger than from the sheet adjacent to the fluid pipes.

Tables of the "fin factor" efficiency for conduction in thin sheets can be found in the handbook published by the American Society of Heating, Refrigerating, and Airconditioning Engineers (Jordan, 1967).

11.28 DIFFERENTIAL THERMAL EXPANSION

Differential thermal expansion is a serious problem for solar collectors and related systems. The basic problem arises from the daily cycling of temperature. Each day, the collector and portions of the system rise from night temperature to midday operating temperature. This rise in temperature leads to thermal fatigue and also to dimensional changes that must be allowed for in the mechanical design.

The linear expansion of a solid is given by

$$\Delta L = kL\Delta T, \tag{11.85}$$

where k is the coefficient of thermal expansion, values of which are given for common materials in Table 11.6. (A more complete table is provided in Appendix P.) Taking a typical expansion, for steel, for a temperature rise of 500 C° applicable for a photothermal power collector, a collector module 10 m long would have a daily expansion of 65 mm. It is obvious that the absorber of such a system could not be rigidly attached to a cooler part of the structure at more than a single point.

Table 11.6 Coefficients of Thermal Expansion

Material	*Temperature range, °C*	*Linear expansion coefficient, k/C°, (10⁻⁶)*
Aluminum	20–100	23.8
Brass	0–100	18.8
Copper	25–100	16.8
Steel	0–100	10.5
Stainless steel	20–100	9.6
Plastics	0–100	17–160
Concrete		10–14
Brick		9.5

Thermal expansion requires that parts of a collector be designed so that the expansion can occur without buckling of any part. It is further important that this expansion occur silently, since a "popping" collector will be objectionable to people nearby. In a flat-plate collector the parts that must be free to expand are the absorber and the windows within the frame, and the frame with respect to the mounting structure.

Differential expansion is especially serious in regard to some plastics. On a flat-plate collector a cover window of polyethylene or another of the high-expansion plastics will be taut at night temperature and very loose at midday. A low-expansion plastic, like Mylar, closely matches the mechanical structure and does not greatly change during the day.

The change in volume with temperature is also important. Liquids in general expand more rapidly than the container, so significant changes in volume can occur for large tanks. The further change of volume when a liquid freezes is significant, as indicated by the seriousness of freezing in a water system. The volume change coefficients for some common liquids are given in Table 11.7, where α and β are related to temperature difference by

$$\delta V = V(1 + \alpha\Delta T + \beta\Delta T^2). \tag{11.86}$$

Table 11.7 Volume Expansion of Liquids

	Volume expansion coefficient	
Liquid	*α/C°, (10^{-3})*	*β/C°, (10^{-6})*
Water, 0–33°C (phase change = 1.11)	−0.064	+ 8.5
Organics	+1.0–1.4	+2–4
Oil	+0.9	+1.4
Mercury	+0.18	+0.008

11.29 PROBLEMS

11.1. How many seconds would it take to heat an element of molten salt in a collector having an aperture 20 times the diameter of the circular pipe containing the salt, when the solar flux is 0.700 W/m^2, the transmittance of the system is 0.80, and the absorptance of the pipe surface is 0.90—up to a temperature difference of 200 C°? Assume the mean loss to be 0.50.

11.2. If the fluid were water under pressure, how many seconds would it take if the geometry of Problem 11.1 prevailed?

11.3. If the fluid were air under 10 atmospheres pressure, how many seconds would it take if the geometry of Problem 11.1 prevailed?

11.4. What would the pressure drop be in the above three cases if the pipe diameter were 5.0 cm and the fluid were caused to flow a distance of 100 m in the appropriate times? Assume that no compressibility effects occur.

11.5. What would be the power expended in pumping the fluid under the above cases? Neglect the v^2 term. What are your conclusions concerning the choice of heat-transfer fluids? Calculate the ratio of useful power generated to power expended for these three cases.

11.6. Derive the expressions for Eq. (11.1) in terms of collector module area. Show how the length and total surface of each pipe size change when a finite gap is provided between collector rows. If the fluid is incompressible, what is the relationship between D_1 and D_2 for 100 modules in a square array?

11.7. An air collector has a cross section of 0.5 m^2 and an absorbing surface 1.50 m wide. How many seconds will air be in the collector for a rise in temperature of 50 C° when the collector efficiency over this interval is 0.50? If the collector is 10 m long, what will be the velocity of the airflow and the volume per unit time passing through the collector?

11.8. If the above collector were placed vertically and natural convection alone provided the driving force for the airflow, what would be the approximate velocity of flow? How hot would the emerging air be compared to the entering air?

11.9. If a collector had a mean efficiency of 0.60 and used air as the working medium, what collector area would be required to match a fan delivering 1500 ft^2/min through the collector when the desired rise in temperature is 45 C°, the solar flux is 0.700 W/m^2, the absorptance is 0.95, and the transmittance of the collector covers is 0.90?

11.10. In the collector described in Problem 11.9, what would be the flow of water required to deliver the same increase in temperature of the fluid?

11.11. Two solar collectors are being compared. Both are evacuated so that radiation loss is the principal loss. The first has a flux concentration X of unity and the second a flux concentration of 3.0. The first collector has a selective surface having a value of $\alpha/\epsilon = 12$ and the second a value of 8. What will be the stagnation temperature of each absorber surface?

11.12. Derive the variation of efficiency of the two collectors in Problem 11.11 as a function of relative distance along the collector when the input fluid temperature is 40°C and the output temperature is 150°C. What will be the relative lengths of these two collectors for equal output temperature of the fluid? State the solar flux value and system parameters you use.

11.13. Given that the two collectors in the above case are to drive a thermodynamic cycle having 0.70 of Carnot efficiency, plot the curves of net efficiency versus relative distance along the collector. Has either collector reached its maximum efficiency? If not, what would be the approximate length for maximum efficiency?

11.14. Calculate the temperature drop between the center of an absorber sheet where the sheet is 0.50 mm thick and made of steel and the fluid pipes are 20 cm apart, assuming that the thermal impedance between the sheet and the fluid is zero. The solar flux is 0.650 W/m^2 incident on the surface and the absorptance is 0.90.

11.15. If the cross section between the sheet and the pipe is 0.4 cm and the pipe wall thickness is the same as the sheet, what additional temperature drop between the sheet and the pipe is needed to deliver the heat flux of Problem 11.14 into the fluid?

Chapter 12
Flat-Plate Collectors

12.0 INTRODUCTION

The flat-plate collector is one of the most important types of solar collector because it is the simplest and has a wide range of important potential applications. In this chapter we discuss this collector, how to calculate its performance under various conditions, and how its performance is altered when such things as selective surfaces are used.

The flat-plate collector is basically a black surface that is placed at a convenient angle to the daily motion of the sun, and provided with a transparent cover and appropriate insulation around the sides and rear. The heat transfer fluid is generally water, but air is often used.

Considerable effort has been expended over the years to improve the efficiency of performance and the output temperatures of flat-plate collectors. This effort has been made with several goals in mind. One is to store heat more efficiently for use during nights and cloudy days: when stored as latent heat a high temperature contains more calories per gram of matter. A second goal is to increase the temperature so that other tasks than simply providing hot water are possible. One application of current importance is the use of heat to operate a gas-absorption cycle refrigerator for cooling applications. Temperatures close to the boiling point of water are desirable, and even low-temperature steam would be useful in optimizing the refrigeration units, which in traditional use operate on steam at about 1 kg/cm^2 (15 psi). One would even like to use flat-plate collectors to provide a heat-transfer fluid hot enough to operate thermodynamic cycles, either directly through steam or via an intermediate organic vapor.

We will first examine the heat balances of a flat-plate collector, using a procedure simple enough that anyone, with a little effort, can calculate the efficiency of a collector under a specific set of operating conditions. We will examine how this basic performance changes with technological additions.

12.1 BASIC COLLECTOR CONFIGURATIONS

Let us examine some of the many forms in which a flat-plate collector can be found. This list is not intended to be comprehensive, as many variants can be devised. For example, in the space of a single decade more than 100 patents were issued in Japan for solar hot-water heaters, most of them involving some sort of flat plate or the optical equivalent thereof.

The basic collector, consisting of a flat plate suitably blackened, could lose significant amounts of heat to the surrounding air, especially when there is wind. One or more windows then become part of the collector, the window providing thermal protection for the absorber and also keeping rain and dirt from the blackened surface. The window also can act to increase the temperature of the absorber through the "greenhouse" effect when the window is opaque to thermal infrared. Glass is opaque, as is most plastic glazing. Plastic film material is not opaque, as shown in Section 8.3.

In Fig. 12.1(a) we show cross-sectional diagrams for fluid heaters, the absorber being shown with a single pipe for the fluid. The types shown range from type A, a bare absorber, to D, a collector with three cover windows. In the following sections we present calculations of the efficiencies for A, B, and C. In practice each of these configurations would have insulation around all the faces except where the sunlight enters the collector.

In Fig. 12.1(b) we show cross-sectional diagrams for air heaters. An air-heater solar collector is basically a duct, one face of which absorbs sunlight. The simplest duct heater is A, in which the outer surface of the duct is made absorbent of sunlight.

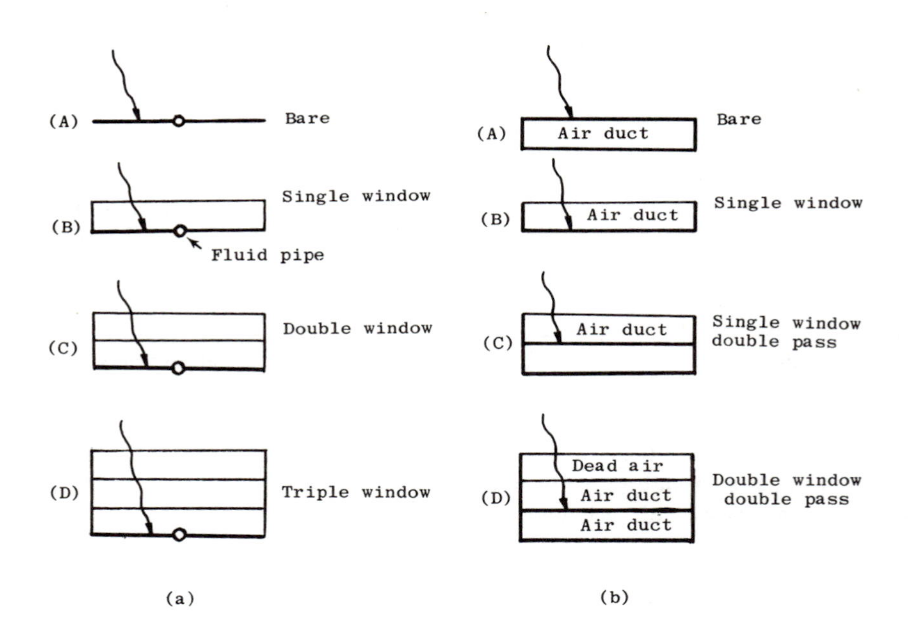

Fig. 12.1 Cross sections through several typical solar flat-plate collector configurations. Liquid heaters on the left show (A) a bare collector, (B) one cover window, (C) two cover windows, and (D) three cover windows. Air heaters on the right show (A) a top surface duct, (B) a bottom surface duct, one cover window, (C) a dual duct, one cover window, and (D) a dual duct with two cover windows.

It is a very simple design but, as we will show, its efficiency is low compared to that of fluid heaters and the other forms of air heater. In heater B the outer wall of the air duct is made transparent, allowing sunlight to reach the absorber. Heaters C and D have dual flow, drawing heat from both sides of the absorber.

In Fig. 12.2 we show perspective diagrams of four types of dual-duct air heaters. Case A is a simple model, with the air flowing parallel down the ducts on both sides of the absorber. A slightly more efficient way is shown in case B, in which entering cool air flows on the outside of the absorber, returning in the opposite direction on the rear of the absorber where the higher-temperature air is protected from heat losses by contact with the window. Case C shows the Lof overlapping-plate air heater, which uses a series of glass plates staggered across the duct. Half of each glass plate is clear, to allow sunlight to pass through it, and the other half is blackened. The clear portion acts to reduce losses from the blackened portion of the plate beneath. Case D shows a blackened porous absorber that allows air to flow through the absorber, the heated air being on the rear of the duct. Of these four cases the Lof configuration is the most efficient because it also restricts loss of thermal infrared (TIR) by means of the overlapping portion of glass plate. Configuration D is very easy to build, using metal shavings or blackened filter pads from evaporative air coolers, the fibrous nature of the absorber eliminating the problem of differential thermal expansion encountered with the rigid plate absorbers of heaters A and B.

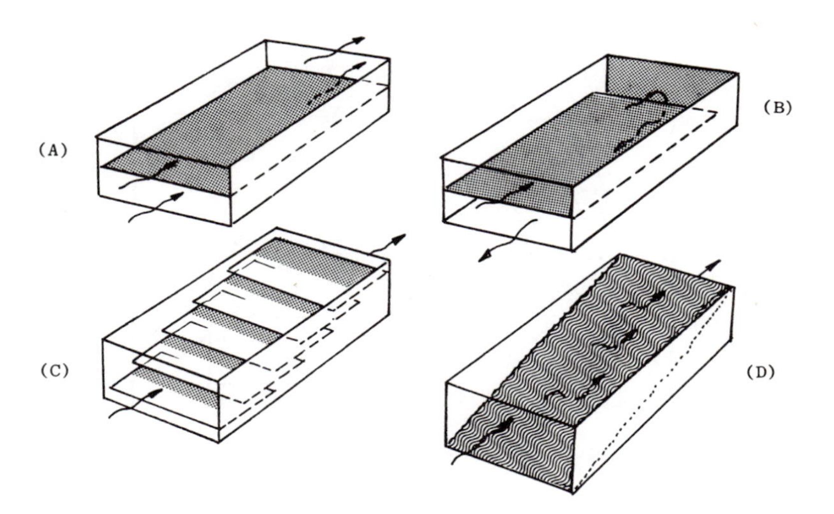

Fig. 12.2 Schematic perspective drawings for several forms of air heaters. The diagrams show (A) a simple dual duct with parallel flow on both sides of the absorber, (B) counterflow, with inlet air on the window-side duct, returning through the rear duct, (C) Lof overlapping plates, and (D) a porous mesh absorber.

12.2 DIURNAL TEMPERATURE PROFILE

The flux on a fixed flat-plate collector varies with the cosine of the angle that the sun is off the normal to the collector; hence the temperature of a black surface under glass covers also shows a wide variation during the day. To give the reader a general idea of flat-plate collector performance before we discuss the details of performance calculations and modifications, we will examine the performance of a typical collector.

In Fig. 12.3 we show the measured temperatures for a standard flat-plate collector with black paint for the absorbing surface, approximating a blackbody at all wavelengths. The collector was tilted normal to the sun at noon. The solid line curves are for the collector with two cover windows. The dashed curve is for the same collector with one cover window. In neither case was fluid circulated through the collector, so the temperatures are *stagnation temperatures.* If one were to extract energy from the collector, one would begin fluid flow at some desired temperature. The other option would be to flow fluid through the collector as soon as the fluid

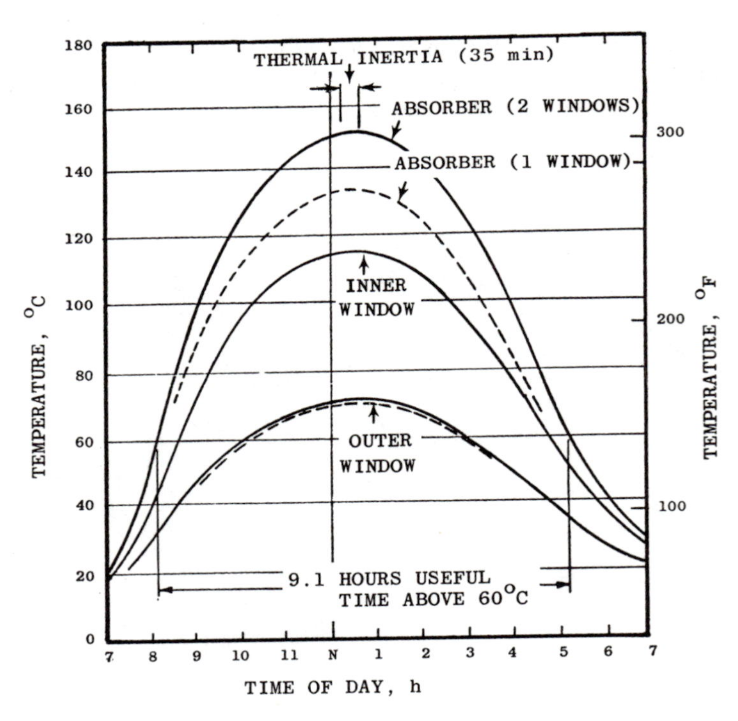

Fig. 12.3 Record of actual observed temperatures (May) for a flat-plate collector at Tucson, Arizona, for one cover window (dashed lines) and two cover windows (solid lines) of plain glass. Note the shift of 35 min after noon (MST), indicating about 15 min of thermal inertia. Temperatures are stagnation temperatures.

began to heat, in which case the operating temperature of 60°C (140°F) would be reached more slowly. A thermosiphon system (Section 10.12) would tend automatically to change the rate of fluid flow to match the heating effect of the sun.

We also show in Fig. 12.3 the temperature readings for the cover windows in addition to the absorber. The noon temperature was 35°C, which means that the outer-cover temperature rose approximately 35 C° above the ambient temperature. Under stagnation conditions all the solar energy entering the collector must also leave via the outer cover window, so its temperature will rise until convection and radiation can carry that amount of energy. Therefore the outer-cover temperature is exactly the same for the one-cover and the two-cover cases. When useful energy is extracted from the collector this simple relationship between one- and two-cover collectors changes in favor of the two-cover collector.

The curves shown in Fig. 12.3 are asymmetrical about local noon. Three effects contribute to this shape. First, Tucson is not on the central meridian of the time zone, being about 15 min west, so solar noon is after local noon. Second, the real sun is not always on the mean sun position, resulting in the "equation of time" correction between apparent noon and solar noon. Third, the collector has thermal inertia; its maximum temperature or hour of symmetry for the entire curve is shifted past noon, depending on the mass of the absorber and associated structure. The arrows denoting "thermal inertia" represent the approximate contribution from thermal inertia.

The temperature curves show that it takes the collector approximately 1.5 h after sunrise to reach the operating temperature of 60°C, and that its useful time, during which it is providing a fluid at or above the desired temperature, is 9.1 h.

12.3 THERMAL INERTIA

Thermal inertia is the term used to refer to the fact that a solar collector has a certain heat capacity, set by the mass and types of materials used to fabricate a square meter of collector. Since a collector will, in general, cool to the environmental temperature during the night, one must each day put in enough thermal energy to bring this mass up to operating temperature. The amount of energy required is given by

$$Q/A = \rho c_p \Delta T. \tag{12.1}$$

Thermal inertia can be expressed in the number of grams of material per square centimeter of collector. This number is of some interest in another regard, that of cost, because one is buying a certain number of grams of material per square centimeter. Typical collector "densities" are in the range of 0.1–10 g/cm^2 of collector. In Table 12.1 we tabulate the time at half the noon solar flux required to raise the collector 50 C° and 150 C° above ambient temperature, assuming a mean efficiency of $\eta = 0.60$ for the 50° rise and $\eta = 0.40$ for the 150° rise. Solar flux is assumed to be 400 W/m^2 absorbed.

Table 12.1 Thermal Inertia for a Flat-Plate Collector

Collector density, g/cm²	*Specific heat* c_p	*Time of rise, min*	
		50 C°	*150 C°*
0.1	0.25	2.4	7.2
0.2	0.25	4.8	14.4
0.5	0.25	12.0	36.0
1.0	0.25	24.1	72.3
2.0	0.25	48.2	144.6
5.0	0.25	120.	360.

The effect of thermal inertia is shown in the stagnation temperature curves of Fig. 12.3 by the fact that the curve is symmetrical about a point some minutes after solar noon. This displacement is approximately equal to the thermal inertia in minutes.

12.4 *U*-FACTOR

The term *U*-factor is often used to denote the rate of heat loss. U is defined by

$$Q/A = U\Delta T, \tag{12.2}$$

where ΔT is the temperature difference between the two sides of the system through which the heat is being transmitted. It is a particularly useful form for engineering of structures. In this book we do not make use of the *U*-factor, but where it does appear one can easily identify it and its relationship to other thermal loss quantities. In solar collectors the *U*-factor refers to the loss of the actual heat *absorbed* by a surface, which is the available sunlight multiplied by the system transmission and absorber absorptance. In our notation

$$U = 1 - P, \tag{12.3}$$

or

$$U = 1 - \eta/\alpha t, \tag{12.4}$$

where η is the efficiency of the collector, α the absorber absorptance, and t the system transmission.

12.5 COLLECTOR HEAT BALANCES

The determination of the heat balance for a flat-plate collector design is essential to the prediction of the efficiency with which it injects solar heat into the heat-transfer fluid. This determination is basically simple, and can be readily translated into computer programming language. But even if the reader has access to a computer program it is essential to know the elements of the calculation. In fact, the calculation is simple enough that three graphs and a slide rule are sufficient to quickly calculate the thermal balance of simple collector designs.

A convenient step-by-step procedure is presented in the evaluation form of Figs. 12.5 and 12.6 (pages 423–424), where the inputs and outputs of each surface in the collector are examined sequentially. We will take each step in turn. Before the balance calculations can be performed, the quantities listed at the left-hand side of the evaluation form need to be determined. We will briefly define each quantity.

Sun Date. The date is of use in defining the solar flux normal to the collector and for noting the time of year when one is evaluating the annual performance of a collector.

Solar Flux. The solar flux is the actual flux normal to the collector on the given date and at a particular time of day. This quantity is the basic resource on which the collector operates. It is also entered at the bottom of the page, above the place for the calculation of the collector extraction efficiency. It is important to show this quantity adjacent to the efficiency calculation to avoid the ambiguity that often enters into a stated "efficiency."

Reflection Loss. The solar flux is attenuated in traversing the collector, and because this energy is retrodirected it is lost to the collector, except for secondary reflections of the outgoing rays. Places are provided for listing the reflection losses for each window and for the absorber.

Net Loss. The net loss is shown because it determines the effective flux injected into the absorber.

Net Solar Flux. The net solar flux is the solar flux less the reflection losses. It is entered on the evaluation form in the boxes for "Sun." Each surface has an entry for "Sun," permitting the inclusion of internal absorption of sunlight passing through the windows. In the evaluation form the absorber is generally taken as surface 0 and the net solar flux value is entered in this box.

Ambient Air Temperature. This quantity is the outside air temperature, and is the basis for the flux balances.

Environmental Thermal Infrared (ETIR). This quantity is the hemispherical flux arriving at the absorber from the total environment. The first box is for ETIR(B), which is the blackbody flux for the ambient temperature. The environment, however, does not function like a blackbody, but emits less than this amount, as discussed in Chapter 2. The effective ETIR flux is entered in the second of the two boxes marked ETIR, generally 0.6(B) for a dry climate and 0.9(B) for a humid climate.

Wind Velocity. Wind velocity is recorded to determine the effective heat-transfer coefficient for the existing conditions. If wind velocity is zero, then the quantity shown is zero, and the appropriate box in the calculation columns is used to show the convection-loss term (from Table 12.2, page 425) instead of the calculated wind-loss term.

Wind (h_w). This quantity is the actual heat transfer coefficient value, read for the appropriate velocity from Fig. 12.4.

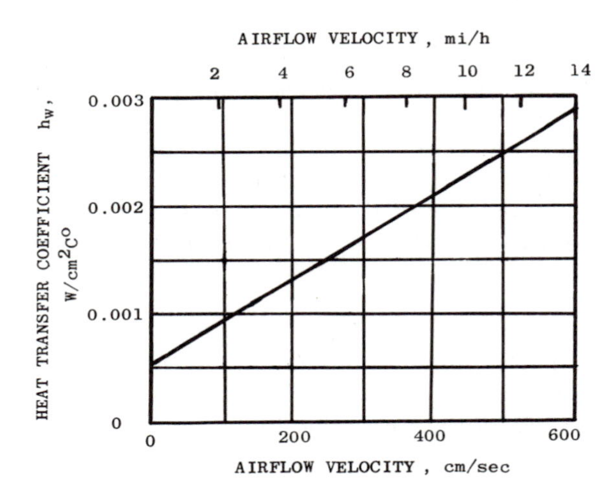

Fig. 12.4 Variation of heat-transfer coefficient h_w with airflow velocity.

Fluid Temperature. If a fluid is used for heat transfer, the temperature of the fluid *in the collector* is shown here. It can generally be assumed for fluids that the surface temperature of the absorber is close to the fluid temperature. This point is a detail that can be specifically evaluated when the thermal conductivity of the absorber plate and the heat-transfer coefficient of the fluid are known. For calculations of the level of accuracy described here, and where the general thermal performance of a collector is desired, this refinement is not necessary.

Airflow Temperature. In an air heater type of collector one specifically needs to know the air temperature in the collector, and to use the heat-transfer coefficients appropriate for the velocity of airflow.

Airflow Velocity. The heat-transfer coefficient is a significant function of the velocity of flow, so this value must be defined and recorded.

Airflow (h_w). The heat-transfer coefficient for a given velocity is obtained from Fig. 12.4, which yields values in agreement with those of Lof and of Hottel and Woertz. When the velocity is appreciable the heat transfer from surfaces is larger than that by TIR radiation, so it is important to define the quantity used for this heat transfer coefficient. Airflow and wind velocity are interchangeable.

Surface Properties. The absorptance α and emittance ϵ must be determined for each surface involved. The boxes show places for values for the top and bottom surfaces of each element. This is because one might wish to evaluate the behavior of a window, for example, where a selective coating is used on one surface and not the other. In the calculations we generally assume that TIR is emitted according to the expression $\epsilon(B)$, but that in absorption the TIR is fully absorbed by the facing surfaces. In reality, an amount B of TIR is reflected from the adjoining surface, but because the reflected TIR from both facing surfaces is trapped and cannot escape, the assumption of $\epsilon = 1.00$ for reabsorption introduces a negligible error. It is, however, important to use the value of ϵ when calculating the TIR loss *from* a surface.

It is important to remember that the flux quantities entered in the boxes in the calculation columns have the blackbody fluxes given in Table 12.2 (page 425) modified by the actual emittance of the surface in question.

Yields. These are the flux quantities, from the calculation columns, that enter the heat-transfer fluid, either the fluid flowing in the absorber plate, always taken as surface 0, or the airflow on one or both sides of the absorber, depending on whether the flow is restricted to one side or is on both. It should be remembered that energy can also enter the airflow from the inner surfaces of the adjoining surfaces 1 and −1, as noted in the sample calculations (Section 12.6).

Efficiency (η). The efficiency η is the total yield divided by the solar flux *normal to the collector.* We use η to denote this net efficiency, including in η the reflection and absorption losses between the aperture of the collector and the absorber as well as the absorptance α of the absorber. In the preceding chapter we also used P to denote efficiency, but P was defined as the ratio of useful energy extracted to the amount of energy absorbed. The term P does not include α, as does η.

One may encounter different definitions of efficiency, depending on which measure of the solar resource is chosen. On occasion one may see efficiency taken with regard to the energy arriving on a horizontal plane, since pyranometer data are widely available. One can then get efficiencies greater than 100% if the actual collector happens, for example, to be normal to the noon sun in winter. We prefer to have students use the definition of efficiency as given for η—the ratio of sunlight entering the collector at the actual angle at which it is used to the energy extracted from the collector.

Collector Orientation. The collector orientation determines the intensity of sunlight reaching the absorber. The appropriate value must be determined from the collector type and orientation, using the graphs and tables from Chapter 3. This value is entered in the box marked "Solar Flux."

12.6 SAMPLE CALCULATION

To illustrate how to perform the energy balance calculations we show cases for two types of collectors. In Fig. 12.5 we show a simple water-heating collector having a single cover window of glass. Water is flowing through the collector at 80°C and we calculate the efficiency of the collector at this temperature, assuming the absorber surface also to be at 80°C. This approximation is quite good for water, as a very small temperature difference is sufficient to cause the solar flux quantity to flow into the water.

The second case, shown in Fig. 12.6, is for an air heater having two cover windows of glass. Because heat transfer to flowing air is not efficient, the absorber is at a much higher temperature than the entering airflow, 20°C for this example, so this case is more complicated than the case with water. For the surface-air heat transfer, we take the same coefficient of velocity as we use for wind loss. The output energy in this case is the sum of four inputs to the airstream, as shown on the left-hand side of the computation sheet.

To begin a sample calculation one needs to estimate the temperatures of the several surfaces. In doing this for the first time, one may wonder how to start. In reality, one can make a poor guess at the start, but the sign of the temperature change to correct the first guess will become obvious from noting the residuals. If more heat is leaving a surface than entering, the surface temperature must be lowered to reduce the outflow, and so forth. The problem is further complicated because other surfaces interact with adjacent ones, so an imbalance might be caused not by an error in the temperature of the surface itself but by errors in the fluxes leaving the facing surfaces.

A good guideline to use in finding a process that converges to the desired answer in three or four iterations is to note the magnitude of change induced by a change in surface temperature. When a wind is blowing or an airflow is present, the change in flux for a given temperature difference is almost equal to the corresponding TIR flux, the convection term being the smallest contributor, changing more slowly with a change in temperature.

One has arrived at a satisfactory set of temperatures and fluxes when the residuals are within 5–10 W/m^2. Further refinement will change the surface temperatures only 1 or 2 C° and change the efficiency only in the second decimal place.

Table 12.2 gives the value for the convection loss for a horizontal upward-facing flat plate. To change this table to fit the actual geometry of the collector being evaluated, one must apply a conversion factor. To aid in the conversion we have provided Table 12.3, which lists the factors generally applicable to geometries other than

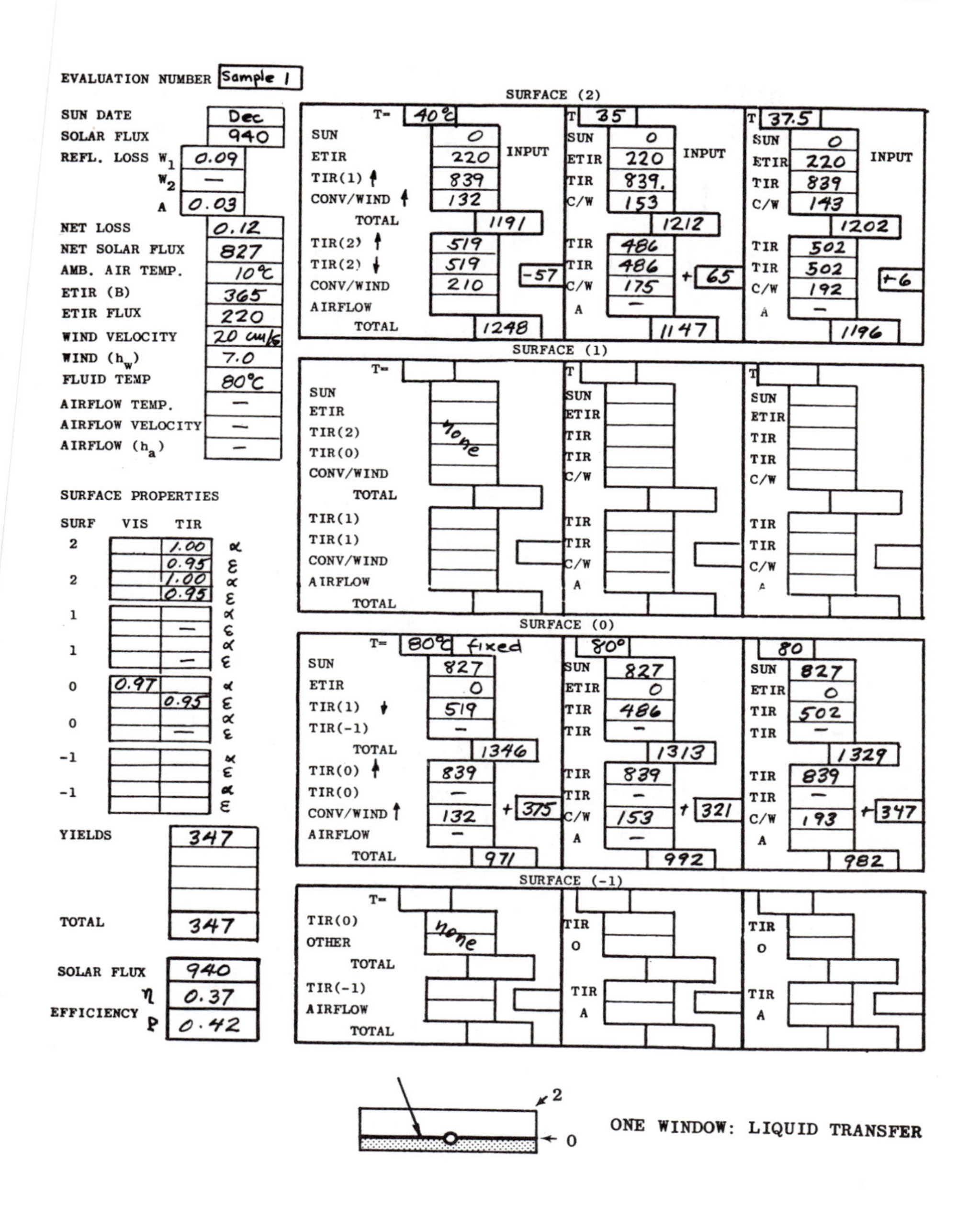

Fig. 12.5 Sample heat balance computation for a water-heater collector. The imbalance between input and loss for the absorber is the useful output.

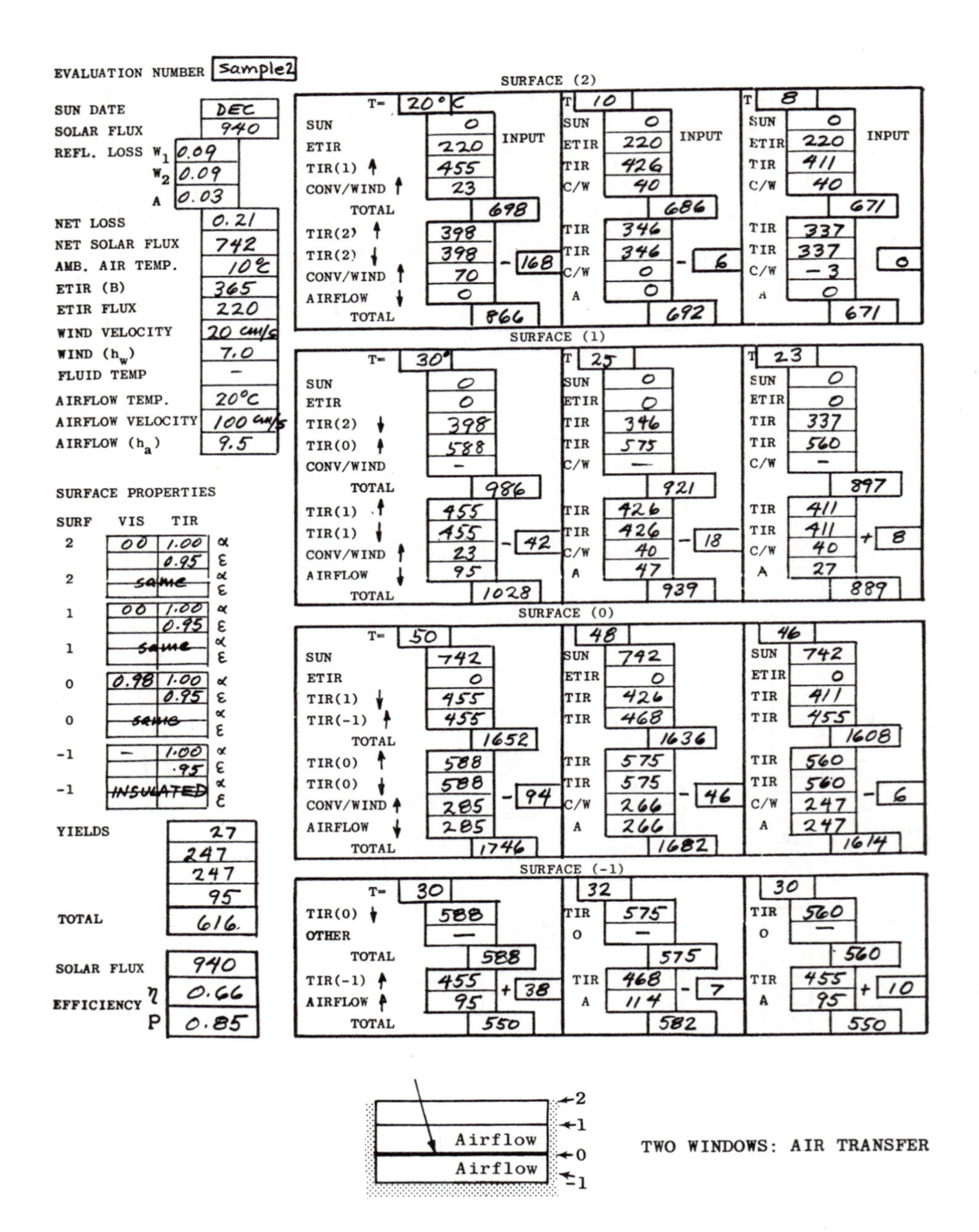

Fig. 12.6 Sample heat balance computation for an air-heater collector with air flowing on both sides of the absorber. The first guess at temperature was far from balance, but two iterations were sufficient to reach the approximate values.

Table 12.2 Radiation and Convection Fluxes for Heat Balance Calculations

English units

Radiation losses

Temp., °F	Flux, Btu/ft²h Value	Diff.
0	76	
10	83	7
20	91	8
30	99	8
40	107	8
50	116	9
60	125	9
70	135	10
80	145	10
90	156	11
100	168	12
110	180	12
120	193	13
130	207	14
140	222	15
150	237	15
160	253	16
170	269	16
180	287	18
190	305	18
200	325	20
210	345	20
220	366	21
230	388	22
240	411	23
250	435	24
260	460	25
270	487	27
280	514	27
290	542	28
300	571	29
310	602	31
320	634	32

Convection losses

Temp. diff., F°	Flux, Btu/ft²h Value	Diff.
0	0	
10	4	4
20	8	4
30	14	6
40	20	6
50	26	6
60	33	7
70	40	7
80	47	7
90	55	8
100	63	8
110	71	8
120	79	8
130	87	8
140	96	9
150	105	9
160	114	9
170	123	9
180	132	9
190	141	9
200	150	9

Metric units

Radiation losses

Temp., °C	Flux, W/m² Value	Diff.
-20	233	
-15	252	19
-10	263	20
- 5	293	21
0	316	23
+ 5	340	24
10	365	25
15	391	26
20	419	28
25	448	29
30	479	31
35	512	33
40	546	34
45	581	35
50	619	38
55	658	39
60	699	41
65	742	43
70	787	45
75	834	47
80	883	49
85	933	50
90	987	54
95	1043	56
100	1101	58
105	1160	59
110	1223	63
115	1288	65
120	1356	68
125	1426	70
130	1499	73
135	1575	76
140	1654	79

Convection losses

Temp. diff., C°	Flux, W/m² Value	Diff.
0	0	
5	10	10
10	23	13
15	39	16
20	56	17
25	73	17
30	92	19
35	112	20
40	132	20
45	153	21
50	174	21
55	196	22
60	219	23
65	242	23
70	266	24
75	290	24
80	314	24
85	339	25
90	364	25
95	390	26
100	415	25

upward-facing flat plates. The precise values for convection-loss coefficients depend in a complicated way upon the geometries and the absolute size of the collector, so these values and those in Table 12.2 should be considered only as approximations useful in deriving first-order computations of collector performances. For a more detailed discussion of convection the student is referred to Tabor (1955) or Fishenden and Saunders (1950).

Table 12.3 Geometrical Shape Factors for Convection Calculations

Orientation	*Factor*
Horizontal upward-facing flat plate	1.00
Horizontal downward-facing flat plate	0.40
Pair of horizontal flat plates	1.00
Vertical flat plate	0.8
Pair of vertical flat plates	0.8
Horizontal cylinder	0.7
Concentric horizontal cylinders	0.8

12.7 SURFACE TEMPERATURES

There are two points to be noted in regard to the surface temperatures in Fig. 12.6. First, at the low temperature of input air (20°C), the outer cover actually drops below the ambient air temperature. This is due primarily to the fact that the ETIR was not sufficient to make up for the TIR radiated upward by the window. One can actually utilize this imbalance between collector and ETIR to achieve significant night cooling, sufficient to be used as a means of recovering fresh water from saline on the Atacama Desert of Chile through production of ice.

The second point is that, for the lower temperature, the inner cover, surface 1, contributes heat to the airflow, but at the higher temperature surface 1 extracts heat from the airflow. The greater heat loss through the collector has in this case heated the outer window to above ambient temperature.

12.8 EFFICIENCY-VERSUS-TEMPERATURE CURVES

A person wishing to calculate the efficiency of a proposed collector needs one particular graph to make this readily possible. This is the graph of efficiency of heat extraction η versus temperature of the working fluid above ambient temperature. The curve for the example of the air-heating collector cited in Fig. 12.6 is shown in Fig. 12.7. We determined two points in the example, which are sufficient for a good determination of the η-versus-ΔT curve because the functions are almost straight lines. (The lines curve downward only near the high-temperature limit of performance, where one seldom uses a collector because its yield is so poor.)

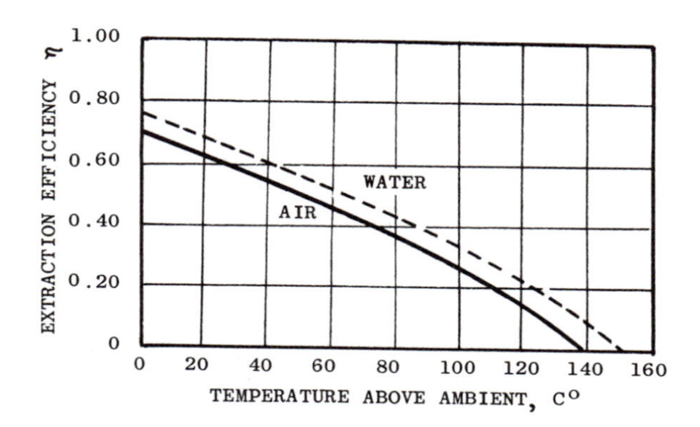

Fig. 12.7 Comparative performance of an air-heating collector having two cover windows versus a water-heating collector, for 10°C ambient temperature and 20-cm/sec wind velocity. The air heater has lower efficiency mainly because the surface of the absorber must operate at about 12 C° above the air flowing through the collector, resulting in larger heat losses.

The procedure we recommend for determining these curves is basically what we have shown: (1) take a fluid temperature close to ambient to establish the best efficiency of the collector, and (2) take a temperature above the anticipated operating temperature to establish the worst efficiency to be encountered. Draw a straight line between these two points, or a gentle curve as shown in Fig. 12.7.

One generally plots η versus temperature difference above the ambient ΔT. In reality the efficiency also depends upon the actual ambient temperature. This means that the curve applied to the case when the ambient temperature is 40°C is not precisely the same as when the ambient temperature is 0°C. In practice the difference is not large, so it is generally ignored in light of the fact that these calculations are not precise enough to show actual differences between real collectors. This is basically because the actual heat losses in a real collector are different from those assumed in model calculations. One can, however, take real collector data and work the question backward to find the heat losses between surfaces necessary to yield the observed efficiency.

The basic difference between air as the heat-transfer medium and water is that for the case of water the heat-transfer coefficient between the absorber walls and the fluid is so high that only a small temperature difference is generated. When air is used the heat-transfer coefficient is small and velocity dependent, so the absorber rises above the airstream temperature. Calculations of efficiency are therefore somewhat simpler for a water system because the temperature (of element 0) is stabilized and equal to the water temperature. In Fig. 12.7 the performance of a water-heating collector is compared to that of an air-heating collector for the conditions cited in the example of Fig. 12.5. The shift of efficiency to lower values at a given temperature simply reflects the fact that the absorber surface is hotter in the case of the air-heating collector, so the losses are higher.

12.9 GENERAL PROPERTIES OF AN η-VERSUS-ΔT CURVE

The general properties of the η-versus-ΔT curve are important to understand. A simplified drawing is shown in Fig. 12.8 to illustrate the salient points.

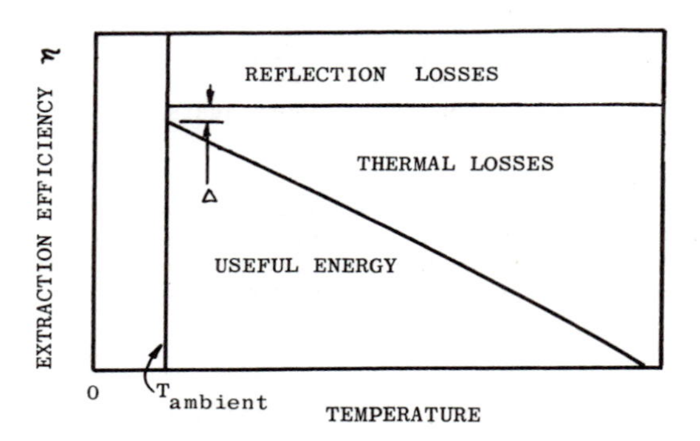

Fig. 12.8 Basic diagram illustrating the functional parts of the η-versus-ΔT diagram. Note that even at $\Delta T = 0$ there is a small thermal loss, indicated by the symbol Δ. Thus, whereas the reflection loss of 0.21 should result in an η of 0.79 at $\Delta T = 0$, the actual value of η is only 0.70.

The curve begins at $\Delta T = 0$ at a value well below 100% efficiency. This is because some solar flux is lost by reflection and never reaches the heat-transfer fluid portion of the collectors. The curve slopes downward with increasing ΔT because heat is lost to the environment as soon as the collector temperature rises above the environmental temperature.

Differences between collectors can be twofold. First, there can be differences in the amount of solar flux reflected back out of the collector. This quantity can be altered by changing the absorptance of the absorber, but since this is generally already close to unity, there is little to be gained here. Some selective surfaces have absorptances of 0.85–0.90, and their η-versus-ΔT curves are immediately penalized by this factor. The use of nonreflection surfaces on the windows can significantly raise the curve at $\Delta T = 0$.

The second difference between collectors is in regard to heat losses. If the losses are reduced, the slope of the curve is reduced, so gains can be realized at large ΔT values even if there were no differences between the curves at $\Delta T = 0$. This is where selective surfaces and convection suppression become important. Both can do much to flatten the curve. The net tradeoff is shown in Fig. 12.9, where a curve of lower net absorptance (higher reflectance) has a lower slope. Beyond a certain temperature the collector with lower losses will become more efficient. It is therefore important to know the desired operating temperature differential before deciding whether selective coatings, convection suppression, or even multiple windows are needed to give significant gains over the simplest collector with a blackbody absorbing surface.

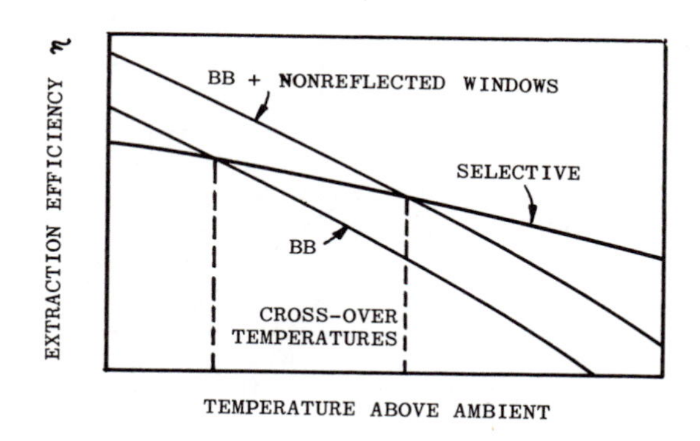

Fig. 12.9 Diagram illustrating the types of performance curves, showing the crossover temperatures defining temperature regions where each type of collector is best. A reduction of reflection losses shifts the entire curve upward; lowering of thermal losses flattens the curve, as shown by the curve marked "selective."

12.10 THE BARE COLLECTOR

The simplest form of collector, shown in Fig. 12.10, is an absorbing plate having no cover windows. Such a collector could have integral or attached pipes to carry the water. If we assume no thermal impedance between the pipes and the absorber surface (not always the case when the lines are simply clamped to the absorber plate), and if we further assume that the rear surface has perfect insulation so that all heat losses are from the front face, we obtain the curves shown in Fig. 12.11.

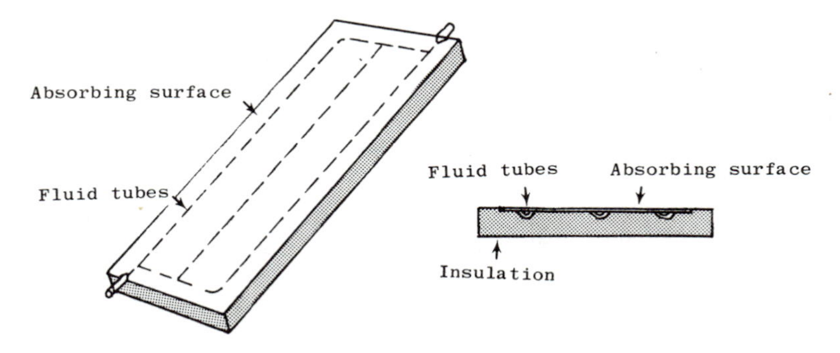

Fig. 12.10 Schematic diagram of a bare solar collector for liquid heat-transfer medium. This type of collector can be used for both heating and cooling (cooling via sky radiation at night), but it is sensitive to wind losses.

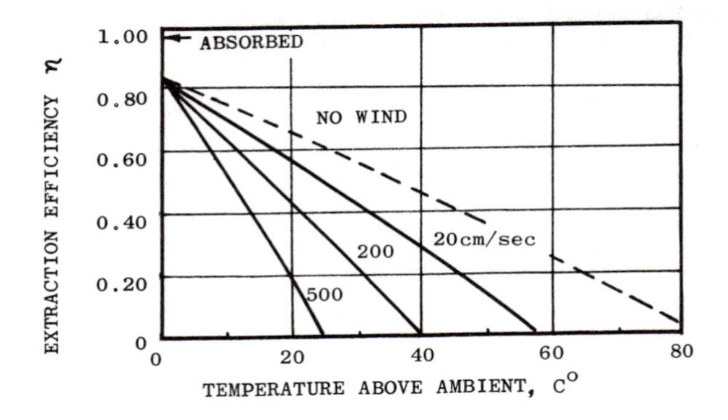

Fig. 12.11 Variation of efficiency of extraction η as a function of temperature above ambient and wind velocity for an absorber with *no* windows. Note that the efficiency at possible operating temperatures drops rapidly with increasing wind.

A bare collector is especially susceptible to wind loss of heat. As shown by the dashed line in Fig. 12.11, for the case of zero wind the efficiency is rather reasonable considering the low potential cost of such a collector. Bliss (1961) used this type of collector for the roof of a solar house in Tucson, Arizona, where the collector was also used to radiate heat to the sky at night during the summer. Since most windows are opaque to thermal infrared, a cover would have prevented efficient sky radiation cooling, so the tradeoff was for both low cost and a cooling mode of operation.

Wind rapidly degrades the performance of a bare collector, the reduction in η being in about equal steps for the several velocities shown in Fig. 12.11. The wind statistics during periods when solar heat is desired are therefore essential to achieving the best performance of a bare collector.

The area of bare collector needed to yield a given temperature rise ΔT can be calculated from these curves and the equation

$$A = \rho c \Delta T V / \eta I, \tag{12.5}$$

where A is the area of solar collector required, V is the volume rate of flow of fluid through the collector, and η is from the above curves.

Another form of bare collector is the simple blackened tube carrying a flow of air, as is used in crop drying, in which the air contacts the inner wall of the absorbing surface. The efficiency of such a collector is lower, but again, when costs are important, an inexpensive collector of modest efficiency may be more cost effective for the assigned task than a more efficient and sophisticated unit. Two forms of air-heating collector are shown in Fig. 12.12. The efficiency curves are shown in Fig. 12.13.

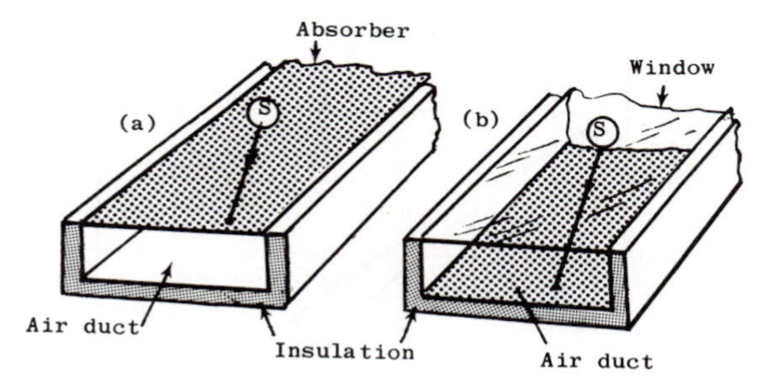

Fig. 12.12 Schematic arrangement of two air-duct solar heaters, showing (a) the upper surface as the absorber and (b) the inner surface as the absorber.

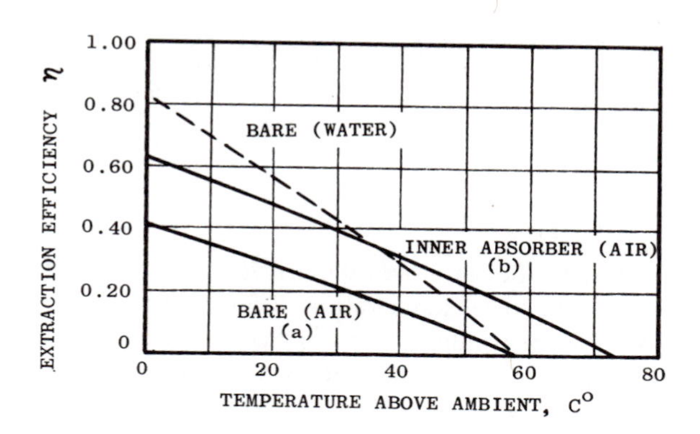

Fig. 12.13 Performance of a bare air-heating collector compared to an air collector in which the sunlight is allowed to reach the inner surface of the air duct, the front surface of the duct becoming a window to protect the absorber from wind losses. For comparison the performance of a bare water-heating collector (from Fig. 12.11) is also shown. In all cases a 20-cm/sec wind is assumed.

12.11 SINGLE-WINDOW COLLECTOR

When a cover window is placed over a bare collector, the efficiency at higher temperatures is greatly improved, even though some sunlight is lost by reflection. In Fig. 12.14 we show the performance of a collector with a single cover for two different wind velocities. We can see that the sensitivity to wind has been greatly reduced and that efficiency has been improved. In the cases cited, the cover temperature drops below the ambient temperature when the collector fluid is at temperatures less than 20 C° above ambient. This results in what appears at first to be an ambiguity: the collector is more efficient under high winds than under low winds. This is because the cover window *acquires* heat from the ambient air under these conditions. As soon as the cover temperature rises above ambient ($\Delta T > 20$ C°), the collector loses efficiency with higher wind velocities.

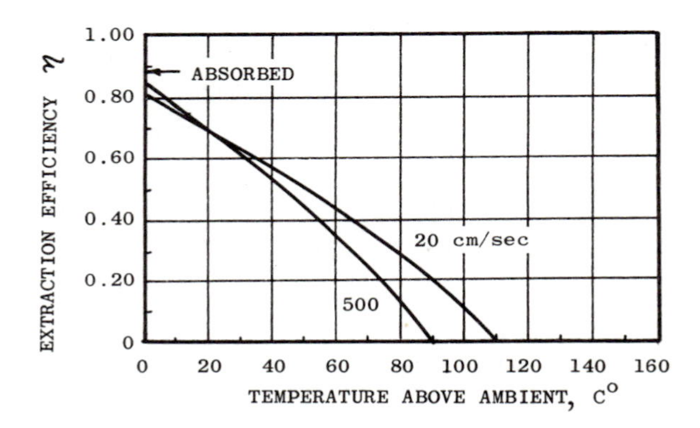

Fig. 12.14 Variation of efficiency of extraction η as a function of temperature above ambient and wind velocity for an absorber having a single cover window. Efficiency actually improves with wind when ΔT is less than 20 C°, since a window normally cools below the environmental temperature for low ΔT and wind helps warm the window. For ΔT greater than 20 C° the window is above ambient and wind cools it.

12.12 DOUBLE-WINDOW COLLECTOR

A second cover window further improves the efficiency of a collector for the greater temperature differences. At lower temperature differences a double-window collector is less efficient than a single-window collector because of the reflection loss of the second window. Examples of zero, one, and two windows are shown in Fig. 12.15. Note that the curve for two windows is flatter, representing lower heat losses, but that the $\Delta T = 0$ crossing is lower than for a single window. The point where a two-window collector becomes more efficient is at about 30 C° above ambient, so if the net requirement is of this magnitude a single-window collector is better, especially considering the cost saving in eliminating the second window.

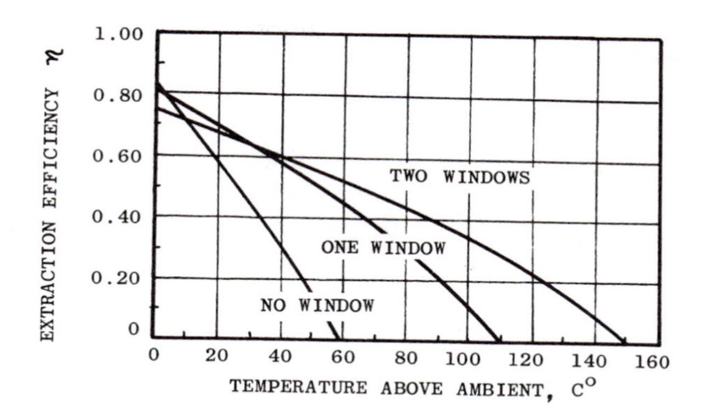

Fig. 12.15 Comparison of a flat-plate collector with gray-body absorber ($\alpha = 0.98$) having zero, one, and two window covers, for a 20-cm/sec wind and a 10°C ambient temperature.

The equation for calculating the area of collector required to provide a given ΔT under conditions of a volume rate of flow of fluid through the collector V is also given by Eq. (12.5), using the appropriate values for η.

12.13 IMPROVEMENT OF PERFORMANCE

Thus far we have discussed the performance of a simple collector having an essentially blackbody absorber ($\alpha = 0.98$). There are a number of potential ways to improve the performance of a flat-plate collector, but each exacts a cost. To determine whether a given way of improving performance is cost effective we need to determine to what extent it improves the efficiency of energy extraction η. There are a number of observations one can make from examining the sample calculation, but the major one is that the radiation terms are generally the largest, being several times the magnitude of the convection terms. The ways to reduce losses and increase efficiency of extraction of solar heat are:

1. reduce convection losses by
   (a) inverting the absorber and other geometrical means,
   (b) using honeycombs,
   (c) reducing the temperature difference between surfaces,
2. reduce radiation losses by using
   (a) selective absorbers,
   (b) selective windows,
   (c) both combined.

Let us examine some aspects of each of these ways.

12.14 GEOMETRICAL SUPPRESSION OF CONVECTION

Convection occurs when a lower surface is hotter than the free air and where buoyancy effects can move the heated air upward. If the heated surface faces down, for example, convection cannot occur because the air layers are stable up against the heated surface. The loss is then by conduction alone (in the absence of wind to move the heated air). One can thus place the collector absorber and window facing downward and use a mirror to reflect sunlight up into the collector. Although this procedure does eliminate convection, it suffers from the additional reflectivity loss by the required mirror. This loss will be about 15–20% of the incident sunlight or 150–200 W/m^2, somewhat larger than the convective losses one is trying to avoid.

Geometrical suppression of convection may be useful in cases where the reflection loss is small compared to the convective loss, or where a mirror already is used to concentrate sunlight on the absorber.

In calculating flat-plate collector efficiencies when convection is suppressed, the appropriate value is one-fourth the value in Table 12.2 to account for the remaining loss due to conduction.

Convection loss is also a function of the shape of the heated surface and the tilt of a flat-plate surface. The largest convection loss is that from a horizontal flat plate, and is the value tabulated. For vertical surfaces the loss is about 0.8 that for a horizontal surface, as shown in Table 12.1.

Reduction of convection through use of honeycombs was discussed in Section 11.24 and will not be repeated here.

12.15 WINDOW TEMPERATURES

During the heat balancing process one notes that the addition of a selective absorber adversely affects one element of the balances: the convective exchange between selective surface and the nearest cover window. The temperature difference is increased and hence the convection is increased because the window is at a lower temperature than with a blackbody absorber. The reason for this effect is simple: less energy is flowing out through the windows.

In Table 12.4 we show the cover temperatures with and without a selective absorber for the cases of one cover and two covers. The convective term for each case is shown by the figures in parentheses. Note the rise in convection because of the increased temperature difference. This effect indicates that we need some way of heating the cover next to the selective surface. One way of accomplishing this further tuning of the collector would be, for example, to inhibit some of the TIR emission *upward* from the inner cover. In other words, one would like to have a selective transmitting surface placed on the *top* of the cover to raise its temperature and decrease the convective exchange with the absorber. One might term this addition a *selective coating to control convection loss* by reducing the temperature difference between surfaces.

Table 12.4 Surface Temperatures

	Smaller ΔT		*Greater* ΔT	
	Blackbody	*Selective*	*Blackbody*	*Selective*
ONE-WINDOW COLLECTOR				
	5°C	2°C	38°C	18°C
	(39)	(44)	(143)	(225)
	20°C	20°C	80°C	80°C
TWO-WINDOW COLLECTOR				
	4°C	0°C	25°C	15°C
	12°C	5°C	55°C	35°C
	(19)	(39)	(73)	(153)
	20°C	20°C	80°C	80°C

All cases assume 10°C ambient temperature and 20-cm/sec wind.

In the next sections we will first examine the effects on flat-plate collector performance of a single selective absorbing surface and then examine comparative performance for selective windows. We also examine cases where selective absorbers and selective windows are used simultaneously.

12.16 EFFECT OF SELECTIVE ABSORBER SURFACES

To explore the effect of selective absorbers, let us assume that there is no emission from such a surface. In reality there will be at most a few percent emission, perhaps as much as 10%, but to see the *best* performance we assume no emission. We will, however, take into account that selective surfaces do not have as good an absorptance as $\alpha = 0.98$. As a realistic value for the absorptance we will assume the selective surface is a good one and has an absorptance of $\alpha = 0.90$.

Bare Collector

The change from a 0.98 gray body to a selective absorber having $\alpha = 0.90$ causes the absorber to reject the ETIR component, as the selective surface is also a perfect mirror for the thermal infrared. Even so, the gain through the reduction of the TIR loss is appreciable, and at all operating temperatures such a collector is superior to the plain absorber, as shown in Fig. 12.16. These curves are for a liquid-medium collector.

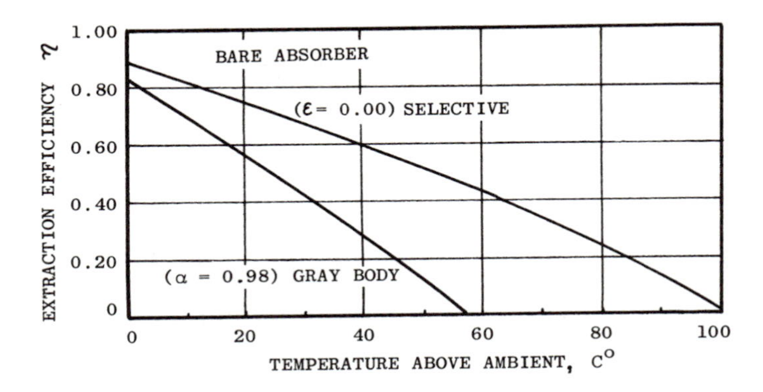

Fig. 12.16 Change in efficiency of extraction of solar heat η for a bare collector using liquid heat-transfer medium in going from an $\epsilon = \alpha = 0.98$ gray-body absorber to a perfect selective absorber having $\alpha = 0.90$ for 10°C ambient temperature. Flux actually absorbed is 846×10^{-4} W/cm².

Single-Window Collector

The change from a 0.98 gray body to a 0.90 selective absorber also provides a distinct advantage for the collector having a single cover window. The slightly lowered absorptance of the selective surface lowers performance only for small temperature rises above ambient. One factor to note in doing the flux balances using the example of Fig. 12.5 is that the absorber is nearly a perfect mirror for the TIR flux from the window, reflecting the TIR back to the window. The window therefore acts as though it loses flux only toward the environment. The change in efficiency caused by the addition of a selective absorber is shown in Fig. 12.17.

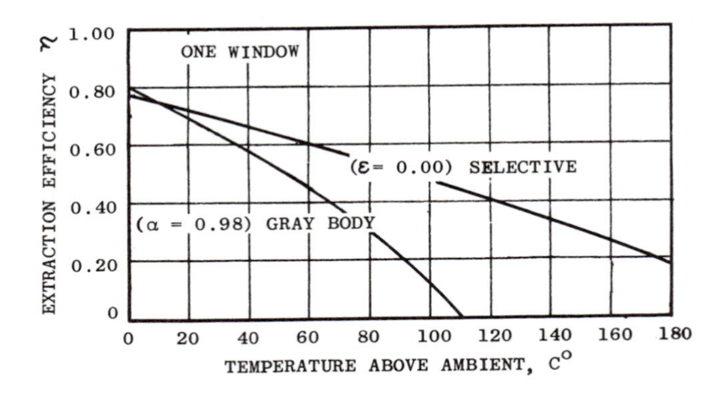

Fig. 12.17 Change in performance for a collector having *one* window cover when the absorber is changed from an $\epsilon = \alpha = 0.98$ gray-body absorber to an $\epsilon = 0.00$ selective absorber having $\alpha = 0.90$, for 10°C ambient temperature. Flux actually absorbed is 770×10^{-4} W/cm².

Double-Window Collectors

The addition of a selective absorber to a double-cover flat-plate collector improves performance as shown in Fig. 12.18. Performance is slightly poorer at small temperature rises because of the lower absorptance of the selective surface, but is better at higher temperature rises.

The effect when the absorber is not fully selective is between these two curves: a coating with emittance of 0.10 is approximately 10% of the way to the blackbody curve from the selective curve shown.

Because a selective coating adds a significant cost to the absorber, one must operate the collector in a region where the selective curve lies significantly above the blackbody curve.

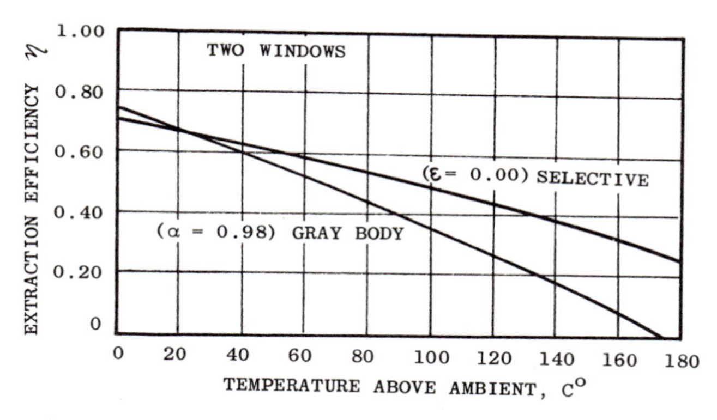

Fig. 12.18 Change in efficiency of extraction η for a fluid collector having two cover windows in going from a 0.98 gray-body absorber to a perfect selective absorber. Flux actually absorbed is 680×10^{-4} W/cm^2.

12.17 SELECTIVE WINDOWS

The simplest case of a selective window is represented by a single window plus an absorber. Let us place a selective coating on the window facing the absorber. Since the window now has high reflectance for the TIR, the thermal infrared loss from the absorber will be reflected back to the absorber. In Fig. 12.19 we show how this reflective surface modifies performance, comparing it to the case of no selective surfaces and a selective absorber.

In the case where the flux absorbed by the absorber is the same, 770 W/m^2, and where the selective reflective layer is perfect, the curve lies exactly on top of the curve for a selective absorber. The best transmitting selective coating, the indium oxide plus 7% tin oxide described in Chapter 9, is about 90% reflecting of the TIR. The performance curve for this coating on the inner face of the window is shown by the dashed line in Fig. 12.19.

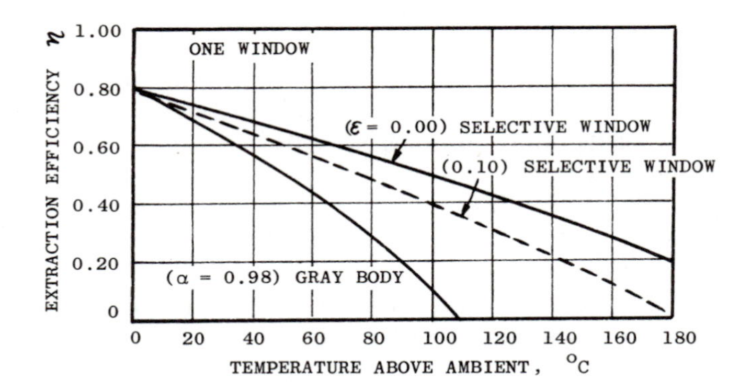

Fig. 12.19 Behavior of a collector having a 0.98 gray-body absorber with a perfect TIR reflecting coating on the inner face of the window, and a 0.90 TIR reflecting coating of indium oxide and tin oxide.

12.18 FACING SELECTIVE SURFACES

Selective windows aid significantly in returning TIR to the absorber when the absorber is essentially a blackbody radiator. The case is not as immediately obvious when the absorber itself is selective. If you add selectivity to the absorber so that its TIR reflectance exceeds that of the window, then the effect of the window coating is quickly diminished. The basic reason is that the TIR returned to the absorber is in turn reflected back to the window. At each reflective pass the window extracts more from the TIR beam than does the absorber. A diagram showing the attenuation of the TIR beam upon repeated reflections from the window and the absorber is given in Fig. 12.20.

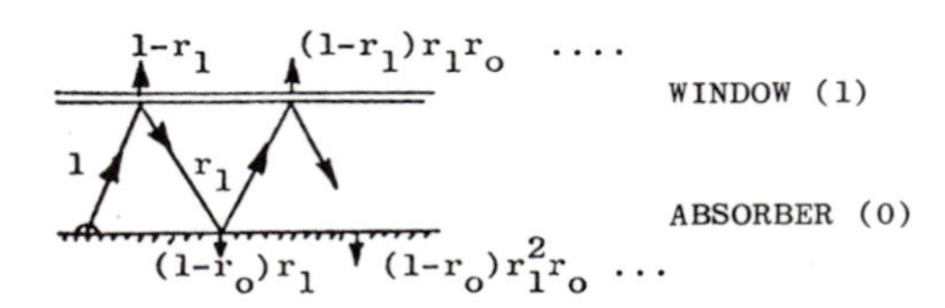

Fig. 12.20 Diagram showing the attenuation of the TIR beam upon repeated reflections from the window and the absorber.

The successive reflection inputs to each surface can be summed conveniently:

$$E \text{ into surface } 1 = (1 - r_1)(1 + r_1 r_0 + r_1^2 r_0^2 + \dots),$$

or, summing the infinite series,

$$E \text{ into surface } 1 = (1 - r_1)/(1 - r_1 r_0) \quad (12.6)$$

and

$$E \text{ into surface } 0 = (1 - r_0) r_1 / (1 - r_1 r_0). \quad (12.7)$$

In Fig. 12.21 we show how the energy balance changes for a given value of window reflectivity of $r_1 = 0.80$ as the absorber TIR reflectivity changes.

If we show the same data in terms of absorber selectivity, we emphasize the reduction in effectiveness of the selective window as the absorber selectivity increases. For the curves shown in Fig. 12.22 we assume that the absorptance of the selective absorber is $\alpha = 0.90$ and that the emittance of the absorber is $\epsilon_0 = 1 - r_0$.

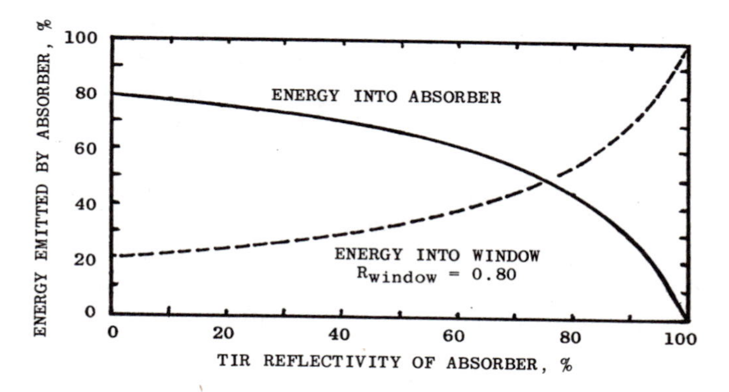

Fig. 12.21 Dependence of thermal infrared (TIR) absorbed into the absorber and window as a function of reflectivity of the absorbing surface.

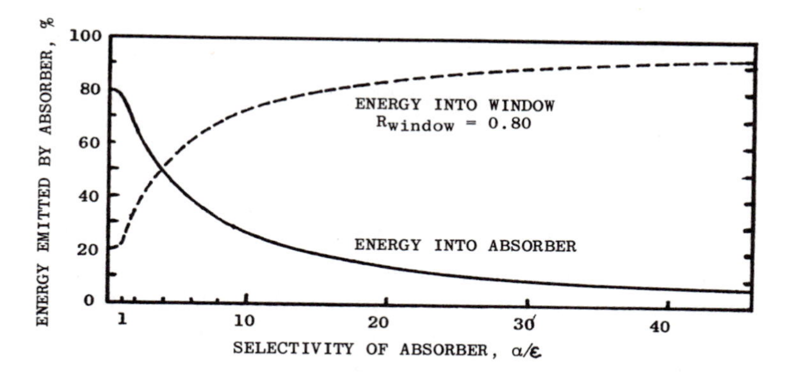

Fig. 12.22 Dependence of thermal infrared (TIR) absorbed into the absorber and window as a function of selectivity (α/ϵ) of the absorbing surface.

12.19 COMBINATION OF ABSORBER AND SELECTIVE WINDOWS

The role of selective windows has generally been restricted to that of acting as a reflector of the thermal infrared emitted from the absorber, returning that energy otherwise lost into the window. As noted above, there is a second role for selective windows, that of preventing TIR emission from the upper sides of windows, thus controlling the convection between surfaces.

A number of materials that have intrinsic selectivity suitable for windows have been used in experiments over a number of years. The usual ones are either tin oxide (SnO_2) or indium oxide (InO_2), singly or in combination, and sometimes with traces of antimony (Sb) in the layer. One problem has been that these coatings are also more reflective and absorbant of solar flux than is an uncoated window. The net result has been that the gains have been small, as reported for example by Solomon (1963), leading to the conclusion that these coatings provide no net gain. In order to understand the limitations and advantages of selective windows, we will consider several cases of selective combinations under *idealized* conditions to see whether more detailed studies are warranted. We will begin by examining some aspects of radiative exchange and then examine the general theory of energy losses before presenting some numerical examples and the conclusion that can be drawn.

To illustrate the effect of multiple selective surfaces, we take the situation where a high-temperature flat-plate collector is required. A high temperature is ideal for illustrating the effects because the usual flat-plate collector has very low η values at a temperature of 120 C° above ambient. In each case we assume two cover windows, and we shall explore the effects of various placements of selective surfaces on these windows.

There are many ways to combine selective surfaces. We present in Figs. 12.23–12.26 the results of an extensive investigation using a computation program we have developed. We plot the results in graphical form of η versus emittance, which shows the maximum difference in collector performance between a blackbody absorber and a selective absorber. The arrangement of selectivity on the covers is denoted by four symbols, *S* being for selective and *P* being for plain. In all cases the emittance of the selective coating on the windows is 0.20. We further assume that no solar energy is absorbed in the selective window coatings, holding this question until after we have determined the nature of the effects gained from selective combinations.

Two Windows with Upper Surfaces Selective (*SPSP*)

In Fig. 12.23 we compare a two-window collector having selective coatings on the upper surface of each window (*SPSP*) with a collector having two plain windows (*PPPP*). The improvement possible with selective coatings is obvious. Also, it is apparent that selective windows give the largest gain when the absorber is a blackbody, but they still add useful performance when the absorber is a perfect selective surface. This gain at 0.00 emittance is caused by the reduction in convection resulting from the heating of the windows by the selective transmitting surface on the top surface of the windows, even though this surface on the outer window causes it to reject the ETIR energy component. T_0 is the absorber temperature and T_a is the ambient temperature.

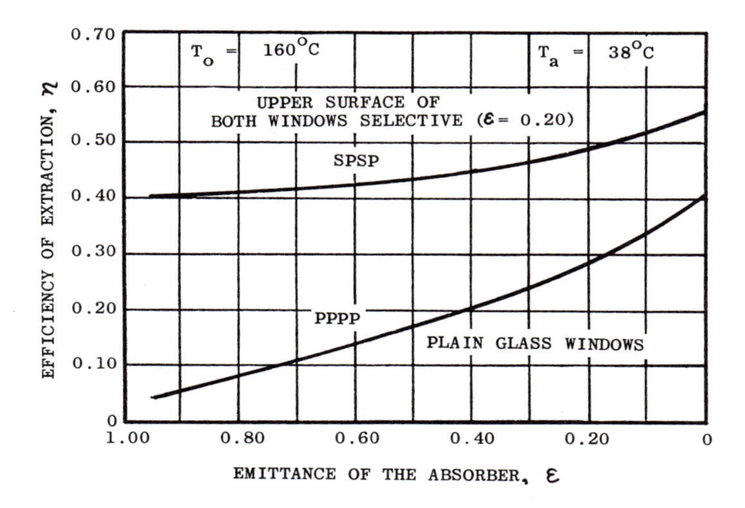

Fig. 12.23 Comparison of the efficiency of two collectors, one having the upper surface of each window selective and the other with plain glass windows.

Selective Surface on One Window (*PPSP, PPPS*)

In Fig. 12.24 we show a comparison involving placement of a single selective window surface. The upper curve shows the effect when the surface is placed on top of the inner window. The lower curve shows the effect when the surface is placed in the traditional manner facing the absorber. When the absorber is a blackbody the traditional placement is better, but when the absorber is also selective the unconventional placement, on the top side, is better, adding 12% to that afforded by the *PPPS* configuration.

Reversal of Window Selective Layer (*SPSP, PSPS*)

To further explore the placement of the selective window coatings we show in Fig. 12.25 two versions of both windows coated on one side, the coatings facing the absorber in one case and the coatings facing the environment in the other. Note that both arrangements are approximately equal for a blackbody absorber, but again the upward placement of the selective window is superior when the absorber is selective.

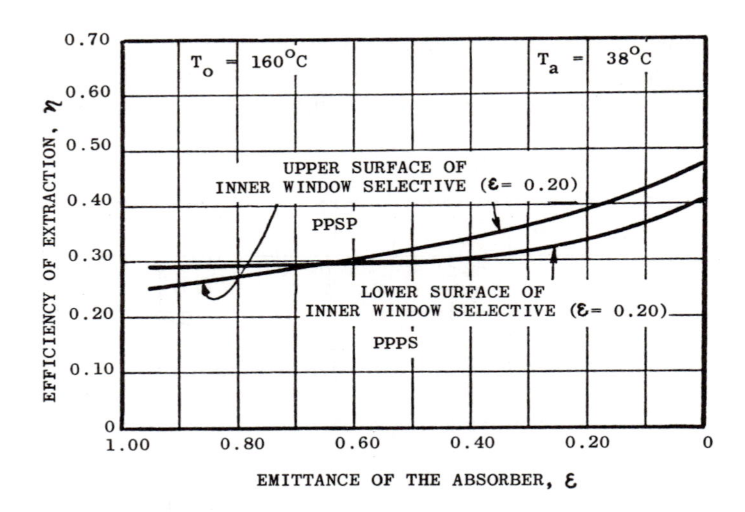

Fig. 12.24 Comparison of efficiency of two collectors with a selective transmitting surface on only the inner window. One inner window has the selective coating on its upper surface; the other has the coating on its lower surface. The outer windows have both surfaces plain.

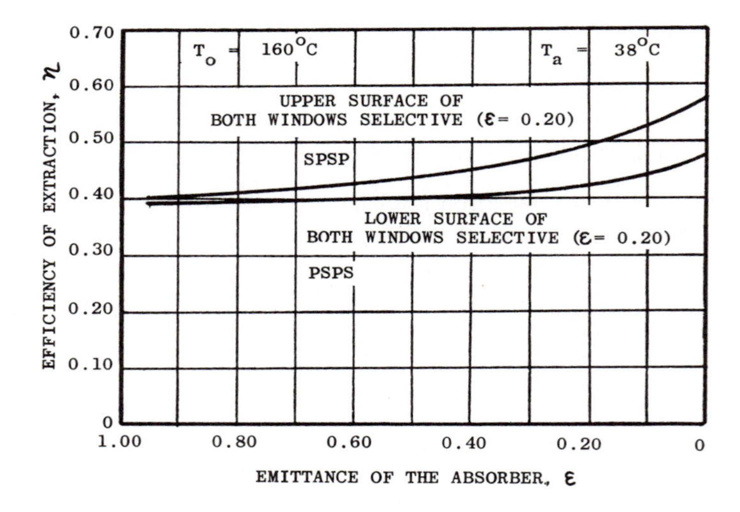

Fig. 12.25 Comparison of efficiency of two collectors with double glass windows. In one case the upper surface of each window has the selective coating; in the other case the lower surface of each window has the coating.

Placement of Two Selective Layers (*SPSP, SSPP*)

The preceding cases, in which each window has one selective surface, raise the question of whether both surfaces could be placed on a single window. The advantage would be that the coatings could be applied in one operation on one piece of glass, as most selective transmitting surfaces are applied either by dip followed by a bake cycle or by an aerosol spray on a heated surface. We show this comparison in Fig. 12.26 and note that there is little difference between the two cases. We can therefore conclude that it is satisfactory to place the selective coatings on both sides of the upper cover window.

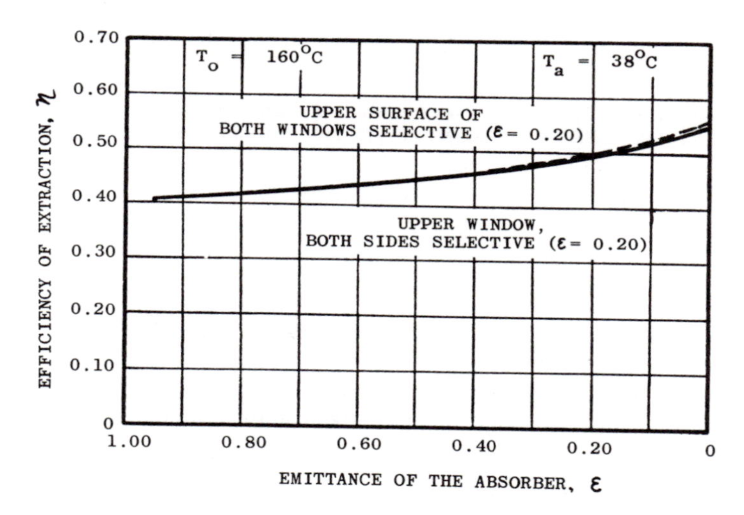

Fig. 12.26 Comparison of efficiency of two collectors with two windows and two selective surfaces. In one case, each window has a selective coating on its upper surface; in the other case, the upper window has selective coatings on both sides.

12.20 COMPARISON OF THERMAL BEHAVIOR FOR SELECTIVE WINDOWS

Now that we have established the relative performances of the several possible combinations of selective surfaces, let us examine how they compare with regard to operating temperature of the collector. In Fig. 12.27 we compare three cases, the lower curve being the standard blackbody absorber with plain windows (*PPPPB*); the notation *B* refers to a blackbody absorber. The middle curve is when only the absorber is made selective (*PPPPS*.0); the notation *S* refers to a selective absorber and the number indicates the value of its emittance. The upper curve is for the maximum effect of selective windows, when two surfaces are selective (*SPSPS*.0). The upper curve would also apply, as we have seen, to the case *SSPPS*.0.

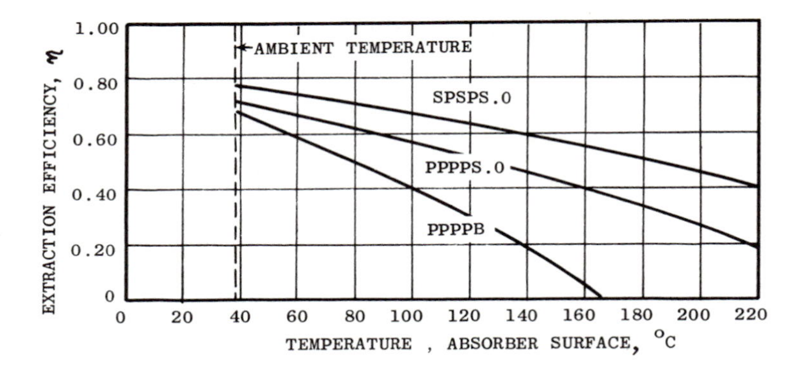

Fig. 12.27 Comparison of efficiency gains for selective coating on the absorber and on absorber and windows, for 725 × 10^{-4} W/cm^2 input in each case, with 38°C ambient temperature and no wind.

12.21 WINDOW ABSORPTION

The major practical objection to the use of selective windows is that the coatings absorb sunlight and have higher reflectance than normal glass. The newer indium oxide coatings when doped with tin oxide appear to have little internal absorptance, and their high index of refraction lends them to effective use as nonreflection coatings. Let us examine the role of absorptance in the coatings. Table 12.5 shows the results of 5% additional absorptance per coating.

Table 12.5 Effect of Window Absorption on Collector Performance

	Temperatures, °C			*Extracted energy, W/cm²*	*Efficiency η*
Configuration	W_1	W_2	*Abs.*		
PPPPB	40.4	71.9	99.0	416 × 10^{-4}	0.44
P_aPP_aPB	43.9	74.5	99.0	380 × 10^{-4}	0.40
SPSPS.2	44.1	71.9	99.0	618 × 10^{-4}	0.65
S_aPS_aPS.2	49.5	78.5	99.0	578 × 10^{-4}	0.61
$SPS_{aa}PS$.2	47.6	81.7	99.0	594 × 10^{-4}	0.63
SPSPS.0	41.7	64.9	99.0	646 × 10^{-4}	0.68
S_aPS_aS.0	47.5	71.7	99.0	600 × 10^{-4}	0.63

Absorptance = 0.05 per surface; subscript *a* indicates window surfaces with an assumed 5% absorption increment. T_{amb} = 38°C; wind = 67 cm/sec.

The results can be summarized quite simply. Absorption in a cover results in about *half* the absorbed energy appearing as useful output, and half being lost. This equipartition of the absorbed energy is a consequence of heating the windows and thus reducing the thermal gradient from the absorber to the outside environment, impeding the loss of energy from the absorber itself. In a sense its physical effect is the same as that of the selective windows, namely raising the window temperature to reduce the convection loss term.

This equipartition of absorbed energy also can be applied to the arguments concerning the type of glass to be used for the collector windows. It is often stated that "water white" glass is needed to minimize internal absorption. Two factors make the increased cost of such glass not worth the performance gain. The first factor is that half of the absorption is useful to the collector, reducing the apparent difference between glasses. The second factor relates to the term "water white" as applied to glass. Glass normally has a slight greenish tint when viewed edge on. Some water-white glasses are made *apparently* white, not by the reduction of the absorption producing the green tint but by the *addition* of absorption to the other parts of the spectrum to reduce the apparent color. In reality the total absolute absorption of such water-white glass is slightly higher than for ordinary float glass. We conclude that ordinary glass is quite satisfactory for solar collector windows.

12.22 NONREFLECTION COATED WINDOWS

Nonreflection coated windows are possible, and produce significant gains in collector efficiency by permitting utilization of the solar energy that otherwise would never reach the absorber. In Fig. 12.9 we show the improvement in a water-heating collector having two windows, each reflection-coated on both sides. The effect is simply to shift the curves upward by the amount of energy gained via the nonreflection coating, about 9%. The types and nature of nonreflection coatings are discussed in Chapter 8.

12.23 VARIATION OF EFFICIENCY WITH SOLAR FLUX

In the graphs shown in this section the solar flux has been taken as the approximate midday flux. We are interested in knowing the efficiency changes with solar flux during the day. The transformation of a curve for a given solar flux to another value is quite simple.

The performance of a collector is generally taken with a constant value of the output temperature. In this case the heat loss U remains independent of the solar flux, so the change in solar flux is directly subtracted from the heat output of the collector. This situation is as shown in Fig. 12.28.

We have the new efficiency η' as given by

$$\eta' = (\alpha t I' - L)/I', \tag{12.8}$$

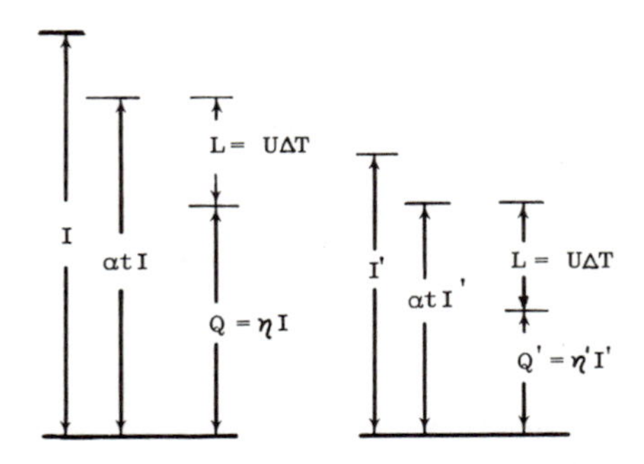

Fig. 12.28 Diagram illustrating the change in efficiency η with changing solar flux when the temperature of the collector remains constant.

but because

$$L = \alpha t I - \eta I, \tag{12.9}$$

we obtain

$$\eta' = \eta I/I' - \alpha t(I/I' - 1). \tag{12.10}$$

If, on the other hand, we use the efficiency with which the absorbed heat is extracted, P, we obtain a simpler relationship. In this case

$$L = I - I(1 - \alpha t) - P\alpha t I \tag{12.11}$$

and also

$$L = I' - I'(1 - \alpha t) - P'\alpha t I', \tag{12.12}$$

in which case we have

$$P' = (I/I')(1 - P) - 1. \tag{12.13}$$

The effect of change in solar flux is shown graphically in Fig. 12.29 using a type of diagram introduced by Tabor (1955). The lower portion of the daily intensity curve required just to equal the heat loss of the collector at operating temperature is shown by the shaded area. The useful energy is the portion above the shaded area, shown as crosshatched. This type of representation is useful in determining the minimum time in the morning required for the collector to heat up to operating temperature. If the collector has additional lag due to thermal inertia, the time will be appropriately longer.

If the day is cloudy, one can use the Tabor graph to show when the collector will fail to reach minimum operating temperature.

The situation displayed in Fig. 12.28 can be shown in graphical form, as in Fig. 12.30. The relationship between η and ΔT for normal solar flux is denoted by

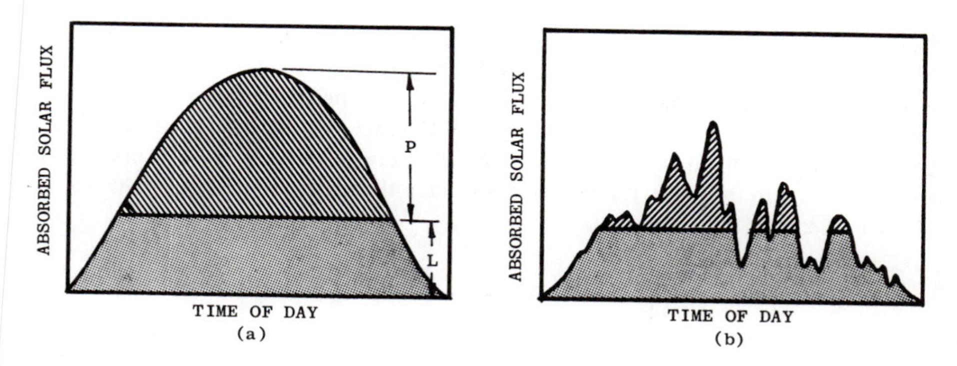

Fig. 12.29 Separation of the daily flux into the amount needed to maintain the heat loss ($L = U\Delta T$) and the amount usefully extracted from the system. For a system having negligible thermal inertia the useful output ceases when the solar flux drops below L.

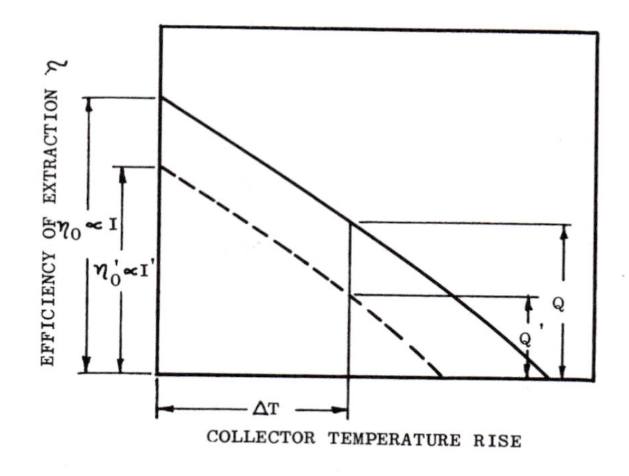

Fig. 12.30 Relationship between the value of the system efficiency at reduced solar flux for a typical curve for a flat-plate collector.

the solid line curve, a typical curve for a flat-plate collector, for example. This curve is essentially a straight line, a fact we will use below. The relationship for reduced solar flux is shown by the dashed curve. The ordinates are proportional to the solar intensity I. The change in energy output with change in flux is indicated by Q for normal solar flux and Q' for reduced solar flux. The value of Q' is obtained by simply sliding the normal solar flux curve horizontally to the left until the ordinate is I'.

The diagram in Fig. 12.30 can be further simplified for use under different solar fluxes by normalizing the curves until a single curve results. Note that the solid and dashed curves form approximately similar triangles. If the η-versus-ΔT curve were a straight line the similarity would be exact. Neglecting the curvature, which means neglecting terms other than ΔT in the full equation for heat balances, one can normalize the curves by dividing the temperature difference by the solar flux, $\Delta T/I$. The resulting curve is shown in Fig. 12.31. The procedure to arrive at a normalized graph when there is a finite rise in temperature is to take the mean temperature value for the collector absorber, $(T_{in} - T_{out})/2$.

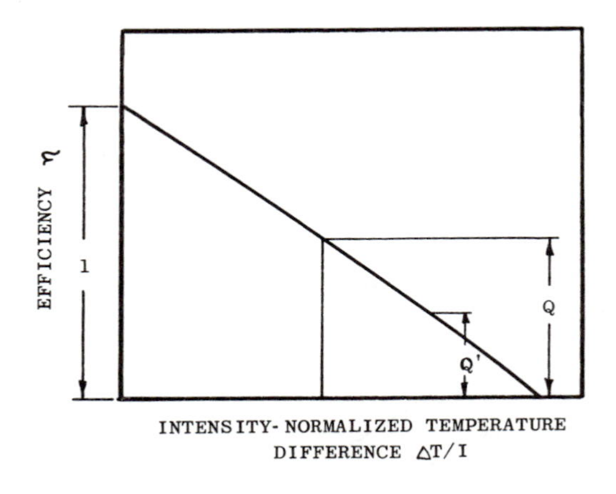

Fig. 12.31 Relationship between efficiency for a collector and the temperature difference normalized by dividing the value of ΔT by the actual solar flux intensity prevailing at the time the collector performance is to be evaluated.

In Fig. 12.32 we show some experimental data points from Gupta and Garg (1967) for an air heater having a single cover window. For comparison, the dashed lines are the calculated performance for a water heater with a single cover window, taken from Fig. 12.14. The two cases are not exactly comparable, but the trend of η versus ΔT is shown by the experimental data. The slope of the data points is affected by heat losses, such as back insulation, and the $\Delta T = 0$ limit is affected by cover transmission and absorber absorptance. The experimental data show considerable scatter. This scatter can be produced by at least three factors: (1) variation of wind velocity, (2) variation of the ETIR flux, and (3) nonlinear terms approximated in the normalization through dividing ΔT by I.

For comparison of collector performance Hall and Kusada (1975) recommend using the normalized temperature functions, as in Fig. 12.32. This procedure permits approximate comparisons, useful for days having a considerable variation in solar flux I, and for the determination of an experimental calibration of any particular type of collector, these data then being of primary importance to the evaluation of perform-

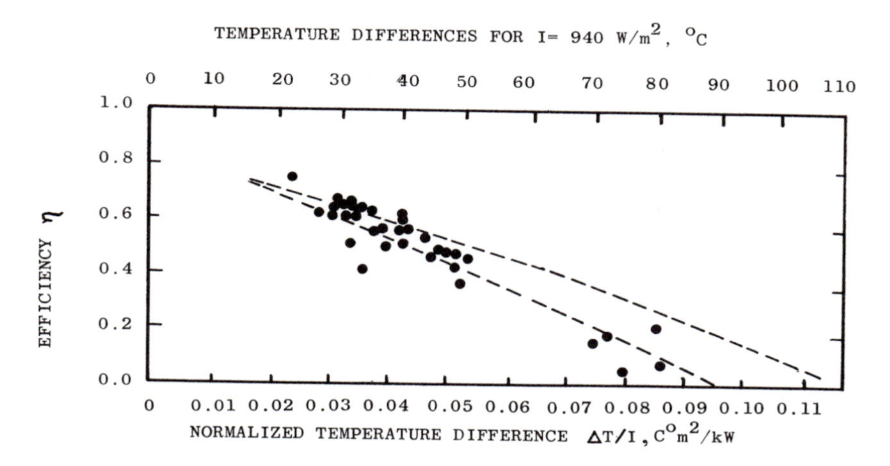

Fig. 12.32 Observed efficiencies versus normalized temperature difference for an air-heating collector with one glass cover window, by Gupta and Garg (1967). The dashed curves are for a single-window collector, from Fig. 12.14.

ance of a proposed solar collector system. If we take a collector in which the temperature rise is ΔT, the energy output Q under normal solar flux I will be given by

$$Q = I\eta(\Delta T), \tag{12.14}$$

where $\eta(\Delta T)$ indicates that η is evaluated at ΔT. Under reduced solar flux I' the energy output Q' is given by

$$Q' = I'\eta'(\Delta T), \tag{12.15}$$

or from Fig. 12.30,

$$Q' = Q - [\eta(0) - \eta'(0)]I, \tag{12.16}$$

where η is evaluated at $\Delta T = 0$, and where we neglect the curvature of the lines. If curvature exists, as is schematically shown, then η at a given ΔT will be slightly greater than the difference as evaluated at $\Delta T = 0$. Since $\eta'(0)$ is equal to $(I'/I)\eta(0)$, we can write

$$Q' = I\eta(\Delta T) - (I - I')\eta(0), \tag{12.17}$$

where Q is actually an amount of energy per unit collector area Q/A. We will use this procedure for evaluating performance of a solar collector with thermal storage subsystem in Section 13.19.

12.24 EVACUATED COLLECTORS

The performance of a flat-plate collector can be further improved as the air is removed from the space between the absorber and windows. In this manner the convective loss can be suppressed. The dependence of loss on residual gas is treated in detail in Chapter 11. The basic problems involved in evacuating a collector are how to seal the system for reliable lifetime operation and how to alleviate the mechanical pressures resulting from the surrounding atmosphere.

If radiation loss is suppressed by use of a selective surface and convection loss is eliminated by evacuating the collector, the limiting performance is set by the incursion of the TIR emission curve into the absorptive region of the selective surface, a limit shown in Fig. 14.27.

Two designs for nonconcentrating flat-plate evacuated collectors are shown schematically in Fig. 12.33. The upper design is based on a patent by Speyer (1965) in which a strip of flat plate containing the incoming and outgoing heat-transfer fluid lines is enclosed in a glass evacuated tube, sealed at one end to the metal absorbing structure. The original Speyer design used a glass tube that had the lower half aluminized to redirect some of the solar flux that missed the absorbing tube structure. Note that the ratio of tube diameter to length is not to the scale of the units constructed to date.

In the other evacuated design in Fig. 12.33 the vacuum region is confined to the space between two glass tubes, the inner one of which is blackened to act as the absorber. A separate tube for injecting the heat-transfer fluid is located in the center of the inner glass tube. In this design the vacuum region can be between glass components, avoiding the problem of vacuum-tight glass-metal seals. When a nonselective absorbing surface is used, the inner wall of the inner tube can receive the black surface. If a selective absorber is to be used, the absorber must be on the outer surface of the inner tube. The alternative is to use a nonselective absorber and place a TIR reflecting selective layer on the inner wall of the outer glass tube. The gap between tubes, which are spaced approximately an absorber tube apart, is backed with a white scattering surface. In this way some of the sunlight is usefully scattered back to the absorbing tubes, increasing the collector efficiency.

In principle these two evacuated designs with selective coatings can yield temperature differences ΔT well above 100 C° with efficiencies in the range of 50–60%. They suffer from the problem posed by the loss of heat-transfer fluid, since their temperatures can rise to a point where selective surfaces can fail by overheating, a problem noted in Chapter 9. An alternative use of these evacuated tube assemblies would be as absorbers in low-concentration collectors, where they would yield very high efficiency of extraction of absorbed solar flux. The problem of overheating could be avoided by providing a means of directing the collector away from the sun should coolant flow failure occur. The major technical problem with evacuated absorber assemblies is in maintaining vacuum integrity under handling and over a lifetime of 10–20 years. The major economic problem is in reducing the cost of these assemblies, possibly through automated fabrication techniques used for other products in the glass industry.

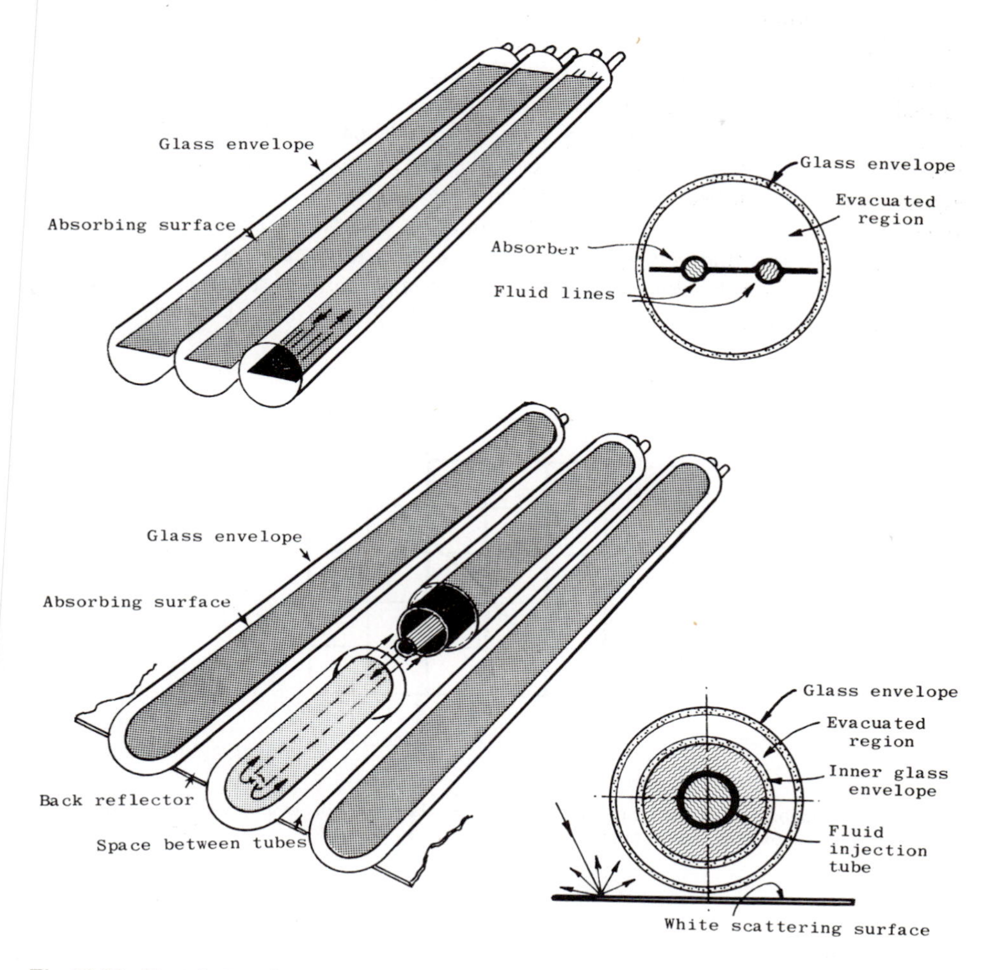

Fig. 12.33 Two designs for nonconcentrating evacuated solar collectors. The top configuration is used by Corning and the bottom by Owens-Illinois.

12.25 RADIATIVE COOLING

In the process of performing flux balance computations the student will become aware that quite often a window surface will drop below the ambient temperature even with sun shining on the collector. The natural result is to ask: How much can the radiative cooling effect be deliberately enhanced? The goal would be to make a collector that would provide cooled fluid during the day and, especially, at night.

The basis for radiative cooling is that the sky does not radiate in the infrared as does a blackbody at the ambient temperature. The nature of the environmental thermal infrared (ETIR) spectrum is discussed in detail in Chapter 2. The major effect is caused by the deficiency of TIR radition between 8 and 12 μm, where the atmosphere is most transparent. The temperature drop of a single glass cover, backed by a perfect mirror, is shown in Fig. 12.34. In this case the temperature of the surface drops until equilibrium is established between the incoming ETIR and convection (conduction) transfer from warm air into the surface and the outgoing TIR from the surface. Figure 12.34 shows that very significant drops can be obtained when the ETIR flux is well below that of a blackbody at the ambient temperature, a situation prevailing in dry climates. Note that at ETIR = 40% the surface will drop from 40–0°C.

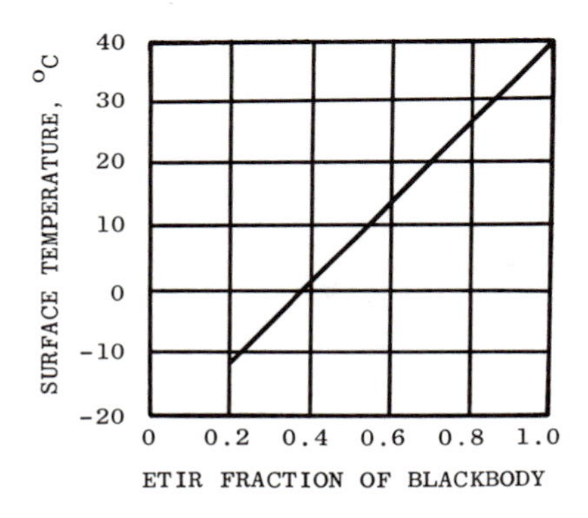

Fig. 12.34 Dependence of the temperature of a single glass cover backed by a perfect mirror on its lower surface upon the environmental thermal infrared flux, for ambient temperature 38°C (100°F).

The large drop in the ETIR between 8 and 12 μm has led researchers to examine whether a selective emitting surface could accentuate this cooling of surfaces by nocturnal radiation, or even make it effective during daytime also. This possibility has led researchers, Silvestrini (1975) in particular, to propose enhancing the ETIR effect by making the collector selectively radiate in the 8- to 12-μm region and thus reject its TIR radiation into space, at the same time reflecting the downcoming sky radiation in the 3- to 8-μm and 12- to 20-μm regions.

We can express the radiative cooling effect in two ways. The first is to define the temperature achieved by a surface and the second is to define the number of watts per square centimeter that can be radiated outward in excess of the number received. In

Fig. 12.34 we show the temperature that a single glass surface would reach as a function of what fraction the ETIR is of a blackbody at the ambient temperature. In most climates the ETIR fraction ranges from 0.50 (dry, high climate) to 0.95 (wet, sea level). Note that the condition for this temperature is that the glass surface be backed by a perfect reflective insulator.

In Fig. 12.35 we show the net radiative flux lost to the sky as a function of the difference between the surface temperature and the ambient temperature when the ambient temperature is 38°C. The upper curve is for the surface at ambient temperature and the other two are for surfaces at 11 C° and 22 C° below ambient temperature respectively.

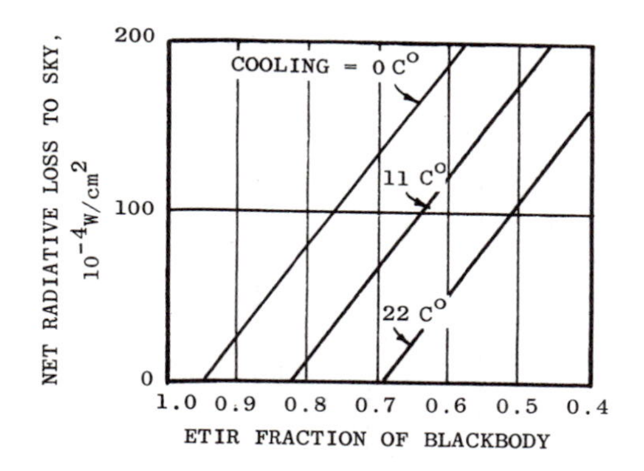

Fig. 12.35 Net loss to the sky of a fully transparent blackbody backed by a perfect reflector and insulation as a function of the ETIR and different amounts of net cooling, for 38°C ambient temperature.

The curves in Fig. 12.35 show that for cooling of 11 C° (20 F°) below ambient temperature the collector can reject about 100 W/cm^2 to the sky in a dry climate. This rate is only about 15% of the typical input energy of a heating solar collector, but is potentially important. It seems clear that more attention should be given to passively cooling air and water through radiative means in view of the complication and cost of actively cooling through use of thermodynamic cycles such as lithium bromide gas absorption systems. Noteworthy examples of radiative cooling for domestic applications are the University of Arizona laboratory building designed by Bliss (no longer existing) and the Hay Skytherm homes in Arizona and California. The Bliss design used uncovered collectors colored dark green to heat during the day and cool during the night; the Hay design uses the radiation to the night sky of a water surface contained in a TIR-transparent plastic.

12.26 SELECTIVE RADIATIVE COOLING

The radiative selective cooling studied by Silvestrini and his colleagues at Naples (Catalanotti *et al.*, 1975) is based upon the natural opacity of some types of plastics, like Tedlar (see Fig. 8.5). The infrared regions where the plastic molecules have absorptance are also those where the material has high thermal emittance. The opacity of such a plastic is shown schematically in Fig. 12.36(a), where the absorption profile is replaced by a step function having highest emittance in the 8- to 12-μm region. The curve in Fig. 12.36(b) is a schematic emittance curve for the atmosphere, showing the dip in flux intensity in the 8- to 12-μm region where the atmosphere is relatively transparent. The net result is that this type of emitter selectively radiates its thermal radiation out through the atmosphere, and because the plastic is not absorptive of the wavelength where the atmosphere does radiate ETIR, the plastic surface cools well below ambient temperature.

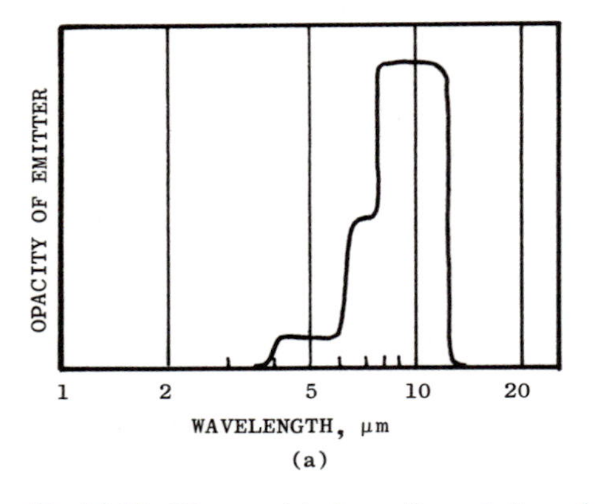

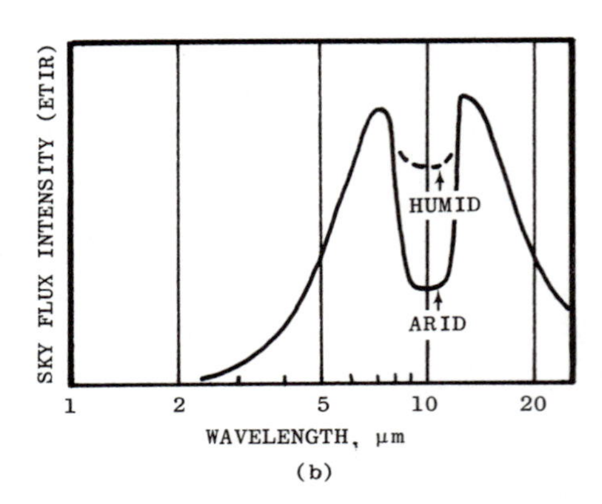

Fig. 12.36 Diagram (a) shows the variation of opacity with wavelength for a plastic suitable for use as a selective emitter, losing radiation to space through the 8- to 12-μm atmospheric window. Diagram (b) is a spectrum of the environmental thermal infrared (ETIR) showing the dip in the 8- to 12-μm region where the atmosphere is transparent.

The magnitude of the selective cooling effect depends upon the amount of water vapor in the atmosphere. The Catalanotti study assumed a rather wet atmosphere, taking Naples to be similar to Cocoa Beach, Florida (see Chapter 2). Even so, it reported experimental cooling rates on the order of 40 W/m^2 for a surface 10 C° below ambient temperature. The type of relationship between radiated energy and surface temperature difference is shown in Fig. 12.37, where a humid climate is assumed. Note that the black emitter is better when the surface is only slightly below ambient, but that the selective emitter is better when a large temperature difference is maintained, as would be the case if significant cooling of an airflow over the emitting surface were desired. The reason the selective surface is poorer at low temperatures is

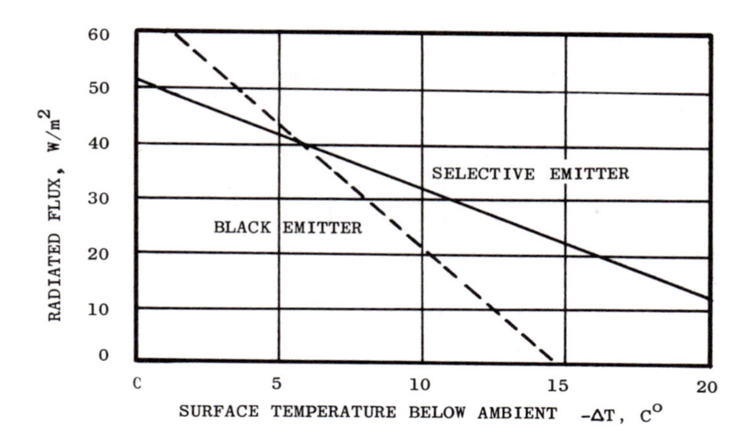

Fig. 12.37 Relationship between emitted flux and surface temperature below ambient for a selective plastic film emitter compared to a black emitter, for an atmospheric humidity as at Cocoa Beach, Florida. For a dry or desert climate the cooling curves are almost twice as high in terms of radiated flux.

that it is not as "black" as a black surface in the 8- to 12-μm region; this is basically the same reason that a selective absorber is poorer than a black surface in a heating collector, the absorptance in this case being less than that possible with a black absorber. Polyethylene is then used as a window because it has no significant absorption in this spectral region, thus protecting the emitting surface from wind losses and dirt.

12.27 CYLINDRICAL COLLECTOR STRUCTURES

In the preceding sections we have discussed the heat balances of flat-plate collectors. The geometry is simple and the experimental data on which heat transfer curves are obtained are extensive. In some cases of practical interest we have cylindrical surfaces for which similar heat balances are required. An example is the parabolic trough collector, in which the absorber is a cylinder. This cylindrical absorber may also be surrounded with cylindrical envelopes that modify the heat losses of the absorber. A key question in the calculation of performance is the temperature of these envelopes and the flux of energy flowing from one envelope to the next.

For concentric cylinders there are two effects that differ from the case of the plane-parallel plates. First, when radiation flows *outward,* its density decreases with increasing radius according to

$$(Q/A)_1 = (Q/A)_0 \, (r_0/r_1), \tag{12.18}$$

where r_0 and r_1 are as shown in Fig. 12.38. Since

$$(Q/A)_0 = \epsilon_0 \sigma T_0^4, \tag{12.19}$$

we have

$$(Q/A)_1 = (r_0/r_1)\epsilon_0 \sigma T_0^4. \tag{12.20}$$

On the other hand, when radiation flows *inward,* its density is constant throughout the volume. The net exchange passing between surface 0 and surface 1 is therefore

$$\underset{\substack{\text{net loss}\\ \text{0 to 1}}}{(\Delta Q/A)_{01}} = \underset{\substack{\text{outward}\\ \text{emitted}}}{(r_0/r_1)\epsilon_0 \sigma T_0^4} - \underset{\substack{\text{inward}\\ \text{absorbed}}}{\alpha_0' \epsilon_1 \sigma T_1^4}, \tag{12.21}$$

where α_0' is the infrared absorptance of surface 0.

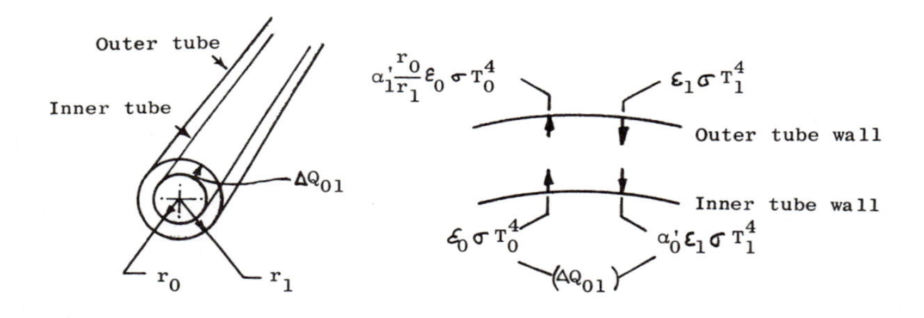

Fig. 12.38 Diagram of the geometry and fluxes for concentric cylinders.

The calculation procedure involves making a table like Fig. 12.5, listing each loss term and each gain term for each surface involved. The temperatures of each surface are then adjusted so that the input and output are balanced, the desired equilibrium situation. The difference between the input to the absorber and its output loss is the amount of energy that can be usefully extracted. The net radiative exchange to space is given by

$$(\Delta Q/A)_{1\infty} = \epsilon_1 \sigma T_1^4 - I_{\text{TIR}}, \tag{12.22}$$

where I_{TIR} is the infrared background flux from the environment.

To calculate the convective loss term we need to determine the value of the Nusselt number (Nu) for the geometry and physical conditions that apply to the design under study. In Section 10.6 we show the experimental data for the variation of

Nu with Rayleigh number (Ra) for flat plates and for a horizontal cylinder. The curve for the horizontal cylinder lies above the curve for the parallel plates. We are not sure that this difference is real, since the difference is approximately π. If the diameter D were used in the reduction of experimental data rather than πD, the difference would largely disappear. We are unable to answer this question in searching the literature, but recommend that the student use the curve for parallel plates 45° upward for calculations of cylinders. This value would be approximately the mean for all possible orientations of parallel plates. We then have the satisfying case of concentric cylinders of small spacing converging to the same answer as parallel plates, a logical conclusion.

The procedure to calculate the combined convection and conduction loss for linear absorbers and for concentric structures is to calculate the Rayleigh number, enter the graph for Nu versus Ra, Fig. 10.11, and add the conduction term. The loss between surfaces is then given by

$$(Q/A)_{01} = (r_0/r_1)\epsilon_0 \sigma T_0^4 - \epsilon_1 \sigma T_1^4 + (\mathrm{Nu} + 1)(k/S)(T_0 - T_1) \ldots, \quad (12.23)$$

where S is a shape factor. These shape factors for some cases of interest are given to a reasonable approximation by

plane parallel plate	$S = L,$
concentric cylinders	$S = r_0 \ln(r_1/r_0),$
free cylinder	$S = 2D_0 = 4r_0.$

12.28 SOLAR PONDS

The simplest type of flat-plate collector, combining both the collection and the thermal storage subsystems, is the solar saline pond, developed by Bloch (1948) and Tabor *et al.* (1959). The solar saline pond is based upon the observation by Kalecsinski (1940) that natural saline lakes can have a steep temperature gradient with depth. This was found to be associated with a chemical gradient in the dissolved solids, resulting in a density gradient sufficient to suppress thermal convection. A number of field experiments have been conducted to establish the feasibility of producing power (and salt) from the operation of such a pond. Important reports were published by Tabor (1963, 1965), Weinberger (1964), and Hirschmann (1970) that detailed the physics and practical problems of solar ponds.

The solar saline pond should be a very low-cost type of collector, as it consists merely of a pond about 1–2 m deep with earthen retaining barriers. This simple picture is complicated, however, by practical problems of managing the saline gradient and of preventing wind stirring, bacterial growth, outgassing of the ground, leakage into the local aquifer, and so forth. New experiments with solar ponds are currently being initiated, and their results may help solve some of these practical problems. In principle, the solar pond nevertheless remains the simplest collector for large-scale uses and is worth careful consideration.

In Fig. 12.39 we show a schematic diagram of a solar saline pond. A typical pond is 1–2 m deep and covers several hundreds to thousands of hectares. The water at the surface has a low salinity, usually that of the feed water used to make up for pond evaporation and for maintenance of the proper saline gradient. If no water were added, the salinity gradient would tend to lessen with time, but if fresh surface water is added and the concentrated effluent is removed from the bottom, the required gradient can be maintained. We also show a typical temperature-density profile through the pond. The surface temperature has a small maximum near the water surface. This is because the infrared portion of the solar spectrum, beyond 0.9 μm, is totally absorbed in the top few centimeters of water, heating this upper layer. The rest of the spectrum is absorbed in the bulk of the pond and in its bottom. The energy absorbed in the top layer is not lost to the system, as this temperature increase impedes the outward flow of energy from the bottom of the collector, acting like a heated front cover window of a flat-plate collector. In a dry climate the surface temperature rise is eliminated by the cooling effect of surface evaporation, especially under windy conditions.

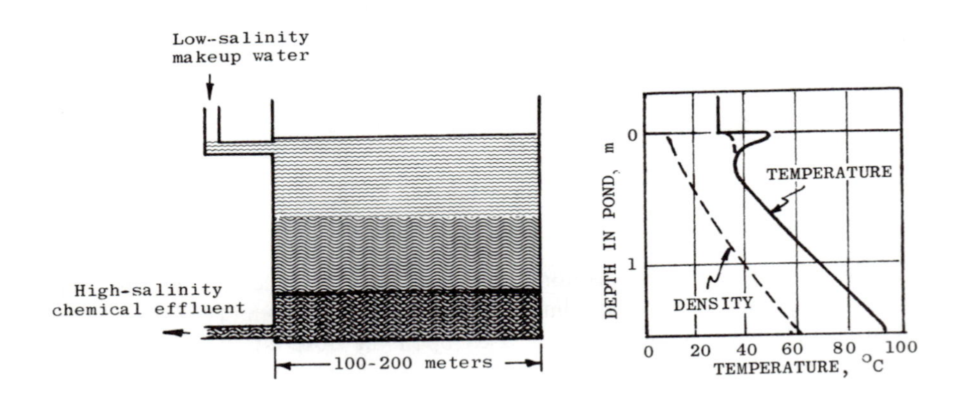

Fig. 12.39 Schematic cross section through a saline solar pond, and a typical temperature-density profile through the pond. The surface temperature rise is due to infrared solar energy absorbed by the water. Low-salinity water is required to make up for evaporation of the pond.

A solar pond can be created from a nonsaline pond wherein one or more plastic covers are used to eliminate evaporation and convective mixing from the absorption zone to the surface zone. Such collectors are in reality only flat-plate collectors in which cost reductions are sought by the floating of one window on the water and the supporting of the upper window with inflated air pressure.

12.29 PROBLEMS

12.1. Prepare, for an air-heater collector, a sample heat balance sheet like that shown in Fig. 12.6, changing the solar flux input to 740 W/m^2. What is the efficiency η?

12.2. Prepare the same sample heat balance sheet as in Problem 12.1, changing the solar flux input to 540 W/m^2. What is the efficiency η?

12.3. Plot Problems 12.1 and 12.2 along with data from Fig. 12.6 to show the variation of collector efficiency with reduction in solar flux. Also plot this variation as a function of time of day, assuming this collector to be a fixed flat plate tilted at 40° latitude, for winter solstice.

12.4. Prepare for a liquid collector a sample heat balance sheet like that shown in Fig. 12.5, but with solar flux input at 740 W/m^2.

12.5. Prepare the same sample heat balance sheet as in Problem 12.4, but change the solar flux to 540 W/m^2.

12.6. Plot Problems 12.4 and 12.5 along with data from Fig. 12.5 to show the variation of collector efficiency with reduction in solar flux, and with time of day under the assumptions of Problem 12.3.

12.7. For the air-heater collector of Fig. 12.6, change the air velocity from 20 cm/sec to 100 cm/sec. Calculate the change in collector efficiency.

12.8. In Fig. 12.6, the optical properties of the surfaces involved are not listed. From the entries in the table portion, calculate the absorptance and emittance of each surface.

12.9. Calculate the change in collector efficiency for the sample in Fig. 12.6 when the absorptance of the absorber is 0.93 and the emittance is 0.10.

12.10. The collector shown in Fig. 12.5 is for operation during the winter, as indicated by the ambient temperature. The same collector is to be used in summer to operate a gas absorption refrigeration system. The fluid temperature is now to be 95°C and the ambient temperature 40°C. Calculate the efficiency and the temperature of the surfaces involved.

12.11. The collector in Problem 12.10 is modified to have a selective absorber with an absorptance of 0.93 and emittance of 0.10. Calculate the system efficiency.

12.12. Repeat Problem 12.10 for a collector having two cover windows of glass.

12.13. Repeat Problem 12.11 for a collector having two cover windows of glass.

12.14. Calculate the winter performance of the collector of Problem 12.12 with the same conditions as in Fig. 12.5 but with a selective absorber.

12.15. Based upon the examples done for collectors with one and with two cover windows and water as the working fluid, what decisions would you recommend for a collector for only winter use? For only summer use? Include the cost factor in your discussion.

12.16. For an overlapping-plate air heater of the Lof type, discuss the question of the length of the overlapping portions in terms of the separation of the plates, assuming that the full temperature rise is to be obtained on a single pass of air through the collector. Outline the steps in the calculation and the relevant equations.

12.17. Take a set of values and perform the numerical integration to derive the air temperature as a function of selected values for the plate separations and lengths.

12.18. Calculate the heat loss from the collector of Problem 12.14 when a cloud passes over the collector. If the collector had a mass of 0.1 g/cm^3 with the effective heat capacity of water, how fast would it cool?

12.19. What would be the equilibrium temperature of the absorber of the collector of Problem 12.14 at night when the ambient temperature has dropped to 0°C? Assume a dry climate model for the ETIR.

12.20. A bare collector having a blackbody absorbing surface is exposed to the night without cover. Calculate the heat balance diagram and the efficiency of heat rejection. Assume an ambient air temperature of 30°C and a dry climate model.

12.21. Repeat Problem 12.20, but with a wet model atmosphere.

12.22. Calculate Problem 12.20 but with a single cover of 6-mil Tedlar for the window. What is your conclusion concerning Problems 12.20–12.22?

Chapter 13
Energy Storage

13.0 INTRODUCTION

In the preceding chapters we discussed aspects of the solar collector portion of the solar energy system. In this chapter we briefly touch upon some of the system aspects and examine some of the other subsystems that together with the solar collector comprise the using system. Energy storage is a key aspect of making a solar collector into a useful system. This topic alone is deserving of an entire book, so this presentation will be only an introduction into some aspects of energy storage. We also discuss some medium-temperature thermodynamic cycle systems that can utilize the hot water produced by solar collectors. The high-temperature Rankine cycle systems, of the type used with high-concentration collectors, are essentially identical to conventional steam or nuclear turbine systems.

To begin this chapter let us look at what comprises a solar thermal system.

13.1 BASIC SYSTEM DIAGRAM

The basic solar energy system has certain subsystems that can be easily defined for subsequent examination. A diagram of such a system is shown in Fig. 13.1.

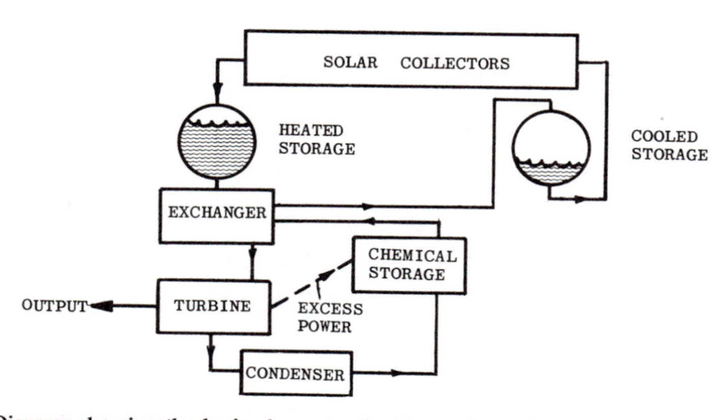

Fig. 13.1 Diagram showing the basic elements of a thermodynamic system utilizing solar heat as an input. The stored energy supply refers to a standby chemical fuel or a fuel created from excess solar energy during the summer.

Solar Collection Area

The solar collection area, where the energy is gathered, is the largest subsystem in both extent and cost. The ultimate economics of the system depend to a large degree on the costs of the collector subsystem. The heat-transfer fluid lines convey cool fluid to the collector and collect heated fluid to be conveyed to a storage location.

Thermal Storage

Each solar energy system has to some degree thermal energy storage, either deliberately provided as a place to store enough energy to smooth out input flux fluctuations, or through the thermal inertia of the extensive system of collectors and heat-transfer fluid. Thermal storage, however, is useful only for a short time, measured in hours or at best a few days, which is generally not enough to carry a system through the worst winter weather. In principle one could make the thermal storage capacity large enough, but the cost is prohibitive except in the special case of the saline solar pond, which is both a solar collector and a thermal storage tank of weeks of heat capacity.

Thermal Exchanger

In general there will be an exchanger between the solar heat-transfer fluid and the using system working fluid. This exchanger also serves as the interface between the energy collection and storage subsystems and the using subsystems.

Turbine

The turbine subsystem is shown as representative of a using subsystem. Such a subsystem could be a house to be heated as well as a turbine. This subsystem provides the customer with the desired output of the solar collector system.

Condenser

The condenser subsystem is necessary whenever a thermodynamic cycle is employed. A turbine requires one, as does a gas absorption refrigeration system. The condenser enters into any calculations of environmental effects of the system, since all of the thermodynamically unused heat must be rejected into the environment at this point in the system. If these condenser temperatures are high enough, one can further add to the using subsystems here by utilizing the waste heat for heating and refrigeration tasks, or even for the operation of enclosed environmental farms.

Chemical Storage

Because the thermal storage subsystem is in general not large enough to permit the system to stand alone, some sort of supplemental backup energy must be injected into the system. We show a subsystem where the excess summer energy output of the system is converted into a storage chemical fuel, such as hydrogen or granular aluminum, for use during the winter when short days and cloudy weather combine to produce problems for a solar energy system. One could inject energy from oil or coal at this point, but we feel such alternatives are not feasible in the long run, such as decades or a century into the future.

In this chapter we discuss some aspects of systems and subsystems. We will not be comprehensive, but will give a glimpse of the nature and complication of the system problems that one encounters. The fundamental problem of integrating solar energy into the national or world energy mix is even more complex, and we do not intend to enter this serious problem area except in the following regard.

13.2 PEAKING EFFECT OF BACKUP DEMANDS

The point we do wish to touch is that it is easy for the solar energy engineer or manager to dismiss the long-range consequences of what appear to be easy solutions to difficult problems encountered in solar energy. Backup energy is one problem that frequently threatens. Most solar energy applications, such as home heating and cooling, rely on an electrical backup for periods of insufficient solar energy. This solution is delightful when only you and a few other people have solar energy systems in your homes. But imagine what would happen if 80% of the people in your city had solar units and all of you wanted the local utility to pick up the load on the fourth cloudy day. If utilities are seriously concerned about peak demand problems today, think what the peaking demand will be if this easy way out of the solar problem is allowed to become general in the systems built in the decades ahead. Either someone will have to pay for the unused generating capacity that is waiting for that "fourth cloudy day," or the sudden demand will pull down the entire regional electrical network.

If the solar energy system has its own backup fuel of a storable nature, like a tank of oil, then the surge problem is reduced to the more manageable problem of simply pulsing the resupply network for the stored fuel. Ultimately, however, if solar energy is to become important enough to affect the net energy usage of the country, some way will have to be found to make the solar energy system independent of a backup energy supply. At present the only candidate for this honor is the lowly solar pond.

13.3 ENERGY STORAGE

Energy storage is critically important to the success of any intermittent energy source in meeting demand. This problem is especially severe for solar energy because storage is needed most when the solar availability is lowest, namely in winter.

Energy storage complicates solar energy utilization in two ways. First, the energy storage subsystem must be large enough to carry the system over the periods of inadequate sunshine. The alternative is to have a backup energy supply, which implies both a capital investment that sits idle until the backup period comes, and a long-range supply of alternative energy. In the short run, solar energy can and probably should be integrated into a system that hypothecates an alternative energy supply, such as coal or oil. However, solar energy cannot meet its destined future until it can stand alone, free from any alternative energy supply.

The second major problem imposed by energy storage is that the primary collecting system must be large enough to rebuild the supply of stored energy during the sunny periods between cloudy periods. This additional collecting area means an additional capital investment. If one examines a typical month of sunshine records, as illustrated in Chapter 2, it is apparent that even in the desert the periods of cloudy and clear weather are about equally spaced, a few days of one followed by a few days of the other. Partly cloudy days appear, in analysis, to make the difference between practical and impractical energy storage. If one can extract the total energy of a partly cloudy day, then the periods requiring energy storage are greatly reduced. This, then, is a systems management problem. Concentrating systems must cope with the on-and-off nature of direct sunlight on a cloudy day. This resolves into a problem of absorber/boiler design; the design must be of major complexity to avoid the burnout problem when sun suddenly returns with full brilliance. A nonconcentrating system has the fundamental problem of trying to provide high enough efficiency at its medium temperatures to yield energy output at reasonable cost.

Since energy storage costs must be reasonable, the problem becomes the cost of the storage medium compared to the cost of the material of the storage subsystem. Let us look briefly at the possible storage options.

13.4 HYDROSTORAGE

Hydrostorage is a technique widely used in the power industry to store off-peak power for peak load periods. It simply utilizes a dam that has sufficient hydrostatic head to be a hydroelectric power plant. Water is pumped back into the reservoir in off-peak periods and drawn out during peak periods. The basic requirement is a dam with extensive water at its foot, or two dams with a height difference between them.

Hydrostorage is reasonably efficient. The energy required to pump water uphill is recovered with about 65–75% efficiency because no thermodynamic cycle is involved. Early solar power plants may in fact be built around existing hydrostorage facilities in the Southwest. Adding new facilities presents serious problems in the sunny areas because of water scarcity, but proposals such as the Montezuma Project (Salt River Project and Arizona Public Service Company of Arizona) to use groundwaters, even unpotable saline groundwaters, in the hydrostorage project in the desert Sierra Estrella mountains opens a new source of hydrostorage potential.

Hydrostorage is ideal for solar power storage after production of the power. The solar plant produces power at the maximum rate during the day and is on standby during the night, maintaining only system temperatures so that it is ready to turn out power the next day as soon as the collector subsystem reaches operating temperature (see Section 12.3).

It is staggering to even conceive of storing a significant fraction of the United States' power demand overnight because of the amount of water required for hydrostorage. One kWh of power is equivalent to lifting 1000 kg (one metric ton) of water 367 m, or 2655 lb of water 1000 ft. To lift and regain 1 kWh of electrical energy one would need to store 1150 kg of water at 314 m, consuming 1.3 kWh of input power. For comparison, 1 kWh, which is 3.67×10^{10} g·cm of energy and requires lifting one metric ton 367 m, is also equivalent to heating the same mass of water 0.86 C°.

How much water would have to be pumped into hydrostorage every day if the entire power needs of the United States had to be stored for 12 h of use, based upon the 1975 average demand of 250,000 MW? During 10 h of sunshine, 3×10^{12} Wh of energy would have to be stored, which means lifting 13.5 million metric tons of water 100 m. This is about equal to emptying and filling each day all the reservoirs behind the largest dams in the western United States.

13.5 CHEMICAL BATTERIES

Some chemical changes are readily reversible upon the application of a voltage and thus make chemical batteries possible. Batteries have been the hope of solar energy since experimentation in the late 19th Century, but even then they were not quite ready for immediate application. One reads in turn-of-the-century articles the phrase "cheap and long-lived batteries are just around the corner." Unfortunately that corner has not yet been turned, but they are still claimed to be "almost here."

The batteries that come closest to having the lifetime characteristics and energy storage capacity needed are the sodium or lithium batteries recently under intensive development at a number of laboratories. They must be maintained at a high temperature (several hundred degrees Celsius) to operate, but they can store, in principle, more than 200 Wh/kg, and nickel-cadmium about 120 Wh/kg. In addition to having storage capacity, a battery for utility functions must stand repeated, deep charge-discharge cycles. One thousand such cycles is a major challenge today, yet the capacity for 10,000 cycles would be desirable. A comparison of storage capacity and lifetime cycles for different types of storage batteries is given in Table 13.1.

Table 13.1 Battery Capacities

Type	*Wh/kg*	*Lifetime cycles*
Lead-acid (auto)	33	300- 1000
Lead-acid (commercial)	29	1600
NiCd	24	2000–3000
NiFe (Edison cell)	23	2000
AgZn	185–220	200
AgCd	110–165	500
NaS (250–350°C)	220	?
$LiCl_2$ (600°C)	440	?

13.6 FLYWHEELS

Flywheels have received attention as potential energy storage devices. They have found practical application for storing energy temporarily dumped from the large magnets of synchrotrons, the energy then being withdrawn within a few seconds to re-energize the magnets. One can manage the energy system of this type simply by means of a switch. One polarity uses the dynamo as a motor, speeding up the flywheel; the other uses the dynamo as a generator, drawing energy from the kinetic energy of the flywheel.

The magnitude of energy stored in a flywheel is small. One Wh of energy is equivalent to 1.8 kg of mass on a 2-m-diameter flywheel rotating at 600 rpm. New super materials, such as carbon fiber composites, can stand large centrifugal forces, and at very high rotational speeds they store much more energy than can steel.

The energy stored in a flywheel is given by

$$E = \tfrac{1}{2} I \omega^2, \tag{13.1}$$

where I is the moment of inertia and ω is the angular velocity of rotation. The equivalent way of looking at the energy stored in a flywheel is as the energy in the "spring" formed by the tension created in the rim of the flywheel by the centrifugal force, which slightly expands the diameter of the flywheel.

As in the case of any energy storage means, the potential of the stored energy to do accidental damage is appreciable.

13.7 CHEMICAL STORAGE

Production of a storable chemical from the excess electricity to be stored, or even from the heat before conversion, is a technical possibility. The simplest chemical to be produced to the conventional way of thinking would be hydrogen. Hydrogen can be produced by electrolysis of water or by direct chemical reactions through multistage processes, as proposed by Marchetti and others. This hydrogen then becomes the fuel for extracting the energy at a later date.

Use of hydrogen as an energy storage fuel suggests use of the *fuel cell.* Fuel cells have proven useful in manned space missions and have demonstrated reliability when hydrogen is the fuel. Cells using pipeline natural gas have not had as good a reliability record to date. The reconversion of hydrogen by means of a fuel cell would be a possible teammate for a solar energy system if it were not for its prohibitive cost. In the first place the hydrogen produced from solar electricity would be several times as costly as the hydrogen used today, which is re-formed from hydrocarbons. To burn it in a fuel cell would add the cost of this process to the basic cost of the fuel. So although the technical possibility is not in question, the economics are.

Another possibility, which does not involve as new a technology as fuel cells, is direct burning of the hydrogen. Burning in a thermal cycle means that one must pay the thermodynamic losses a second time. Hydrogen can be burned in a Brayton topping cycle and yield a combined thermodynamic cycle efficiency in the vicinity of

60%, not much worse than that of a fuel cell, but again the capital costs of the reserve system and the basic cost of the solar hydrogen raise cost barriers rather high.

We have explored the possibility of other chemical fuels as the storable fuels to be produced with excess solar electricity and find the field to be full of potential candidates. One that has some interesting possibilities is aluminum. Three electrons can be placed in each aluminum atom, so a lot of energy can be stored in a small mass of aluminum, which is exactly the problem in producing aluminum metal from its ores. The aluminum in granular form could easily be stored in open piles, not needing the cryogenic tanks for liquid hydrogen or tanks for other hydrogen and carbon chemical intermediates. The aluminum could be readily burned in air in a fluidized-grate burner, like powdered coal. The combustion product, aluminum oxide, is also a solid and can be recovered from the stacks with high efficiency. The aluminum oxide can be stored until it is subsequently reprocessed into aluminum. The cycle is therefore close to ideal. The only problem, again, is the economics.

13.8 COMPRESSED AIR

Significant amounts of energy could be stored in the form of compressed air in underground caverns. Early studies indicate costs comparable to those of hydrostorage, but the requirement of a large cavern limits the usefulness of this approach to regions where natural caverns exist, or where caverns can be easily formed, as in salt domes. The air in such a storage facility is compressed by a turbine, which later serves as the generator.

13.9 BIOLOGICAL STORAGE

The storage of energy in chemical form by means of biological processes is an important method of storage for long periods of time. In this book we do not discuss bioconversion, as it is a subject for separate and extensive treatment. The student should be aware, however, that if the quantum efficiency of biological processes can be increased by a factor of 10, bioconversion takes on a much different perspective than when its efficiency is on the order of 1%.

13.10 THERMAL STORAGE

All of the preceding storage processes are applicable to energy after conversion to electrical energy. Thermal storage offers the possibility of storing the energy before conversion. As mentioned above, storing one kWh can be done by lifting one metric ton of water 314 m *or* heating one metric ton of water 0.86 C°. On the other hand, one could choose to heat 10 kg of water 86 C°, an attractive possibility.

Thermal storage can be accomplished in two basic ways: by means of *sensible heat* and by means of *latent heat.* Sensible heat utilizes the specific heat capacity of a material c and a temperature rise ΔT. Latent heat utilizes the heat associated with a phase change of the material, occurring at a constant temperature. We discuss each separately, although a given thermal storage system may utilize both forms.

13.11 SENSIBLE-HEAT STORAGE

Sensible-heat storage involves a material that undergoes no change in phase over the temperature domain to be encountered in the storage process. The basic equation for the amount of heat stored in a mass of material is given by

$$Q/m = c\,\Delta T, \tag{13.2}$$

where c is the specific heat of the material at constant pressure and ΔT is the temperature rise above some minimum temperature for the system. If we have a specific volume for the container for the material, we obtain an equation that has the familiar product ρc:

$$Q/V = \rho c \Delta T, \tag{13.3}$$

where V is the volume of the container. The ability to store sensible heat in a given container therefore depends on the value of the product ρc. Water has the highest value but other materials are within a factor of 2 of water. Some values of materials of potential interest for thermal storage applications are shown in Table 13.2. The material must be inexpensive and have a good heat capacity factor ρc. Water is cheap but, being liquid, must be contained in a better-quality container than a solid. In our home solar heating system we used water as the thermal storage medium for an air-transfer unit, the water being contained in 1000 one-gallon polyethylene bottles stacked so that air could flow between them. They worked satisfactorily until some desert pack rats invaded the storage bin, making nests of the insulation and chewing holes in the water bottles.

Rock is another good thermal storage material from the standpoint of cost, but its capacity is only about half that of water. The rock storage bin used by George Lof in his home air-heating system in Fort Collins, Colorado, is practical. The advantage of rock over water is that it can easily be used for heat storage at above 100°C.

The rate at which heat can be injected and extracted is also important in thermal storage using sensible heat. The ability of a substance to store heat is therefore also a function of the thermal diffusivity $k/\rho c$, where k is the coefficient of thermal conductance. Values of thermal diffusivity are also shown in Table 13.2. Note that by this criterion iron shot is an excellent thermal storage medium, having both high heat capacity and high thermal conductance.

For high-temperature thermal storage, up to several hundred degrees Celsius, iron or iron oxide (red) is as good as water per unit volume of storage. The cost is moderate for either pellets of the oxide (taconite spherules ready for smelting are ideal) or metal

Table 13.2 Heat Storage Capacity

Material	*Density* ρ	*Specific heat* c	*Volume heat capacity* ρc	*Heat conductivity* k	*Thermal diffusivity* $\alpha = k/\rho c$
	g/cm^3	$cal/g\,C°$	$cal/cm^3 C°$	$cal/cm\,C° sec$	cm^2/sec
Water	1.00	1.00	1.00	0.0014	0.0014
Iron (cast)	7.60	0.11	0.84	0.112	0.134
Fe_2O_3	5.20	0.18	0.94	0.0070	0.0074
Granite	2.70	0.19	0.52	0.0065	0.0127
Marble	2.70	0.21	0.57	0.0055	0.0097
Concrete	2.47	0.22	0.54	0.0058	0.0107
Al_2O_3	4.00	0.20	0.80	0.0060	0.0075
Brick	1.70	0.20	0.34	0.0015	0.0044
Dry earth	1.26	0.19	0.24	0.0006	0.0025
Wet earth	1.70	0.50	0.86	0.0060	0.0070

balls. Since iron and its oxide have equal performance, the slow oxidization of the metal in a high-temperature liquid or air system would not degrade performance.

The high heat storage capacity of water makes it doubly attractive as a combined heat transfer and heat storage medium. In Chapter 14 we will discuss this combination under thermodynamic cycle considerations, where the maximum pressure of water in storage sets the system parameters.

The amount and volume of thermal storage material needed to contain a given amount of energy are given by

$$M = Q/c\Delta T$$

and

$$V = Q/\rho c\Delta T. \qquad (13.4)$$

To illustrate the amounts these equations imply, let us take the case of water, heated 100 C° above the minimum temperature. We then have

$$M = 8.62 \text{ kg/kWh}$$

and

$$V = 8.62 \text{ l/kWh}.$$

If a utility system desired to store energy in blocks measured in MWh, the amount of water required to store 1000 MWh_t (thermal) would be

$$M = 8.62 \times 10^3 \text{ m}^3 \text{ of water} = 2.12 \text{ acre-ft.}$$

The volume of 2.12 acre-ft is that contained in a sphere 20.5 m (67 ft) in diameter. If the using system had a net conversion efficiency for heated fluid of 35%, typical of a 500°C working fluid, the tank would be 27.8 m in diameter to store water needed to produce 10^3 MWh_e of electrical energy. The resulting utility-sized storage systems get large, on the order of size of oil storage tanks, but underground thermal storage tanks of this size would not be much out of scale with existing structures found at the average utility.

13.12 LATENT-HEAT STORAGE

The latent heat involved in phase changes is an important potential way of storing heat. There are several types of phase changes that have large enough quantities of heat involved to be useful. The largest phase change is that from water to steam, storing 548 cal/g, or less, depending on the temperature at which the boiling change occurs. Steam, however, is a difficult medium to contain, and attempts to store steam in gigantic pressure vessels have been made and abandoned as impractical and dangerous. In general we want a phase change from solid to liquid rather than from liquid to gas.

Phase changes from solid to liquid require less energy than those from liquid to gas, but some solid-to-liquid changes still provide useful amounts of storage potential. In Table 13.3 we list some materials having high heat of fusion that might be useful.

The temperature at which the phase change occurs is important because it must be compatible with the system temperatures in which the subsystem is to be integrated. It is therefore useful to list heat of fusion also as a function of temperature. As shown in the table, most of the phase-change temperatures are too high for most potential solar energy applications. The only ones that appear of interest are LiOH, NaOH, B_2O_3, KNO_3, and Al_2Cl_6.

An interesting suggestion about how to construct a phase-change thermal storage bin is to encapsulate the phase-change chemicals in "beer cans" suspended by strings throughout the tank. The heat-transfer fluid flows freely through the bin filled with cans, but never mixes with the phase-change materials. The heat conductivity through each can is rapid whether the contents are solid or liquid.

Phase-change materials involving water of hydration have long been experimented with in solar energy. Maria Telkes, one of the pioneers of this approach, discovered that some compounds work well and others fatigue with repeated cycling. The basic problem is that some compounds melt and freeze congruently and other incongruently. Congruent melting means that all phases behave uniformly and there is no chemical separation within the container of chemicals. The observed consequence of incongruent behavior is that a fresh tank of materials absorbs and releases the expected

Table 13.3 Heat of Phase Change

Material	*Phase change*	*Transition point, °C*	*Heat of change, cal/g*
H_2O	Liquid→Gas	100	540
$BeCl_2$	Solid→Liquid	547	310
NaF	Solid→Liquid	992	168
NaCl	Solid→Liquid	803	123
LiOH	Solid→Liquid	462	103
$LiNO_3$	Solid→Liquid	264	88
KCl	Solid→Liquid	776	82
B_2O_3	Solid→Liquid	449	76
Al_2Cl_6	Solid→Liquid	190	63
$FeCl_3$	Solid→Liquid	306	62
NaOH	Solid→Liquid	318	40
H_3PO_2	Solid→Liquid	26	35
KNO_3	Solid→Liquid	337	28
$Na_2SO_4 \cdot 10H_2$ (Glauber's salt)	Solid→Liquid	32	56*
$CaCl_2 \cdot 6H_2O$	Solid→Liquid	30	41*

*These are congruent melting materials, so the value of heat of change depends on the degree of "aging" of the solid-solution mixture.

amount of energy when first cycled. After a few cycles, less and less energy is exchanged and the storage function is degraded until the tank is rested or otherwise reconditioned. Attempts to use additives such as gelatine to inhibit separation of the phases have not solved the problem. Major new research on the use of phase-change salts for use in storing heat or cold for home applications is being done at the Center for Energy Management, University of Pennsylvania, Philadelphia.

13.13 SALT EUTECTICS

Salt eutectics offer the possibility of lowering the temperature of a phase change to below the normal melting points of any of the compounds forming the mixture. The term *eutectic* is generally used to describe a mixture of chemicals having a desired temperature behavior; strictly, however, the term refers to that particular mixture where the melting point is lowest. Some of the lowest melting point eutectics are useful as heat-transfer media and are discussed in Chapter 10. The properties and temperature-mixture relationships for two eutectics are shown below and in Figs. 13.2 and 13.3. Note that the minimum temperature is a sharp function of mixture. The data on heat of fusion of mixtures at their eutectic points are difficult to find in the literature, and we have no values to quote for these two mixtures.

In examining the tables for heat of fusion for different compounds and eutectics it becomes apparent that, with few exceptions, high heats of fusion are associated with

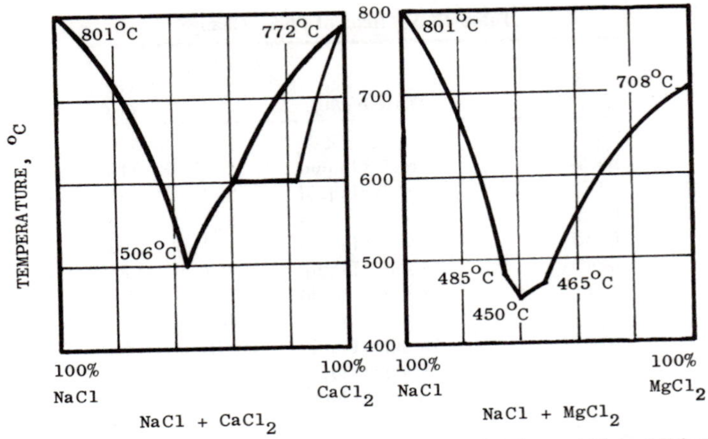

Fig. 13.2 Two eutectic mixtures of common salt with inexpensive additives.

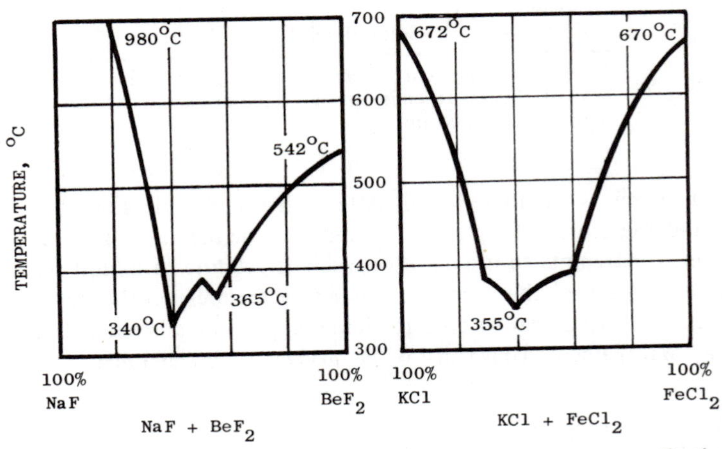

Fig. 13.3 Two high-energy eutectic mixtures for medium-temperature applications.

low atomic weights of the constituents. Most of the materials with high heat of fusion are, unfortunately, expensive materials such as beryllium and lithium compounds. If we are restricted to inexpensive materials of high heat of fusion and low temperature, we find few candidates. Common materials like sodium hydroxide (NaOH) or potassium nitrate (KNO_3) have reasonable heat of fusion but poor chemical handling properties. Aluminum chloride (Al_2Cl_6) eutectic with common salt (NaCl) looks like an attractive candidate except that the aluminum chloride is volatile and sublimes. Encapsulating it would solve the vaporization problem, but it would also add cost.

13.14 ZONED THERMAL STORAGE FLUID TANK

For small thermal storage systems one can utilize a single tank to store both the heated and the cooled fluid. The principle is based on gravity separation of fluids of different density, the hot water being of a lower density than the cooled water. A typical thermal storage tank of this type is shown in Fig. 13.4. The requirement for good performance is that the cooled water being injected into the bottom of the tank not be dynamically mixed with the overlying hot water, indicated by a baffle in the illustration. The second requirement is that the storage cycle time be short enough to prevent the static thermal pulse from the cool water from traveling upward through the hot water.

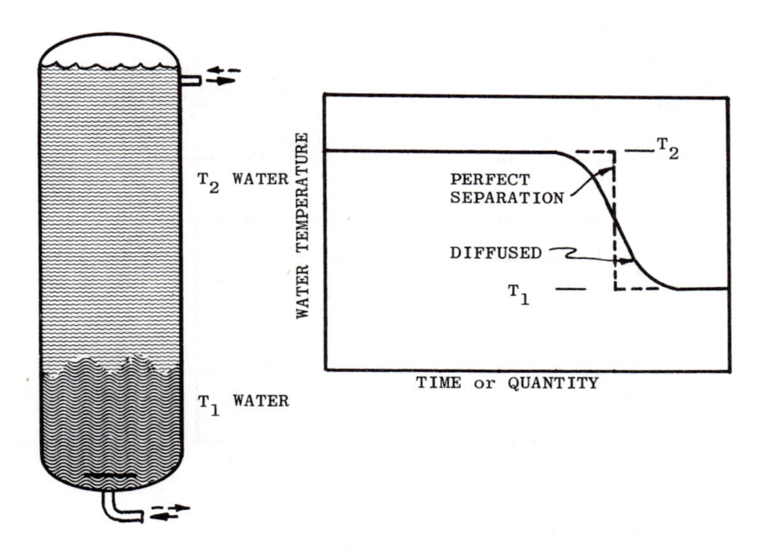

Fig. 13.4 Thermal storage tank for pressurized water using natural gravity separation of the hot and cool water. Direction of flow depends on whether freshly heated water is being added or stored hot water is being removed.

The temperature-time profile for a typical single-tank unit is also shown in Fig. 13.4. If the separation were perfect, the outflowing water would be at a constant temperature T_2 until all of the hot water were withdrawn. In reality the temperature will remain constant for some time, then drop as the interface is approached, as indicated by the curve marked "Diffuse."

One could place a floating barrier between the two quantities of water, but the cost effectiveness of such an addition is open to question.

13.15 ROCK THERMAL STORAGE TANK

To store heat from an air heater a horizontal or vertical storage bin can be used, as gravity separation is not especially important. A horizontal bin is illustrated in Fig. 13.5. The hot inflowing air transfers its heat to the rocks, leaving the tank cooled. When the tank is at full thermal capacity the entire length is at the input temperature. When air is to be heated from storage, the direction of flow is reversed and the incoming cool air draws heat from the rocks, exiting at the storage temperature until most of the heat is withdrawn from storage. A typical time-temperature profile for such a tank is also shown in Fig. 13.5.

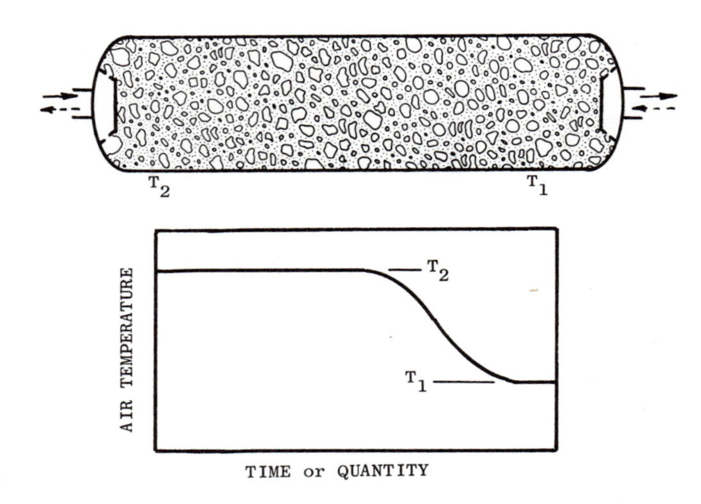

Fig. 13.5 Schematic diagram for a horizontal rock-filled thermal storage bin. For a long bin the output temperature can be held constant at T_2 for more than half of the total energy held in storage. Unlike the rocks in the schematic diagram, the actual materials should *not* vary widely in size, as packing must be avoided in order to ensure free flow of the air through the bin.

In the illustration we show the rock bin as being filled with material of different sizes. This would not be the case for a well-engineered bin. The rocks should be of reasonably uniform size. This prevents the phenomenon of *packing,* which can occur when particles of smaller size fill the gaps between the larger rocks. The ideal rock size is between 1 and 3 in. If the rocks are too small they will impede the airflow and necessitate a higher horsepower fan. If the rocks are too large the time needed to inject and remove the heat becomes long enough to broaden the transition in temperature from T_2 to T_1.

A bin filled with water bottles also can provide a time-temperature profile as illustrated in Fig. 13.5. We have used such a bin for our home air heater, where an exterior adobe-walled bin is filled with one-gallon polyethylene bottles filled with water. The bottles are stacked regularly but spaced so that air can flow readily through them. The fluid filling ensures that heat will flow readily into and out of each bottle. The water-bottle bin allows one to approximately double the heat storage in a given size bin, as indicated by the relative ρc values shown in Table 13.2.

13.16 THERMAL STORAGE TANK FARM

In contrast to the single-tank storage of hot fluid discussed above, the tank farm approach allows perfect separation of the hot and cool fluids. In this case one of the tanks is empty, to permit management of the fluids. As hot fluid is drawn from one tank and processed through the using system, the cooled fluid is pumped into the empty tank. When that tank is filled with cooled fluid, the tank originally holding the hot fluid is empty and ready to receive cooled fluid drawn from a fresh tank.

When only two tanks are provided the use factor is not good, because half of the storage capacity is empty at all times. In such a situation it is better to allow some diffusion and use the single-tank approach described in Section 13.14. If the tank farm has a multiplicity of tanks, n, then the utilization factor approaches unity because

$$U = (n-1)/n. \tag{13.5}$$

In Fig. 13.6 we show schematically the management of a thermal storage tank farm. The light-colored waves represent hot water and the dark ones cooled water. The tank farm approach provides for both good management of a large utility-type solar energy system and the expansion of the storage facility with time, as costs permit and as the availability of the normal backup heat source diminishes.

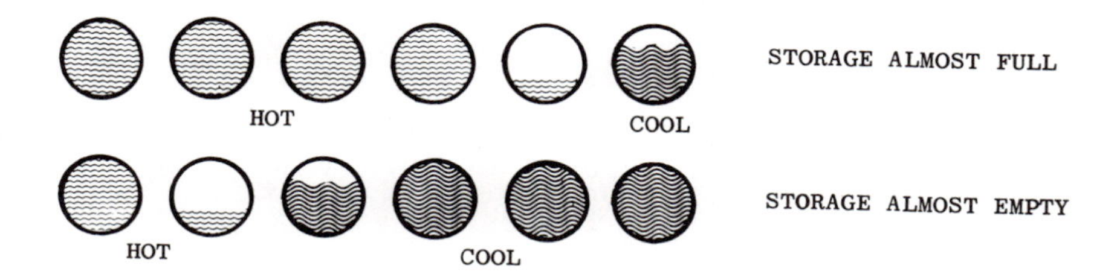

Fig. 13.6 Management of a thermal storage tank farm requires one tank to be empty (equivalent) to permit perfect separation of the hot and cool water in storage. The tank farm approach allows for expansion of the facility with time, thus reducing backup demands.

13.17 HEAT MANAGEMENT WITH AND WITHOUT PHASE CHANGE

The injection and extraction of heat from thermal storage presents some problems not immediately obvious. These difficulties flow from the fact that heat is transferred only from a warmer body to a cooler one. A phase-change material, while operating at a constant temperature for heat injection and heat release, requires (1) that the fluid supplying the heat be at a higher temperature and (2) that the fluid receiving the heat be at a lower temperature than the phase change. A corollary statement is that the heat-transfer fluid in a solar collector supplying heat to a phase-change storage system must operate at all points in the collector *above* the phase-change temperature. For sensible-heat storage these points are not required. The heat-transfer fluid can enter the collectors at less than the system operating temperature.

To illustrate the problems that arise when one wishes to use thermal storage, we present several cases. The first set of factors determining the characteristics of the exchange is the heating profile of the heat-transfer fluid. There are basically three profiles, as illustrated in Fig. 13.7. The simplest system is case (a), in which a fluid without phase change in the temperature range of use rises from some inlet temperature to a final outlet temperature linearly with regard to heat input Q. This type of heat-transfer fluid is to be preferred for solar collectors mainly because there are no changes of volume, density, gas flow rate, and so forth, which complicate the phase-change situation. In case (b) the heat-transfer fluid changes to a gas and exits at the boiling temperature of the fluid. In case (c) some superheat is added to the heat-transfer fluid gas. In each case we denote the same temperature rise and output temperature.

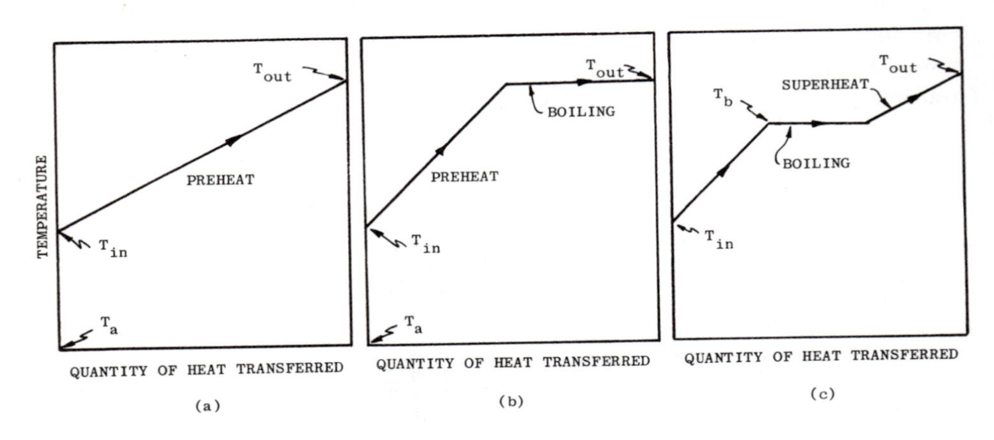

Fig. 13.7 Schematic representation of the three basic heating cycles inside a solar collector. The simplest arrangement is (a), where the heat-transfer fluid never reaches boiling point. In arrangement (b), the output is vapor at essentially the boiling temperature, and in (c) the vapor is superheated above the boiling point. These curves are for the case where there is a significant temperature rise from inlet temperature to outlet temperature.

An alternative case, in which the heat-transfer fluid has a phase change, is shown in Fig. 13.8. This type of change is better adapted to heat storage with a phase-change material, but the small increase in temperature raises a problem of efficiency. This problem is shown in Fig. 13.9, where the typical curve of efficiency of extraction η is plotted versus fluid temperature. Two cases are shown having the same outlet temperature. In one case the increase of temperature in the collector is small, a rise of ΔT_2. The average efficiency is $\bar{\eta}_2$. If, on the other hand, a larger temperature rise is permitted by the system, where the inlet temperature is lower, the average efficiency $\bar{\eta}_1$ is considerably higher. We therefore would like to have the collector operate over as wide a value of ΔT as possible for best use of the collector.

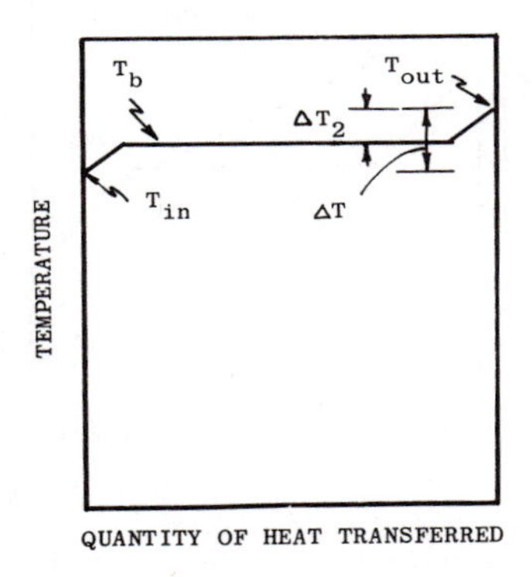

Fig. 13.8 Schematic representation of the case where most of the energy transfer into the fluid is by means of phase change, and where a small temperature increase (ΔT_2) occurs in the collector.

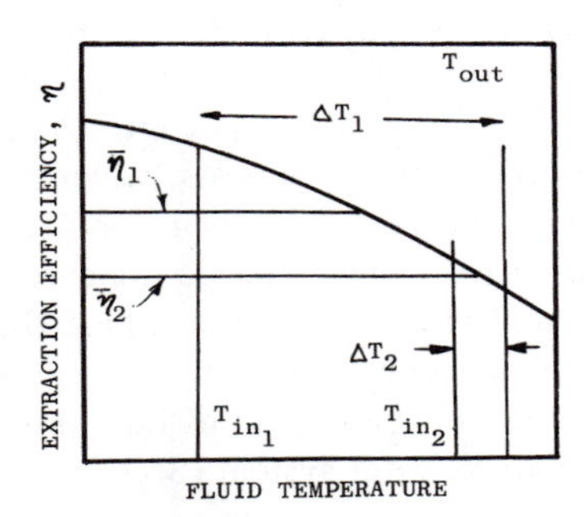

Fig. 13.9 Schematic representation of the significance of a small or a large temperature increase in the collector. For a given solar input Q, the efficiency of conversion into a hot working fluid is higher for the wider temperature range ΔT_1. Average values for the efficiency are denoted by $\bar{\eta}$.

When we examine the influence of a thermal storage material having a phase change, we encounter a situation where the collector ΔT must be small and therefore less efficient. In Fig. 13.10 we show three cases. In case (a) a heat-transfer fluid having no phase change is transferred to the phase-change storage material. The transfer fluid quickly cools to a point close to the phase-change temperature and ΔT is so low that transfer essentially ceases. A new stream of fluid is then required to further heat the storage material. As a consequence there is only a small drop in temperature in the collector heat-transfer fluid, resulting in a poor value for η.

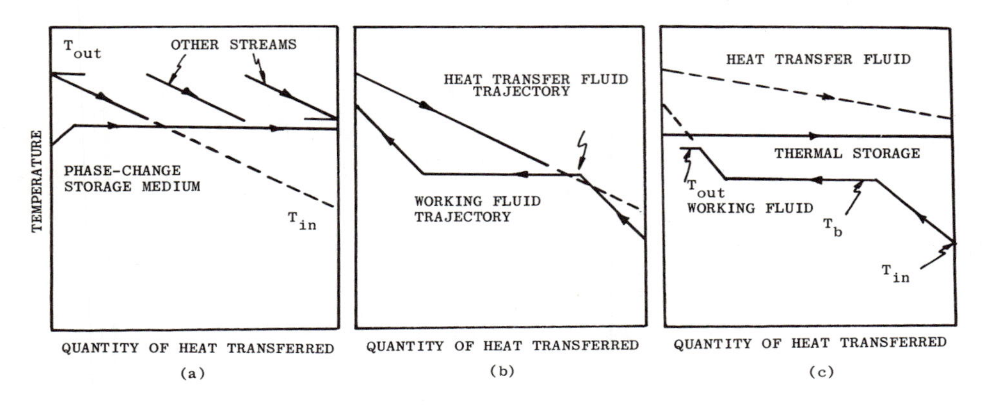

Fig. 13.10 Three illustrative trajectories for flow of energy from heat-transfer fluid to thermal storage medium and to the working fluid to show problems encountered because of the requirement that heat must flow from a higher to a lower temperature.

The problem is different when the heat-transfer fluid both has no phase change and acts as its own thermal storage medium. In case (b) we show the fluid coming from storage and heating the working fluid of the turbine. The turbine fluid must have a phase change because we wish to send a vapor into the turbine. We can imagine a counterflow heat exhanger-boiler for the process, where fluid enters hot at one end and vaporizes the other fluid, the two paths being indicated by arrows in case (b). Since the turbine fluid must have a phase change, its curve has the typical knee and horizontal portions as shown. There is a situation where the heat-transfer fluid cools too fast and the knee crosses the temperature curve, as shown. This situation chokes the heat-transfer process. A further illustration of how this is avoided is shown when we discuss a medium-temperature water system (Section 14.10). Note, however, that barring the choke problem, we can operate the turbine at almost the temperature of the fluid arriving from the solar collector, the ΔT value here being small and set by the difference required in the heat exchanger. Thus a wide operating temperature difference is possible, its advantage being in regard to efficiency η.

In case (c) we show the problem posed by a phase-change thermal storage medium. The phase-change temperature must be considerably lower than the outlet temperature from the solar collector so that enough heat can be transferred before the fluid temperature drops to where transfer ceases, following the trajectory shown by the dashed line. Now, when we want to use the phase-change material to heat the working vapor for the turbine, the turbine fluid must always be at a lower temperature than the thermal storage medium, so the entire temperature range for the turbine working fluid is depressed relative to what it could be if a sensible-heat thermal storage medium were used.

The above arguments indicate that the apparent advantages of having a phase-change material for the thermal storage medium are not realized. For this reason we are inclined to the type of system we discuss in Section 14.10, in which pressurized water is used for both the heat transfer and the thermal storage medium, this heat eventually being transferred to an organic working vapor for the turbine.

13.18 THERMAL INERTIA

Page (1975) has stressed that, in thermal storage by means of sensible heat, the thermal inertia of a building is an important contributor, with the potential for maximization. Whereas most buildings have their mass on the exterior and insulation on the interior, one could greatly increase the usefulness of the mass if the reverse could be done: insulation on the outside and mass on the inside. The storage capacity thus obtained could be larger than could be economically added by means of, for example, a rock storage bin.

If massive structures are involved in augmenting thermal storage, the velocity of adding or extracting the heat is important. The velocity of propagation of a heat pulse is given by

$$v = 2(\pi\alpha/P)^{1/2} = 2(\pi k/\rho c P)^{1/2}, \tag{13.6}$$

where α is the thermal diffusivity $k/\rho c$ and P is the period of the impulse. As applied to solar thermal storage effects, the period is 24 h. The velocity for a 24-h wave in several materials is given in Table 13.4.

Other properties of a wall in regard to a periodic heat source are the amplitude of the wave inside the wall, the lag in temperature at a particular depth, the wavelength of the thermal impulse in the wall, and the amount of heat transferred in a wave. These quantities are

$$\text{Amplitude:} \quad T_r = 2\Delta T_0 \exp[-x(\pi/\alpha P)^{1/2}], \tag{13.7}$$

where x is the depth below the surface of the wall and ΔT_0 is the amplitude of the

Table 13.4 Velocity of a 24-h Wave

Material	*Velocity, cm/h*
Taconite	10.
Granite	6.5
Concrete	3.3–3.7
Brick	2.6–2.8
Timber	1.7
Copper	46.2

temperature wave at the surface, the total range of the temperature disturbance being $2T_0$,

$$\text{Lag in temperature:} \quad t = (x/2)(P/\pi\alpha)^{1/2} \tag{13.8}$$

$$\text{Wavelength:} \quad \lambda = 2(\pi\alpha P)^{1/2} \tag{13.9}$$

$$\text{Amount of heat:} \quad Q/A = k\Delta T_0(2P/\pi\alpha)^{1/2}, \tag{13.10}$$

where heat flows into the wall during one half of the cycle and out of the wall for the other half of the cycle.

Heat flow into and out of rocks in a storage bin is a more complicated problem in heat conduction, but the equations given by Ingersoll (1948) are useful. The mathematical model is a sphere surrounded by a medium at a given temperature T_S. We are interested in the temperature at point r inside the sphere, its central temperature as a function of time, and the average temperature of the sphere. The equations are

Temperature at r:

$$\frac{T-T_s}{T_0-T_s} = \frac{2R}{\pi r}\left[\sin\left(\frac{\pi r}{R}\right)\exp\left(\frac{-\pi^2\alpha t}{R^2}\right) - \tfrac{1}{2}\sin\left(\frac{2\pi r}{R}\right)\exp\left(\frac{-4\pi^2\alpha t}{R^2}\right) + \ldots\right], \tag{13.11}$$

which simplifies for the central temperature to

$$\frac{T_C-T_s}{T_0-T_s} = 2\left[\exp\left(\frac{-\pi^2\alpha t}{R^2}\right) - \exp\left(\frac{-4\pi^2\alpha t}{R^2}\right) + \ldots\right]. \tag{13.12}$$

For the average temperature we have

$$\frac{T_a-T_s}{T_0-T_s} = \frac{6}{\pi^2}\left[\exp\left(\frac{-\pi^2\alpha t}{R^2}\right) + \tfrac{1}{4}\exp\left(\frac{-4\pi^2\alpha t}{R^2}\right) + \ldots\right]. \tag{13.13}$$

13.19 CALCULATION OF DETAILED PERFORMANCE

Calculation of the detailed performance of a solar collector and the status of energy in storage is not complicated when first-order accuracy is acceptable. The need for this type of calculation is evident when we examine the sunshine availability and also the demand for energy during the heating or cooling season. A hypothetical demand for winter heating on a 10-day average is shown in Fig. 13.11. If the solar collector is sized so that only a portion of the maximum winter demand is met, then the system will be working at design capability only for a limited time. Earlier and later in the season the system will collect more energy than is needed; it is not proper to count this unusable energy when evaluating the economics of the collector.

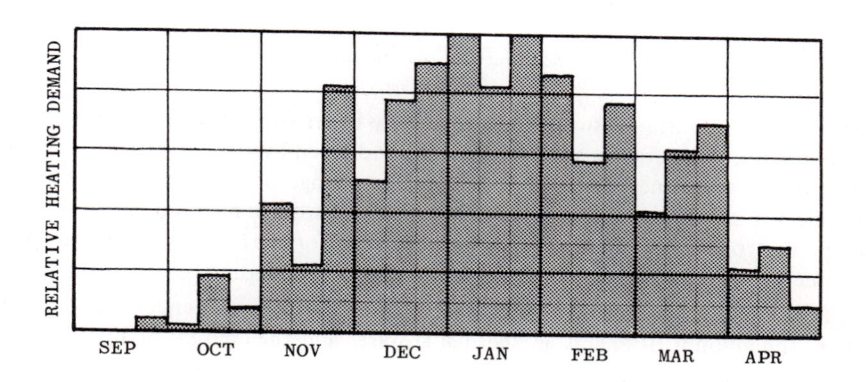

Fig. 13.11 Annual heating demand curve for a typical year normalized to unity for the maximum demand.

For better evaluation of actual situations one would like to have daily averages of the solar energy availability because weather cycle effects on energy in storage are apt to be distorted since weather cycles in records we have examined are shorter than 10 days. The solar energy available is further modified by the characteristics of the collector. A thermal collector is not a linear transducer of solar flux, as is the solar cell. If the curve of efficiency versus temperature difference, η versus ΔT, has been determined for any given value of the solar flux, such as 300 Btu/ft^{2}h, the efficiency at any other flux value can be readily determined. If the collector output temperature is maintained, the heat losses will remain the same; hence, reduced flux is directly translated into a reduced output. This fact was used in Section 12.23 to derive the procedure to be used for evaluating the detailed performance of a solar collector. We will use this derivation now to calculate the behavior of a solar collector when combined with a thermal storage subsystem under conditions of changing solar flux.

To calculate the detailed performance, one begins with the particular η versus ΔT curve determined for the collector under study, obtained by performing the energy balance calculations described in Sections 12.5 and 12.23. This curve can then be shifted laterally so that the ordinate at ΔT is proportional to the actual solar flux. This quantity is to be denoted by $Q_{\text{available}}$, and is indicated by Q' in Fig. 12.30.

Since the amount of heat demanded, denoted by Q_{demand}, is also known from the record under study, the amount of solar heat available to be stored is

$$\Delta Q_{\text{storage}} = Q_{\text{available}} - Q_{\text{demand}}. \tag{13.14}$$

The energy in storage is then the running sum

$$Q_{\text{storage}} = \sum \Delta Q_{\text{storage}}. \tag{13.15}$$

This simple relationship assumes that the energy losses from storage are not a function of the amount of energy in storage, generally indicated by the temperature of storage. This procedure also neglects effects such as the change of temperature of fluid from the collector as compared to the temperature of the storage medium, but it is accurate enough that one can evaluate system performance for the determination of approximate energy value and costs. These will be examined in Chapter 16.

In Table 13.5 we show a sample calculation of the solar energy collected by a system, the amount of energy in thermal storage, and the amount of auxiliary energy required. The input data are for a fictitious month of December, where the solar flux data are taken from the record for Tucson, Arizona, as representative of the sequence of clear and partly cloudy days. The ambient temperatures are a typical set for the amount of cloudiness and the presence of cold and warm air masses associated with frontal systems. The demand for heat is made proportional to the ambient temperature. If this calculation were being done for a real dwelling, we would need to know the actual demand for the dwelling. The efficiency of the collector is taken from a typical graph, as in Fig. 12.30, of η versus ΔT, where $\eta = 0.80$ at $\Delta T = 0$ and $\eta = 0$ at $\Delta T = 180$ C°, typical of a flat-plate collector having two cover windows and a selective surface with emittance of 0.10. This curve is then shifted vertically so that the value of η at $\Delta T = 0$ is proportional to the actual flux for the day. The appropriate value for η' is then read off this shifted curve for the actual ΔT value between the ambient temperature for that day and the system output temperature of 70°C. The value of the collector area to form the product $\eta' A$ is taken as 2.0 in the arbitrary units used for this illustration. This value gives us an interesting case where auxiliary energy is required, and happens to give a reasonable division between solar and auxiliary energy. The student can readily see from this case what happens when the collector area is increased or decreased over the value used herein.

Table 13.5 Sample Calculation for Solar Energy Collected and Stored and Auxiliary Energy

	Date (December)																	
	10	*11*	*12*	*13*	*14*	*15*	*16*	*17*	*18*	*19*	*20*	*21*	*22*	*23*	*24*	*25*	*26*	
Fractional solar I'/I	98	100	100	45	97	71	99	69	42	58	39	73	66	98	74	85	37	M
Ambient temp. T_a, °C	11	12	15	10	2	5	8	10	7	3	9	6	2	12	10	11	5	M
Fluid temp. T, °C	70	70	70	70	70	70	70	70	70	70	70	70	70	70	70	70	70	S
Temp. rise ΔT, C°	59	58	55	60	68	65	62	60	63	67	61	64	68	58	60	59	65	C
Efficiency η'	0.48	0.48	0.51	0.15	0.44	0.30	0.48	0.29	0.11	0.22	0.10	0.32	0.25	0.47	0.33	0.39	0.08	C
$Q_{available}$ η'–A	96	96	102	30	88	60	96	58	22	44	20	64	50	94	66	78	16	C,S
Q_{demand}	75	70	60	80	95	90	82	80	85	95	81	88	96	74	81	77	90	M
ΔQ to storage	+21	+26	+42	−50	−7	−30	+6	−22	−63	−51	−61	−24	−46	+20	−15	+1	−74	C
Energy stored	21	47	89	39	32	2	8	0	0	0	0	0	0	20	5	6	0	C
Auxiliary demand	0	0	0	0	0	0	0	14	63	51	61	24	46	0	0	0	68	C
Accum. solar	96	192	294	324	412	472	568	626	648	672	712	775	826	920	985	1064	1080	
Accum. auxiliary	0	0	0	0	0	0	0	14	77	128	189	213	259	259	259	259	327	

Fraction of demand from solar = 76.8%

Fraction of demand from auxiliary = 23.2%

M = measured quantities
S = specified quantities
C = calculated quantities

13.20 PROBLEMS

13.1. A solar heating system uses a 10,000-l water tank for thermal storage. The maximum temperature of fluid from the collector is 90°C. The minimum temperature of water to the heat exchanger in the house is 35°C. How much heat energy can be stored in (a) kJ and (b) Btu? If the value of the heat is $4.00/MBtu, what is the value of the heat in storage?

13.2. If the house has a heat demand of 40,000 Btu/h, how many hours will the tank in Problem 13.1 last?

13.3. If the monthly demand for heat for a particular year is as shown in the curve in Fig. 13.11, and the solar collector can provide 60% of the maximum winter demand, make a graph of the dependence on a backup energy supply. Is your graph representative of peak demand during the winter? If not, what additional information would be required to evaluate peak demand? Assume that a maximum of two days energy is contained in storage.

13.4. A location has weather that on a statistical basis has cloudy periods averaging three days in length and subsequent clear periods of five days. If 100% of the energy demand is desired in these average periods, how much additional solar collector area will be required to meet demand on a clear day in order for the thermal storage tank to be up to full temperature when the next cloudy period arrives?

13.5. What are the dimensions of a rock bin filled with granite stones occupying 70% of the volume in order to have the same thermal storage capacity as in Problem 13.1? Assume that the length is twice the lateral dimensions, which are equal.

13.6. If a 5000-l bin is filled with calcium chloride hexahydrate and the upper limit of temperature is 50°C, what would be the heat storage capacity if the bin is 60% filled by the chemical when the operating range is between 50 and 29°C?

13.7. A typical gas absorption cooling system can operate with a water temperature above 80°C. The maximum water storage temperature is 90°C. How many kJ and Btu can usefully be stored in an 80,000-l water tank? If the coefficient of performance (COP) of the system is 0.55, how much cooling can be produced?

13.8. Discuss the tradeoffs between basement water storage tanks for a heating and gas-absorption cooling system where the energy is stored solely as hot water and where a cold water tank is used in summer.

13.9. If the water tanks for a gas absorption cooling system can operate at 120°C, how much more thermal energy can be stored than in Problem 13.7? Would this allow more money to be spent on collectors that would provide this temperature? What factors would tend to increase the cost of such collectors?

13.10. Approximately how much of an increase could be justified on the collector of Problem 13.9 before a breakeven point is reached?

13.11. What mixtures of salt and magnesium chloride would be necessary to make a thermal storage system for high-temperature operation at 600°C? Which mixture would be preferable and why?

13.12. What would be an acceptable operating temperature for a system to use the storage tank with the mixture in Problem 13.11? What factors need to be taken into account in arriving at this answer?

13.13. Calculate the amount of heat that can be injected into a rock wall of a building during a day when the effective temperature difference between the airflow from a solar collector and the wall is 30 C°. Define the factors that determine the wall surface temperature and estimate what the wall surface temperature would be. If the wall surface temperature were 30°C, what would the amount of heat be?

13.14. Recalculate Table 13.5 for the case where the area factor is increased from 2.0 to 2.2.

13.15. Recalculate Table 13.5 for the case where the area factor is reduced from 2.0 to 1.8.

13.16. Calculate the effectiveness of the total collector area in supplying energy in the case of Table 13.5 and Problems 13.14 and 13.15. Why does the effectiveness of the collector diminish as the fraction of solar heat is increased?

Part 4
Utilization and Applications

Chapter 14
Thermodynamic Utilization Cycles

14.0 INTRODUCTION

In this chapter we discuss conversion of solar heat into work by means of three thermodynamic cycles: the *Rankine cycle,* the *Brayton cycle,* and the *Stirling cycle.* The Rankine cycle is familiar as the cycle on which most steam turbines operate, and the Brayton cycle is familiar as the cycle of the aircraft jet engine. The fundamental difference between the two is that the Rankine cycle accomplishes the *compression* stage with the working medium in the liquid state, the heat causing the fluid to go through a phase change before or as it enters the expansion stage of a turbine. The Brayton cycle, on the other hand, always works with a gaseous medium, compressing it, adding heat, and expanding it through the turbine. The Stirling cycle is not widely used, but it is widely discussed as a possibility for solar applications because it is basically efficient and uses an external heat source.

In the course of our discussion we use two types of diagram: (1) the temperature-entropy (*T-S*) diagram, and (2) the enthalpy-entropy (*Q-S*) diagram, also called the Mollier diagram.

14.1 RANKINE CYCLE

The entropy and enthalpy curves for a water-steam Rankine cycle are shown schematically in Fig. 14.1. In the *T-S* diagram the path of the cycle is shown by numbers 1 through 7. Path 1 is heating the water at constant pressure to the boiling point, where it intersects the equation of state for water. Path 2 is vaporizing water at constant pressure and temperature. The saturated steam is then superheated beyond the equilibrium condition, along path 3, becoming superheated to the maximum temperature of the system T_2. The steam is next expanded adiabatically along path 4 until it reaches saturation, whereupon it is reheated to a superheated condition along path 5 to T_2, but now at a lower pressure than along path 3. In the final turbine stage the steam is expanded until it has some degree of saturation, dropping below the saturated steam line. The spent steam is then condensed at constant temperature along path 7, at T_1. The equivalent Carnot cycle is indicated by the dotted line, being simply a rectangle with its top at T_2 and its bottom at T_1.

The *Q-S* diagram, Fig. 14.1(b), generally known as the *Mollier diagram,* is more commonly used to describe and evaluate a Rankine system performance than is the

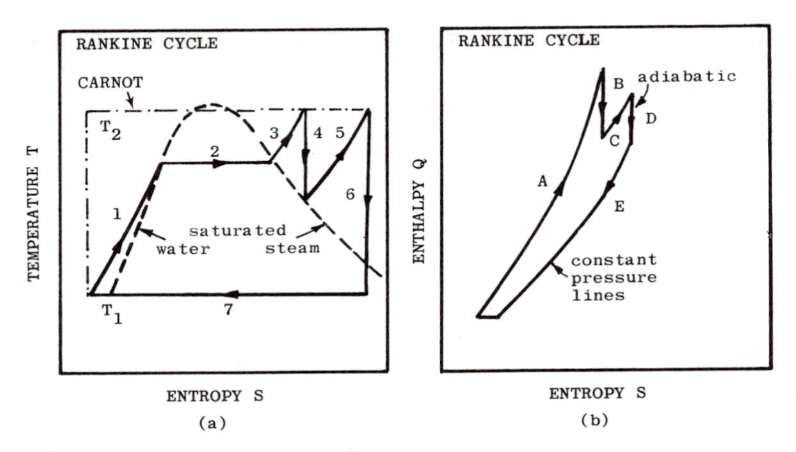

Fig. 14.1 (a) Temperature-entropy (*T-S*) diagram for water-steam Rankine cycle, showing a double superheating cycle. The equivalent Carnot cycle is shown by the dotted line. (b) Enthalpy-entropy (*Q-S*) diagram (Mollier diagram) for water-steam Rankine cycle. Cycle efficiency is the ratio of the *Q*-values for paths *B*+*D* divided by paths *A*+*C*.

T-S diagram. The useful work is obtained from the Mollier diagram by summing the ΔQ-values for the expansions *A* and *B*. The work put into heating the working vapor is the increase in *Q* from its starting point to its maximum value. The Rankine cycle when plotted in the *Q-S* diagram does not show the boiling plateau, but since the quantity of heat is plotted as enthalpy the curve changes slope only slightly at the boiling point.

In the *Q-S* diagram the input energy is the energy added to the water-steam fluid along path *A* plus the additional energy added along path *C.* The useful energy output is that along paths *B* and *D.* The unavailable energy is that along path *D.* The efficiency of the Rankine cycle is therefore

$$\eta_R = (B + D)/(A + C). \tag{14.1}$$

14.2 BRAYTON CYCLE

The *T-S* and *Q-S* diagrams for a typical Brayton cycle, shown in Fig. 14.2, are rather similar in shape because no boiling plateau is present for a fluid that is gaseous at all times. The path in the *T-S* diagram is also simple, the adiabatic compression along path 1 being followed by a heating at constant pressure along path 2. The expansion stage is also adiabatic along path 3. The heat rejection path, 4, is generally in free air and should be a dotted rather than solid line because paths 1 and 3 are not physically connected. In a closed-cycle Brayton system the gas does go through a cooling stage, and path 4 has physical reality.

In the *Q-S* diagram the input energy is the sum of the increases in *Q* for paths *A* and *B*. The useful energy output is path *C*. The unavailable energy is path *D*. The input energy is represented by the work done in compressing the gas in *A* and the energy added to heat the gas in *B*. The efficiency of the Brayton cycle is therefore

$$\eta_B = C/(A + B). \tag{14.2}$$

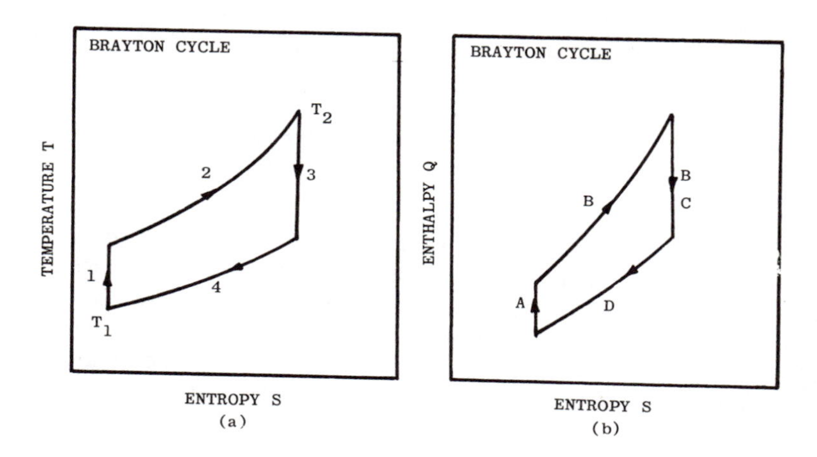

Fig. 14.2 (a) Temperature-entropy diagram for a gas turbine Brayton cycle. In an open cycle, as in a jet engine, path 4 is not physically connected to path 1. (b) Enthalpy-entropy diagram (Mollier diagram) for a gas turbine Brayton cycle. Cycle efficiency is the ratio of the *Q*-values for path *C* divided by paths *A*+*B*. Path *D* represents unavailable energy.

14.3 STIRLING CYCLE

The Stirling cycle is one of the several hot-air cycles. Its basic characteristic is that it uses a constant mass of gas in a totally sealed system, alternately adding and removing heat from this volume of air. The *T-S* diagram for the Stirling cycle is shown in Fig. 14.3. The modern forms of this cycle have two moving parts per cylinder, a working piston attached to the crankshaft and a transfer piston that alternately allows the gas to be exposed to the hot surface and the cold surface. The Stirling cycle tends to approach the Carnot cycle better than other cycles, as can be seen in Fig. 14.3. The *T-S* diagram has vertical paths in the constant-volume portion that are more nearly vertical than the constant-pressure paths of the Brayton cycle, and hence approach the rectangular *T-S* plot of the Carnot cycle.

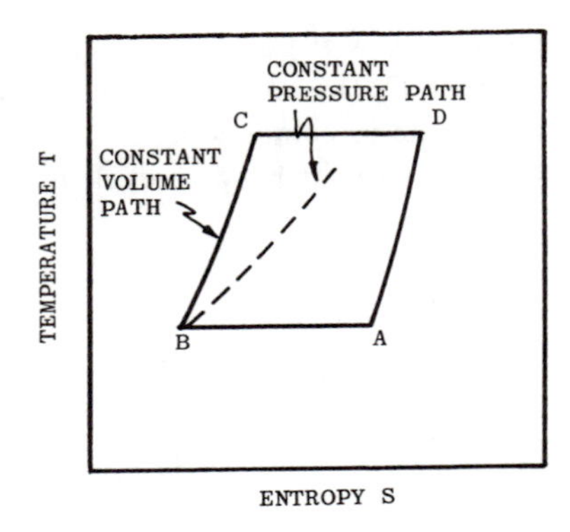

Fig. 14.3 Typical *T-S* diagram for a Stirling cycle.

Although simple in principle, the mechanical embodiment of the Stirling cycle is complicated. Two factors are especially troublesome: (1) the relative motions of the two pistons and (2) sealing the gas charge. Philips-Eindhoven and Ford-USA have made experimental engines. The Philips engine is shown schematically in Fig. 14.4. The gas is compressed in the low-temperature portion of the cylinder, whereupon the transfer piston moves downward, allowing the gas to come into contact with the hot dome of

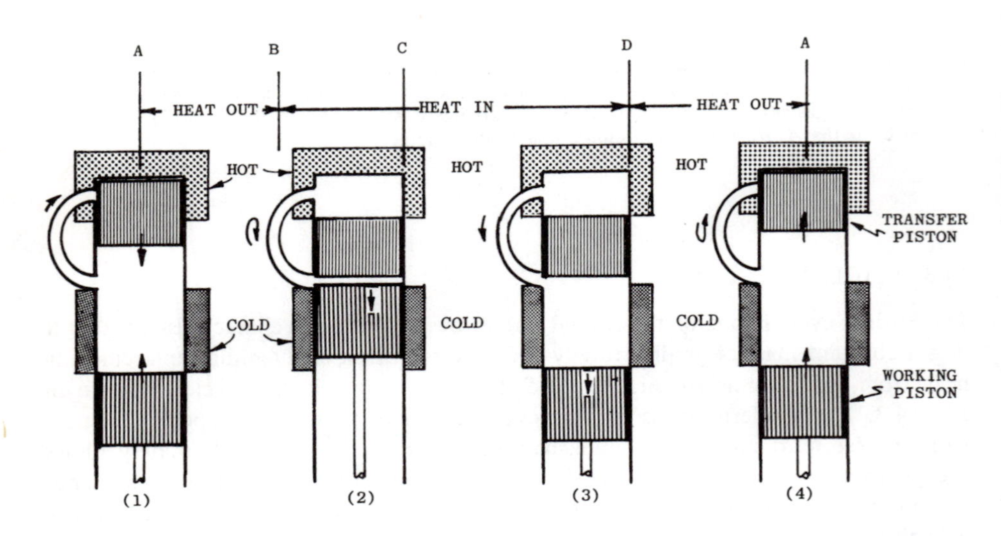

Fig. 14.4 Schematic Stirling cycle. Step 1 is charging the hot zone; step 2 is the transfer of heat at constant volume; step 3 is the transfer of heat at constant temperature; and step 4 is the rejection of heat to the cold sink. A regenerator in the transfer loop to conserve heat is not shown.

the cylinder. The gas then expands between (2) and (3), driving the working piston downward. The floating transfer piston is then replaced against the hot surface (4) and the gas cools, causing the working piston to rise.

The gas in the engine can be varied from air. The Ford system uses hydrogen for its high thermal conductivity. Helium has also been used, but when gases other than air are used the sealing problem is accentuated.

14.4 ERICSSON CYCLE

The Ericsson hot-air cycle may have some applicability as a means of utilizing solar energy for mechanical applications. The cycle is similar to the Stirling cycle in that heat is added at constant volume. A form of this cycle is shown in Fig. 14.5. The charge of air is compressed by a cylinder (not necessarily the same cylinder as will extract the work) and injected into a separate chamber where heat is added to the air. The heated air is then allowed to enter the working chamber and expand at somewhere between a constant temperature and an adiabatic curve, depending on the rate of transfer to the gas as it works on the cylinder. The expanded gas is then discharged through a recuperator to add some of its heat to the next charge. In a sense this cycle is like a Brayton cycle, except that the charge is heated at constant volume in the Ericsson cycle and at constant pressure in the Brayton cycle.

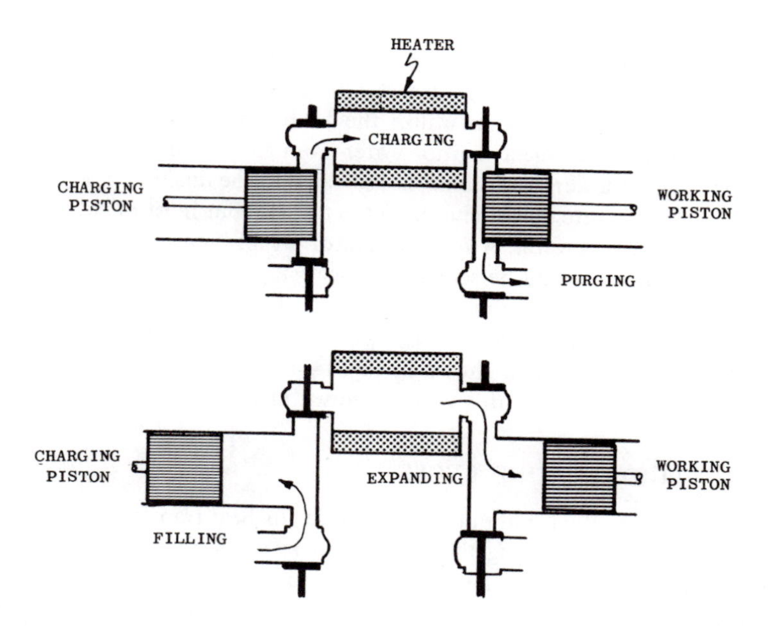

Fig. 14.5 Schematic diagram of an Ericsson hot-air engine cycle. Note that valves are required to contain the charge during the heating at constant volume phase.

The basic factor limiting use of these hot-air engines in solar energy development is the fact that they do not exist in the commercial market. People are therefore justifiably reluctant to combine an untried engine with a novel heat source, the sun, feeling that a double innovation is at least twice as likely to develop unanticipated troubles. It is hoped that the renewed surge of interest in solar energy will see some development of potential new types of heat cycle engines that are better suited to solar heat input than the standard Rankine cycle.

14.5 HOT-WATER RANKINE CYCLES

Adapting a solar energy collecting system to a Rankine cycle steam turbine presents a number of problems. The fundamental one is how to handle boiling directly in the solar collector. One can design a suitable boiler, but the wildly fluctuating energy inputs on a partly cloudy day lead to the threat of *burnout* of the boiler. One can, however, utilize a heat-transfer fluid that does not change phase, such as sodium or molten salts, in which case the transfer to water and subsequent utilization is identical to that for any other type of heat source used to drive turbines. In the following sections we examine the performance of a Rankine cycle using hot water as the input working fluid.

14.6 DIRECT INJECTION OF WATER

There is a distinct advantage to keeping the heat-transfer fluid from changing phase in the collectors, but when the heat-transfer fluid is also the working fluid for the turbines one is faced with how to utilize the heated working fluid effectively. The simplest way would be to inject the hot water from the thermal storage tank directly into the first stage of a suitable turbine, as indicated in the diagram in Fig. 14.6. Some of the water will turn into steam and drive the turbine, but most of the fluid will exit from the turbine stage as water. A water separator would then separate the water from the spent steam, which would be sent to a reheat stage to be brought back up to the original inlet temperature (less the exchanger ΔT). The reheated steam would then go to the second stage of the turbine, a low-pressure stage. The Mollier diagram for a hot-water turbine with reheat is shown in Fig. 14.7. A representative set of operating temperatures and pressures for this cycle is shown in Fig. 14.8.

14.7 DUAL-FLUID RANKINE CYCLE

In a system using one fluid without phase change for heat transfer and a second fluid with phase change for operating the turbine, there are a wide variety of choices of fluids but always the same problems. As an example of such a dual-fluid system, we show in Fig. 14.9 the system proposed by Rotoflow Corporation of Los Angeles for the San Diego Light and Power Company for a 10-MWe geothermal application. In this case hot water is used to vaporize pressurized isobutane, the water being at 205°C

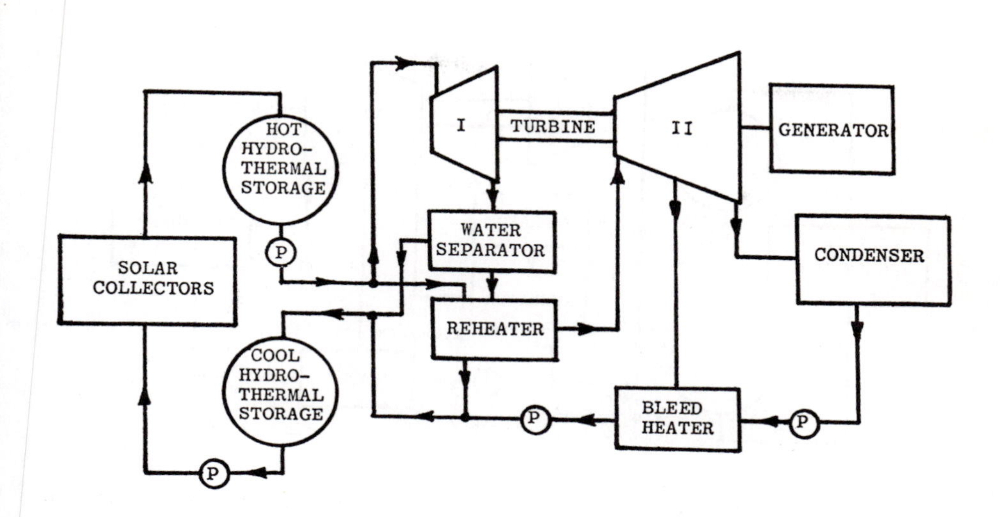

Fig. 14.6 Schematic diagram of a thermodynamic cycle where hot water is injected directly into the first stage of a two-stage turbine, for use where a common heat-transfer fluid and working fluid are used.

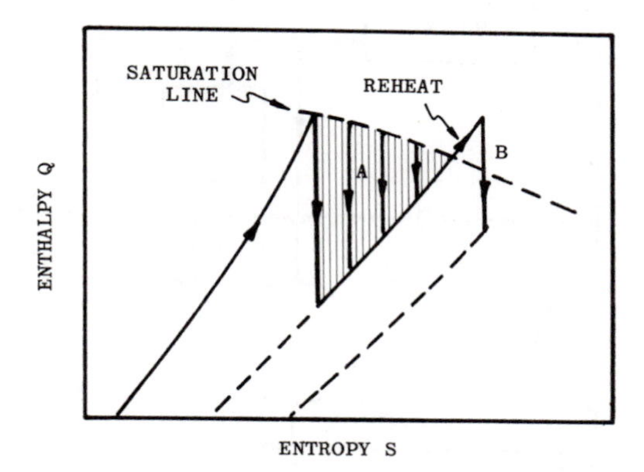

Fig. 14.7 Mollier diagram for a two-stage turbine cycle starting with direct injection of water into the first stage. The steam from stage *A* is separated from the water, reheated at 15 psi, and expanded in stage *B*.

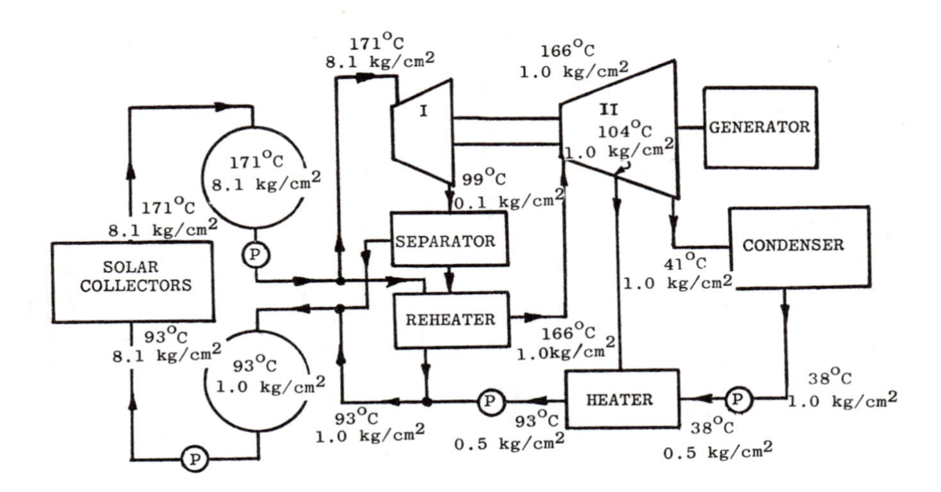

Fig. 14.8 Representative set of operating temperatures and pressures for a two-stage turbine using water injected directly into the first stage. The spent steam at 15 psi is separated from the water issuing from the first stage and is reheated before injection into the low-pressure stage of the turbine.

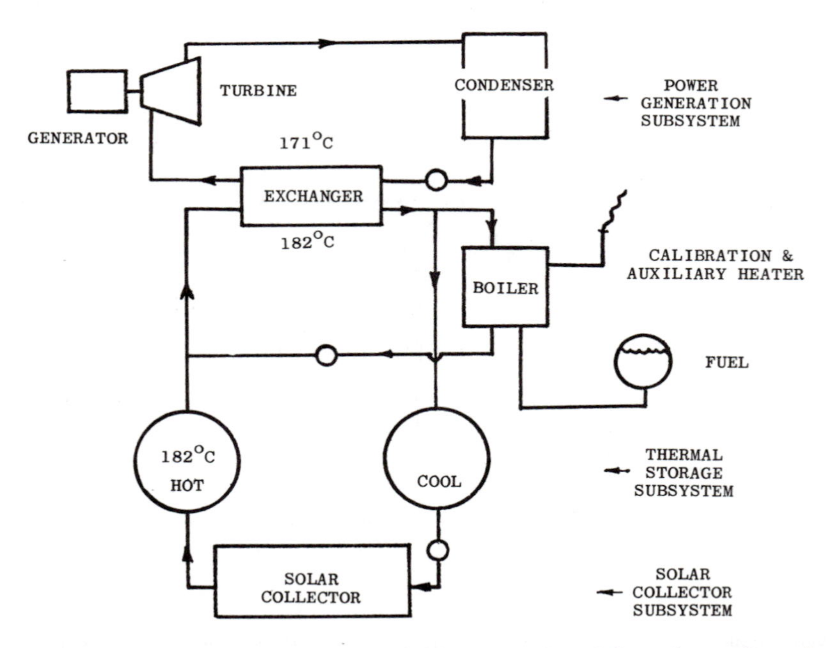

Fig. 14.9 Basic schematic diagram of a dual-fluid system adapted from the geothermal power plant designed for the San Diego Light and Power Company, using isobutane as the working fluid.

(400°F) and the isobutane at 170°C (340°F). Note the provision for oil-fired hot water to be used for system calibration and as a standby.

The working fluid must have a lower boiling point than the heat-transfer fluid because heat transfer depends on there being a significant temperature difference between the two. The heat transferred to the working fluid is only that heat stored *above* the temperature of the boiler. The difference must be large in order for each gram of hot water to be used efficiently. This means that the secondary working vapor must be significantly superheated, and the net efficiency of the cycle depends upon the *T-S* diagram for the fluid under study. Some fluids do not yield good performance from the additional energy transferred in superheating. In the case of solar application, superheating is always the most efficient way of using the stored energy in the hot water.

To illustrate the type of heat-transfer cycle encountered in a dual-fluid system we show such a cycle schematically in Fig. 14.10. The shaded portion represents a *T-Q* diagram for temperature as a function of heat transferred into the working fluid. The coordinate system at the left side of the figure illustrates the relative temperatures encountered. The working fluid is first raised from the working vapor, then returned to the solar collector at a temperature slightly above the boiler temperature T_b. The second stream now transfers its energy load into the boiler, not affecting the temperature of the working fluid in the boiler but producing vapor. It too is returned to the

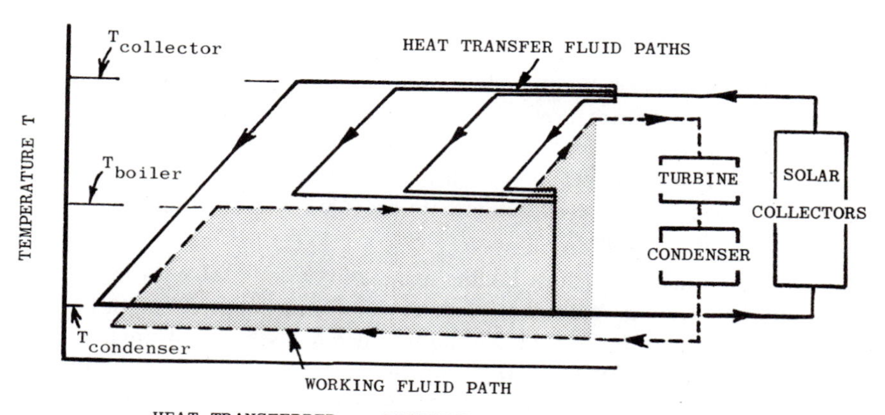

Fig. 14.10 Basic system diagram for a power system using different heat-transfer fluid and working fluid. The fluid paths in the shaded area approximately represent temperature on the ordinate and quantity of heat transferred on the abscissa. Note that four heat-transfer fluid loops are needed for one working fluid loop, with three of the heat-transfer fluid loops returning to the collector at substantially the boiler temperature and only one of the heat-transfer fluid loops returning at the condenser temperature.

solar collector at a temperature that is above the boiler temperature only to the degree of the temperature difference inside the heat exchanger. The third stream is also used to boil working fluid, and it then returns to the solar collector. The fourth stream also boils some working fluid, but it continues down in temperature because it can be used to preheat the working fluid before it enters the boiler. The four streams are then recombined to enter the solar collector and be reheated up to T_C.

An important question in this case is: What is the optimum temperature for the boiler in relation to the collector? The difference must be large enough to extract the maximum number of calories from each gram of hot water in thermal storage, but the spike of the superheated vapor portion of the *T-S* diagram, Fig. 14.1(a), leads to a degradation of the theoretical Rankine efficiency as compared to the theoretical Carnot efficiency.

14.8 PHASE CHANGE IN THE COLLECTOR

In a collector in which a phase change occurs, boiling is a potentially serious problem. It is serious because it *chokes* the fluid flow in the collector. In principle, a phase change is desirable because the fluid enters as a liquid and exits as a vapor ready to be sent to the turbine, without the efficiency or temperature losses that would be encountered in a separate phase-change separator outside the collector. Although the vapor can be used directly in a turbine, there is the problem of how it can be stored for later use, since the vapor occupies a larger volume than a heated fluid when the vapor is below its critical point.

The change of phase in a collector means that the collector must operate as a boiler. A standard boiler has rather short boiler tubes. Water and steam are violently propelled by the expanding steam out the upper end of the boiler tubes into the steam drum. The steam drum separates the steam from the water, the steam going to the superheater tubes while the water is returned to the bottom of the boiler tubes. The solar collector must therefore be designed to provide these functions within the constraints of the collector design.

A schematic diagram of how boiling in a relatively long collector would function is shown in Fig. 14.11. At a number of points along the collector length the accumulated steam would have to be bled off and returned in a separate steam line. The flow volume would diminish along the length of the collector, as indicated by the number of arrows shown. The returning flow of steam could also be within the solar collector so that the steam could be superheated in the process of returning to the using system.

The alternative to separation of the steam from the water would be to make the duct containing the water large enough to handle the flow of steam. If you carefully examine the Shuman-Boys collector (Fig. 1.9), you will note that the cross section of the absorber is trapezoidal, being larger at the top, apparently to provide the duct for easy removal of steam from the boiling water.

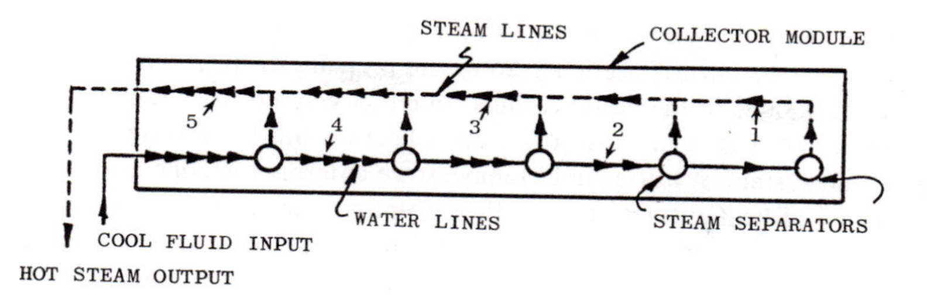

Fig. 14.11 Schematic diagram for a long module in which a phase change occurs during the passage of the heat-transfer fluid down the collector. Periodic steam separator drums are provided to collect the steam and direct it back along another path through the module to superheat the steam, the water being circulated onward through the collector to the end.

14.9 EXPANSION BOILER PLUS REHEAT

The second way to utilize hot, pressurized water is shown schematically in Fig. 14.12. In this case the hot water is conveyed to a boiler, where a pressure drop allows the water to vaporize until the water temperature has dropped to the new equilibrium temperature at the lower pressure. The steam is drawn off from this boiler and reheated (superheated) before going to the turbine, which in this case would probably be a single-stage turbine. The cooler water is then returned to the solar collectors for reheating.

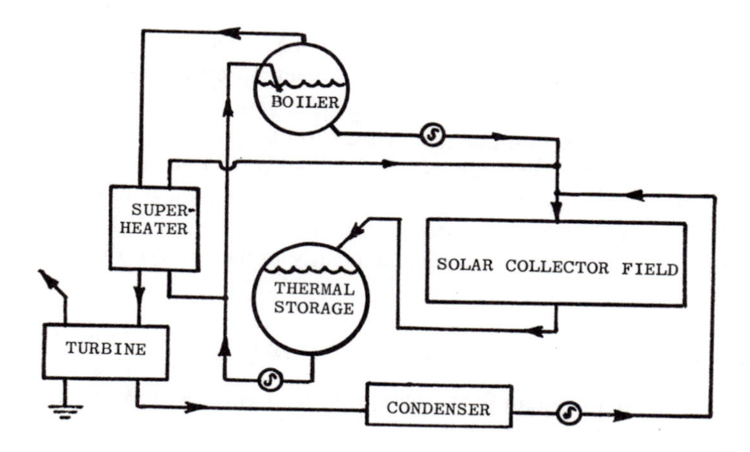

Fig. 14.12 Basic system diagram for a low-temperature power system using a heat-transfer fluid without phase change, where the heat-transfer fluid is also the working fluid of the turbine. Since the boiler operates at a lower pressure than the solar collector field and thermal storage tank, it is advantageous to reheat (superheat) the steam exiting from the boiler.

The tradeoff between a liquid turbine system and a separate boiler system involves the fact that in the second system the pressure drop in the boiler is less than that in the first stage of the liquid turbine. In either case no work is extracted from the hot water before it is returned to the solar collectors, but the returning water is hotter in the second system. A hotter fluid temperature entering the collectors means a lower net efficiency of heat extraction from the collectors, but the system is simpler because a standard turbine could be used. The liquid turbine system would be more efficient because the fluid temperature returning to the collector is lower, but it involves development of new types of turbine.

When thermal storage is included in the system considerations, the fluid turbine appears even more desirable. This is because more heat is extracted per gram of stored water, so the mass rate of consumption of water in storage is minimized.

A practical matter to consider when building a pressurized hot-water collector and storage system is that one could begin operations with the simpler but less efficient separate boiler system and later switch to the liquid turbine mode simply by changing the turbine and adding a steam separator.

In the expansion boiler arrangement the steam has lost a significant amount of heat in the expansion process. Further expansion of this steam in a turbine would not yield much useful energy. The steam would, moreover, always be supersaturated. The proper way to use the steam from the expansion boiler is to reheat it to the temperature of the stored hot water, as is discussed in Section 14.10. This reheating superheats the steam and sets the stage for multiple reheatings and expansions to gain the maximum amount of energy from a given quantity of stored hot water.

The basic diagram needed to evaluate the performance of a steam thermodynamic cycle is the Mollier diagram, as shown in Fig. 14.13. The Mollier diagram plots enthalpy versus entropy. An adiabatic expansion or compression is a vertical line. Heating at constant pressure is an inclined line. The curves most steeply inclined in Fig. 14.13 are lines of constant pressure. The nearly horizontal curves at the top of the diagram are lines of constant temperature for the vapor. The curves inclined slightly downward to the right are curves of constant moisture percentage in the portion below the saturation line and curves of constant superheat temperature above the saturation line. We will refer to the Mollier diagram frequently in the analysis that follows.

14.10 SINGLE EXPANSION TO A WET CONDITION

The schematic Mollier diagram for a single expansion cycle is shown in Fig. 14.14. The expansion goes directly from an initial p_c,T to a final p_o by a vertical path. One then finds the final temperature by tracing the p_o curve up to the saturation line and reading the temperature of the horizontal curve to T_o. For ease of analysis we show the temperature, pressure, and mass fraction along each of the fluid paths in the system diagrams (see Figs. 14.15–14.18).

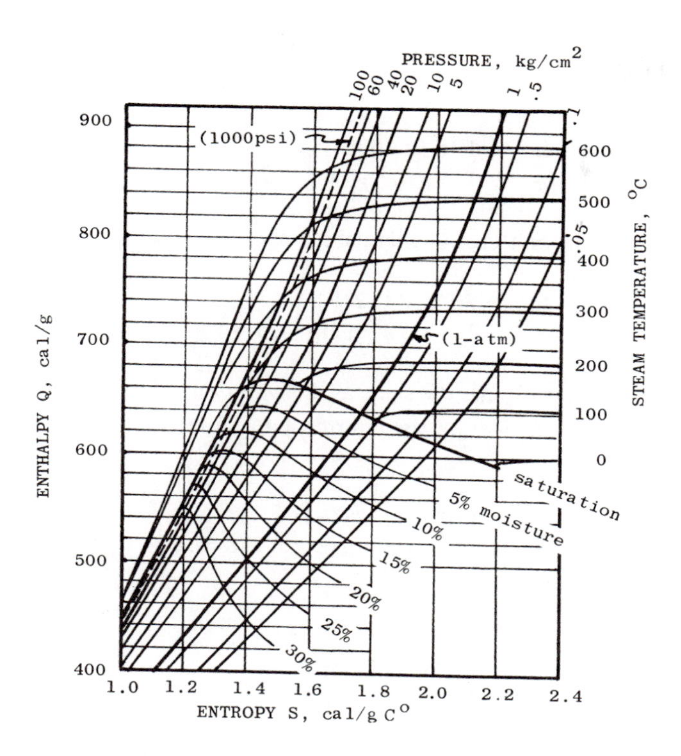

Fig. 14.13 Mollier diagram for steam, with metric pressure and temperature lines.

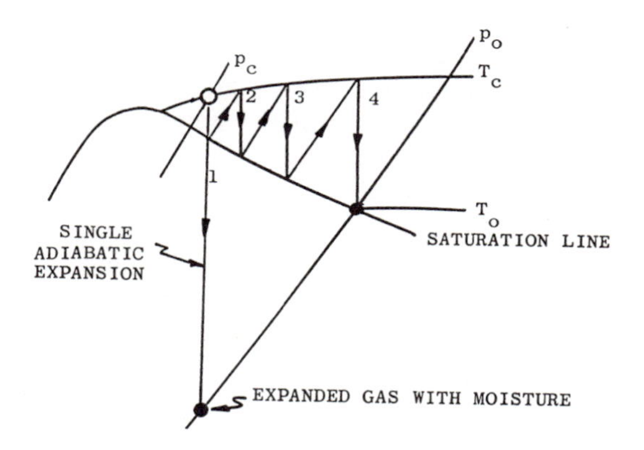

Fig. 14.14 Schematic Mollier diagram, simplified to show only the relevant pressure-temperature lines. Path 1 shows the single expansion path from a superheated T_c to an oversaturated T_o,p_o. A more efficient path is shown by four separate adiabatic expansions to the saturation line, followed by reheating of the steam to T_c at intermediate pressures.

In each case we have the same initial and final conditions of temperature and pressure:

$$T_c = 195°\text{C} \quad (380°\text{F}),$$
$$p_c = 14 \times 10^4 \ \text{g/cm}^2 \quad (203 \ \text{psi})$$

for the water out of the collector or thermal storage tank, and

$$T_o = 50°\text{C} \quad (122°\text{F}),$$
$$p_o = 1.2 \times 10^2 \ \text{g/cm}^2 \quad (1.8 \ \text{psi})$$

for the steam out of the turbine, the same as the water out of the condenser.

The three cases evaluated are for intermediate temperatures and pressures of the flash boiler:

Case 1: $T_b = 170°\text{C} \quad (338°\text{F})$
$p_b = 8.1 \times 10^4 \ \text{g/cm}^2 \quad (118 \ \text{psi}),$

Case 2: $T_b = 150°\text{C} \quad (302°\text{F})$
$p_b = 4.9 \times 10^4 \ \text{g/cm}^2 \quad (69 \ \text{psi}),$

Case 3: $T_b = 125°\text{C} \quad (255°\text{F})$
$p_b = 2.4 \times 10^3 \ \text{g/cm}^2 \quad (34 \ \text{psi}).$

To illustrate these three cases we show in Figs. 14.15–14.18 the calculated temperatures and pressures at each point in the system. At the bottom of each figure is a summary of the calculations relating to efficiency.

The heat input to the steam would at first appear to be the heat transferred to the steam in the flash boiler plus the heat transferred in the superheater. These are only part of the heat input because the role of the condenser is not included. It is more correct to calculate the temperature of the water flowing back into the solar collector and see how much heat is added to bring the water back up to the desired inlet temperature. It is this quantity of heat that we will use in calculating the efficiency of each cycle.

For the three cases we have the following:

Case 1:			
	Water returned from boiler	=	19.3 g @ 170°C
	Steam from condenser	=	1.0 g @ 50°C
	Water from superheater	=	0.5 g @ 175°C
	Net return to collector	=	20.8 g @ 164°C
	Reheating returns		20.8 g @ 195°C
	requiring the addition of		32.1 cal/g_w
	for a total of		670 cal/g_s

CASE 1

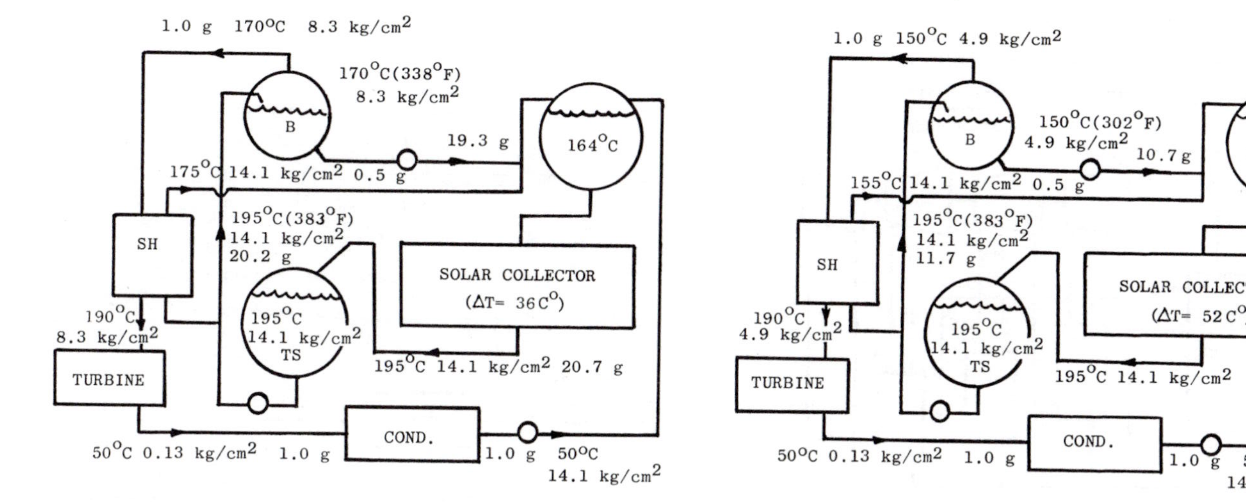

Boiler temperature drop of 25 C°, releases 25.9 cal/g.
Production of steam in boiler absorbs 497 cal/g;
therefore 1.0 g of steam requires 19.2 g (water).

Work output of the adiabatic expansion from the Mollier diagram is

$$T_b, P_b^{*} \qquad H_b = 675 \text{ cal/g}$$
$$T_o, P_o \qquad H_o = 517 \text{ cal/g @ 17\% moisture}$$
$$\Delta H = 158 \text{ cal/g (285 Btu/lb)}$$

Water to steam = 4.9%; water returned to storage = 95.1%.

*Note that the pressure to be used for H_b is the boiler pressure and T_b is the temperature out of the superheater, denoted by SH.

Fig. 14.15 Thermodynamic quantities for case 1.

CASE 2

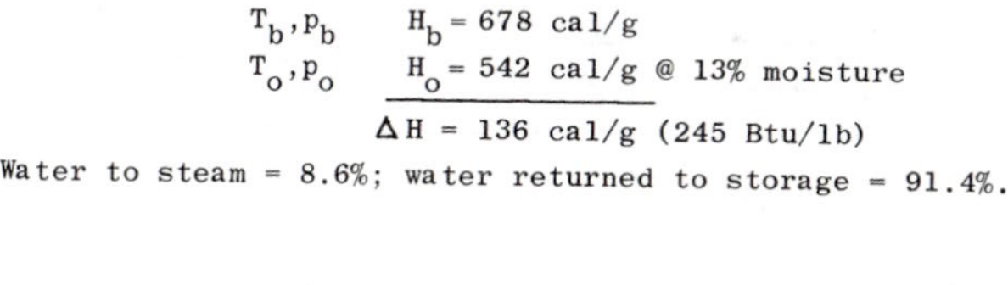

Boiler temperature drop of 45 C°, releases 47 cal/g.
Production of steam in boiler absorbs 504 cal/g;
therefore 1.0 g of steam requires 10.7 g (water).

Work output of the adiabatic expansion from the Mollier diagram is

$$T_b, P_b \qquad H_b = 678 \text{ cal/g}$$
$$T_o, P_o \qquad H_o = 542 \text{ cal/g @ 13\% moisture}$$
$$\Delta H = 136 \text{ cal/g (245 Btu/lb)}$$

Water to steam = 8.6%; water returned to storage = 91.4%.

Fig. 14.16 Thermodynamic quantities for case 2.

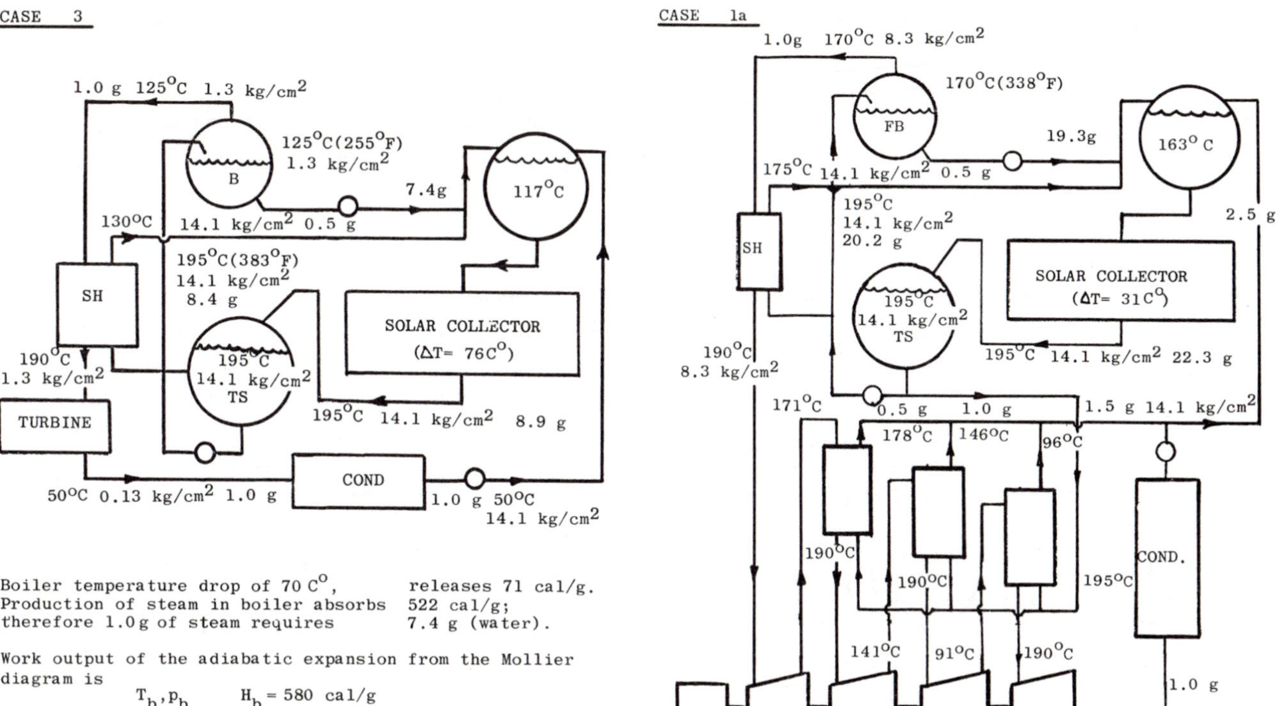

Boiler temperature drop of 70 C°, releases 71 cal/g.
Production of steam in boiler absorbs 522 cal/g;
therefore 1.0 g of steam requires 7.4 g (water).

Work output of the adiabatic expansion from the Mollier diagram is

$$T_b, p_b \qquad H_b = 580 \text{ cal/g}$$
$$T_o, p_o \qquad H_o = 480 \text{ cal/g @ 7\% moisture}$$
$$\Delta H = 100 \text{ cal/g (180 Btu/lb)}$$

Water to steam = 11.9%; water returned to storage = 88.1%.

Fig. 14.17 Thermodynamic quantities for case 3.

Boiler temperature drop of 25 C°, releases 25.9 cal/g.
Production of steam in boiler absorbs 497 cal/g;
therefore 1.0 g of steam requires 19.2 g (water).

Work output of the four adiabatic expansions to saturation from the Mollier diagram is found to be ΔH = 188 cal/g (340 Btu/1b).

Fig. 14.18 Thermodynamic quantities for case 1a.

Case 2:	Water returned from boiler	=	10.7 g @ 150°C
	Steam from condenser	=	1.0 g @ 50°C
	Water from superheater	=	0.5 g @ 155°C
	Net return to collector	=	12.2 g @ 143°C
	Reheating returns		12.2 g @ 195°C
	requiring the addition of		53.6 cal/g_w
	for a total of		682 cal/g_s

Case 3:	Water returned from boiler	=	7.4 g @ 125°C
	Steam from condenser	=	1.0 g @ 50°C
	Water from superheater	=	0.5 g @ 130°C
	Net return to collector	=	8.9 g @ 117°C
	Reheating returns		8.9 g @ 195°C
	requiring the addition of		80.1 cal/g_w
	for a total of		711 cal/g_s

Case 1a is a more elaborate system employing three reheats and four adiabatic expansions to show what gains can be obtained from a four-stage turbine:

Case 1a:	Water returned from boiler	=	19.3 g @ 170°C
	Steam from condenser	=	1.0 g @ 50°C
	Steam from stages	=	1.5 g @ 139°C
	Water from superheater	=	0.5 g @ 175°C
	Net return to collector	=	22.3 g @ 163°C
	Reheating returns		22.3 g @ 195°C
	requiring the addition of		33.4 cal/g_w
	for a total of		744 cal/g_s

The results of these model calculations are presented in Table 14.1 and Fig. 14.19. Figure 14.19 shows that the theoretical Rankine efficiency drops as the temperature drop in the expansion boiler is increased. If one were using hot water directly from the solar collector one would, therefore, like to use a small temperature drop, limited only by the increased size of the return burden of water to the solar collector when only a small amount of energy is extracted per pass through the boiler. If, on the other hand, one is using stored hot water, as from a thermal storage tank, the most effective use of the water is for a rather appreciable drop in temperature in the boiler, about 55 C°.

Table 14.1 System Thermodynamic Behavior

Case	*Enthalpy change H* Btu/lb	cal/g_S	*Work output,* cal/g_W	*Heat inputs,* cal/g_S	*System efficiency,* %
1	285	158	8.2	670	23.7
2	245	136	11.1	682	20.0
3	180	100	11.2	711	14.1
1a	340	188	8.4	744	25.3

The column for heat inputs is the sum of heat inputs into water in the collector to raise the temperature back to 195°C.

Case 1a uses four adiabatic expansions, each to the saturation line, with reheating to 190°C in each step. Note that the improvement is less than 2%, which may not be cost effective for the additional turbine stages and three reheaters.

The column for work output in cal/g_W shows the efficiency of using stored hot water (g_W). When the cost of thermal water storage for energy is appreciable, this factor may weigh heavily against the higher efficiency of the cycle for cases 1 and 1a.

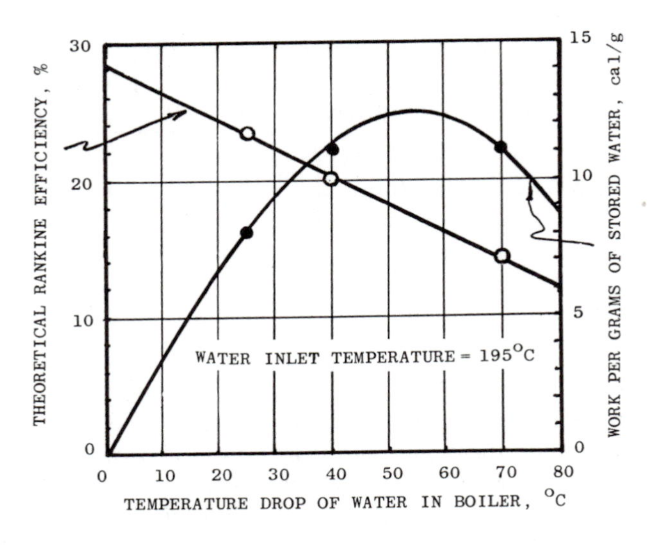

Fig. 14.19 Summary graph of system performance for cases 1, 2, and 3 for theoretical Rankine efficiency and for the work output from the cycles in terms of the efficiency of use of stored hot water. Note that the best usage of water in thermal storage is where the net cycle efficiency is almost half its maximum value of $\Delta T = 0$.

The other option, not yet discussed, is what determines the temperature of the hot water for this type of steam. In the example we took a temperature of 195°C at the collector. The vapor pressure of water at this temperature is approximately 13 kg/cm² (190 psi). This pressure is low enough to be readily contained in the system. A higher temperature would increase the efficiency of extraction of energy, but it would increase the pressure because it is rising rapidly in this temperature domain. The practical question is that of the tradeoffs among the operating pressure, the cost of the containment of the water, and the resultant efficiency. We think that 13 kg/cm² is near the top of the practical range for this type of using system, but we have not explored the tradeoff between efficiency and temperature. A graph showing the change of water pressure with temperature is given in Fig. 14.20.

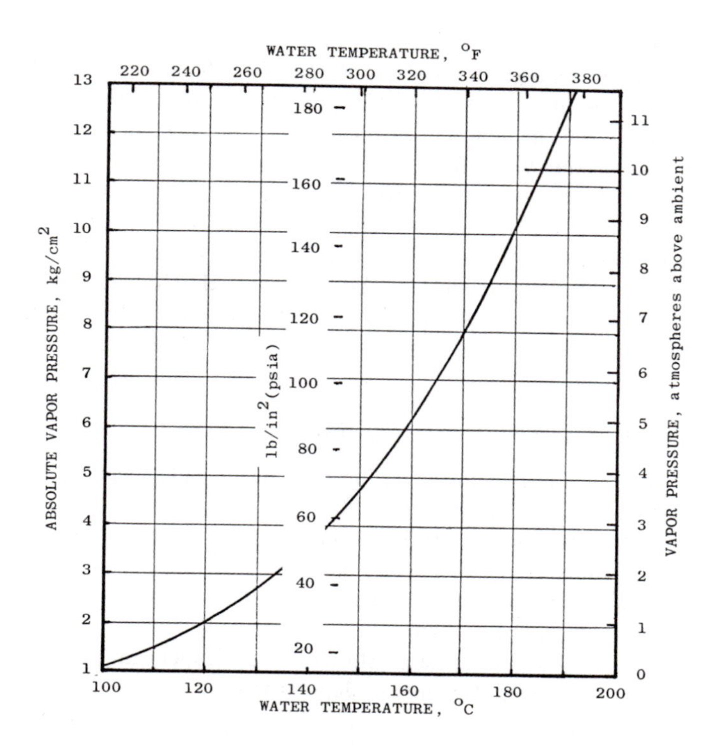

Fig. 14.20 Graph of the vapor pressure of water as a function of temperature.

Another option is to use the steam coming directly from the hot thermal storage tank, eliminating the expansion boiler. The result would be that the initial conversion efficiency would be better than in cases 1–3, but the storage tank would cool as energy is extracted via the steam. The efficiency would therefore drop with time rather than remain constant as in the cases cited. The decision on whether to elect the constant-temperature cycles or a decreasing-temperature cycle requires more information than we present here.

14.11 MULTIPLE REHEATS AND EXPANSIONS

In case 1a, a multiple-stage turbine is used in order to prevent the expanded steam from reaching supersaturated conditions and to permit injection of additional heat into the steam before each successive stage of expansion. The cycle with temperatures is shown in Fig. 14.18.

Although the efficiency of this particular system is not much greater than with the single adiabatic expansion, case 1, it is important to note that the last expansion ends with the steam exiting at 68°C (155°F) rather than 50°C (122°F). This additional temperature is important if we wish to use the spent steam for domestic or industrial heating. One then can count the waste heat as useful heat and also not be required to use open-loop water for cooling the steam.

The Carnot efficiency for the thermodynamic cycle between a maximum system temperature of 195°C and a condenser temperature of 50°C is 30%. We thus see that the theoretical Rankine efficiencies above are reasonably good. The actual Rankine efficiency depends on the efficiency of the turbine used. This efficiency ranges from 50% for small turbines to close to 90% for modern 1000-MW steam turbines. A 10-MW_e turbine, as an example, would be close to the 50% efficiency range.

14.12 BRAYTON CYCLE SOLAR CONFIGURATIONS

The Brayton cycle is generally considered to be efficient only when very high temperatures are involved, between 1500 and 1800°C for the gas entering the turbine stage. These high temperatures may be attainable in very high-concentration solar collectors, but the materials problems involved appear to be very serious. The question arises: Can the Brayton cycle be used at lower temperatures? If so, what are the operating conditions and theoretical efficiencies involved?

The theoretical Brayton cycle is shown on a temperature-entropy diagram in Fig. 14.21. The actual cycle, also shown, departs from the theoretical one in a significant way. The adiabatic compressions and expansions are not without loss of energy to the environment surrounding the gas.

The compressor cycle begins with gas at pressure p_1 and compresses it quasi-adiabatically to p_2. The actual compression path is the inclined line rather than the adiabatic line, which is shown as dashed. As a consequence, the input energy is greater than the adiabatic energy by a factor of $1/e_c$:

$$Q_c = Q_{ad_1}/e_c\,, \tag{14.3}$$

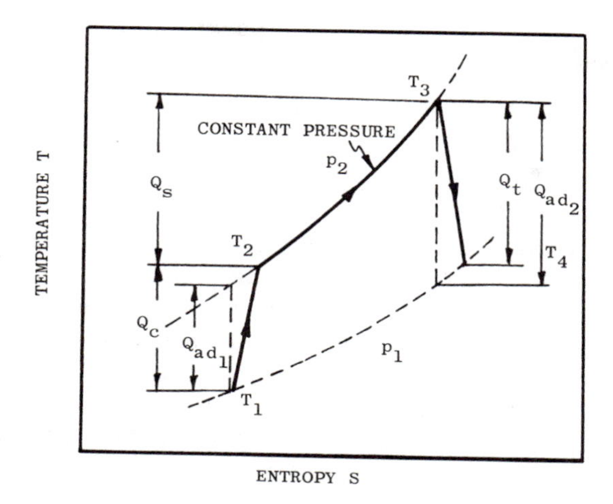

Fig. 14.21 Temperature-entropy diagram for an actual Brayton cycle (solid line) compared to a theoretical cycle (dashed line) used for the derivation of the efficiency for a typical cycle.

where

$$Q_{ad_1} = c_p T_1 (y-1)/e_c, \tag{14.4}$$

where

$$y = (p_2/p_1)^{(\gamma-1)/\gamma} = T_2/T_1, \tag{14.5}$$

and where γ is the ratio of specific heat at constant pressure to specific heat at constant volume, which for air is approximately 0.25 at the temperatures and pressures of interest.

The energy yielded in the turbine path is

$$Q_t = Q_{ad_2} e_t, \tag{14.6}$$

where

$$Q_{ad_2} = c_p T_3 (1-z), \tag{14.7}$$

and

$$z = (p_3/p_4)^{(\gamma-1)/\gamma} = T_3/T_4. \tag{14.8}$$

The actual work input in the compression stage is therefore the theoretical adiabatic work done in compressing a gas from initial to final pressure *divided* by the compressor efficiency. The actual work output in the turbine stage is equal to the theoretical adiabatic work *multiplied* by the turbine efficiency. The system efficiency is consequently given by the ratio of the solar energy input in raising the gas from its initial temperature at compressor output pressure to the final temperature at the turbine inlet.

The amount of solar heat added is given by

$$Q_s = c_p(T_3 - T_2). \tag{14.9}$$

Hence the system efficiency is

$$\eta = (Q_{ad_2} - Q_{ad_1})/Q_s. \tag{14.10}$$

In Fig. 14.22 we show the results of some model calculations of a low-temperature Brayton cycle. The turbine inlet temperature for the two curves is indicated; it is still rather high for a solar collector using ordinary metals. Note that the compressor pressure ratios are low. A typical aircraft turbine operates in the range of 6–12. The reason the solar Brayton is effective at low pressures is simple. When the gas is compressed adiabatically it is heated. At a high compression ratio, the gas is in fact heated almost to the maximum temperature of the collector; hence little additional thermal energy can be injected into the compressed gas.

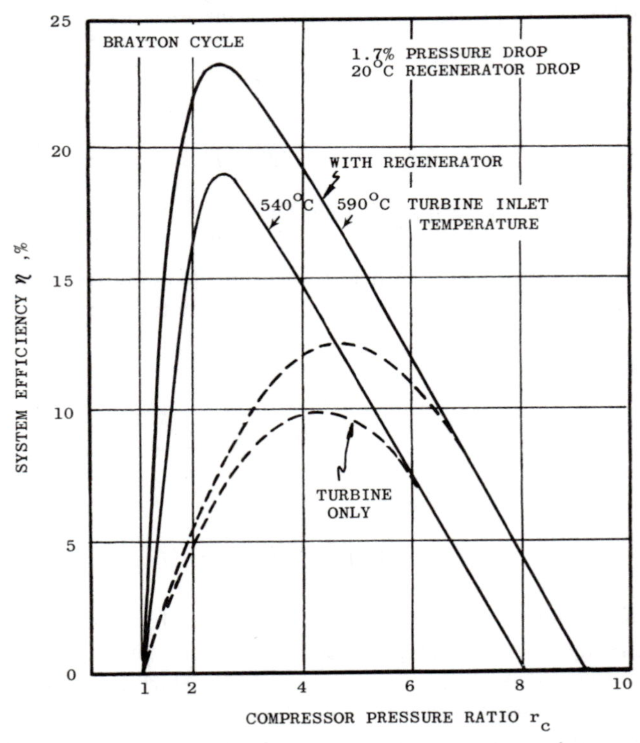

Fig. 14.22 Performance of a Brayton cycle for air temperatures in the range accessible to solar collectors. Calculations include a compressor efficiency of 0.82 and a turbine efficiency of 0.86, typical values for a gas turbine in the aircraft power rating size. No collector losses are included.

A second operational characteristic of concern is that when the compressor ratio is small, the gas does not cool enough in expanding back to atmospheric pressure and is rejected to the air at a high temperature. One therefore finds it attractive to use some of this waste heat to preheat the entering air by means of a refrigerator. In the models cited the actual energy output for a turbine only was almost zero when losses were considered, but rose to a peak of 15% when the regenerator was added.

The regenerator is especially useful at low compressor pressure ratios. This is because the heated compressed gas does not expand enough to cool significantly, but the hot exhaust is hot enough to effectively transfer its residual heat to the gas before it enters the compression stage. The curves shown in Fig. 14.22 are only for losses in the Brayton cycle system. Additional losses in the solar collector further reduce efficiency. A typical concentrating collector having 60% efficiency would result in a Brayton system with a net efficiency of about 15%, compared to a Carnot efficiency of 56% for 590°C input and 100°C output. This 15% efficiency is rather low, but the using system is simple and requires no water to help dispose of the waste heat, which is blown directly into the air.

14.13 LIMITS TO SYSTEM EFFICIENCY

In the course of our studies of several systems it became apparent that there is a well defined upper limit to the performance of a solar thermodynamic system. In Fig. 14.23 we show the variation of efficiency as a function of the upper temperature

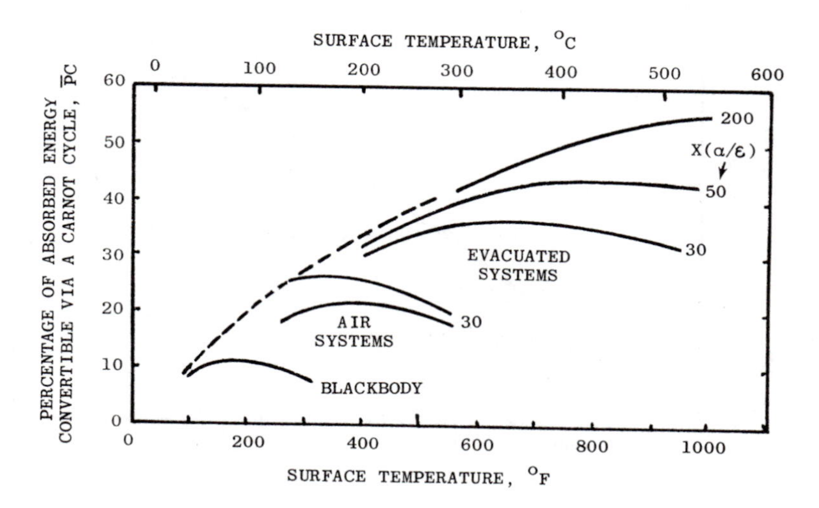

Fig. 14.23 Variation of theoretical conversion efficiency by means of a Carnot cycle of the absorbed solar energy for several system temperatures and types.

of the system for six different systems, ranging from a blackbody absorber in air to a relatively high-concentration system with absorber in a vacuum. It is apparent that these curves tend to define an upper limit, even though some had different condenser temperatures than others. In the graph we have taken the efficiency measure as the percentage of the *absorbed* energy convertible by means of a Carnot cycle. If the percentage of available energy were used there would, however, be little significant change in the limit these curves define.

In Fig. 14.24 we have indicated the locus of the tangent to the curves of Fig. 14.23 as a heavy line, in order to denote the upper limit of possible system performance. We have also taken the data for systems studied by Tabor (1955) and formed a tangent to his family of curves, essentially defining the same curve as in our study. For interest we plot the published efficiencies of a number of actual solar energy systems constructed early in the 20th Century. Although these early published efficiencies must be taken with caution, it is interesting to note that they all lie on the "permitted" side of the limit curve. The Shuman flat-plate collector constructed at Philadelphia in 1909 appears to be close to the limit. The point for Abbot (1943) is a point he calculated (the system was never constructed), but it agrees well with the limit curve.

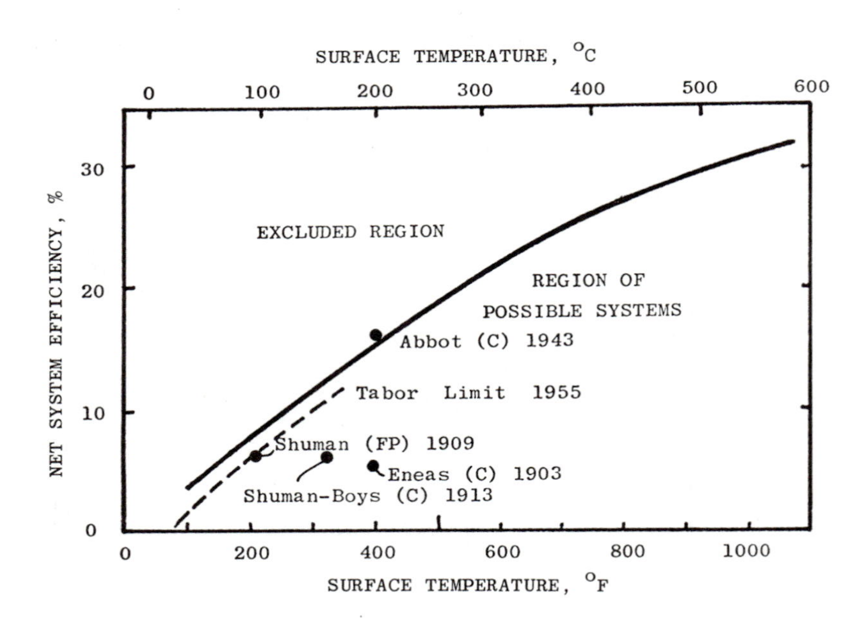

Fig. 14.24 Historical system conversion efficiencies as compared to the theoretical maximum conversion efficiency. ***FP*** **denotes flat-plate and** ***C*** **denotes concentrating collectors.**

The limits to performance can be plotted in a different way, as shown in Fig. 14.25. Here we show the limit obtainable with a flat-plate collector as a function of the complication accepted in the design of the unit. The blackbody absorber with two cover glasses in air defines the lower limit of performance. The limit for an absorber in air when the absorber emittance is zero is shown by the curve at the upper boundary of the shaded area. If the air is removed from the collector and selective surfaces are used, then the limit is quite high and is set by the shift of the TIR emission into the passband of the selective surface. No combination of improvements can yield a flat-plate collector above the upper line into the dotted region.

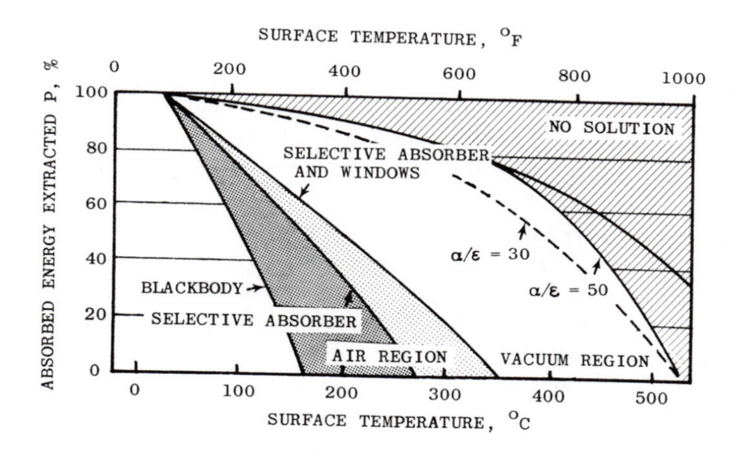

Fig. 14.25 Limits to performance for flat-plat collectors using two cover windows, as a function of changes in the heat losses of the collector.

The limit curve in Fig. 14.24 is useful in that it also sets the upper limit to the relative cost of collectors for the different temperature zones. If a solar saline pond, deriving temperatures on the order of 90°C through the stabilizing effects of a strong saline gradient with depth, costs one unit, it will operate with a top efficiency of about 6%. A medium-temperature pressurized water collector system operating at 150°C will yield a top value of 12%. The medium-temperature collector will be less cost effective than the lowly solar pond if the collectors cost more than twice the pond. Since a pond can in principle be made rather cheaply, the cost constraint on the medium-temperature collector is quite severe. If we go to a high-temperature power-tower system, operating at 500°C, the top efficiency will be about 28%, and the cost-effective breakpoint is about 4.5 times the cost of the solar pond. For tracking heliostats this limit also looks severe.

14.14 HEAT REJECTION FROM THERMODYNAMIC CYCLES

A basic problem with solar thermodynamic cycles, often dismissed with little comment, is heat rejection to the environment. The problem is on two levels, the first being simply the disposal of the heat and the second being the maintenance of the heat balance of the environment.

Heat rejection involves disposing of all of the heat not converted into power. Low-temperature solar cycles, therefore, need to reject between 60% and 90% of the solar energy collected because of lower thermodynamic efficiency. Higher operating temperatures mean less heat to be rejected into the environment. In either case a wet cooling tower requires that the condenser evaporate enough grams of water to have the latent heat of the phase change of water absorb the waste heat. Water in large quantities is therefore a necessity, but the sunniest portions of the world, where solar power conversion would be most attractive, are also the regions where water resources are scarce. Wet cooling towers are most desirable when maximum turbine efficiency (minimum heat rejection) is the goal because they can work close to the ambient temperature. In Arizona, the power plant cooling towers operate within 5–10 C° of the ambient temperature.

Dry cooling towers are less efficient because they require a temperature high enough above that of the environment to dispose of their heat by convection and wind or induced drafts. This temperature difference is necessary to keep the total surface area of the hot side of the cooling tower exchanger small yet still dispose of the required heat per unit of time.

Some very large exchange areas have been constructed for present power plants. The one shown in Fig. 14.26 is a cooling pond that was extended in size to aid in loss of heat by radiative and convective action to the atmosphere, with evaporation being small because the plant was situated in a humid climate.

Solar power plants will have very extensive areas for the collectors, and it would be nice if the same large area could be used for the heat rejection. In principle this could be done, but the economics do not appear satisfactory.

One possibility for reducing the disposal problem would be to use the waste heat constructively. Such a use would be for an integrated energy system for a community, where the waste heat could be used for heating and cooling functions. Because a normal power plant with wet cooling tower rejects heat so close to the environmental temperature, this type of waste heat is quite useless. The integrated plant would need to operate the heat rejection system at a high enough temperature to operate the heating and cooling cycles. Such a temperature would be close to 100°C rather than the 40–50°C currently widely used. This means that the basic power cycle would work with a lower conversion efficiency. It also means that low-temperature thermodynamic cycles would be at a disadvantage in such a system because the upper and lower temperatures would come too close together. For high-temperature systems operating in the 400–600°C range, the increase of condenser temperature from 50°C to 100°C would not cause a major reduction in system efficiency.

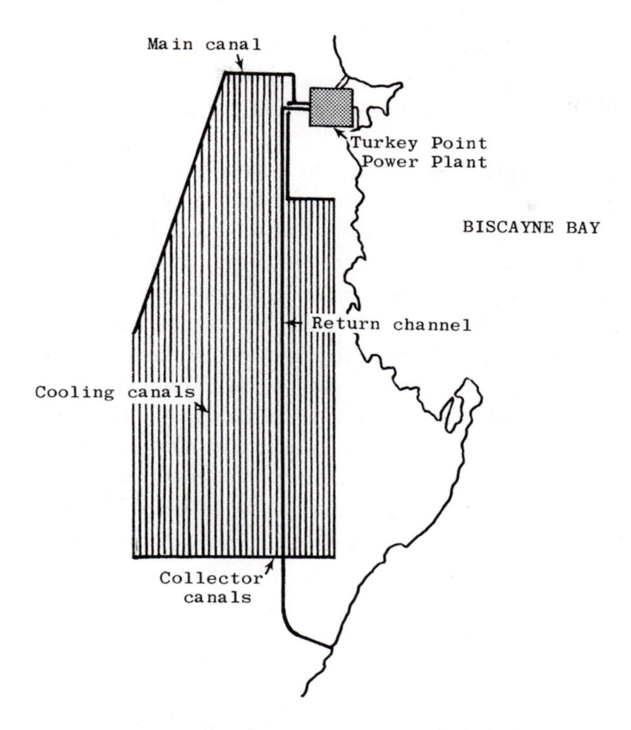

Fig. 14.26 Plan for a 1600-hectare (4000-acre) radiator developed for the Turkey Point electrical power plant, Florida, to dispose of the waste heat. The plan uses 270 km of water canals 60 m wide and 1 m deep. This large area begins to compare with land use for solar collection for a similar power plant.

14.15 ENVIRONMENTAL IMPACT FACTORS

The topic of waste heat rejection also raises some more general environmental questions. These involve, in addition to the simple rejection of waste heat, the broader topics of environmental heat balance and impact from land utilization, and of societal secondary impact from large-scale solar power farming.

Let us first look at the question of environmental heat balance. By heat balance we mean the general question of how much the consumption of energy affects the global temperature. The creation of every new energy or heat source on the earth requires a very small increase in the temperature of the earth in order to have the earth reject that heat into space, the final dumping place for energy. If large amounts of energy are released in one area, like the Los Angeles metropolitan area in California, as is now the case, the effect is about equal to the sun's brightening 1 C°. This increased input of energy raises the average temperature of the region. One can foresee the day when, if the energy consumption increases of the past three decades are extrapolated into the future, the entire world's temperature could be increased by this amount. The possible consequences for the world climate and the stability of the polar ice masses become of concern.

The increase in societal sensitivity to excessive energy consumption makes it necessary to examine the potential impact of solar energy consumption. Many say it is nonpolluting energy, but is that assessment fully correct? To a first approximation one can say that the conversion of solar energy into the many possible forms of heat and power would not alter the global heat balance, simply because the sunlight is coming into the environment whether we use it or not. Its ultimate fate is to be converted into heat, heat we enjoy because it keeps our planet comfortable. But as we have already pointed out, we do modify this normal dissipation of solar energy in the environment in two regards: (1) we make the collecting areas darker than the environment (increase the *albedo* of the region), and (2) we may export some energy from one area to another, as in the case of solar power farms, where energy might be collected in the desert areas and delivered to the cloudy areas of the globe.

In regard to disturbing the natural albedo, covering the total surface required to meet the entire national energy needs would still make less of an impact than the areas of blacktop now used for roads and parking lots. Further, the plowing of the Midwest each spring and the harvest each fall cause a larger net perturbation of the albedo of the United States than would all the solar collectors in the foreseeable future.

But the observation that "we do worse things anyway" is really not a valid answer. If it were determined that the regional albedo change would be deleterious to the climate, one could balance the albedo change by whitening the spaces near or between the collectors or mirrors. Thus, in principle, one could keep the albedo of the collecting area unchanged. At the other end (the energy-consuming area), the injection of solar power would still modify the microclimate of the cities, but in general the warming of the city climate only reduces the severity of winter temperatures, while summer temperatures, which do not block roads with snow or water, can be handled by cooling systems. Even the city heat loads could be compensated for if one were to regularly whiten the structures and areas within the city to reduce the input load of direct solar energy. Such white gleaming cities might be quite attractive compared to the darkened aged buildings we so often have today.

The concept of Meinel and Meinel (1971) of conversion of solar energy into electrical power through large-scale solar power farms brings to the fore energy balance and other environmental problems. A model of one of our solar power farm concepts is shown in Fig. 14.27, where an array of long module collectors (Section 11.2) is deployed over a land area of 1 km by 1 km. Note the sun reflection in the cover windows of the collectors. This farm would produce on the order of 60 MW_e average output, and hence a large land area would be required to equal the output of a modern nuclear power plant. One possible environmental objection to such large-scale farms is that they would require large areas of the deserts of the earth. On a global scale there are vast amounts of desert, but few close to load centers. If one could locate solar plants in remote deserts, as in the Sahara or Arabian deserts, and produce a transportable fuel, such as hydrogen, then the world could be supplied with fuel from sunlight indefinitely without significant impact on the deserts themselves. Because there are deserts near major load centers in the United States and Mexico, they will become the prime areas for the first solar power farms. The possible damage to prime new desert land areas is of some concern. We feel, however, that a recent observation by a member of the Audubon Society is a good example of progressive thinking in this regard. The comment was that one should look to land in the arid regions now being taken out of pumped agricultural production because of receding water table problems. The same land would be used for a new crop: energy. In this manner the entire United States' electrical power demand could be met without requiring a single hectare of new desert land. The new income from energy would also replace the income from the abandoned crop.

Fig. 14.27 Model of a solar power farm using long module collectors. (Artist: Don Cowen.)

Perhaps the major environmental impact from large-scale solar power farms will be from the new population associated with the construction and operation of the farms. If a single *national solar power facility* were established, it would probably be desirable to distribute the generating areas over several states in order to minimize the population impact. Secondary effects, such as industrial use of waste heat and power from a national solar power facility, would imply new industry in the regions of the farms. Thus the impact of indirect effects could be more severe than that of the direct effect of placing the solar collectors on several thousand square miles of arid lands. We are sure that this question will get much study if and when the day comes for such a project to be implemented, perhaps by the turn of the 21st Century.

14.16 COMMUNITY HEATING AND COOLING SYSTEMS

Community heating and cooling systems may find easier application than individual home heating and cooling. The basic reason is that this approach removes many of the economic and aesthetic constraints imposed by individual systems. One community-type system for the Southwest is shown schematically in Fig. 14.28.

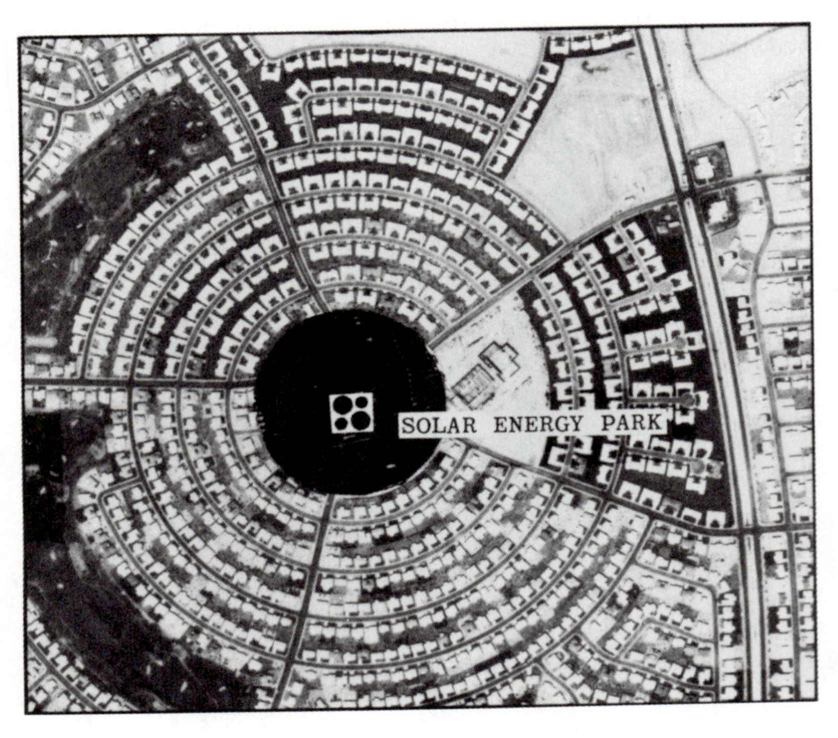

Fig. 14.28 Architect's drawing of a possible community arrangement about a central energy park where hot and chilled water are generated for distribution to homes.

The community system uses a central area for the collectors and thermal storage tanks. The tanks can be located underground. Both hot and chilled water would be generated at this central location and distributed to the homes in the surrounding community through buried pipes. Communities of 50 to 5000 homes appear to be in the range of possibility, a figure based solely upon energy losses in the transmission of the fluids to the homes.

One attractive feature of the underground distribution system is that it provides considerable natural thermal storage. In regions where the ground is relatively dry, the ground itself is a good insulator for the pipes, so no other insulation would be required. The ground will be heated or cooled within a diameter of several feet about each pipe, storing considerable reserve heat or cold for the system and thus leveling fluctuations in input temperature at the central location.

The advantage in operational aspects is considerable for the community system. For new subdivisions the piping is placed in the ground when the subdivision is started, and the central facility is operated by the local community or utility, with its own maintenance and operating staff. For existing subdivisions one needs about one "lot" for each 100 homes. If vacant land is not available in this area, then an urban renewal program would need to acquire the land. Adding underground uninsulated pipes is readily done by utilities and should pose no serious problems.

The community heating and cooling system is one step toward the community total energy system of the future. When solar power plants of community size are developed, the community provides a place to deploy the waste energy from power production. The total energy package still has some problems to be solved—for example, how to dump the excess energy in seasons when neither heating nor cooling is required. Nevertheless, the community total energy center based upon solar energy utilization appears to be the thing of the future. The individual can then have a home of any architectural style, unencumbered by solar collectors, placed in any direction with regard to the sun, with surrounding trees, and without threat from the neighbor's trees.

14.17 SOLAR WATER PUMPING

Water pumping is an application where demand can be modified to meet availability. In Southwestern irrigated agriculture one currently pumps water 24 hours a day, distributing it through the farming area as needed. To replace present natural gas or diesel pumps with solar pumps entails either a change in procedures to follow sun availability or the additional cost of water or thermal energy storage.

In Fig. 14.29 we show a schematic diagram for a solar water-pumping system with storage. This system is a dual-fluid system, which uses pressurized water for energy gathering and storage and organic fluid for the working vapor for the turbine. An auxiliary boiler is shown that supplies hot water to the heat exchanger when solar-heated water is not available. The turbine condenser is cooled by water pumped from the well.

Solar water pumps of a single Veinberg parabolic cusp collector, as shown in Fig. 14.30, can be deployed along access roads through the farming area or in strips

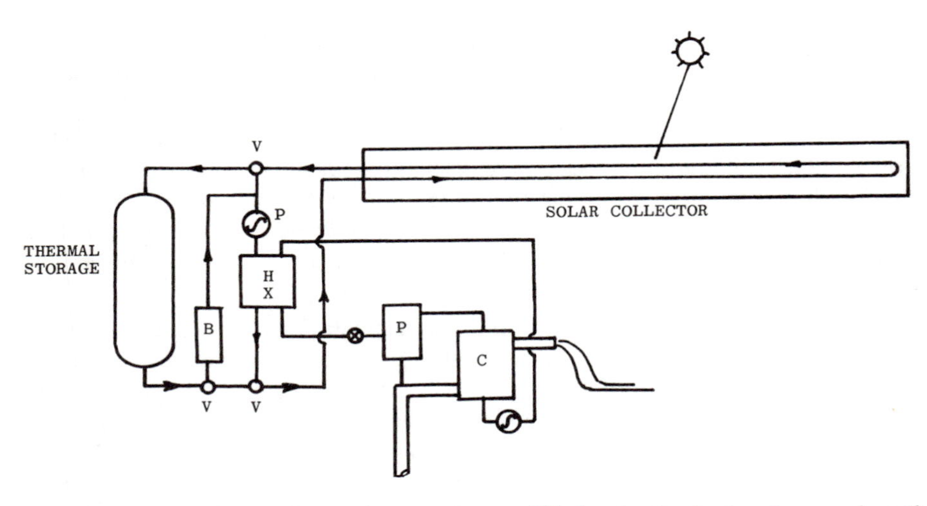

Fig. 14.29 System diagram for a solar water pump. *HX* denotes the heat exchanger where the organic vapor for the turbine is vaporized; *B* denotes a standby heater for supplementing the solar hot water; *P* is the pump-turbine unit; and *C* is the condenser cooled by the pumped water. The design temperature and pressure for this system are 150°C and 10.5 kg/cm^2.

Fig. 14.30 Drawing of a solar water pump using a single Veinberg parabolic cusp with fixed absorber, as designed for installation in Willcox, Arizona, by Helio Associates. (Artist: Don Cowen.)

along the irrigation canals. In this manner the pumps can be integrated with the present bore holes. Their location close to the farm area may, however, have disadvantages: the pumps might interfere with farming operations, or the collectors might become dirty or damaged. The alternative is to place the solar collectors in an area away from the wells or farms, attach the turbine to a generator instead of directly to the pump, and deliver the energy to electric motors at the well heads.

14.18 SOLAR GAS-ABSORPTION REFRIGERATION

There are several cycles for converting heat into refrigeration, classified as *intermittent* or *continuous.* The basic refrigeration process is termed *absorption*; the other two are *compression,* in which a mechanical compressor causes the gas to make the transition to the liquid state, and *adsorption,* in which a gas is adsorbed on a solid substrate, such as aluminum oxide. The transition involved is generally between a gaseous state and a liquid state, $l \rightleftharpoons g$, but some refrigeration cycles based on gas to gas have been developed, such as when compressed air is the refrigerant.

In the intermittent cycle, heat and cold are alternately applied to a solution containing the refrigeration medium, relying on the changes in solubility of the solute with temperature for the generation of a gas of sufficient pressure that it can be liquefied by the use of cooling water in a heat exchanger. The intermittent type of cycle needs no pumps for achieving the required gas compression, and hence is good for locations where there is no electrical power.

In the continuous cycle a pump is used to pressurize the solution containing absorbed gas, so that when heat is applied to the solution the gas is released at sufficient pressure to be liquefied in the condenser stage.

A number of working gases can be used with a gas-absorption refrigeration system. Ammonia is used, the ammonia having a widely varying solubility in water as a function of temperature. The solar absorption refrigeration systems currently being used have water as the refrigerant and lithium bromide as the solute. Solid lithium bromide, like lithium chloride, is hygroscopic, forming a brine in which the water content can be varied with temperature.

In Fig. 14.31 we show a continuous-cycle lithium bromide refrigeration system. Cold-water vapor proceeds to the absorber, which contains a water solution of lithium bromide. The cool solution absorbs the cool vapor, diluting the brine. The brine is pressurized and sent to the stripper, where heat is applied. The heated brine evaporates some water vapor, which flows to the condenser, where it is liquefied. The liquid then proceeds to an expansion valve, where it turns into gas, cooling in the process. For further information on the varieties of cycles and data the reader is referred to a recent handbook on refrigeration, such as those published by the American Society of Heating, Refrigerating, and Airconditioning Engineers (ASHRAE).

Ammonia mixed with water or with a solution of sodium thiocyanate (NaSCN) also makes a good solar-type gas-absorption system, and was studied by Sargent and Beckman (1968).

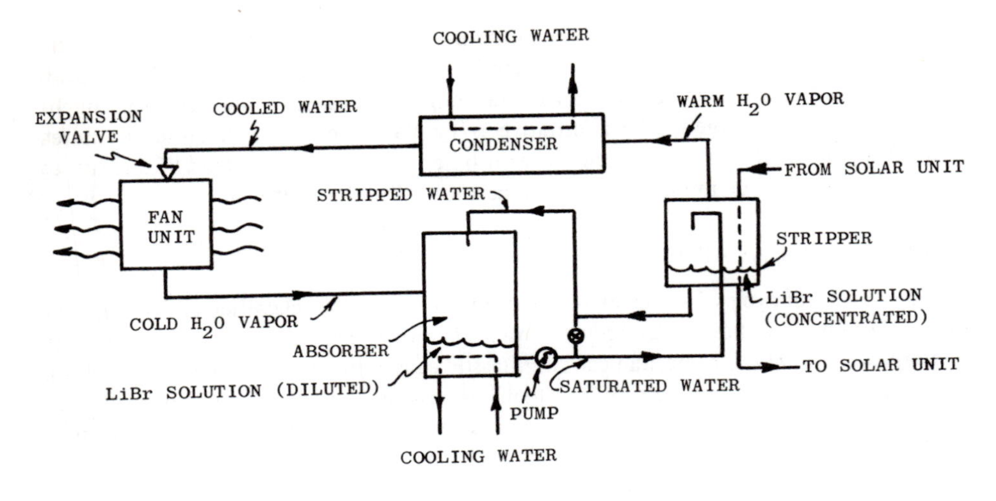

Fig. 14.31 Schematic thermodynamic cycle for a gas absorption refrigeration system using water vapor as the working gas and a solution of lithium bromide in water as the absorbant.

14.19 MEC COOLING SYSTEM

The Munter Environmental Control (MEC) cooling system is based upon a very simple principle: the cooling of air when water is evaporated into it. It is the basis for the old swamp-box coolers of the desert Southwest. The addition that Munter introduced was to *first* dry the ambient air so that *after* evaporation the humidity is restored to a comfortable level along with the cooled air. The key to the success of the MEC system is the efficient drying of the air and regeneration of the dryer.

The basic component of the Munter drying wheel is a thin pancake made of rolled asbestos, corrugated so that air can flow through the small corrugations along the thin dimension of the pancake. The asbestos corrugations are covered with lithium chloride (LiCl), which is hygroscopic and stable, absorbing moisture when cold and releasing it when heated. Such dehydration wheels were first extensively used to reduce the humidity in missile silos. An improved form of the wheel now uses specially treated anodized aluminum as the adsorption surface for moisture and has a longer use lifetime than the original lithium chloride wheels.

The basic system is as shown in Fig. 14.32. Two counterrotating wheels are provided, *A* being the dehydration wheel and *B* the heat exchanger wheel. Two airstreams are used, one of outside air and a second of inside air to be cooled. Wheel *A* rotates slowly but wheel *B* rotates rapidly. A simplified version of the system developed by the Institute of Gas Technology in Chicago has the wet air from the room (plus makeup air from outside) pass through the drying wheel, where water is extracted. This airstream is heated in the process of removal of water from the air, and the second wheel, rotating more rapidly, brings the temperature back down to ambient outside air temperature. The dried air, now at outside temperature, is then cooled by passage through a water screen or evaporative cooler pad where water is added to the air, cooling it to about 10°C (50°F) and raising its humidity back to about 50%.

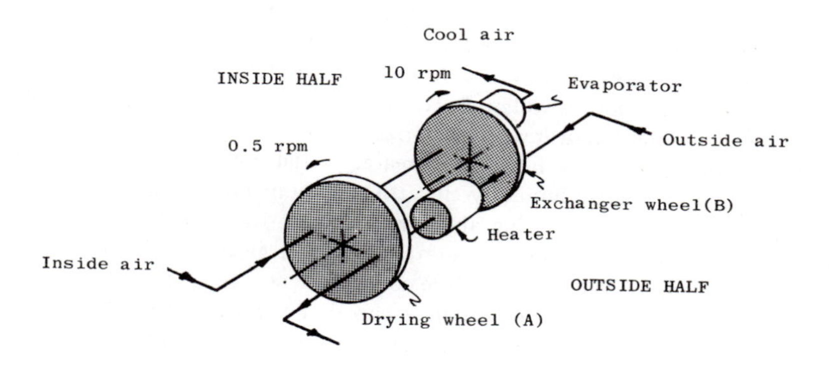

Fig. 14.32 Schematic diagram of the MEC cooling/heating system. Humid air is first dried and then cooled by evaporation of water. The coefficient of performance (COP) for a Houston climate is 0.31 for 35°C ambient air at 25°C wet bulb.

In the outside airstream, the airflow enters the rapidly rotating exchanger wheel, where it picks up some heat while cooling the exchanger wheel. This warmed air is further heated by a gas flame, electrical heat, or solar heat in the space between the two wheels, so that the air entering the dehydration wheel A is at 110–120°C (230–248°F), which is hot enough to drive off the water adsorbed on the wheel, conditioning it for readsorption of water. This wheel rotates slowly because it must be cooled to ambient when it enters the position to readsorb water.

The MEC system has the added advantage that one can use it for heating in winter by reversing the airflow direction in the outside air loop and turning off the evaporation stage.

Typical performance for the MEC unit in a 2.7-ton size is as follows:

T_{ambient}	$T_{\text{wet bulb}}$	T_{output}	COP
27°C (80°F)	20°C (67°F)	16°C (61°F)	0.73
35°C (95°F)	25°C (77°F)	16°C (61°F)	0.31
38°C (100°F)	28°C (83°F)	16°C (61°F)	0.16

Note that the humidities are very high, as indicated by the high wet-bulb temperatures. In a dry climate the operating temperatures for equal coefficients of performance (COP, defined as the ratio of watts of cooling output to watts of heat input) are higher than indicated above.

14.20 TWO-STAGE EVAPORATIVE COOLING

Evaporative cooling is effective in a dry climate. It is also simple and inexpensive to operate and modest in price. It is not effective in a humid climate since the degree of cooling depends on the addition of water vapor to the entering air.

In a single-stage evaporative cooler, as discussed in Section 16.9, a fan pulls outside air through a porous wet pad. The airflow picks up some water vapor, in the process absorbing the latent heat of vaporization of water that has been evaporated. The result is that the airflow now has a greater partial pressure of water vapor and a significantly higher relative humidity since the air temperature has also been lowered. At low humidity for the entering air more grams of water per kilogram of air are evaporated and a lower air temperature results for a given final humidity. At high humidity, less water is evaporated and the cooling effect is less, so the comfort index is very poor. The variant of the evaporative cooler, known as a two-stage evaporative cooler, avoids this problem by injecting the cooled but wet air into a heat exchanger rather than into the building. The air for the building is then cooled by contact with the heat exchanger, thus adding no additional water burden to the inside air. Since the first stage of such a cooler cannot reach the dew point of the local air, one cannot dry the air passing through the second stage, but nevertheless the system is better for marginal climates than the single-stage system. Several variants of the two-stage evaporative cooler have been tested in Arizona, but few are currently in use.

14.21 PROBLEMS

14.1. Steam at 40 kg/cm^2 and 300°C is expanded adiabatically by means of the Rankine cycle to saturation at 1 kg/cm^2. What is the amount of energy extracted from a gram of fluid? How much heat went into heating the fluid to its initial temperature and pressure? What is the efficiency of the process? What is the Carnot efficiency for the same change in states? What percent of Carnot is achieved by this Rankine cycle change of state?

14.2. Steam at 40 kg/cm^2 and 300°C is expanded adiabatically by means of the Rankine cycle to 15% saturation and reheated to 300°C, then expanded in a second stage to 0.05 kg/cm^2 pressure. Calculate the same quantities as requested in Problem 14.1.

14.3. Explain why the stage *A* expansion in Fig. 14.7 is represented by an area rather than a single path as for stage *B*.

14.4. Calculate the theoretical Rankine efficiency for an expansion boiler plus reheat, with 195°C and 14.1 kg/cm^2 pressure water and a temperature drop in the expansion boiler of 55 C°. What is the pressure of the steam drawn from the boiler? Calculate through the cycle as in Fig. 14.16. How much work is extracted per gram of water entering the boiler?

14.5. Calculate the theoretical Rankine efficiency if steam were taken directly from the thermal storage reservoir of Fig. 14.15 when the tank is initially at 195°C and 14.1 kg/cm^2. What fraction of the stored water in the tank can be used before the temperature of the remaining water drops to 125°C? What is the theoretical Rankine efficiency of the system at this temperature and pressure? Compare the average efficiency over the same conditions shown in Fig. 14.21.

14.6. What is the Brayton cycle efficiency for a compressor pressure ratio of 3.0 when the inlet temperature is 550°C for the conditions cited in this chapter? What would be the theoretical Rankine efficiency for steam at this temperature and 60 kg/cm^2 pressure? What is the factor between these two turbines? What additional factors would influence a decision to be made toward either the Brayton or Rankine cycle?

14.7. A solar thermodynamic cycle is to operate at 200°C. What is the limit of efficiency for an evacuated collector system and for an air system with $X(\alpha/\epsilon)$ of 30? If the air system collector costs \$100 per square meter, what is the maximum allowable cost for the evacuated system? Repeat the calculations for an operating temperature of 300°C.

14.8. In case 1 (Fig. 14.15), how many kilograms of water will be required by a wet cooling tower to dissipate the heat load from a 1-kW$_e$ output? If a river is handy and its temperature is 10°C, how many liters per second would be required to dissipate the heat load from a 1-kW$_e$ output, assuming the river water discharge can be heated only 2 C° for environmental limitations?

14.9. Draw a diagram illustrating the environmental heat balance for a solar power plant operating at 200°C and one operating at 500°C. Indicate by the lengths of arrows the magnitude of each heat component or power component. State how you arrive at the approximate efficiencies and heat loads. What could be done to balance each system at the collecting sites, and also the points where the energy load is delivered?

Chapter 15
Direct Conversion to Electricity

15.0 INTRODUCTION

Converting sunlight into electricity without the complications of gathering the solar heat and passing it through a thermodynamic cycle has great appeal for its simplicity and, ultimately, its low cost. The achievement of low-cost solar cells, however, has been an exceedingly elusive goal. In 1975 a cost reduction on the order of 100–500 would be required for solar cells to compete with utility electric power. The vision of large energy farms using passive solar cells, on the earth or in space, or of cells on the home rooftop has ensured a major effort toward this cost reduction goal.

Direct conversion to electricity depends upon the *photoelectric effect,* which has been discovered in various forms over the past century. The basic photoelectric effect is the release of an electron when a sufficiently energetic photon of light is absorbed by certain materials. Two additional effects can produce electricity directly, but through the avenue of heat rather than the direct energy of the photon. The first, termed the *thermoelectric effect,* results when heat is applied to the junction between two dissimilar metals. The second, termed the *thermionic effect,* results when a metal inside a vacuum enclosure is heated to very high temperature. We thus have three principal ways to convert sunlight directly into electricity: (1) photoelectric, (2) thermoelectric, and (3) thermionic. The photoelectric effect is embodied principally in the form known as the solar cell.

In this chapter we discuss aspects of each relating to solar energy conversion. We will not go into the same level of detail as we have with thermal conversion, but we will supply enough information to provide an introduction to the field, and to give operating characteristics and assess the application potential of each method.

In operational practice solar cells have an advantage over thermal conversion modes in that a solar cell responds linearly to the flux. As the solar intensity fluctuates, the output stays in step. In a properly designed system this change in intensity does not change the voltage but only the current. There also is no inertia to a solar cell system: it immediately produces its output at the level appropriate to the solar intensity. Thermal systems, including thermoelectric and thermionic, need time to reach operating temperature.

15.1 DIRECT CONVERSION BY MEANS OF SOLAR CELLS

There are two types of photoelectric effect to be noted. In the case of the *external photoelectric effect* free electrons are emitted from a surface by the absorption of energetic photons. In the case of the *internal photoelectric effect* charge carriers are freed within the bulk of a material by the absorption of energetic photons.

The external effect is observed principally with metals inside a vacuum enclosure. The opacity of metals and some compounds to photons in the visible and ultraviolet is high, so the photon is absorbed very close to the surface. If the photon energy is above the surface work function of the absorber, the electron is ejected with sufficient energy to break free from the surface. The acceleration of these freed electrons by an external voltage applied to the absorber and an anode creates a photocurrent. The efficiency of the external effect is generally less than 50% since absorption events can occur deep enough in the material that the electron cannot escape. One cannot make a solar cell by use of the external photoelectric effect since no self-generated field is present to cause a current to flow.

The internal effect is observed principally with semiconductors. The opacity of these materials is lower than that of metals, so the principal effect occurs at a depth where very few electrons can escape from the surface. The process is one in which the absorption of an energetic photon creates a pair, an electron and a hole. When the photon has an energy greater than the band gap of the material, the charge carriers are free to move in the conduction band for a short time, after which they can recombine. The mean time between creation and recombination of a pair is termed the *lifetime* and the distance traveled in this time the *diffusion length.*

The condition for making a solar cell using the internal photoelectric effect is that a self-generated electric field must be present. This field is created by forming what is termed a *junction* between semiconductor materials of different types. This junction can be provided by using different materials or one material in which different dopant atoms are used. The junction causes the electrons to drift one way and the holes the opposite way. If a conductor is attached to the opposite sides of the junction a current is observed.

Efficiencies of Solar Cells

The efficiency of the internal effect is 100% for photon energies above the band-gap energy. The net efficiency of a solar cell is much lower owing to several effects to be described below.

The materials used for solar cells are summarized, along with their efficiencies, in Table 15.1. The approximate band-gap voltage is also indicated. For several materials the effective band gap is not that of the pure material, so these values are only for order-of-magnitude orientation.

Table 15.1 **Efficiencies of Solar Cells**

Material	*Band-gap voltage, eV*	*Efficiency, %*	*Status*
Si (wafer)	1.1	12–18	Commercial
Si (thin film)		2–5	Experimental
GaAs/GaAlAs	1.4	16–20	Experimental
CdS/Cu_2S	2.3	5–8	Advanced development
CdTe	1.4	5–6	Experimental
SiC		1–3	Experimental
GaP	1.9	1–3	Experimental
InP	1.3	2–5	Experimental

15.2 SILICON CELLS

The most widely used and technically developed type of solar cell is the silicon cell. Its popularity stems not from its scientific excellence but from the fact that it builds on the extensive solid-state technology and manufacturing experience of the semiconductor industry. Silicon is chemically stable and yields cells of long lifetime potential in the earth's environment. Most commercial cells yield 10% conversion efficiency; some now approach 15% in reliable quantities. The cost of solar silicon cells is, however, so high that their use as an energy supply for terrestrial applications is limited to specialized remote installations where the cost of power is minor compared to other costs, or where the service is essential, as with remote telephones or remote instrumentation sites.

The silicon cell is shown in cross section in Fig. 15.1. Single-crystal silicon of ultra-high purity is doped through its bulk with arsenic to produce *n-type* silicon. The surface of a wafer is subsequently doped with boron to produce *p-type* silicon. This type of cell is called a *pn-junction solar cell.* One can also reverse the types, yielding an *np-junction solar cell.*

Silicon itself responds to photons of solar wavelength, the absorption producing an electron-hole pair. If the silicon is sufficiently pure and structurally perfect, this electron-hole pair exists for a considerable time before finally spontaneously recombining. If, however, there are impurity centers and structural imperfections, as between different crystals of a multicrystalline silicon wafer, there is a high probability that the electrons and holes will recombine quickly. For this reason, high purity and crystalline perfection are necessary to make a high-efficiency silicon cell.

The additional factor necessary to get the electrons to flow and produce an electric current is an electric field. One can attach an external potential and get a photocurrent, as in ordinary photocells, but then one is supplying the field with external energy and no net yield is obtained. The doping, however, does provide an

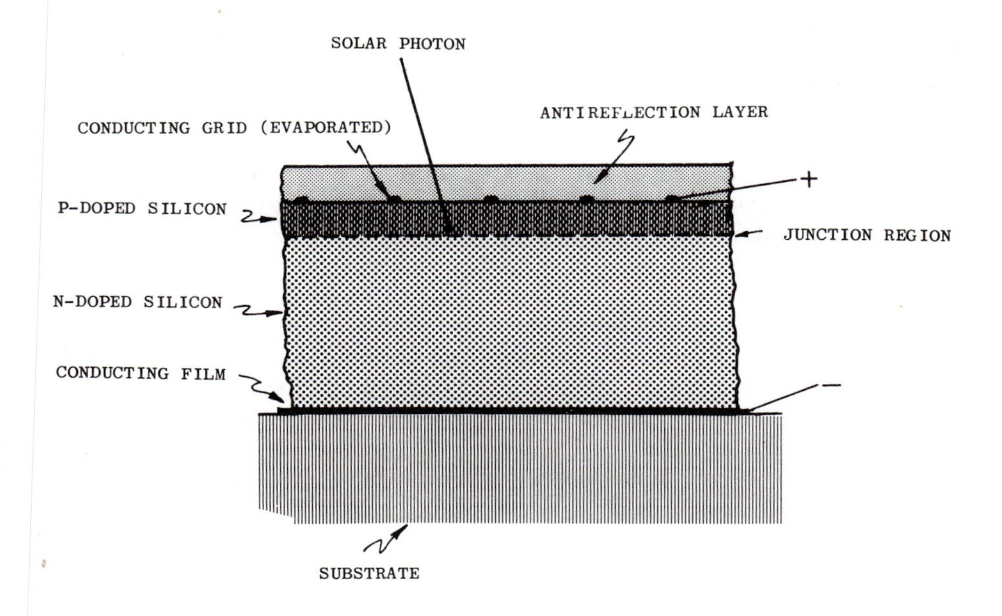

Fig. 15.1 Schematic cross section of a silicon (*pn*) solar cell.

internal field. The *pn* or *np* junction provides a strong field in the immediate region of the junction, sweeping the electrons to one side of the junction and the holes to the opposite side. This field is permanent and provides the charge concentrations that yield a spontaneous photocurrent when conductive leads are attached to the two sides of the junction.

The junction and its field occupy a very small linear thickness and affect the electron-hole pairs formed within a short distance of the junction, called the *diffusion length.* The diffusion length is the average distance a pair can travel before being recombined with a recombination center. The depth of the upper *p* layer is one or two micrometers and the diffusion length a few tenths of a micrometer (0.1–0.3 μm). To yield an electron to the photocurrent, the absorption event in the silicon must occur within a diffusion length of the junction. If the absorption occurs deeper, the electron-hole pair will not reach the junction. For this reason the actual yield of current is less than is theoretically possible, since every photon of enough energy will create a pair. The difference between silicon and other semiconductors is their absorption. One wants high optical opacity to stop the photons at a short distance from where the junction is placed.

Silicon cells are constructed with a grid of fine conducting paths on the face to pick up the photocurrent. The fineness of this grid is determined by the flux of radiant energy for which the cell is designed. There is no problem in extracting the current for solar fluxes, but if one uses flux brightness concentration C, care is required in the design of the conducting grid spacing and pattern.

The variation of sensitivity with wavelength for silicon solar cells is determined by a number of competing effects. The exact cutoffs in the infrared and the blue can be adjusted in manufacture by varying appropriate physical parameters. The opacity of silicon changes rapidly with wavelength, and consequently the depth of placement of the junction below the surface of the silicon has much to do with the spectral sensitivity. If the junction depth and diffusion length are beyond the region of absorptance for blue light, the photon simply produces heat because the pairs recombine before they reach the junction. In the infrared, on the other hand, the opacity is lower than in the visible, and the photon penetrates too deeply for the pair to diffuse to the junction, the result also being heat production rather than electron production. In Fig. 15.2 we show the variation of opacity with depth in silicon for the solar spectrum. Most of the shorter wavelengths are absorbed in the top layers, whereas the infrared wavelengths penetrate deeply.

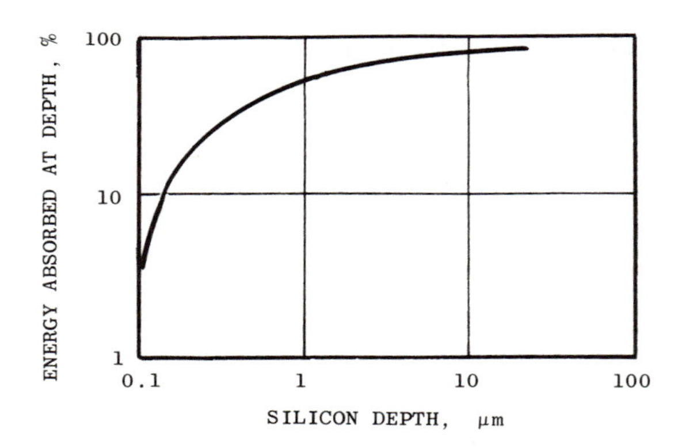

Fig. 15.2 Variation of the amount of the solar spectrum absorbed in silicon as a function of depth in the silicon.

In Fig. 15.3 we show the net effect on the solar spectrum, following Seraphin (1974). At the wavelength of optimum efficiency the photons are statistically absorbed at the junction. Those absorbed progressively on either side of the optimum wavelength are in part lost to the junction, and the efficiency drops.

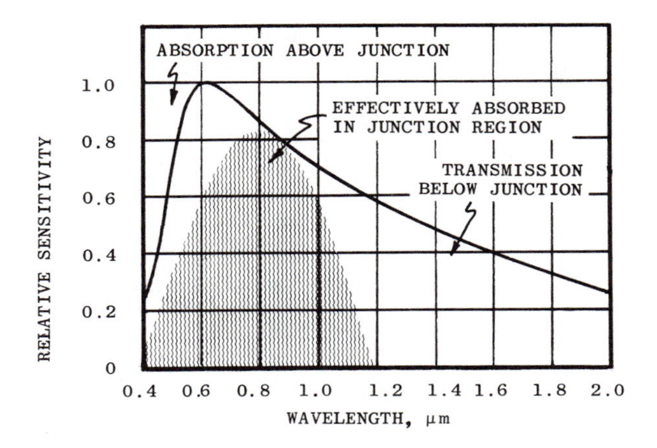

Fig. 15.3 Functional description of the wavelength sensitivity of a silicon solar cell, after Seraphin (1974). The shaded area represents that portion of the spectrum where the photons are absorbed within the diffusion length of the junction. All other photon events produce only heat.

The variation of sensitivity with wavelength for silicon solar cells is shown in Fig. 15.4. The shaded curve shows the range in which one finds the absolute sensitivity of cells, the exact curve depending upon the particular cell. The user should determine the exact curve for the type of silicon cell he plans to use before calculating other aspects of the system. In general a silicon cell has no sensitivity longer than 1.1 μm and shorter than 0.4 μm. Because the solar intensity curve is skewed with regard to the

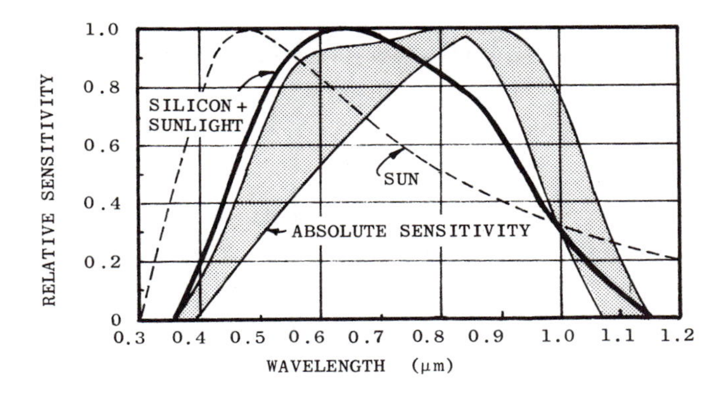

Fig. 15.4 Sensitivity as a function of wavelength for silicon solar cells. The shaded area represents the range of values for sensitivity to a source at constant brightness. The heavy curve is for the sensitivity to sunlight.

silicon cell sensitivity curve, the resulting solar cell sensitivity for sunlight is shifted to the blue of the absolute curve. A typical solar sensitivity curve for a silicon cell is shown in Fig. 15.4.

The characteristic curve for a silicon cell is shown in Fig. 15.5, where cell voltage is plotted against cell current. The point of maximum power output, *Va,* is near the knee of the curve. This characteristic curve is also variable from cell to cell, depending on the conditions of manufacture. If the electrical system behavior is to be adequately designed, the characteristic curve for the cell to be used should be utilized. The characteristic curve for a cell is also temperature dependent, the drop in voltage for a given current being the major effect, as discussed below.

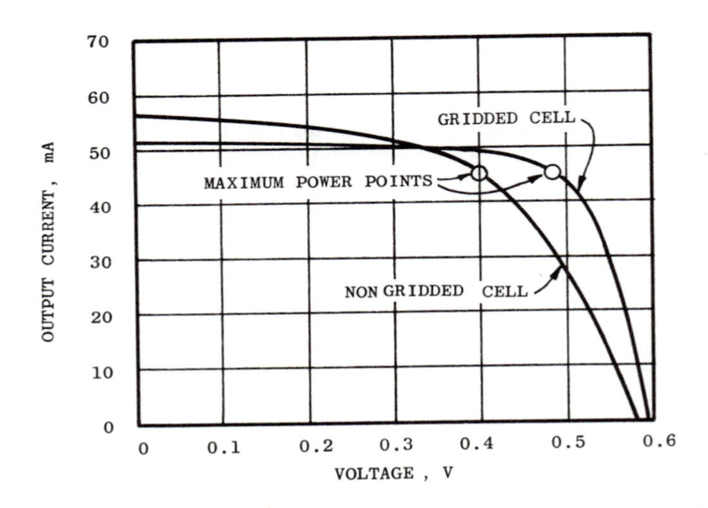

Fig. 15.5 Characteristic curve for a silicon cell with collection grids and without. The points of maximum power output are indicated.

The characteristic curve also depends on the design of the grids and the intensity of the sunlight, the variation being caused by changes in the efficiency of collection of the photocurrent. Closer grid spacing is required for cells designed to use flux concentration.

The variation of silicon solar cell output with solar flux intensity is shown in Fig. 15.6. When the cell is shorted, so that the current is maximum and the voltage zero, the cell is linear in response. When a constant voltage is maintained, the current output is approximately linear. The locus of maximum power output is shown as a dashed line. The cell output along this line is very close to linear.

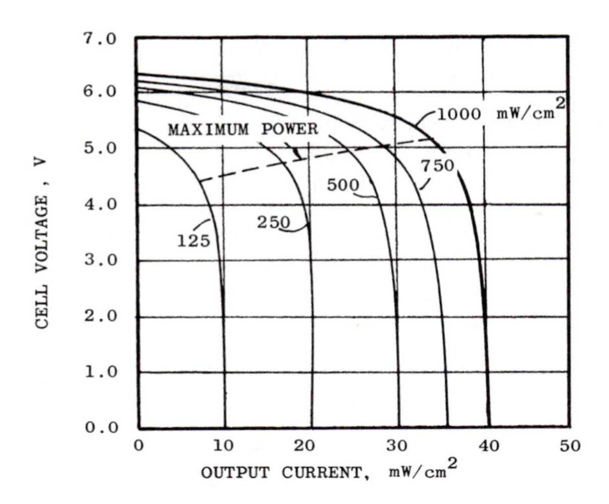

Fig. 15.6 Variation of output with solar flux brightness for a typical silicon solar cell. The cell output is linear with intensity when shorted and approximately linear at maximum output wattage.

15.3 MANUFACTURE OF SILICON CELLS

The high cost of silicon solar cells is attributable not to the intrinsic cost of the basic materials but to the cost of processing and energy. Silicon is one of the most abundant elements in the crust of the earth, but it is never found in native form and must be separated from its compounds. The raw material for metallurgical grade silicon is silica sand, SiO_2. This is reduced by heating with carbon (coke) in an electric furnace, yielding a purity of about 96–98%. Further purification, to 99.7% and better, can be obtained by repeated leaching.

The purity required for solar cells is extremely high. To obtain high purity, the commercial grades can be chemically converted to silicon tetrachloride ($SiCl_4$), which is reduced to silicon by reaction with metallic zinc, yielding a material 99.97% pure. This material is further purified by growing a crystal by the Czochralski method, whereby a boule of silicon is slowly drawn out of a melt of silicon. The purity of silicon at this point is seldom referred to in percentage, but instead in terms relevant to its final use–its resistivity in ohms-centimeters. Good silicon for solar cells is in the range of 0.2 Ω cm. The alternative processes, currently favored, are pyrolytic decomposition of silane (SiH_4) or hydrogen reduction of trichlorosilane ($SiHCl_3$).

The traditional method of making solar cells requires a high-purity silicon crystal boule to which a dopant is added in the crystal-making process to make the boule *n*-doped. Six additional steps are then performed, each of which has several substeps:

1. The boule is sliced into wafers as thin as can be handled in subsequent processing. A typical wafer is 75 mm in diameter and 0.5–1.0 mm thick.

2. The wafer is lapped and polished to remove defects in the surface structure caused by sawing, and the wafer is then cleaned.

3. Doping atoms of a different type are diffused into the wafer to produce a surface layer p-doped, establishing the *pn junction* for typical cells. For *np* cells the wafer is originally doped p and the diffused surface layer is doped n.

4. The wafers are carefully cleaned and placed in a vacuum chamber, where a conductive grid is formed by evaporating metal onto the wafer surface. A mask keeps the metal from being deposited on the portions of the wafer surface that are to be the active regions.

5. The wafer is then sintered or soldered to the mounting substrate contacts. At this point the wafer has become a solar cell.

6. Testing of the completed wafers is necessary to provide quality control of the final product. Cells meeting the specifications are then assembled into the final array, consisting of tens to thousands of wafers.

For terrestrial uses the solar silicon cell wafers are round, the diameter of the original boule. For space applications, where weight is critical and structures must be compact, the solar cell wafers are generally rectangular for dense packing.

A significant part of the cost is the energy involved in the above steps. Iles (1974) has given an energy breakdown for the production of 2×10^6 cells per year, along with the number of years required for the finished cell to "repay its debt," as shown in Table 15.2. This table shows cells of different types. Note that when the efficiency is low, as for polycrystalline film cells, the payback time becomes very long in terms of the energy required in their production.

Table 15.2 Debt Times for Five Sample Cell Production Examples

Cell type[a]	*Cell size, cm²*	*Efficiency, %*	*Annual operating hours*	*Power for annual cell production, kW*	*kWh/yr*	*Debt time, yr*
A Space	4	11	8.75×10^3	120	1.05×10^6	2.1[b]
B Space	12	11	8.75×10^3	360	3.15×10^6	0.7[b]
C Terrestrial (single crystal)	20	10	1.8×10^3	400	7.2×10^5	3.05[b]
D Terrestrial (polycrystalline film)	50	2	1.8×10^3	200	3.6×10^5	5.1[c]
E Terrestrial (polycrystalline film)	50	1	1.8×10^3	100	1.8×10^5	10.2[c]

[a] Assume 2×10^6 cells per year manufactured.
[b] With annual energy usage of 22×10^5 kWh.
[c] With reduced annual energy usage of 18.5×10^5 kWh.

The goal of inexpensive solar cells calls for different approaches. One approach is to stick with the familiar Czochralski process and make the cost breakthrough in the automation of the production of the cells. A second approach is to make the cells by thin-film processes, like sputtering or chemical vapor deposition (CVD), upgrading the resulting polycrystalline material to have very large polycrystals, and raising its efficiency from the order of 1% to as much as 5–10%. A third approach is to grow the silicon crystal in the final shape to be used, for example, ribbon silicon.

15.4 EFG RIBBON SILICON

One of the most promising techniques of reducing the cost of silicon cells is through the production of EFG silicon, or ribbon-growth silicon. This technique was first developed by Stephanov for growing ribbons of germanium, and hence is known alternatively as the *Stephanov process.* In this process the stages of crystal growth, cutting, and wafer production are shortened by growing the crystalline silicon as a thin ribbon of a single crystal. The term EFG stands for *edge-defined film growth.* The formation of a ribbon of crystalline silicon is illustrated in Fig. 15.7. The silicon is pulled directly from a melt of high-purity silicon through a die the approximate shape of the final ribbon. Surface tension causes the liquid silicon to adhere to the seed crystal and the edges of the die. When the rate of pulling is properly adjusted, the silicon emerges and quickly solidifies into a ribbon of the dimensions of the die,

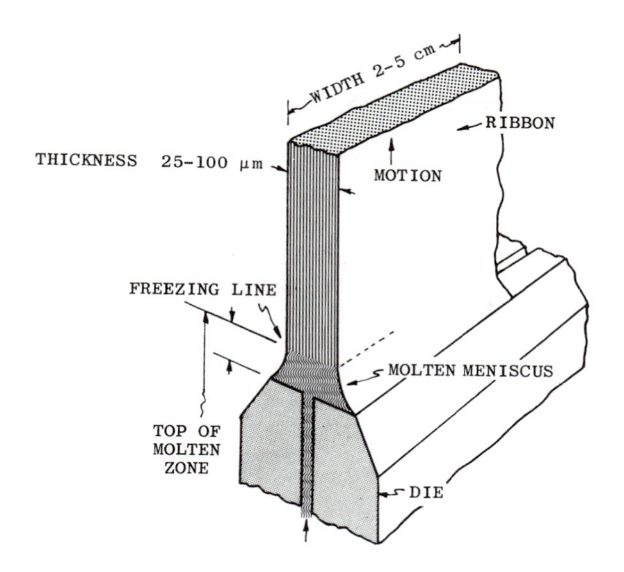

Fig. 15.7 Schematic cross section through the die and ribbon of the Tyco EFG silicon ribbon-growing system. The molten silicon flows up through the die, forming a meniscus the shape of the top of the die, solidifying to form the ribbon.

approximately 100 μm in thickness and 2–4 cm in width. The length depends on how long the process can be stabilized and how long the purity of the emerging silicon remains within the strict limits necessary for long pair lifetime. Silicon is almost the perfect solvent, rivaling water. The result is that it is difficult to find a die material that will not inject harmful impurity atoms into the silicon ribbon. Progress by Tyco Corporation in the EFG process offers encouragement that ribbon-grown silicon will bring about a major reduction in the cost of silicon solar cells of 10–15% conversion efficiency.

15.5 POLYCRYSTALLINE SILICON CELLS

The cost of producing silicon cells from single-crystal silicon has led to considerable effort to find ways of using polycrystalline material. It is generally thought that if the crystal size can be made large compared to the required diffusion length, the cell will act reasonably like a single-crystal cell.

Polycrystalline silicon can be deposited in a number of relatively inexpensive ways, including vacuum evaporation, sputtering, and chemical vapor deposition. The attendant problem of reducing cost is to find an inexpensive substrate upon which to deposit the silicon. Chu (1974) has reported experiments with steel, sapphire, and graphite for substrate materials. Both spinel and sapphire have lattice constants reasonably close to silicon and are easily formed into single-crystal ribbon, or slabs, on which the silicon is deposited. Unfortunately the cost of such substrates is high and the goal of low cost could not be achieved even if good cell performance were obtained. At present the best lifetimes for silicon on sapphire are in the range of 10^{-8}–10^{-9} sec, a factor of at least 100 shorter than necessary for efficient solar cells.

Steel substrates are inexpensive; so is aluminum as used by Fang *et al.* (1974). Steel has bad impurities for silicon; these impurities and iron diffuse readily into the silicon. A barrier layer of evaporated or sputtered silica (SiO_2) or borosilicate glass was found by Fang to isolate the silicon. However, problems are still serious enough to lead Fang to discount the prospect of obtaining good silicon solar cells from steel/borosilicate substrates. Experiments by Fang using graphite for the substrate did yield polycrystalline silicon cells having efficiencies for sunlight of 1.5%. Some properties of substrate materials of importance to thin-film solar cells are listed in Table 15.3.

15.6 CADMIUM SULFIDE SOLAR CELLS

Cadmium sulfide (CdS) cells were developed concurrently with silicon cells for early space applications, but soon lost the race to silicon. One of the problems with cadmium sulfide cells was the unreliability of performance of manufactured cells and their propensity to degrade in the terrestrial environment. The fact, however, that they can be produced by inexpensive thin-film deposition techniques has enhanced their potential value for terrestrial applications.

A cross section of a cadmium sulfide cell is shown in Fig. 15.8. The cadmium sulfide cell was named thusly when it was thought that the cadmium sulfide was the

Table 15.3 Thermal Expansion and Lattice Parameters

Material	*Crystal structure*	*Lattice parameters,* Å	*Thermal expansion* $k/C°$, (10^{-6})
Silicon	Diamond		3.6
Sapphire	Rhombohedral	a = 4.76 c = 12.99	8.4
Spinel	Cubic	8.08	7.45
Steel	–	2.87	9.6–11.2
Aluminum	–	4.09	23.1–30.0
Carbon	–	2.46	0–2.0

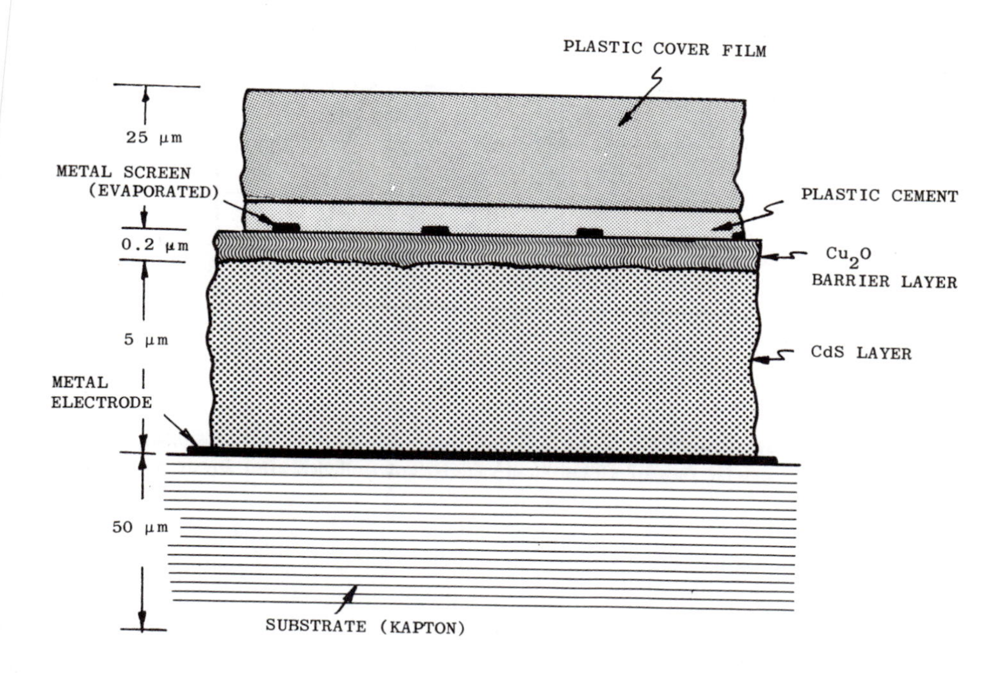

Fig. 15.8 Schematic cross section through a cadmium sulfide heterojunction solar cell. The cover can also be evaporated glass and the substrate metal foil.

active material. The importance of the copper oxide and the junction between the two materials is now recognized, and the term *heterojunction* is now applied to this type of cell, the junction between two chemical species being the active region. The proper designation for this cell would be $CdS/Cu_{1.98}O$, but for brevity heterojunction cells are designated by one or the other of the two materials, the choice being somewhat the choice of the researchers first working with the materials. In Fig. 15.8 we denote the copper oxide layer as being Cu_2O. The actual ratio of copper to sulfur must depart from 2:1 for an efficient cell. The nominal composition is $Cu_{1.96}$ to $Cu_{1.98}$ per sulfur atom. The several mineral forms of copper oxide are: chalcocite ($Cu_{2.00}S$), djurlite ($Cu_{1.96}S$ to $Cu_{1.93}S$), diginite ($Cu_{1.80}S$), anilite ($Cu_{1.75}S$), and covellite ($Cu_{1.00}S$).

The characteristic curve for a typical cadmium sulfide solar cell is shown in Fig. 15.9. The maximum power output point is at 0.75 A and 0.37 V. This curve changes with temperature, as discussed in Section 15.9. The efficiency of this cell was measured to be 5.1%.

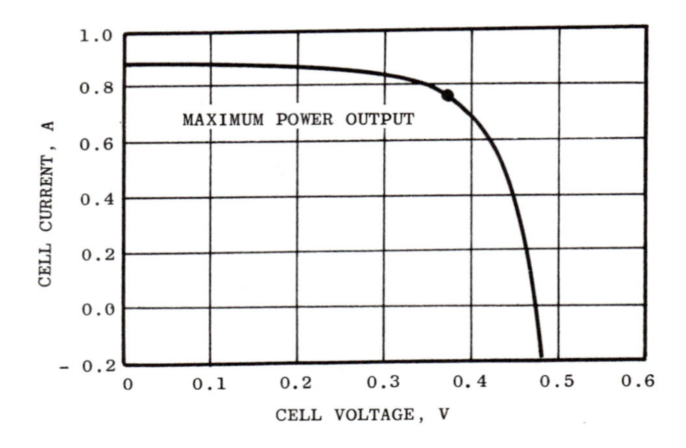

Fig. 15.9 Characteristic curve for a cadmium sulfide solar cell having 53-cm² area yielding an efficiency of 5.1% for a solar flux of 100 mW/cm² equivalent sunlight, after Brody and Shirland (1974).

The basic problem with the cadmium sulfide cells is that they are especially sensitive to water vapor. These cells must therefore be carefully encapsulated to survive in terrestrial applications. In addition the toxicity of cadmium is high enough to raise questions about the safety of wide utilization of these cells. A less toxic material such as zinc selenide (ZnSe) is a possible substitute.

15.7 MANUFACTURE OF CADMIUM SULFIDE CELLS

Cadmium sulfide cells are manufactured by the successive evaporation of the materials onto a substrate. This process requires very small amounts of the materials, and therefore can use considerably more expensive materials than the starting material for silicon cells. Brody and Shirland (1974) have outlined the steps in making cadmium sulfide cells on either a plastic (Kapton) or metal foil substrate, as given in Table 15.4. In the case of the metal foil substrate the Cu_2S layer and contact grid are covered with an evaporated layer of SiO_2 or glass.

Table 15.4 Fabrication of Cadmium Sulfide Solar Cells (Kapton Base)

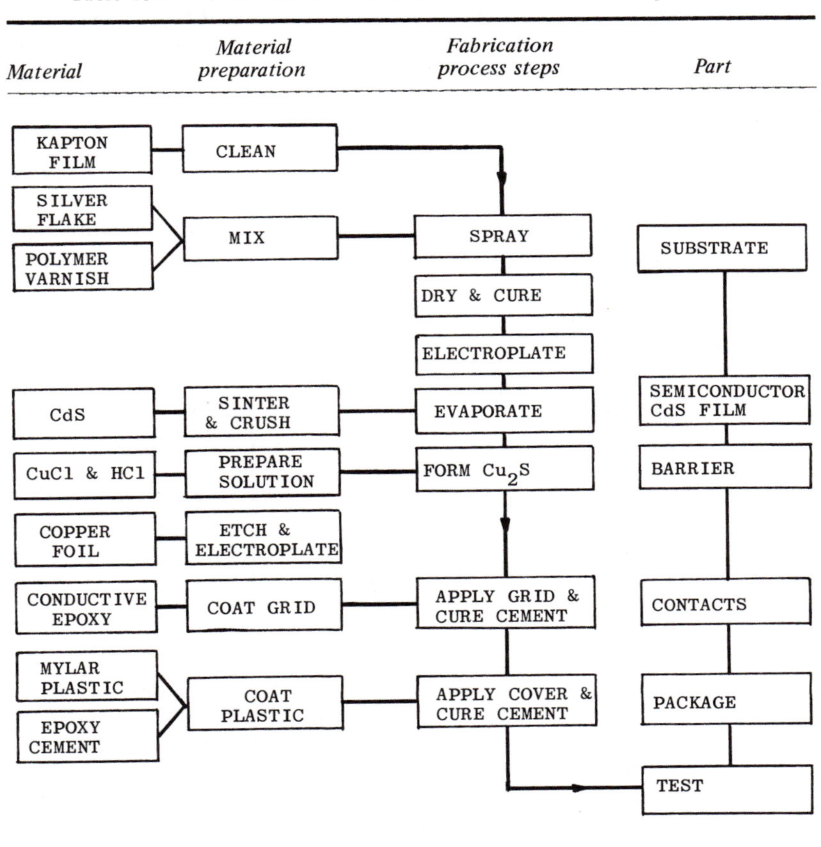

15.8 GALLIUM ARSENIDE SOLAR CELLS

Gallium arsenide (GaAs) cells are also the heterojunction type, the junction being between gallium aluminum arsenide and gallium arsenide. Their potential advantage is higher absorptance, allowing more solar photons to be absorbed within the diffusion length of the junction. Their further advantage is that the larger band gap allows this material to function at higher operating temperatures than silicon or cadmium sulfide.

Gallium arsenide cells are still very expensive, but they can yield 20–28% efficiency. Fabrication methods appear to include CVD deposition as well as standard crystalline processes.

15.9 THERMAL BEHAVIOR OF SOLAR CELLS

The thermal behavior of solar cells is very important to their application for terrestrial uses. The high cost of cells could be overcome significantly if they could be used with optical concentration. The area of the cell would thus be $1/X$ times the concentration, and since mirrors are presumably cheaper per square meter than solar cells, a net reduction in system application cost would be effected.

The basic problem with the use of optical concentration is that the solar cells heat under solar flux. This heating arises from the absorption of solar photons and subsequent pair recombination, which yields only phonon stimulation, yielding heat. All solar cells degrade with temperature in that their output is temperature dependent. In Fig. 15.10 we show the behavior of several types of cells as a function of temperature. The data used in this figure are from Wysocki and Rappaport (1960), but are here reploted with temperature as the abscissa.

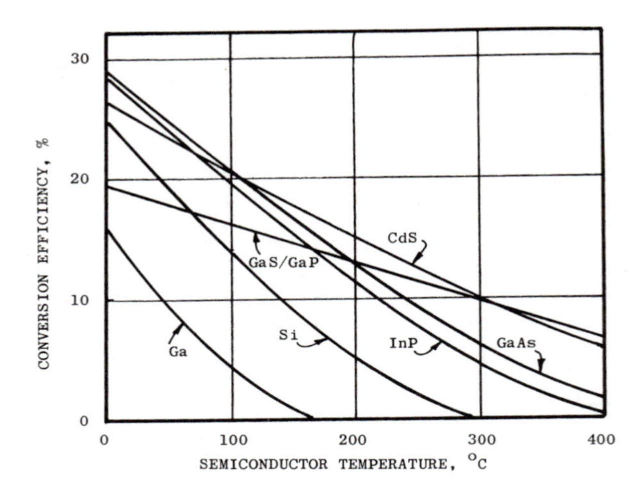

Fig. 15.10 Variation of efficiency of conversion of a semiconductor used as a solar cell as a function of material (bandgap) and temperature.

We note from Fig. 15.10 that gallium arsenide should be only slightly less temperature sensitive than silicon. In laboratory experiments gallium aluminum arsenide appears significantly better, giving yields at over 200°C. This theoretical curve therefore is not absolute when it comes to heterojunction cells. For example, cadmium sulfide cells should have better temperature performance than gallium arsenide, but in fact they have poor thermal performance curves. The difference is apparently that the heterojunction devices do not act as high-purity semiconductors should act. One of the goals is therefore to develop the full thermal potential of the wide-band-gap materials.

Since solar cells are generally sensitive to only the near infrared and the red side of the visible solar spectrum, the incident solar flux load can be reduced by coating the cells to reflect the unwanted wavelengths, and at the same time antireflecting the wavelengths effective in producing the photocurrent.

The change in cadmium sulfide solar cell output with temperature is shown in Fig. 15.11. At a constant cell voltage the output current drops rapidly with temperature. At constant cell current, the voltage drop with rising temperature is not as serious. To maintain maximum power output, such a cell must have its operating voltage changed with temperature. The locus for both constant voltage and maximum power output is shown in Fig. 15.11.

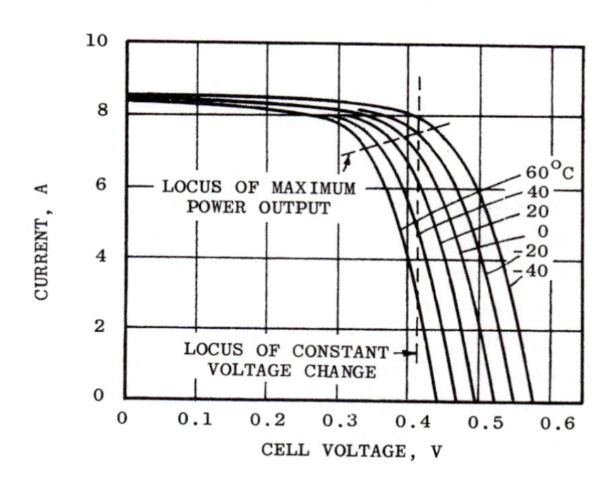

Fig. 15.11 Variation of the characteristic curve for cadmium sulfide cells as a function of temperature. The change in cell efficiency for constant voltage and maximum power output is shown in Fig. 15.12.

The temperature dependence of silicon solar cells (Wichner, 1974) is shown in Fig. 15.12 for a barrier height of 0.8 eV and an ideal *pn* junction cell having no recombination current, compared with experimental values for a standard diffused junction silicon cell.

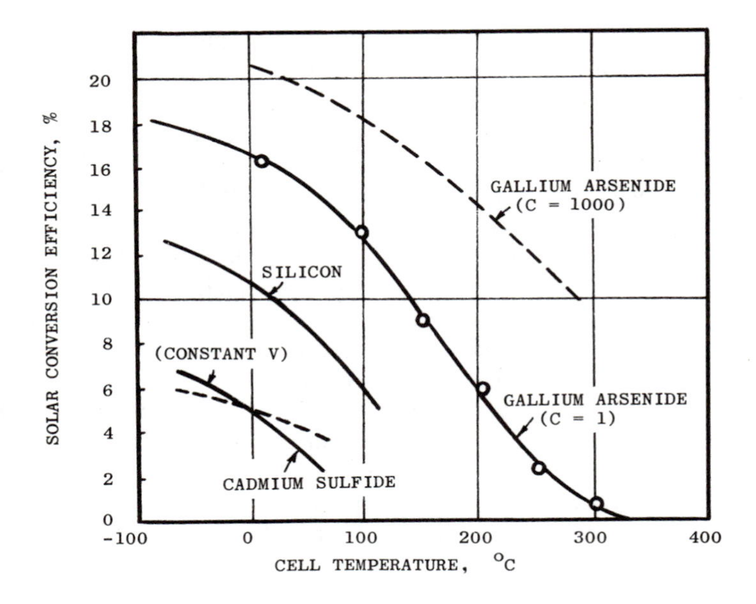

Fig. 15.12 Variation of efficiency with cell temperature for the major types of solar cells.

The temperature dependence of a gallium arsenide heterojunction cell (Bell, James, and Moon, 1974) is also shown in Fig. 15.12. The efficiency of solar cells is dependent upon flux concentration, improving significantly at higher temperatures, as illustrated for gallium arsenide, for unit flux concentration and 1000 concentration.

The curves for variation of cell efficiency with temperature, shown in Fig. 15.12, are rather similar in shape but different in height. These curves are for a particular cell sensitivity, but it is apparent that one can safely take the sensitivity of the actual cell to be used and estimate its temperature behavior simply by drawing a curve of the form of the curves shown, through the value of the cell sensitivity to be used.

The reason for the increase in efficiency with temperature predicted by Bell *et al.* is obtained from the equations for the voltage and current at optimum power:

$$V_p = V_{oc} - (nkT/q)\ln(qV_pI_s/nkTi_p) \tag{15.1}$$

and

$$i_p/I_s = 1 - \exp[q(V_p - V_{oc})/nkT], \tag{15.2}$$

where V_p and i_p are the values of the voltage and current at optimum power loading, V_{oc} is the open circuit voltage, nkT is the Boltzmann factor, and I_s is the solar intensity. These two equations must be solved by iteration, and the product $i_p V_p$ is formed to calculate the efficiency as a function of I_s. The projected gain is, however, subject to the inherent adverse temperature effects due to the construction of the cell.

The long-range advantage of a cell of the high-temperature gallium arsenide type is that such a cell could be used in a combined cycle system. The cell would produce (1) direct power output and (2) heat, the heat being transferred to a coolant at approximately 400°C to operate a standard thermodynamic cycle. The combined net conversion efficiency could be as high as 40%, that is, 15% for the gallium arsenide cells and another 25% for the steam cycles.

15.10 COOLED SOLAR CELLS FOR CONCENTRATING SYSTEMS

One approach to reducing the cost of solar cells is to use optical concentration, thereby reducing the area of solar cell for collection of a given amount of solar flux. Concentrators for use with solar cells can be any of the many types discussed in Chapters 5, 6, and 7. Even plane mirrors set on either side of a row of solar cells can double or triple the flux on the cells. The basic problem with concentration is the decrease of cell efficiency as the cell temperature increases, as shown in Fig. 15.11. It is therefore necessary to find ways to reduce the amount of heating by either rejecting unusable solar flux or removing heat rapidly from the cell.

The operating temperature of a solar cell under optical concentration can be calculated in the simple manner described in Section 12.5 for calculating the heat balances for a flat-plate collector: by balancing the heat inputs and outputs. The input flux can be reduced by means of optical coatings to reflect sunlight in spectral regions where the cell is insensitive, thus allowing only light that produces the cell current to enter the solar cell. Alternatively, one could add a coating that would be highly emissive in the thermal infrared to help reject heat by radiative means. Whether one could combine the reflective filter for shorter wavelengths with the emissive filter for the thermal infrared is not certain.

Silicon cells are being studied as the best candidate for a low-cost concentrating solar cell system. There are two problems to be faced: (1) the control of cell temperature and (2) collection of the generated electrical carriers. The first problem is met by conducting the heat from the cell surface to a copper block or heat pipe, dispersing the heat to a large surface where natural convection can carry off the required heat load. The second problem is met by using fine grid spacing on the cell surface. A number of designs are being explored that are reputed to allow flux concentrations on the order of 500–1000 with temperature rises on the order of 25–50 C° above ambient.

Active cooling of solar cells under flux concentrations up to 280 was done by Beckman, Schoffer, Hartman, and Lof (1966). With close-gridded cells they were able to hold the temperature rise to 90 C°. A diagram of the configuration used is shown in Fig. 15.13. The cell efficiency was reduced from about 7% at room temperature to 6% at 90°C, with an output power of 2 W/cm^2 of solar cell. The temperature rise of the cell surface over that of the cooling water under the assumption of one-dimensional

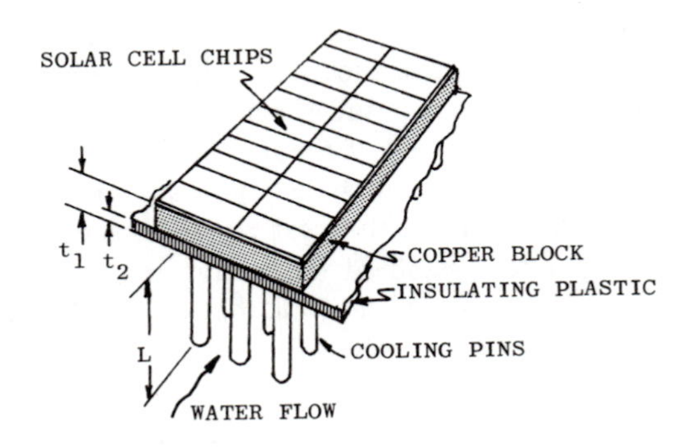

Fig. 15.13 Schematic diagram of a water-cooled silicon solar cell array used by Beckman *et al.* (1966) with a 280 concentration parabolic concentrator.

heat conduction is given by

$$T = \frac{4\alpha t I A t_2}{\pi k N D^2}\left(1 + \frac{1}{t_2 m \tanh mL} + \frac{t_1 N D^2}{4 t_2 A}\right). \qquad (15.3)$$

where

N = number of pins in the array,
D = diameter of the pins,
L = length of the pins,
A = total solar cell area,
$\alpha t I$ = absorbed solar flux,
k = thermal conductivity of pins,
h = heat transfer coefficient,
t_1 = thickness of electrical insulation adjacent to the pins,
t_2 = thickness of the copper block, and
$m = (4h/kD)^{1/2}$.

The heat-transfer coefficient used by Beckman *et al.* was determined from an empirical correlation

$$h = (k/D)(1.11\,C)(\mathrm{Re})^n(\mathrm{Pr})^{1/2}, \qquad (15.4)$$

where the two constants are to be taken as appropriate from Grimison (1938).

The test system used eighteen 1 × 2-cm silicon solar cells with 20 grid lines per centimeter and produced 50 W using a parabolic mirror concentrator 167 cm in diameter. The rate of water cooling was 90 l/h.

15.11 THERMOELECTRIC SOLAR CELLS

A thermoelectric solar cell is one in which a current is generated as the result of the voltage appearing at the junctions of two dissimilar metals when one set of junctions is maintained at a different temperature than the other. The ability of metals to generate this voltage is given by the Seebeck coefficient α. Any two metals produce a Seebeck coefficient, but by careful selection one can greatly increase the effect.

Thermoelectric cells, like silicon solar cells, yield a system with no moving parts; unlike solar cells, however, they require high temperature and thus concentrating collectors, to attain usable efficiency. The voltage such a device generates is linearly proportional to the temperature difference between the hot and the cold junctions. A diagram of a basic thermojunction is shown in Fig. 15.14. The thermal energy is fed into the unit through a plate of material best matching the expansion coefficients of the two thermoelectric materials, so as to minimize the problem of soldering the materials together. One side of the junction is a metal with as negative a value of the Seebeck coefficient as possible, while the other side is as positive as possible, to make the largest difference. The contacts are then attached to the cold end of the unit. By interconnecting the cold ends of one junction with the opposite sign end of the adjacent thermoelectric element, one can cascade the voltages up to the desired level.

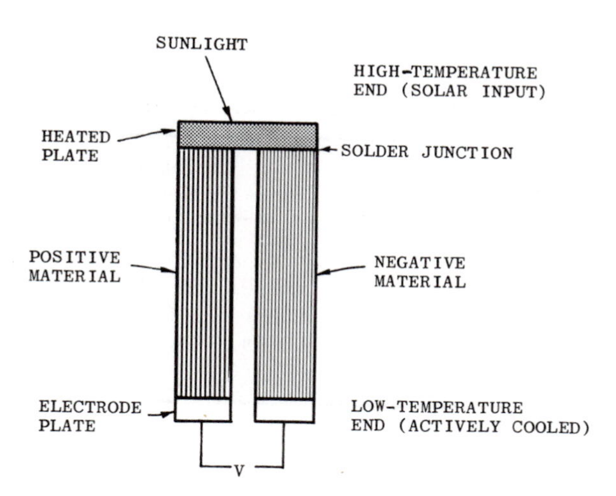

Fig. 15.14 Basic configuration of a thermojunction device combining two materials of widely different Seebeck coefficients.

For a good presentation of the theory of the thermoelectric effect the reader is referred to the excellent text by Kettani (1974). There are two forms used to summarize the efficiency equations for thermoelectric effects (Gaibnazarov, Malevskii, and Rezgol, 1969). The first is

$$\eta_{\text{TEmax}} = \underbrace{\frac{T_2 - T_1}{T_2}}_{\text{Carnot term}} \underbrace{\left(\frac{X-1}{X + T_2/T_1} \right)}_{\text{Heat conduction loss term}}, \tag{15.5}$$

where X is the ratio of the external load resistance to the internal load resistance. The first term is the inescapable Carnot term, limiting efficiency of a perfect system. The second term is the irreversible heat conduction and Joule effect. The second equation uses different input quantities:

$$\eta_{\text{max}} = \eta_C (X-1) / [\alpha(X+1)/\alpha' - \eta_C (X-1)], \tag{15.6}$$

where η_C is the Carnot factor, α is the total Seebeck coefficient of a cascaded system, α' is the coefficient for the hot junction, and X is the ratio of the external to internal load resistances.

Thermoelectric materials having high Seebeck coefficients and high operating temperatures are listed in Table 15.5.

Table 15.5 Properties of Thermoelectric Materials

Material	*Upper temperature limit, °C*	*Type*
0.8 Bi_2Te_3 + 0.2 Bi_2Se_3	200	*n*
PbSe	450	*n*
0.75 PbTe + 0.25 SnTe	600	*n*
0.8 Sb_2Te_3 + 0.14 Bi_2Te_3 + 0.06 Te	200	*p*
0.95 GeTe + 0.05 Bi_2Te_3	600	*p*

Thermoelectric elements are generally used in combination to yield usable voltages. They can be cascaded as multistage elements or as composite elements. A very basic problem with such thermopile systems is how to attach them together and how to attach junctions. The principal problem is that they often have very different thermal expansion coefficients. Also, one must find a solder that is compatible with the diverse elements. Plating the thermoelectric materials with an alloy of tungsten and cobalt has been done in the USSR, and a solder made of nickel and tin has been used as well. A bismuth-tin-lead solder is also used in conjunction with the contact areas of the thermoelectric material plated with palladium. (Typical voltage output as a function of temperature difference between the hot and cold junctions is shown in Fig. 15.16.)

The problem of the thermal expansion of the material is also complicated by the need to maintain a large thermal difference between the hot and cold junctions, and to keep these junctions close together. Water cooling is essential, but it adds to the sealing problem.

A typical variation of the voltage as a function of the temperature is shown for a germanium-tellurium thermojunction device measured by Baranova, Malevskii, and Saplizhenko (1969). Small changes in the content of tellurium affect the Seebeck coefficient, and the range of values is indicated by the swath in Fig. 15.15. Values for the Seebeck coefficient can be as high as 230 μV/C° for a *p*-type Bi_2Te_3-Sb_2Te_3 compound at 200°C. The actual voltage output as a function of temperature difference is shown in Fig. 15.16 for a relatively simple compound of lead and selenium.

Thermoelectric devices have not found widespread utilization in solar energy conversion because of a combination of technical difficulties and cost in terms of conversion efficiency. Conversion efficiency is low, ranging from 1.8% at a temperature difference of 195 C° to 2.5% at 350 C°.

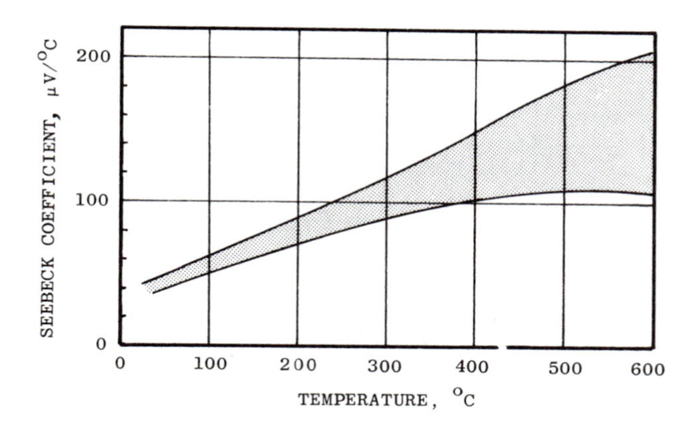

Fig. 15.15 Variation of the Seebeck coefficient for germanium with from 62–65% tellurium added as a function of material temperature.

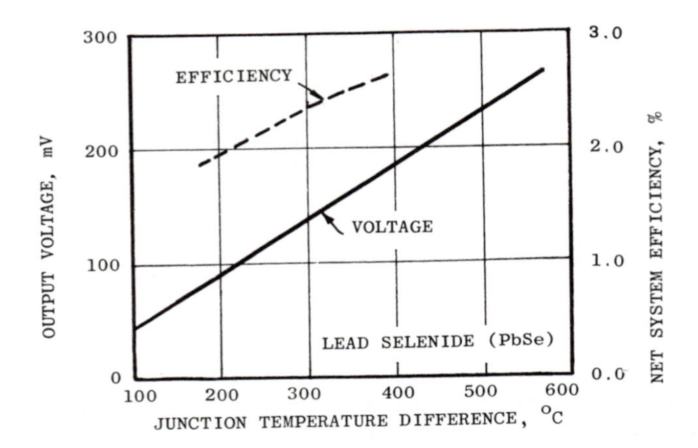

Fig. 15.16 Variation of cell output voltage as a function of temperature for a lead-selenium thermojunction device, by Alatyrtsev *et al.* (1969).

15.12 THERMIONIC SOLAR CELLS

A heated metal surface in a vacuum enclosure will emit electrons, forming a space charge about the heated surface. If a second plate of metal at a lower temperature is placed in the same vacuum enclosure, an electric current will flow from the hot to the cold plate. The thermionic cell is therefore a vacuum tube designed to accentuate this effect to a point where practical amounts of power can be extracted from the device. The basic elements of such a device are shown in Fig. 15.17.

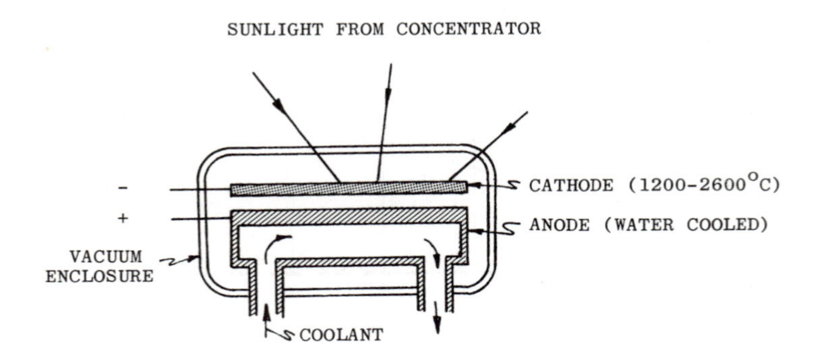

Fig. 15.17 Schematic diagram of a thermionic solar cell. Electrons emitted from a very hot cathode plate travel to the cold metal plate, generating the electric current.

The basic problem in making a practical thermionic device is the high temperature needed to get high rates of electron emission per square centimeter of cathode, being in the range of 2000–2700°C for pure metals and above 1200°C when easily ionized atoms such as cesium are added to the vacuum space. These high temperatures lead to high radiation losses and evaporation of the cathode material, which is then deposited on the cool window of the device.

The spacing between the cathode and the anode is generally very small—otherwise objectionable space-charge effects reduce the net current. The spacing in some experimental units has been as small as 10–20 μm, but heat transfer is so high across the small gap that it is difficult to maintain a high temperature difference between the cathode and anode. A high temperature difference is required to maximize the current.

The objectionable space-charge effects can be reduced by adding cesium vapor to the space, or by other methods outlined by Kettani (1974).

The efficiency of some bare metal cathode thermionic tubes has been reported as 4% at 2600°C by the USSR and 5% at 2400°C by the United States. Brosens (1965) reported that with a cesium thermionic tube one could expect from 10% conversion at 1200°C to 15% at 2200°C.

Because of their high operating temperatures, thermionic solar cells require high optical concentration, in the vicinity of 1000. This high value requires a parabolic mirror of considerable optical precision. Because of the cost of the collector, the problems of maintaining a high vacuum, and heat losses, thermionic cells appear to be cost effective only for space applications, losing out to simple silicon cells for simplicity and reliability. Further development of thermionic solar devices appears promising, but presents a major challenge to materials science.

15.13 PHASE-CHANGE THERMAL DIRECT CONVERSION

A particularly interesting novel thermodynamic cycle involving a phase change between liquid and gaseous sodium has been studied by Weber (1974), as diagrammed in Fig. 15.18. The fluid is pumped through a high-temperature zone to heat the sodium to 500–800°C. The hot sodium is then passed through a porous electrode of beta alumina (solid). A large pressure drop occurs on the two sides of the alumina electrode, so there is liquid sodium on the high-pressure side and sodium vapor on the low-pressure side. When conducting electrodes are placed on the two sides of the alumina plug, a voltage is developed and a current generated. The sodium vapor is then condensed at a cold region (condenser), repressurized by a pump, and returned to the hot region. The system is therefore analogous to a standard thermodynamic cycle.

The fact that the Weber cycle has no moving parts other than the sodium pump makes it attractive for development. Conversion efficiencies are high, being 20% at 600°C. As in many solar energy areas, technical problems are in the field of materials properties, in particular with regard to the production and lifetime of the beta alumina electrode.

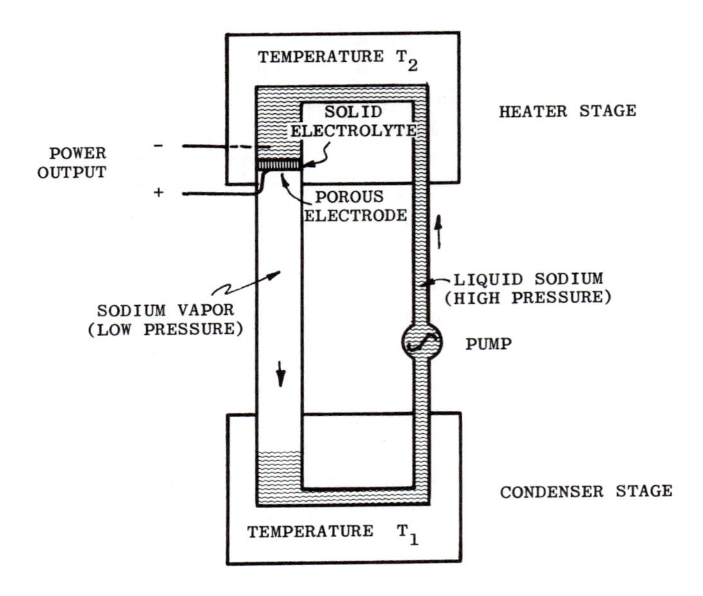

Fig. 15.18 Schematic diagram of a thermoelectric cycle based on liquid sodium and a beta alumina solid electrolyte, after Weber (1974).

The pressure at the hot side is equal to or slightly greater than the vapor pressure of sodium at T_2. A typical pressure is 0.60 atm at 800°C. Pressures of this magnitude raise practical problems of the support of the porous electrode over the required surface areas. The pressure at the cold side is also equal to the vapor pressure, or 5×10^{-5} atm at 200°C. At 805°C the test unit delivered 5.0 kW/m^2 at 0.50 V, with an overall efficiency of 28%.

15.14 PROBLEMS

15.1. A silicon solar cell converts 11% of the incident sunlight to electrical energy. The cell is not overcoated, so part of the sunlight is reflected. If there is no wind, what will be the temperature rise above ambient of the solar cell? If the 10% figure is measured at 20°C, what will be the percentage reduction in conversion efficiency for this cell?

15.2. A solar cell of 1 m^2 optical cross section is overcoated so that half the solar spectrum is rejected but the efficiency is unchanged. If the cell is placed at the focus of a concentrating collector producing uniform brightness over the cell area of 100 solar fluxes, and if convection and radiation are as applied to a flat-plate collector, what will be the energy load to be carried off by water flow to keep the temperature rise below 20 C°? How many liters per second is this flow when the water enters at ambient temperature? Assume that the water temperature is the same as the cell surface temperature.

15.3. What is the maximum fill factor for a solar cell system where the panels use circular silicon wafers and where the collectors must have no sun shadowing at 9 a.m. at winter solstice at 45° latitude?

15.4. If a silicon solar cell is exactly linear in response to solar flux variation and has 11% conversion efficiency, what is the value of the electrical energy output for a collector fixed at 60° tilt at 45° latitude at winter solstice? At summer solstice? What is the allowable cost per 75-mm-diameter solar cell chip? Assume the power being displaced has a value of 50 mills/kWh.

15.5. If a cadmium sulfide solar cell has a conversion efficiency of 6%, and if the cells have a fill factor in an assembly of 90%, what is the allowable cost per square foot for the sun conditions given in Problem 15.4?

15.6. What is the difference between the curve for the efficiency of cadmium sulfide given in Fig. 15.10 and actual cadmium sulfide cells?

15.7. If a gallium arsenide cell has the efficiency curve shown in Fig. 15.12 and is used at an ambient temperature of 20°C, what is the maximum total energy yield of a hybrid system where flux concentration is used and the cell is allowed to rise in temperature? The excess temperature is used to drive a thermodynamic cycle that has an efficiency 50% that of a perfect Carnot cycle. Plot the variation of combined photovoltaic and thermal power output as a function of increasing temperature. Is there a maximum operating temperature, and what is the percentage gain to be had from this hybrid system compared to simple operation of the gallium arsenide cell at ambient temperature? At the temperature of the maximum combined performance? Assume a flux concentration of 1000.

15.8. If the conversion efficiency of a thermoelectric-effect solar cell is 4%, what is the allowable cost per square centimeter of a junction device for a fully tracking system operating all year at maximum output? The cost of the displaced power is 50 mills/kWh.

15.9. A solar thermal system is to use thermoelectric devices to generate auxiliary and startup power requirements for the fluid system. If the junction temperature difference is 400 C° for a lead-selenium device, and the collector is a central receiver system with an area of 1000 m^2, how many kilowatts of electrical auxiliary power will be available one hour after sunrise at the equinox at 35° latitude?

15.10. A thermionic cell is to operate at a cathode temperature of 1200°C, and has the efficiency as projected by Brosens (Section 15.12). If the anode coolant operates at a temperature of 600 C° less and operates a thermodynamic cycle of 70% Carnot, what will be the combined efficiency when the temperature of the condenser is 50°C?

Chapter 16
A Survey of Applications

16.0 INTRODUCTION

In this chapter we consider a number of potential applications of solar energy. We present a first-order evaluation of each application in order to give guidance to persons interested in the order-of-magnitude values for the collector areas involved and in the value of solar energy as compared to conventional energy used for the same purpose.

Each typical application is accompanied by a sample calculation. All of the input information is available in this book. Of particular importance are the graphs and tables for the daily output and annual output for different collector configurations. These graphs can be used to obtain the energy income values for a particular latitude and configuration.

For the examples we have linearized the equations so that performance is obtained by simply multiplying the individual performance factors. The problem then is reduced to the choice of values for these performance factors. The terms generally required for a calculation are as follows:

E_{in} = solar input (energy income) for the selected geometry and time interval (hour, day, season, or year), per unit area,
η_1 = percentage of possible sunshine, of the type accepted by the collector,
η_2 = collector efficiency at operating temperature or mean value over the temperature range involved; this factor is the same as η,
η_3 = utilization factor, indicating the fraction of time the collector output can be utilized, including time out for maintenance,
η_4 = using system efficiency for heat delivered from the collector system,
η_5, η_6, etc., are other efficiency factors appropriate in given situations, and
E_{out} = energy output (yield) delivered to the user in final form for the design application, per unit collector area.

We then have

$$E_{out} = E_{in}(\eta_1 \cdot \eta_2 \cdot \eta_3 \cdot \eta_4 \cdot \eta_5 \; ...). \tag{16.1}$$

The values for collector efficiency η_2 are estimates from the η-versus-ΔT curves in this book. For cases not covered by a graph in this book we have estimated reasonable values based on our experience and reported by other groups.

The utilization factor η_3 is not adequately treated in this book; it involves system studies of such length that we had to omit it. The utilization factor involves the fact that at some times of the year the collector will provide more energy than can be utilized by the system and at other times it cannot provide enough energy. One aspect of this question is covered in Section 13.19. The utilization factor depends, therefore, on the tradeoffs in the design of the overall system. We estimate some typical utilization factors in the sample calculations in order to draw attention to the existence of this factor and its importance in the overall system behavior.

If the energy required by the user is Q, then the area of the collector required is

$$A = Q/E_{out}. \tag{16.2}$$

The value of the energy delivered to the customer is a most important factor in his deciding whether solar energy is economic for his application. If the cost of the competing form of energy is taken as C, then the value of a unit area of collector is

$$V = E_{out} \cdot C, \tag{16.3}$$

where V is per unit time and E_{out} is for the same unit time, and where E_{out} and C are in the same energy units (kWh or kJ).

We calculate the *value* of the solar energy delivered to the user by the system in terms of typical energy costs. This type of calculation does not involve speculation about how much a given collector design will *cost,* a matter on which there can be wide differences of opinion. The value of the energy delivered from a given collector design or system can be calculated with reasonable certainty, knowing the cost of alternative energy supplies. One can then readily assign a *cost limit* to each square meter of a given collector on the simple basis of how much energy it will deliver. Whether or not such a collector can be manufactured, distributed, and installed within that limit is up to the builder. It should be noted that in translating value into allowable collector cost one must take into account the interest rate on the capital investment involved. We do not consider this question in the examples that follow.

16.1 WATER DISTILLATION

Water distillation is treated in many professional papers, with many types of collectors and details of performance being presented. Our example represents the simplest type of distillation system, the *single-effect* system. In a single-effect system the energy input is exactly equal to the heat of vaporization of water (540 cal/g), the heat lost in condensation being unavailable to the system except for maintaining the temperature of the hot water. In a *multiple-effect* system some of the heat of vaporization is recovered and less energy is required per gram of water produced.

The distillation of water does not imply "boiling" in the usual sense. In the presence of a cold surface water evaporates from the surface of the warm water and condenses on the cold surface. The simplest water-distillation system, therefore, consists of a shallow pool of water in an enclosure having a transparent window, as shown in Fig. 16.1. The window, being exposed to the environment, will be cooler than the air inside the enclosure. The water will be even warmer than the enclosed air. The vapor pressure of the water will tend to saturate the air in the enclosure at equilibrium with the water temperature. The window, on the other hand, will tend to limit the saturation of the air to that appropriate to the temperature of the window. The result is a net transfer of water from the hot water to the cool window.

The rate of water production for a single-effect still is exactly the incoming energy at a rate of 540 cal/g of water multiplied by the efficiency of the collectors η. A typical set of values is shown in Fig. 16.1.

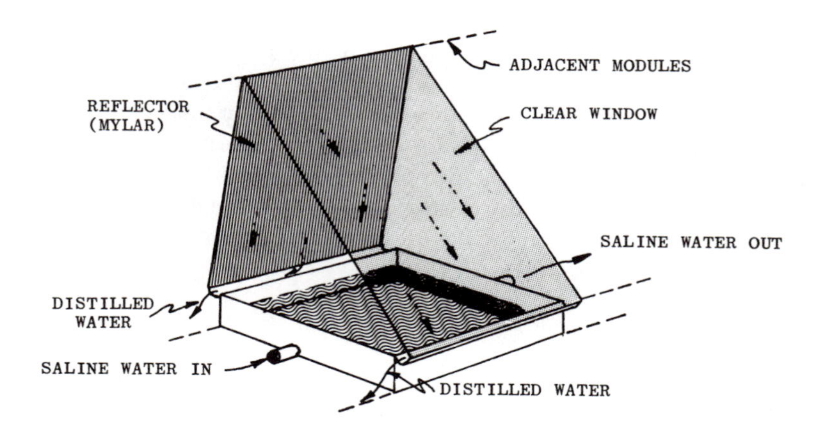

WATER DISTILLATION

Latitude	30	deg
Average annual energy income (tilted 30°)	8.1×10^6	kJ/m^{2}y
Percentage of possible sunshine	84	%
Collector efficiency	0.40	-
Average energy yield to water, per year	2.7×10^6	kJ/m^{2}y
Heat of vaporization (540 cal/g)	2.2×10^3	kJ/kg
Water yield per year	1.2×10^3	kg/m^{2}y
Water yield per day	3.29	kg/m^{2}d
Value of water at $2/1000 gal (3785 kg)	1.7×10^{-3}	$/m^{2}d
Value of water at $2/1000 gal (3785 kg)	0.62	$/m^{2}y
Value of collector	0.62	$/m^{2}y

Fig. 16.1 Schematic diagram of a single-effect water still, and sample calculation.

The rate of water production is dependent on how cool the window is, since η is a function of the window temperature. If there is a wind, heat is removed rapidly from the window, cooling it and making the transfer of water to the window more efficient.

An efficient multiple-effect solar distillation system was designed by Hodges *et al.* (1966) at the Environmental Research Laboratory of the University of Arizona. A pilot plant constructed at Puerto Peñasco, Sonora, Mexico, produced 19,000 liters/day (5000 gal/day) from a simple solar collector of 1000 m^2 (10,400 ft^2). A schematic diagram of this plant is shown in Fig. 16.2. The solar collector used by Hodges consisted of a series of long shallow ponds of water covered by a single layer of plastic. The plastic was stretched tight because the air pressure inside the collector was slightly above atmospheric pressure. The multiple effect in this system was obtained not by pressure gradients but by simply reusing the airstream that was dehumidified by the cooling water. After a first pass through the hot water in the tower, the water vapor was condensed out by the cooling water system, this "dried" air being again passed through the warm water in a lower section of the tower and rehumidified for subsequent condensation in the cooling portion of the system. The net result was a multiple-effect system that effectively used one-eighth of the heat of vaporization to distill one gram of water. This means that the value listed in the sample calculation in Fig. 16.1 would be multiplied by eight to yield \$4.96/m^2year.

Multiple-effect evaporation can also be had by pressure-staging the evaporator-condenser. If the water vapor from one stage is condensed at a higher pressure than the next stage evaporator, the heat of vaporization from the condensing water can be transferred into evaporating water in the stage at next lower pressure.

16.2 HOT-WATER HEATING

Hot-water heating is one of the most popular of the first-generation uses of solar energy. At this point it is necessary to note that there are two general types of solar application. The more demanding type is one that must be deliverable at any time, depending on the wishes of the user. This we term a *demand* application. A less demanding type of solar application is one that need not be delivered at a moment's notice but can be delivered when sunlight is available. This we term a *when-available* application. Solar hot-water heating is a demand application since one wants hot water whenever it is needed, day or night. The utilization of hot water for domestic applications is reasonably independent of the season, yet solar availability varies widely with season. One then faces a conflict with regard to the collector: it gathers more heat than is needed in summer and probably has a significant deficiency in winter. The solar collector for hot-water heating has different efficiencies η of operation in summer and in winter. The efficiency of a typical collector in summer is about $\eta = 0.60$ because the temperature difference between the collector water flow and the environment is small. In winter a typical collector efficiency needed to produce 50°C (122°F) water would be $\eta = 0.40$. Incoming water would be about 25°C in summer and about 15°C in

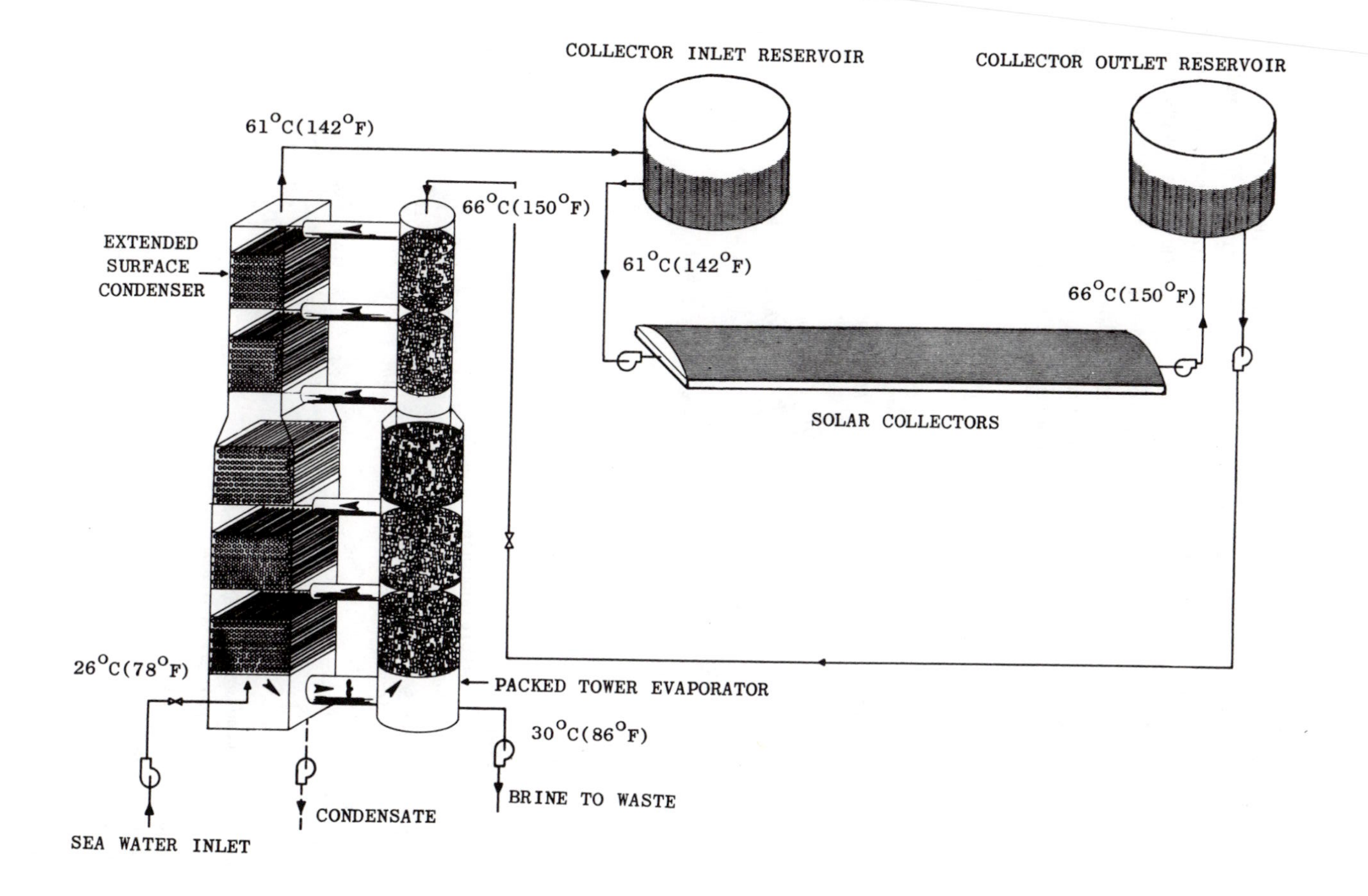

Fig. 16.2 Schematic diagram of the University of Arizona/University of Sonora solar desalination multiple-effect plant at Puerto Peñasco, Sonora, Mexico.

winter, so the amount of heating required is larger in winter, but in winter the collector is less efficient and sunshine less abundant.

Solar water heaters cool rapidly at night, and except in the warmest climates they will freeze on winter nights. They therefore must be drained, or a trickle of hot water must be continued through them at night. An antifreeze could also be used, in which case the solar portion must be permanently sealed from the domestic water supply and a heat exchanger used. All of these complications tend to increase the cost of solar water heaters to a point where their basic advantages can be lost to the consumer.

A typical hot-water heater system using either a pump or domestic water pressure to circulate the water through the collector is shown in Fig. 16.3. Sample calculations are shown for both summer and winter.

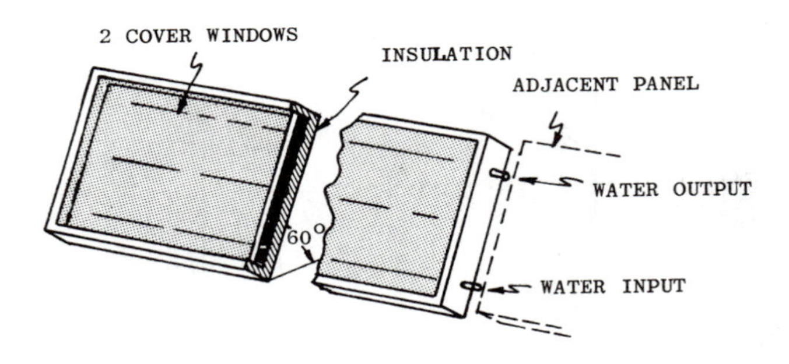

HOT–WATER HEATING

	Summer	Winter	
Latitude	45	45	deg
Seasonal average energy income ($\lambda+15^o$)	2.8×10^4	1.4×10^4	kJ/m^2d
Percentage of possible sunshine	80	50	%
Collector efficiency	0.60	0.40	-
Seasonal daily collector yield	1.3×10^4	2.8×10^3	kJ/m^2d
Required rise in water temperature to 50°	25	35	C°
Energy required at 4.2 kJ/kgC°	1.05×10^2	1.47×10^2	kJ/kg
Hot water yield per day	124 (33)	19 (5)	kg/m^2d gal/m^2d
Average home use per day	400 (100)	400 (100)	kg/d gal/d
Area of collector required for 100%	3.2	21.0	m^2
Percentage supply from 3.2 m^2	100	20	%
Value of hot water at \$0.05/kWh	0.18	0.18	\$/d
Value of hot water at \$0.05/kWh	66.00	66.00	\$/y
Value of collector	20.62	20.62	$\$/m^2y$

Fig. 16.3 Schematic diagram of a hot-water heater, and two sample calculations.

There are many designs for solar hot-water heaters and many different materials. Since lifetime is one factor of importance to cost, collectors made of glass windows and copper absorber panels are best, but they are expensive. Aluminum panels that have been corrosion protected with a diffused zinc layer are reasonably satisfactory for most water types. Plastics do not react with water but do present potential problems because of their low softening point. If for any reason water does not flow through a plastic absorber panel it can overheat, causing permanent damage or loss of the unit.

In warm climates one can utilize a collector with a single window cover and get the same performance as with a double cover at 45°N latitude. In a desert climate one can even dispense with cover windows entirely and get reasonable performance.

The value of the solar hot water, if the cost of heating by electricity is assumed to be 5 mills/kWh, is approximately $\$20/\text{m}^2\,\text{yr}$. This allowable cost must include the solar collector and also the storage tank and interconnections with the existing domestic hot-water supply, with appropriate controls.

16.3 SWIMMING-POOL HEATING

Swimming pools are frequently heated with natural gas, one of the prime fuels to disappear first from the energy scene. The problem of using solar energy instead is compounded by the large energy demands of swimming pools. They lose energy by radiation and convection to the atmosphere, plus a large amount through evaporation in the desert regions where pools are very popular. The magnitude of the loss from a pool becomes apparent by examining the heating ratings on typical pool heaters. They run between 200,000 and 400,000 Btu/h, whereas the gas heater in the home beside the pool may be rated only between 100,000 and 150,000 Btu/h.

Let us examine the heat loss for a desert heated swimming pool. The surface area of a typical pool is 40 m^2 (430 ft^2). The weekly evaporation loss is approximately 2.5 cm (1 in.), resulting in a loss of 10^3 kg of water per week, or 142 kg per day. In addition, we have convection heat loss and radiation to the sky, which also can be appreciable during the night in a dry climate. For heat input to the pool we have the sunshine falling on the pool, plus the additional heating we wish to obtain from a solar pool heater. Sample calculations for a pool with external solar heater are given in Fig. 16.4.

Since the surface losses from the pool are very high (so high that it is difficult to heat the pool without having a solar heater larger than the pool), a second approach is to combine a solar heater with a pool cover. A pool cover does an excellent job of reducing heat losses, but most owners object to the appearance and to the nuisance of having a pool cover. The alternative approach is to *combine* the pool heater and cover. A number of such floating heaters are being marketed. Let us examine a solar pool heater consisting of floating discs of plastic. These discs partially cover the pool surface, reducing evaporation loss. They could also be made so that they reduce radiation loss, but we will consider only evaporation loss. If these floating covers are kept dry on their upper surface (no evaporation loss) and if they are a dark color so that 60–80% of the incident sunlight is absorbed, they can significantly heat the pool. The sample calculation for this type of pool heater is given in Fig. 16.5.

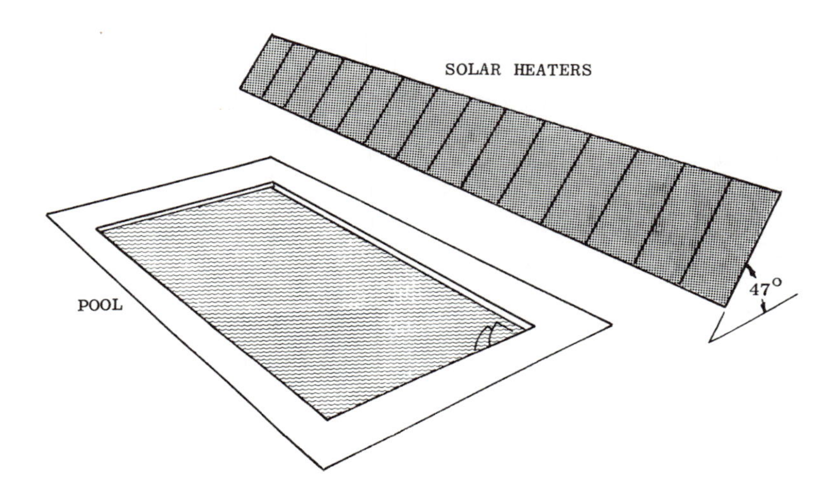

SWIMMING-POOL HEATING	Horizontal external heater	$\lambda + 15^o$ external heater	
Latitude	32	32	deg
Winter average energy income	$1.28x10^4$	$2.04x10^4$	kJ/m^2d
Percentage of possible sunshine	70	70	%
Collector efficiency	0.90	0.90	-
Daily collector yield	$8.1x10^3$	$1.29x10^4$	kJ/m^2d
Pool surface area	40	40	m^2
Evaporation loss	142	142	kg/d
Evaporation energy loss	$-3.1x10^5$	$-3.1x10^5$	kJ/d
Radiation imbalance loss at 50 W/m^2	$-1.8x10^5$	$-1.8x10^5$	kJ/d
Convection and wind loss	$-0.1x10^5$	$-0.8x10^5$	kJ/d
Normal solar input to pool at $\alpha = 0.30$	$1.1x10^5$	$1.1x10^5$	kJ/d
Net daily energy loss from pool	$-3.9x10^5$	$-3.9x10^5$	kJ/d
Area of collectors required	48	30	m^2
Value of heat at \$2/MBtu = \$2/MkJ	0.77	0.77	\$/d
Value of heat at \$2/MBtu = \$2/MkJ (200 d)	155.00	155.00	\$/season
Value of collector	3.23	5.17	$\$/m^2y$

Fig. 16.4 Schematic diagram of an external swimming-pool heater, and sample calculation.

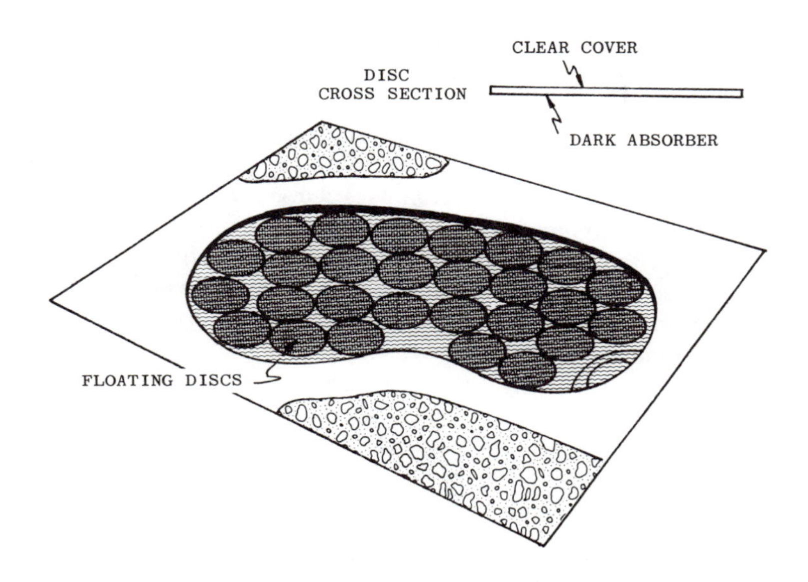

SWIMMING-POOL HEATING	Floating heater	
Latitude	32	deg
Winter average energy income	1.28×10^4	kJ/m^2d
Percentage possible sunshine	70	%
Collector efficiency	0.80	-
Daily collector yield	7.2×10^3	kJ/m^2d
Pool surface area	40	m^2
Pool surface covered with collectors	25	m^2
Evaporation energy loss	-1.1×10^5	kJ/d
Radiation imbalance loss	-0.9×10^5	kJ/d
Convection and wind loss	-0.0×10^5	kJ/d
Solar input to pool through gaps	0.4×10^5	kJ/d
Net daily energy loss	-1.6×10^5	kJ/d
Area of collector required	22	m^2
Value of heat at \$2/MBtu = \$2/MkJ	0.32	\$/d
Value of heat at \$2/MBtu = \$2/MkJ (200 d)	63.00	\$/season
Value of collector	1.38	$\$/m^2y$

Fig. 16.5 Schematic diagram of a swimming-pool heater combined with a cover, and sample calculation.

16.4 CROP DRYING

Crop drying with solar energy can be divided into two general categories, each having several subdivisions. These two categories are (1) drying of grains and (2) drying of leafy crops or crops of high moisture. Grains in general have a low moisture content, on the order of 20–30% at the time of harvest; drying is for the purpose of bringing the moisture down to 12–14%, at which point the grains can be stored for the required length of time without further biological degradation. Leafy crops and moist crops like fruit require drying from high moisture contents, above 50%, in order to be stored or further processed into relatively dry feed products for animal consumption.

The role of solar energy in crop drying is to change the equilibrium vapor pressure of the crop to accelerate the diffusion of the moisture into a flowing airstream. The solar heat is to increase the temperature of that airstream.

A problem with crop drying is that one must keep the added heat either very low or very high. A temperature in between can accelerate spoilage. Low-cost solar collectors appear inadequate for the high-temperature processing mode, so the low-temperature, high-volume flow mode is of more potential importance.

To take a specific example, let us consider a grain like wheat, corn, or sorghum, where we want to lower the moisture from 25% at harvest to 12% within 20 days. For this work we want to add only 3–6 C° (5–10 F°). Let us assume the grain bin is filled to produce a backpressure of about 50 mm of water so that we can maintain a volume rate of flow through the grain of 1.0 liter/sec (2 cfm) per bushel (about 9 kg of corn). If our fan provides under these conditions a flow of 944 liter/sec (2000 cfm), how many square meters of collector will we need in order to heat the air exiting from the collector by 6 C° at noon on a clear day?

The type of collector we will assume is a single tube of black polyethylene 1 m in diameter, placed north-south for the fall equinox season. The circular cross section means that the optical cross section is constant during the day, so it acts like a north-south (NS) horizontal, east-west (EW) tracking collector even though nothing moves. On the other hand, the circular cross section is less efficient in absorbing sunlight, since some sunlight is incident at large angles near the tangent point of the tube. The efficiency of the collector is low because it has neither a protective envelope nor curved sides, so we take η as being about 0.25 when the air is heated only 6 C° above the ambient temperature.

An alternative to low-heat grain drying is to use electricity. The fan typically used to aerate grain raises the air temperature 0.5–1 C° from the energy it dissipates into the airstream from the motor. If one were to use electricity for the additional degrees of heat, the appropriate rate would be, or soon will be, about $0.05/kWh. We use this figure in the sample calculation.

A sketch of such a solar collector and the sample calculation are shown in Fig. 16.6.

It is interesting to note that the value during a single drying period of 20 days would more than pay back the cost of a seamless polyethylene black tube. If the same

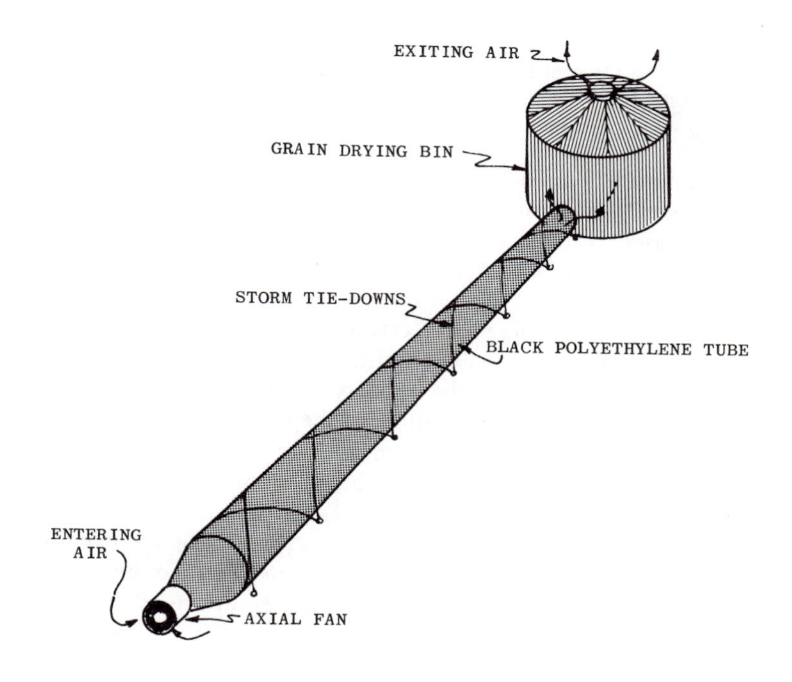

GRAIN DRYING

Latitude	45	deg
Energy income	2.6×10^4	kJ/m^2d
Energy income (noon hour)	2.4×10^3	kJ/m^2h
Percentage of possible sunshine	80	%
Utilization factor	100	%
Collector efficiency	0.25	-
Collector yield	5.2×10^3	kJ/m^2d
Collector yield (noon hour)	0.6×10^3	kJ/m^2h
Mass of air to be heated	4.3×10^3	kg/h
Heat required for $6C^o$ rise ($1.04\ J/gC^o$)	2.7×10^4	kJ/h
Collector area required	45	m^2
Tube length (1.0 m diameter)	45	m
Equivalent energy from collector	1.09×10^2	kWh/d
Value of heat at $0.05/kWh	3.25	$/d
Value of heat in 60-day season	195.00	$/season
Value of collector	4.34	$\$/m^2y$

Fig. 16.6 Schematic diagram of a grain dryer, and sample calculation.

dryer tube can be used for several crops in succession (wheat in October, corn in November, sorghum in December), the cost looks reasonable.

16.5 THE BIOCONVERSION WOODLOT

We have not given proper attention to the very large topic of bioconversion of sunlight into fuels and chemical feed stocks. Many ways of utilizing bioconversion are being explored under current Federal support. Bioconversion is limited, however, by the maximum natural rate at which a plant or tree can convert sunlight into biomass. When a plant is small it does not fully utilize the sunshine falling on the land. When it is mature it may be converting close to 1% efficiency. For the best crops at the peak of their growing season, generally a small part of the year, the conversion rate can exceed 1% and approach 2%. However, for the ordinary plant or tree the rate is well below 1%, averaging close to 0.5% over the lifetime of the plant.

Bioconversion solves one basic problem of solar energy utilization in that the collecting subsystem is simple to deploy and tend: it requires only planting of the seed or seedling and occasional cultivation. In some regions watering is also necessary, and some of the most efficient crops require intensive watering for optimum growth.

A practical question arises in regard to an elementary form of bioconversion: the woodlot. The woodlot was widely used in the past century in the midwestern and northern states as a place to grow the trees to provide fuel for the farm and home. It may again become important as the conventional fuels—natural gas, oil, and coal—become too expensive for casual uses. The question is: How big a woodlot do I need in order to grow the fuel for my own needs?

The heating demand for a home in the upper Midwest is about 100 million Btu per year. Some homes may require twice this amount, but we will assume for this example 100 MBtu of demand.

The available solar energy per square meter of land during the six months of the growing season is 3.6×10^6 kJ of energy. If the area receives 80% of the possible sunshine, if tree conversion efficiency is 0.5% over its lifetime, and if 50% of the tree is usable, then the annual yield of wood (energy equivalent) is 7250 kJ of energy per square meter of land. Since the annual heating demand is 100×10^6 kJ, this means that 13,790 m^2 of land (1.38 ha or 3.41 acres) will be required for the woodlot.

In terms of trees, one fast-growing tree would mature at about 200 kg (440 lb), equivalent to a log 28 cm (11 in.) in diameter and 600 cm (24 ft) long. Since wood has a heat content of 20,000 kJ/kg, one needs 5000 kg (5 metric tons) of wood per year. At 200 kg per tree, this means harvesting 25 trees per year.

If the tree grows to maturity in nine years, the woodlot dimensions are as shown in Fig. 16.7. A total area of 31 ha (76.7 acres) is required. This is a large woodlot for one home, illustrating the growth of energy needed for an individual home over the past century. The woodlot of a century ago was only several hectares. Of course, the standard mode then was to heat only one or two rooms when occupied, with much of the heat generated from a wood-burning (or cob-burning) kitchen stove.

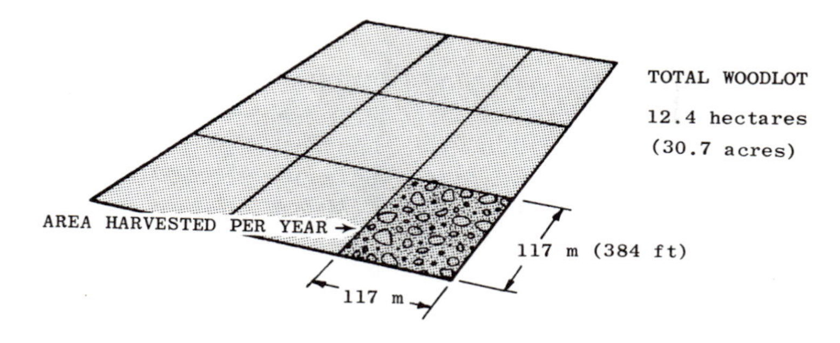

BIOCONVERSION "WOODLOT"

Latitude	45	deg
Maximum possible energy income (180 d)	3.6×10^6	kJ/m^2
Percentage of possible sunshine	80	%
Tree conversion efficiency	0.5	%
Usable fraction of tree	50	%
Annual wood yield, equivalent energy	7.25×10^3	kJ/m^2
Annual home heating demand	1.0×10^8	kJ/season
Area of woodlot to be harvested	1.38×10^4	m^2/y
	(3.41)	(acres/y)
Weight of wood harvested at 2×10^4 kJ/kg	5×10^3	kg/y
Wood per tree at 9 years age	2×10^2	kg/tree
Number of trees harvested	25	/y

Fig. 16.7 Schematic diagram of a bioconversion woodlot for one home, and sample calculation.

Wood growing would be acceptable for the conversion of solar energy if the land, water, and fertilizer could be provided. Improvement in the efficiency we have used in the example could come from using a crop that covers a maximum area as soon as possible, and that has a better efficiency of conversion. If one could raise the net efficiency to 1% for the growing season, the land use would shrink from 30.7 acres per home to 15.3 acres, which, although still large, is more reasonable.

Burning biomass for thermal energy will inevitably become of major importance to the chemical industry, which today relies almost exclusively on petrochemicals as the basic feed stock. Plants produce less-complicated hydrocarbon molecules than those that form the basis of the petrochemical industry, but these plant molecules are a far better starting point than water and carbon dioxide when economics are important. For example, alcohol is produced in large quantity from oil (methanol), but plants also produce alcohol (ethanol). The difference is significant for internal consumption, but in the chemical process field the difference is unimportant.

16.6 HOUSE HEATING

One of the solar energy applications enjoying great popularity at the moment is house and space heating. Hundreds of designs and dozens of manufacturers, some large but most small, are beginning to enter this field. The question of how large a home solar heater must be often gets an ambiguous answer because the question is incomplete. The size is determined by how much of the winter demand of the home one expects to fill by means of the solar collector. If the answer is 100%, one is in for a disappointment because the collector needed for most climates will be prohibitively large. The proper compromise is a collector that sometimes cannot deliver all the heat it gathers because the house is heated and the thermal storage bin is full of heat and that, an equal number of times, cannot supply heat because there is no sun and the thermal storage bin is empty. Thus there are two important efficiencies: (1) the collector efficiency η and (2) the utilization efficiency η_3. For a typical collector having double windows, η is approximately 0.60. For a reasonably optimum collector size, η_3 is approximately 0.60–0.80. In our example we will assume that 70% of the collected solar heat can be used in the home. This means that 30% of the demand for the home must be supplied by a backup source.

We will take a vertical south-facing collector, which is somewhat different from most solar house heaters but practical in the sense that it can fit existing building situations rather well. It can be attached to the south side of the house, or detached, as a patio wall, with the storage bin forming the junction with the house. We will take the same latitude and house heat demand, 45°N and 100 MBtu per heating season, as for the woodlot example. The sample calculation is given in Fig. 16.8.

In examining the cost, we note that if electricity is used for "resistance" heating the value of the solar heat is rather high, but if lower-cost oil is used the allowable solar cost for equal season costs is much lower. The practicality of solar heating is therefore highly sensitive to the form of energy presently used in one's area.

Solar house heating can be done by either of two types of collector system: (1) hot water and (2) hot air. The system diagrams for these two classes are shown in Figs. 16.9 and 16.10. The practical choices between the two are several. The hot-water system is more efficient, by about 10%, but water collectors freeze at night and must be drained or have some heat supplied during the night to keep the water above freezing. The alternative is to use antifreeze in the collector. Antifreeze in the hot-water storage tank would be prohibitively expensive, so a heat exchanger is required between the collector/antifreeze system and the plain-water thermal storage tanks.

A hot-air collector is less efficient but simpler to build and maintain. A rock-storage bin or bin filled with bottles of sealed water is required for heat storage. The control system for air is more expensive than for water because the duct size is large and mechanical dampers or separate fans are necessary to divert the air to be heated to the proper heat source (collector storage bin or backup heater). We have used both the water and the air type at our home in Tucson, and find the air system to be freer of maintenance and simpler to construct.

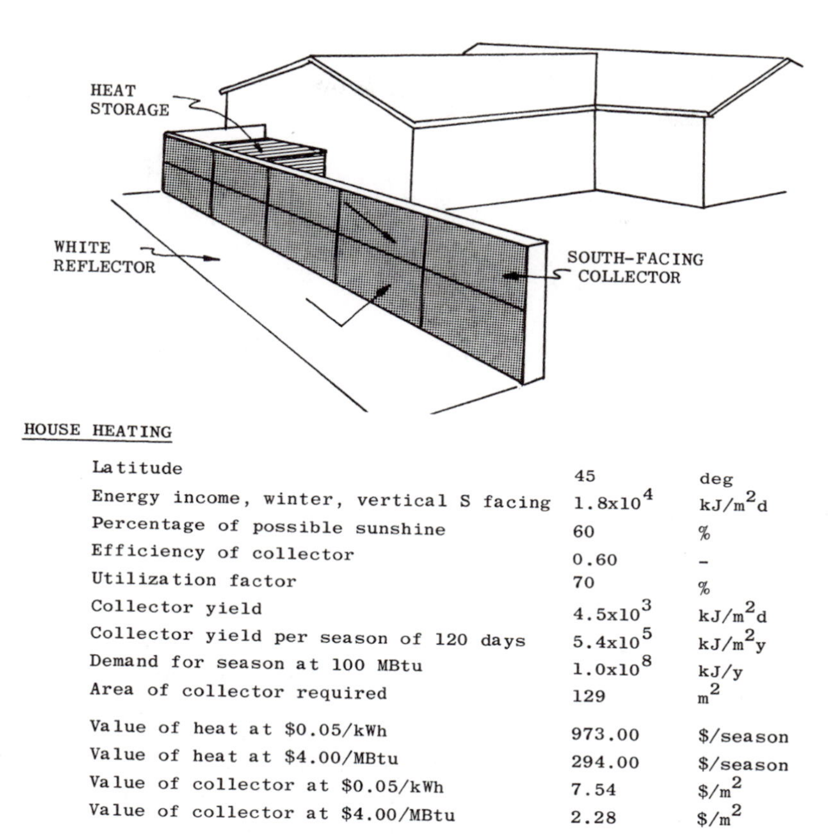

HOUSE HEATING

Latitude	45	deg
Energy income, winter, vertical S facing	1.8×10^4	kJ/m^2d
Percentage of possible sunshine	60	%
Efficiency of collector	0.60	-
Utilization factor	70	%
Collector yield	4.5×10^3	kJ/m^2d
Collector yield per season of 120 days	5.4×10^5	kJ/m^2y
Demand for season at 100 MBtu	1.0×10^8	kJ/y
Area of collector required	129	m^2
Value of heat at $0.05/kWh	973.00	$/season
Value of heat at $4.00/MBtu	294.00	$/season
Value of collector at $0.05/kWh	7.54	$\$/m^2$
Value of collector at $4.00/MBtu	2.28	$\$/m^2$

Fig. 16.8 Schematic diagram of a house heating system, and sample calculation.

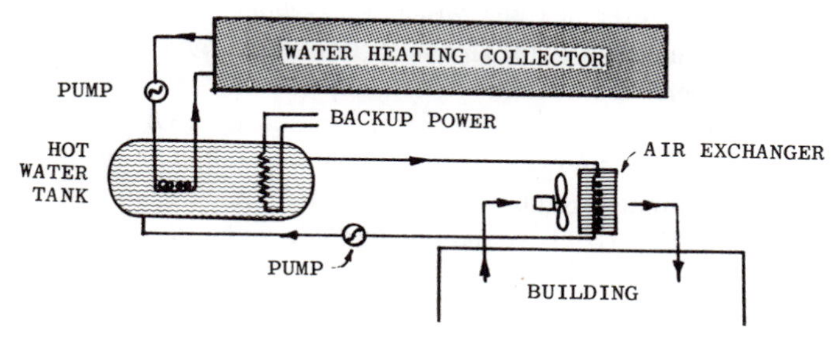

Fig. 16.9 Basic diagram for a hot-water solar collector system with thermal storage tank and backup power.

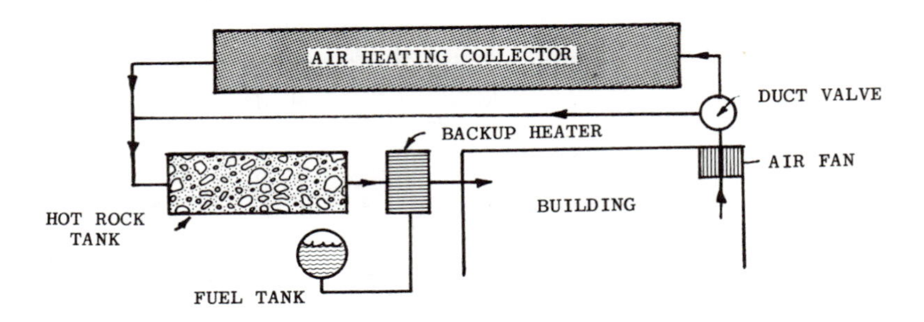

Fig. 16.10 Basic diagram for a hot-air solar collector system with thermal storage tank and backup heater.

16.7 GAS-ABSORPTION COOLING

The basic system diagram for a gas-absorption refrigeration system is shown in Section 14.18. Such a system would generally be used in combination with a solar heating system for winter use. The choice of collector geometry is therefore affected by the need to balance energy input to the collector for both use seasons. This means that the collector is placed facing more directly upward than would be the case for heating only. If we examine the seasonal daily energy yield data in Table 3.1, we note that we can change the ratio of summer to winter collection by adjusting the tilt (Fig. 3.26). The proper tilt then depends upon the required ratio of summer energy to winter energy. The use of gas-absorption refrigeration for summer cooling entails an additional energy loss over winter heating. This loss is represented in the *coefficient of performance* (COP) of the refrigeration system. Typical COP values for solar adapted units are 0.40–0.60, depending upon the exact operating conditions. In our sample calculation we assume a COP of 0.50. This means that half of the heat delivered to the unit is rejected: 1 Btu of cooling requires 2 Btu of heat.

The choice of angle of the collector further depends upon the relative amounts of summer cooling versus winter heating, which vary widely with region in the United States. In the desert Southwest twice as many cooling energy units are required as for winter heating. The reverse is true for the north-central regions, where summer cooling is still not widely used. In view of the demand factor and the COP factor, let us assume that twice the summer solar energy output will be required. The proper angle for the collector is then approximately equal to the latitude.

In the example in Fig. 16.11 we assume that approximately 1 MBtu of cooling is required for a peak summer day, and that the solar refrigeration portion will be 70%. This means that solar energy would be able to handle most of the cooling demand earlier and later in the season. We will assume the "effective" cooling season at latitude 30°N to be 120 days.

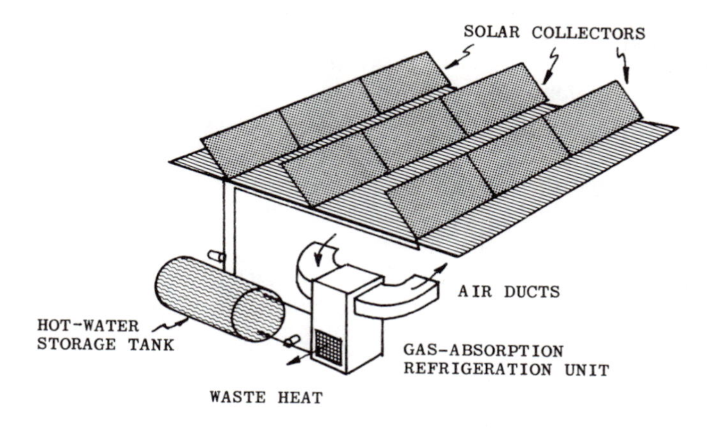

GAS-ABSORPTION COOLING

Latitude	30	deg
Energy income, summer, λ-tilted	2.3×10^4	kJ/m^2d
Percentage of possible sunshine	80	%
Efficiency of collector	0.60	-
Utilization factor	80	%
Collector yield of heat	8.9×10^3	kJ/m^2d
System COP	0.50	-
Daily cooling yield	4.4×10^3	kJ/m^2d
Daily maximum cooling demand	1.0×10^6	kJ/d
Fraction of demand from solar heat	70	%
Daily solar cooling demand	0.7×10^6	kJ/d
Area of collector required	159	m^2
Value of cooling at $0.05/kWh	9.72	$/d
Value of cooling per season of 120 days	1166.00	$/season
Value of collector	7.33	$\$/m^2y$

Fig. 16.11 Schematic diagram of a gas-absorption refrigeration system, and sample calculation.

16.8 NIGHT RADIATION COOLING

Night radiation cooling employs a "collector" that has a net radiation exchange *to* the night sky. Water in such a collector, for example, will then be chilled below the ambient night temperature. The amount of cooling depends upon the infrared nature of the collector. For a significant effect to be obtained, such a collector cannot have glass covers. A collector with no cover can cool, as was demonstrated by Bliss with the University of Arizona house, but such a cover is not efficient because a night wind can warm the surface, reducing its net cooling effect. The selective radiator of Silvestrini (Section 12.26) is much better but more complicated.

For the sample calculation we assume that the collector radiator has a net rejection of 50 W/m^2 (50 J/sec) or 0.18×10^3 kJ/m^2h, which would be approximately 1.4×10^3 kJ/m^2 for the night. The collector to be used is the same as for winter heating. Let us assume the area to be 130 m^2 (1400 ft^2). How much cooling would this collector provide? The sample calculation, given in Fig. 16.12, shows that only 18% of the demand would be met in this manner. This means that the solar collector in the night cooling mode would be of only marginal effectiveness in a hot climate; some additional means of cooling would be required for ideal comfort.

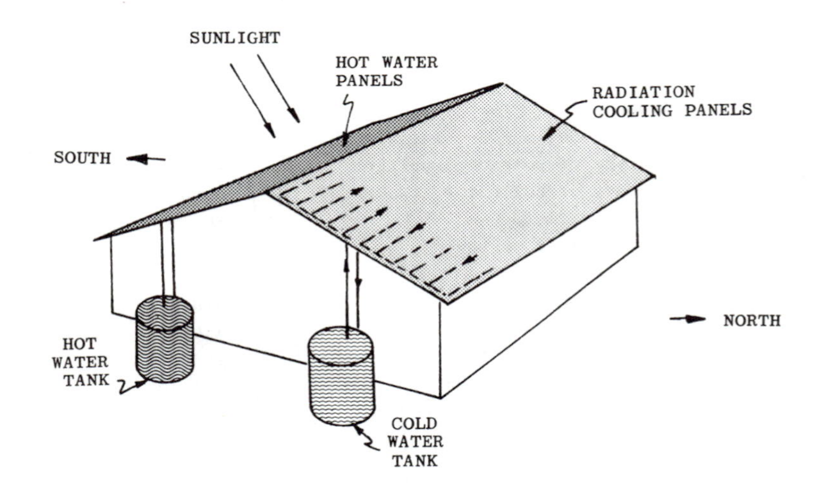

NIGHT RADIATION COOLING

Latitude	35	deg
Cooling income per night	1.4×10^3	kJ/m^2n
Total cooling income for 130 m^2 area	1.82×10^5	kJ/n
Average cooling over 24 hours	7.6×10^3	kJ/h
Average cooling demand, previous example	1.0×10^6	kJ/d
Average cooling demand, previous example	4.2×10^4	kJ/h
Fraction of cooling demand met	18	%
Value of cooling at $0.05/kWh	2.53	$/d
Value of cooling per season of 200 days	505.00	$/y
Value of cooling panels	3.89	$/$m^2$y

Fig. 16.12 Schematic diagram of a night radiation cooling system, and sample calculation.

In the illustration we show two water tanks, one for the daytime heated water and the other for the nighttime chilled water. One can do as Bliss did, and combine both the hot-water and cold-water storage in a single vertical tank, the hot water being separated by density from the cold in the bottom of the tank. When mixing is prevented in the tank, the hot and cold zones remain well separated. This arrangement also allows programing the tank to hold mostly hot water in winter and mostly cold water in summer. A floating insulating disc can be added to further augment gravity in keeping the hot and cold waters separated.

The diagram also shows only the north roof supplying chilled water. In reality the collector could be on the south slope and collect hot water during the day and cold water at night. Since cooling is less efficient than heating, one actually may want to use both the north and south roof slopes for night cooling to get enough chilled water for use during the day for cooling the house.

16.9 EVAPORATIVE COOLING

Summer cooling in a dry climate can be done efficiently with an evaporative cooler. This type of cooler functions by evaporating water into the airstream flowing from the outside into the house. The amount of cooling is dependent upon the humidity of the air before the evaporation stage and the maximum tolerable humidity after evaporation. When the entering air is already loaded with water vapor, the cooling will be slight and the humidity unbearably high after evaporation. In Arizona, where evaporative coolers were used long before refrigeration, they function well in the early summer months, but when the monsoon arrives in July they cease to yield desirable cooling.

The use of two-stage evaporative coolers has often been proposed but has never gained acceptance. With a two-stage cooler, the saturated cooled air is not injected into the house, but exchanges its coolness with air circulated from the house.

The amount of cooling effect is set by the heat of vaporization of water and the number of grams of water evaporated into each cubic meter of air. In the example in Fig. 16.13 we assume a simple case, where we know how much water is evaporated into the air. In a real situation the amount of water evaporated depends upon airflow velocity and cooler pad dynamics.

The input air has a temperature of 40°C and a relative humidity of 15%. The cooler causes 6.5 g of water to be evaporated into a cubic meter of flowing air. The temperature drop and the final humidity are presented in the sample calculation.

The final temperature is still rather high. If one were to improve the efficiency of evaporation so that 10 g of water were evaporated, the temperature would drop to 23°C (73°F), but the humidity would rise to more than 80%.

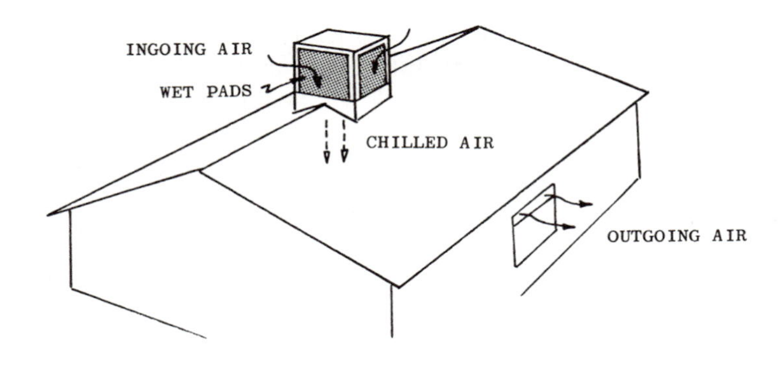

EVAPORATIVE COOLING

Latitude	not relevant	
Outside air temperature	40	°C
Outside air humidity	15	%
Mass of water evaporated into air	6.5	g/m^3
Mass of water in outside air	7.5	g/m^3
Mass of air per cubic meter	$1.29x10^3$	g/m^3
Heat capacity of a cubic meter of air	$0.32x10^3$	cal/C^om^3
Heat released by 6.5 g of water	$3.5x10^3$	cal
Temperature drop of airstream	10.9	C^o
Final air temperature	29.1	°C
Final air humidity	52	%

Fig. 16.13 Schematic diagram for an evaporative cooling system, and sample calculation.

16.10 WATER PUMPING

Water pumping with solar energy is a return to one of the first high-technology uses of solar energy at the turn of the 20th Century, but with a difference. In olden times water from such pumps could be accepted on a "when-available" basis. Modern irrigation farming is on a 24-hour commercial basis. Wells represent a capital investment that must be worked 24 hours a day to maximize the return on the investment. The older convenience of solar energy with its occasional interruptions no longer applies. This means that solar water pumping requires either a backup energy supply or enough energy storage to make the system sufficient for the practical task it faces.

A solar water pump can be attached directly to the well pump, or at some distance if electrical energy is the means of coupling the solar unit to the well. The difference is that the electrical mode allows the pump to rely on electrical power at night and at off-peak power rates. Direct coupling means that the solar collector must be adjacent to the wellhead, and this often means that expensive land would be taken out of production. In addition, the solar collectors would be in a dusty environment part of the year and an inconvenience to operations on the farm land.

In the example in Fig. 16.14 we assume that the solar unit is directly coupled to the well pump. The case of the generation of electrical energy (see Section 16.11) would apply to the other case of an electric well pump, the electric motor efficiency being the only difference between the two cases.

The well depth and pumping rate are from a typical case for the irrigated farms in southern Arizona.

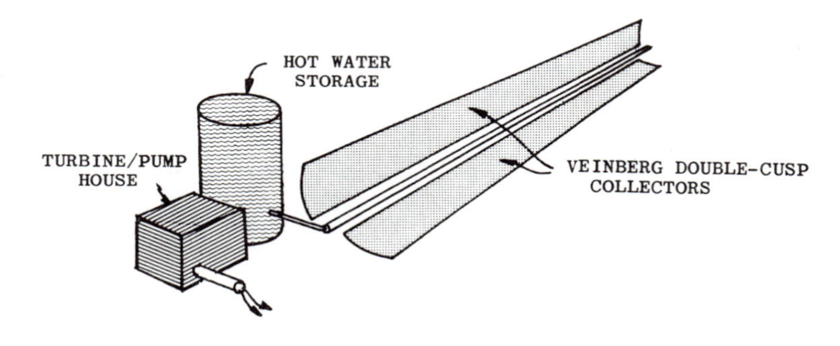

WATER PUMPING

Latitude	35	deg
Energy income (EW:NS tracking, Equinox)	2.38×10^4	kJ/m^2d
Percentage of possible sunshine	80	%
Utilization factor	86	%
Collector efficiency	0.65	-
Turbine thermodynamic efficiency	20	%
Pump efficiency	80	%
Energy output in 12 hours	1.7×10^3	kJ/m^2d
Pumping output (2.55 acre-ft/day)	3.15×10^6	kg/d
Pumping lift (500 ft)	150	m
Well pump power	67	kW
Pumping energy requirement	2.45×10^5	kJ/h
Area of collector for 12 hour usage	1727	m^2
Area of collector for 24 hour usage	3646	m^2
Value of energy at $0.05/kWh (12 h)	40.20	$/d
Value of energy at $0.05/kWh (250 days)	10,050.00	$/season
Value of collector	5.82	$\$/m^2y$

Fig. 16.14 Schematic diagram of a water-pumping system supplying 2.55 acre-ft/day, and sample calculation.

The value of pumped water varies widely with end use. In many parts of the world the assured production of a crop depends upon pumped irrigation. For this water to be affordable the pumping cost must be low. One way of estimating the value of the water pumped is presented in Table 16.1, which shows the amount of water needed to produce a kilogram of produce.

Table 16.1 Agricultural Water Consumption

	Water consumed		*Product value**	
Product	*Liters/kg*	*Gal/lb*	*$/kg*	*$/lb*
Alfalfa (dry)	2128	219.2	0.081	0.037
Barley	1900	195.7	0.264	0.132
Sorghum	1930	198.8	0.231	0.105
Wheat	1890	194.8	0.165	0.075
Lettuce	1100	113.8	0.660	0.300
Cotton	4115	423.8	0.572	0.236
Copper (mined)	182	18.8	1.386	0.630

*1974 Commodity prices.

16.11 ELECTRIC POWER GENERATION

In this example we consider the collector area required to provide electrical power for a community of 1000 homes. The type of collector is a high-temperature system like the "power tower" or the "power bowl." The system is assumed to produce the energy needed for 70% of the demand with enough thermal storage to carry the producing period four hours past the time the collector would normally cease to function. Power during the off-peak period would be provided by a baseload power plant as part of the community energy supply.

We will consider a power demand per home of 1200 kWh/mo, which is equal to $60/mo at 50 mills/kWh. If the home uses electrical heating, a heat pump, or refrigeration, the monthly demand will be approximately double this figure. The average power demand for the month for the community is therefore 1.2×10^6 kWh/mo. If the peak demand is during the day when the solar power unit is on line, it will be approximately twice the average, or 3.33×10^3 kW of power (3.33 MW). We assume that the solar plant will supply a peak of 2.3 MW_e, the balance to come from the baseload plant. The calculations are shown in Fig. 16.15.

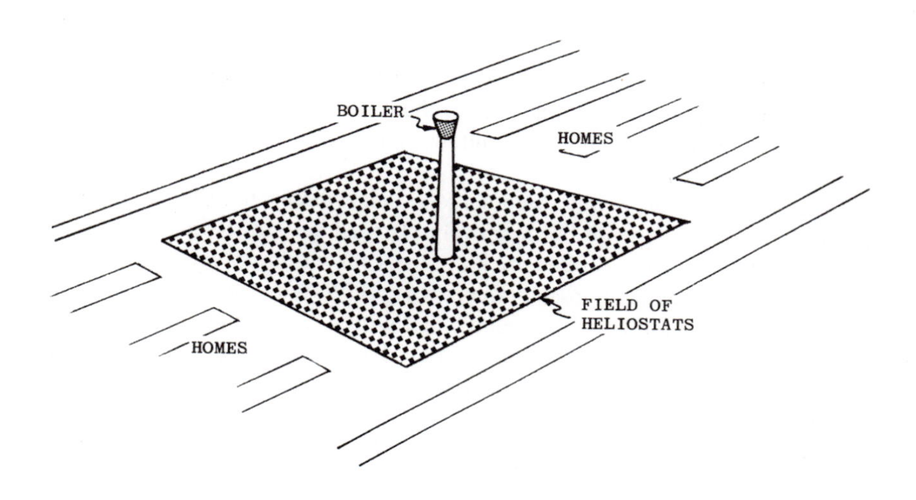

PHOTOTHERMAL POWER

Latitude	40	deg
Energy income, winter, fully tracking	$1.8x10^4$	kJ/m^2d
Percentage of possible sunshine	60	%
Utilization factor	90	%
Collector efficiency	0.70	-
Turbine thermodynamic efficiency	33	%
Generator efficiency	90	%
Energy yield	$2.04x10^3$	kJ/m^2d
Demand, 1000 homes at $2.0x10^4$kWh/d each	$7.2x10^7$	kJ/d
Demand to be met with solar heat	$4.8x10^7$	kJ/d
Area of collector required (winter)	$2.47x10^4$	m^2
Average annual energy income	$2.6x10^4$	kJ/m^2d
Average annual power output	$6.9x10^7$	kJ/d
Annual total power output	$2.5x10^{10}$	kJ/y
Value of energy, \$0.05/kWh = $1.4x10^{-5}$\$/kJ	351,000	\$/y
Value of energy, \$0.05/kWh = $1.4x10^{-5}$\$/kJ	14.21	$\$/m^2y$

Fig. 16.15 Schematic diagram of an electrical power system for 1000 homes, and sample calculation.

16.12 PHOTOVOLTAIC POWER

It has been proposed that the cost of photovoltaic cells could be reduced by using them with concentrating collectors. In this manner a small piece of cell could process solar energy arriving on an area much larger than the cell. The basic problem with this idea is that the high flux of sunlight would heat the silicon cell, reducing its efficiency, and destroying it if it got too hot.

The method of using water to cool a solar cell is discussed in Section 15.10. This approach involves the expense of an actively pumped system. RCA has explored the possibility of using passive cooling to keep the cell within operating limits. A schematic diagram of the RCA approach (Rappaport, 1975) is shown in Fig. 16.16. In this design the cell is placed in good thermal contact with a large copper block. The copper conducts the heat rapidly to the extremities of the block, where convection and radiation dissipate the heat. This approach is claimed to be effective at flux concentrations approaching 1000.

Let us now look at the value derived from this type of system and determine the amount that can be paid for the solar cells in this application. The energy income is taken to be that of a fully tracking collector. For flux concentrations of 1000 it is necessary to have a precision optical system that accurately tracks the sun, but these units can be small, on the order of a square meter for a cell having an area of 10 cm^2.

The result of the sample calculation is that a reasonable cost for the solar cells is \$8000/kW for 10% conversion efficiency. If the cell efficiency can be raised to 15% at the operating temperature, the allowable cost rises to \$12,000. The 1975 price for silicon solar cells is approximately \$20,000. We thus see that photovoltaic power through flux concentration is becoming reasonably close to the cost of photovoltaic cells.

16.13 COMMUNITY-SCALE TOTAL ENERGY SYSTEMS

The application of solar energy to the individual home or other small units is impeded by a number of problems, including (1) appearance, (2) compatibility with the building, (3) the difficulty of retrofitting to existing buildings, (4) maintenance complexity of heating and cooling systems, and (5) high capital cost affecting mortgage factors. Many of these problems could be alleviated or eliminated by means of larger installations capable of serving a community of users. The basic advantage of a community application is that the system would be large enough for professional management, both financial and operational. The size would lend itself to some form of "utility," providing the equipment and the service desired. In this case the solar collectors would no longer need to be on the buildings. A centrally located *energy park* could be provided, with power lines and hot- and cold-water lines distributing the energy to the individual homes.

The application of solar energy to dense communities, such as high-rise apartments or office buildings, would be complicated by the lack of adequate space in

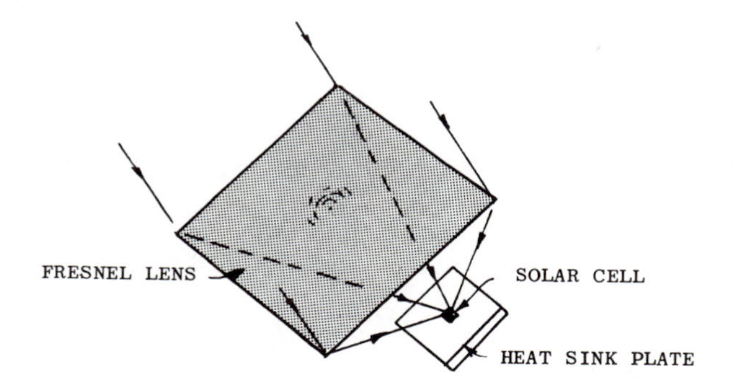

PHOTOVOLTAIC POWER

Latitude	40	deg
Energy income, fully tracking	1.04×10^7	kJ/m^2y
Percentage of possible sunshine	70	%
Utilization factor	90	%
Conversion efficiency	10	%
Energy output	6.5×10^5	kJ/m^2y
Average demand, 3650 h/y	1.0	kW
Average demand, 3650 h/y	1.31×10^7	kJ/y
Area of collector required	20.1	m^2
Value of energy at $0.05/kWh	182.00	$/y
Value of energy, 7 years	1,280	$
Value of system, 7 years	65.00	$\$/m^2$
Area of cells for 500 concentration	400	cm^2
If half the value is assigned to cells,		
Value of cells at 10% efficiency	8.00	$/W
Value of cells at 10% efficiency	8,000	$/kW

Fig. 16.16 Schematic diagram for photovoltaic power, and sample calculation.

populated urban areas. Roof surfaces could be used but would be inadequate for more than two levels in the building. Vertical faces, especially south-facing ones, would be good energy collectors for winter heat, but shadowing by adjacent tall buildings could seriously affect this placement of collectors. Therefore, solar energy applications for high-density high-rise buildings will probably be limited to supplemental heating of hot water. Other applications of solar energy in dense urban areas will be forced to await the delivery of solar electric energy from the countryside beyond the city.

The application of solar energy to low-density high-rise buildings may provide the opportunity for more complete usage of the exterior walls of the building for solar energy collection. Land areas between buildings could also accommodate solar collection areas. In this case the distribution of hot and chilled water for heating and cooling would be practical.

The best areas for application of community solar energy are the suburbs, where the density of homes is low. In this case some land could be used for the energy park, with resulting energy products being distributed to the community.

The basic attraction for community energy parks is the possibility of using the maximum amount of collected sunlight by cascading the uses. Electrical power by thermal conversion results in a significant fraction of the energy appearing as waste heat. If this waste heat is rejected at a high enough temperature, on the order of 100–120°C, it can be used to provide gas-absorption cooling, space heating, and water heating.

The energy balances for a total energy system are shown in Fig. 16.17. The problem with a community total energy system is that the waste-heat utilization is seasonal. In midwinter the entire load of waste heat can be absorbed for heating. In fact, as the example shows, there is not enough waste heat from the electrical power needed by the community. In midsummer we have a similar situation, that of absorbing all of the waste heat and requiring more. Between these seasons relatively little heating or cooling is demanded, so the power station needs cooling towers capable of disposing of this waste-heat load.

The mismatch between electrical power demands and waste-heat availability can be met. In summer the long days produce more electricity than is needed for normal home uses. The gap between waste heat availability and cooling demand can be made up by using the excess electrical power for compressor refrigeration.

A full exploration of the options for community and total energy systems would require a separate book. We hope, in the interim, that these points and examples will be sufficient to launch the student on further exploration of these options.

It should be noted that all the values in Fig. 16.17 refer to the *installed* cost of the collector and heat transfer system, divided by the area of the collector. To arrive at an allowable budget for the collector before it is assembled into a system one needs to analyze the fraction of the system cost represented by the collector alone. In the examples cited the cost for heating, cooling, and electrical applications would be between two and three times the collector cost alone, and hence the values cited would have to be divided by this factor to yield a reasonable cost for the collector.

TOTAL ENERGY SYSTEM

Power production alone		
Latitude	40	deg
Energy income, winter, fully tracking	1.8×10^4	kJ/m^2d
Energy income, year, fully tracking	1.05×10^7	kJ/m^2y
Percentage of possible sunshine	60	%
Collector efficiency	0.70	-
Utilization factor	90	%
Thermodynamic efficiency	30	%
Distribution efficiency	95	%
Energy yield, winter	1.94×10^3	kJ/m^2d
Energy yield, year	1.13×10^6	kJ/m^2y
Demand, 1000 homes at 2.0×10^4kWh/d each	7.2×10^7	kJ/d
Fractional demand met by solar power	4.8×10^7	kJ/d
Power demand per year	1.75×10^{10}	kJ/y
Area of collectors, winter demand	2.47×10^4	m^2
Annual solar power production	2.8×10^{10}	kJ/y
Value of energy delivered, $0.05/kWh	388,000	$/y
Value of collector	15.70	$\$/m^2y$
Winter heating added		
Waste thermodynamic heat at 2.3 x power	1.1×10^8	kJ/d
Waste heat, 120-day season	1.3×10^{10}	kJ/y
Demand, 1000 homes at 100 MBtu/season	1.0×10^{11}	kJ/y
Fraction of heating demand met	13	%
Added value of heat at $0.05/kWh	181,000	$/y
Added value of collector	7.32	$\$/m^2y$
Summer cooling added		
Solar income, summer	3.5×10^4	kJ/m^2d
Power output, summer	3.8×10^3	kJ/m^2d
Power output from collector, summer	1.13×10^8	kJ/d
Waste heat produced at 2.3 x power	2.3×10^8	kJ/d
Waste heat, 120-day season	2.7×10^{10}	kJ/y
COP to produce cooling	0.60	-
Cooling from waste heat	1.6×10^{10}	kJ/y
Demand, 1000 homes at 1.5×10^4kWh/y	1.5×10^7	kWh/y
Demand, 1000 homes at 1.5×10^4kWh/y	5.4×10^{10}	kJ/y
Fraction of cooling demand met	44	%
Added value of cooling at $0.05/kWh	334,000	$/y
Added value of collector	13.51	$\$/m^2y$
Total value of collector	36.53	$\$/m^2y$

Fig. 16.17 Sample calculations for a total energy system.

16.14 COMMUNITY HEATING AND COOLING

In the example in Fig. 16.18 we calculate the efficiency for supplying heating and cooling on a community basis. A central solar energy park contains the collectors and water storage tanks for the entire community. Hot water and chilled water are produced in this park and distributed to the homes in the community through underground piping. The water is continuously circulated through the mains and feeder up to the exterior of the house. These pipes are buried in the ground without any insulation other than the soil. We assume a low soil moisture typical of the Southwest because it has good thermal insulating values compared to saturated soils of other regions.

The buried pipes create a surrounding heated zone or cooled zone, contributing significantly to the thermal inertia of the system. The central water storage tanks are considered large enough that with the added inertia of the ground around the pipes this system provides heating and cooling at all times.

The question we consider in this example is: What fraction of the collected solar energy will be lost in the distribution system? We will assume that 1000 homes are serviced, and each home is on a one-acre lot.

In this example we are not interested in how many square meters of collector will be required for heating and cooling. The amount per home is taken as shown in Fig. 16.18, plus the additional area for the losses in the distribution system. This example therefore deals with the losses incurred in delivering hot water and chilled water to a community of 1000 homes.

The layout of the homes is taken to be along a single line, with houses on both sides. Each lot is 200 ft wide, so the 1000 homes are distributed along a line 100,000 ft long. Since the water distribution also has return lines, the main lines will total 200,000 ft for the hot water and an equal length for the chilled water. We also assume that an equal length of smaller service pipe will connect to each house, again each being a double line, totaling 200,000 ft of service lines. In the calculation we assume a loss per foot as indicated.

The sample calculation indicates that 32% of the heat will be lost from the hot-water distribution system, probably the upper limit that can be tolerated. To reduce the losses the solar energy park could be designed to serve fewer houses, the houses could be grouped in a more compact arrangement, or the lot shape could be altered to reduce distribution-line length.

The loss from the chilled-water lines is less, 22%. This lower value is due to the fact that the temperature difference between the chilled-water line and the ground is much less than in the case of the hot water. Again, this loss can be reduced by better design of the community.

These figures are intended only to indicate the magnitude of the loss and to show that it is close enough to practical values that the community distribution of hot and chilled water could be an alternative to placing the solar collectors on the homes themselves. This option frees the homeowner from having his home architecture dominated by the solar collectors, it allows trees and so forth to be placed without regard for shadows that might be cast on the owner's or neighbor's solar collector, and it frees the homeowner from the acquisition and maintenance of the solar collectors–all factors that would indicate that this option deserves further study.

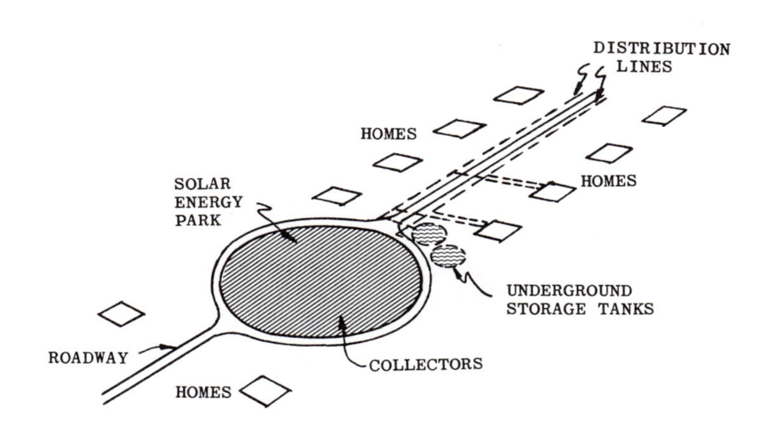

COMMUNITY HEATING AND COOLING

Latitude	40	deg
Energy income	not relevant	
Hot water service		
Total length of mains for 1000 homes	6.1×10^4	m
Total length of service lines, 1000 homes	6.1×10^4	m
Heat loss, mains	2.7×10^2	kJ/mh
Heat loss, service lines	0.9×10^2	kJ/mh
Total energy loss, 12.2×10^4 m	2.2×10^7	kJ/h
House demand at 1.67×10^6 kJ/d	7.0×10^7	kJ/h
Loss of heat energy	32	%
Cold water service		
Total length of mains for 1000 homes	6.1×10^4	m
Total length of service lines, 1000 homes	6.1×10^4	m
Heat gain, mains	1.3×10^2	kJ/mh
Heat gain, service lines	0.5×10^2	kJ/mh
Total energy loss, 12.2×10^4 m	1.1×10^7	kJ/h
House demand at 9.8×10^5 kJ/d	4.1×10^7	kJ/h
Loss of cold energy	22	%

Fig. 16.18 Schematic diagram of a community heating and cooling system for 1000 homes, and sample calculation.

16.15 INDUSTRIAL PROCESS HEAT

A considerable amount of fossil fuel is consumed in heating water for industrial processing. Where this water is used in the 50–80°C range, solar water heating is possible with simple collectors. An example of this application is the study made by Lawrence Berkeley Laboratories for Exxon on heating water for processing uranium in New Mexico. The collector consists of a long "waterbed" of vinyl, black on the bottom and with a clear top layer. This bag is filled with water to be heated. A second clear cover is provided for insulation and is inflated by means of air-pressure loading. This collector is shown schematically in Fig. 16.19(a).

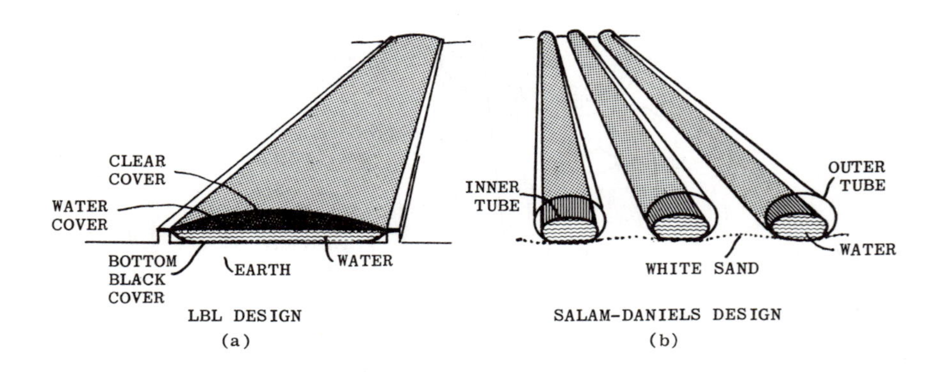

INDUSTRIAL PROCESS HEAT

Latitude	35	deg
Energy income, horizontal flat plate	6.7×10^6	kJ/m^2y
Percentage of possible sunshine	82	%
Utilization factor	70	%
Collector efficiency	0.60	-
Collector yield	2.3×10^6	kJ/m^2y
Demand of 100 MBtu per day average	1.0×10^8	kJ/d
Solar portion of demand	50	%
Area of collector required	8.4×10^3	m^2
Value of heat at $5.00/MBtu	91,200	$/y
Value of collector	10.88	$\$/m^2y$

Fig. 16.19 Schematic diagram of an industrial water heating system, and sample calculation.

The mode of operation could be by either of two procedures. One could flow water through the collectors during the day, extracting the hot water, or one could use a "batch" approach, where only enough water is placed in the collector for it to reach the desired temperature early in the afternoon. The heated water is then pumped out and placed into storage, where it can be drawn upon during the subsequent 24 hours.

The tradeoffs between the continuous flow and the batch are several, and a decision on which to follow requires a careful study of each case.

In this example we take the solar availability as for a horizontal flat plate. This means that winter performance will be low, but a side mirror booster could be used to increase winter performance, as the cost effectiveness ratio is favorable. One could also use tubular absorbers, like the grain dryer, to increase the optical cross section during winter.

Very simple collectors can be designed for medium-temperature industrial heat applications. Salam and Daniels (1959) designed a concentric tube collector consisting of an inner clear tube 4.85 cm in diameter partly filled with water containing a dye, and an outer tube twice the diameter that acts as a transparent window. This collector is shown schematically in Fig. 16.19(b). Air is circulated over the warm-water surface, and the humid air is then cooled to produce distilled water. Thus a single collector could produce distilled water and hot water. Comparable performance can be obtained if the inner tube is black plastic and the outer tube clear. The clear plastic should be opaque to the TIR to derive the maximum "greenhouse" effect.

These industrial heat collectors generally differ from the standard solar water-heater collectors in that their thermal inertia is very large. In Fig. 16.20 we show the diurnal temperature rise for a collector of low thermal inertia and one of high thermal inertia. In the latter case the temperature of the water continues to rise after noon until it reaches the temperature of a zero thermal inertia collector at the same hour. At this point the water batch would be pumped into storage.

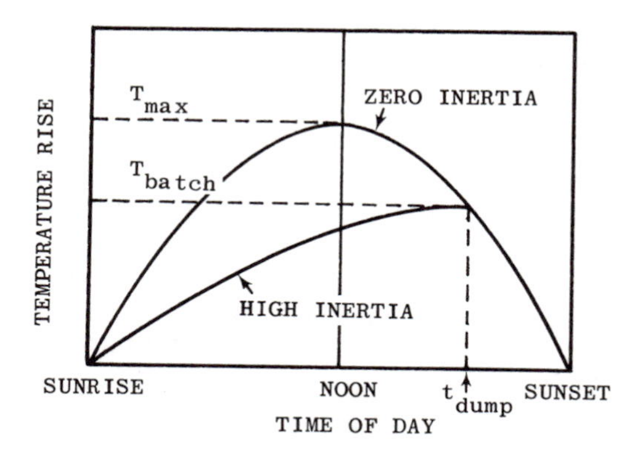

Fig. 16.20 Diurnal temperature rise for collectors of high and low thermal inertia.

As in other applications, the goal is not simply heated water, but water heated at a price that is competitive with that of other fuels. This means that the collector must be simple and have sufficient lifetime to repay the investment. The field is so extensive and the regions for potential applications so widespread that several types of collectors will undoubtedly be developed.

In the preceding examples we have taken only a first examination of the questions involved. For example, in several cases we have examined the performance only on an annual average basis or for a single season. It is obviously important to examine the behavior of any of these applications in more detail. We hope that the student will select one or more of these applications for a closer examination.

A particularly important tradeoff that requires more detail than we present is the optimum portion of demand that is to be met with solar energy. Although one might aspire to the goal of complete dependence on solar energy for any application, the hard fact of economics rules against this. The reason is quite simple. When annual demand covers a considerable range, and peak demand occurs infrequently, making a stand-alone solar application means that a portion of that capital investment is idle most of the year. Optimizing the solar portion then involves a decision on whether to use a backup energy supply, or tolerate the discomfort or inconvenience of the system's not being able to meet the full demand on these particular occasions. One also must weigh the consequences of having *other* energy-supplying alternatives idling until the time when the solar application needs this backup energy. The questions are quite involved and take detailed numerical studies for rational choices.

16.16 COST LIMITS

Up to this point we have said little about the economics of solar energy conversion. In the preceding examples we have stopped at the point of evaluating the economic value of the energy displaced by the solar energy, based on the assumption of certain costs for the energy being displaced. We can use these examples to determine the upper limit to the cost of the solar collector systems if we do not exceed the cost of retaining use of the energy to be displaced.

An important factor in future judgments is the instability of energy costs with traditional fuels, which are driven up by the growing shortages of petroleum. The cost limit as of the time of writing this textbook may change as fuel costs escalate. One must be very cautious about making the simplistic statement that "as the cost of traditional fuels rises, solar energy will become economic." The basic reason for caution is that energy costs are quickly reflected in the cost of every product manufactured with that energy, and in the labor costs incurred through the escalation of wages driven by the need of the worker for a larger wage to permit him to afford the more expensive energy. Recently the *average* cost of basic materials like steel, glass, and aluminum has risen at about the same rate as the *average* cost of energy used in the industrial community.

The reduction of the cost of solar energy conversion systems is urgent and is the basic problem in this field. Design, materials, and manufacturing methods all have the potential for cost reductions. One also must face the large increment of final system cost that must be attributed to the distribution and installation of systems, a cost factor that approximately doubles the FOB factory cost of a solar collector system.

It is interesting to note that present-day cost estimates for solar collectors actually built *as a system,* as contrasted to home-built components, show them to be about five times more expensive than the competing energy source. When we remember that in 1890 the French government also reached the conclusion that solar energy was a factor of 5 more expensive than competing energy sources (see Chapter 1), and when we recognize that the dollar value for a ton of coal is approximately 100 times more in 1975 than in 1890 (in monetary units, not in GNP values), we see that this feedback of general cost escalations has not bettered the economics of solar energy.

An answer to economic solar energy is to either "do it yourself" and avoid the normal operations of the marketplace, or find new points of leverage for cost reductions at the factory so that solar energy systems can reach the consumer in return for an acceptable expenditure of his annual income.

To return to the examples cited in the preceding sections of this chapter, we can divide the value of the energy displaced by the area of the solar collector in each example. This number is the value one can pay for the collector system and still break even with the cost of the fuel displaced. In reality the system is made up of other components in addition to the collector, but since the collector is the most expensive component the number we obtain gives some feeling for the allowable cost per square meter of the collector. We note that the values are rather low compared to the typical cost of a collector, which is $70–100/m^2 FOB factory (1975), and $200–400/m^2 installed.

As a final point, it should be noted that once solar energy systems are installed the "energy equivalent" cost of the solar power is stabilized, while other energy costs could rise by large amounts during the years of lifetime of the solar collector. This fact encourages us that solar energy can have a significant stabilizing effect on the escalation of energy costs, once a reasonable fraction of any demand section in the economy has elected the solar option.

There are two meaningful ways to compare costs. The first is to compare the cost of solar means with the traditional cost for the product. The second way is to compare the cost on a common basis of cost per unit of alternative energy. In Table 16.2 we show the cost relative to the traditional cost, taking the values shown in the sample calculations for each application. We see a wide range in allowable cost for the collector. We present the cost per year of operation and for a period of years corresponding to an extended lifetime for the collector. With the current capital rate charge of 15%, the effective lifetime for a capital investment is therefore seven years. If a lower capital rate charge or other investment credits are appropriate, the effective value lifetime will increase accordingly. Under any such conditions one should be cautious about using a value lifetime of more than 10 years. One reason for this conservatism is that achieving more than a 10-year operating lifetime for solar collectors is complicated by the need to use inexpensive materials and construction processes and the unreliability of inexpensive materials. Even with expensive materials there is no proof of lifetime this early in the course of the development of a new energy option. There is some evidence that hot-air collectors can give long lifetime, as reported by Lof. In the case of hot-water heaters the evidence indicates that corrosion is a continuing problem, but solvable if inert closed-loop coolants are used in lieu of water for the heat transfer role.

Because Table 16.2 mixes different base comparison costs it is hard to place the different applications on a common basis. The comparison is further complicated by the different use periods during the year, from only 60 days for the grain dryer to 365 days for power generation (subject to the utilization factor reduction in the effective number of days). For comparison on a common base we show Table 16.3, which gives the daily output of each collector type.

From Table 16.2 we note that the highest breakeven cost is for water heating. This is because the system is fully used all year and the system efficiency (see Table 16.3) is the highest of the simple applications. Water distillation and grain drying have the lowest value, but in these cases the collector can be simple and have a limited lifetime. If a permanent grain dryer installation were made, the allowable cost would increase from \$2.42 to \$16.94/m^2 (7 yr). It is interesting to note that the allowable budget for photothermal power and total energy systems is reasonable; this is due to

Table 16.2 Breakeven Collector Costs

Application	*Value per yr, \$/m2*	*Value/lifetime (7 yr)*		*Basis*
		\$/m2	*\$/ft2*	
Water distillation	0.62	4.34	0.47	\$2/1000 gal
Water heating	20.62	144.37	13.41	\$0.05/kWh
Swimming pool				
(exterior heater)	3.23	22.60	2.10	\$2/MBtu
(integral heater)	2.86	20.16	1.87	\$2/MBtu
Grain drying	4.34	2.42	0.22*	\$0.05/kWh
House heating	7.54	42.70	3.97	\$0.05/kWh
	2.28	12.88	1.20	\$4/MBtu
House cooling (LiBr)	7.33	51.31	4.77	\$0.05/kWh
Radiation cooling	3.89	27.86	2.59	\$0.05/kWh
Water pumping	5.82	40.74	3.78	\$0.05/kWh
Photothermal power	14.21	99.47	9.24	\$0.05/kWh
Photovoltaic power	9.00	63.00	5.85	\$0.05/kWh
Total energy system	36.53	255.71	23.76	\$0.05/kWh
Industrial heat	10.88	76.16	7.07	\$0.05/kWh

*Based on 60 days usage in lifetime and total tube surface rather than optical cross section.

Table 16.3 Comparison of System Yields

Application	*Yield per day-m², MJ/m²day*	*Use season, days*	*Yield per season, MJ/m²day*
Water distillation	7.3	365	2.66×10^3
Water heating	9.6	365	3.50×10^3
Swimming pool			
(exterior heater)	8.1	200	1.62×10^3
(integral heater)	7.2	200	1.44×10^3
Grain drying	5.2	60	0.32×10^3
House heating	3.6	120	0.44×10^3
House cooling (LiBr)	8.9	120	1.07×10^3
Radiation cooling	1.4	200	0.28×10^3
Water pumping	1.7	250	0.42×10^3
Photothermal power	2.0	365	0.73×10^3
Photovoltaic power	1.8	365	0.65×10^3
Total energy system	9.8	365	3.59×10^3
Industrial heat	6.3	365	2.30×10^3

their utilization over 365 days, which offsets the lower efficiency involved in the use of a thermodynamic cycle.

We would recommend that the designer or buyer do a similar evaluation of breakeven costs. If reasonable values are taken for the different terms in the sample calculations in this section, the output of energy from the proposed system can be derived. The allowable cost for the solar option can then be established on the basis of reasonable assumptions about the future costs of the energy alternatives.

16.17 SOLAR COLLECTOR COSTS

It is generally conceded that solar energy systems yield useful energy at costs above those of most alternative energy sources. The closest competing solar option is house and water heating in localities that use electrical resistance heating. To be effective in the national energy future, solar energy will need to become competitive on a broader basis, but this will require cost reductions. If we look into the future, what will the costs be in terms of 1975 dollars? If we assume a continuation of the present inflationary trend, then apparent prices of solar collector systems will certainly rise. The problem is how to normalize future apparent prices to a common basis. We think that the critical factor for each consumer of energy is the fraction of his gross personal income that is expended for the services rendered by energy. With this leveling factor it is probable that conventional energy costs will show a net rise, but what will solar energy costs show?

There is so little commercial experience at hand that predicting future solar collector costs is difficult. In the absence of an adequate record, two techniques can be applied: (1) the Delphi process and (2) cost sensitivity analysis.

The Delphi process is a simple technique of surprising accuracy in some situations. It involves the preparation of a set of questions defining precisely what aspect of the future is being assessed. These questions are submitted as a questionnaire to a broad selection of people who are knowledgeable about one or more aspects of the subject, asking for their best guess. For various reasons only a limited number of persons asked will respond. Some refuse to guess. Others know much about the topic but think the questions are inadequately phrased, pointless, or unanswerable. The respondents will generally be those who have a definite opinion, either optimistic or pessimistic. The results may therefore be widely variant with only a few estimates near the mean of the variance. In social, political, and economic matters the Delphi process has notable success when conditions remain normal. If unusual new realities suddenly affect any given situation the Delphi process is not accurate, as was the case in trying to predict unemployment just before the 1973–1974 Arab oil embargo disturbed the existing economic situation. The Delphi process is probably most successful when mass opinion has a direct influence on the subject being explored. A broad sample can then determine what people think, and what they think can be a self-fulfilling prophesy. In the case of solar energy costs this process may be less reliable, the guessing being more a hope than a reality.

The second way of estimating future costs is the sensitivity analysis method. In it one takes the present costs for existing designs and estimates within the cost breakdown where significant changes could occur. This method is more firmly based on fact, but it still involves considerable uncertainty in what changes might occur that could affect the cost elements. In the case of solar energy one is dealing on the whole with a technology the elements of which are quite conventional insofar as thermal applications are concerned. Solar cells, on the other hand, may evolve considerably away from using conventional semiconductor technology, increasing the uncertainty of future costs. In areas of conventional technology the major cost reduction to be looked for is the effect of mass production. In the area of new technology the reductions can be due to lowered cost of materials and processing energy requirements, as well as to mass production technology.

A set of future cost projections based upon the method of sensitivity analysis is presented in Figs. 16.21–16.23. The curves are fixed by 1975 costs of solar collectors or, where no commercial units have been sold, by estimates. The shaded area represents what we estimate to be the limits to future cost, and the solid line represents the most probable cost. In each case we consider a commercial, turn-key installation. Each collector type also includes the heat collection and transfer system to deliver the heat (electrical power in the case of direct conversion) to the using system. This means that the system is directly comparable to systems using other energy supplies, the solar collector and its transfer system simply replacing the alternative energy source. Neither energy storage nor a backup energy system is included since these costs vary widely.

The equivalent power cost for each application is shown at the right-hand edge of the graph. These costs are from the examples summarized in Table 16.2 and will shift upward or downward if different input quantities and interest factors are selected. This cost is where the solar collector system would break even with commercial electrical power to provide the same function. If commercial power costs are higher than the indicated equivalent cost, the solar energy is cheaper.

All of the costs are reduced to a constant factor of 1975 gross personal income. As this income level changes in the future, the costs of collector systems and commercial power must be multiplied by the ratio of the 1975 income to the actual income on the date of examination in order to plot the costs on these graphs.

In the estimation of future costs there are several factors influencing costs upward and downward, the interplay between these factors determining the final cost. We have already mentioned volume of production and sales as being a major factor for downward adjustment of costs and prices. The problem then becomes one of estimating the growth of volume. In the case of many new products the market does not develop rapidly even though the technology does. People wait for others to test the technology and find out whether it is attractive and reliable. When the technology is finally accepted as being economic and desirable the volume can rapidly build to higher levels. For this reason the curves in Figs. 16.21–16.23 do not immediately swing downward in the first decade.

It will be noted that some of the curves actually move upward during the first few years. This is because early experience may indicate the need for changes in materials and for correcting small problems before a satisfactory product evolves. These changes generally tend to increase prices. In addition, companies often accept smaller profit margins when introducing a new product to a mass market, correcting this situation as sales improve.

In the more distant future a price rise may be forced by the fact that materials costs will rise more rapidly than personal income owing to scarcity of natural resources and of fossil energy to process these resources into solar eollector systems. If solar energy still is only a minor contributor to the energy economy by the year 2010 and if no other exotic energy source replaces the inexpensive fuels of the mid 20th Century, it seems that this price rise will inevitably occur. This factor is responsible for the rise shown in our estimates, which becomes significant beyond the year 2000. The magnitude of the rise is dependent upon the type of materials required in each specific application and the amount of energy required to produce these materials.

Flat-Plate Collectors

There are two types of flat-plate collector to be considered, the air heater and the liquid heater, as shown in the curves in Fig. 16.21. In general the air heater is cheaper but less efficient for heat transfer and storage. There are also two modes of installation that affect the cost. Retrofitting existing buildings is, in general, more expensive than new construction. We estimate very little cost reduction for retrofitting but significant reduction for new construction, where the inclusion of solar heating and cooling is planned in the architecture and construction methods. We assume a lag time of 10 years, which means that cost reductions will come slowly in the 1980s but rapidly in the 1990s, the magnitude of the reduction depending on the size of the potential market and on the rate of escalation of materials costs above income. We predict that the decades of maximum cost advantage for solar energy installations will be 1990–2010. If solar energy or another exotic energy option can become a major source for industry, then a cost rise beyond 2000 might not occur, but time seems rather short for such a major transition to a stable low-cost energy supply.

Concentrating Collectors

We group all the concentrating collectors into two types, cylindrical and central focus, as presented in Fig. 16.22. We do not see significant differences in cost between the variants of each class, especially when the heat transfer subsystem is included in the cost. In both cases we see that early complications and design evolution will lead to increased costs before the phase of cost reduction begins. We also show a more rapid rise in cost beyond the year 2000 because more energy-intensive materials are used for these collectors. We note that the equivalent electrical power cost to break even is high for agricultural water pumping because the capital investment is generally not used the entire year. The equivalent cost is most favorable for the integrated energy system application since it is used all of the year, and also waste heat from the power cycle can be used much of the year for either heating or cooling.

Direct Conversion

Solar cells could be subject to large reductions in cost if new methods of production prove successful. It should be noted that the current cost spread of solar cells is small, indicative of a mature technology. This means that there must be a breakthrough in production and cost levels if the prediction curve is to move downward. Because of this unknown factor, future estimates for solar cells have a wider spread, as shown in Fig. 16.23, than thermal conversion. We show the minimum of the estimated cost of solar cells in the year 2000 as somewhat higher than others predict because we take the cost of a fully installed system and because it costs more to market, install, and protect cells in the real environment than the cell itself costs at the end of the production line. Concentrating solar cell systems yield a lower cost prediction than planar systems but are still limited by the cost of the concentrating portion of the system.

We hope that these several graphs will be useful in providing a running guideline of where various solar applications costs might be in future years and where the equivalent cost goal is. The confidence level for these curves is not high because the actual factors that affect costs are highly complex. Further studies to accurately define major points of cost sensitivity may prove useful so that research and development can focus on these points of maximum leverage on final system costs.

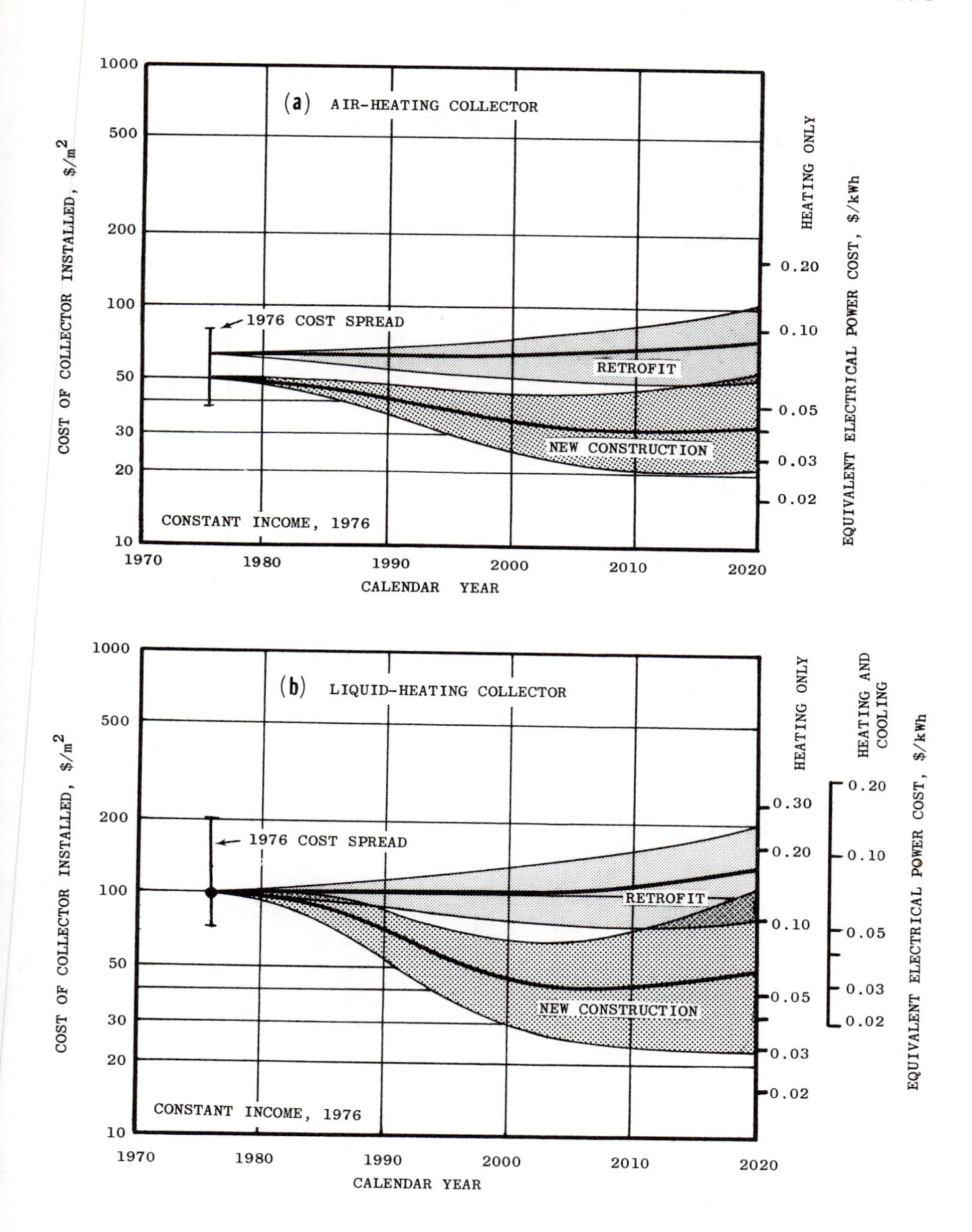

Fig. 16.21 Estimated future costs for air-heating and liquid-heating collectors and the equivalent electrical power cost for a heating application only.

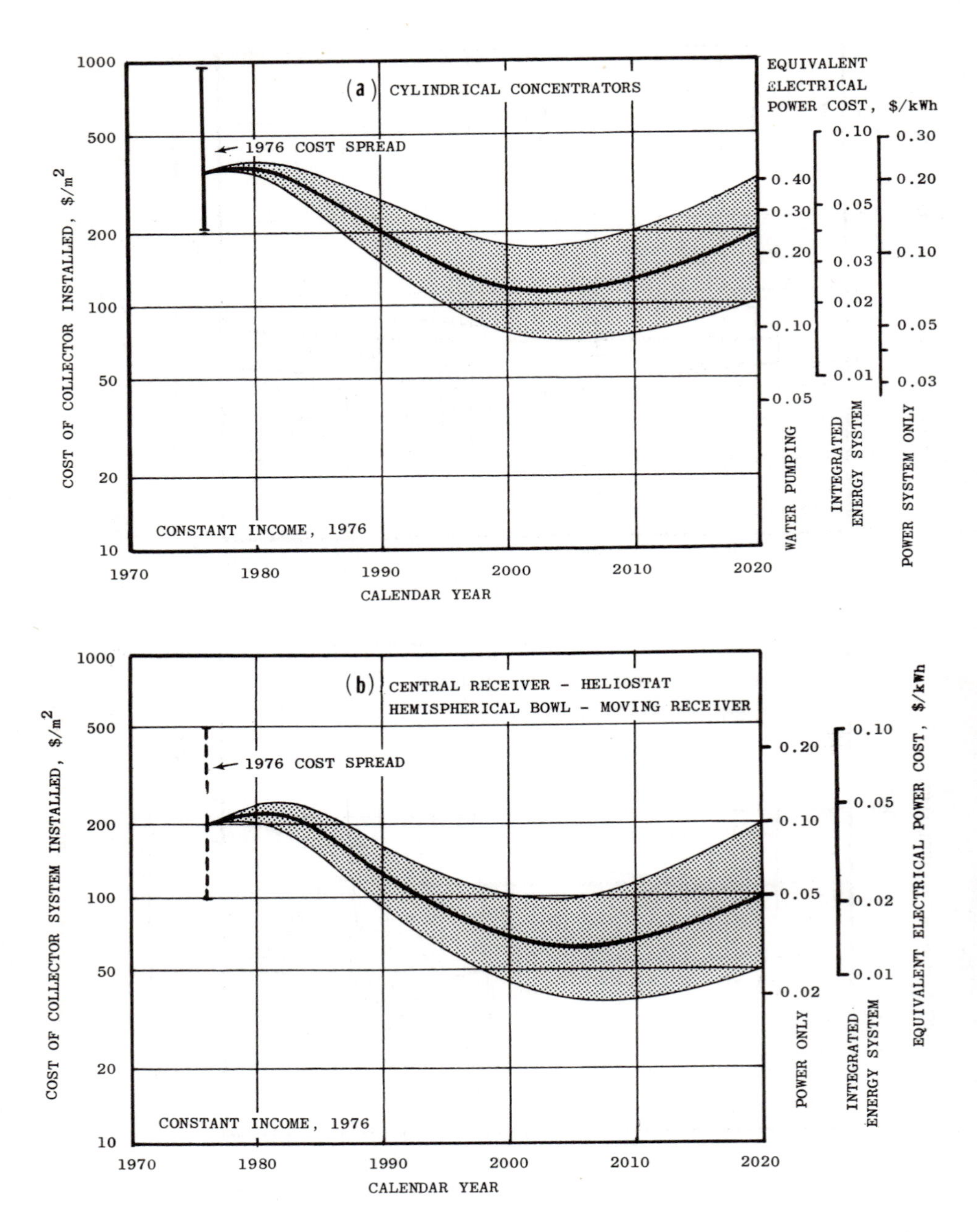

Fig. 16.22 Estimated future costs for cylindrical concentrating collectors and central receiver-heliostat and hemispherical bowl-moving receiver systems, and the equivalent electrical power cost for production of electricity.

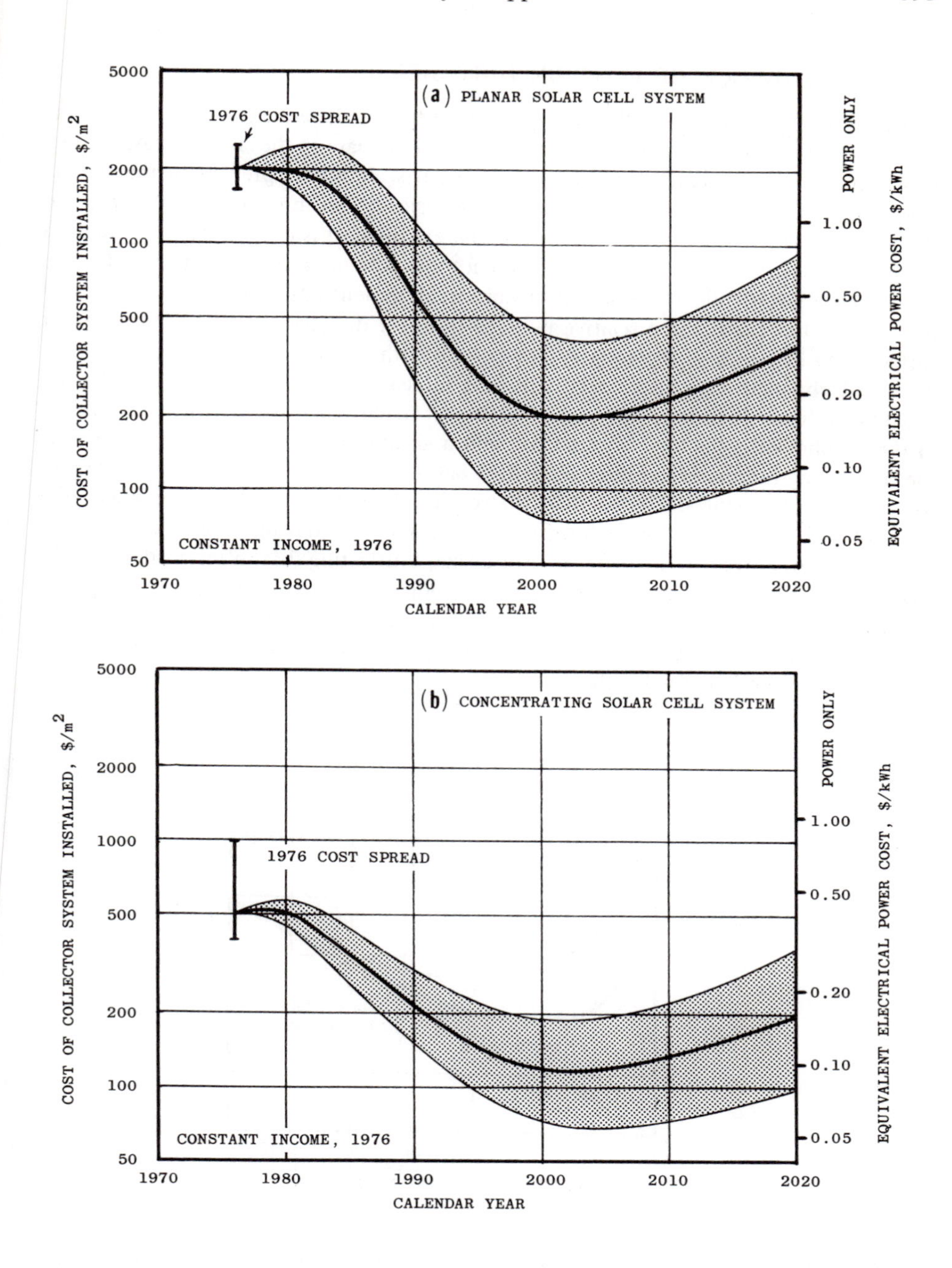

Fig. 16.23 Estimated future costs for planar and concentrating solar cell systems, and the equivalent electrical power cost for production of electricity.

16.18 DEPENDENCE OF COST ON UNIT SIZE

The cost of solar collectors relative to energy output is dependent on the size of the collector. Generally the ratio of output to cost will increase as the size decreases, down to a limit where incidental costs having different scaling laws prevent further reduction. This scaling dependence for collectors is basically because the output is proportional to the surface area of the collector whereas the cost is proportional to the mass of the collector, which in turn is proportional to its volume. It is therefore readily seen that when a given collector is scaled upward in all dimensions, the volume of material increases as the cube of the relative increase in linear dimensions, while the collector surface area increases as the square. The costs of other items, such as labor in assembly, field connections, and so forth, do not increase as the cube, and generally are constant or linear functions. When the collector is made small, there will be an increase in the number of collectors required to yield a given output. In this case the effect of the nonscaling factors is to increase the cost per unit output because of the increase in the number of modules. The result is that each collector type will show a curve like those of Fig. 16.24. Here curve *B* differs from curve *A* in having a minimum ratio of cost to energy output at a smaller aperture size, illustrating that different designs will have different optimum aperture sizes. A minimum cost per unit energy output generally will be obtained. The optimum size is determined by analyzing the individual cost elements for the design under study and then applying the appropriate scaling laws for each component of the final system cost.

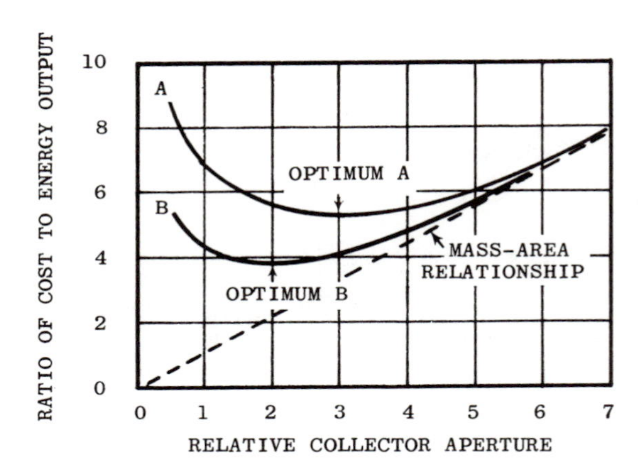

Fig. 16.24 Dependence on collector aperture of the ratio of cost to energy output for two hypothetical designs, illustrating the existence of an optimum size collector. Design *B* differs from design *A* in having a minimum ratio of cost to energy output at a smaller aperture size.

16.19 ENERGY COST STABILITY

We have shown that the figures on energy values at today's energy costs make the economic projections for most solar energy applications look bleak. The gap between present fuel energy costs and solar energy costs is a factor of about 2 to 5. There is, however, another way of looking at solar energy costs that alters the picture more favorably toward solar applications. We will only touch upon the essence of this point of view.

If the rise in energy costs continues into the indefinite future, a new phenomenon enters the economic picture. In the past several years we have seen the start of a general rise in energy costs that exceeds the rate of inflation. The inevitable growing shortage of petrofuels will continue to put pressure on the costs of all fuels. The cost of solar energy, on the other hand, will remain approximately constant, at least in regard to the capital investment portion. At the end of the lifetime of the facility its contribution to the energy cost is the same as the day the facility first produced energy.

In Fig. 16.25 we show the change in relative energy costs of solar versus alternative fuels based upon three values for the cost escalation of the alternative fuels. We show the case where the cost of solar energy was twice that of competing fuels on the day the solar installation was placed in operation. Energy costs are rising rapidly and show little sign of slowing down because of pressures on petrofuel prices and inflation. We note in Fig. 16.25 that even at a 10% cost escalation rate, modest in terms of

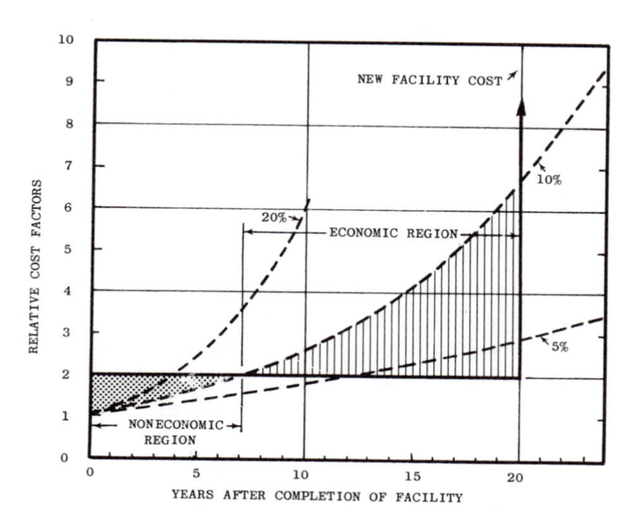

Fig. 16.25 Model of the cost of solar power in terms of original facility cost compared to alternative fuel costs where these fuel costs rise at an annual rate of between 5% and 20%. This shows that a facility that originally is more expensive than alternative fuels becomes economic during the lifetime of the facility.

present trends, solar energy from the facility becomes economic after about seven years and a real bargain by the end of the facility lifetime, assumed in this case to be 20 years. If the rate of energy cost escalation is 20%, the solar facility becomes economic in four years.

When the solar facility is replaced at the end of its lifetime, the replacement cost will be at the new market level, and if inflation in capital goods costs has continued at 10%, the cost for the replacement will make the original price look extremely low. Utilities are today faced with the fact that new facilities cost 5–10 times as much per kilowatt of generating capacity as 30 years ago, and in addition that the rate of inflation was much lower during those years.

One must be careful to use the same basis in comparing relative costs for a solar versus fuel option. As an example, a total energy system appears to have a good cost-value potential, as shown in Table 16.2, as a result of using the energy resource in an optimum way throughout the year. The solar option therefore looks reasonable in comparison to utility energy service, when it is not of the total energy type. If one then compares the economics of a similar coal or nuclear total energy system, it will show a clear advantage under present economic conditions.

There is another aspect of the cost picture that currently can be applied to narrow the gap between solar and fuel energy costs. There are various state and Federal legislative tax credits that apply only to renewable energy utilization options. These must be investigated by the prospective customer to see which ones apply in his area and what their impact is on the cost-value relationship.

16.20 COST EFFECTIVENESS OF IMPROVEMENTS

In Chapter 12 we discussed the efficiency of extraction of energy by a collector, presenting a wide variety of η-versus-ΔT curves. We also discussed how collector performance is affected by selective surfaces on the absorber, on the windows, and on combinations of both. The question naturally arises: What is the cost effectiveness of a change in performance? To illustrate the procedure we will consider the specific case of the cost effectiveness of superselective absorbing surfaces.

The important aspect of selective surfaces is their cost effectiveness, which is increased gain versus increased cost. If the net gain is exactly the same as the cost of additional ordinary collectors of lower efficiency with equal energy output, then there is no point in using selective surfaces. The net gain will therefore be the difference between the cost of additional ordinary collectors and the cost of the collector with the selective surface.

The efficiency of extraction of available sunlight η is a function of both operating temperature T and temperature rise ΔT. The system output then can be written in linear form where η is the average value appropriate to the operation of the system. Then

$$Q = I\eta, \tag{16.4}$$

where I is the solar flux. For simplicity in evaluating cost effectiveness in this example we will take the value of η at the listed temperatures. The energy gain resulting from a change in η because of a selective surface is

$$\Delta Q = I\,\Delta\eta. \tag{16.5}$$

If the cost per unit surface area of the standard collector is C for a given output Q, the breakeven extra cost allowed for a selective surface yielding improved performance $\Delta\eta$ is

$$\Delta C/C = \Delta\eta/\eta. \tag{16.6}$$

The reduction in collector area for the same energy output is

$$\Delta A/A = \Delta\eta/\eta. \tag{16.7}$$

For a specific case let us take the improvement in selective surfaces of the type by McKenney, Beauchamp, and Harper (MBH), similar to that shown in Fig. 9.43, compared to an electroplated black chrome coating. The efficiency curves for these two types of coating are shown in Fig. 16.26. Both coatings have absorptances of 0.95, but different emittances. The change in the performance curves is small but can be translated directly into values. In Table 16.4 we show the incremental efficiencies for the MBH coating relative to black chrome and a blackbody.

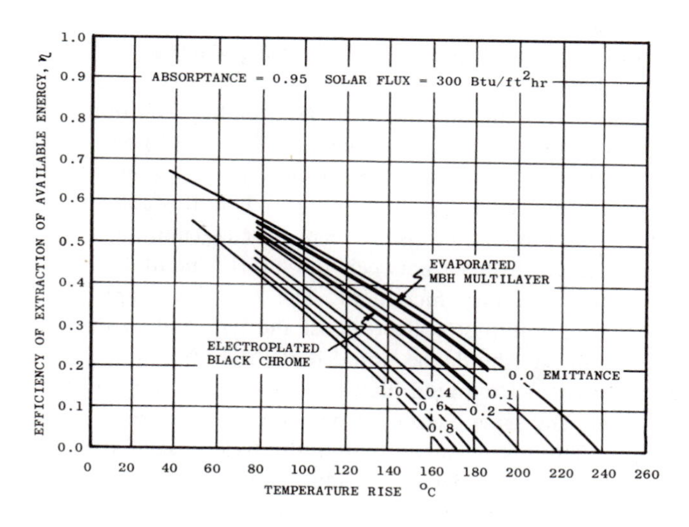

Fig. 16.26 Approximate relative performances for electroplated black chrome and MBH evaporated multilayer selective surfaces for normal solar flux and for a surface absorptance of 0.95.

Table 16.4 Incremental Efficiencies

			Blackbody		Black chrome	
Operating temperature, °*C*	*Tempera-ture rise,* *C*°	*Efficiency* η	*Incremental efficiency* $\Delta\eta$	*Incr. ratio* $\Delta\eta/\eta$	*Incremental efficiency* $\Delta\eta$	*Incr. ratio* $\Delta\eta/\eta$
150	150 (winter)	0.38	0.200	1.67	0.060	0.190
	115 (summer)	0.50	0.175	0.63	0.040	0.090
110	110 (winter)	0.55	0.145	0.35	0.045	0.089
	75 (summer)	0.60	0.110	0.22	0.030	0.053

The translation of incremental changes in efficiency into incremental value can be made when the cost of the normal collector is known. In Table 16.5 we show the incremental values for several base collector costs. Note that the incremental value of a selective coating is large compared to a blackbody absorber and still significant for the difference between the two types of selective surfaces. The last column in Table 16.5 shows the added value, taking into account the difference in cost estimated for the production of the two selective surfaces. We take the added cost of making a black chrome selective surface as $\$8.25/m^2$ ($\$0.75/ft^2$) and the added cost of making the MBH selective surface as $\$16.00/m^2$ ($\$1.45/ft^2$). This difference in cost is because the black chrome surface is produced by electroplating whereas the MBH surface is produced by vacuum deposition.

The cost saving with the MBH surface over the black chrome surface is shown in Fig. 16.27. In this graph we show no abscissa. This type of presentation is commonly used when a multivariate distribution is displayed. It is constructed by arbitrarily spacing the several groups of data points equidistant. In this case the point for the cost savings for the 115 C° temperature rise for each base collector cost is arbitrarily spaced uniformly. The abscissa distance between the 115 C° point and the 150 C° point is also spaced a similar arbitrary increment. The resulting graph clearly displays the curves as a function of both base collector cost and temperature rise.

From this sample it should be noted that the base collector cost is that of the collectors installed with all required plumbing. Thermal storage is not included, since both collectors yield the same energy output and hence presumably require the same storage subsystem. The result shows that if the base cost of the collector system is close to $\$50/m^2$, the added efficiency produces little cost saving. If the cost is closer to $\$150/m^2$, still low by actual experience as of this date, then the cost saving with the MBH surfaces is significant. We leave it to the student to explore the cost savings of solar collectors when other aspects affecting collection efficiency are considered.

Table 16.5 Net Added Value of an Improvement in Selectivity

Base collector cost	*Operating temperature*	*Temperature rise*	*Added value*
$/m²	°C	C°	$/m²
50	150	150	1.75
50	150	115	−3.25
50	110	110	−3.30
50	110	75	−5.10
100	150	150	11.25
100	150	115	1.25
100	110	110	1.15
100	110	75	−2.45
150	150	150	20.75
150	150	115	5.75
150	110	110	5.55
150	110	75	0.20
200	150	150	30.25
200	150	115	10.25
200	110	110	10.05
200	110	75	2.85

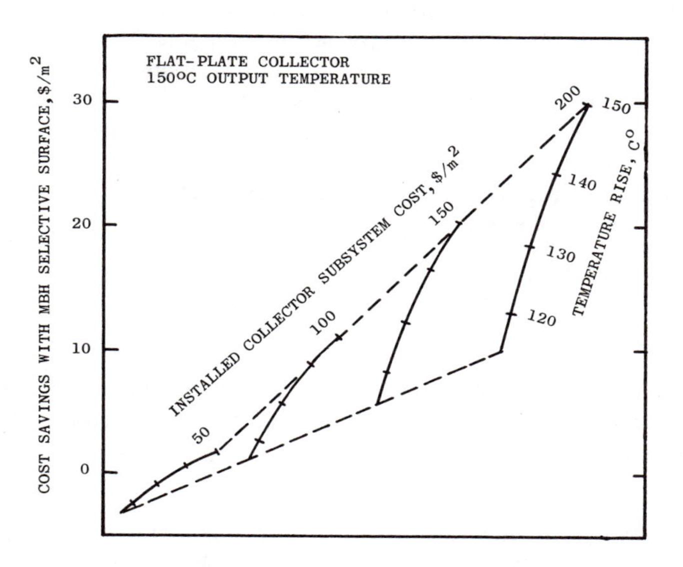

Fig. 16.27 Cost savings for a flat-plate collector geometry as a function of collector subsystem cost and temperature rise for an evaporated MBH selective surface and an electroplated black chrome selective surface.

16.21 PROBLEMS

16.1. If solar energy is used with a multiple-effect still having an effect factor of 4, how does the allowable cost change for the solar collectors? If the factor is 20, how does the cost change? What other economic factors arise when one goes from a single-effect to a multiple-effect still?

16.2. What would the seasonal variation in water output be for a single-effect still as discussed in Fig. 16.1?

16.3. In Table 16.2 we show the allowable cost for water heating based upon summer yield. What is the allowable cost based upon the energy delivered by the solar heater over the entire year? What would be the average fraction of demand over the year if the collector provides 100% of the water heating demand for the summer?

16.4. How much additional energy would be delivered if the solar hot-water heater of Fig. 16.2 were tilted to optimize for winter sun part of the year and summer sun the rest of the year?

16.5. What practical problems have to be faced with night cooling of a solar hot-water heater? Itemize the possible avenues to solve these problems. What are the economic consequences of these avenues?

16.6. In the example in Fig. 16.4 for the external pool heater, what would be the collector size (approximately) to maximize the solar heat input for March and April, where the pool owner wants to start the swimming season two or three months early but does not want to use it in the depth of winter? If he does not want the pool to overheat in summer, how many months can he use the heater for a pool temperature of approximately 30°C? Assume a location and ambient temperature for your answer. How does the economic factor change for this partial season use?

16.7. If the pool heater is plastic, what problems are faced in regard to system durability in winter? In summer? What are the practical answers to these problems?

16.8. What is the maximum fraction of a rectangular pool that can be covered with circular floating pool heaters? Assume (a) a perfect match in diameters and (b) the maximum mismatch.

16.9. A flat-plate plastic collector for grain drying is made of a single tube of black polyethylene. What is the relative energy yield for this amount of material being deployed flat on the ground with the tube flattened so that it has small thickness compared to the width of the tube compared to the circular case? When compared to the optimum tilt of the flat-plate deployment?

16.10. How many metric tons of wood will be produced from a woodlot per year in Minnesota? With how many tons of coal will this compare? Per hectare? Per acre?

16.11. If a crop of cane can be grown in three months, and three crops per year are possible, what would be the approximate yield per hectare per year?

16.12. What would be the value of heat delivered for house heating per year at latitude 30°? State the values used in this problem. How does the allowable cost per square meter of collector compare with a site at 45° latitude?

16.13. If the heating system of Problem 16.12 were combined with the gas-absorption refrigeration system of Fig. 16.11, what would be the allowable cost per square meter for a home at 30° latitude? State the quantities selected and used in this problem.

16.14. If the radiating roof of Fig. 16.12 were facing south, how much solar heat could be gathered on a winter day at 40° latitude if the roof pitch were 15°? How much night cooling could be obtained on a summer night from this same roof? The roof has no cover windows and there is no wind. The heat is to be delivered at 40°C and the outside daytime temperature is 15°C.

16.15. If the wind were 3 m/sec, how much heat could be delivered at 40°C in the above problem?

16.16. If you wished to gain radiative cooling 24 hours a day by placing the radiating surface beneath a honeycomb baffle, what would be the optical properties of the honeycomb so that sunlight would not reach the radiator but TIR emitted from the radiator would escape over a 2π solid angle? What would be the alternative in regard to the radiator itself?

16.17. A house requires 150,000 Btu/day of cooling. If an evaporative cooler is used, how many cubic feet per minute will be required of the fan in the evaporative cooler when the conditions are as stated in Fig. 16.13?

16.18. If alfalfa is 1% efficient in converting solar energy to biomass, how many hectares of alfalfa will the water pump in Fig. 16.14 supply? How many kilograms of alfalfa will be produced in a three-month growing cycle? What will be the added cost to a metric ton of alfalfa when this solar pump is used?

16.19. Assume that the electrical power demand for a community is two units during the period of 0900 to 1900 h and one unit at other times. If the solar electric system just meets the demand of 2.3 MW_e at noon, what fraction of the total demand must still be supplied by a backup system at the winter solstice (40°N) and at summer solstice? Use the system as described in Fig. 16.15.

16.20. If the power plant of Problem 16.19 is to store enough solar energy in the form of heat to make the system self-sufficient at winter solstice, and if sensible heat at 1 cal/g C° is used and the allowable temperature rise (or drop) is 200 C°, what is the diameter of a spherical tank of this capacity? What is the value of the energy stored?

16.21. If the power plant of Problem 16.20 is to have the capacity to store 48 h of additional energy to the overnight energy reserve, how many tanks of the size as stated in Problem 16.19 will be required?

16.22. If the weather is such that there are five clear days after a period of stormy weather (when all heat in the storage system is exhausted), how much additional solar collection area will be required to bring the storage system up to full storage at the end of the five days?

16.23. What must be the diameter of the solar energy park in Fig. 16.18, if the collector fill factor is 50%, to supply 1000 homes with the winter energy demand described in Fig. 16.8? How large would a spherical underground water storage tank have to be to have a reserve thermal energy supply for ten days for this community?

16.24. Prepare a diagram like Fig. 16.27 for the cost savings for a black chrome selective surface over a blackbody having equal absorptance. Use the efficiency graph in Fig. 16.26.

16.25. If you assume the annual heating demand curve of Fig. 13.11 and the conditions in Problem 13.3, what will be the utilization factor for the solar heating collector?

Concluding Remarks

In the main portion of this textbook we have presented the case for technological conversion of solar energy into heat and work. There are many ways this can be done. If you now go back and reread the historical introduction, you may wonder what really is new about solar energy conversion. What is new that was never tried before the turn of the century? Other than the photovoltaic cell, whose beginnings antedate the 20th Century, there is little. Hopefully, we understand more about the functioning of systems and subsystems and have a wider range of materials at our disposal. But is this enough?

The fundamental question for solar energy is how much it costs. If solar energy does not compete economically with other fuels, which often are much more versatile and convenient to use, it faces the same decision it received late in the 19th Century when the French government examined Mouchot's and Pifre's devices and found the energy to be too expensive for further development.

One can say that the ultimate exhaustion of fossil energy was not a factor then. Ultimate exhaustion is not necessarily the driving question today, as there are energy alternatives that can postpone the time of crisis for a few decades or a century. The real problem is whether, at any future cost for fossil fuels, solar energy will be preferred. This question and its answer revolve about the coupling of energy costs into products, such as collectors, made using that energy.

The central problem and challenge for solar energy applications is whether the costs can be reduced from their present levels sufficiently to find applications. If any points of market penetration can be found, such as hot-water production (one of the more cost-effective applications), then will other applications follow? If we look at the predictions of the sensitivity analysis exercise and compare them with those in Table 16.2, we note a gap today and a gap even in the year 2020, based on the average costs. When one further considers that our estimates are for the total system, and the sensitivity analysis predictions refer only to the cost of the solar collectors, the gap widens. Whether this gap can be narrowed is a challenge not only for the engineer and scientist of tomorrow; it is a great challenge to manufacturing ingenuity and marketing innovation. We hope this book can make a small contribution toward these goals.

Concluding Remarks

Appendixes

APPENDIX A. CONVERSION OF UNITS

From	To	Multiply by	From	To	Multiply by
acre-ft	gal	3.26×10^5	cal/g	Btu/lb	1.80000
"	liters	1.233×10^6	"	Wh_t/kg	1.1622
"	m^3	1233.5	cal/sec	Btu/h	14.286
acre	ha	0.4047	"	W	4.184
Angstrom (Å)	μm	1.--- $\times 10^{-4}$	"	J/sec	4.184
atm	kg/m^2	1.0332×10^4	"	Btu/sec	3.9684×10^{-3}
"	kg/cm^2	1.0332	"	ft-lb/sec	3.0860
"	g/cm^2	1.0332×10^3	"	g-cm/sec	4.2665×10^4
"	lb/in^2	14.70	cal/cm^2sec	Btu/ft^2h	1.3272×10^4
"	$tons/ft^2$	1.058	"	W/cm^2	4.184
bar	atm	0.9869	"	W/m^2	4.184×10^4
"	$dyne/cm^2$	10^6	"	J/m^2sec	4.184×10^4
"	kg/m^2	1.020×10^4	"	kW/m^2	41.84
"	lb/in^2	14.5038		MJ/m^2h	150.62
Btu	J	1.0548×10^3	cal/cm^2	L (langley)	1.0000
"	kJ	1.0548	"	Wsec	4.184
"	cal	251.996	cal/cm^2h	kW/m^2	1.1622×10^{-2}
"	kWh_t	2.928×10^{-4}	"	W/cm^2	1.1622×10^{-3}
"	ft-lb	778.3	cal/cm^2min	kW/m^2	6.9732×10^{-1}
Btu/h	ft-lb/sec	0.2162	"	L/min	1.0000
"	cal/sec	0.0700	$cal/cm^2C^osec/cm$	$Btu/ft^2F^oh/ft$	0.067197
"	W_t	0.2931	"	$Btu/ft^2F^oh/in$	0.80636
Btu/min	kW_t	0.01757	"	$J/cm^2C^osec/cm$	4.184
"	W_t	17.57	"	$W/cm^2C^o/cm$	4.184
Btu/lb	J/kg	2.326×10^{-3}	"	$W/m^2C^o/cm$	4.184×10^4
"	cal/g	0.55555	cm	in	0.39370
Btu/ft^2min	W/m^2	18.914	"	ft	0.03281
"	cal/cm^2sec	4.521×10^{-3}	cm^2	in^2	0.15500
Btu/ft^2h	cal/cm^2sec	7.5344×10^{-5}	"	ft^2	1.0764×10^{-3}
"	J/cm^2sec	3.1524×10^{-4}	cm^3	in^3	6.1023×10^{-2}
"	W/m^2	3.1524	"	ft^3	3.5315×10^{-5}
"	kJ/m^2h	11.348	cfm	liters/sec	0.4720
"	cal/cm^2h	0.27125	day	sec	86,500.0
"	cal/cm^2min	4.5208×10^{-3}	dyne	J/sec	10^{-7}
"	L/min	4.5208×10^{-3}	$dyne\text{-}cm^2$	bar	10^{-6}
Btu/ft^2F^oh	cal/m^2C^osec	1.3573	erg	Btu	9.4845×10^{-11}
"	W/m^2C^o	5.6783	"	cal	2.3901×10^{-8}
$Btu/ft^2F^oh/ft$	$cal/cm^2C^oh/cm$	14.881	"	g-cm	1.0197×10^{-3}
"	$J/m^2C^osec/m$	1.73073	"	kWh_t	2.7778×10^{-14}
"	$cal/m^2C^osec/cm$	4.1336	ft	cm	30.480
"	$W/m^2C^o/cm$	17.294	"	m	0.30480
$Btu/ft^2F^oh/in$	$cal/cm^2C^oh/cm$	1.2401	ft(water)	atm	0.0295
"	$kJ/m^2C^oh/cm$	100.45	"	lb/in^2	0.4335
"	$cal/m^2C^osec/cm$	3.4447	"	kg/m^2	304.8
"	$cal/cm^2C^osec/cm$	3.4447×10^{-4}	ft^2	cm^2	929.03
"	$W/m^2C^o/cm$	14.412	"	m^2	0.092903
"	$W/m^2C^o/m$	0.14412	ft^3	cm^3	2.8317×10^4
cal	g·cm	42664.9	"	m^3	2.8317×10^{-2}
"	J	4.184	"	gal(U.S.)	7.48052
"	Btu	3.9683×10^{-3}	"	liters	28.3106
"	kWh_t	1.1622×10^{-6}	ft/sec	cm/sec	30.48
"	kJ	4.184×10^{-3}	"	km/h	1.09728
"	Wh_t	1.1622×10^{-3}	"	mi/h	0.68182

From	To	Multiply by
ft-lb	Btu	1.2841×10^{-3}
"	cal	0.32405
"	J	1.35582
gal(U.S.)	cm^3	3785.41
"	lb(water)	8.34517
"	kg(water)	3.7852
"	ft^3	0.13368
"	acre-ft	3.0689×10^{-6}
"	liters	3.7852
g	lb	2.2046×10^{-3}
g/cm^2	lb/in	1.4223×10^{-2}
"	atm	9.6784×10^{-4}
g/cm^3	kg/m	10^3
"	lb/in^3	3.6126×10^{-2}
"	lb/ft^3	62.4280
g-cm	Btu	9.3011×10^{-9}
"	J	9.8066×10^{-5}
g-cm/sec	cal/sec	2.3438×10^{-5}
hp	Btu/min	42.4356
"	kW	0.74570
"	kJ-sec	0.74570
"	cal/min	1.06936×10^4
in	cm	2.54000
in(water)	lb/in^2	3.6126×10^{-2}
"	kg/m^2	25.3993
"	atm	2.458×10^{-3}
in^2	cm^2	6.4516
"	ft^2	6.9444×10^{-3}
in^3	cm^3	16.3871
"	m^3	1.6387×10^{-5}
"	ft^3	5.7870×10^{-4}
in(Hg)	atm	3.3421×10^{-2}
"	ft(water)	1.1330
J	Btu	9.48451×10^{-4}
"	kWh	2.7778×10^{-7}
"	cal	0.239006
"	erg	10^7
"	ft-lb	0.737562
"	Wh	2.7778×10^{-4}
J/cm^2	cal/cm^2	0.239006
J/sec	W	1
"	Btu/min	5.6907×10^{-2}
"	cal/min	14.3404
"	hp	1.34102×10^{-3}
kJ	kWh	2.7778×10^{-4}
kJ/m^2	Btu/ft^2	8.8111×10^{-2}
"	cal/m^2	239.006
"	cal/cm^2	0.02390
"	$kcal/m^2$	0.23901
$kcal/m^2 day$	$MJ/m^2 day$	4184.0
"	$kJ/m^2 day$	4.1840
kg	lb	2.20462
kg/cm^2	atm	0.96784
"	ft(water)	32.8093
"	lb/ft^2	14.2233

From	To	Multiply by
km	mi	0.62137
"	ft	3280.84
kcal	J	4.1840×10^3
km^2	acres	247.105
"	mi^2	0.38610
km/h	ft/sec	0.9113
"	mi/h	0.62137
knots	cm/sec	51.444
"	ft/sec	1.6878
"	mi/h	1.1508
"	km/h	1.852
kW	Btu/min	56.8253
"	Btu/h	3414.43
"	hp	1.3410
"	cal/min	1.43197×10^4
"	cal/sec	238.662
"	J/sec	10^3
"	kJ/h	3600
kWh	Btu	3410.08
"	cal	8.59326×10^5
"	kcal	8.59326×10^2
"	kJ	3600
"	lb(H_2O evap.)	3.53
kW/m^2	$Btu/ft^2 h$	317.21
"	W/cm^2	0.10000
"	$cal/cm^2 sec$	0.23901
"	$kJ/m^2 h$	3600
L (langley)	cal/cm^2	1.00000
"	j/cm^2	4.184
"	kJ/m^2	41.84
"	Btu/ft^2	3.6866
L/min	$cal/cm^2 min$	1.0000
"	$cal/cm^2 sec$	1.6667×10^{-2}
"	$Btu/ft^2 h$	1.3272×10^4
liter	in^3	61.0254
"	gal	0.26418
"	ft^3	3.53157×10^{-2}
"	cm^3	1000
lb/in^2	g/cm^2	70.3070
"	kg/m^2	703.070
lb/in^3	g/cm^3	27.6799
"	kg/m^3	2.76799×10^4
lumen	W	1.4706
m	mi	6.2137×10^{-4}
"	ft	3.2808
"	in	39.3701
m^2	ft^2	10.7639
"	in^2	1550.00
"	ha	10^{-4}
"	acres	2.47105×10^{-4}
m^3	ft^3	35.3147
"	acre-ft	8.1071×10^{-4}
"	in^3	6.10237×10^4
"	gal (U.S.)	264.172

From	To	Multiply by	From	To	Multiply by
MBtu	kJ	1.0548×10^6	therm	kWh	29.28
"	MkJ	1.0548	"	MJ	1.0548×10^2
"	kWh	292.875	"	kcal	2.5200×10^4
MkJ/m^2	$kBtu/ft^2$	88.11	W	Btu/h	3.41443
mi	yds	1760	"	cal/sec	0.239045
"	m	1609.34	"	hp	1.34102×10^{-3}
"	km	1.60934	"	J/sec	1.00000
mi^2	km^2	2.58999	W/cm^2	cal/cm^2sec	0.239045
"	acres	640.000	"	Btu/ft^2h	31.721
"	ha	258.9988	"	kW/m^2	10.0000
N (newton)	dynes	10^5	"	cal/cm^2min	14.3310
poundal	lb	0.03108	"	L/min	14.3310
"	g	14.0981	"	kJ/m^2h	3.60000×10^4
lb	g	453.5924	Wh	cal	859.184
"	kg	0.453592	"	ft-lb	2655.22
therm	Btu	10^5	"	kg-m	367.098
"	MBtu	0.10000	"	kJ/m^2	3.6000×10^4

APPENDIX B. USEFUL SOLAR FLUX QUANTITIES

Useful Solar Flux Quantities

Quantity	kW/m^2	MJ/m^2h	Btu/ft^2h
Extraterrestrial solar flux (Solar Constant)	1.353	4.871	429.2
Desert sealevel, noon, direct (D)	0.97	3.49	308.
Desert sealevel, noon, direct + scattered (D+S)	1.05	3.78	334.
Standard sealevel, noon, (D)	0.93	3.35	295.
Standard sealevel, noon, (D+S)	1.03	3.71	327.
Urban, typical, noon, (D)	0.61	2.16	193.
Urban, typical, noon, (D+S)	0.81	2.92	257.
24h average, Desert, fully tracking, (D+S)	0.40	1.43	149.
24h average, Desert, fixed λ tilt, (D+S)	0.24	0.87	76.

Ratio, noon/24h average, Desert, fully tracking, (D)	2.66
Ratio, noon/24h average, Desert, fully tracking, (D+S)	2.57
Ratio, noon/24h average, Desert, fixed λ tilt, (D)	4.30
Ratio, noon/24h average, Desert, fixed λ tilt, (D+S)	4.20
Ratio, extraterrestrial/24h average, Desert, FT, (D+S)	3.40
Ratio, extraterrestrial/24h average, Desert, Fλ, (D+S)	5.55

APPENDIX C. MAXIMUM ANNUAL COLLECTOR PERFORMANCE

Maximum Possible Annual Energy Yields

Configuration	*Latitude, degrees*	*Energy yields* 10^3 *kWh/m*^{2}*yr*	10^6 *kJ/m*^{2}*yr*	*kBtu/ft*^{2}*yr*
Fully tracking (*D*)	45	2.85	10.3	910
	30	3.11	11.3	985
Horizontal plate (*D*)	45	1.64	5.9	520
	30	1.92	6.9	610
Horizontal plate (*D* + *S*)	45	1.92	6.9	610
	30	2.26	8.1	710
Fixed (+15°) (*D* + *S*)	45	2.03	7.3	642
	30	2.25	8.1	713
Polar EW tracking (*D*)	45	2.80	10.0	886
	30	3.00	11.0	952
Horizontal EW:NS tracking (*D*)	45	2.30	8.3	728
	30	2.45	8.8	775
Horizontal NS:EW tracking (*D*)	45	1.70	6.1	537
	30	2.51	9.0	790

APPENDIX D. COMPARATIVE DAILY COLLECTOR PERFORMANCE

Geometry		Energy collected				Percent of fully tracking collector		
		Summer	Equinox	Winter		Sum.	Equi.	Win.
Fully tracking (*D*)	(45°)	11.47	7.89	4.40	kWh/m^{2}day	100	100	100
		41.3	28.4	15.7	MJ/m^{2}day			
		3.63	2.50	1.39	kBtu/ft^{2}day			
	(30°)	10.82	8.85	6.24	kWh/m^{2}day	100	100	100
		39.0	32.0	22.4	MJ/m^{2}day			
		3.44	2.81	1.97	kBtu/ft^{2}day			
Fully tracking (*D+S*)	(45°)	12.67	8.59	5.10	kWh/m^{2}day	110	110	110
		46.0	30.9	17.9	MJ/m^{2}day			
		4.02	2.73	1.57	kBtu/ft^{2}day			
	(30°)	11.96	9.60	6.50	kWh/m^{2}day	110	110	110
		43.0	34.4	23.8	MJ/m^{2}day			
		3.79	3.05	2.10	kBtu/ft^{2}day			
Horizontal flat plate (*D*)	(45°)	7.54	4.31	1.23	kWh/m^{2}day	64	54	27
		27.0	15.5	4.5	MJ/m^{2}day			
		2.38	1.36	0.42	kBtu/ft^{2}day			
	(30°)	7.85	5.53	3.00	kWh/m^{2}day	72	63	46
		28.5	20.0	10.8	MJ/m^{2}day			
		2.50	1.75	0.95	kBtu/ft^{2}day			
Horiz. flat plate (*D+S*)	(45°)	8.55	4.90	1.62	kWh/m^{2}day	73	64	37
		31.0	18.0	5.8	MJ/m^{2}day			
		2.70	1.57	0.51	kBtu/ft^{2}day			
	(30°)	8.90	6.34	3.50	kWh/m^{2}day	82	72	56
		32.0	22.9	12.7	MJ/m^{2}day			
		2.82	2.01	1.10	kBtu/ft^{2}day			
Latitude tilt (*D*)	(45°)	6.48	6.10	3.43	kWh/m^{2}day	56	76	78
		23.3	21.8	12.3	MJ/m^{2}day			
		2.05	1.94	1.09	kBtu/ft^{2}day			
	(30°)	6.47	6.33	4.87	kWh/m^{2}day	59	72	79
		23.3	22.8	17.6	MJ/m^{2}day			
		2.05	2.00	1.53	kBtu/ft^{2}day			
Latitude tilt (*D+S*)	(45°)	7.60	7.00	3.95	kWh/m^{2}day	64	86	88
		27.3	25.0	14.2	MJ/m^{2}day			
		2.40	2.21	1.25	kBtu/ft^{2}day			
	(30°)	7.63	7.28	5.56	kWh/m^{2}day	69	82	89
		27.5	26.2	20.0	MJ/m^{2}day			
		2.41	2.30	1.76	kBtu/ft^{2}day			
Latitude tilt + 15° (*D*)	(45°)	4.63	6.16	3.56	kWh/m^{2}day	40	79	81
		16.7	22.2	12.8	MJ/m^{2}day			
		1.47	1.95	1.13	kBtu/ft^{2}day			
	(30°)	4.79	6.49	5.10	kWh/m^{2}day	44	74	81
		17.4	23.4	18.4	MJ/m^{2}day			
		1.52	2.06	1.62	kBtu/ft^{2}day			
Latitude tilt + 15° (*D+S*)	(45°)	5.80	7.04	4.08	kWh/m^{2}day	49	89	91
		20.7	25.3	14.6	MJ/m^{2}day			
		1.83	2.23	1.29	kBtu/ft^{2}day			

Geometry		Energy collected Summer	Equinox	Winter		Percent of fully tracking collector Sum.	Equi.	Win.
	(30°)	6.00	7.45	5.80	kWh/m²day	54	84	91
		21.4	26.7	20.8	MJ/m²day			
		1.89	2.35	1.84	kBtu/ft²day			
Vertical (*D*)	(45°)	1.65	3.98	3.57	kWh/m²day	14	52	82
		5.95	14.3	12.9	MJ/m²day			
		0.52	1.26	1.14	kBtu/ft²day			
	(30°)	0.36	3.22	4.40	kWh/m²day	3	36	71
		1.3	11.6	15.8	MJ/m²day			
		0.12	1.03	1.40	kBtu/ft²day			
Vertical (*D*+*S*)	(45°)	2.40	4.82	4.40	kWh/m²day	22	62	92
		8.8	17.3	15.9	MJ/m²day			
		0.78	1.53	1.40	kBtu/ft²day			
	(30°)	1.27	4.16	5.17	kWh/m²day	12	46	81
		4.6	14.9	19.1	MJ/m²day			
		0.40	1.32	1.68	kBtu/ft²day			
Polar axis (*D*)	(45°)	11.20	7.90	3.97	kWh/m²day	94	100	93
		40.0	28.4	14.3	MJ/m²day			
		3.51	2.50	1.26	kBtu/ft²day			
	(30°)	10.20	8.83	5.91	kWh/m²day	93	100	93
		36.7	31.8	21.6	MJ/m²day			
		3.24	2.80	1.88	kBtu/ft²day			
NS horiz.; E-W track. (*D*)	(45°)	11.21	6.04	1.99	kWh/m²day	95	76	46
		40.2	21.8	7.2	MJ/m²day			
		3.54	1.92	0.63	kBtu/ft²day			
	(30°)	10.80	8.03	4.45	kWh/m²day	97	90	71
		39.0	28.9	16.1	MJ/m²day			
		3.43	2.55	1.41	kBtu/ft²day			
EW horiz.; season (*D*)	(45°)	6.65	6.07	3.72	kWh/m²day	58	77	86
		23.9	21.9	13.4	MJ/m²day			
		2.11	1.93	1.18	kBtu/ft²day			
	(30°)	6.92	6.41	5.50	kWh/m²day	64	73	88
		24.9	23.1	19.8	MJ/m²day			
		2.20	2.03	1.71	kBtu/ft²day			
NW horiz.; N-S track (*D*)	(45°)	8.54	6.11	3.92	kWh/m²day	74	77	91
		30.7	21.4	14.1	MJ/m²day			
		2.69	1.88	1.24	kBtu/ft²day			
	(30°)	8.18	6.50	5.72	kWh/m²day	75	72	91
		29.4	23.4	20.6	MJ/m²day			
		2.59	1.97	1.82	kBtu/ft²day			

APPENDIX E. RADIATION AND CONVECTION FLUXES FOR HEAT BALANCE CALCULATIONS

Radiation and Convection Fluxes for Heat Balance Calculations

English units

Radiation losses

Temp., °F	*Flux, Btu/ft²h* Value	*Diff.*
0	76	
10	83	7
20	91	8
30	99	8
40	107	8
50	116	9
60	125	9
70	135	10
80	145	10
90	156	11
100	168	12
110	180	12
120	193	13
130	207	14
140	222	15
150	237	15
160	253	16
170	269	16
180	287	18
190	305	18
200	325	20
210	345	20
220	366	21
230	388	22
240	411	23
250	435	24
260	460	25
270	487	27
280	514	27
290	542	28
300	571	29
310	602	31
320	634	32

Convection losses

Temp. diff., F°	*Flux, Btu/ft²h* Value	*Diff.*
0	0	
10	4	4
20	8	4
30	14	6
40	20	6
50	26	6
60	33	7
70	40	7
80	47	7
90	55	8
100	63	8
110	71	8
120	79	8
130	87	8
140	96	9
150	105	9
160	114	9
170	123	9
180	132	9
190	141	9
200	150	9

Metric units

Radiation losses

Temp., °C	*Flux, W/m²* Value	*Diff.*
-20	233	
-15	252	19
-10	263	20
- 5	293	21
0	316	23
+ 5	340	24
10	365	25
15	391	26
20	419	28
25	448	29
30	479	31
35	512	33
40	546	34
45	581	35
50	619	38
55	658	39
60	699	41
65	742	43
70	787	45
75	834	47
80	883	49
85	933	50
90	987	54
95	1043	56
100	1101	58
105	1160	59
110	1223	63
115	1288	65
120	1356	68
125	1426	70
130	1499	73
135	1575	76
140	1654	79

Convection losses

Temp. diff., C°	*Flux, W/m²* Value	*Diff.*
0	0	
5	10	10
10	23	13
15	39	16
20	56	17
25	73	17
30	92	19
35	112	20
40	132	20
45	153	21
50	174	21
55	196	22
60	219	23
65	242	23
70	266	24
75	290	24
80	314	24
85	339	25
90	364	25
95	390	26
100	415	25

APPENDIX F. COLLECTOR HEAT BALANCE COMPUTATION SHEET

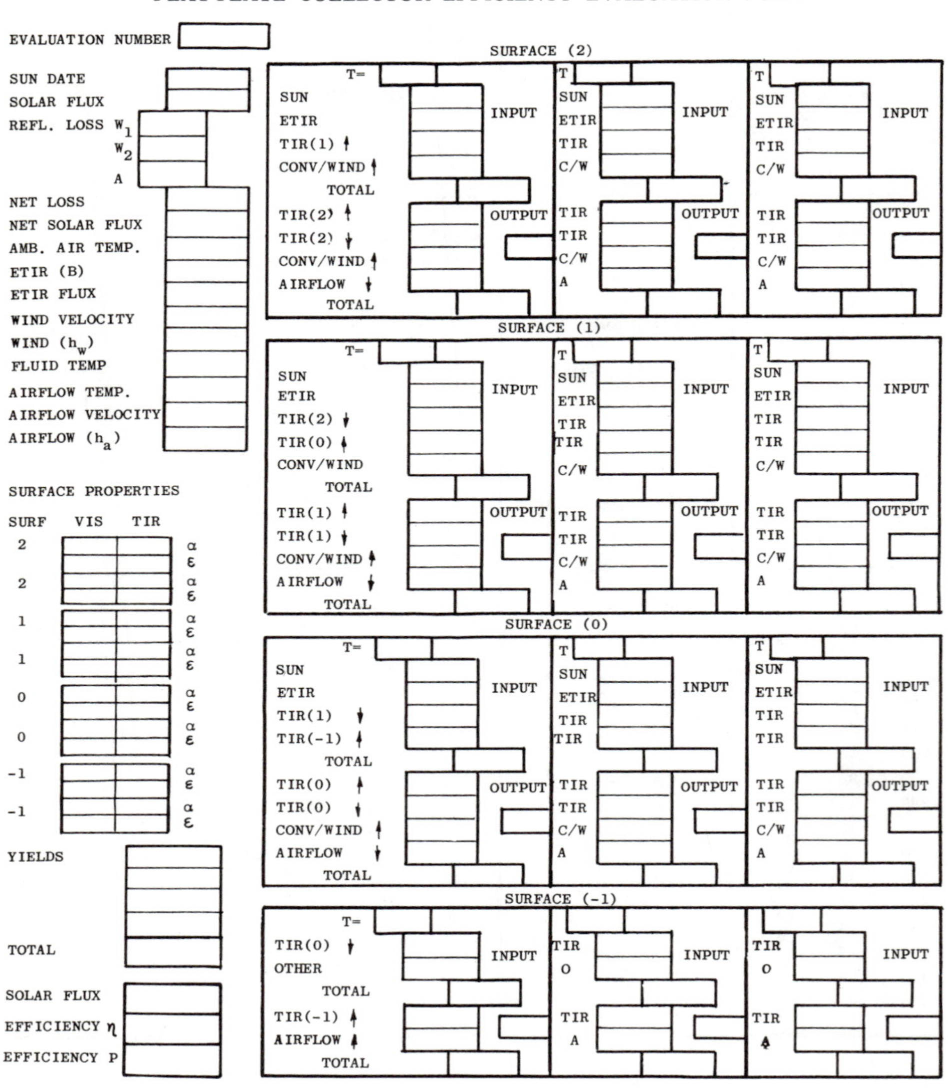

FLAT-PLATE COLLECTOR EFFICIENCY EVALUATION FORM

EVALUATION NUMBER

SUN DATE
SOLAR FLUX
REFL. LOSS W_1
W_2
A
NET LOSS
NET SOLAR FLUX
AMB. AIR TEMP.
ETIR (B)
ETIR FLUX
WIND VELOCITY
WIND (h_w)
FLUID TEMP
AIRFLOW TEMP.
AIRFLOW VELOCITY
AIRFLOW (h_a)

SURFACE PROPERTIES

SURF	VIS	TIR	
2			α
			ε
2			α
			ε
1			α
			ε
1			α
			ε
0			α
			ε
0			α
			ε
-1			α
			ε
-1			α
			ε

YIELDS

TOTAL

SOLAR FLUX
EFFICIENCY η
EFFICIENCY P

SURFACE (2)
T=
SUN
ETIR
TIR(1) ↑
CONV/WIND ↑
TOTAL
INPUT
TIR(2) ↑
TIR(2) ↓
CONV/WIND ↑
AIRFLOW ↓
TOTAL
OUTPUT
T
SUN
ETIR
TIR
C/W
INPUT
TIR
TIR
C/W
A
OUTPUT
T
SUN
ETIR
TIR
C/W
INPUT
TIR
TIR
C/W
A
OUTPUT

SURFACE (1)
T=
SUN
ETIR
TIR(2) ↓
TIR(0) ↑
CONV/WIND
TOTAL
INPUT
TIR(1) ↑
TIR(1) ↓
CONV/WIND ↑
AIRFLOW ↓
TOTAL
OUTPUT
T
SUN
ETIR
TIR
TIR
C/W
INPUT
TIR
TIR
C/W
A
OUTPUT
T
SUN
ETIR
TIR
TIR
C/W
INPUT
TIR
TIR
C/W
A
OUTPUT

SURFACE (0)
T=
SUN
ETIR
TIR(1) ↓
TIR(-1) ↑
TOTAL
INPUT
TIR(0) ↑
TIR(0) ↓
CONV/WIND ↑
AIRFLOW ↓
TOTAL
OUTPUT
T
SUN
ETIR
TIR
TIR
INPUT
TIR
TIR
C/W
A
OUTPUT
T
SUN
ETIR
TIR
TIR
INPUT
TIR
TIR
C/W
A
OUTPUT

SURFACE (-1)
T=
TIR(0) ↓
OTHER
TOTAL
INPUT
TIR(-1) ↑
AIRFLOW ↑
TOTAL
TIR
O
INPUT
TIR
A
TIR
O
INPUT
TIR
A

APPENDIX G. CONVERSION OF CENTIGRADE TO FAHRENHEIT

Temp °C	0	1	2	3	4	5	6	7	8	9
-50	-58.0	-59.8	-61.6	-63.4	-65.2	-67.0	-68.8	-70.6	-72.4	-74.2
-40	-40.0	-41.8	-43.6	-45.4	-47.2	-49.0	-50.8	-52.6	-54.4	-56.2
-30	-22.0	-23.8	-25.5	-27.4	-29.2	-31.0	-32.8	-34.6	-36.4	-38.2
-20	- 4.0	- 5.8	- 7.6	- 9.4	-11.2	-13.0	-14.8	-16.6	-18.4	-20.2
-10	+14.0	+12.2	+10.4	+ 8.6	+ 6.8	+ 5.0	+ 3.2	+ 1.4	- 0.4	- 2.2
- 0	32.0	30.2	28.4	26.6	24.8	23.0	21.2	19.4	+17.6	+15.8
+ 0	32.0	33.8	35.6	37.4	39.2	41.0	42.8	44.6	46.4	48.2
+10	50.0	51.8	53.6	55.4	57.2	59.0	60.8	62.6	64.4	66.2
20	68.0	69.8	71.6	73.4	75.2	77.0	78.8	80.6	82.4	84.2
30	86.0	87.8	89.6	91.4	93.2	95.0	96.8	98.6	100.4	102.2
40	104.0	105.8	107.6	109.4	111.2	113.0	114.8	116.6	118.4	120.2
50	122.0	123.8	125.6	127.4	129.2	131.0	132.8	134.6	136.4	138.2
60	140.0	141.8	143.6	145.4	147.2	149.0	150.8	152.6	154.4	156.2
70	158.0	159.8	161.6	163.4	165.2	167.0	168.8	170.6	172.4	174.2
80	176.0	177.8	179.6	181.4	183.2	185.0	186.8	188.6	190.4	192.2
90	194.0	195.8	197.6	199.4	201.2	203.0	204.8	206.6	208.4	210.2
100	212.0	213.8	215.6	217.4	219.2	221.0	222.8	224.6	226.4	228.2
110	230.0	231.8	233.6	235.4	237.2	239.0	240.8	242.6	244.4	246.2
120	248.0	249.8	251.6	253.4	255.2	257.0	258.8	260.6	262.4	264.2
130	266.0	267.8	269.6	271.4	273.2	275.0	276.8	278.6	280.4	282.2
140	284.0	285.8	287.6	289.4	291.2	293.0	294.8	296.6	298.4	300.2
150	302.0	303.8	305.6	307.4	309.2	311.0	312.8	314.6	316.4	318.2
160	320.0	321.8	323.6	325.4	327.2	329.0	330.8	332.6	334.4	336.2
170	338.0	339.8	341.6	343.4	345.2	347.0	348.8	350.6	352.4	354.2
180	356.0	357.8	359.6	361.4	363.2	365.0	366.8	368.6	370.4	372.2
190	374.0	375.8	377.6	379.4	381.2	383.0	384.8	386.6	388.4	390.2
200	392.0	393.8	395.6	397.4	399.2	401.0	402.8	404.6	406.4	408.2
210	410.0	411.8	413.6	415.4	417.2	419.0	420.8	422.6	424.4	426.2
220	428.0	429.8	431.6	433.4	435.2	437.0	438.8	440.6	442.4	444.2
230	446.0	447.8	449.6	451.4	453.2	455.0	456.8	458.6	460.4	462.2
240	464.0	465.8	467.6	469.4	471.2	473.0	474.8	476.6	478.4	480.2
250	482.0	483.8	485.6	487.4	489.2	491.0	492.8	494.6	496.4	498.2
260	500.0	501.8	503.6	505.4	507.2	509.0	510.8	512.6	514.4	516.2
270	518.0	519.8	521.6	523.4	525.2	527.0	528.8	530.6	532.4	534.2
280	536.0	537.8	539.6	541.4	543.2	545.0	546.8	548.6	550.4	552.2
290	554.0	555.8	557.6	559.4	561.2	563.0	564.8	566.6	568.4	570.2
300	572.0	573.8	575.6	577.4	579.2	581.0	582.8	584.6	586.4	588.2
310	590.0	591.8	593.6	595.4	597.2	599.0	600.8	602.6	604.4	606.2
320	608.0	609.8	611.6	613.4	615.2	617.0	618.8	620.6	622.4	624.2
330	626.0	627.8	629.6	631.4	633.2	635.0	636.8	638.6	640.4	642.2
340	644.0	645.8	647.6	649.4	651.2	653.0	654.8	656.6	658.4	660.2
350	662.0	663.8	665.6	667.4	669.2	671.0	672.8	674.6	676.4	678.2
360	680.0	681.8	683.6	685.4	687.2	689.0	690.8	692.6	694.4	696.2
370	698.0	699.8	701.6	703.4	705.2	707.0	708.8	710.6	712.4	714.2
380	716.0	717.8	719.6	721.4	723.2	725.0	726.8	728.6	730.4	732.2
390	734.0	735.8	737.6	739.4	741.2	743.0	744.8	746.6	748.4	750.2
400	752.0	753.8	755.6	757.4	759.2	761.0	762.8	764.6	766.4	768.2
410	770.0	771.8	773.6	775.4	777.2	779.0	780.8	782.6	784.4	786.2
420	788.0	789.8	791.6	793.4	795.2	797.0	798.8	800.6	802.4	804.2
430	806.0	807.8	809.6	811.4	813.2	815.0	816.8	818.6	820.4	822.2
440	824.0	825.8	827.6	829.4	831.2	833.0	834.8	836.6	838.4	840.2
450	842.0	843.8	845.6	847.4	849.2	851.0	852.8	854.6	856.4	858.2
460	860.0	861.8	863.6	865.4	867.2	869.0	87.08	872.6	874.4	876.2
470	878.0	879.8	881.6	883.4	885.2	887.0	888.8	890.6	892.4	894.2
480	896.0	897.8	899.6	901.4	903.2	905.0	906.8	908.6	910.4	912.2
490	914.0	915.8	917.6	919.4	921.2	923.0	924.8	926.6	928.4	930.2
500	932.0	933.8	935.6	937.4	939.2	941.0	942.8	944.6	946.4	948.2
510	950.0	951.8	953.6	955.4	957.2	959.0	960.8	962.6	964.4	966.2
520	968.0	969.8	971.6	973.4	975.2	977.0	978.8	980.6	982.4	984.2
530	986.0	987.8	989.6	991.4	993.2	995.0	996.8	998.6	1000.4	1002.2

APPENDIX H. BLACKBODY FLUX

Blackbody Flux

Temperature			*Blackbody flux*			
°*K*	°*C*	°*F*	*W/cm²*	*J/m²sec*	*cal/cm²sec*	*Btu/ft²h*
250	-23	-8	0.0222	222	0.00531	70.4
260	-13	+9	0.0260	260	0.00621	82.5
270	- 3	27	0.0302	302	0.00722	95.8
280	+ 7	45	0.0349	349	0.00834	110.7
290	17	63	0.0402	402	0.00961	127.5
300	27	81	0.0460	460	0.01099	145.9
310	37	99	0.0524	524	0.01252	166.2
320	47	117	0.0600	600	0.01434	190.3
330	57	135	0.0674	674	0.01611	213.8
340	67	153	0.0759	759	0.01814	240.8
350	77	171	0.0852	852	0.02036	270
360	87	189	0.0954	854	0.02280	302
370	97	207	0.1065	1065	0.02545	338
380	107	225	0.1184	1184	0.02830	376
390	117	243	0.1314	1314	0.03140	417
400	127	261	0.1454	1454	0.03475	461
410	137	279	0.1605	1605	0.03836	509
420	147	297	0.1768	1768	0.04225	561
430	157	315	0.1942	1942	0.04641	616
440	167	333	0.2128	2128	0.05086	675
450	177	351	0.2328	2328	0.05564	738
460	187	369	0.2542	2542	0.06075	806
470	197	387	0.2771	2771	0.06623	879
480	207	405	0.3015	3015	0.07206	956
490	217	423	0.3274	3274	0.07825	1038
500	227	441	0.3549	3549	0.08482	1126
520	247	477	0.4152	4152	0.09923	1317
540	267	513	0.4829	4829	0.11541	1532
560	287	549	0.5585	5585	0.13348	1771
580	307	585	0.6426	6426	0.15358	2038
600	327	621	0.7360	7360	0.17590	2334
620	347	657	0.8392	8392	0.20056	2662
640	367	693	0.9527	9527	0.22769	3022
660	387	729	1.078	10780	0.2576	3419
680	407	765	1.215	12150	0.2903	3854
700	427	801	1.364	13640	0.3260	4327
720	447	837	1.527	15270	0.3650	4844
740	467	873	1.703	17030	0.4070	5401
760	487	909	1.895	18950	0.4529	6011
780	507	945	2.102	21020	0.5024	6668
800	527	981	2.326	23260	0.5559	7378

APPENDIX I. RADIATIVE FLUX INTENSITIES

Radiative Thermal Flux from a Blackbody

Temperature			λ_{max},	Flux at λ_{max},	Hemispherical total flux		
°K	°C	°F	μm	W/cm²μm	W/cm²	kW/m²	Btu/ft²h
5800	5523	9974	0.50	9500.	6420.0	64200.	2.03x10^{6}
1200	923	1693	2.4	3.209	11.80	118.0	37400.
1000	723	1333	2.9	1.290	5.679	56.79	18010.
800	523	973	3.6	0.423	2.326	23.26	7378.
700	423	793	4.1	0.217	1.364	13.64	4327.
600	323	613	4.8	0.100	0.736	7.36	2335.
500	223	433	5.8	0.0403	0.355	3.55	1126.
400	123	253	7.2	0.0132	0.145	1.45	460.
300	23	73	9.6	0.0031	0.046	0.46	146.
273	0	32	10.5	0.0019	0.030	0.30	96.
Solar flux, sea level, desert					0.097	0.97	308.

APPENDIX J. SOLAR FURNACE RADIATIVE FLUX TEMPERATURES

Solar Furnace Radiative Flux Temperatures

Radiative flux, W/cm²	Temperature, °K	Radiative flux, W/cm²	Temperature, °K
7360	6000	673.5	3300
6350	(Sun)	595.5	3200
5197	5500	524.5	3100
4365	(Sun)*	460.0	3000
3549	5000	401.7	2900
2328	4500	349.1	2800
1454	4000	301.8	2700
1314	3900	259.5	2600
1184	3800	221.8	2500
1065	3700	188.5	2400
953.8	3600	159.0	2300
852.2	3500	133.1	2200
758.9	3400	110.5	2100
		90.86	2000

*Through one air mass, standard atmosphere.

APPENDIX K. DISTORTED-WAVELENGTH GRAPH

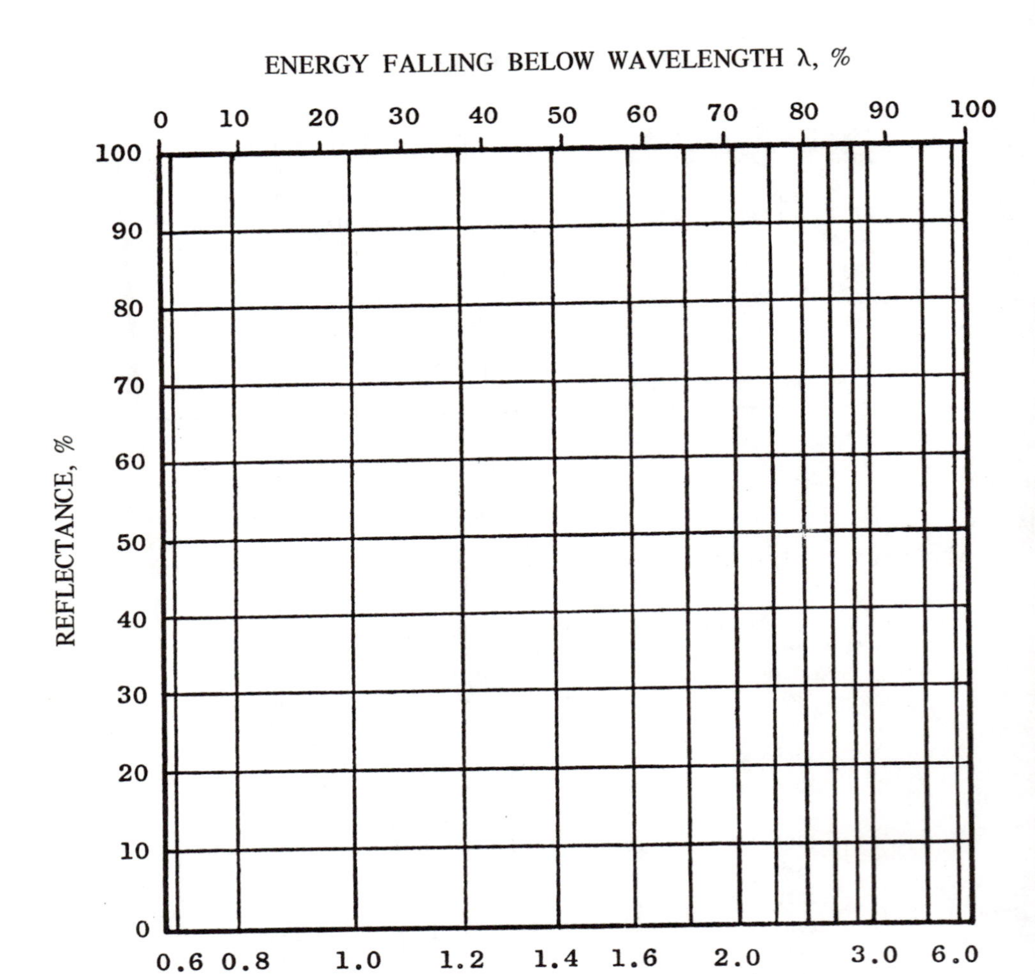

APPENDIX L. THERMAL CONDUCTIVITY OF MATERIALS

Thermal Conductivity of Materials

Material	*Temperature, °C*	*Thermal conductivity*		
		cal cm/sec C°cm²	*W cm/C°m²*	*Btu in./h F°ft²*
Metals				
Aluminum	100	0.49	20500	1130
	200	0.55	23000	1590
	400	0.76	31800	2210
	600	1.01	42300	2932
Brass	0	0.25	10400	720
Copper	20	0.934	39300	2711
	100	0.908	38100	2640
	300	0.89	37200	2580
Magnesium	20	0.37	15400	1070
Steel	18	0.115	4800	330
	100	0.107	4500	310
Nonmetals				
Asbestos fiber	500	0.00019	8.0	0.55
Brick	20	0.0015	62	4.3
Concrete	20	0.00071	30	2.0
Cotton wool	20	0.000043	1.8	0.12
Diatomaceous earth	20	0.00013	5.4	0.38
Granite	20	0.0045	188	13
Glass	20	0.0017 –0.0025	71 –100	5.0 –7.2
Glass wool	20	0.000081	3.4	0.24
Infusorial earth	100	0.00034	14.0	1.0
	300	0.00040	17.0	1.2
Magnesia brick	500	0.0050	210	14.0
Plaster of Paris	20	0.00070	29.0	2.0
Sand, dry	20	0.00093	39.0	2.6
Soil, dry	20	0.00033	14.0	0.96
wet	20	0.0010	42.0	2.9
Wood, across grain	20	0.00010 –0.00030	4.2 –12.6	0.29 –0.87
Gases				
Air	20	0.000057	2.4	0.13
Carbon dioxide	20	0.000031	1.3	0.07
Helium	20	0.000339	14.2	0.79
Hydrogen	100	0.000369	15.4	0.86

APPENDIX M. SPECIFIC HEAT OF MATERIALS

Specific Heat of Materials

Element		*Specific heat, cal/g*	*Compound*		*Specific heat, cal/g*
Hydrogen	H_2	3.41	Water	H_2O	1.00
Helium	He	1.24	Lithium hydride	LiH	0.6
Lithium	Li	0.85	Lithium fluoride	LiF	0.37
Berillium	Be	0.436	Lithium chloride	LiCl	0.28
Sodium	Na	0.293	Sodium fluoride	NaF	0.28
Nitrogen	N_2	0.249	Sodium chloride	NaCl	0.22
Neon	Ne	0.246	Calcium fluoride	CaF_2	0.21
Boron	B	0.245	Potassium fluoride	KF	0.20
Magnesium	Mg	0.243	Sodium hydroxide	NaOH	0.2
Oxygen	O_2	0.219	Potassium hydroxide	KOH	0.2
Aluminum	Al	0.215	Magnesium chloride	$MgCl_2$	0.19
Fluorine	F_2	0.197	Calcium chloride	$CaCl_2$	0.16
Phosphorus	P	0.181	Potassium clhoride	KCl	0.16
Potassium	K	0.180	Iron chloride	$FeCl_3$	0.15
Sulfur	S	0.175	Zinc chloride	$ZnCl_2$	0.14
Silicon	Si	0.168	Copper chloride	$CuCl_2$	0.14

APPENDIX N. HEAT STORAGE CAPACITY

Heat Storage Capacity

	Density ρ	*Specific heat* c	*Volume heat capacity* ρc	*Thermal conductivity* k	*Thermal diffusivity* $k/\rho c$
Material	*g/cm*^{3}	*cal/g C°*	*cal/cm*^{3}*C°*	*cal cm/cm*^{2}*C°sec*	*cm*^{2}*/sec*
Water	1.00	1.00	1.00	0.0014	0.0014
Iron shot	7.86	0.13	1.02	0.160	0.157
Oils	1.00	0.60	0.60	0.00036	0.0006
Rock	2.50	0.20	0.50	0.0060	0.0120
$MgCO_3$	3.0	0.20	0.60	0.0025	0.0042
$MgCO_3 \cdot 6H_2O$	1.7	0.38	0.65	0.0030	0.0046

APPENDIX O. HEAT OF FUSION OF MATERIALS

Heat of Fusion of Materials

Material		*Heat of fusion, cal/g*	*Melting temperature, °C*
Beryllium chloride	$BeCl_2$	310	547
Sodium fluoride	NaF	168	992
Lithium hydride	LiH	139	680
Nickel chloride	$NiCl_2$	139	1030
Sodium metaborate	$NaBO_2$	135	966
Beryllium fluoride	BeF_2	128	800
Sodium chloride	$NaCl$	123	803
Potassium fluoride	KF	117	860
Calcium silicate	$CaSiO_3$	115	1512
Magnesium chloride	$MgCl_2$	109	708
Lithium hydroxide	$LiOH$	103	462
Magnesium fluoride	MgF_2	95	1396
Lithium fluoride	LiF	93	896
Sodium silicate	$NaSiO_3$	84	1087
Iron chloride	$FeCl_2$	82	677
Potassium chloride	KCl	82	776
Lithium silicate	$LiSiO_3$	80	1201
Boric oxide	B_2O_3	76	449
Lithium chloride	$LiCl$	76	613
Manganese chloride	$MnCl_2$	71	708
Chromium chloride	$CrCl_2$	66	814
Aluminum chloride	Al_2Cl_6	63	190
Sodium bromide	$NaBr$	61	742
Potassium bromide	KBr	59	730
Sodium hydroxide	$NaOH$	38	318
Lithium bromide	$LiBr$	34	547
Potassium nitrate	KNO_3	28	337
Copper chloride	$CuCl$	25	430

APPENDIX P. COEFFICIENTS OF LINEAR THERMAL EXPANSION

Coefficients of Linear Thermal Expansion

Material	*Temperature range, °C*	*Linear thermal expansion coefficient k*	
		1/C°, (10⁻⁶)	*1/F°, (10⁻⁶)*
Aluminum	20-100	23.8	13.2
	20-500	25.4	14.1
Brass	0-100	18.8	10.4
Copper	25-100	16.8	9.3
	25-300	17.8	9.9
Iron, cast	40	12.1	5.7
Steel	0-100	10.5	5.8
	200-300	13.0	6.1
	500-600	16.0	8.9
Lead	0-100	25.1	13.9
Molybdenum	0-100	4.9	2.7
Stainless steel	20-100	9.6	5.3
	20-600	11.2	6.2
Silver	20	17.0	9.4
Silicon	40	7.6	4.2
Zinc	10-100	26.3	14.6
Bakelite	20- 60	22.	12.
Tedlar	20	50.4	28.0
Mylar	20	17.0	9.4
Kapton	20	20.0	11.1
Acrylic	20	80.0	44.0
Methacrylate	20	80.0	44.0
Acetate	20	50-160	28-89
Vinyl	20	86.	48.
PVC	20	69.	38.
Polystyrene	20	68.	37.
Brick	-	9.5	5.3
Concrete	-	10-14	5.5-7.8
Glass, plate	0-100	8.9	4.9
Rock	-	7-12	4-7
Wood, along grain		4-6	2-3
across grain		30-50	17-28

APPENDIX Q. LOGARITHMIC RATIOS

Logarithmic Ratios

r_2/r_1	$\ln(r_2/r_1)$	$1/\ln(r_2/r_1)$	$2d/r$	$\ln(2d/r)$	$1/\ln(2d/r)$
10.00	2.303	0.434	200	5.298	0.189
8.0	2.079	0.481	100	4.605	0.222
6.0	1.792	0.558	80	4.382	0.228
4.0	1.386	0.722	60	4.094	0.244
2.0	0.693	1.443	40	3.689	0.271
1.8	0.588	1.792	20	2.996	0.334
1.6	0.470	2.13	18	2.890	0.346
1.4	0.336	2.98	16	2.773	0.361
1.2	0.182	5.49	14	2.639	0.379
1.10	0.095	10.5	12	2.485	0.402
1.05	0.049	20.4	11	2.398	0.417
1.00	0.000	-	10	2.303	0.434

APPENDIX R. FRACTIONAL POWERS

Fractional Powers

	Power	
Number	*0.250*	*1.250*
1	1.000	1.00
2	1.189	2.43
3	1.316	3.95
4	1.414	5.66
5	1.495	7.48
10	1.788	17.78
20	2.115	42.30
30	2.340	70.22
40	2.515	100.62
50	2.659	132.97
60	2.783	166.99
70	2.892	202.49
80	2.990	239.26
90	3.080	277.21
100	3.162	316.21

APPENDIX S. CONVECTIVE PROPERTIES OF GASES

Convective Properties of Gases

Gas	Molecular weight M	Density ρ	Specific heat c_p	Thermal expansivity y	Viscosity (poises) μ	Thermal conductivity k	Modulus a	Modulus ratio a/a_{air}	Convectivity $ka^{1/4}$
	g/mole	$\frac{g}{cm^3}$, (10^{-3})	cal/g C°	cm³/cm³C°, (10^{-3})	g/cm sec, (10^{-6})	cal/cm C°sec, (10^{-6})	1/cm³C°		cal/cm²C°, (10^{-6})
H_2	2	0.066	3.39	3.66	98	490	1.1	0.23	499
He	4	0.130	1.25	3.66	228	360	0.9	0.19	352
CO	28	0.915	0.25	3.67	210	74	48	1.02	194
Air	29	0.946	0.24	3.67	220	75	47	1.00	195
Ar	40	1.306	0.125	3.68	269	52	55	1.17	141
CO_2	44	1.447	0.199	3.72	186	54	151	3.21	189
SO_2	64	2.142	0.152	3.90	161	35	477	10.1	163
Cl_2	71	2.352	0.11	3.90	168	27	513	10.9	128
CS_2	76	2.479	0.157	3.90	126	26	1130	24.0	150
Kr	84	2.740	0.09	3.68	260	28	334	7.1	120
Xe	131	4.316	0.07	3.68	280	26	644	13.7	131
CCl_3F	154	5.02	0.12	3.9	110	34	3090	65.7	254
Br_2	160	5.22	0.055	3.9	188	22.5	1320	28.1	136
CCl_4	166	5.41	0.115	3.9	120	24	4470	95.1	196
$SiCl_4$	170	5.54	0.132	3.9	100	19	7640	123.0	178

All data at 100°C.

APPENDIX T. GASES OF LOW CONDUCTIVITY

Gases of Interest for Convection Suppression

Gas	Molecular weight M g	Boiling point T_b °C	Pressure $p(-10°C)$ g/cm²	Heat of vaporization H cal/g
CO	28	−192		
CO_2	44	−78		
CS_2	76	+46	66	19
Ar	40	−186		
Cl_2	71	−34		
Br_2	160	+59		44
$SiCl_4$	170	+96	13	11
Si_2H_6	62	−14		
SiF_4	104	−86		
$SiClF_3$	120	−70		
SF_6	193	−34		
SO_2F_2	102	−55		
SeF_6	193	−34		
COS	60	−48		
CCl_4	159	+77	13	11
CCl_3F	137	+24	144	10
ClF_3	92	+11	295	14
POF_3	104	−40		
NOF	49	−56		
CCl_2F_2 (F12)	121	−30		
$CClF_3$ (F13)	104	−81		
$CHClF_2$ (F22)	86	−41		
C_2ClF_5	154	−38		
C_2NF_7	171	−35		
$GeCl_2F_2$	181	−3		
$GeClF_3$	165	−21		
C_2F_6	138	−79		

APPENDIX U. MOLLIER DIAGRAM FOR STEAM

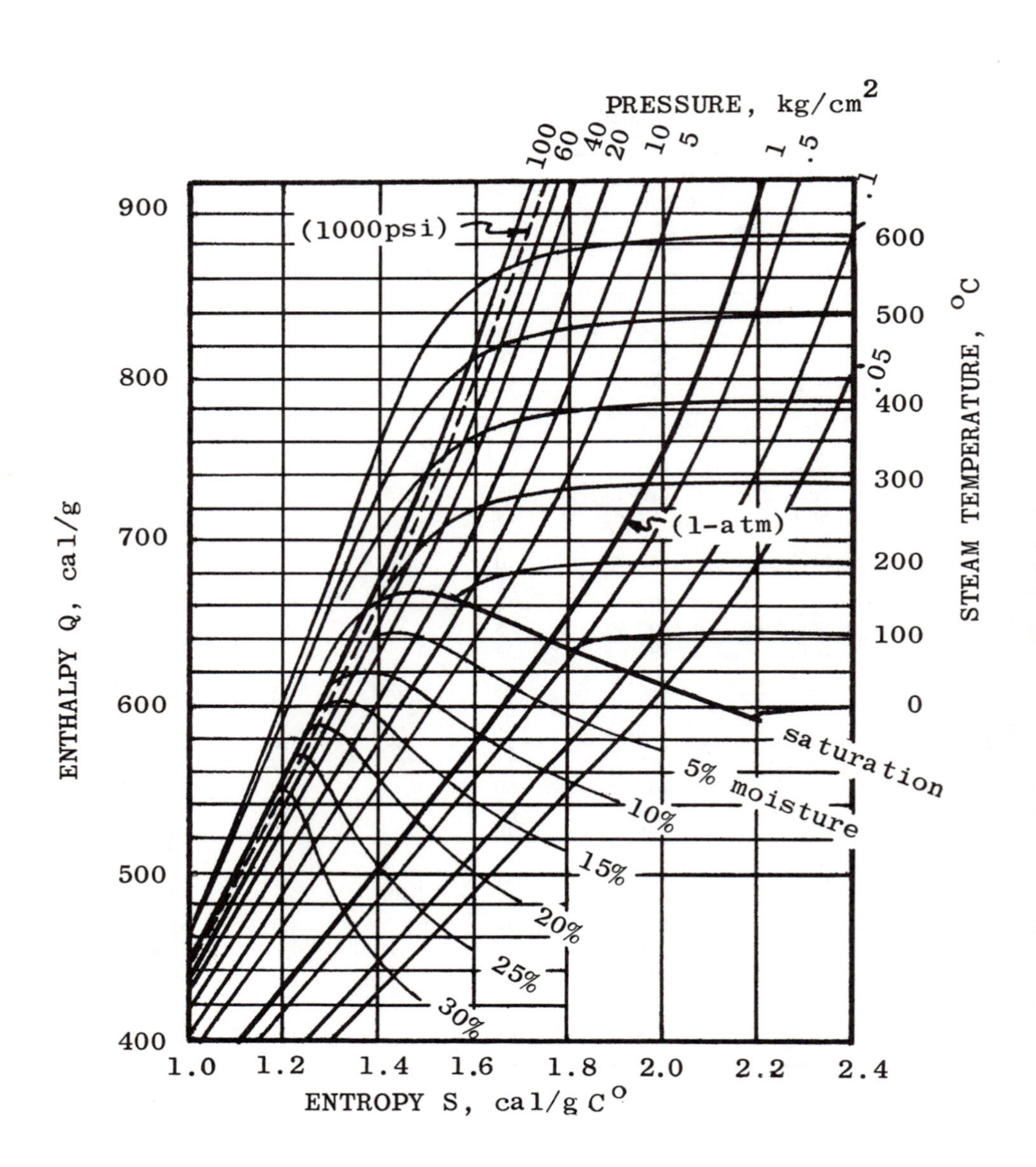

APPENDIX V. SYMBOLS AND DEFINITIONS

a	modulus of air
A	collector area
Å	Angstrom unit = 10^{-4} μm
$b(T)$	flux contained in a finite solid angle
$B(T)$	source brightness in unit solid angle
c_p	specific heat at constant pressure
C	brightness flux concentration in the solar image
D	entrance aperture diameter
D	fluid pipe diameter
f	f-ratio of an optical system
F	focal length of an optical system
f	friction factor for a pipe
g	acceleration of gravity
h	heat-transfer coefficient
H	effective hydrodynamic height
i,r	angles of incidence, refraction
I	total solar flux per unit area at earth's surface, zenith
I_d	direct solar flux intensity (D)
I_s	scattered (diffuse) flux intensity (S)
I_0	extraterrestrial solar flux
k	coefficient of heat conductivity
K	coefficient of thermal expansion
L	characteristic length
L	collector fluid path length
L/D	ratio of pipe length to diameter
m	air mass
M	mass
n	index of refraction
Δp	pressure drop
P	efficiency defined as useful energy out/energy absorbed
P	power
Q/A	quantity of heat flowing through area A
r,R	radii of curvature
r	reflectance
t	transmittance
T	temperature
ΔT	temperature difference
U	heat loss factor; $Q/A = UT$
v	velocity
V	volume rate of flow
w	width of cylindrical collector aperture

W	energy accumulation factor
X	radiation balance flux concentration (aperture area/absorber area)
y	volumetric expansion of air
z	zenith distance angle
Gr	Grasshof number
Nu	Nusselt number
Pr	Prandtl number
Ra	Rayleigh number
Re	Reynolds number
α, β	solid angles in descriptions of radiant flux in a solid angle
α	absorptance over a finite wavelength interval
β	angle of incidence of sun on collector
γ	ratio of specific heats
δ	solar declination
δ	incremental angle error
ϵ	emittance over a finite wavelength interval
η	collector efficiency
η	pump, turbine, etc., efficiency (with appropriate subscript)
θ, ϕ	angles of incidence with respect to an optical axis
θ	collector tilt measured normal to plate from zenith
λ	wavelength
λ	latitude
μ	dynamic viscosity (g cm/sec)
ν	kinematic viscosity
ρ	density
σ	incremental angle error
σ	Stefan-Boltzmann constant
σ	root-mean-square angular or surface error
$d\Omega$	incremental solid angle

References

GENERAL REFERENCES

The student and researcher is referred to the ERDA publication *Solar Energy–A Bibliography,* which is a comprehensive survey of solar energy literature of the past 30 years. It comes in two volumes. The first, TID-3351-R1P1, is citations, and the second, TID-3351-R1P2, is an index by name and subject. The two volumes, at $13.95 each, are obtainable from the Technical Information Center, U.S. Department of Commerce, Springfield, Virginia 22161.

A second general bibliography is the *Solar Energy Bibliography,* published by NTIS (National Technical Information Service), P.O. Box 2409, Grand Central Station, New York, N.Y. 10017. References are arranged in broad subject categories. Within each category the references are arranged chronologically. Three indexes are also provided: by author, by subject, and by NTIS reporting number.

As a third general index, the student and researcher is referred to the *Solar Energy Index,* published in IBM printout form by the Arizona State University Science/Engineering Reference Service, University Library, Tempe, Arizona 85281. This index is approximately 300 pages long and is updated periodically. References are listed by topic and by author, the author index being arranged chronologically. References cited in this book by author and year can be quickly located.

Abstracts of NSF/RANN Research Reports are contained in NSF 75-6, RANN Document Center, Washington, D.C. 20550, covering the period from October 1970 through December 1974. Additional abstract compilations should be available from time to time through NTIS. Most of the documents cited in these summaries are available from NTIS, Document Sales, U.S. Department of Commerce, Springfield, Virginia 22161. Another source of abstracts and documents on solar energy is the Energy Information Center, Holifield National Laboratory (formerly Oak Ridge National Laboratory), Oak Ridge, Tennessee 37830.

Solar Energy Journals

Solar Energy, published by the International Solar Energy Society (ISES), c/o Smithsonian Radiation Biology Laboratory, 12441 Parklawn Drive, Rockville, Maryland 20852.

Applied Solar Energy (Geliotekhnika), the Soviet solar energy journal published in English translation by Allerton Press, 150 Fifth Avenue, New York, N.Y. 10011.

Bulletin, Cooperation Méditerranéenne pour l'Energie Solaire (COMPLES), 32 Cours Pierre-Puget, 13006 Marseilles, France.

Textbooks

Brinkworth, B. J. (1972), *Solar Energy for Man,* Compton Press, Salisbury, England.

Daniels, F. (1964), *Direct Use of the Sun's Energy,* Yale University Press, New Haven, Conn.; Paperback Edition (1973), Ballantine Books.

Daniels, F., and Duffie, J. A. (1955), *Solar Energy Research,* University of Wisconsin Press, Madison, Wis.

Duffie, J. A., and Beckman, W. A. (1974), *Solar Energy Thermal Processes,* Wiley, New York.

Jordan, R. C., Ed. (1967), *Low Temperature Engineering Application of Solar Energy,* American Society of Heating, Refrigerating, and Airconditioning Engineers (ASHRAE), 345 E. 47th St., New York, N.Y. 10017.

Zarem, A. M., and Erway, D. D. (1963), *Introduction to the Utilization of Solar Energy,* Extension Series, McGraw-Hill, New York.

CHAPTER REFERENCES

The references given below cover those cited in the text plus some additional references deemed important in rounding out the list.

Chapter 1. History of Applications

References to some of the literature between 1637 and 1782 are given in Ackerman, A. S. E. (1915), *Annual Report, Smithsonian Institution,* p. 141, and Benveniste, G. (1956), *Sun at Work* **1**, No. 2, p. 4.

Veinberg, V. B. (1959), *Optics in Equipment for the Utilization of Solar Energy,* State Publ. House, Moscow, Transl. AEC-tr-4471, USAEC, OTS, Dept. Commerce, Washington, D.C.

Chapter 2. Solar Flux and Weather Data

Allen, C. W. (1955), *Astrophysical Quantities,* Athlone Press, London.

Bos, P. B. (1974, 1975), *Solar Thermal Conversion Mission Analysis,* Vols. I–V, Summary Report, NSF/RANN-74-017.

Bos, P. B., Kammer, W. A., and Blond, E. (1974), Aerospace Corp. Rept. No. ATR-74(7417-16)-2, Vol. 1.

Duffie, J. A., and Beckman, W. A. (1974), *Solar Energy Thermal Processes,* Wiley, New York, p. 30.

Haurwitz, B. (1948), *J. Meteorol.* **5,** No. 3, p. 110.

Kondratyev, K. Ya. (1954), *Radiant Solar Energy,* Publ. Leningrad, pp. 264, 330, 592. (Cited by Robinson, 1966.)

Laue, E. G. (1970), *Sol. Energy* **13,** No. 1, p. 43.

Liu, B. Y. H., and Jordan, R. C. (1960), *Sol. Energy* **4,** No. 3, p. 1.

McKenney, D. B. (1974), cited by McKenney *et al.* (1974).

McKenney, D. B., Beauchamp, W. T., and Meinel, A. B. (1974), *Solar Microclimatology,* NASA Final Report, NASW-274.

Meinel, A. B., and McKenney, D. B. (1975), cited by Meinel *et al.* (1975).

Meinel, A. B., McKenney, D. B., and Beauchamp, W. T. (1975), NSF/RANN/SE/GI-41895/FR/75/7, NTIS, U.S. Dept. Commerce, Washington, D.C.

Moon, P. (1940), *J. Franklin Inst.* **230,** No. 5, p. 583, cited by Valley (1965).

Oetjen, R. A., Bell, E. E., Young, J., and Eisner, L. (1960), *J. Opt. Soc. Am.* **50,** No. 12, p. 1308.

Robinson, N. (1966), *Solar Radiation,* Elsevier, London.

Valley, S. L., Ed. (1965), *Handbook of Geophysics and Space Environments,* Air Force Cambridge Research Laboratories (AFCRL), Bedford, Mass.

Chapter 3. Solar Availability

Veinberg, V. B. (1959), *Optics in Equipment for the Utilization of Solar Energy,* State Publ. House, Moscow, Transl. AEC-tr-4471, USAEC, OTS, Dept. Commerce, Washington, D.C.

Winston, R. (1974), *Sol. Energy* **16,** No. 2, p. 89.

Chapter 4. Luminance of Collector Optics

Chapter 5. Refractive Collector Optics

Lomonsov, M. V. (1741), cited by Veinberg, V.B. (1959).

Meinel, W. B. (1972), unpublished research.

Moreau, M. Ya. (1924), cited by Veinberg, V. B. (1959).

Newton, I. (1722), cited by Veinberg, V. B. (1959).

Veinberg, V. B. (1959), *Optics in Equipment for the Utilization of Solar Energy,* State Publ. House, Moscow, Transl. AEC-tr-4471, USAEC, OTS, Dept. Commerce, Washington, D.C.

Chapter 6. Mirror Collector Optics

Aparisi, R. R., Kolos, Ya. G., and Shatov, N. I. (1968), *Semiconductor Solar Energy Converters,* Ed. Baum, V. A., p. 101, Transl., Consultant's Bureau, New York, 1969.

Mouchot, A. (1875), cited by Benveniste, G. (1956), *Sun at Work* **1,** No. 2, p. 4.

Shuman, F. (1911), cited by Blake, A. D. (1911), *Power* **34,** No. 13, p. 506.

Tabor, H. (1966), *Sol. Energy* **10,** No. 3, p. 111.

Taylor, C. S. (1975), *Proceedings, Solar Energy Conference, Dhahran, Saudi Arabia, 1975,* Ed. Kettani, M. A., University of Petroleum and Minerals, Dhahran (1976).

Veinberg, V. B. (1959), *Optics in Equipment for the Utilization of Solar Energy,* State Publ. House, Moscow, Transl. AEC-tr-4471, USAEC, OTS, Dept. Commerce, Washington, D.C., p. 100.

Chapter 7. Fixed-Mirror Collectors

Baranov, V. K. (1966), *Geliotekhnika* **2,** No. 3, p. 11.

McKenney, D. B. (1975), unpublished data, EPRI Contract RP548-1, Helio Associates, Inc., Tucson, Ariz.

McKenney, D. B., and Meinel, A. B. (1975), *Air Stable Selective Surfaces,* NSF/RANN/SE/GI-41895/FR/74/7, Helio Associates, Inc., Tucson, Ariz.

Meinel, A. B. (1972), cited by McKenney and Meinel (1975).

Meinel, A. B. (1974), cited by McKenney and Meinel (1975).

Russell, J. L. (1974), Tech. Rept., Gulf General Atomics Corp.

Steward, W. G. (1974), cited by Steward and Kreith (1975).

Steward, W. G., and Kreith, F. (1975), *Appl. Opt.* **14,** No. 7, p. 1509.

Tabor, H. (1958), *Sol. Energy* **2,** Nos. 3–4, p. 27.

Tabor, H., and Zeimer, H. (1962), *Sol. Energy* **6,** No. 2, p. 55.

Trombe, F. (1957), French patent (No. prov. P.Y. 681855), expired.

Veinberg, V. B. (1959), *Optics in Equipment for the Utilization of Solar Energy,* State Publ. House, Moscow, Transl. AEC-tr-4471, USAEC, OTS, Dept. Commerce, Washington, D.C., p. 118.

Winston, R. (1965), cited by Winston (1974).

Winston, R. (1974), *Sol. Energy* **16,** No. 2, p. 89.

Chapter 8. Optical Surfaces

Cox, J. T., and Hass, G. (1964), *Physics of Thin Films,* Vol. 2, Ed. Hass, G., Academic Press, New York, p. 239.

Catalanotti, S., Cuomo, V., Piro, G., Ruggi, D., Silvestrini, V., and Troise, G. (1975), *Sol. Energy* **17,** No. 2, p. 83.

McKenney, D. B. (1973), unpublished analysis for Smithsonian Astrophysical Observatory, Mt. Hopkins, Ariz.

Meinel, W. B. (1973), unpublished data, Helio Associates, Inc., Tucson, Ariz.

Silvestrini, V. (1975), cited by Catalanotti *et al.* (1975).

Tabor, H. (1967), Ch. IV in Jordan, R. C., Ed., *Low Temperature Engineering Application of Solar Energy,* ASHRAE, New York.

Chapter 9. Selective Surfaces

Baum, V.A., Ed. (1969), *Semiconductor Solar Energy Converters,* Transl., Consultant's Bureau, New York.

Bennett, H. E. (1974), *Proceedings of the Symposium on the Material Science Aspects of Solar Energy Conversion, Tucson, Arizona, 1974,* published as NSF/RANN Rept. GI-43795, Ed. Seraphin, B. O. (1975), p. 145.

Bennett, H. E., and Bennett, J. M. (1970), *Physics of Thin Films,* Vol. 4, Ed. Hass, G., Academic Press, New York, p. 70.

Blandenet, G., Lagarde, Y., and Spitz, J. (1975), *Proceedings of the Chemical Vapor Deposition Conference,* Ed. Blocher *et al.,* Electrochemical Society, Inc., Princeton, N.J.

Burrafato, G., Giaquinte, G., Mancini, N. A., and Pennisi, A. (1975), *Proceedings, Course on Solar Energy Conversion, Procida,* Ed. Mancini, N. A., and Quercia, I. F., University of Catania, p. 391.

Drummeter, L. F., and Hass, G. (1964), *Physics of Thin Films,* Vol. 2, Ed. Hass, G., Academic Press, New York, p. 305.

Edwards, D. K., Gier, J. T., Nelson, K. E., and Roddick, R. D. (1961), *Proceedings of the UN Conference on New Sources of Energy,* United Nations, New York, Vol. 4 (1964).

Goldner, R. B., and Haskal, H. M. (1975), *Appl. Opt.* **14**, No. 10, p. 2328.

Hass, G. (1964), cited by Drummeter and Hass (1964).

Hass, G., Schroeder, H. H., and Turner, A. F. (1956), *J. Opt. Soc. Am.* **46**, No. 1, p. 31.

Hottel, H. C., and Unger, T. A. (1959), *Sol. Energy* **3**, No. 3, p. 10.

Kauer, E. (1966), U.S. Patent 3,288,625.

Kokoropoulos, P., Salam, E., and Daniels, F. (1959), *Sol. Energy* **3**, No. 3, p. 10.

Langley, R. C. (1974), *Proceedings of the Symposium on the Material Science Aspects of Solar Energy Conversion, Tucson, Arizona, 1974,* published as NSF/RANN Rept. GI-43795, Ed. Seraphin, B. O. (1975), p. 321.

Mancini, N. A. (1975), cited by Burrafato *et al.* (1975).

Masterson, K. D., and Seraphin, B. O. (1975), *Inter-Laboratory Comparison of the Optical Characteristics of Selective Surfaces for Photo-Thermal Conversion of Solar Energy,* NSF/RANN Rept. GI-36731X.

McKenney, D. B., *et al.* (1974), see Meinel *et al.* (1974).

McKenney, D. B., and Beauchamp, W. T. (1974, 1975), see Meinel *et al.* (1974).

McMahon, T. J., and Jasperson, S. N. (1974), *Appl. Opt.* **13**, No. 12, p. 2750.

Meinel, A. B., McKenney, D. B., and Beauchamp, W. T. (1974), NSF/RANN/SE/GI-41895/PR/74/4, NTIS, U.S. Dept. Commerce, Washington, D.C.

Moon, P. (1940), *J. Franklin Inst.* **230**, No. 5, p. 583, cited by Valley (1965).

Rekant, N. B., and Sheklein, A. V. (1969), cited by Baum (1969), p. 200.

Schmidt, R. N., and Park, K. C. (1965), *Appl. Opt.* **4**, No. 8, p. 917.

Seraphin, B. O. (1974), *Chemical Vapor Deposition Research for Fabrication of Solar Energy Converter,* Tech. Rept. for NSF/RANN Grant SE/GI-36731.

Seraphin, B. O., and Meinel, A. B. (1975), *Optical Properties of Solids–New Developments,* Ed. Seraphin, B. O., North-Holland, Amsterdam, Ch. 17, p. 929.

Sievers, A. J. (1973), Materials Research Council Summer Conference, La Jolla, Calif., sponsored by ARPA, Washington, D.C., Vol. II, p. 111.

Tabor, H. (1955), *Bull. Res. Counc. Isr.* **5C**, No. 1, p. 5.

Tabor, H. (1956), *Bull. Res. Counc. Isr.* **5A**, No. 2, p. 119.

Tabor, H., cited by Wolfe, W. L., Ed. (1965), *Handbook of Military Infrared Technology,* U.S. Government Printing Office, Washington, D.C., p. 353.

Valley, S. L., Ed. (1965), *Handbook of Geophysics and Space Environments,* Air Force Cambridge Research Laboratory (AFCRL), Bedford, Mass.

Veinberg, V. B. (1959), *Optics in Equipment for the Utilization of Solar Energy,* State Publ. House, Moscow, Transl. AEC-tr-4771, USAEC, OTS, Dept. Commerce, Washington, D.C., p. 142.

Watson-Munro, C. N., and Horwitz, C. M. (1975), *Solar Energy,* Ed. Messel, H., and Butler, S. T., Pergamon International Library, New York, Ch. 8, p. 293.

Wehner, G. K. (1975), cited by Seraphin and Meinel (1975), p. 966.

Welty, J. R. (1974), *Engineering Heat Transfer,* Wiley, New York, p. 316.

Williams, D. A., Lappin, T. A., and Duffie, J. A. (1963), *Trans. ASME J. Eng. Power* **85**, No. 3, p. 213.

Chapter 10. Basic Elements of Heat Transfer

Duffie, J. A., and Beckman, W. A. (1974), *Solar Energy Thermal Processes,* Wiley, New York, p. 80.

Gaudenzi, P. (1975), Lectures, International College of Applied Physics, Nerano, Italy, unpublished.

Hottel, H. C., and Woertz, B. B. (1942), *Trans. ASME,* **64**, No. 2, p. 91.

Kreith, F. (1973), *Principles of Heat Transfer,* 3rd Ed., International Textbook Company, Scranton, N.Y.

McAdams, W. H. (1954), *Heat Transmission,* 3rd Ed., McGraw-Hill, New York, p. 259.

Pope, C. H. (1903), cited by Ackermann, A. S. E. (1915), *Annual Report, Smithsonian Institution.*

Tabor, H. (1955), *Bull. Res. Counc. Isr.* **5C**, No. 1, p. 5.

Tabor, H. (1956), *Bull. Res. Counc. Isr.* **5A**, No. 2, p. 119.

Tabor, H. (1958), *Bull. Res. Counc. Isr.* **6C**, No. 3, p. 155.

Trombe, F. (1974), private communications to the authors.

Welty, J. R. (1974), *Engineering Heat Transfer,* Wiley, New York, p. 316.

Chapter 11. Heat Transfer in Solar Collectors

Buchberg, H., Lalude, O. A., and Edwards, D. K. (1971), *Sol. Energy* **13**, p. 193.

Charters, W. W. S., and Peterson, L. F. (1972), *Sol. Energy* **13**, p. 353.

Francia, G. (1961), cited by Francia, G. (1968), *Sol. Energy* **12**, p. 51.

Hollands, K. G. T. (1965), *Sol. Energy* **9**, No. 3, p. 159.

Hollands, K. G. T., and Edwards (1967), cited by Charters and Peterson (1972).

Hottel, H. C. (1927), cited by McAdams, W. H. (1954), *Heat Transmission,* 3rd Ed., McGraw-Hill, New York.

Jordan, R. C., Ed. (1967), *Low Temperature Engineering Application of Solar Energy,* ASHRAE, New York.

Kemme, J. E., *et al.* (1969), Los Alamos Special Contribution LA-4221-MS; also *IEEE Trans. Electron Devices* **ED-16,** No. 8, p. 717.

Speyer, E. (1965), *Trans. ASME J. Eng. Power* **87**, No. 3, p. 270.

Veinberg, B. P., and Veinberg, V. B. (1929), cited by Veinberg, V. B. (1959), *Optics in Equipment for the Utilization of Solar Energy,* State Publ. House, Moscow, Transl. AEC-tr-4471, USAEC, OTS, Dept. Commerce, Washington, D.C., p. 144.

Chapter 12. Flat-Plate Collectors

Bliss, R. W. (1961), *Proceedings of the UN Conference on New Sources of Energy, Rome, 1961,* United Nations, New York, Vol. 5 (1964).

Bloch, R. (1948), cited by Tabor (1963).

Catalanotti, S., Cuomo, V., Piro, G., Ruggi, D., Silvestrini, V., and Troise, G. (1975), *Sol. Energy* **17**, No. 2, p. 83.

Fishenden, M., and Saunders, O. A. (1950), *An Introduction to Heat Transfer,* Clarendon Press, Oxford.

Gupta, C. L., and Garg, H. P. (1967), *Sol. Energy* **11**, No. 1, p. 25.

Hall, J. E., and Kusada, T. (1975), National Bureau of Standards NBSIR 74-635.

Hirschmann, J. R. (1970), *Sol. Energy* **13**, No. 1, p. 83.

Jordan, R. C., Ed. (1967), *Low Temperature Engineering Application of Solar Energy,* ASHRAE, New York.

Kalecsinski, V. A. (1940), cited by Tabor (1963).

Silvestrini, V. (1975), see Catalanotti *et al.* (1975).

Solomon, R. A. (1963), Solar Energy Lab Report, University of Arizona, Tucson, Jan.

Speyer, E. (1965), *Trans. ASME J. Eng. Power* **87**, No. 3, p. 270.

Tabor, H. (1955), *Bull. Res. Counc. Isr.* **5C**, No. 1, p. 5.

Tabor, H. (1959), cited by Tabor (1963).

Tabor, H., *et al.* (1959), cited by Tabor (1963).

Tabor, H. (1963), *Sol. Energy* **7**, No. 4, p. 189.

Tabor, H. (1965), *Sol. Energy* **9**, No. 4, p. 177.

Weinberger, H. (1964), *Sol. Energy* **8**, No. 2, p. 45.

Chapter 13. Energy Storage

Brumleve, T. D. (1974), Internal Report by Sandia Corp., Albuquerque, N.M., Contract AT(29-1)-789.

Daniels, F. (1962), *Sol. Energy* **6**, No. 3, p. 78.

Ingersoll, L. R. (1948), *Heat Conduction*, McGraw-Hill, New York, p. 47.

Page, J. K. (1975), Lectures, International College of Applied Physics, Nerano, Italy.

Chapter 14. Thermodynamic Utilization Cycles

Abbot, C. G. (1943), *Mil. Eng.* **35**, No. 208, p. 70.

Babcock and Wilcox Co., Publ. (1955), *Steam, Its Generation and Use*, 37th Ed., Babcock and Wilcox Co., New York.

Jordan, R. C. (1963), *Introduction to the Utilization of Solar Energy*, Ed. Zarem, A. M., and Erway, D. D., McGraw-Hill, New York, p. 125.

Meinel, A. B., and Meinel, M. P. (1971), *Bull. At. Sci./Science & Public Affairs*, **27**, No. 8, p. 32.

Sargent, S. L., and Beckman, W. A. (1968), *Sol. Energy* **12**, No. 2, p. 137.

Tabor, H. (1955), *Bull. Res. Counc. Isr.* **5C**, No. 1, p. 5.

Chapter 15. Direct Conversion to Electricity

Alatyrtsev, G. A., Baum, V. A., Malevskii, Yu. N., and Milevskaya, N. G. (1969), *Semiconductor Solar Energy Converters*, Ed. Baum, V. A., Consultant's Bureau, New York, p. 3.

Baranova, R. Kh., Malevskii, Yu. N., and Saplizhenko, N. F. (1969), *Semiconductor Solar Energy Converters,* Ed. Baum, V. A., Consultant's Bureau, New York, p. 49.

Beckman, W. A., Schoffer, P., Hartman, W. R., Jr., and Lof, G. O. G. (1966), *Sol. Energy* **10,** No. 3, p. 132.

Bell, R. L., James, L. W., and Moon, R. L. (1974), *Proceedings of the Symposium on the Material Science Aspects of Solar Energy Conversion, Tucson, Arizona, 1974,* published as NSF/RANN Rept. GI-43795, Ed. B. O. Seraphin (1975), p. 232.

Brody, T. P., and Shirland, F. A. (1974), *Proceedings of the Symposium on the Material Science Aspects of Solar Energy Conversion, Tucson, Arizona, 1974,* published as NSF/RANN Rept. GI-43795, Ed. B. O. Seraphin (1975), p. 170.

Brosens, P. J. (1965), *Trans. ASME J. Eng. Power* **87**, No. 3, p. 281.

Chu, T. L. (1974), *Proceedings of the Symposium on the Material Science Aspects of Solar Energy Conversion, Tucson, Arizona, 1974,* published as NSF/RANN Rept. GI-43795, Ed. B. O. Seraphin (1975), p. 300.

Fang, P. H., Ephrath, L., and Nowak, W. B. (1974), *Proceedings of the Symposium on the Material Science Aspects of Solar Energy Conversion, Tucson, Arizona, 1974,* published as NSF/RANN Rept. GI-43795, Ed. B. O. Seraphin (1975), p. 351.

Gaibnazarov, M., Malevskii, Yu. N., and Rezgol, I. A. (1969), *Semiconductor Solar Energy Converters,* Ed. Baum, V. A., Consultant's Bureau, New York, p. 9.

Grimison, E. D. (1938), cited by McAdams (1954), p. 273.

Iles, P. A. (1974), *Proceedings of the Symposium on the Material Science Aspects of Solar Energy Conversion, Tucson, Arizona, 1974,* published as NSF/RANN Rept. GI-43795, Ed. B. O. Seraphin (1975), p. 37.

Kettani, M. A. (1974), *Direct Energy Conversion,* Addison-Wesley, Reading, Mass.

McAdams, W. H. (1954), *Heat Transmission,* 3rd Ed., McGraw-Hill, New York, p. 273.

Seraphin, B. O. (1974), *Proceedings of the Symposium on the Material Science Aspects of Solar Energy Conversion, Tucson, Arizona, 1974,* published as NSF/RANN Rept. GI-43795, Ed. B. O. Seraphin (1975), p. 7.

Weber, N. (1974), *Energy Convers.* **14,** No. 1, p. 1.

Wichner, R. (1974), *Proceedings of the Symposium on the Material Science Aspects of Solar Energy Conversion, Tucson, Arizona, 1974,* published as NSF/RANN Rept. GI-43795, Ed. B. O. Seraphin (1975), p. 223.

Wysocki, J. J., and Rappaport, P. (1960), *J. Appl. Phys.* **31,** No. 3, p. 571.

Chapter 16. A Survey of Applications

Hodges, C. N., Thompson, T. L., Groh, J. E., and Frieling, D. H. (1966), *Solar Distillation Utilizing Multiple Effect Humidification,* Environmental Research Laboratory, University of Arizona, Tucson, Ariz.

Rappaport, P. (1975), Lectures, International College of Applied Physics, Nerano, Italy.

Salam, E., and Daniels, F. (1959), *Sol. Energy* **3**, No. 1, p. 19.

Index

SURNAME INDEX

SUBJECT INDEX